Environmental Economics & Management

Theory, Policy, and Applications

Scott J. Callan • Janet M. Thomas

Bentley College

Australia · Canada · Mexico · Singapore · Spain · United Kingdom · United States

Environmental Economics and Management: Theory, Policy, and Applications, 3d ed.

Scott J. Callan and Janet M. Thomas

Vice President / Editorial Director:
Jack Calhoun

Vice President / Editor-in-Chief:
Mike Roche

Publisher of Economics:
Michael B. Mercier

Senior Acquisitions Editor:
Peter Adams

Developmental Editor:
Vicki Hunter Ross

Marketing Manager:
Janet Hennies

Production Editor:
Margaret M. Bril

Manufacturing Coordinator:
Sandee Milewski

Compositor:
G&S Typesetters

Printer:
Phoenix Color
Hagerstown, Maryland

Internal Designer:
Bethany Casey

Cover Designer:
Bethany Casey

Cover Photographer:
PhotoDisc, Inc.; Artville, LLC.; and Digital Vision, Ltd.

Printed in the United States of America
1 2 3 4 5 06 05 04 03

For more information contact South-Western, 5191 Natorp Boulevard, Mason, Ohio 45040.
Or you can visit our Internet site at:
http://www.swlearning.com

Library of Congress Control Number:
2002116031

ISBN: 0-324-17181-1

Preface

Few contemporary issues have so influenced consumer behavior, corporate strategy, and public policy as environmental concerns. In the last few decades, we have witnessed changes in product design, capital investment practices, tax policies, product packaging, and technology—all because environmental issues have been integrated into private and public decision making. New industries have emerged in environmental products and services. National and international policies have been developed to preserve natural resources and ecosystems. Firms have redefined their business strategies in response to new regulations and the changing demands of more environmentally conscious consumers. As a society, we have come to recognize that economic activity and the natural environment are inexorably linked, and this profound relationship is at the core of environmental economics and management.

There is no question that environmental economics is a dynamic field. Hence, we have made every effort to keep the content of this text current, accessible to students, and lively. In this third edition of our book, we had the opportunity to integrate many good suggestions offered by our review panel, insightful comments made by our adopters, plus some new ideas of our own. We still believe that teaching environmental economics is an exciting opportunity to show students the broad applicability of economic thinking. Students are more environmentally literate than they were a decade ago, and most are eager to understand how the market process can help explain and even solve environmental problems. It is, to say the least, an energizing challenge to present this evolving field to what typically is a diverse audience of students.

What hasn't changed is our underlying purpose in writing this text. We wrote *Environmental Economics and Management* to offer undergraduate students and certain MBA and other graduate students a clear perspective of the relationship between market activity and the environment. Although we generally assume that students have been exposed to principles of microeconomics, we offer a good review of basic microeconomic fundamentals in Chapter 2 and the major concepts of public goods and externality theory in Chapter 3, both of which should provide students with the necessary foundation for the course.

Our general approach is to illustrate in a practical manner how economic analytical tools such as market models and benefit-cost analysis can be used to assess environmental problems and to evaluate policy solutions. Along with traditional discussions, we incorporate many contemporary examples of business practices that are part of environmental decision making—a vantage point often overlooked in other textbooks. The presentation does not compromise economic theoretical concepts. Rather, it complements the theory with timely real-world applications. In so doing, seemingly abstract concepts are given relevance through actual cases about consumers, industry, and public policy.

Content: A Modular Approach

Organizing the vast amount of material that an environmental economics course attempts to cover is a challenge at best. Mindful of the usual time constraint in a one-semester course and the fact that the student audience can be highly varied, we devised a modular structure for the text. This approach not only organizes the presentation by major topic but also provides a format that facilitates customizing the material to suit a variety of course objectives. At the instructor's discretion, certain chapters within a given module can be omitted or covered less thoroughly without loss of continuity in the overall presentation. Likewise, the order in which the modules are covered can be varied to suit instructor preferences or student interests.

The first three modules form the foundation for the course. These are:

Part 1. Modeling Environmental Problems: a three-chapter module illustrating how environmental problems are modeled from an economic perspective. Primary topics are the materials balance model, a review of market theory and price determination in an environmental context, and the market failure of pollution using both a public goods model and externality theory.

Part 2. Modeling Solutions to Environmental Problems: a two-chapter module on environmental regulatory approaches—one on the command-and-control approach and one on the market approach. Allocative efficiency and cost-effectiveness are used to analyze these, and models are developed to study various control instruments such as technology-based standards, pollution charges, deposit/refund systems, and tradeable pollution permits.

Part 3. Analytical Tools for Environmental Planning: a four-chapter module introduced by an in-depth investigation of risk assessment, risk management, and benefit-cost analysis. Included is a thorough presentation of benefit estimation procedures such as the contingent valuation method and the averting expenditure approach.

There are also three media-specific modules, which are actually comprehensive case studies of major environmental problems and policy solutions. Using economic modeling and analytical tools, each module assesses the associated environmental risk, evaluates the policy response, and presents a benefit-cost analysis of major U.S. legislation. These three modules can be covered in any sequence following the foundational material covered in the first half of the text.

Part 4. The Case of Air: a four-chapter module assessing major air pollution problems and the policy initiatives aimed at controlling them. The first chapter lays the groundwork by discussing air quality policy in general and the standard-setting process. The next two chapters cover policy implementation—one aimed at stationary sources and the other at mobile sources, with accompanying discussions of urban smog and acid rain. The last chapter in this module deals with global air quality, specifically ozone depletion and climate change, as well as domestic and international policies that address these problems.

Part 5. The Case of Water: a three-chapter module covering the problems of groundwater and surface water contamination and specific policies aimed at point and nonpoint polluting sources. Two chapters are devoted to an economic analysis of the Clean Water Act, and a third conducts an analogous investigation of the Safe Drinking Water Act.

Part 6. The Case of Solid Wastes and Toxic Substances: a three-chapter module analyzing the solid waste cycle and the use of pesticides and other toxic substances. Among the primary topics discussed are risk management of the hazardous waste stream, the Superfund controversy, market solutions to controlling municipal solid waste, and risk-benefit analysis in pesticide control.

Each of these media modules utilizes the analytical tools presented in Parts 1 through 3, such as economic models, risk management, and benefit-cost analysis.

The concluding module covers topics in global environmental management. While we continue to integrate international issues throughout the text, we have rewritten this final module to focus on environmental objectives, policies, and strategies that involve the global community. We focus here on sustainable development, international trade and environmental protection, industrial ecology, and pollution prevention.

> **Part 7. Global Environmental Management:** a two-chapter module examining sustainable development as a worldwide objective and various efforts underway to achieve it. The first chapter in the module addresses the effect of economic growth on environmental quality, international agreements aimed at transboundary pollution, and the effect of environmental protection on international trade. In the second chapter, the focus is on approaches, specifically industrial ecology and pollution prevention. Beyond explaining these concepts at a fundamental level, we also illustrate how these ideas are put into practice through various programs and partnerships in nations around the world.

In essence, this module "closes the loop" of the text by revisiting the materials balance model introduced originally in Chapter 1. Here, we use this model to illustrate the importance of long-run environmental planning and global policy initiatives that go beyond command-and-control abatement efforts.

New to the Third Edition

Because our reviewers and adopters have been pleased with the organization and writing of the text, we have maintained its basic style, structure, and underlying motivation. This allows existing adopters to face minimal adjustment in transitioning from the second edition to the third edition. In particular, we have maintained the overall content and completeness of each module as well as the mobility among them. We also continue to offer supporting examples and data that are as current as possible and to emphasize the important connections between theory and reality through our boxed applications. That said, we did make some important changes in direct response to recent events in the news, various policy reforms, and comments offered by reviewers and adopters.

More International Coverage

Part 7 has been completely rewritten, largely in response to reviewer and adopter requests. While we continue to offer an integration of international issues throughout the text, we also recognize that some instructors want to devote a class or two strictly to international concepts in environmental economics. Accordingly, Part 7 has been revised to focus on global environmental management. This module has been expanded to two chapters—one dedicated to sustainable development, international trade, and agreements on transboundary pollution, and the other committed to sustainable approaches, specifically industrial ecology and pollution prevention. A quick scan of the table of contents will convey the depth of coverage and the interesting content in supporting applications.

Other Structural Changes

Two additional structural changes have been made to the third edition. First, our discussion of the public policy development process has been significantly reduced, since most reviewers said there was insufficient time in a semester to cover this material. This change necessitated the elimination of what was Chapter 6 in the previous edition, though selected public policy topics have been integrated into other chapters. A second change involved splitting what was formerly Chapter 12 in the air module into two shorter, more focused chapters—one aimed at stationary sources and acid rain and the other focusing on urban smog and mobile sources. The content is essentially the same, but the split accommodates the way most instructors cover the material—in two separate classes.

New Topics, New Applications, and Other Updates

We searched diligently for updated data, examples, and applications to assure that instructors and students have access to the most current information. Outdated examples and data were deleted and, where possible, replaced with more contemporary examples or insights. We also added a number of **new topics** to this edition, including the following:

- Environmental Kuznets Curve
- Remanufacturing
- Recent developments surrounding the Kyoto Protocol
- International trade and environmental protection
- President Bush's Clear Skies Initiative
- The ongoing debate about the New Source Review
- Green Chemistry
- ISO 14000 series
- EPA's prospective analysis of the Clean Air Act
- The new arsenic standard
- Environmental costs of the September 11 attacks
- Extended Product Responsibility
- Hybrid vehicles
- World Trade Organization

All boxed applications, which have been well received by adopters and their students, were reviewed, updated, and in some cases replaced with more current cases. Overall, we deleted 10 applications, added 16 new ones, and updated the remainder for a total of 77 applications scattered throughout the text. The **new applications** deal with recent issues that have affected, or are continuing to influence, policy decisions or business practices. These new titles are as follows:

- G.E. Ordered to Dredge the Hudson River: Do the Benefits Justify the Costs?
- The United States Must Play Catch-Up in Hybrid Vehicles
- Why the U.S. Withdrew from the Kyoto Protocol
- The Energy Star Program
- Incremental Environmental Costs of the September 11, 2001, Terrorist Attacks
- Can the Profit Motive Help the Environment?
- Is Standard Setting Under the Clean Air Act Unconstitutional?
- Developing a Regional NO_X "Cap and Trade" Plan
- Should the New Source Review for Stationary Sources Be Revoked?
- Debating the Merits of President Bush's Clear Skies Initiative
- Trading Programs at the State Level: California's RECLAIM
- The Nonpoint Source Program: A Collaborative Effort
- Regulatory Impact Analysis (RIA) for the New Lead Standard in Drinking Water
- Economic Growth and Environmental Quality: The Environmental Kuznets Curve
- ISO 14000 International Standards on Environmental Management
- Remanufacturing: A Lucrative Approach to Pollution Prevention

A Simpler Presentation

We conducted a complete review of the entire text, noting places where we could simplify the presentation without harming the integrity of the content. In addition to general editing, we also moved some of the more detailed tables to a newly added **Appendix.** The motivation was to simplify the text of each chapter but still offer the depth and detail to those wishing to use it in classroom presentations, assignments, or course projects. We simplified many of the data tables and converted others to easier-to-read pie charts or other graphics. We also added more subheadings to help students organize their thinking and more readily learn the material when presented in smaller segments.

Updated Internet Links and Icons

As in the second edition, we continue to offer students and instructors **Internet links** to Web sites that support or enhance the text presentation. Every effort has been made to update all URLs and to add new Web sites that have come to our attention. These links are integrated within the text and in footnotes and are also summarized in a useful list at the end of each chapter. New to this edition are icons placed next to each link within the main text, which points them out more clearly to instructors and students.

Pedagogical Features

Our text has a number of features designed to help instructors prepare lectures and class materials and to make the material interesting and accessible to students. There are both end-of-chapter and end-of-text pedagogical tools, including lists of key concepts, chapter summaries, review questions, relevant Web sites, and an extensive list of references. In every chapter, important definitions are given in the margins, key concepts are shown in boldface, and an extensive offering of real-world applications are provided in shaded boxes.

Applications

Seventy-seven boxed applications complement the text presentation by illustrating the relevance of economic theory, environmental risk, and public policy. The content has been drawn from many sources, including the business press, government reports, economic research, and the environmental science literature. Topics range from corporate strategies to international policy formulation. These real-world cases motivate learning because they illustrate fundamental concepts in relevant, contemporary settings. They also may stimulate more in-depth study in a term paper or course project. In addition to the new applications listed previously, other titles include:

- Conflict Between Economic Development and the Environment: China's View
- Inequities of Air Pollution: Who Suffers More?
- BMW's Design for Recycling
- How NAFTA Was Affected by the National Environmental Policy Act
- Market Incentives to Phase Out Lead Emissions
- Searching for Alternatives to CFCs: The Corporate Response
- Mexico City's Serious Smog Problem
- Who Regulates the Quality of Bottled Water?
- Taxing Gasoline Consumption: An International Comparison
- Germany's Effluent Charge System
- Industrial Symbiosis in Kalundborg, Denmark: When a Bad Becomes a Good

Margin Definitions

In each chapter, **Margin Definitions** of terms and relationships are placed adjacent to the associated text presentation. This feature calls attention to important points in the text, helps familiarize students with new terminology, and assists them in reviewing and self-testing their comprehension.

End-of-Chapter Learning Tools

Each chapter concludes with a **Summary** to help students review and assimilate what they have read. Instructors also may find these summaries valuable in organizing the course and in preparing lectures. There is also a list of **Key Concepts,** which includes all the terms that appear in the chapter as margin definitions. We also provide conceptual and analytical **Review Questions** that can be used for regular assignments, in-class discussions, or sample test questions. Solutions are provided in an Instructor's Manual available through the publisher. We also offer a selection of **Additional Readings** (beyond those cited in the chapter), which are useful for supplementing reading assignments or supporting student projects. A list of **Related Web Sites** is offered to give students and instructors easy access to online information pertaining to each chapter. Many new readings and Web sites have been added to this third edition. We also provide a reference list of commonly used **Acronyms** at the end of the chapters in the media-specific modules, i.e., Parts 4, 5, and 6.

End-of-Text Learning Tools

At the end of the text is a list of **References,** which gives complete information on all sources cited in short form throughout the book. Both instructors and students should find this collection of resources helpful in conducting independent investigations of selected topics. There is also a convenient **Glossary** of all defined terms given in the chapters. New to the third edition is an **Appendix,** which includes a small selection of exhibits to specific chapters, each of which gives greater detail on a particular topic than is covered in the chapter itself.

Text Ancillaries

The following ancillary materials are available to users of the text:

- An **Instructor's Manual,** by the text authors, is available to instructors only, and includes quantitative solutions and suggested responses to all end-of-chapter Review Questions. The Instructor's Manual is also available online.
- A collection of **PowerPoint®** slides for key figures from the text is available on the text Web site to facilitate classroom presentations and aid in student note-taking.
- The text Web site at **http://callan.swlearning.com** contains links to Internet addresses from the text, InfoTrac, PowerPoint slides, and more.
- Economic Applications **(e-con @pps)** at **http://econapps.swlearning.com**, is the site for South-Western's dynamic Web features: EconNews Online, EconDebate Online, and EconData Online. Organized by topic, these features deepen understanding of theoretical concepts through hands-on exploration and analysis of the latest economic news stories, policy debates, and data, and are updated on a regular basis. The Economic Applications Web site is complimentary to every new book buyer via an access card packaged with the book. Used-book buyers can purchase access to the site at **http:econapps.swlearning.com**.
- An **InfoTrac® College Edition** password is packaged with every new copy of the textbook. InfoTrac, at **http://www.infotrac-college.com**, is a fully searchable online university library containing complete articles and their images. Its database allows access to hundreds of scholarly and popular publications—all reliable sources, including magazines, journals, encyclopedias, and newsletters. Many of the Key Concepts at the end of each chapter can serve as useful keywords when searching this database.

Acknowledgments

No text is ever produced without the help and support of many individuals. In our case, we owe a debt of gratitude to the administration at Bentley College. Its ongoing support of this project and our related research in environmental economics is much appreciated. We also have been fortunate to hear from some of our adopters over the years, giving us the opportunity to integrate their ideas into our revision plans. Similarly, we are grateful for input offered directly by students, as they used the text and worked with the review problems, boxed applications, and other pedagogical features.

We also wish to acknowledge the many economists who contributed to this text at various stages of the revision process. In preparing this third edition, we learned much from the suggestions given by our most recent review panel, which included the following individuals: Craig A. Gallet, California State University, Sacramento; Richard D. Horan, Michigan State University; Brian J. Peterson, Manchester College; George D. Santopietro, Radford University; Hilary A. Sigman, Rutgers, The State University of New Jersey; Douglas D. Southgate Jr., Ohio State University; and David Yoskowitz, Texas A&M International University. All of their comments were carefully considered, and many of their suggestions were integrated into this new edition. To each of them, we extend our appreciation for their important contribution to this work.

We also are most grateful to our review panel for the second edition, which included Mark Aldrich, Smith College; Douglas F. Greer, San Jose State University; Darwin C. Hall, California State University; Stanley R. Keil, Ball State University; and Warren Matthews, Houston Baptist University. Likewise, we are indebted to the reviewers of the first edition: Bill Ballard, College of Charleston; Laurie Bates, Bryant College; John Braden, University of Illinois (Urbana); Michelle Correia, Florida Atlantic University; Warren Fisher, Susquehanna University; Joyce Gleason, Nebraska Wesleyan College; Sue Eileen Hayes, Sonoma State University; Charles Howe, University of Colorado; Donn Johnson, University of Northern Iowa; Supriya Lahiri, University of Massachusetts; Donald Marron, University of Chicago; Arden Pope, Brigham Young University; H. David Robison, La Salle University; Richard Rosenberg, Pennsylvania State University; Jeffrey Sundberg, Lake Forest College; David Terkla, University of Massachusetts; John Whitehead, East Carolina University; and Keith Willett, Oklahoma State University.

To all these teachers and scholars, we offer our sincere appreciation for their supportive commentary, constructive criticism, and many useful ideas. By incorporating many of their suggestions along the way, we believe our text has been strengthened in every dimension.

We also wish to thank our entire project team at South-Western for their support and excellent work in developing, designing, and producing this edition. This group of fine professionals includes Peter Adams, Senior Acquisitions Editor; Mike Roche, Vice President/Editor-in-chief; Mike Mercier, Publisher of Economics; Susan Smart, Senior Developmental Editor; Vicki Hunter Ross, Developmental Editor; Janet Hennies, Senior Marketing Manager; Vicki True, Media Technology Editor; Peggy Buskey, Media Developmental Editor; Casey Gilbertson, Designer; Marge Bril, Production Editor; Pam Wallace, Media Production Editor; and Sandee Milewski, Manufacturing Coordinator. In particular, we are indebted to Peter Adams, Marge Bril, and Vicki Ross, each of whom offered invaluable guidance throughout the preparation of this edition. Their enthusiasm, experience, and genuine interest in this project at every phase were a great source of encouragement throughout the revision process.

Finally and most important we thank our respective spouses, Karen and David, for their continued support of our work on this project and all our professional endeavors.

Scott J. Callan
Janet M. Thomas

About the Authors

Professor Scott J. Callan is a professor of economics at Bentley College. Professor Callan received his M.S. and Ph.D. degrees from Texas A&M University in 1985. Prior to joining the Bentley College faculty in 1987, Professor Callan was a member of the business faculty at Clarkson University in Potsdam, New York. His teaching areas of interest focus on quantitative methods and applied microeconomic topics, such as environmental economics and managerial economics. He has taught courses in environmental economics and the economics of natural resources at both the undergraduate and graduate levels.

In addition to his textbook, Professor Callan is the author of numerous applied microeconomic articles that have been published in a variety of economic journals, including the *Southern Economic Journal, Advances in the Economics of Energy and Resources,* the *Review of Industrial Organization,* and the *Journal of Business and Economic Statistics.* His recent environmental economics research investigates the demand and supply characteristics associated with the market for municipal solid waste (MSW). Demand-side topics examined include the impact of pay-as-you-throw programs on waste generation, disposal, and recycling activities. Supply-side issues have focused on the extent of economies of scale and scope in the provision of MSW services. His research findings have appeared in *Land Economics* and *Environmental and Resource Economics,* among other academic journals. In addition to his many publications, Professor Callan has reviewed numerous scholarly articles for academic publications as well as research grant proposals for the U.S. Environmental Protection Agency. Professor Callan is a member of several professional organizations, including the American Economics Association, the Southern Economic Association, and the National Association of Business Economists.

Professor Janet M. Thomas earned her M.A. and Ph.D. degrees in economics at Boston College. After completing her doctorate, she was appointed to the faculty of Bentley College in 1987. She is currently a full professor at Bentley, teaching at both the undergraduate and graduate levels. In addition to environmental economics, Professor Thomas teaches intermediate microeconomics, industrial organization, principles of microeconomics, and principles of macroeconomics. She has been actively involved in course and curriculum development in environmental economics and served as coordinator of the MBA – Environmental Management Concentration Program at Bentley.

Professor Thomas also is an active researcher in environmental economics, industrial organization, and other fields in applied microeconomics. Working with data provided by the Massachusetts Department of Environmental Protection, her present environmental economics research centers on municipal solid waste markets, studying such issues as economies of scale and scope, demand determinants, unit pricing, and the influence of policy on recycling efforts. In addition to her textbook, she has published her research results in such academic journals as *Land Economics, Southern Economic Journal, Environmental and Resource Economics, Review of Industrial Organization, Eastern Economic Journal,* and the *Journal of Transport Economics and Policy.* She is a member of the American Economics Association and the Industrial Organization Society and has served as a reviewer for a number of academic journals and textbook publishers. Professor Thomas was named Faculty Member of the Year by Bentley's Student Government Organization in 1991. In 1993, she received the Gregory H. Adamian Award for Teaching Excellence, and in 1996 she received the Bentley College Scholar of the Year Award.

Brief Contents

Contents

Environmental Economics & Management

PART 1

Modeling Environmental Problems

Environmental problems, although not new, have taken on a more significant role in business decisions and corporate planning in the past two decades. The world has become more aware of the natural environment and more sensitive to the implications of ecological damage. People are altering their consumption patterns to incorporate environmental responsibility into their market decisions. Many are reordering their preferences in favor of biodegradable detergents, non-ozone-depleting products, and recyclable packaging. Similarly, governments are responding on both a local and a national level by passing environmental laws and establishing pollution monitoring networks to protect the ecology. Firms are adding environmental concerns to their list of business priorities. This corporate response is necessary, not only to comply with regulations on production and product design, but also to remain competitive in a marketplace where many consumers are seeking environmentally responsible producers.

To comprehend this changing marketplace, it is necessary to understand the fundamentals of how markets work and the relationship between market activity and nature. Economic analysis uses models to explain the strategic decision making and economic conditions that define the marketplace. By eliminating unnecessary detail, models allow economists to test theories about economic relationships and to make predictions about behavioral and institutional reactions to changes in market conditions. For these reasons, modeling is a fundamental tool in environmental economics.

In this first module of the text, we devote three chapters to basic models that are useful for understanding environmental issues. In Chapter 1 we develop the materials balance model, which illustrates the linkages between the circular flow of economic activity and nature. In so doing, we show how environmental damage and resource depletion occur as a result of market decision making. In Chapter 2 we focus on the fundamentals of the market process, reviewing such key concepts as supply and demand, economic efficiency, and measures of social welfare. All of this lays the foundation for Chapter 3, in which we explain how environmental problems occur when the market fails. Here, too, we use models to illustrate the sources of market failure and the conditions under which such an outcome arises.

Chapter 1

The Role of Economics in Environmental Management

As society moves through the twenty-first century, it faces an important challenge—to protect and preserve the earth's resources as society continues to develop economically. The rapid growth and advancing technology that began in earnest with the Industrial Revolution did take a toll on the natural environment. Mass transport, manufacturing processes, telecommunications, and synthetic chemicals are responsible both for the highly advanced lifestyle that society enjoys *and* for much of the environmental damage it now faces. With 20/20 hindsight, we now recognize that the trade-off between economic growth and environmental quality was significant.

An important objective is to understand the critical relationship between economic activity and nature and to use that knowledge to make better and wiser decisions. Of course, there will always be some amount of trade-off—precisely what economic theory conveys. We cannot expect to have perfectly clean air or completely pure water, nor can we continue to grow economically with no regard for the future. But there *is* a solution, although it is a compromise of sorts. We first have to decide what level of environmental quality is acceptable and then make appropriate adjustments in our market behavior to sustain that quality as we continue to develop as a society.

The adjustment process is not an easy one, and it takes time. As a society, we are still learning—about nature, about market behavior, and about the important relationships that link the two together. What economics contributes to this learning process are analytical tools that help to explain the interaction of markets and the environment, the implications of that relationship, and the opportunities for effective solutions.

In this chapter, we support these assertions with a simple but powerful model that illustrates the link between economic activity and nature. As we will discover, the underlying relationships motivate economic analysis of environmental issues, which is formally defined through two disciplines: natural resource economics and environmental economics, the latter of which is our focus in this book. With this model as a foundation, we lay the groundwork for our course of study, starting with an introduction of basic concepts. From there, we identify and discuss major objectives in environmental economics and then present an overview of public policy development and the role of economics in that process.

Economics and the Environment

One of the most pervasive applications of economic theory is that it logically explains what we observe in reality. For example, through microeconomic analysis we can understand the behavior of consumers and firms and the decision making that defines the marketplace. This same application of economic theory can be used to analyze environmental problems—why they occur and what can be done about them. Stop to consider how pollution or resource depletion comes about—not from a sophisticated scientific level, just from a fundamental perspective. The answer? Both arise from decisions made by households and firms. Consumption and production draw on the earth's supply of natural resources. Furthermore, both activities generate by-products that can contaminate the environment. This means that the fundamental decisions that comprise economic activity are directly connected to environmental problems. To illustrate this relationship, we begin by presenting a basic model of economic activity. Then, we expand the model to show exactly how this connection arises.

The Fundamental Model of Economic Activity: The Circular Flow Model

The basis for modeling the relationship between economic activity and the environment is the same one that underlies all of economic theory—the **circular flow model,** which is shown in Figure 1.1. Typically, this is the first model students learn about in introductory economics.

circular flow model
Illustrates the real and monetary flows of economic activity through the factor market and the output market.

First, consider how the flows operate, holding all else constant. Notice how the **real flow** (i.e., the nonmonetary flow) runs counterclockwise between the two market sectors, households (or consumers) and firms (or producers). Households supply resources or factors of production to the **factor market,** where they are demanded by firms to produce goods and services. These commodities are then supplied to the **output market,** where they are demanded by households. Running clockwise is the **money flow.** The exchange of inputs in the factor market generates an income flow to households, and that flow represents costs incurred by firms. Analogously, the money flow through the output market shows how households' expenditures on goods and services are revenues to firms.

Now, think about how the volume of economic activity and hence the size of the flow are affected by such things as population growth, technological change, labor productivity, capital accumulation, and natural phenomena such as drought or floods. For example, holding all else constant, technological advance would expand the productive capacity of the economy, which in turn would increase the size of the flow. Similarly, a population increase would lead to a greater demand for goods and services, which would call forth more production and lead to a larger circular flow.

Notice that by analyzing how the flows operate and how the size of an economy can change, we can understand the basic functioning of an economic system and the market relationships between households and firms. However, the model does not explicitly show the linkage between economic activity and the environment. To illustrate this interdependence, the circular flow model must be expanded to depict market activity as part of a broader paradigm called the materials balance model.

Figure 1.1 *Circular Flow Model of Economic Activity*

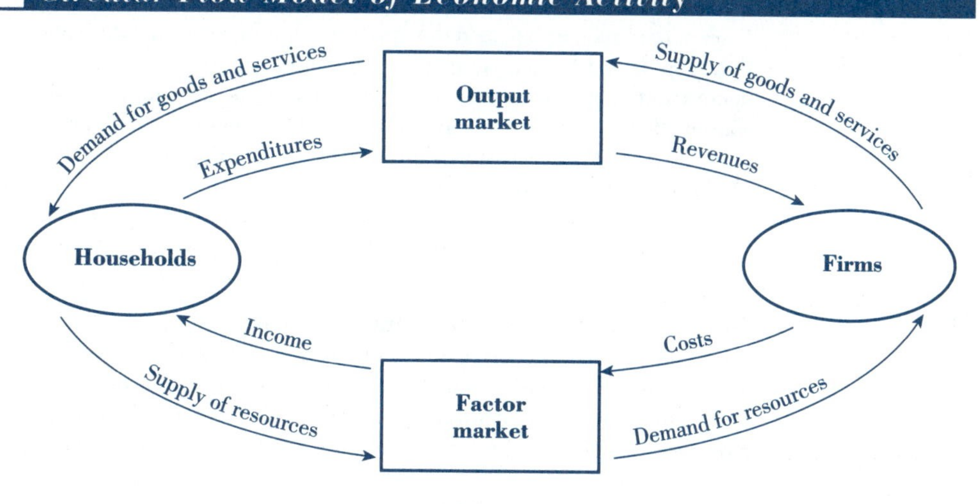

Figure 1.2 *Materials Balance Model: The Interdependence of Economic Activity and Nature*

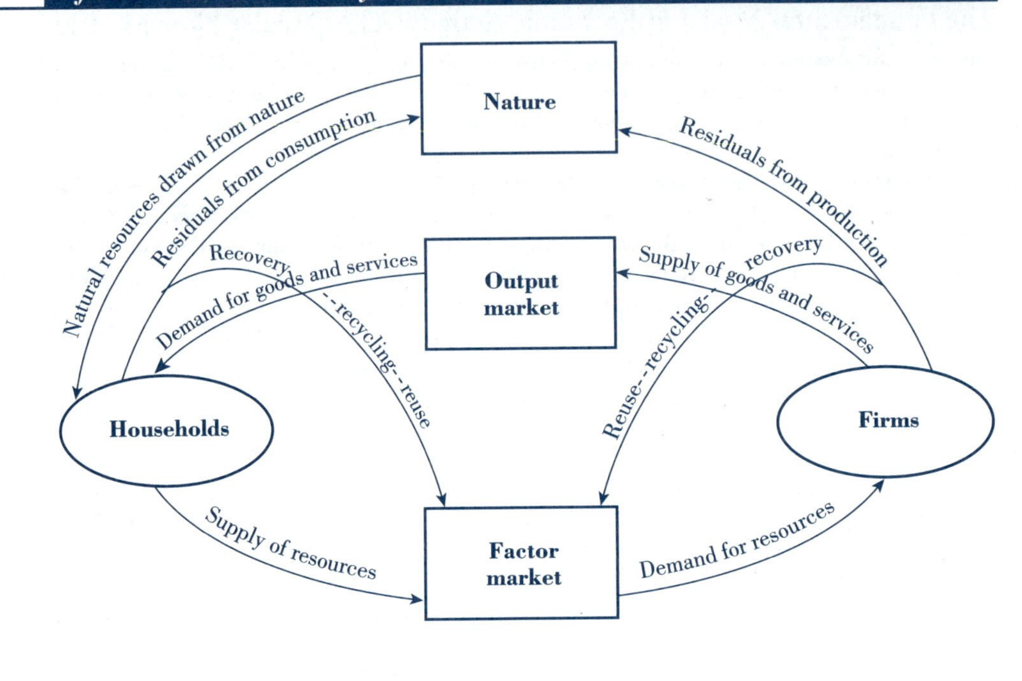

Source: Adapted from Kneese, Ayres, and D'Arge (1970).

materials balance model
Positions the circular flow within a larger schematic to show the connections between economic decision making and the natural environment.

natural resource economics
A field of study concerned with the flow of resources from nature to economic activity.

residual
The amount of a pollutant remaining in the environment after a natural or technological process has occurred.

The Materials Balance Model

The explicit relationship between economic activity and the natural environment is illustrated by the **materials balance model,** shown in Figure 1.2.[1] Notice how the real flow of the circular flow model is positioned within a larger schematic to show the connections between economic decision making and the natural environment.

Flow of Resources: Natural Resource Economics

Look at the linkages between the upper block representing nature and the two market sectors (households and firms), paying particular attention to the direction of the arrows. Notice that one way an economic system is linked to nature is through a flow of materials or natural resources that runs *from* the environment *to* the economy, specifically through the household sector. (Recall that, by assumption, households are the owners of all factors of production, including natural resources.) This flow describes how economic activity draws on the earth's stock of natural resources, such as soil, minerals, and water. It is the primary focus of **natural resource economics,** a field of study concerned with the flows of resources from nature to economic activity.

Flow of Residuals: Environmental Economics

A second set of linkages runs in the opposite direction, *from* the economy *to* the environment. This flow illustrates how raw materials entering the system eventually are released back to nature as by-products or **residuals.** Most residuals are in the form of gases released into the atmosphere, and in the short run, most are not harmful. In fact, some are absorbed naturally

[1] Kneese, Ayres, and D'Arge (1970). See p. 9, chart 1, in this source for a more detailed depiction of this model.

through what is called the **assimilative capacity** of the environment. For example, carbon dioxide emissions from the combustion of fossil fuels (i.e., oil, coal, and natural gas) can be partially absorbed by the earth's oceans and forests. Other released gases are not easily assimilated and may cause harm, even in the short term. There are also liquid residuals, such as industrial wastewaters, and solid residuals, such as municipal trash and certain hazardous wastes—all of which are potential threats to health and the ecology.[2] Notice in Figure 1.2 that there are two residual outflows, one leading from each of the two market sectors, meaning that residuals arise from both consumption and production activity. This set of flows is the chief concern of **environmental economics.**

environmental economics
A field of study concerned with the flow of residuals from economic activity back to nature.

It is possible that the flow of residuals back to nature can be delayed, though not prevented, through **recovery, recycling,** and **reuse.** Notice in the model that there are inner flows running from the two residual outflows back to the factor market. These inner flows show that some residuals can be recovered from the stream and either recycled into another usable form or reused in their existing form. For example, Application 1.1 discusses how Germany's Bavarian Motor Works (BMW) has made advances in automobile design to facilitate recycling once a vehicle has reached the end of its economic life.

Although recycling efforts are important, keep in mind that they are only short-term measures, because even recycled and reused products eventually become residuals that are returned to nature. Indeed, what the materials balance model shows is that all resources drawn from the environment ultimately are returned there in the form of residuals. The two flows are balanced, a profound fact that is supported by science.

Using Science to Understand the Materials Balance

According to the **first law of thermodynamics,** matter and energy can neither be created nor destroyed. Applying this fundamental law to the materials balance model means that in the long run the flow of materials and energy drawn from nature into consumption and production must equal the flow of residuals that run from these activities back into the environment. Put another way, when raw materials are used in economic activity, they are converted into other forms of matter and energy, but nothing is lost in the process. Over time, all these materials become residuals that are returned to nature. Some arise in the short run, such as waste materials created during production. Other resources are first transformed into commodities and do not enter the residual flow until the goods are used up. At this point, the residuals can take various forms, for example, carbon monoxide emissions from gasoline combustion or trash disposed in a municipal landfill. Even if recovery does take place, the conversion of residuals into recycled or reused goods is only temporary. In the long run, these too end up as wastes.

first law of thermodynamics
Matter and energy can neither be created nor destroyed.

There is one further point. Because matter and energy cannot be destroyed, it may seem as though the materials flow can go on forever. But the **second law of thermodynamics** states that nature's capacity to convert matter and energy is not unlimited. During energy conversion, some of the energy becomes unusable. It still exists, but it is no longer available to use in another process. Consequently, the fundamental process on which economic activity depends is finite.

second law of thermodynamics
Nature's capacity to convert matter and energy is not without bound.

These scientific laws that support the materials balance model communicate important, practical information to society. First, we must recognize that every resource drawn into economic activity ends up as a residual, which has the potential to damage the environment. The process can be delayed through recovery but not stopped. Second, nature's ability to convert resources to other forms of matter and energy is limited. Taken together, these assertions provide a comprehensive perspective of environmental problems and the important connections between economic activity and nature. It is the existence of these connections that motivates the discipline of environmental economics.

[2] Kneese, Ayres, and D'Arge (1970), pp. 11–12.

Application

1.1 BMW's Design for Recycling

Although BMW is known primarily for its well-engineered automobiles, it is also recognized for another distinction. The German carmaker is committed to developing and building BMWs in an environmentally responsible manner. For some time, the company's objective has been to build its automobiles from 100 percent reusable or recyclable parts. Today, new BMWs are almost completely recyclable. In fact, recycling was an important factor in the design of the company's reintroduced MINI Cooper.

Thanks to disassembly analyses conducted at its Recycling and Disassembly Center in Lohhof, Germany, BMW is developing what it calls "solutions for environmentally and economically sensible recycling." These analyses determine the time and resources needed to dismantle a vehicle at the end of its useful life; this information is then integrated into vehicle construction plans. Such an approach is called "Design for Disassembly (DFD)"—a manufacturing method aimed at building a product specifically to facilitate end-of-life recycling. Along with other major corporations such as Volkswagen, 3M, and General Electric, BMW is investigating ways to manufacture a DFD product that is economically competitive and that stands up to the company's high standards for quality engineering.

BMW's research in DFD is part of a long-term commitment to environmentally responsible production decisions. The company has been recycling its catalytic converters since 1987. A year later, it introduced its limited-production Z1 roadster. The two-seater is totally recyclable and is considered the first DFD product ever made. Subsequently, BMW built a pilot plant in Bavaria dedicated solely to researching the DFD approach to manufacturing. Teams of workers at the facility systematically dismantle cars, beginning with the fluids and oils and ending with the removal and sorting of interior materials.

A major objective is to build an automobile that can be dismantled at a relatively low cost. Long hours for disassembly elevates costs, which ultimately forces up car prices and reduces competitiveness—an outcome no carmaker can afford. Another key goal is to ensure that parts can be readily sorted. This is particularly critical for plastics, which are more complex to recycle but are increasingly used to lower vehicle weight and to improve fuel efficiency. Today, 162 kilograms of plastics are used in the BMW 3-series model, representing a 15 percent increase over the previous model year. Of this amount, close to 90 kilograms can be economically recycled. BMW has also been establishing a network of recycling centers in Germany and the United States to make it more convenient for BMW owners to dispose of their cars.

BMW has been a leader in advancing the DFD approach, although other manufacturers have begun to follow suit. Japan's Nissan Motor Company has launched a variety of new research programs; some are aimed directly at the DFD approach, others at using more recycled materials and fewer plastics in production. In the United States, manufacturers are labeling plastic parts to facilitate sorting for recycling, and they have established a consortium with suppliers and recyclers. All of this suggests that increasing the recyclability of automobiles is a worthwhile pursuit.

Sources: Protzman (July 4, 1993); Nussbaum and Templeman (September 17, 1990); U.S. Congress, OTA (October 1992), p. 11; Knepper (January 1993); BMW Group (2001).

pollution
The presence of matter or energy whose nature, location, or quantity has undesired effects on the environment.

Fundamental Concepts in Environmental Economics[3]

Environmental economics is concerned with identifying and solving the problem of environmental damage, or **pollution,** associated with the flow of residuals. Although pollution is defined differently in different contexts, such as within various laws, it can be defined gener-

[3] Although it is not necessary to master the rigors of environmental science, it is important to become familiar with the basic concepts used to identify environmental damage. A useful reference is the EPA's Web site, which gives a glossary of environmental terms at **http://www.epa.gov/OCEPAterms.**

ally as the presence of matter or energy whose nature, location, or quantity has undesired effects on the environment. Virtually any substance can cause pollution solely on the basis of a single characteristic, such as its fundamental constituents, its location, or its quantity. What this implies is that finding solutions to environmental damage depends critically on identifying the causes, sources, and scope of the damage.

Identifying the Causes of Environmental Damage: Types of Pollutants

How do we identify which substances are causing environmental damage? One way is to distinguish them by their general origin—that is, whether they are **natural pollutants** arising from nature or **anthropogenic pollutants** resulting from human activity.

- **Natural pollutants** arise from nonartificial processes in nature, such as particles from volcanic eruptions, salt spray from the oceans, and pollen.
- **Anthropogenic pollutants** are human induced and include all residuals associated with consumption and production. Examples include gases from combustion and chemical wastes from certain manufacturing processes.

natural pollutants
Contaminants that come about through nonartificial processes in nature.

anthropogenic pollutants
Contaminants associated with human activity, including polluting residuals from consumption and production.

Of the two, anthropogenic pollutants are of greater concern to environmental economists, particularly those for which nature has little or no assimilative capacity.

Identifying the Sources of Pollution: Classifying Polluting Sources

Once pollutants have been identified, the next step is to determine the sources responsible for their release. Polluting sources are many and varied, ranging from automobiles to waste disposal sites. Even a seemingly pristine setting such as farmland can be a polluting source if rainwater transports chemical pesticides and fertilizers to nearby lakes and streams. Because polluting sources are so diverse, they are usually classified into broadly defined categories that are meaningful to policy development. Depending on the environmental media (air, water, or land), sources of pollution are generally grouped by (1) their mobility or (2) how readily they can be identified.

Sources Grouped by Mobility

Whether or not a polluting source remains in a single location affects how the pollution can be controlled. The relevant categories are designated simply as **stationary sources** and **mobile sources.**

- A **stationary source** is a fixed-site producer of pollution, such as a coal-burning power plant or a sewage treatment facility.
- A **mobile source** refers to any nonstationary polluting source, such as an automobile or an airplane.

stationary source
A fixed-site producer of pollution, such as a building or manufacturing plant.

mobile source
Any nonstationary polluting source, including all transport vehicles.

This distinction is commonly used to characterize air pollution sources, since each requires a different form of control.

Sources Grouped by Identifiability

In some contexts, the identifiability of a polluting source is an important factor, in which case a distinction is made between **point sources** and **nonpoint sources.**

- A **point source** refers to any single identifiable source from which pollutants are released, such as a factory smokestack, a pipe, or a ship.
- A **nonpoint source** is one that cannot be identified accurately and degrades the environment in a diffuse, indirect way over a relatively broad area.

point source
Any single identifiable source from which pollutants are released, such as a factory smokestack, a pipe, or a ship.

nonpoint source
A source that cannot be identified accurately and degrades the environment in a diffuse, indirect way over a relatively broad area.

This set of classifications is most commonly used in water pollution control policy because nonpoint sources are so significant to this particular problem. As these definitions suggest,

Figure 1.3 Comparison of Pollutants Contributing to Urban Air Pollution in Major Cities

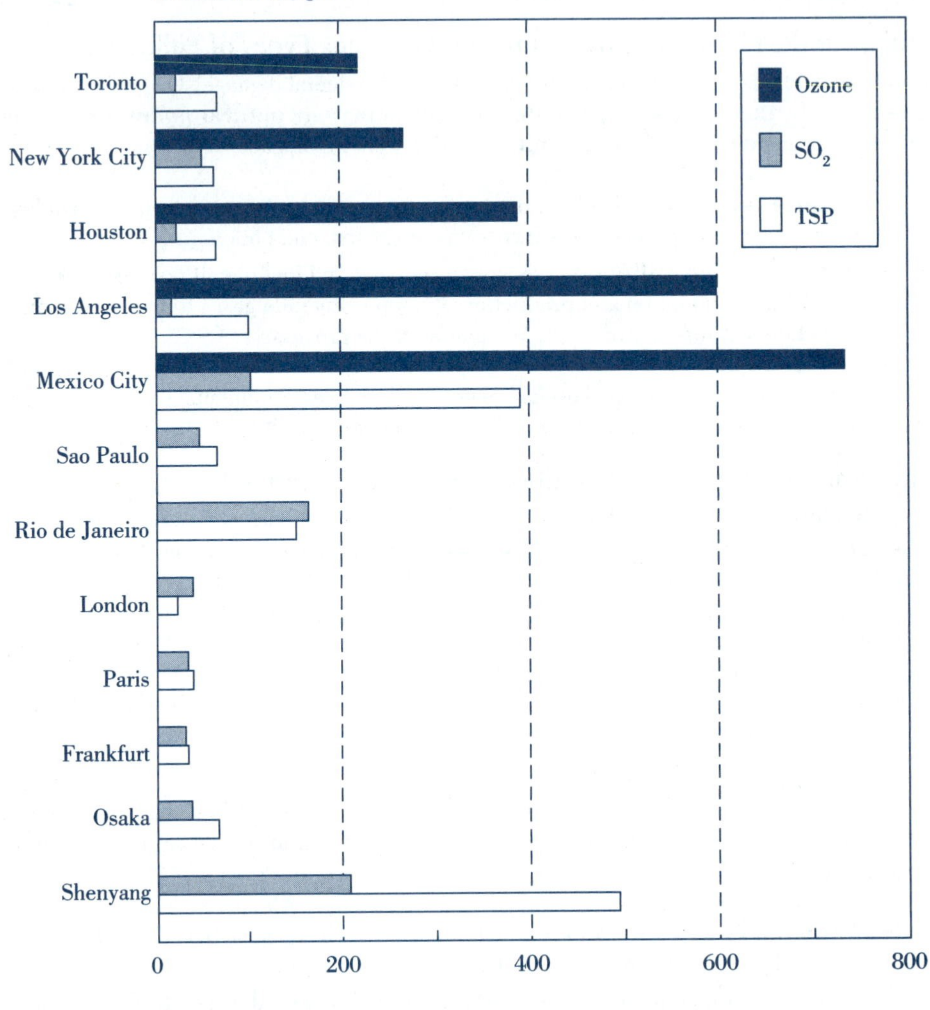

Source: U.S. EPA, Office of Air Quality Planning and Standards (October 1992), pp. 6–7, fig. 6-5, which is drawn from numerous sources.

Note: Ozone refers to ground-level ozone; SO_2 is sulfur dioxide, the condensation of which contributes significantly to particulate matter in the air; and TSP refers to total suspended particles such as dust, soot, smoke, and other contaminants. Concentrations are given in $\mu g/m^3$ (micrograms per cubic meter).

pollutants released from nonpoint sources are more difficult to control than those released from point sources.

Identifying the Scope of Environmental Damage: Local, Regional, and Global Pollution

Although environmental damage is a universal concern, some types of pollution have detrimental effects that are limited to a single community, whereas others pose a risk over a large geographic region. The point is, the extent of the damage associated with pollution can vary

Table 1.1 ***Per Capita Municipal Solid Waste Generation and Gross National Product (GNP) in Selected Countries***

Country	Waste (kg per capita)	1998 GNP ($ per capita)
United States	720	29,240
Australia	690	20,640
Iceland	650	16,180
Norway	600	34,310
France	590	24,210
Denmark	560	33,040
Ireland	560	18,710
Canada	500	19,170
Italy	460	20,090
Japan	400	32,350
Greece	370	11,740
Poland	320	3,910
Czechoslovakia	310	5,150
Mexico	310	3,840

Sources: OECD (1999b), as cited in U.S. Census Bureau (2001); World Bank, as cited in U.S. Census Bureau (2000).
Note: Waste figures are for 1998 or latest available year.

considerably—an observation vitally important to policy formulation. Consequently, environmental pollution is often classified according to the relative size of its geographic impact as local, regional, or global.

Local Pollution

Local pollution refers to environmental damage that does not extend very far from the polluting source and typically is confined to a single community. Although the negative effects are limited in scope, they nonetheless pose a risk to society and can be difficult to control. A common local pollution problem is **urban smog.** Visible as a thick yellowish haze, smog is caused by a number of pollutants that chemically react in sunlight. It is particularly severe in major cities such as Los Angeles and Mexico City, as Figure 1.3 suggests.

local pollution
Environmental damage that does not extend far from the polluting source, such as urban smog.

Another local pollution problem that is receiving increasing attention is **solid waste pollution.** Poor waste management practices can allow such contaminants as lead and mercury to leach into soil and nearby water supplies. Beyond measures aimed at improving waste management are efforts to reduce the amount of waste being generated in the first place. Look at the estimates of municipal waste generation and gross national product (GNP) per capita for selected countries given in Table 1.1. Note that these data generally suggest a positive relationship between waste generation and industrialization.

Regional Pollution

Environmental pollution that poses a risk well beyond the polluting source is called **regional pollution.** An important example is **acidic deposition,** which arises from acidic compounds that mix with other particles and fall to the earth either as dry deposits or in fog, snow, or rain. Acidic deposition is commonly referred to as "acid rain." Acid rain is characterized as regional pollution because the harmful emissions can travel hundreds of miles from their source.

regional pollution
Degradation that extends well beyond the polluting source, such as acidic deposition.

Global Pollution

global pollution Environmental effects that are widespread with global implications, such as global warming and ozone depletion.

Some environmental problems have effects so extensive that they are called **global pollution.** Global pollution is difficult to control, both because the associated risks are widespread and because international cooperation is needed to achieve effective solutions. Consider, for example, the problem of **global warming.** Also known as the "greenhouse effect," global warming occurs as sunlight passes through the atmosphere to the earth's surface and is radiated back into the air where it is absorbed by so-called greenhouse gases (e.g., carbon dioxide). Although this warming process is natural, such activities as fossil-fuel combustion add to the normal level of greenhouse gases, which in turn can raise the earth's natural temperature. These climate disruptions could affect agricultural productivity, weather conditions, and the level of the earth's oceans—all effects that are worldwide in scope.

Similarly widespread are the risks of **ozone depletion,** a thinning of the earth's ozone layer. The ozone layer protects the earth from harmful ultraviolet radiation, which can weaken human immune systems, increase the risk of skin cancer, and harm ecosystems. In 1985, scientists discovered that a previously observed thinning of the ozone layer over the Antarctic region had become an "ozone hole" the size of North America. Ozone depletion is caused mainly by a group of chemicals known as chlorofluorocarbons (CFCs), which had been commonly used in refrigeration, air conditioning, packaging, insulation, and aerosol propellants. Although national governments have controlled CFC usage, the main policy thrust has arisen through international agreements because of the global nature of the problem.

Identifying Environmental Objectives

Just as fundamental environmental problems are universal, so too are the overall objectives. However, articulating the specifics of these objectives and accepting the trade-offs that such goals imply is a process that is not without debate. Indeed, such is the substance of environmental summits, where national leaders, industry officials, and environmentalists gather to exchange ideas about appropriate objectives and to garner cooperation from one another. A case in point was the 2001 United Nations climate conference in Morocco at which negotiations continued on the Kyoto Protocol.[4] A more comprehensive environmental agenda was addressed at the United Nations Conference on Environment and Development (UNCED) held in Rio de Janeiro, Brazil, in 1992. Known as the Earth Summit, the event marked the twentieth anniversary of the first worldwide environmental conference in Stockholm and was attended by 6,000 delegates from more than 170 countries.[5]

Although the objective-setting process has been difficult at times and is often immersed in political debate, it is nonetheless moving forward. Today, virtually every environmental decision is guided by what have become worldwide objectives: **environmental quality, sustainable development,** and **biodiversity.**

Environmental Quality

environmental quality A reduction in anthropogenic contamination to a level that is "acceptable" to society.

Given the pervasive problems of local, regional, and global pollution, few would debate including **environmental quality** among the world's objectives. However, there is a lack of consensus about how to define this concept in practice. Most of us consider environmental quality to mean clean air, water, and land. However, when environmental quality is being defined to guide policy, we have to decide just how clean is clean.

The debate usually begins with asking why environmental quality should not mean the absence of all pollution. The answer is that such an objective is impossible, at least in a pure sense. Recall that some pollution is natural and therefore not controllable. Furthermore, the absence of all anthropogenic pollutants could be achieved only if there were a prohibition on

[4]For more information on the Kyoto conference, visit the Web site for the United Nations Framework Convention on Climate Change at **http://unfccc.int/.**
[5]Council on Environmental Quality (January 1993), p. 140.

1.2 *Conflict Between Economic Development and the Environment: China's View*

Application

As China moves toward developing its economy into an industrialized power, the quality of its air, water, and land resources is deteriorating. Despite a $43.5 billion environmental protection program for the 1996–2000 period, much of the ecological damage is going virtually unchecked. As is sometimes the case in developing economies, environmentalism is perceived as an obstacle to economic advance that can frustrate progress toward industrial development. Many observers believe that unless officials change their perspective toward environmental issues, the response to China's pollution problems will be insufficient. Indeed, in a recent speech to environmental planners, China's premier, Zhu Rongji, argued that all levels of government must participate in this effort.

China's air pollution problems can be traced primarily to the country's heavy reliance on coal as an energy source. In fact, during the 1980s, its coal consumption increased from 620 million tons to more than 1 billion tons. By 1996, China had become the largest producer and consumer of coal, and according to most experts, its dependence on coal is expected to at least double by 2020. Industrial centers, such as Chongqing, already face severe acid rain problems because of the use of high-sulfur coal. Documented damages range from erosion of buildings to the destruction of crops and other plant life. Reportedly, trees along city streets have had to be replaced 3 times in the last 30 years. China's water resources are also at risk. Sewage treatment is often inadequate, even nonexistent in some locations, and industrial wastes are contaminating many of the country's rivers and streams. These are nontrivial problems, but they are particularly serious for a nation with hundreds of cities facing water shortages.

According to China's environmental protection committee, economic losses from environmental pollution are about 3 percent of China's GNP, or approximately $11.5 billion per year. The country now spends about 0.7 percent of its GNP on environmental protection, which apparently is not nearly enough to address the extent and seriousness of China's pollution problems. Nonetheless, Chinese officials are reluctant to allocate resources away from economic development and toward environmental cleanup and protection, accepting pollution as a necessary sacrifice to achieve industrial growth. Instead, China is relying on financial support from other countries and international organizations. Denmark is helping Chongqing build a modern sewage treatment plant, and the World Bank is providing loans to fund major environmental efforts, including a multi-million-dollar project to clean up Beijing. Notwithstanding the significance of these international endeavors, the most important effort has to come from China itself. And that can follow only if this vast nation learns to balance growth with resource preservation and the needs of future generations with those of the present.

Sources: Biers (June 16, 1994); Radin (January 2, 1995); "Coal Dependence Has China Looking for Pollution Source" (November 3, 1996); "China's Environmental Outlook Bleakens" (January 10, 2002).

virtually all the goods and services that characterize modern living. This means that a more rational perception of environmental quality is that it represents a reduction in anthropogenic contamination to a level that is "acceptable" to society. This "acceptable" level of pollution will, of course, be different for different contaminants, but in each case, certain factors are considered in making the determination. Among these factors are the gains to human health and ecosystems, expenditures needed to achieve the reduction, availability of technology, and the relative risk of a given environmental hazard.

Although the world has made progress toward achieving environmental quality, there is still work to be done. In some parts of the world, such as Eastern Europe and developing nations, environmental pollution is extreme, and progress toward reducing the effects is slow. Application 1.2 discusses some of the environmental problems China is dealing with as it strives to advance economically.

Recognizing environmental quality as a worldwide objective has triggered an awareness of its importance over a longer time horizon. Society has begun to realize that pursuing economic growth could so adversely affect the natural resource stock that the productive capacity and overall welfare of future generations would be threatened. The potential of such an *intertemporal* trade-off has prompted a sense of obligation to the future that has materialized into two related objectives: **sustainable development** and **biodiversity.**

Sustainable Development

sustainable development
Management of the earth's resources such that their long-term quality and abundance is ensured for future generations.

Economic growth is defined as an increase in real gross domestic product (GDP). Although growth is a favorable outcome, there are long-term environmental implications, as the materials balance model suggests. Achieving an appropriate balance between economic growth and the preservation of natural resources is the essence of the objective known as **sustainable development,** which calls for managing the earth's resources to ensure their long-term quality and abundance.[6] This reminds us that the circular flow of economic activity cannot be properly understood without recognizing how it fits into the larger scheme of the natural environment. Yet only in the recent past have economists and society at large begun to accept this broader and more realistic view. For example, new methods have been proposed to capture the ecological effects of growth in macroeconomic performance measures—an issue discussed in Application 1.3.

On a much broader scale, the *Rio Declaration,* drafted at the Rio Summit, outlines 27 principles to act as guidelines for global environmental protection and economic development. Similar commitments are given in the summit's 40-chapter document, *Agenda 21,* an international agenda of comprehensive environmental goals. *Agenda 21* is dedicated in large part to sustainable development, with an emphasis on regions where achieving this objective is particularly critical, such as in developing nations.[7] In celebration of the Rio Summit's tenth anniversary, another worldwide event was held in South Africa—the Johannesburg Summit 2002. As in Rio, thousands of people attended, including heads of state, national delegates, and business leaders, all coming together to discuss issues relating to the goal of sustainable development.

Biodiversity

biodiversity
The variety of distinct species, their genetic variability, and the variety of ecosystems they inhabit.

Another environmental objective that addresses the legacy left to future generations is **biodiversity.** This refers to the variety of distinct species, their genetic variability, and the variety of ecosystems they inhabit.[8] There is still much that scientists do not know about the diversity of life on earth. Although approximately 2.5 million species have been identified, most estimates suggest that the actual number may be at least 5–10 million, and some biologists believe there may be as many as 100 million.[9] Although there are many unknowns, the consensus within the scientific community is that the variety of species on earth is important to the ecology. Beyond the preservation of a species for its own sake, it is also the case that all life on earth is inexorably connected. Hence, the loss of one species may have serious implications for others, including human life.

The longevity of any biological species can be directly threatened by exposure to pollutants or by other human actions, such as commercial or sport hunting. The major threat to biodiversity, however, is natural habitat destruction, which affects entire ecosystems. Population growth, poverty, and economic development are primarily responsible for this destruction, which includes the harvesting of tropical forests and the engineered conversion of natural landmasses into alternative uses.[10] For example, by the mid-1970s, literally half of the 200 million acres of wetlands within the contiguous United States had been lost to such uses as

[6] Council on Environmental Quality (January 1993), p. 135.
[7] Parson, Haas, and Levy (October 1992).
[8] Council on Environmental Quality (January 1993), p. 135.
[9] Raven, Berg, and Johnson (1993), p. 144.
[10] Raven, Berg, and Johnson (1993), pp. 262–63, 353–58.

1.3 *National Income Accounting and the Bias Against Environmental Assets*

Application

Assessing a nation's macroeconomic performance is an important and complex undertaking. The objective is to monetize total production and use the result to assess and monitor a nation's growth. Measures of economic activity are also useful in making international comparisons, provided that the accounting methods used are standardized. Currently, the accepted method follows the System of National Accounts (SNA) endorsed by the United Nations, a universal accounting framework for calculating such performance measures as gross domestic product (GDP) and national income.

As defined in most introductory economics texts, GDP is the monetized value of all final goods and services produced in a country each year. Although designed to be a comprehensive measure of productive activity, GDP is admittedly flawed. Difficulties in attempting to capture the value of nonmarketed goods, such as the value of "do-it-yourself" projects and the so-called underground economy, necessarily bias the measure. Beyond these well-known flaws, other inherent biases have caused some economists and government officials to question the validity of GDP as a measure of economic welfare. Among these is the absence of consideration for ecological damage and natural resource depletion associated with economic activity.

Recognizing this shortcoming, some economists have argued in favor of an environmentally adjusted national income measure that explicitly values natural resources as economic assets. The justifications for this proposal are based on the following considerations. Just as the SNA allows for the depreciation of physical capital, so too should it recognize the devaluation of natural resources associated with economic activity. If pesticides leach into underground springs, then the damage to drinking water should be monetized and recognized by the GDP measure. Likewise, if a forest is destroyed to make way for urban development, then that loss should be recorded. In both cases, assets would be depreciating, and that depreciation should be captured in the SNA to avoid a serious bias.

Proponents of environmentally adjusted performance measures also make a broader-based argument. They claim that ignoring the economic value of natural resources perpetuates society's failure to acknowledge the effect of economic growth on future generations. In fact, in some cases, current accounting practices falsely record environmental deterioration as a contribution to economic welfare. A case in point is increased medical spending associated with the effects of a toxic chemical leak, which ironically *elevate* a nation's GDP. An explicit accounting of resource depletion and environmental damage would correct this outcome and restore accuracy to national product and income accounts.

Officials in many nations are aware of this environmental bias in national income accounts, and some are compiling internal accounts that explicitly value natural resources and the negative effects of environmental pollution. In the United States, the Department of Commerce's Bureau of Economic Analysis (BEA), which is charged with measuring the nation's output, began to investigate this issue during the early 1990s. In 1994, Congress halted the BEA's investigation and directed the National Research Council to form a panel to independently study the matter and assess the work to date of the BEA. The panel concluded that the BEA should continue to develop measures that will better incorporate nonmarketed environmental assets into national product accounts.

Despite such efforts in the United States and abroad, there is still no universal method to correct the bias against environmental assets in the SNA. In general, the official response at the United Nations has been limited to developing guidelines on how individual nations should prepare their own natural resource accounts. Many believe that such a position is inadequate so long as the SNA is the universally accepted approach to measuring a nation's economic well-being, and so the issue remains unresolved.

Sources: Repetto (September 1992); Repetto (June 1992); Nordhaus (November 1999).

Table 1.2 ***Threatened and Endangered Species in the United States as of July 2002***

	Endangered	Threatened	Total
Mammals	65	9	74
Birds	78	14	92
Herptiles (reptiles, amphibians)	26	31	57
Fish	71	44	115
Invertebrates (snails, clams, crustaceans, insects, arachnids)	148	31	179
Plants	597	147	744
Total	985	276	1,261

Source: U.S. DOI, Fish and Wildlife Service (July 31, 2002), as cited in *Information Please Almanac*, **http://www.infoplease.com/ipa/A0004713.html.**

agriculture and urban development.[11] Biodiversity can also be jeopardized by habitat alteration, which is often attributed to environmental pollution. For example, acid rain has been linked to changes in the chemical composition of rivers and lakes as well as to forest declines in Europe and North America. Such disturbances in the natural conditions to which biological life has become adapted can pose a threat to the longevity of these species.

The extent of biodiversity loss is not known. There are, however, indications that concern about diversity loss is warranted. As of 2002, over 1,250 plant and animal groups have been classified as endangered or threatened in the United States, as shown in Table 1.2[12] That this issue is one of global concern is evidenced by the *Convention on Biodiversity,* executed by 153 nations at the Rio Summit. Among the convention's mandates are devising measures that identify which species are in decline and discovering why the declines have occurred.

Collectively, the goals of environmental quality, sustainable development, and biodiversity set an ambitious agenda. As such, all of society must work toward developing effective environmental policy initiatives. Central to this effort is a planning process in which public officials, industry, and private citizens participate. In the context of environmental problems, this process involves a series of decisions about assessing environmental risk and responding to it, as the following overview explains.

Environmental Policy Planning: An Overview

An important observation to make about environmental policy planning is that it involves the interdependence of many segments of society, including government agencies, private industry, the scientific community, and environmentalists. Each group of participants, albeit from a different vantage point, plays a significant role in formulating policy, and each offers much-needed expertise to the outcome.

Policy Planning in the United States

We can categorize the individuals who are instrumental to environmental policy planning in many ways. Using the broadest classification, there are two major groups: the public sector and the private sector. In the United States, the Environmental Protection Agency (EPA) acts

[11] U.S. EPA (August 1988), p. 57.
[12] U.S. DOI, Fish and Wildlife Service (July 31, 2002).

Figure 1.4 ***Parties Involved in Environmental Policy Planning***

Source: Adapted from Vaupel (1970), p. 75, fig. 5.3.

as a sort of liaison between the various constituents of each sector, as shown in Figure 1.4. Established in 1970, the EPA was created by President Nixon from various components of existing federal agencies and executive departments.[13] Today, the EPA operates as an independent agency headed by a president-appointed administrator who oversees its vast infrastructure.

As Figure 1.4 implies, some environmental issues fall under the jurisdiction of federal agencies other than the EPA. For example, policies aimed at occupation-specific environmental hazards, such as worker exposure to asbestos, are assigned to the Occupational Safety and Health Administration (OSHA). Issues relating to the contamination of foods or the use of food additives such as saccharine are the responsibility of the Food and Drug Administration (FDA).

The National Environmental Policy Act

The National Environmental Policy Act (NEPA) of 1969 directs the integration of effort across administrative agencies, executive departments, and branches of government. This legislation guides the formulation of all U.S. federal environmental policy and also requires that the environmental impact of public policy proposals be formally addressed. Among NEPA's provisions is a requirement for an Environmental Impact Statement (EIS) on proposals for legislation or major federal actions, which according to law must entail the following:

- **(i)** The environmental impact of the proposed action,
- **(ii)** Any adverse environmental effects which cannot be avoided should the proposal be implemented,
- **(iii)** Alternatives to the proposed action,
- **(iv)** The relationship between local short-term uses of man's environment and the maintenance and enhancement of long-term productivity, and
- **(v)** Any irreversible and irretrievable commitments of resources which would be involved in the proposed action should it be implemented.

[13] The EPA's mission statement and strategic plan can be accessed at **http://www.epa.gov/epahome/epa.html.**

Since the NEPA was passed in 1969, literally thousands of EISs have been completed by a variety of federal agencies. During 1997 alone, 498 EISs were completed by major federal agencies and executive departments.[14]

In the United States and around the world, environmental policy planning relies on careful research and analysis, which in turn depends on individuals with expertise in a variety of disciplines, among them biology, chemistry, economics, law, and medicine. Input from these and other fields is used to evaluate data and make decisions that ultimately lead to specific policy prescriptions. The underlying tool that guides this policy planning process is **risk analysis,** which consists of two decision-making procedures—**risk assessment** and **risk management.**

Risk Assessment

At any point in time, a number of environmental objectives must be met with a limited amount of economic resources. This means that as problems are identified, they have to be prioritized. In general, this is done through scientific assessment of the relative risk to human health and the ecology of a given environmental hazard—a procedure known as **risk assessment.** The assessment must determine whether a causal relationship exists between the identified hazard and any observed health or ecological effects. If causality is determined, scientists then attempt to quantify *how* the effects change with increased exposure to the hazard. These findings are critical, because they determine whether or not a policy response is necessary, and, if so, how immediate and how stringent that policy should be.

risk assessment
Qualitative and quantitative evaluation of the risk posed to health or the ecology by an environmental hazard.

Risk Management

Assuming that the risk assessment findings warrant it, the planning process enters its next phase, called **risk management.** This refers to the decision-making process of evaluating and choosing from alternative responses to environmental risk. In a public policy context, risk responses refer to various types of control instruments, such as a legal limit on pollution releases or a tax on pollution-generating products. The objective of risk management is clear—to choose a policy instrument that reduces the risk of harm to society. What is less obvious is how public officials determine the level of risk society can tolerate and on what basis they evaluate various policy options.

risk management
The decision-making process of evaluating and choosing from alternative responses to environmental risk.

Policy Evaluation Criteria

A number of risk management strategies have been devised to guide these important policy decisions. These strategies use criteria to evaluate policy options. In general, these criteria are based on measures of risk, costs, or benefits—either singularly or in comparison to one another. Two criteria that are economic in motivation are **allocative efficiency** and **cost-effectiveness.**

allocative efficiency
Requires that resources be appropriated such that the additional benefits to society are equal to the additional costs incurred.

cost-effectiveness
Requires that the least amount of resources be used to achieve an objective.

- **Allocative efficiency** requires that resources be appropriated such that the additional benefits to society are equal to the additional costs incurred.
- **Cost-effectiveness** requires that the least amount of resources be used to achieve an objective.

In practice, the choice of criteria is mandated, or at least implied, by law. For example, some provisions in U.S. legislation prohibit cost considerations in policy formulation, implicitly blocking either an efficient or a cost-effective outcome—a result that is frequently debated in the literature and one that we will analyze in later chapters.

Although both efficiency and cost-effectiveness are rooted in resource allocation, another criterion, called **environmental equity,** has a different perspective. It considers the fairness

environmental equity
Concerned with the fairness of the environmental risk burden across segments of society or geographic regions.

[14] Council on Environmental Quality (1997), pp. 51, 534–45; Council on Environmental Quality (1998), pp. 355–58.

of the risk burden across geographic regions or segments of the population. Policy officials have begun to place an increasing emphasis on environmental equity. In fact, during his administration, President Clinton signed Executive Order 12898, which specifically directed all federal agencies to incorporate environmental justice into their policy- and decision-making activities. Pollution problems such as exposure to hazardous waste sites and urban smog are locationally oriented, and there is some evidence to suggest that income and cultural factors are linked to the locational decisions of certain population groups. Consequently, public officials are trying to rectify existing inequities and to incorporate the equity criterion into future decision making.[15]

Government's Overall Policy Approach

An important element of risk management is the regulatory approach selected by government. Policies calling for direct regulation of polluting are indicative of a **command-and-control approach.** This form of regulation uses rules or standards to control the release of pollution. In practice, standards either set a maximum on the amount of residuals polluters may release or designate an abatement technology that all sources must use. In either case, polluters have little or no flexibility in deciding how they comply with the law. Every polluter must meet the same standard regardless of its location, access to resources, existing emission levels, or technology. The command-and-control approach has been the predominant approach used in the United States over the past several decades, but in recent years more economic incentives have been integrated into strategic policy plans, suggesting a shift toward a **market approach** to policy.

command-and-control approach
A policy that directly regulates polluters through the use of rules or standards.

The **market approach** is incentive based, meaning it attempts to encourage conservation practices or pollution reduction strategies rather than force polluters to follow a specific rule. Many policy instruments can achieve this result, such as a charge on pollutant releases or a tax levied on pollution-generating commodities. What all such instruments have in common is that they tap into natural market forces so that polluters' optimizing decisions will benefit the environment.

market approach
An incentive-based policy that encourages conservation practices or pollution reduction strategies.

Think about how private firms are influenced by profit. What a market approach does is strategically use this information to design environmental policy. For example, if a profit-maximizing firm were discharging a chemical into a river, a market approach might be to charge that polluter a fee for every unit of chemical released. In so doing, the firm would have to pay for the damage it caused, which would erode its profits. This tactic is sometimes called the **"polluter-pays principle."** The expected outcome is that a rational, profit-maximizing firm will reduce the amount of chemicals it releases, using the least-cost method available. The favorable outcomes are that society enjoys the benefit of a cleaner environment and that the associated costs to achieve that gain are minimized.

That such market-based strategies can be effective is validated by experience all over the world. In the United States and other industrialized nations, more market incentives are being integrated within the conventional command-and-control policy approach. In fact, in a 1998 survey of 24 nations, the Organisation for Economic Co-operation and Development (OECD), which supports the polluter-pays principle, found that approximately 270 economic instruments were in use across these countries.[16]

Setting the Time Horizon

Another element of risk management decision making is determining the most effective time plan for policy initiatives. One approach is to target policy at more immediate or *short-term* problems. These types of initiatives are referred to as **management strategies,** since their purpose is to manage an existing problem. Here, the intent is *ameliorative.* In terms of the materials balance model, such strategies attempt to reduce the damage from the flow of residuals back to the environment.

management strategies
Methods that address existing environmental problems and attempt to reduce the damage from the residual flow.

[15] For a collection of articles on the issue of environmental equity, see Heritage (March/April 1992).
[16] OECD (1999a), appendix 1, pp. 100–108.

Application

1.4 *The Coalition for Environmentally Responsible Economies:* Valdez *Principles*

http://

In the fall of 1989 the Coalition for Environmentally Responsible Economies (CERES) was formed. Its purpose is simple and direct: to encourage the corporate sector to assume full responsibility for the environmental consequences of its actions. The coalition is composed of 14 environmental organizations, including the Sierra Club and the National Wildlife Association, plus the 325-member Social Investment Forum (SIF), which is a national trade association of investment professionals concerned with social issues. To accomplish its goal, CERES embarked on a plan to draft a formal pledge document through which private firms would commit to environmental objectives.

The membership collaborated to define specific standards to which signatories of the compact would be held accountable. Biodiversity, sustainable development, and pollution prevention were among the issues considered for the final document. Ultimately, the group settled on a set of 10 guidelines, originally called the "*Valdez* Principles," after the infamous March 1989 Alaskan oil spill.

The next step was critical. The CERES membership invited thousands of corporations to sign the environmental pledge. With endorsements from groups that collectively control $150 billion in investment funds, CERES believed its financial clout would be effective in garnering and rewarding corporate participation. Despite a strong effort, however, response has been less than enthusiastic. As of 2001, only 60 companies have agreed to commit formally to what are now known as the "CERES Principles," although some of these are major corporations such as Ford Motor Company and Nike, Inc. To learn more about which firms have made this commitment, visit **http://www.ceres.org/about/endorsing_companies.htm.** The 10 guide-lines that make up the CERES Principles are itemized in the following list, and more detail is available at **http://www.ceres.org/our_work/principles.htm.**

The CERES Principles

1. Protection of the biosphere
2. Sustainable use of natural resources
3. Reduction and disposal of waste
4. Energy conservation
5. Risk reduction
6. Safe products and services
7. Environmental restoration
8. Informing the public
9. Environmental directors and managers
10. Audit and reports

Sources: Ohnuma (March/April 1990); Coalition for Environmentally Responsible Economies (1989), as reported in the Canadian Institute of Chartered Accountants (1992), pp. 7–8, table 2.1; Parrish (February 4, 1994).

pollution prevention
A long-term strategy aimed at reducing the amount or toxicity of residuals released to nature.

An alternative approach deals with the potential of future deterioration and is therefore preventive in purpose. This *long-term* strategy is referred to as **pollution prevention,** and it is implemented by reducing the flow of residuals back to nature and/or by minimizing the harmful components of residuals, such as toxic chemicals. Preventive strategies are becoming more prevalent in U.S. policy. In fact, the United States made a formal commitment to pollution prevention when Congress enacted the Pollution Prevention Act of 1990. There also have been important private initiatives that support preventive strategies. One such effort is exemplified in a corporate pledge to guidelines originally known as the "*Valdez* Principles," the subject of Application 1.4.

Conclusions

Concerns about the risks of pollution and the threat of natural resource depletion have been expressed by private citizens, the business community, and governments all over the world. In large part, this perspective has come from a growing awareness of the delicate balance between nature and economic activity. The materials balance model illustrates the strength of this relationship and the consequences of naive decision making that ignores it.

Recognizing the implications, many nations have made measurable progress in identifying problems and setting an agenda to address them. As that work continues, comparable efforts are under way to develop and implement solutions. To that end, policy reform, collaborative arrangements between government and industry, and international summits are being aimed at resolving environmental problems. Research scientists are working to learn more about the ecology, the diversity of species, and environmental risks. Laws are being changed to incorporate more preventive measures as well as incentive-based instruments that encourage pollution reduction and resource conservation. As this process evolves, society is changing the way it thinks about the earth's resources, the long-term consequences of its decisions, and its obligation to the future.

Economics has much to contribute to this evolution, in large part because of the interdependence between market decisions and nature. The fundamental concepts of price and optimizing behavior can be used to analyze the effectiveness of environmental policy and to develop alternative solutions. As we explore the discipline of environmental economics, we will use these same concepts to study the effects of environmental pollution as well as the public policy and private responses to the associated risks.

As this chapter suggests, there is a lot of ground to cover, but the importance and relevance of the issues justify the effort. Environmental economics, much like the problem it examines, presents both a challenge and an opportunity—a characterization drawn from the opening address to the Rio Summit:

> The Earth Summit is not an end in itself, but a new beginning. . . . The road beyond Rio will be a long and difficult one; but it will also be a journey of renewed hope, of excitement, challenge and opportunity, leading as we move into the 21st century to the dawning of a new world in which the hopes and aspirations of all the world's children for a more secure and hospitable future can be fulfilled.[17]

Summary

- The circular flow model is the basis for modeling the link between economic activity and nature.
- The relationship between economic activity and the natural environment is illustrated by the materials balance model.
- The first law of thermodynamics asserts that matter and energy can be neither created nor destroyed. The second law of thermodynamics states that the conversion capacity of nature is limited.
- Pollution refers to the presence of matter or energy, whose nature, location, or quantity produces undesired environmental effects. Some pollutants are natural; others are anthropogenic.
- Sources of pollution are sometimes grouped into mobile and stationary sources. Another common classification is to distinguish point sources from nonpoint sources.
- Local pollution problems are those whose effects do not extend far from the polluting source.
- Regional pollution has effects that extend well beyond the source of the pollution.
- Global pollution problems are those whose effects are so extensive that the entire earth is affected.

---

[17] Maurice F. Strong, UNCED Secretary-General, in his opening address to the UN Conference on Environment and Development, Rio de Janeiro, Brazil, June 3, 1992, as cited in Haas, Levy, and Parson (October 1992), p. 7.

- Among the most critical environmental objectives are environmental quality, sustainable development, and biodiversity.
- The National Environmental Policy Act (NEPA) guides the formulation of U.S. federal environmental policy and requires that the environmental impact of *all* public policy decisions be formally addressed.
- The underlying tool that guides policy planning is risk analysis, which consists of two decision-making procedures: risk assessment and risk management.
- Risk assessment is a scientific evaluation of the relative risk to human health or the ecology of a given environmental hazard. Risk management refers to the process of evaluating and selecting an appropriate response to environmental risk.
- Two economic criteria used in risk management are allocative efficiency and cost-effectiveness.
- Another criterion, environmental equity, considers the fairness of the risk burden across geographic regions or across segments of the population.
- A command-and-control policy approach uses limits or standards to regulate environmental pollution.
- A market approach uses economic incentives to encourage pollution reduction or resource conservation.
- Management strategies have a short-term orientation and are ameliorative in intent.
- Pollution prevention strategies have a long-term perspective and are aimed at precluding the potential for further environmental damage.

Key Concepts

circular flow model
materials balance model
natural resource economics
residual
environmental economics
first law of thermodynamics
second law of thermodynamics
pollution
natural pollutants
anthropogenic pollutants
stationary source
mobile source
point source
nonpoint source
local pollution
regional pollution
global pollution
environmental quality
sustainable development
biodiversity
risk assessment
risk management
allocative efficiency
cost-effectiveness
environmental equity
command-and-control approach
market approach
management strategies
pollution prevention

Use the Key Concepts listed above to begin your search for additional articles and information using the InfoTrac College Edition® database.

Review Questions

1. a. State how each of the following factors affects the materials balance model: (i) population growth; (ii) income growth; (iii) increased consumer recycling; (iv) increased industrial recycling; (v) increased use of pollution prevention technologies.
   b. Assume that stringent pollution controls are placed on the flow of residuals released into the atmosphere. According to the materials balance model, what does this imply about the residual flows to the other environmental media and/or the flow of inputs into the economy?
2. Faced with the oil crisis of the mid-1970s, the U.S. Congress instituted Corporate Average-Fuel Economy (CAFE) standards. (For an overview of the CAFE standards, visit the Web site **http://www.ita.doc.gov/td/auto/cafe.html** at the U.S. Department of Commerce, International Trade Administration.) These standards were intended to increase the miles per gallon (MPG) of automobiles.
   a. Briefly describe the expected environmental effect of increasing the MPG of automobiles, holding all else constant.
   b. Serious criticism has been lodged against the CAFE standards because U.S. automakers responded by using more plastics in automobiles (to make the cars lighter in weight) to meet the more restrictive standards. Explain how the use of this technology affects your answer to part (a). Are there any other relevant issues associated with this manufacturing decision?

http://

3. Use your knowledge of economic principles to discuss how the market premise operates under the "polluter-pays principle."
4. Reconsider the problem of U.S. wetlands loss and the implications for biological diversity. Briefly contrast how a command-and-control policy approach to this problem would differ in intent and implementation from a market approach.
5. In your view, should employment in the forest industry be sacrificed to save the northern spotted owl? If so, by how much, and how should the extent of employment loss be determined? If not, why not?

Additional Readings

Bullard, Robert D. "Overcoming Racism in Environmental Decisionmaking." *Environment* 36 (May 1994), pp. 11–20, 39–44.

Commoner, Barry. "Economic Growth and Environmental Quality: How to Have Both." *Social Policy* (summer 1985), pp. 18–26.

Cruz, Wilfrido, Mohan Munasinghe, and Jeremy Warford. "Greening Development: Environmental Implications of Economic Policies." *Environment* 38 (June 1996), pp. 6–11, 31–38.

Fialka, John J. "Kyoto Treaty Moves Ahead Without U.S." *Wall Street Journal,* November 12, 2001, p. A4.

Haas, Peter M., Marc A. Levy, and Edward A. Parson. "Appraising the Earth Summit: How Should We Judge UNCED's Success?" *Environment* 34 (October 1992), pp. 6–11, 26–33.

Hahn, Robert W. "The Impact of Economics on Environmental Policy." *Journal of Environmental Economics and Management* 39 (May 2000), pp. 375–99.

Landy, Marc K., Marc J. Roberts, and Stephen R. Thomas. *The Environmental Protection Agency: Asking the Wrong Questions: Nixon to Clinton.* New York: Oxford University Press, 1994.

Metrick, Andrew, and Martin L. Weitzman. "Conflicts and Choices in Biodiversity Preservation." *Journal of Economic Perspectives* 12 (summer 1998), pp. 21–34.

Munasinghe, Mohan, ed. *Macroeconomics and the Environment.* Northampton, MA: Elgar, 2002.

Nordhaus, William D. "New Directions in National Economic Accounting." *American Economic Review, Papers and Proceedings* 90 (May 2000), pp. 259–263.

Polasky, Stephen. *The Economics of Biodiversity Conservation.* Burlington, VT: Ashgate, 2002.

Raustiala, Kal, and David G. Victor. "The Future of the Convention on Biological Diversity." *Environment* 38 (May 1996), pp. 16–20, 37–45.

Smith, Emily T. "Growth vs. Environment: In Rio Next Month, a Push for Sustainable Development." *Business Week,* May 11, 1992, pp. 66–75.

Solow, Robert M. "Sustainability: An Economist's Perspective." In Robert N. Stavins (ed.), *Economics of the Environment: Selected Readings,* pp. 131–38. New York: Norton, 2000.

Spofford, Walter O., Jr. "Chongqing: A Case Study of Environmental Management During a Period of Rapid Industrial Development." *Resources* 123 (spring 1996), pp. 13–16.

Stavins, Robert N. *Environmental Economics and Public Policy.* Northampton, MA: Elgar, 2001.

Related Web Sites

Coalition for Environmentally Responsible Economies (CERES)	**http://www.ceres.org**
Council on Environmental Quality	**http://www.whitehouse.gov/CEQ**
Council on Environmental Quality, NEPAnet (includes access to the full text of NEPA)	**http://ceq.eh.doe.gov/nepa/nepanet.htm**
President's Commitment to Environmental Protection	**http://www.whitehouse.gov/infocus/environment/**
United Nations Environment Programme (UNEP)	**http://www.unep.org**
United Nations Framework Convention on Climate Change	**http://unfccc.int/**
U.S. Department of Commerce, International Trade Administration CAFE standards	**http://www.ita.doc.gov/td/auto/cafe.html**
U.S. EPA Mission Statement and Strategic Plan	**http://www.epa.gov/epahome/epa.html**
U.S. EPA Office of Environmental Justice (OEJ)	**http://www.epa.gov/compliance/environmentaljustice/**
U.S. EPA Terms of Environment	**http://www.epa.gov/OCEPAterms**
The World Bank Group (information on economic development and the environment)	**http://www.worldbank.org/environment**

Chapter 2

Modeling the Market Process: A Review of the Basics

According to the materials balance model, environmental problems are directly linked to market activity. The basic decisions made by consumers and firms affect both the abundance and the quality of the earth's natural resource stock. Because environmental economics is concerned with resource damage, we need to develop a thorough understanding of how market activity gives rise to polluting residuals and why market forces cannot solve the problem. From an economic perspective, environmental pollution is characterized as a market failure. Hence, environmental economics uses market failure models to analyze the problem and to identify solutions, but these models rely on a solid understanding of the market process itself.

To that end, in this chapter we review the essential components of a market and the basic concepts used in microeconomic analysis. Essentially, we focus on the operation of the circular flow model, which is central to the materials balance paradigm. The fundamentals of supply and demand are reviewed to reestablish a good grasp of market behavior, the motivations for consumer and firm decision making, and price determination.

The context for our review is a hypothetical market for bottled water in which competitive conditions are assumed. Through an analysis of market equilibrium, we discuss the economic criterion of allocative efficiency, a notion used throughout the study of environmental economics to evaluate public and private responses to pollution problems. From there, we develop welfare measures, which are useful in evaluating the effect of environmental policy on society. All the analytical and modeling tools presented here will serve as the foundation for our study of market failure in Chapter 3.

Market Models: The Fundamentals

Defining the Relevant Market

market
The interaction between consumers and producers to exchange a well-defined commodity.

In economic analysis, the concept of a market is given a broader definition than in everyday usage. Specifically, a **market** refers to the interaction between consumers (or buyers) and producers (or sellers) for the purpose of exchanging a well-defined commodity. This more theoretical definition is purposefully general and abstract, since an economic market is meant to refer to the process of exchange and the conditions underlying that exchange for a broad range of economic activities. For example, this definition is as relevant to the purchase and sale of labor in the factor market as it is to the exchange of groceries at a supermarket. As we will discover in upcoming chapters, it can even be applied to an analysis of pollution control in the "market for environmental quality." Hence, one of the more critical steps in economic analysis is defining the market context for the good or service under investigation.

Specifying the Market Model

Once the relevant market has been defined, a model of that market and its characteristics must be specified. The form of the model varies with the objective of the prospective study and its level of complexity. Simple qualitative relationships among economic variables can be modeled using a two-dimensional graph. To quantify these relationships, models are refined through the use of equations or functions.[1] Formal testing is accomplished through empirical analysis of these theorized relationships using real-world data.[2]

The Model of Supply and Demand: An Overview

By definition, a market exchange for any commodity contains two sets of independent decision makers: buyers and sellers. Each is motivated by different objectives, and each is influenced and even constrained by different economic factors. The decisions of sellers (or producers) are modeled through a **supply** function, while consumers' decisions are modeled through a **demand** function. When considered simultaneously, the resulting market model of supply and demand determines equilibrium output and price.

Purpose of the Model

The primary objective of the supply and demand model is to facilitate an analysis of market conditions and any observed changes in price. An investigation of price movements can discern the presence of shortages and surpluses, the existence of resource misallocations, and the economic implications of government policy initiatives. For example, environmental economists can use supply and demand models to investigate the effectiveness of a gasoline tax on reducing gas consumption as a way to help improve urban air quality. By studying the associated changes in market conditions and movements in gasoline prices, economists can determine how consumption patterns are affected, how the tax burden is shared between the consumer and the producer, and how income distribution is affected.

More complex analysis is necessary when the market system fails to operate properly. In these instances, the conventional model of supply and demand must be modified to account for those conditions that weaken the operation of market forces. Economic theory suggests that the persistence of environmental problems such as urban smog and water pollution are the result of failures or breakdowns in the market system. Of course, to understand why this is so and to begin to formulate solutions, we must have a good command of the market process, the underlying supply and demand conditions, and the mechanisms of the price system.

Building a Basic Model: Competitive Markets for Private Goods

To develop a basic model of supply and demand, we make a number of assumptions. First, a competitive *goods* market is assumed, which is characterized primarily by (1) a large number of independent buyers and sellers with no control over price, (2) a homogeneous or standardized product, (3) the absence of entry barriers, and (4) perfect information. Second, the market for *resources* also is assumed to be competitive. This implies that the individual firm has no control over input prices—a result that will simplify the model. Finally, the output being exchanged in the market is assumed to be a **private good.** A **private good** is a commodity that has two characteristics—rivalry in consumption and excludability. This means that consumption of the good by one person precludes that of another, and the benefits of consumption are exclusive to that single consumer. This assumption of a private (as opposed to a public[3]) good is critical to a conventional analysis of quantity and price determination.

private good
A commodity that has two characteristics, rivalry in consumption and excludability.

[1] For the most part, when we use equations in the text, we specify linear relationships for simplicity.
[2] In later chapters, we examine the results of some important empirical studies in environmental economics.
[3] Public goods are excluded from our analysis because they represent a type of market failure. These are commodities such as national defense and clean air, whose consumption is both nonrival and nonexclusive. That is, once a public good is provided to one consumer, it is difficult or very costly to prevent others from sharing in its consumption. A more formal development of the theory of public goods as it pertains to environmental issues is presented in Chapter 3.

Application

2.1 *Consumer Demand and Environmental Issues: What Really Matters?*

According to a 1991 survey of 410 New York adults, environmental safety has become an important influence on consumer decision making. About 93 percent of the respondents in the survey believe that protecting the environment is a "very important factor" in their purchasing decisions. However, although this influence on consumer demand is a viable one, only 1 in 10 survey respondents places environmental safety as a top priority. Product price and quality continue to be the most significant determinants of consumers' buying decisions. These findings have been supported by other sources. A 1994 national poll, for example, indicates that over 80 percent of Americans are in favor of more stringent environmental laws, yet only 54 percent are willing to incur the economic costs of achieving improved environmental quality. However, attitudes apparently are changing. A 2001 Gallup poll shows that almost 75 percent of Americans would be willing to pay higher product prices to cover the added costs associated with more stringent emission standards.

Notwithstanding the influence of price and quality on American buyers, the survey of New York adults *does* indicate that consumers will switch brands in favor of environmentally safer products. This change in preferences implies a shift in market demand toward these products and away from less safe substitutes. Some of the nondurable consumer product groups affected by this phenomenon are detergents, diapers, aerosol sprays, and cleaners. Overall, 44.6 percent of all survey participants say they have switched brands for environmental reasons, with a higher proportion reported for the 31- to 45-year-old age group. Gender differences also exist with respect to this switching phenomenon. Some 49.4 percent of females in the survey express a willingness to substitute toward more environmentally safe brands, whereas the comparable proportion for males is 37.9 percent.

Recognizing the market opportunity, manufacturers are responding to this trend by promoting their brands as "environmentally safe." The survey provides some information about the relative success of this effort across competitive commodities. Brand names such as Clorox, Arm & Hammer, Tide, L'Oreal, and Revlon are among those perceived by consumers as associated with environmentally superior products. This brand recognition suggests that the advertised commitment to environmental concerns of certain manufacturers is perceived as more credible than that of other firms.

Just as demand theory dictates, consumer decisions depend on a variety of market conditions and product characteristics. As these survey data show, price continues to be a major determinant of demand, particularly in an economic slowdown. Nonetheless, environmental concerns have some effect on consumers' buying habits, at least for some nondurable products. In response, manufacturers are adjusting production and marketing strategies to meet these demand changes. Of course, firms will have to monitor the degree of market responsiveness to environmental issues to determine if the trend continues.

Sources: Manly (March 23, 1992); Fialka (August 27, 1997); Dunlap and Saad (April 16, 2001).

Market Demand

Demand refers to the market response of consumers, who adjust their purchasing decisions to maximize their satisfaction, or what economists term utility. There are many factors that influence consumers' decisions. However, because a key objective of market analysis is price determination, the demand function is specified as the relationship between quantity demanded and price, holding constant all other variables that influence this decision. Economists use the Latin phrase *ceteris paribus*, abbreviated *c.p.*, to mean "holding all else con-

Table 2.1 *Single Consumer's Demand Data for Bottled Water*

Price P	Quantity Demanded (bottles/month) $q_d = -4P + 20$
\$0.50	18
1.00	16
1.50	14
2.00	12
2.50	10
3.00	8
3.50	6
4.00	4
4.50	2
5.00	0

stant." Hence, **demand** is formally defined as the quantities of a good the consumer is willing and able to purchase at some set of prices during some discrete time period, *c.p.* The consumer's "ability to pay" refers to the income constraint that limits consumer choice. The "willingness to pay" is the value or benefit the consumer expects to receive from consumption of the commodity. In fact, this willingness to pay, or **demand price,** is considered a measure of the **marginal benefit (*MB*)** associated with consuming another unit of the good.

demand
The quantities of a good the consumer is willing and able to purchase at a set of prices during some discrete time period, *c.p.*

The key economic variables that are held constant when specifying demand are the wealth and income of the consumer, the prices of related goods (i.e., substitutes and complements), preferences, and price expectations. A change in any of these variables alters the entire price-quantity relationship, which represents a *change in demand.* This is distinct from the effect on consumption of a change in price, which causes only a *change in quantity demanded.* For example, consumers typically buy more of a good when it goes on sale. This is simply a change in the quantity demanded of the product in response to a price change. On the other hand, if, for example, consumers' tastes change such that they desire more of a product at *all* possible prices, the result would be a change in demand. A case in point is the observed shift in consumer preferences toward "environmentally friendly" products at all possible prices. Application 2.1 discusses survey data that speak to this phenomenon and its impact on market demand.

The Law of Demand

Under conventional circumstances, the relationship between quantity demanded and price is an inverse one, which is referred to as the **Law of Demand.** This means that a *rise* in price is associated with a *fall* in quantity demanded, *c.p.*, and the converse is true. This is a highly intuitive theory, because we expect consumers to view price as an obstacle that limits consumption, given their income constraints.[4]

Law of Demand
There is an inverse relationship between price and quantity demanded of a good, *c.p.*

Modeling Individual Demand

To illustrate the Law of Demand, Table 2.1 presents hypothetical data for an individual's demand for one-liter bottles of water. The values show the utility-maximizing quantity decisions in a one-month period for a set of prices ranging from \$0.50 to \$5.00 per bottle. Notice how the inverse relationship between quantity demanded and price holds throughout. Although

[4]There are more advanced explanations of the Law of Demand, specifically, the income effect and the substitution effect. Interested readers can refer to any microeconomics text for a discussion of these concepts.

Figure 2.1 *One Consumer's Demand (d) for Bottled Water*

This model of one consumer's demand (d) uses the *inverse* form of the function $P = -0.25q_d + 5$, where -0.25 is the slope and $+5$ is the vertical intercept. The negative slope illustrates the Law of Demand.

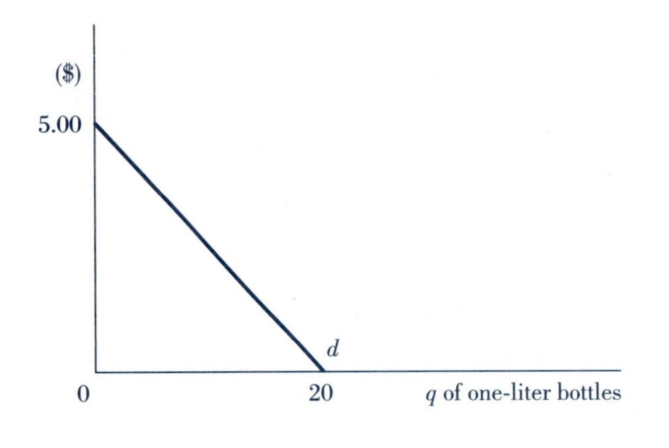

these data give only a limited sampling of price-quantity pairs, an equation of the same relationship models all possible price-quantity responses for the consumer. In this case, the demand function is the simple linear relationship $q_d = -4P + 20$, where q_d signifies the quantity demanded of a single individual and P is the dollar price per one-liter bottle. Notice that if any of the P values from Table 2.1 are substituted into the right-hand side of the equation, the corresponding q_d values can be obtained algebraically. For example, if $P = \$3$, then $q_d = -4(3) + 20 = 8$.

By convention, the graphical depiction of demand uses the *inverse* form of the function, that is, $P = f(q_d)$. Solving the equation for P in terms of q_d gives $P = -0.25q_d + 5$. The graph of this single consumer's demand curve (d) is shown in Figure 2.1.

Deriving Market Demand from Individual Demand Data

market demand for a private good
The decisions of all consumers willing and able to purchase a good, derived by *horizontally* summing the individual demands.

For most applications in economics, the collective decision making of all consumers in a given market is more relevant than that of a single consumer. Thus, the more appropriate concept is **market demand,** representing all consumers who are willing and able to purchase the commodity. Market demand is derived by summing over the individual demand data. For private goods, this summing is done over the quantity levels at each demand price and is therefore referred to as "horizontal summing," because quantity is conventionally plotted on the horizontal axis.[5] Even though price is the same for all consumers, quantity decisions are variable because of differences in other factors, such as income, wealth, tastes, and expectations.

For simplicity, we first illustrate this summing procedure by adding the demand of only one other consumer to the model specified previously. Individual demand data for two hypothetical consumers, Consumer 1 and Consumer 2, are given in Table 2.2. Notice that each consumer makes unique decisions about quantity because each has a unique income level, stock of wealth, preference ordering, and so forth. The aggregate demand for these two individuals is found by summing the two quantity columns at each price level, the result of which is shown in the far right-hand column of the table. The same method can be applied to an algebraic

[5] Horizontal summing is characteristic of **private goods.** Because consumption of such goods is excludable, each individual is able to choose his or her own quantity. As we discuss in Chapter 3, this is *not* true for public goods, an outcome with important environmental, economic implications.

Table 2.2 *Combined Demand Data for Bottled Water for Two Consumers (Consumer 1 and Consumer 2)*

Price P	Quantity Demanded by Consumer 1 (bottles/month) $q_{d1} = -4P + 20$	Quantity Demanded by Consumer 2 (bottles/month) $q_{d2} = -2P + 10$	Combined Quantity Demanded by Both Consumers (bottles/month) $q_{d(1+2)} = (q_{d1} + q_{d2}) = -6P + 30$
$0.50	18	9	27
1.00	16	8	24
1.50	14	7	21
2.00	12	6	18
2.50	10	5	15
3.00	8	4	12
3.50	6	3	9
4.00	4	2	6
4.50	2	1	3
5.00	0	0	0

Table 2.3 *Market Demand Data for Bottled Water*

Price P	Quantity Demanded by Consumers 1 and 2 (bottles/month) $q_{d(1+2)} = -6P + 30$	Quantity Demanded by Consumers 3 through 100 (bottles/month) $q_{d(3,...,100)} = -94P + 1{,}120$	Market Demand (bottles/month) $Q_D = -100P + 1{,}150$
$0.50	27	1,073	1,100
1.00	24	1,026	1,050
1.50	21	979	1,000
2.00	18	932	950
2.50	15	885	900
3.00	12	838	850
3.50	9	791	800
4.00	6	744	750
4.50	3	697	700
5.00	0	650	650

model by adding each pair of corresponding terms in the two demand equations. This is shown by.[6]

$$\begin{array}{lrl} & \text{Demand for Consumer 1:} \quad q_{d1} &= -4P + 20 \\ + & \text{Demand for Consumer 2:} \quad q_{d2} &= -2P + 10 \\ \hline & \text{Demand for Consumers 1 and 2:} \quad q_{d(1+2)} = (q_{d1} + q_{d2}) &= -6P + 30 \end{array}$$

To derive the market demand, we use the same approach to aggregate across *all* consumers. Maintaining the characteristics of a competitive market, we assume there are 100 consumers in the market for bottled water. The hypothetical data are shown in Table 2.3. For each price,

[6] As an exercise, verify that the price-quantity pairs given in Table 2.2 satisfy each corresponding algebraic expression.

Figure 2.2 *Market Demand (D) for Bottled Water*

This model of the market demand (*D*) for bottled water uses the *inverse* market demand function $P = -0.01Q_D + 11.5$, which represents the aggregate decisions of all consumers in the market. The market demand curve is derived using a horizontal summing procedure across all individual demand curves. Thus, at each price level, the horizontal distance between the vertical axis and *D* is exactly equal to the sum of the analogous horizontal intervals for each of the 100 individual demand (*d*) curves.

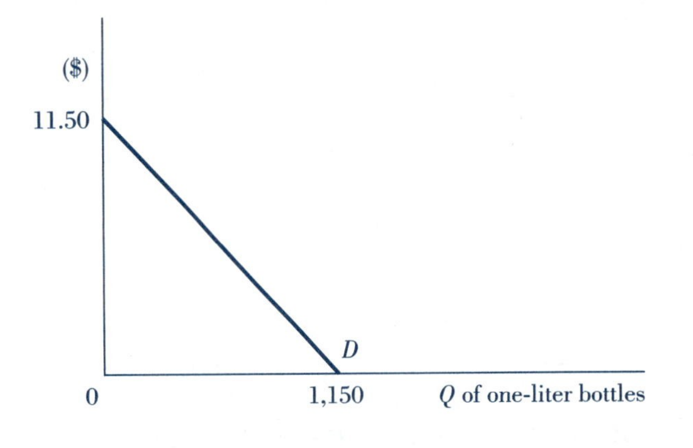

we show the combined quantity values for Consumer 1 and Consumer 2 in the second column, the aggregated data for the remaining 98 consumers in the third column, and the market quantity demanded ($Q_D = \Sigma q_d$) in the fourth column. The corresponding equations are

$$\begin{array}{lrl} \text{Demand for Consumers 1 and 2:} & q_{d(1+2)} & = -6P \quad + 30 \\ + \text{ Demand for Consumers 3 through 100:} & q_{d(3,...,100)} & = -94P \quad + 1{,}120 \\ \hline \text{Market demand:} & Q_D & = -100P + 1{,}150 \end{array}$$

The graph of the *inverse* market demand curve (*D*) is shown in Figure 2.2.[7]

Market Supply

supply
The quantities of a good the producer is willing and able to bring to market at a given set of prices during some discrete time period, *c.p.*

On the opposite side of the market, we derive a supply relationship based on the decisions of producers who are motivated by profit. Each firm's supply decision is modeled as a function of price, even though this decision is influenced by many other variables. Hence, we say that **supply** refers to the quantities of a good the producer is willing and able to bring to market at a given set of prices during some discrete time period, *c.p.* Among the variables that potentially affect the price-quantity response of a firm are production technology, input prices, taxes and subsidies, and price expectations. Analogous to the demand side of the market, changes in these determinants affect the entire price-quantity relationship, causing a *change in supply*, whereas a movement in price is associated with a *change in quantity supplied.*

Law of Supply
There is a direct relationship between price and quantity supplied of a good, *c.p.*

The Law of Supply

The relationship between quantity supplied and price is generally a positive one, which is referred to as the **Law of Supply.** This means that a *rise* in price is associated with a *rise* in quantity supplied, *c.p.*, and the converse holds as well. The conventional assumption that

[7] The algebraic expression for the inverse market demand function is $P = -0.01Q_D + 11.5$.

Application

2.2 *Can the Profit Motive Help the Environment?*

For a time, the common perception of how environmental goals would affect the business sector was that such efforts added only to corporate costs, which in turn reduced profitability. Although it is true that meeting new environmental standards and regulations can be costly, it is also the case that today's companies are finding ways to reduce pollution *and* enhance profits. How? The approach varies widely across firms and across industries. A few examples will explain.

One obvious way that firms are enjoying profits tied to environmentalism is through recycling. A commonly cited example is the use of sludge from electric utilities' smokestacks to produce gypsum, an input in the manufacture of such products as wallboard. Another is the recycling and reusing of wastewater from production processes, an effort used by many manufacturers across the economy. Although corporate recycling may seem like old news, what *is* new is that such efforts are becoming more pervasive, showing up in entirely new contexts. An innovative example is the use of old picture tubes from televisions and computer monitors to make a glass block product that protects technicians from X-rays. Another is McDonald's experimental use of a Big Mac package made from discarded juice boxes. In every case, the environment is saved from excess waste and pollution, and the company enjoys lower input costs and a better bottom line.

A fairly new corporate approach to "profitable environmentalism" is something called remanufacturing. Unlike recycling, where used products or wastes are recovered and formed into alternative raw materials, remanufacturing involves the collection, reconditioning, and reselling of the same product over and over again. A case in point is the collaborative effort of Eastman Kodak Company and Fuji Photo Film Company, two firms that offer so-called disposable cameras to the marketplace. These popular cameras are not immediately discarded but in fact are rebuilt and resold, sometimes as many as 10 times. Another example is the remanufacturing of used printer cartridges, which has grown to a $3 billion industry. This new remanufacturing trend is a win-win situation. The gains to the environment are energy conservation and significantly reduced waste disposal. For the firm, the gain is cost savings, assuming disassembly and reconditioning are cheaper than starting from scratch, where the cost difference is generally 40–65 percent.

Other environmentally based profits arise in an entirely different way. Sometimes, companies launch a project solely for environmental reasons, but it ends up being a lucrative source of new profit that is entirely unexpected. Starbucks Corporation experienced such an outcome in late 1998. The company had received the brunt of negative commentary about coffee production causing the destruction of rain forests. In response, Starbucks subsidized Mexican farmers to find ways to improve the production of coffee beans grown in the shade. Quite unexpectedly, the new beans were a success with consumers, and the coffee-making giant expanded the project to four other countries. In this case, the increased profits were generated not from lower production costs, but from increased demand totally unrelated to the underlying environmental motivation. The new coffee simply tasted good to consumers, and Starbucks enjoyed higher revenues as a result.

No matter the source or the motivation, in every case and in numerous other cases like these, corporations are proving that profits and environmental responsibility need not be mutually exclusive. As economic theory predicts, firms continue to be profit-driven entities even in the face of greater environmental pressures. With new innovation and technology, however, the lure of profit can actually contribute to a better environment.

Sources: Arndt (April 16, 2001); Deutsch (September 9, 2001); Ginsburg (April 16, 2001).

firms are profit maximizers suggests that a higher price should be an incentive for firms to bring more of the product to market. A contemporary example of how powerful the profit motive can be is in the context of green markets, discussed in Application 2.2.

A more formal justification for the Law of Supply is based on the nature of firms' costs as production is carried out. As the firm produces more output (Q), its **total costs (TC)** rise proportionately faster, meaning that the ratio of the change in TC (ΔTC) to the change in Q (ΔQ) is increasing. This ratio ($\Delta TC/\Delta Q$) defines the firm's **marginal cost (*MC*)** of production, the additional cost of producing another unit of output. Because MC rises as Q rises, firms need to charge a higher price for each extra unit of output they produce. Hence, the existence of rising MC supports the positive price-quantity relationship given by the Law of Supply.

Modeling Individual Supply

The Law of Supply is illustrated by using the hypothetical data for a single producer of bottled water given in Table 2.4. The quantity column shows a single firm's profit-maximizing output decisions associated with prices ranging from $0.50 to $5.00. Notice how the positive relationship between the **supply price** and the quantity supplied holds throughout. The linear equation associated with these data is $q_s = 16P - 4$, where q_s is used to signify the quantity supplied by a single firm. Each of the price-quantity pairs given in Table 2.4 satisfies this equation.

The graphical model of the single firm's supply curve (s) is illustrated in Figure 2.3. Again, the convention is to plot price on the vertical axis and quantity on the horizontal axis, using the inverse form of the equation. In this case, the inverse supply function is $P = 0.0625q_s + 0.25$. Given the absence of extremes, the supply curve has the expected positive slope in accordance with the Law of Supply and the underlying theory of rising MC. In fact, under the assumption of competitive output markets, the firm's short-run supply curve *is* its MC curve.

Deriving Market Supply from Individual Supply Data

market supply of a private good
The combined decisions of all producers in a given industry, derived by *horizontally* summing the individual supplies.

To analyze production at an aggregated level, **market supply** of a private good is derived using a *horizontal* summing procedure over all the individual quantity decisions of firms, analogous to what is done on the demand side. However, because the output market is assumed to be competitive, the adding-up process on the supply side is simpler, since all firms are identical under such market conditions.

Just as on the demand side, we begin by illustrating the aggregation procedure using only two firms. Hypothetical data for two representative producers of bottled water and their combined supply decisions are given in Table 2.5. The algebraic counterpart is

$$\begin{array}{lrl} \text{Supply by Producer 1:} & q_{s1} & = +16P - 4 \\ + \ \text{Supply by Producer 2:} & q_{s2} & = +16P - 4 \\ \hline \text{Combined Supply by Producers 1 and 2:} & q_{s(1+2)} = (q_{s1} + q_{s2}) & = +32P - 8 \end{array}$$

To bring the supply relationship up to the market level, we follow the same summing procedure for *all* firms in the industry, assumed to total 25 in number. The hypothetical **market**

Table 2.4 *Single Producer's Supply Data for Bottled Water*

Price P	Quantity Supplied (bottles/month) $q_s = 16P - 4$
$0.50	4
1.00	12
1.50	20
2.00	28
2.50	36
3.00	44
3.50	52
4.00	60
4.50	68
5.00	76

Figure 2.3 One Producer's Supply (s) of Bottled Water

The model shows the supply (s) of one representative producer in the bottled water market. It is based on the inverse supply function $P = 0.0625q_s + 0.25$. Notice that this function has a slope of $+0.0625$ and a vertical intercept of $+0.25$. The positive slope supports the relationship given in the Law of Supply and the theory of rising *MC*. Because the market for bottled water is assumed to be competitive, this supply curve is the firm's *MC* curve.

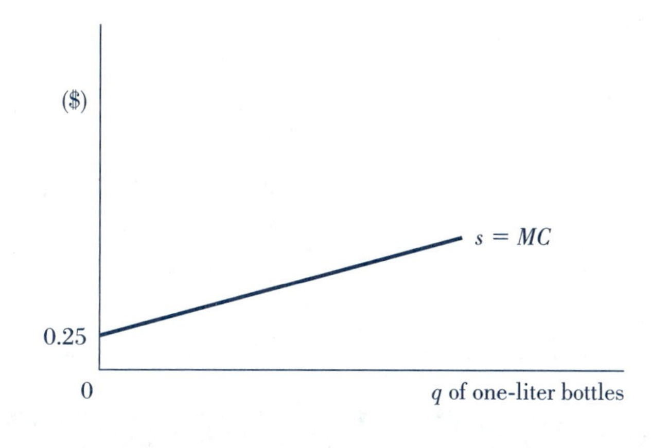

Table 2.5 Combined Supply Data for Bottled Water for Two Producers (Producer 1 and Producer 2)

Price P	Quantity Supplied by Producer 1 (bottles/month) $q_{s1} = +16P - 4$	Quantity Supplied by Producer 2 (bottles/month) $q_{s2} = +16P - 4$	Combined Quantity Supplied by Both Producers (bottles/month) $q_{s(1+2)} = q_{s1} + q_{s2} = +32P - 8$
$0.50	4	4	8
1.00	12	12	24
1.50	20	20	40
2.00	28	28	56
2.50	36	36	72
3.00	44	44	88
3.50	52	52	104
4.00	60	60	120
4.50	68	68	136
5.00	76	76	152

supply data are shown in Table 2.6. For each market price, the combined supply decisions for Producer 1 and Producer 2 are given in the second column, the aggregated data for the remaining 23 producers are given in the third column, and the market quantity supplied ($Q_S = \Sigma q_s$) is given in the last column. The corresponding supply equations are

$$
\begin{array}{lll}
 & \text{Supply by Producers 1 and 2} & q_{s(1+2)} = 32P - 8 \\
+ & \text{Supply by Producers 3 through 25} & q_{s(3,\ldots,25)} = 368P - 92. \\
\hline
 & \text{Market Supply} & Q_S = 400P - 100
\end{array}
$$

Table 2.6 Market Supply Data for Bottled Water

Price P	Quantity Supplied by Producers 1 and 2 (bottles/month) $q_{s(1+2)} = 32P - 8$	Quantity Supplied by Producers 3 through 25 (bottles/month) $q_{s(3,...,25)} = 368P - 92$	Market Supply (bottles/month) $Q_s = 400P - 100$
$0.50	8	92	100
1.00	24	276	300
1.50	40	460	500
2.00	56	644	700
2.50	72	828	900
3.00	88	1,012	1,100
3.50	104	1,196	1,300
4.00	120	1,380	1,500
4.50	136	1,564	1,700
5.00	152	1,748	1,900

Figure 2.4 Market Supply (S) of Bottled Water

This model depicts the market supply curve (S) for bottled water based on the inverse form of the market supply function $P = 0.0025Q_S + 0.25$. The market supply curve is derived by summing all the individual supply curves (s) horizontally. So, at each price level, the horizontal distance between the vertical axis and S is exactly equal to the sum of the analogous horizontal intervals for each of the 25 individual supply (s) curves. Because each individual supply (s) curve is the single firm's MC curve, the market supply (S) represents the horizontal sum of all these MC curves.

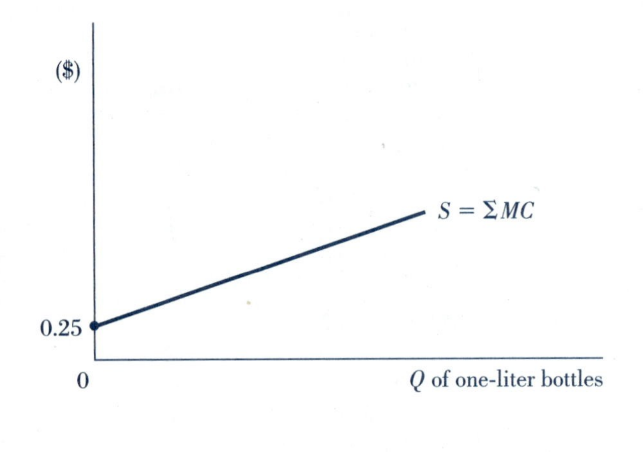

Figure 2.4 shows the graphical model of the market supply curve (S) in inverse form.[8] At each price level, the horizontal distance between the vertical axis and S is exactly equal to the sum of the analogous horizontal intervals for each of the 25 individual supply (s) curves. Furthermore, because the competitive firm's supply (s) curve is its MC curve, the market supply (S) curve is in turn the horizontal sum of each of the individual firm's MC curves.

[8]The corresponding equation for market supply in inverse form is $P = 0.0025Q_S + 0.25$.

Table 2.7 *Market Supply and Demand Data for Bottled Water*

Price P	Market Quantity Supplied $Q_S = 400P - 100$	Market Quantity Demanded $Q_D = -100P + 1{,}150$	Market Surplus or Shortage
$0.50	100	1,100	Shortage = 1,000
1.00	300	1,050	Shortage = 750
1.50	500	1,000	Shortage = 500
2.00	700	950	Shortage = 250
2.50	**900**	**900**	**Equilibrium**
3.00	1,100	850	Surplus = 250
3.50	1,300	800	Surplus = 500
4.00	1,500	750	Surplus = 750
4.50	1,700	700	Surplus = 1,000
5.00	1,900	650	Surplus = 1,250

Market Equilibrium

Thus far, we have considered each side of the market separately to develop distinct models of economic decision making. To generate a model of price determination, we must consider supply and demand *simultaneously* to allow for the interaction of consumers and producers in the marketplace. The formal theory that price is simultaneously determined by supply and demand is one of the most significant in all of economic analysis.[9]

Equilibrium Price and Quantity

Together, the forces of supply and demand determine a unique **equilibrium price (P_E)** at which point the market system "comes to rest" or has no tendency for change. Equilibrium, or market-clearing, price (P_E) is the price at which quantity demanded by consumers (Q_D) is exactly equal to the quantity supplied by producers (Q_S) (i.e., where $Q_D = Q_S$). Only at P_E will the associated **equilibrium quantity (Q_E)** be both the profit-maximizing production level for firms and the utility-maximizing consumption level for consumers.

equilibrium price and quantity
The market-clearing price (P_E) associated with the equilibrium quantity (Q_E), where $Q_D = Q_S$.

Market equilibrium for our hypothetical bottled water market is illustrated in Table 2.7, where market supply and demand data are presented together. Notice that equilibrium price in this market is $2.50, the only price at which quantity supplied (Q_S) is exactly equal to the quantity demanded (Q_D) of 900 units. This result also can be determined algebraically by solving the market demand and market supply equations simultaneously.[10]

Graphically, market equilibrium is modeled by diagramming the market demand (*D*) and the market supply (*S*) curves together. This is illustrated in Figure 2.5. Note that equilibrium is shown as the point where *D* and *S* intersect, or where $Q_D = Q_S$.

Market Adjustment to Disequilibrium

If the prevailing market price is at some level other than the equilibrium level, it must be the case that $Q_D \neq Q_S$, and the market is said to be in **disequilibrium.** As a consequence, consumers and producers have an incentive to make some adjustment that will restore equilibrium to the market. The motivation for this adjustment and the process by which it is implemented depend on whether there is excess demand or excess supply.

[9] This theory is attributable to Alfred Marshall, who first published this hypothesis in his *Principles of Economics* in 1890.

[10] Verify this assertion by substituting the market supply equation, ($Q_S = 400P - 100$) and the market demand equation ($Q_D = -100P + 1{,}150$) into the equilibrium condition ($Q_S = Q_D$).

Figure 2.5 ***Equilibrium in the Market for Bottled Water: Market Supply and Market Demand***

Market demand (D) and market supply (S) determine equilibrium price (P_E) and quantity (Q_E) at their point of intersection.

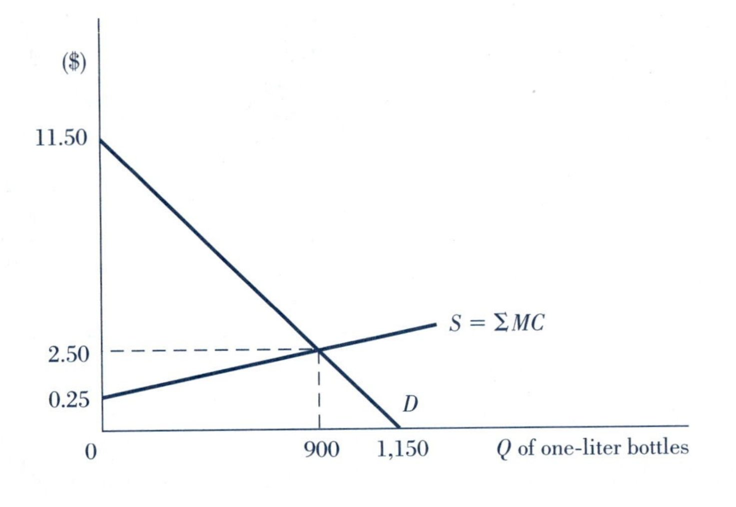

shortage
Excess demand of a commodity, equal to ($Q_D - Q_S$), that arises if price is *below* its equilibrium level.

Shortage

If the actual price P is *below* the equilibrium price P_E, then there is **excess demand,** meaning that Q_D exceeds Q_S at that price. The result is a market **shortage** of the good, equal to ($Q_D - Q_S$). Refer to Table 2.7 to see the shortages in the bottled water market at every price below $2.50. This means that consumers want more of the good at the existing price level than firms are willing and able to provide to the market. In response to this, firms would be willing to increase quantity supplied, moving up the market supply curve and elevating price as they do so. This process would continue until the incentive to change quantity no longer exists, which occurs when price is increased to its equilibrium level.

surplus
Excess supply of a commodity, equal to ($Q_S - Q_D$), that arises if price is *above* its equilibrium level.

Surplus

When the actual price P is *above* the equilibrium price P_E, then firms are unable to sell all their output at the prevailing price level, and they begin to observe their inventories accumulating with unsold stock. In this case, there is **excess supply,** meaning that Q_S exceeds Q_D at the prevailing price level. The result is a market **surplus,** equal to ($Q_S - Q_D$). Look at Table 2.7 to see the surpluses in the bottled water market at all prices above $2.50. Observing unsold supplies, firms would have an incentive to lower price to eliminate the surplus. This price reduction would continue until all the unsold supplies are purchased, which occurs at the equilibrium price.

Notice how price movements serve as a signal that a shortage or a surplus exists, whereas the stability of price suggests equilibrium. The adjustments that occur when the market is in disequilibrium happen because of the market's internal forces. As long as the assumptions of the model hold (and this is an important point), there is no need for government intervention or any other third-party mediation to achieve equilibrium. An example of how the market adjustment process operates is discussed in Application 2.3 in the context of the recycled newspaper market.

Thus far, we have examined the fundamental elements of a competitive market and illustrated how the simultaneous effect of demand and supply decisions determines equilibrium

Application

2.3 *Recycling Efforts and the Volatile Market for Newsprint*

The amount of trash generated in the United States has risen from 88 million tons in 1960 to 230 million tons in 1999. Of this tonnage, 38 percent is paper and paper products. In a logical move, many communities established paper recycling programs in the 1980s. The first step was to encourage individuals and firms to bring paper wastes to collection centers. According to EPA statistics, shown in the accompanying table, this recovery stage has met with some success.

Types of Paper Waste	Percent Recovered			
	1980	1990	1995	1999
Corrugated boxes	37.4	48.0	64.2	65.1
Newspaper	27.3	38.0	53.0	59.0
Books and magazines	8.3	20.9	47.1	40.8
Office papers	21.8	26.5	44.3	52.7

Sources: U.S. EPA, Office of Solid Waste and Emergency Response (May 1997a), p. 65, table 16, and p. 73, table 21; U.S. EPA, Office of Solid Waste and Emergency Response (2000), chap. 2, p. 32, table 4.

Although the data in the table suggest that society responded responsibly to the waste disposal dilemma, they belie a very real problem. In their haste to reduce paper waste in landfills, many communities failed to recognize the need to create a market for recovered materials. The recovery stage of the recycling process generates a supply of used materials, but if there is insufficient demand for these materials, communities face a glut of recovered wastes.

This potential problem became a reality for recovered newspapers, the amount of which grew by 34 percent over a five-year period in the late 1980s. The excess supply sent the price of used newsprint plummeting—a financial dilemma for many American cities and towns. According to a *Business Week* report, towns in New Jersey that used to sell the material for $20 per ton later had to pay as much as $10 a ton to get rid of it. Hence, the price actually fell to a negative value. Similar scenarios were reported by communities throughout the northeastern United States.

To correct the excess supply of recovered newspapers without facing dramatic price declines, it was necessary to stimulate market demand. Virtually all levels of government took an active role. A number of state governments passed laws requiring newspapers to be partly printed on recycled paper. In 1993, President Clinton signed Executive Order 12873, which called for all printing and writing paper to contain at least 20 percent recovered paper. (This amount was subsequently raised to 30 percent in Executive Order 13101.) The EPA established clearinghouses and hotlines to bring together suppliers and demanders of recyclables. Added influences were the thriving domestic economy and the rising demand of developing nations, whose growth required new sources of paper inputs.

Taken together, market demand swamped existing supplies, and, in 1995, there was a shortage of recycled newsprint. Just as predicted by economic theory, the shortage placed upward pressure on price, which rose to between $100 and $200 per ton. The boom in the market was temporary, however. By the following year, still more market fluctuations tipped the scales again, this time in the opposite direction. By 1996, excess supplies and falling demand drove prices back to the $20 per ton level of the early 1990s.

The lessons are clear. First, recycling is a complex process, and collection is just one step in that process. For recycling to be successful, there have to be complete markets—both supply *and* demand. Second, the market for recovered paper can be volatile. Although a supply of recovered paper is ensured by routine recycling practices, the same is not true for demand. With the unpredictability of market demand, price in this market is anything but stable.

Sources: Alexander (July/August 1992); Yang, Symonds, and Driscoll (April 22, 1991); Cahan (July 17, 1989); "Newspaper Recycling Booming" (July 11, 1995); Reidy (July 24, 1996).

price and quantity. However, we need to go beyond this point to understand the implications of a competitive equilibrium—first in terms of the allocation of economic resources and then in terms of the well-being of society.

Economic Criteria of Efficiency

Economic analysis is guided by specific criteria. One of these criteria deals with the proper allocation of resources among alternative productive uses, referred to as **allocative efficiency.** Another is concerned with economizing on resources used in production, called **technical efficiency.** Each of these criteria is relevant to all applied economic disciplines, including environmental economics. We begin with a discussion of allocative efficiency.

Allocative Efficiency

As illustrated by the materials balance model, the way a market system uses resources is critical, not only for production and consumption, but also for the natural environment. But how is the resource allocation to be evaluated? The answer is through a procedure that involves the following two elements:

- an assessment of benefits and costs
- the use of marginal analysis

We have already introduced these elements in our earlier discussion of demand and supply. Now, we use them to draw important conclusions about decision making and equilibrium under competitive conditions, first at the *market* level and then at the *firm* level.

Evaluating Resource Allocation at the Market Level

Competitive markets are considered an ideal, a standard by which other market conditions are evaluated. If we look carefully at what is conveyed by a competitive equilibrium, we can better appreciate this characterization. At equilibrium, we know that market demand intersects market supply. What does this mean in terms of resource allocation? Recall that we argued that prices along a demand curve are measures of marginal benefit (*MB*). Each demand price communicates the value consumers place on the next, or *marginal,* unit of the good based on the added benefit they expect to receive from consumption. On the supply side, there is also a marginal interpretation, but here the prices are measures of economic cost. Because the market supply in a competitive market is the horizontal sum of each firm's marginal cost (*MC*) curve, each supply price represents the additional cost of resources needed to produce another unit of the good. Economic costs include both the **explicit,** or out-of-pocket, costs associated with production and all **implicit** costs based on the highest valued alternative use of any economic resource.

Taking these two interpretations together, we arrive at an important result. At the competitive equilibrium, the value society places on the good is equivalent to the value of the resources given up to produce it, or $MB = MC$. By definition, this result ensures that there is an efficient allocation of scarce economic resources, or that **allocative efficiency** is achieved. **Allocative efficiency** requires that the additional value society places on another unit of the good is equal to what society must give up in scarce resources to produce it. Recognizing this outcome as part of the competitive model is important, but it is just as critical to understand why it arises. For that, we need to focus on the decision making of individual producers under these market conditions.

allocative efficiency
Requires that resources be appropriated such that the additional benefits to society are equal to the additional costs incurred.

Evaluating Resource Allocation at the Firm Level

Starting from a general perspective, the assumed motivation governing firm decision making is profit maximization. We further assume that the choice variable for producers is output. Hence, all firms, regardless of competitive conditions, choose the level of output that will maximize profit. **Total profit (π)** is defined as **total revenue (*TR*)−*total costs* (*TC*).** *TR* is simply the dollar value of the firm's sales, found as the product of market price (P) and the quantity of output sold (q) (i.e., $TR = Pq$). Total costs (TC) include all economic costs associated with producing the output.

total profit
Total profit (π) = total revenue (TR) − total costs (TC).

To find the q that achieves the highest possible π, the firm makes its decisions *at the margin*, considering the relative benefits and costs of producing each additional unit of output. From the firm's perspective, the benefit is measured by *TR* and the cost by *TC*. Thus, it considers the profit implications of each successive unit of output by looking at the associated changes in *TR* and *TC*. If producing the next unit of output adds more to its *TR* than it does to its *TC*, then the firm increases production. If production adds more to *TC* than to *TR*, production is decreased. The process goes on until there is no incentive to continue, or when the *change in TR* (ΔTR) from increasing output is equal to the *change in TC* (ΔTC). At this point, the *change in* π ($\Delta\pi$) from producing the last unit of output is zero, and any further increase in output would cause π to decline. Thus, at this precise point, π is at its maximum.[11]

Notice that the entire decision-making process relies on *changes*, which is the definition of marginal analysis. In this context, the relevant marginal variables are as follows:

- **Marginal revenue (*MR*)** is the change in total revenue (TR) associated with a change in output (q), that is, $MR = \Delta TR/\Delta q$.
- **Marginal cost (*MC*)** is the change in total costs (TC) associated with a change in output (q), that is, $MC = \Delta TC/\Delta q$.
- **Marginal profit (*M*π)** is the change in total profit (π) associated with a change in output (q), that is, $M\pi = \Delta\pi/\Delta q$.

Thus, the firm implicitly makes its profit-maximizing decisions according to the following rules:

The firm increases production as long as $MR > MC$ or as long as $M\pi > 0$.
The firm contracts production as long as $MR < MC$ or as long as $M\pi < 0$.
The firm achieves **profit maximization** at the output level where $MR = MC$ or where $M\pi = 0$.

profit maximization
Achieved at the output level where $MR = MC$ or where $M\pi = 0$.

Notice that the individual firm's optimal output level occurs at precisely the point where the marginal benefit *to the firm* of doing so (*MR*) is exactly offset by the marginal cost of the resources it uses (*MC*). Although this outcome validates the use of benefits and costs at the margin, it does not necessarily result in an allocatively efficient outcome. Why? Because *MR* is the marginal benefit *to the firm*, which is not necessarily equal to the marginal benefit *to society*, which is measured by price (P). In fact, the only way that the firm's optimizing behavior achieves allocative efficiency is if *MR* is equal to *P*, and this equivalency occurs only under competitive conditions. To understand this, we need to reconsider how the firm's profit-maximizing decision is affected by a competitive market.

As a price taker, each firm in a competitive market must accept the market-determined price as a given. It is unable to charge a higher price, because by assumption its product is identical to that of all other firms in the market. Thus, if a firm were to raise its price, consumers would demand none of its output and buy from other suppliers. Moreover, a firm has no incentive to lower its price, because it can sell all it wants at the market-determined price. (Recall that the single firm is small compared to the entire market, so "selling all it wants" means that its output adjustments are, in a relative sense, insignificant.)

[11] This, of course, assumes that in the short run the firm's revenues cover its variable costs. If not, it will shut down. In the long run, the firm produces as long as its total revenues cover all its total costs. If not, it will exit from the market.

Because the competitive firm has no control over prevailing market conditions, it faces a price that is constant. Consequently, each additional unit that the firm sells raises its total revenue by an amount exactly equal to the price of the good. This outcome translates to an important equality that is unique to competitive markets, namely, that $P = MR$. Thus, although competitive firms follow the profit-maximizing decision rules derived earlier, the equilibrium outcome is markedly different in that allocative efficiency is automatically ensured. This is summarized in the following derivation:

Profit maximization requires:	$MR = MC.$
Competitive markets imply:	$P = MR.$
Thus, profit maximization for competitive firms requires:	$P = MC.$

Now, it should be clear why the competitive *market* equilibrium achieves allocative efficiency—because every firm in that market independently produces where $P = MC$.

To illustrate this outcome, Table 2.8 presents selected revenue and cost data for a representative firm in our bottled water market. First, let's confirm that $P = MR$ at all output levels. In accordance with the competitive model, notice that the prices faced by the firm are constant at the market-determined equilibrium price (P_E) of \$2.50. The MR values are found by calculating the change in TR divided by the change in q. So, for example, when output rises from 28 units to 36 units, generating an increase in TR from \$70 to \$90, $MR = \Delta TR/\Delta q = (\$90-\$70)/(36-28) = \$20/8 = \$2.50$. As this calculation is repeated for all changes in q shown in the table, $MR = \$2.50$ in every case—exactly equal to P_E.

Now let's examine how the firm chooses an output level that maximizes profit. As predicted by the competitive model, the MC values shown in the last column of the table are the same as the set of prices associated with the firm's supply schedule. To guide its production decisions, the firm considers MC relative to MR at each output level. In this case, the profit-maximizing equilibrium output occurs at $q_E = 36$, where $MR = MC = \$2.50$. And because $P = MR$ at all q levels, this decision also ensures that $P = MC$, indicating that resources are being allocated to production in an efficient manner. Figure 2.6 illustrates this equilibrium.

Two final observations are worth noting. First, be sure to recognize why alternative output decisions would not maximize profit. Notice in Figure 2.6 that at any output level *below* q_E, the firm's MR exceeds MC, meaning that its $M\pi$ is greater than zero. Hence, the firm can increase π by expanding production. Conversely, at all output levels *above* q_E, the firm's MR is less than MC, meaning that its $M\pi$ is less than zero or that its π is declining. So the firm is better off contracting production. Only at $q = 36$ is $MR = MC$ or $M\pi = 0$, meaning that further additions to π are not possible or that π is at its maximum.

Table 2.8 ***Revenue and Cost Data for a Representative Firm in the Bottled Water Market***

Price P_E	Quantity q	Total Revenue TR	Marginal Revenue MR	Marginal Cost MC
\$2.50	4	\$10.00	\$2.50	\$0.50
2.50	12	30.00	2.50	1.00
2.50	20	50.00	2.50	1.50
2.50	28	70.00	2.50	2.00
2.50	36	90.00	2.50	2.50
2.50	44	110.00	2.50	3.00
2.50	52	130.00	2.50	3.50
2.50	60	150.00	2.50	4.00
2.50	68	170.00	2.50	4.50
2.50	76	190.00	2.50	5.00

Figure 2.6 ***Competitive Firm's Profit-Maximizing Equilibrium***

The profit-maximizing equilibrium for the competitive firm occurs at $q_E = 36$, where $MR = MC = \$2.50$. Because $P = MR$ at all q levels, $P = MC$ at equilibrium, indicating that resources are allocated efficiently. At any q *below* q_E, $MR > MC$, meaning that $M\pi > 0$. Hence, the firm can increase π by expanding production. Conversely, at all output levels *above* q_E, $MR < MC$, meaning that $M\pi < 0$, or π is declining. So the firm is better off contracting production. Only at $q_E = 36$ is $M\pi = 0$, meaning that further additions to π are not possible, and π is at its maximum.

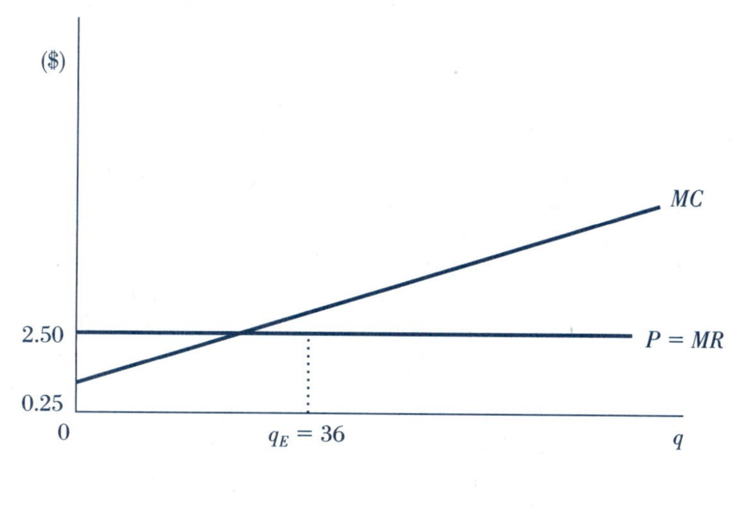

The second observation is that each of the other 24 firms in the bottled water market would follow this same decision-making process, and each would arrive at the same q_E of 36, because competitive firms are identical. Therefore, the market equilibrium quantity (Q_E) is equal to 900 units, the product of the number of firms in the market (25) and the firm level q_E of (36). Notice that this result confirms the equilibrium quantity determined from the market supply and demand model.

Technical Efficiency

Another important economic criterion used in market analysis is **technical efficiency.** This refers to production decisions that generate maximum output, given some stock of resources, or (saying the same thing from a slightly different perspective) decisions to produce a given output level using a minimum amount of resources. In the context of the materials balance model, achieving technical efficiency preserves the stock of natural resources and minimizes the subsequent generation of residuals arising from resource use. Furthermore, given the relationship between production and costs, technical efficiency implies that economic costs are minimized when producing a given level of output. When viewed from this perspective, it becomes apparent that technical efficiency is an application of the more general cost-effectiveness criterion introduced in Chapter 1.

technical efficiency Production decisions that generate maximum output given some stock of resources.

A key point is to understand that market forces can achieve technical efficiency as long as competitive conditions prevail. To remain viable, the competitive firm must minimize costs, because it cannot raise its price to cover the added expense of inefficient production. If it attempted such a strategy, demand for its product would fall to zero, because many other firms can bring the same product to market at a lower price. Recognizing how technical efficiency is achieved under such ideal market conditions helps economists determine *why* it is not being met in some other market context. More important, the magnitude of a technically inefficient decision can be assessed by comparing the resulting costs to what they *would* have been if the market had been allowed to operate freely.

Welfare Measures: Consumer Surplus and Producer Surplus

In economics, an important objective is to assess the gains and losses to society associated with any event that alters market price. The supply and demand model provides the information necessary to perform these types of analyses, using concepts known as **consumer surplus** and **producer surplus.** By comparing these measures before and after a market disturbance, it is possible to quantify how society has been affected.

Consumer Surplus

consumer surplus The net benefit to buyers estimated by the excess of the marginal benefit (*MB*) of consumption over market price (*P*), aggregated over all units purchased.

To get a sense of the logic of consumer surplus, we start with a general working definition. **Consumer surplus** is a measure of net benefit accruing to buyers of a good estimated by the excess of what they are willing to pay over what they must actually pay, aggregated over all units of the good purchased in the market. Notice that consumer surplus depends on two distinct notions of price—one that measures a *willingness to pay,* and one that measures what is *actually paid.* The series of prices consumers are willing to pay for various quantities of a good are those that define the demand curve. As discussed, each demand price is a measure of the marginal benefit (*MB*) associated with consumption. Conversely, the price that consumers must actually pay is the prevailing market price (*P*) determined by both demand and supply.

There are two major differences between these prices. First, they have different determinants. The demand price (*MB*) is determined *solely* by demand, a sort of psychic price based on how consumers value a good. On the other hand, market price (*P*) arises from the forces of supply and demand and is driven by *both* producer and consumer incentives. Second, although there is a whole series of demand prices, there is only one market price charged for *all* units sold. The result? Once the market price is determined, *all* units are sold for that single price, even those for which the demand price is much higher. Hence, consumers receive a surplus benefit for every unit purchased at a demand price that exceeds the market price.

To graphically illustrate consumer surplus, we reproduce in Figure 2.7 the market demand for bottled water derived previously. Added to the diagram is a reference price line drawn horizontally at the equilibrium price level of $2.50. Notice that for every output level up *to* the equilibrium quantity of 900, the demand price is higher than the market price. So each unit purchased yields consumers a surplus benefit over and above what they had to pay for it. For example, for the first bottle of water, consumers are willing to pay a price of $11.49 based on the market inverse demand function,[12] but they actually have to pay only $2.50. That means they receive a net benefit from consuming the first unit of the good equal to the excess of $11.49 over $2.50, or $8.99. Geometrically, this amount is measured as the vertical distance from the demand curve to the price line at $Q = 1$, shown in Figure 2.7 as distance *ab*.

Because consumers receive a net benefit for every unit purchased up to the equilibrium point, this value must be aggregated over all units consumed to derive the measure of **consumer surplus.** Graphically, this is the triangular area above the price line and below the demand curve. In Figure 2.7, the consumer surplus for bottled water is area *WXY.* This makes sense because the entire area under the demand curve is a measure of total benefit from consumption (i.e., the aggregation of the marginal benefit from each unit of the good), and the rectangular area under the price line is the total expenditure on the good. Thus, the difference between the two is the net benefit to the consumer.

[12]The price of $11.49 is found by substituting $Q_D = 1$ into the inverse demand function, $P = -0.01Q_D + 11.5$.

Figure 2.7 **Consumer Surplus in the Competitive Market for Bottled Water**

For the first bottle of water, consumers' demand price is \$11.49, but the market-determined price is only \$2.50, so consumers receive a net benefit from this first bottle of water equal to \$11.49 − \$2.50 = \$8.99, shown as the vertical distance *ab*. Aggregating this net benefit over all units of water consumed yields the measure of consumer surplus, shown as the triangular area *WXY*. The dollar value of this area is ½(900)(\$9.00) = \$4,050.

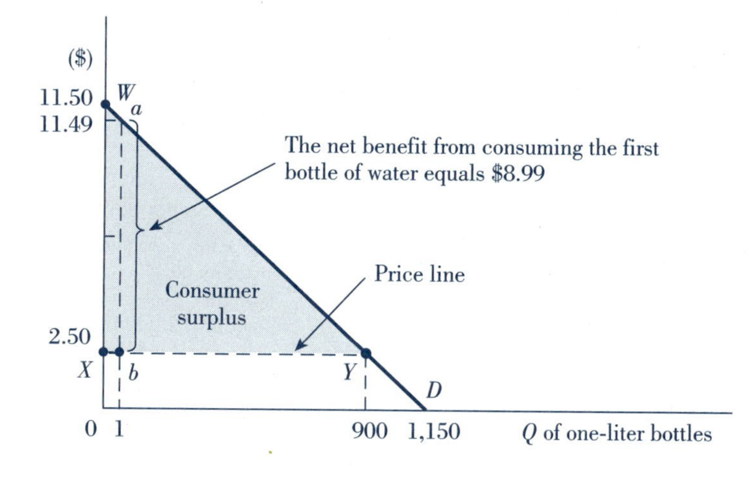

The dollar value of consumer surplus can be found by calculating the area of the triangle that represents it in the graphical model.[13] In Figure 2.7, the base of the *WXY* triangle is the horizontal distance from the vertical axis to the equilibrium quantity, or 900. The height is \$9.00, which is the difference between the vertical intercept of the demand curve (\$11.50) and the price line (\$2.50). Thus, the dollar value of consumer surplus in this market is ½(900)(\$9.00) = \$4,050.

The applicability of consumer surplus stems from the fact that its magnitude is related to equilibrium price and quantity. So any disturbance to market equilibrium will change the size of consumer surplus. Because this surplus is a measure of consumer benefit, any change in its value can be used to assess the associated gain or loss to consumer welfare.

Producer Surplus

On the supply side of the market, the comparable measure of welfare is **producer surplus.** It is a measure of net gain accruing to sellers, estimated by the excess of the market price (*P*) of a product over the marginal cost (*MC*) to produce it, aggregated over all units sold.

producer surplus The net gain to sellers of a good estimated by the excess of market price (*P*) over marginal cost (*MC*), aggregated over all units sold.

Based on our discussion of supply decisions, we know that firms must charge a price for their product that covers *MC*. Further, we know that the competitive market supply curve is the horizontal sum of all firms' *MC* curves. Therefore, because market price is determined by the intersection of market supply and demand, $P = MC$ at a competitive equilibrium. However, at every output level below equilibrium, *MC* is *lower* than *P*. So firms are actually willing to supply these smaller quantities at prices below what is dictated by the market. The

[13] Recall that the formula for finding the area of a triangle is ½ (base)(height).

Figure 2.8 Producer Surplus in the Competitive Market for Bottled Water

For the first bottle of water produced, the *MC* is $0.2525, based on an evaluation of the market supply function at $Q_S = 1$. But because firms can sell this bottle at a price of $2.50, they receive a net gain of $2.50 − $0.2525, or $2.2475, shown as the vertical distance *cd*. Aggregating this gain over all units sold yields the measure of producer surplus, represented by the triangular area *XYZ*. The dollar value of this area is ½(900)($2.25) = $1,012.50.

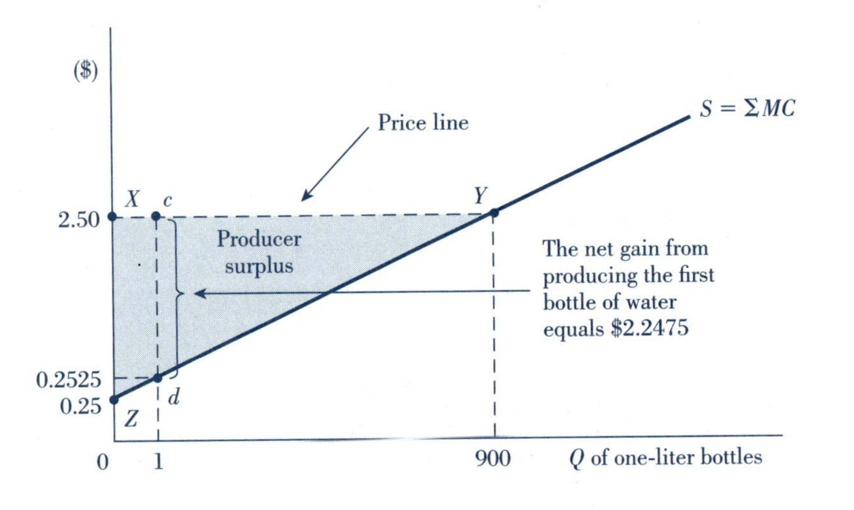

price that firms are "willing to accept" for each output level is their supply price, and it is this price that is reflected in the *MC* curve. Thus, at each quantity below equilibrium, firms accrue a net gain measured by the excess of *P* over *MC*.

This net gain is illustrated in Figure 2.8. The diagram shows the competitive market supply for bottled water, which is also the *MC* curve, and a reference line drawn horizontally at the equilibrium price of $2.50. For every unit of output supplied up *to* the equilibrium point at 900 units, producers receive a surplus equal to the excess of *P* over *MC*. For example, for the first unit of bottled water produced, the *MC* is $0.2525, based on the market supply function.[14] However, firms can sell this first bottle at the market-determined price of $2.50. Thus, the net gain associated with this unit of output is the excess of $2.50 over $0.2525, or $2.2475. Geometrically, this is the vertical distance from the supply curve to the price line at $Q = 1$, labeled as distance *cd* in Figure 2.8.

Just as on the demand side of the market, this net gain must be aggregated over all units sold up to the equilibrium quantity to find the measure of producer surplus. Graphically, this is the sum of all the vertical distances between the *MC* curve and the price line, or the triangular area bounded by the *MC* curve and the price line up to the equilibrium point. In Figure 2.8, this is shown as area *XYZ*.

By using the same method as described for consumer surplus, we can find the dollar value of producer surplus by calculating its representative triangular area depicted in the graphical model. In Figure 2.8, the base of the triangle *XYZ* is 900, and its height is $2.25, found as the difference between the price line at $2.50 and the vertical intercept of the supply curve,

[14]The supply price, or *MC*, of producing the first unit of output is found by evaluating the market supply function at $Q_S = 1$, or $MC = 0.0025(1) + 0.25 = \$0.2525$.

$0.25. Thus, producer surplus at equilibrium in the bottled water market is ½(900)($2.25) = $1,012.50. Notice that the magnitude of producer surplus, just like consumer surplus, is based on equilibrium price and quantity. Hence, any market disturbance will change its value and thus provide a way to assess any associated welfare gain or loss to firms.

The Welfare of Society: Sum of Consumer and Producer Surplus

Economists use the sum of consumer and producer surplus to capture the gains accruing to both sides of the market, or **society's welfare.** By applying this concept to the market for bottled water, we see that society enjoys a surplus valued at $4,050 + $1,012.50, or $5,062.50. Because this surplus is based on a competitive equilibrium, its value is maximized. This is so because of the efficient allocation of resources that characterizes a competitive market. Put another way, it is not possible to reallocate resources to improve society's welfare. By default, this implies that any market outcome that does not meet the criterion of allocative efficiency has a negative effect on society's well-being. In such an instance, the loss can be monetized by comparing the resulting sum of consumer and producer surplus to what it *would* have been if allocative efficiency had been achieved.

society's welfare
The sum of consumer surplus and producer surplus.

Measuring Welfare Changes

By way of illustration, let's consider a hypothetical policy initiative in the bottled water market that forces price above *MC* up to $6.50 per unit. To measure the effect on society's welfare, we need to compare the post-policy level of consumer and producer surplus to the benchmark competitive level of $5,062.50. We begin by reproducing the model of the bottled water market in Figure 2.9, adding to the diagram the policy price and quantity of $6.50 and 500, respectively.[15] Capital letters A through F have been added to facilitate our discussion. The benchmark level of consumer surplus is represented by the area (A + B + C), and producer surplus is represented by the area (D + E + F).

Now, we determine the comparable surplus values under the policy-determined price and quantity. At the $6.50 price level, consumer surplus is reduced to the triangular area A, valued at $1,250. This new value is $2,800 *lower* than it was before the policy, so we know that the policy generates a net *loss* to consumers. The producer surplus at the $6.50 price level is area (B + D + F), valued at $2,812.50, found by summing the area of the rectangle (B + D) and the area of the triangle (F). Because this magnitude is $1,800 *higher* than the original surplus, we see that producers enjoy a net *gain* as a result of the policy.

Finally, consider the overall effect. The total surplus under the new policy is $1,250 + $2,812.50, or $4,062.50, which is $1,000 less than the original value. Although producers enjoy a net gain of $1,800, this gain is outweighed by the loss to consumers of $2,800. Hence, the new pricing policy causes a decline in society's welfare of $1,000. This change can be confirmed geometrically by looking at the areas representing surplus before and after the policy:

	Change in consumer surplus:	(A)−(A + B + C) = − (B + C)
+	Change in producer surplus:	(B + D + F)−(D + E + F) = + (B − E).
	Net loss to society:	= − (C + E)

Area (C + E), valued at $1,000, is referred to as the **deadweight loss to society** because it was once a part of the surplus accruing to producers and consumers under allocatively efficient conditions but, as a result of the policy, it has been lost or unaccounted for. Notice that area B, although a loss to consumers, is a gain to firms. Thus, the value is redistributed from one market sector to another. Some may view this outcome as unfair, but the relevant point is that such transfers are not inefficient, because the amount is captured somewhere

deadweight loss to society
The net loss of consumer and producer surplus due to an allocatively inefficient market event.

[15] At a price of $6.50, $Q_S > Q_D$, so the quantity exchanged in the market is Q_D. Thus, quantity is found by substituting $6.50 into the demand equation: $Q_D = -100(6.50) + 1{,}150 = 500$.

Figure 2.9 ***Deadweight Loss to Society Under a Pricing Regulation in the Bottled Water Market***

At the allocatively efficient equilibrium, area (A + B + C) is the consumer surplus, valued at \$4,050, and area (D + E + F) is the producer surplus, valued at \$1,012.50, for a total welfare measure of \$5,062.50. Under a pricing policy that sets price at \$6.50, consumer surplus falls to \$1,250 (shown as area A) and producer surplus increases to \$2,812.50 (shown as area B + D + F), for a new total welfare measure of \$4,062.50. Consumers incur a net loss of −(B + C), valued at \$2,800, whereas producers gain +(B − E), valued at \$1,800. Thus, as a result of this policy, there is a **deadweight loss to society** of −(C + E), valued at \$1,000.

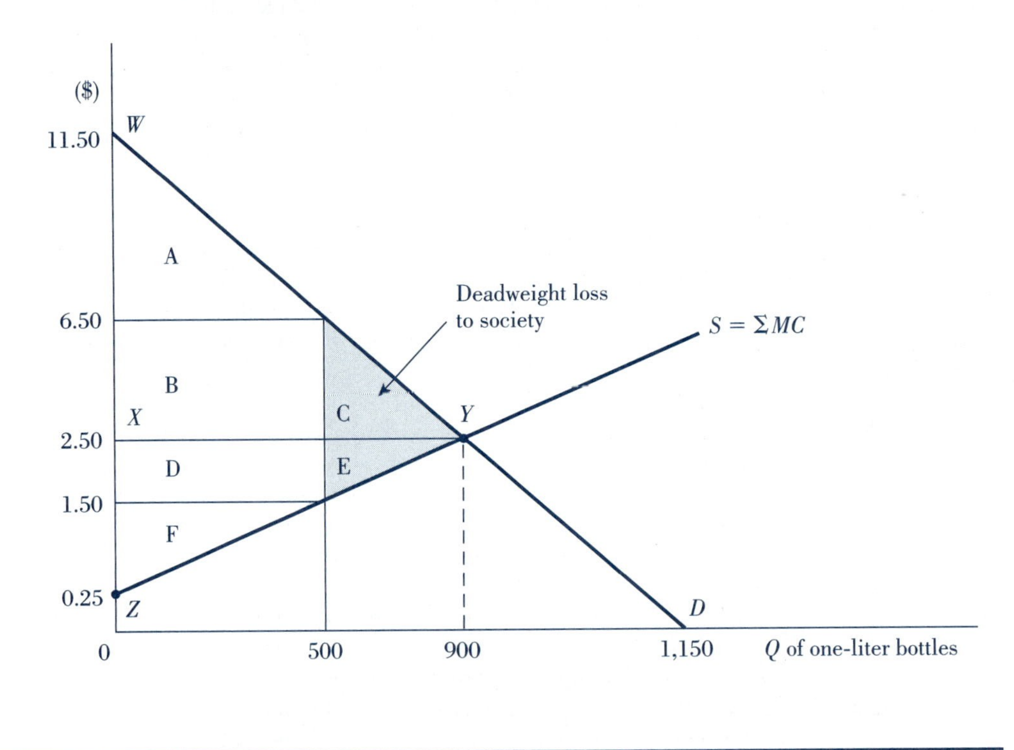

within the market system. What *is* problematic is that the policy generates a loss to society as a whole because it forces P above MC, violating the allocative efficiency criterion.

Conclusions

Understanding the fundamentals of how markets operate is an important basis for the study of environmental economics. To that end, we have limited our discussion in this chapter to the circular flow model—reviewing the mechanics of supply and demand, the signaling mechanism of price, and the importance of marginal analysis—all within the context of a classical competitive market system. Competitive markets establish a benchmark that helps economists evaluate the effects of market failures and market disturbances, both of which are important to environmental economics. From a practical perspective, we can assess the effects of environmental pollution or any policy initiative using allocative efficiency as a criterion. Furthermore, these effects can be measured by quantifying the associated changes in consumer and producer surplus.

The competitive model also illustrates how an economic system operates in the absence of any condition that impedes natural market forces. Recognizing how a fully functioning market performs is necessary to understanding the economic perception of environmental pollution as a market failure—the subject of our next chapter. To make this transition, we will expand our analysis to the full materials balance model, allowing for the interdependence of the circular flow with the natural environment. By using the modeling tools shown in this chapter and expanding on the concepts of marginal benefits and costs, we will develop more elaborate market models that explicitly account for this interdependence. These models will show *how* and *why* the market fails to correct environmental damage, which in turn will suggest approaches to finding effective policy solutions.

Summary

- A market refers to the interaction between consumers and producers for the purpose of exchanging a well-defined commodity.
- A competitive market model is characterized by a large number of independent buyers and sellers with no control over price, a homogeneous product, the absence of entry barriers, and perfect information.
- Demand is a relationship between quantity demanded (Q_D) and price (P), holding constant all other factors that may influence this decision, such as wealth, income, prices of related goods, preferences, and price expectations.
- The Law of Demand posits an inverse relationship between quantity demanded and price, *c.p.*
- Market demand for a private good is found by horizontally summing the individual demands.
- Supply is a relationship between quantity supplied (Q_S) and price (P), holding constant all other supply determinants, such as technology, input prices, taxes and subsidies, and price expectations.
- The Law of Supply states that there is a direct relationship between quantity supplied and price, *c.p.*
- Market supply for a private good is found by horizontally summing the individual supplies.
- The equilibrium or market-clearing price (P_E) is the price at which $Q_D = Q_S$. If a price is above (below) its equilibrium level, there is a surplus (shortage) of the commodity, which puts pressure on the prevailing price to fall (rise) toward its equilibrium level.
- Allocative efficiency requires that the additional value that society places on another unit of a good is precisely equivalent to what society must give up in scarce resources to produce it.
- All profit-maximizing firms expand (contract) output as long as the associated additional revenue (MR) is greater (lower) than the increase in costs (MC). The profit-maximizing output level occurs where $MR = MC$, or where marginal profit ($M\pi$) = 0.
- Competitive firms are price takers. Because $P = MR$ for competitive firms, the profit-maximizing output level at $MR = MC$ is also the point where $P = MC$, the condition signifying allocative efficiency.
- Technical efficiency arises when the maximum amount of output is produced from some fixed stock of resources, or, equivalently, when a minimum amount of resources is used to produce a given output level.
- Consumer surplus measures the net benefit accruing to buyers and is measured as the excess of what consumers are willing to pay (MB) over what they must actually pay (P), aggregated over all units purchased.

- Producer surplus measures the net gain accruing to sellers, estimated as the excess P over MC, aggregated over all units sold.
- Society's welfare is measured as the sum of consumer and producer surplus, which is maximized when allocative efficiency is achieved.
- The deadweight loss to society measures the net change in consumer and producer surplus caused by an allocatively inefficient market event.

Key Concepts

market
private good
demand
Law of Demand
market demand for a private good
supply
Law of Supply
market supply of a private good
equilibrium price and quantity
shortage
surplus
allocative efficiency
total profit
profit maximization
technical efficiency
consumer surplus
producer surplus
society's welfare
deadweight loss to society

Use the Key Concepts listed above to begin your search for additional articles and information using the InfoTrac College Edition database.

Review Questions

1. Suppose $Q_D = 200 - 4P$ and $Q_S = 100$ describe market demand and market supply in a given market.
   a. Algebraically find equilibrium price and quantity, and support your answer graphically.
   b. What is unusual about this market? Give an example of a good or service that might be characterized in this way.

http://

2. In 1995, the Food and Drug Administration published new labeling standards for bottled water. (The full text of the final rule can be found at **http://www.cfsan.fda.gov/~lrd/n095-323.txt**). Prior to that time, bottlers could sell regular tap water under a bottled water label. In fact, the FDA estimated that approximately 25 percent of the supply of bottled water was nothing more than ordinary tap water. Consider how these tougher standards eliminated 25 percent of the supply of bottled water. If market demand is unaffected, what qualitative impact would this labeling change have on equilibrium price and quantity for bottled water? Support your answer with a graphical model.
3. Reconsider the implications of the revised labeling standards discussed in Question 2 in the context of the hypothetical market for bottled water modeled in the text. Recall that the market demand and market supply equations are

$$Q_D = -100P + 1{,}150 \quad \text{and} \quad Q_S = 400P - 100,$$
$$\text{where } P_E = \$2.50 \text{ and } Q_E = 900.$$

   Now, suppose the change in standards results in a new market supply of $Q_S{}' = 400P - 350$, with no change in market demand.
   a. Determine the new $P_E{}'$ and $Q_E{}'$ for bottled water. Do your results agree with your intuitive answer to Question 2?
   b. Graphically illustrate the market for bottled water before and after the change in labeling standards. Be sure to label all relevant points.

c. Compare the values of consumer and producer surplus before and after the change in labeling standards. Is this result expected? Why or why not?

4. a. Describe a real-world government policy that creates a market surplus. Be sure to carefully define the relevant market.

b. Explain the efficiency implications of such a policy. Be specific.

c. In the instance you have described, what is the government's motivation for intervening in the market in this way?

Additional Readings

Cortese, Amy. "Can Entrepreneurs and Environmentalists Mix?" *New York Times*, May 6, 2001, p. 3.

Friedman, Milton. *Capitalism and Freedom.* Chicago: University of Chicago Press, 1962.

Heilbroner, Robert L. *The Worldly Philosophers.* New York: Simon and Schuster, 1980.

Jenkinson, Tim, ed. *Readings in Microeconomics.* New York: Oxford University Press, 1996.

Mankiw, N. Gregory. *Principles of Microeconomics.* 2d ed. Mason, OH: South-Western, 2002.

Nicholson, Walter. *Intermediate Microeconomics and Its Application.* 8th ed. Mason, OH: Dryden, 2000.

Pindyck, Robert S, and Daniel Rubinfeld. *Microeconomics.* 5th ed. Upper Saddle River, NH: Prentice-Hall, 2001.

Related Web Sites

http://

Executive Order 12873: Federal Acquisition, Recycling, and Waste Prevention — **http://es.epa.gov/program/exec/eo12873.html**

Executive Order 13101: Greening the Government Through Waste Waste Prevention, Recycling, and Federal Acquisition — **http://www.epa.gov/fedrgstr/eo/eo13101.htm**

U.S. Food and Drug Administration Final Rule on bottled water — **http://www.cfsan.fda.gov/~lrd/n095-323.txt**

Chapter 3

Modeling Market Failure

According to the circular flow model, free markets provide desired goods and services to the market, resolve shortages and surpluses, and eliminate inefficiency through the pricing mechanism—all without government intervention. This is a remarkable result, given that consumers and producers are not motivated by philanthropic goals but rather are driven by their own self-interest. As first described through Adam Smith's metaphor of the "invisible hand," this market outcome emerges as though consumers and firms are guided to make decisions that enhance the society's well-being.[1] Recognizing the efficiency and welfare implications of a competitive equilibrium underscores what is at stake when something impedes the market process that underlies it. A case in point is the persistence of pollution.

When we consider the circular flow in the fuller context of the materials balance model, we become aware of how economic activity generates residuals that can damage natural resources. However, we need to look further to understand the economics of why pollution persists in the absence of third-party intervention. Why is the market unable to respond to environmental pollution, or can it? The most immediate answer is that pollution is a **market failure** that distorts the classical market outcome.

From an economic perspective, environmental problems persist because they implicitly violate the assumptions of a fully functioning market. The incentive mechanisms that normally achieve an efficient solution are unable to operate, and government has to intervene. However, if the market failure is understood, incentives can be restored through environmental policy. Conceptually, the idea is to ferret out the conditions that cause the pricing system to break down, make the needed adjustments to the underlying conditions, and then let the power of the market work toward a solution.

In this chapter we provide the analytical tools necessary for understanding market failure in the context of environmental problems. Our discussion centers on the development of two economic models. The first is based on the public goods characteristics of environmental quality. The second uses what is known as externality theory to show how market incentives fail to capture the effects of pollution associated with production or consumption. Finally, we link the two models through a discussion of property rights and their role in environmental market failures.

Environmental Problems: A Market Failure

Classical microeconomic theory predicts an efficient outcome given certain assumptions about pricing, product definition, cost conditions, and entry barriers. If any of these assumptions fail to hold, market forces cannot operate freely. Depending on which assumption is violated, the result will be any of a number of inefficient market conditions, collectively termed **market failures.** These include imperfect competition, imperfect information, public goods, and externalities. For example, if we relax the assumption of freedom of entry in the competitive model, some degree of market power will develop. As this occurs, society's welfare declines, and resources are allocated inefficiently.

market failure
The result of an inefficient market condition.

[1] Smith (1937), p. 423 (originally published in 1776).

Economists model environmental problems as market failures using either the theory of public goods or the theory of externalities. Each is distinguished by how the market is defined.

- If the market is defined as "environmental quality," then the source of the market failure is that environmental quality is a **public good.**
- If the market is defined as the good whose production or consumption generates environmental damage, then the market failure is due to an **externality.**

Although each of these models suggests a different set of solutions, the theories are not totally unrelated. In the context of environmental problems, both are exacerbated by a third type of market failure, **imperfect information.** We begin by analyzing the market failure aspect of public goods.

Environmental Quality: A Public Good

Economists distinguish public goods from private goods by examining their inherent characteristics — *not* by whether they are publicly or privately provided.[2] A **public good,** or, more technically, a *pure* public good, is one that possesses the following characteristics: (1) It is **nonrival** in consumption, and (2) its benefits are **nonexcludable.**[3] At the other extreme is a pure private good, which is characterized by rivalness and excludability.

public good
A commodity that is nonrival in consumption and yields benefits that are nonexcludable.

Characteristics of Public Goods

Nonrivalness refers to the notion that the benefits associated with consumption are *indivisible,* meaning that when the good is consumed by one individual, another person is not preempted from consuming it at the same time. Put another way, the marginal cost (*MC*) of another individual sharing in the consumption of the good is zero. Consider, for example, the network television broadcast of the NBA championship finals. The benefits to the existing television audience are completely unaffected when another individual tunes in to view the broadcast. Contrast this result with what happens when a private good, such as a personal computer, is consumed. Once someone is using the computer, that consumption prohibits another person from using it at the same time.

nonrivalness
The characteristic of indivisible benefits of consumption such that one person's consumption does not preclude that of another.

Nonexcludability means that preventing others from sharing in the benefits of a good's consumption is not possible (or prohibitively costly in a less strict sense). An example of a good with this characteristic is a jogging path. It would be virtually impossible to ration the use of the path to a select group of runners. In contrast, consider the inherent excludability of a conventional private good, such as hotel lodging. Exclusive rationing of hotel services to the consumer paying for them is easily accomplished, and the associated benefits accrue solely to that single consumer.

nonexcludability
The characteristic that makes it impossible to prevent others from sharing in the benefits of consumption.

Although nonrivalness and nonexcludability may seem similar, they are not identical. A good way to distinguish them is as follows. Nonrivalness means that rationing of the good is not *desirable,* whereas nonexcludability means rationing of the good is not *feasible.*[4] In fact, it is possible for a good to possess one of these attributes but not the other. Reconsider the examples for each characteristic. Although a televised NBA game is a nonrival good, it can be made excludable by broadcasting it over a cable network only to subscribers who have paid for it and by scrambling the signal to everyone else. Likewise, provision of the jogging path, which is nonexcludable absent prohibitively high costs, does not possess the nonrivalry attribute. Use of the path by more and more runners would lead to congestion, which would affect every consumer of the jogging path.

[2] For a good discussion of this distinction, see Rosen (2002), chap. 4.
[3] Much of the important work in the theory of public goods is credited to Samuelson (1954, 1955, 1958).
[4] Stiglitz (1988), pp. 119–23.

Two classic examples of public goods are a lighthouse and national defense. Contemplate the services provided by these goods to see that the benefits of each are both nonrival and nonexcludable. A more contemporary and, from our perspective, more relevant example of a public good is environmental quality.[5] Just like the lighthouse, cleaner air, for example, is both nonexcludable and indivisible. Consider the futility of trying to restrict the benefits of air quality to a single person. It is unreasonable to think that others could be excluded from the consumption of a cleaner air supply just because another person has paid for it. Moreover, once the air is made cleaner for one person, others could simultaneously enjoy the benefits of breathing healthier air.

So what, then, is the problem? Having accepted the idea that environmental quality is a public good, we still need to explain why this, or any public good for that matter, is a market failure. To do this, we must develop a model of a public good.

Modeling a Public Goods Market for Environmental Quality

Public goods generate a market failure because the nonrivalness and nonexcludability characteristics prevent natural market incentives from achieving an allocatively efficient outcome. To illustrate this assertion, we reintroduce the supply and demand model but redefine the market as the public good, air quality.[6] It turns out that by changing the market definition from a private good to a public good, the conventional derivation of market demand is no longer viable. It is this modeling dilemma that is at the root of the public goods problem.

Allocative Efficiency in the Market for a Public Good

Just as in the private goods case, achieving an allocatively efficient equilibrium in a public goods market depends on the existence of well-defined supply and demand functions. To develop these functions for air quality, we adjust the market definition so that output can be quantified. Air quality can be defined as "an acceptable level of pollution abatement," which for this discussion we assume is some percentage reduction in sulfur dioxide (SO_2) emissions.

Market Supply for Air Quality

Although the market supply of a public good often represents government's decisions about production, as opposed to those of private firms, the general derivation of the supply function is analogous to what we developed in Chapter 2 for a private good. We begin by assuming that there is some number of hypothetical producers, each of which is willing and able to supply various reductions in SO_2 at different price levels, *c.p.* The aggregation of these production decisions gives rise to market supply, which we assume is represented by the data in Table 3.1. Price (P) is measured in millions of dollars, and quantity supplied (Q_S) is measured as a percentage of SO_2 abatement. The algebraic counterpart for these data is

$$\text{Market supply: } P = 4 + 0.75Q_S.$$

Market Demand for Air Quality

On the demand side, the model for a public good is quite different from that for a private commodity. Recall from Chapter 2 that the market demand for a private good is found by *horizontally* summing the demands of individual consumers. It is as though each consumer were asked, "What *quantity* of this good would you consume at each of the following prices?" But this question is not relevant to the demand for a public good because once such a commodity

[5] Alternatively, we could define environmental pollution as a public "bad."

[6] To direct attention solely to the public goods aspect of the model, we continue to assume competitive markets. This prevents confounding the market failure of public goods with that of imperfect competition. Further, the assumption of competitive markets allows the market supply curve to be modeled as the horizontal sum of all producers' marginal cost curves, just as was done for the private goods case in Chapter 2.

Table 3.1 ***Hypothetical Supply Data in the Market for Air Quality, Measured as a Percentage of Sulfur Dioxide (SO_2) Abatement per Year***

Quantity Supplied Q_S (% of SO_2 abatement)	Market Supply Price $P = 4 + 0.75Q_S$ ($ millions)
0	4.00
5	7.75
10	11.50
15	15.25
20	19.00
25	22.75
30	26.50

is provided, it is available at the same quantity to all consumers—a direct consequence of the **nonrivalness** characteristic. How, then, *is* market demand determined for a public good if quantity is not a decision variable?

The key is to recognize that the demand price for a public good is variable, even though the quantity is not. So the relevant question in deriving this demand must be, "What *price* would you be willing to pay for each quantity?" In theory, each consumer should express a unique **willingness to pay (WTP)** for the public good based on the benefits each expects to derive from consumption. The **market demand for a public good** is the aggregate demand for all consumers in the market. It is derived by summing each individual demand *vertically* to determine the market price ($P = \sum p$) at each and every possible market quantity (Q), *c.p.*

market demand for a public good
The aggregate demand of all consumers in the market, derived by *vertically* summing their individual demands.

To illustrate this procedure in the market for SO_2 abatement, we initially focus on only two consumers. The scenario is that we conduct a survey, asking the two consumers how much each would be willing to pay each year for various amounts of SO_2 abatement, *c.p.* The results are given in Table 3.2. The quantity demanded (Q_D) column shows a selection of possible abatement levels to be provided to all consumers. The demand prices, or WTP responses, are

Table 3.2 ***Hypothetical Demand Data in the Public Goods Market for Air Quality, Measured as a Percentage of Sulfur Dioxide (SO_2) Abatement per Year***

Quantity Demanded Q_D (% of SO_2 abatement)	Consumer 1's WTP $p_1 = 10 - 0.1Q_D$ ($)	Consumer 2's WTP $p_2 = 15 - 0.2Q_D$ ($)	Combined Demand Price for Consumers 1 and 2 $p_1 + p_2 = 25 - 0.3Q_D$ ($)
0	10.00	15.00	25.00
5	9.50	14.00	23.50
10	9.00	13.00	22.00
15	8.50	12.00	20.50
20	8.00	11.00	19.00
25	7.50	10.00	17.50
30	7.00	9.00	16.00

Note: WTP = Willingness to pay.

Figure 3.1 ***Combined Demand of Two Consumers for Air Quality (SO_2 Abatement)***

The demand curve labeled ($d_1 + d_2$) represents the combined consumption decisions of two hypothetical consumers for air quality. For any given quantity on this curve, the corresponding price is equal to the vertical sum of the individual price responses for that same quantity as given on curves d_1 and d_2.

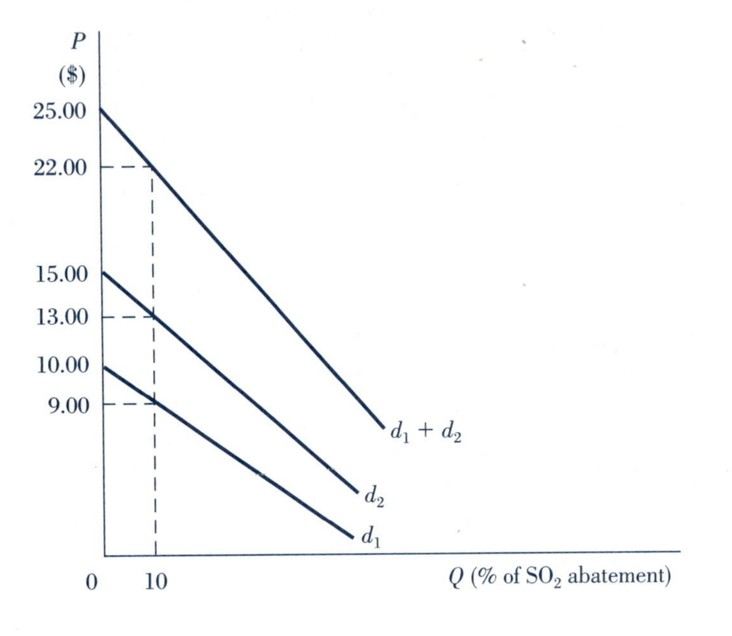

labeled p_1 and p_2 for Consumer 1 and Consumer 2, respectively. These responses are based on the following demand equations, which are expressed in inverse form to signify that price is the decision variable:

Demand for Consumer 1: $p_1 = 10 - 0.1Q_D$,
Demand for Consumer 2: $p_2 = 15 - 0.2Q_D$.

Notice that each consumer's WTP response for a given Q_D is unique. For example, for a 5 percent reduction in SO_2, Consumer 1 is willing to pay $9.50 per year, whereas Consumer 2 is willing to pay $14.00 per year. These responses are different because each consumer has a unique level of income, wealth, preferences, and so forth. In this context, we might expect the two consumers to have distinct preferences for air quality and even disparate views about the associated benefits. For example, perhaps Consumer 2's WTP is higher because he or she is a member of an environmental group and is more aware of the benefits of cleaner air.

To find the combined demand for air quality, the individual price responses at each quantity level are added together. The far right-hand column of Table 3.2 shows the results. The algebraic counterpart is found by summing each pair of corresponding terms on the right-hand side of each equation:

	Demand for Consumer 1:	$p_1 = 10 - 0.1Q_D$
+	Demand for Consumer 2:	$p_2 = 15 - 0.2Q_D$.
	Demand for Consumers 1 and 2:	$p_1 + p_2 = 25 - 0.3Q_D$

The graphical model is shown in Figure 3.1. For any given quantity on the combined demand curve ($d_1 + d_2$), the corresponding measure of price is equal to the vertical sum of the individual price responses for that same quantity based on the demand curves d_1 and d_2.

Having developed the general procedure, we now bring the demand relationship up to the market level by assuming that the combined demand of the two consumers represents the decisions of 1 million consumers. Notice that this conversion means that price (P) is now denominated in millions of dollars, which corresponds to the scale of the market supply function. More formally, the market demand function is specified as

$$\text{Market Demand: } P = 25 - 0.3Q_D.$$

Equilibrium in the Air Quality Market

To find equilibrium in the market for SO_2 abatement, we begin by combining the market demand and supply data in Table 3.3. We observe that Q_E is equal to 20 percent, the point at which both the supply price and the demand price are equal to \$19 million, which is P_E. This result also can be found algebraically by solving the equations for market supply and demand simultaneously.[7] The corresponding graphical model is shown in Figure 3.2, where the intersection of the market supply (S) and market demand (D) curves identifies P_E and Q_E.

Assessing the Implications

Determining P_E and Q_E for the abatement of any kind of pollution is a significant result. As will become apparent in upcoming chapters, it is this result to which economists refer when addressing market-based solutions to environmental pollution. Notice in Figure 3.2 that Q_E represents the efficient or optimal level of *abatement* measured from *left to right* and implicitly the optimal level of *pollution* measured from *right to left.* As the model suggests, this optimal level is not necessarily zero.[8]

From a general perspective, abating at the 100 percent level to reduce pollution to zero involves prohibitive opportunity costs. These include the forgone production and consumption of any good generating even the smallest amount of pollution. Given our present technology, a zero-pollution world would be one without electricity, advanced transportation systems, and virtually all manufactured products. Thus, it makes little sense to argue for the elimination of all pollution in our environment.[9]

Understanding the Market Failure of Public Goods Markets

As we stated at the outset, the achievement of an allocatively efficient outcome in a public goods market depends on the identification of well-defined demand and supply functions. Although we developed both functions for our hypothetical market for air quality, market demand was identified only because we implicitly made one critical assumption—that consumers would reveal their willingness to pay (WTP) for SO_2 abatement. However, without third-party intervention, the nonexcludability of this or any public good makes it difficult, if not impossible, to ascertain such information. If consumers' WTP responses are unknown, market demand cannot be identified, and an efficient outcome cannot be obtained. It is precisely the inability of free markets to capture the WTP for a public good that causes the market failure.

Consider the meaning of demand in the context of private versus public goods. In general, market demand captures the expected benefits associated with consumption. If the good is private (i.e., excludable), its benefits can be obtained only through purchase. Therefore, the consumer's WTP for a private good is a suitable proxy for the anticipated marginal benefits of consuming it. However, in the case of a public good that is nonexcludable, the consumer can

[7] Verify this assertion by setting the market demand equation ($P = 25 - 0.3Q_D$) equal to the market supply equation ($P = 4 + 0.75Q_S$), recognizing that $Q_S = Q_D$ at equilibrium.

[8] Of course, a zero optimal level of pollution is *possible* under certain conditions, although this is the exception rather than the rule.

[9] One of the classic research papers that discusses this economic view of pollution is Ruff (spring 1970).

Table 3.3 ***Hypothetical Market Demand and Market Supply Data for Air Quality, Measured as a Percentage of Sulfur Dioxide (SO_2) Abatement per Year***

Quantity Q (% of SO_2 abatement)	Market Demand Price $P = 25 - 0.3Q_D$ ($ millions)	Market Supply Price $P = 4 + 0.75Q_S$ ($ millions)
0	25.00	4.00
5	23.50	7.75
10	22.00	11.50
15	20.50	15.25
20	**19.00**	**19.00**
25	17.50	22.75
30	16.00	26.50

Figure 3.2 ***Market Supply and Market Demand for Air Quality (SO_2 Abatement)***

Assuming the market for SO_2 abatement is competitive, equilibrium price (P_E) of $19 million and equilibrium quantity (Q_E) equal to 20 percent abatement represent an allocatively efficient solution. Q_E represents the optimal level of *abatement* measured from *left to right* and implicitly the optimal level of *air pollution* measured from *right to left*. Notice that the optimal level of pollution is not zero.

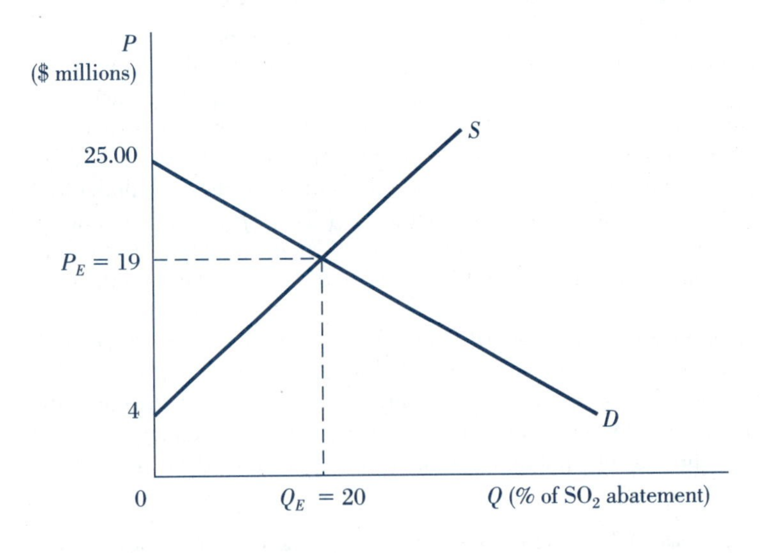

3.1 *Boston Harbor: The Changing Condition of an Environmental Public Good*

During the 1988 presidential campaign, the environmental degradation of Boston Harbor became a highly charged political issue. Wanting to promote an image as the prospective "environmental president," then Vice President George H. W. Bush focused the nation's attention on what he called the "dirtiest harbor in America." Political motivations aside, Bush was, at least in spirit, accurate in his assessment of the famous harbor that for a century has suffered the ill effects of public and industrial waste dumping. Boston Harbor and its nine beaches have been fouled by pollutants ranging from raw sewage to heavy metals, such as lead and mercury.

Why did the condition of this vital natural resource decline to such a deplorable state? The specifics are complex and entangled with bureaucracy and local politics, but the bottom line is that the harbor became seriously polluted because clean waterways are a public good. A cleaner Boston Harbor possesses the requisite characteristics of a public good, namely, nonexclusivity and nonrivalness in consumption. The result is the typical free-ridership problem that characterizes the market failure aspect of all public goods. No one is willing to reveal their preferences for a cleaner harbor and accept a share of the financial responsibility as long as the associated benefits can be enjoyed without having to pay for them. In addition, most harbor users are not fully aware of the benefits of cleaning up the harbor. Therefore, no one accepts responsibility for the preservation and protection of a waterway that is, in a real sense, commonly owned. The unfortunate outcome is that industries have used the harbor as a dumping ground for chemical discharges, and residents have left their trash on the harbor's beaches.

As predicted by theory, third-party intervention—implemented in this case through a federal court order—was necessary to initiate corrective action to save the harbor. The federally mandated plan, spearheaded by the Massachusetts Water Resources Authority (MWRA) (see **http://www.mwra.state.ma.us**), was completed in 2001 at a price tag of $4.1 billion. The monies were spent on an extensive reconstruction of the area's sewage treatment facilities. Although the project has taken years to complete, the results have been impressive. Discharges of toxic metals and sludge have been dramatically reduced, and beach closings are less frequent. However, despite the substantial environmental benefits achieved, more needs to be done to protect Boston Harbor. For example, after a heavy rainstorm, local beaches must be closed because of unhealthy levels of bacteria and other pollutants. These contaminants are entering the harbor's waters from the area's antiquated storm drainage system, a problem the MWRA is beginning to address.

Sources: Allen (September 6, 1992); Dolin (July/August 1992); Doneski (spring 1985); Daley (September 10, 2001).

http://

share in its consumption even when it is purchased by someone else. Thus, there is no incentive for a rational consumer to volunteer a WTP for something he or she can consume without having to pay for it.

Formally, this problem is known as **nonrevelation of preferences,** which in turn is due to the more basic dilemma of **free-ridership.** The rational consumer recognizes that the benefits of a public good are accessible simply by allowing someone else to purchase it. The consumer becomes a free-rider. Individual preferences about the public good remain undisclosed, and thus market demand is undefined. When the public good is environmental quality, the consequence can be serious ecological damage. A case in point is the pollution that threatens Boston Harbor, discussed in Application 3.1.

If we relax the conventional assumption of perfect information, adding more realism to the model, the identification of market demand becomes even more obscure. In many public goods markets, consumers are not fully aware of the benefits associated with consumption. This is certainly the case for environmental quality. Most people are not aware of all the

nonrevelation of preferences
An outcome that arises when a rational consumer does not volunteer a willingness to pay because of the lack of a market incentive to do so.

free-ridership
Recognition by a rational consumer that the benefits of consumption are accessible without paying for them.

Application

3.2 Informing the Public Through Truth in Advertising: Green Marketing Guidelines

Strategic labeling of products to promote their environmental attributes had become a growing business trend in the 1970s and early 1980s. This so-called green marketing was used by businesses to make their products more appealing to an emerging group of environmentally minded consumers. By using claims such as "environmentally safe," "biodegradable," or "made with 100% recycled material" on product packaging, firms could gain market share without having to cut product price. However, as consumers' awareness of environmental issues grew, so too did their demand for accurate green labeling and their skepticism about advertised ecological promises.

An often cited case of misleading advertising was Mobil Corporation's use of a biodegradable label to market its new line of Hefty trash bags in the late 1980s. Unfortunately for Mobil, its decision came in the midst of a controversy about the relevance of degradable products. Scientists had pointed out that the absence of oxygen and sunlight for wastes buried in a landfill would deter the degradation process. In response to this finding, Mobil attempted to preempt any legal attacks by adding a disclaimer to its packaging, but its reaction came too late. In 1990, Mobil was faced with several lawsuits that charged it with deceptive advertising and consumer fraud (Lawrence 1991). Ultimately, Mobil paid $150,000 in damages to six states to settle the lawsuits.

Reacting to the Mobil case as well as to countless other incidents of misleading ecological claims, consumers demanded that environmental claims be accurate and independently substantiated. Would the marketplace resolve the problem on its own? Not likely. No single producer wanted to be the only one to qualify its environmental claims in the name of accuracy, only to run the risk of losing market share. At the same time, the individual consumer lacked the information and the financial resources to take on corporate giants participating in this not-so-truthful eco-advertising. It was obvious that government had to step in to solve the dilemma.

In response to a call for action from consumer, industry, and environmental groups, the Federal Trade Commission (FTC) on July 28, 1992, announced environmental marketing guidelines for industry that were developed in consultation with the Environmental Protection Agency (EPA) and the Office of Consumer Affairs. These guidelines are summarized in the following list. For more detail, visit the FTC's Web site at **http://www.ftc.gov/bcp/grnrule/guides92.htm**.

http://

General Guidelines for Green Marketing

- Product labels that state environmental claims should be clear and sufficiently prominent to avoid deception.
- It should be clear whether the environmental claim refers to the product, its packaging, or both.
- Claims of environmental benefits should not be overstated.
- Manufacturers that make environmental claims through comparison with other products should clearly state the basis for comparison and be prepared to substantiate the claim.

Although the guidelines are voluntary, businesses likely will cooperate to preserve their image as environmentally conscious enterprises. In so doing, accuracy of information about the product and the effects of its consumption on the environment should improve.

Sources: Snyder (1992); Lawrence (January 29, 1991); U.S. EPA, Office of Solid Waste and Emergency Response (fall 1992); "State Your Claim" (July/August 1992).

health, recreational, and aesthetic benefits associated with pollution abatement. So even if consumers could be induced to express their WTP for a cleaner environment, it is highly likely that the resulting demand price would underestimate the true benefits. This added complication is due to **imperfect information,** which is another source of market failure.

What can be concluded from all this? As we initially claimed, market forces alone cannot provide an allocatively efficient level of a public good. This realization helps to explain what we observe in public goods markets — intervention by some third party, typically government. But to what degree should government become involved in the market process, and what tasks should government perform to achieve an efficient solution? These are difficult questions, but the underlying theories that identify the public goods problem suggest approaches to solving it.

The Solution: Government Intervention

In practice, a common means by which government responds to the dilemma of free-ridership and nonrevelation of preferences is through **direct provision of public goods.** Simple observation reveals that many public goods, such as fire protection, parks, and roadways, are provided by government. Similarly, government is involved in the preservation of natural resources and the provision of environmental quality. Among the government agencies whose responsibilities include the environment are the Army Corps of Engineers, the Federal Energy Regulatory Commission, and, of course, the Environmental Protection Agency (EPA).

An alternative government response is the use of **political procedures** and **voting rules** aimed at identifying society's preferences about public goods.[10] For example, members of Congress are responsible for discerning their constituents' views on environmental issues and representing their interests in developing legislation. Other environmental laws, such as bottle bills, are enacted at the state level through majority voting rules on state referenda.

Responding to the problem of imperfect information, governments regularly provide **education** and **public information** to citizens about public goods. For example, the EPA allocates some of its resources to educating the public about the benefits of a cleaner environment, and the Federal Trade Commission (FTC) issued green marketing guidelines to improve the accuracy of advertised environmental claims about products and packaging, an effort discussed in Application 3.2.

Environmental Problems: Externalities

Another way to model environmental problems is through **externality** theory. Instead of defining the market as environmental quality or pollution abatement, this approach specifies the relevant market as the good whose production or consumption generates environmental damage *outside* the market transaction.[11] Any such effect that is *external to* the market is aptly termed an **externality.**

Basics of Externality Theory

Microeconomic theory argues that price is the most important signaling mechanism in the market process. Equilibrium price communicates the marginal value that consumers assign to a good and the marginal costs incurred by firms in producing it. Under ordinary conditions, this theory predicts the realities of the market remarkably well. Sometimes, however, price fails to capture *all* the benefits and costs of a market transaction. Market failures such as these occur when a third party is affected by the production or consumption of a commodity. Such a third-party effect is called an **externality.** If the external effect generates *costs* to a third party, it is a **negative externality.** If the external effect generates *benefits* to a third party, it is a **positive externality.**

externality
A spillover effect associated with production or consumption that extends to a third party outside the market.

negative externality
An external effect that generates costs to a third party.

positive externality
An external effect that generates benefits to a third party.

Although the notion of an externality may seem obscure, it is nonetheless familiar conceptually. If a family purchases an unsightly satellite dish and installs it in their front yard, that action imposes costs to neighbors in the form of declining property values, a negative externality not reflected in the price of the satellite dish. Conversely, if one firm conducts research that advances a production process, there is a benefit to the entire industry, a positive externality not accounted for in the research investment decision.

[10] Such approaches are suggested by public choice theory, a subdiscipline of applied microeconomics. For an overview of public choice theory, see Rosen (2002), chap. 6.

[11] Recall from the materials balance model that residuals are generated by *both* producers and households.

Application

3.3 *Tokyo's Just-in-Time Deliveries Create a Negative Externality*

In an ironic turn of events, Tokyo's traffic congestion and air pollution have been worsened by local merchants' use of an efficiency-driven system known as just-in-time deliveries. The just-in-time (JIT) production system was developed and refined by the Toyota Motor Company during the late 1950s and early 1960s. This efficiency concept is based on the premise that firms should operate with a minimum amount of inventory to avoid waste and minimize costs. Hence, raw materials and product components should arrive only just before they are needed in manufacturing. For the system to work, deliveries of inputs must be made frequently and in a timely manner to achieve maximum efficiency.

Over the years, the use of JIT production allowed Japanese manufacturers to bring their output to market at a substantially reduced cost. The system's success in Japan's industrial sector is partly responsible for the nation's dominant position in many international markets. Following the lead of their industrial counterparts, several Japanese service industries began to institute JIT delivery systems. Thousands of convenience stores, grocery markets, and department stores concentrated in Japan's capital city now maintain only a minimum stock, relying on frequent deliveries from suppliers, sometimes as often as every few hours. The result? Just-in-time delivery in this densely populated city has backfired. Adapted incorrectly, the system designed to save time is perversely wasting time. The parade of delivery trucks through the city streets slows traffic to a crawl, contributes to air pollution, and wastes fuel.

Few would question that appropriate use of the JIT concept generates efficiency gains and resource productivity and thus substantially reduces private costs. However, Japan's local merchants have unwittingly compromised the integrity of the concept, as evidenced by delivery trucks filled to only half their capacity. When the system is misused, as in Tokyo, there are external costs to society. These include the waste of fuel, increased urban air pollution, and longer commutes for workers, none of which is reflected in the price of convenience goods and grocery products. Hence, there is a true negative externality associated with the use of the JIT delivery system by Tokyo's merchants.

Sources: Schrage (March 22, 1992); Goddard (1986).

Common to both examples is a spillover effect that occurs outside the market transaction; this effect is not captured by the price of the commodity being exchanged. If price does not reflect *all* the benefits and costs associated with production and consumption, it is unreliable as a signaling mechanism, and the market fails. An important consequence is that scarce resources are misallocated. If consumption generates external benefits, the market price undervalues the good, and *too little* of it is produced. If there is a negative externality, the market price does not reflect the external costs, and *too much* of the commodity is produced. An example of this latter phenomenon is the congestion and pollution in Tokyo associated with merchants' overuse of the just-in-time delivery system. Application 3.3 gives an inside look at this unusual dilemma.

Environmental Externalities

Of interest to environmental economists are externalities that damage the atmosphere, water supply, natural resources, and the overall quality of life. The classic case is the negative externality associated with *production*. For example, the provision of air transportation causes noise pollution, damages air quality, and reduces the value of nearby residential properties. These are real costs that are not absorbed by airlines or by air travelers. Because these costs are incurred by parties outside the market transaction, they are not captured in the price of airline tickets. Environmental externalities also can be associated with *consumption*. A good example is the cost of waste disposal associated with consuming products with excess or nonbiodegradable packaging. Application 3.4 discusses how this type of externality affected the market for CDs.

3.4 *An Industry Response to a Negative Consumption Externality: CD Product Packaging*

Compact disks (CDs), introduced into the U.S. market in 1982, were originally packaged in two boxes. The outer box was a 6-by-12-inch cardboard package, known in the industry as the long box. Inside the long box was a 5-by-5½-inch jewel box constructed of clear plastic, which housed the CD. The jewel box, which continues in use today, generally is saved by the consumer as a protective container for the CD when it is not in use. However, the long box had no practical value beyond identifying the contents at the point of sale. In fact, most consumers discarded the long box immediately after purchase, so this part of the packaging ended up in landfills or burned in municipal incinerators.

Of the 250 million CDs sold in 1990, an estimated 23 million pounds of CD packaging was discarded. Think about the costs associated with the generation and disposal of this much waste, costs that are external to the purchase and sale of CDs. The producer of the CD does not consider these external costs as part of its production expenditures, so they are not reflected in the CD's price. Who, then, bears the costs? Society as a whole has to pick up the tab, the classic symptom of a negative externality—in this case, one associated with consumption.

The obvious question is to ask why the long box was used at all in packaging CDs. Music company executives defended this packaging with two arguments. First, its large size helped to discourage shoplifting. The second and most important justification was to facilitate CD sales. The long box allowed retailers to use existing display racks originally designed for 12-by-12-inch record albums. According to one industry official, the long box saved retailers an estimated $100 million in redesign costs for their display units.

Despite the industry's arguments, followers of the green movement pressured the industry to find an alternative to the wasteful packaging. In 1991, a "Ban the Box" campaign materialized to eliminate the long box from CD packaging. Environmentally minded recording artists joined the crusade, demanding that a more environmentally friendly package design be developed.

Industry officials searched for a new design that would facilitate retailers' display needs *and* placate environmentalists. After weighing all the options, the recording industry agreed in 1992 to a voluntary ban on the long box, and since April 1993, CDs have been shipped without the objectionable packaging. The standard practice by most American companies has become the use of the shrink-wrapped jewel box.

The CD packaging controversy is an important example of how externalities persist in the marketplace. Wasteful CD packaging was used for a full decade before any action was taken. In this case, the third-party intervention came not from government but from environmentalists who made market participants aware of how their decisions adversely affected the environment.

Sources: Cox (July 25, 1991); Newcomb (May 13, 1991); Block (June 1993).

Positive externalities also can help to explain the persistence of environmental problems. Consider the market for pollution abatement equipment, such as scrubbers. Scrubbers are elaborate systems used to clean emissions from producers' smokestacks. When a producer of electricity, for example, purchases and installs a scrubber system, the benefits of cleaner air accrue to all people living in the nearby area. Because these individuals are not a part of the market transaction, the external benefits are not captured in the scrubber system's price. Resources are misallocated, and too few scrubbers are exchanged in the marketplace.

Notice that there is a qualitative relationship between the external benefits associated with pollution abatement and the external costs of pollution-generating commodities. They are just the inverse of one another. If the market is defined as the abatement equipment industry, there is a *positive* externality, and the external benefits are improved health, natural resources, aesthetics, etc. On the other hand, if the market is defined as electricity production, for example, there is a *negative* externality, and the external costs are the damages to health, natural resources, and aesthetics. Which of the two models is relevant depends solely on which market is specified.

Relationship Between Public Goods and Externalities

Environmental externalities are those affecting air, water, or land, all of which have public goods characteristics. What this implies is that, although public goods and externalities are not the same concept, they are closely related. In fact, if the externality affects a broad segment of society and if its effects are nonrival and nonexcludable, the externality is itself a public good.[12] If, however, the external effects are felt by a more narrowly defined group of individuals or firms, then those effects are more properly modeled as an externality.[13]

Modeling Environmental Damage as a Negative Externality

Having established the basics of what externalities are conceptually, we now develop a formal model of a negative environmental externality. We elect to model a production externality, because this approach addresses the most prevalent source of environmental pollution.

Defining the Relevant Market

As always, one of the first steps in model building is to define the relevant market. In this case, we define the market as refined petroleum products. This is a fitting choice because refined petroleum plants are major water polluters. According to EPA statistics, an estimated 4.1 million pounds of toxic chemicals were released to surface waters by the refined petroleum industry in 1995.[14] Among the associated external costs are serious health risks for people using the rivers and streams.

Modeling the Private Market for Refined Petroleum

To avoid complicating our analysis with the market failure of imperfect competition, we assume that the private market for refined petroleum is competitive. The hypothetical supply and demand relationships for refined petroleum products are

$$\text{Supply:} \quad P = 10.0 + 0.075Q,$$
$$\text{Demand:} \quad P = 42.0 - 0.125Q,$$

where Q is measured in thousands of barrels per day and P is the price per barrel.

Recall from Chapter 2 that supply represents the marginal costs of production and demand represents the marginal benefits of consumption, both based on *private,* or *internal,* decision making. In markets such as this one where there are also external costs, it is necessary to distinguish internal, or private, costs from external costs. To do this, we formally refer to the supply function as the **marginal *private* cost (*MPC*)** of production. For consistency, we also will refer to the demand relationship as the **marginal *private* benefit (*MPB*)** function, even though it is assumed that there are no external benefits in this market. Thus, the two functions are restated as

$$MPC = 10.0 + 0.075Q,$$
$$MPB = 42.0 - 0.125Q.$$

[12]Technically, if the externality provides *benefits* to a large component of society, it is a public *good;* if the opposite is true, the externality is a public *"bad."*

[13]For further discussion on the relationship between externalities and public goods, interested readers are referred to Holtermann (February 1972).

[14]See U.S. EPA, Office of Policy, Planning, and Evaluation (March 1991), p. 4–2; U.S. EPA, Office of Pollution Prevention and Toxics (April 1997), p. 28, table 4-10.

Figure 3.3 ***Competitive Equilibrium in the Market for Refined Petroleum***

The intersection of the *MPB* and *MPC* curves identifies the competitive equilibrium in the refined petroleum market, where P_C = \$22 per barrel and Q_C = 160,000 barrels. This equilibrium ignores the external costs of contaminated water supplies caused by production of refined petroleum products. Therefore, the *MPC* undervalues the true opportunity costs of production, and the competitive output level is too high.

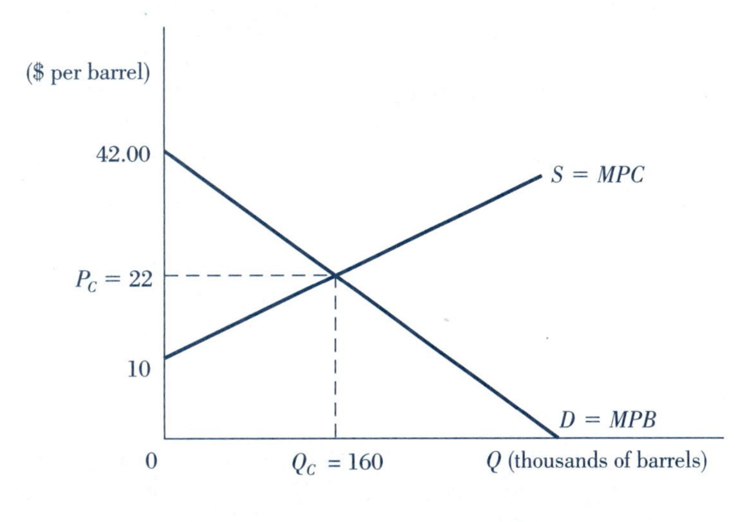

Inefficiency of the Competitive Equilibrium

Given the usual assumptions about the underlying motivations of supply and demand, the competitive market clears where $MPB = MPC$, or, equivalently, where marginal profit ($M\pi$) or $(MPB - MPC) = 0$. Solving the *MPB* and *MPC* equations simultaneously yields a competitive market price (P_C) of \$22 per barrel and a market quantity (Q_C) of 160,000 barrels. The graph is shown in Figure 3.3, where equilibrium occurs at the intersection of the *MPB* and *MPC* curves.

The problem with this equilibrium is that it ignores the external costs to society of contaminated water supplies caused by refined petroleum production. Remember, natural market forces motivate firms to satisfy their own interests, not those of society. The costs of the water pollution are *external* to the market exchange and consequently are not factored into private market decisions. The implications are serious, because allocative efficiency requires that marginal benefits be equal to *all* marginal costs of production. Because the external costs are not included in private decision making, the *MPC* undervalues the opportunity costs of production, and the resulting output level is too high.

From a practical perspective, economists want to identify and monetize these external costs, but assigning a dollar value to negative externalities is difficult. Think about trying to monetize the damage to aquatic life from water pollution or the increased health risks from swimming in a polluted lake. Although there are economic methods that estimate these costs, they are not straightforward. Therefore, we will defer discussion of these methods to later chapters. For now, we will simply argue that these costs do exist and assume for the present that they can be quantified.

Modeling the External Costs

To complete our analysis of the refined petroleum market, we model the hypothetical **marginal external cost (*MEC*)** function as

$$MEC = 0.05Q.$$

Think about the economic interpretation of this equation. Based on the constant slope of 0.05, the *MEC* resulting from water pollution is increasing at a constant rate of 0.05 with respect to oil production. Because Q is measured in thousands of barrels, this value implies that for every additional 1,000 barrels of refined oil produced, the marginal external costs of pollution rise by $0.05 per barrel.

Modeling the Marginal Social Costs and Marginal Social Benefits

marginal social cost (*MSC*)
The sum of the marginal private cost (*MPC*) and the marginal external cost (*MEC*).

To achieve allocative efficiency, we must consider the external costs in determining equilibrium price and quantity. To accomplish this, we must add the *MEC* to the firm's *MPC* to derive the **marginal social cost (*MSC*)** equation:

$$\begin{aligned} MSC &= MPC + MEC \\ &= 10.0 + 0.075Q + 0.05Q \\ &= 10.0 + 0.125Q. \end{aligned}$$

Figure 3.4 ***Comparing Competitive and Efficient Equilibria Using Marginal Benefit and Marginal Cost: Refined Petroleum Market in the Presence of a Negative Externality***

The *MSC* curve is found as the vertical sum of the *MEC* and the *MPC* curves. The intersection of *MSC* and *MSB* identifies the efficient equilibrium point at P_E = $26 and Q_E = 128,000. Notice how this compares to the competitive equilibrium where P_C = $22 and Q_C = 160,000, which corresponds to the intersection of *MPC* and *MPB*. At Q_C, *MSB* is *below* *MSC*, which means that society is giving up more in scarce resources to produce petroleum than it gains in benefits from consuming it.

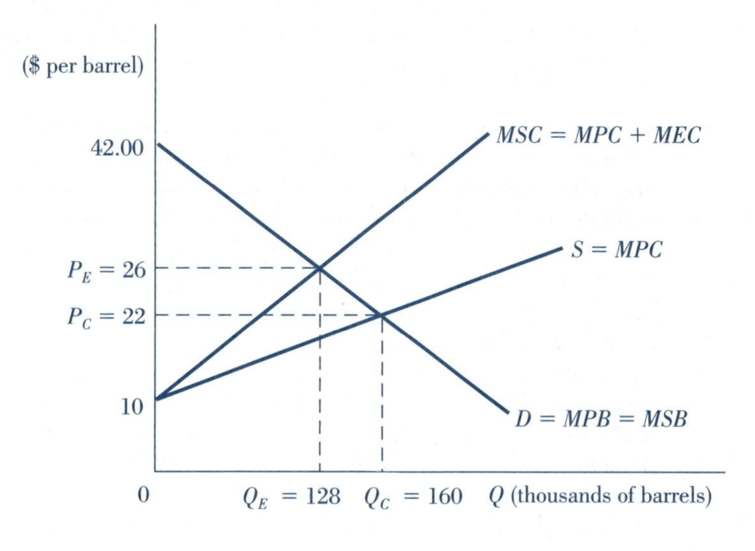

The *MSC* is relevant to production decisions because it captures all the costs of producing refined petroleum, that is, the private costs of production *and* the external costs of environmental damage to society.

On the demand side, there is an analogous benefit relationship, called the **marginal social benefit (*MSB*),** which is the sum of the *MPB* and any marginal external benefit (*MEB*). Because we have assumed that there are no positive consumption externalities, the *MEB* is zero; therefore, the *MPB* equals the *MSB* in this case.

marginal social benefit (*MSB*) The sum of marginal private benefit (*MPB*) and marginal external benefit (*MEB*).

Efficient Equilibrium

Once determined, *MSC* must be set equal to *MSB* to solve for the efficient equilibrium price (P_E) and quantity (Q_E). In this case, the efficient level of refined petroleum products is 128,000 barrels per day sold at a market price of $26 per barrel. Compare this to the competitive equilibrium of $Q_C = 160{,}000$ and $P_C = \$22$. As we asserted, the competitive equilibrium in the presence of a negative externality is characterized by an overallocation of resources to production. Furthermore, the competitive price is too low, because the *MEC* is not captured by the market transaction.

The results are shown graphically in Figure 3.4. Geometrically, the *MSC* curve is the vertical sum of the *MEC* curve and the *MPC* curve. The intersection of *MSC* and *MSB* identifies the efficient equilibrium at $P_E = \$26$ and $Q_E = 128{,}000$. The graph also shows the competitive equilibrium at $P_C = \$22$ and $Q_C = 160{,}000$, which corresponds to the intersection of *MPC* and *MPB*. Notice that at Q_C, *MSB* is *below MSC*. This signifies that society is giving up more in scarce resources to produce petroleum than it gains in benefits from consuming it. To restore the equality of $MSB = MSC$, signifying allocative efficiency, output must be decreased—precisely what is predicted by theory.

An alternative way to analyze the two equilibria is to examine the corresponding levels of $M\pi$. At the competitive equilibrium, we know that $M\pi = 0$, because that point is defined where $MPB = MPC$. At the efficient equilibrium, we know that $MSB = MSC$, which can be reexpressed as $MSB - MSC = 0$, or, equivalently, where $MPB - MPC = MEC$. We conclude that at the efficient equilibrium point, $M\pi = MEC$. These derivations are summarized by

Competitive equilibrium:

$$\begin{aligned} MPB &= MPC, \\ MPB - MPC &= 0, \\ M\pi &= 0. \end{aligned}$$

Efficient equilibrium:

$$\begin{aligned} MSB &= MSC, \\ MPB + MEB &= MPC + MEC, \\ MPB - MPC &= MEC \quad \text{(because } MEB = 0\text{)}, \\ M\pi &= MEC. \end{aligned}$$

competitive equilibrium The point where marginal private benefit (*MPB*) equals marginal private cost (*MPC*), or where marginal profit ($M\pi$) = 0.

efficient equilibrium The point where marginal social benefit (*MSB*) equals marginal social cost (*MSC*), or where marginal profit ($M\pi$) = marginal external cost (*MEC*).

These $M\pi$ relationships are shown in Figure 3.5. There is a direct correspondence between the curves in this graph and those in Figure 3.4. The $M\pi$ function in Figure 3.5 is equivalent to the vertical distance between the *MPB* curve and the *MPC* curve shown in Figure 3.4. The *MEC* function in Figure 3.5 is equivalent to the vertical distance between the *MSC* and the *MPC* curves in Figure 3.4. The efficient equilibrium, where $Q_E = 128{,}000$, occurs where $M\pi$ intersects the *MEC* function. At this point, notice that the *MEC* and $M\pi$ are equal to $6.40 per barrel:

$$\begin{aligned} MEC &= 0.05Q = 0.05(128) = \$6.40, \\ M\pi &= MPB - MPC = 42.0 - 0.125(128) - [10.0 + 0.075(128)] = \$6.40. \end{aligned}$$

The competitive equilibrium, where $Q_C = 160{,}000$, occurs where the $M\pi$ function crosses the horizontal axis or where $M\pi = 0$. However, *MEC* evaluated at this point equals $8 [i.e., $MEC = 0.05(160) = \$8$]. Thus, at the competitive equilibrium, $M\pi \neq MEC$, which means the result is inefficient.

The interpretation of this outcome is as follows. In the presence of a negative externality, efficiency requires that firms set their production levels such that the price covers not only

Figure 3.5 ***Comparing Competitive and Efficient Equilibria Using Marginal Profit and Marginal External Cost: Refined Petroleum Market in the Presence of a Negative Externality***

The efficient equilibrium is shown where $M\pi$ intersects *MEC* at $Q_E = 128{,}000$ barrels. At this point, both $M\pi$ and *MEC* equal \$6.40. The competitive equilibrium where $Q_C = 160{,}000$ occurs where the $M\pi$ function crosses the horizontal axis, or where $M\pi = 0$. The *MEC* evaluated at this point is equal to \$8. Because $M\pi \neq MEC$ at the competitive equilibrium, the result is not efficient.

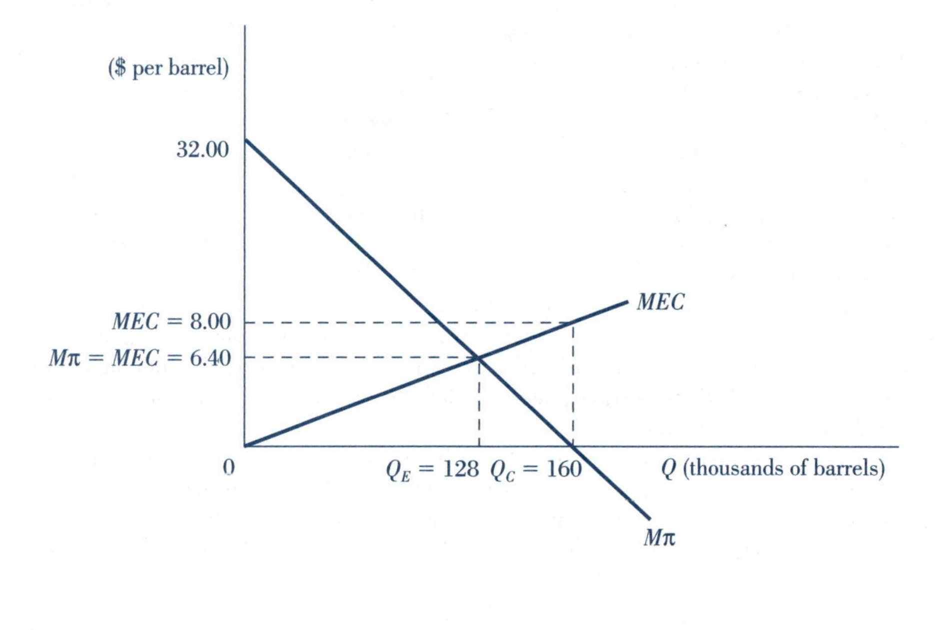

private but also external costs at the margin. In this case, the external costs are the damages to the environment, and these should be accounted for by producers in their profit decisions.

Measuring the Welfare Gain to Society

One important conclusion of the preceding analysis is that efficiency in the market for refined petroleum would improve if output were restricted by 32,000 barrels per day (i.e., 160,000 − 128,000). This output adjustment would increase society's welfare. To illustrate this, the *MPC, MSC,* and *MPB* curves are reproduced in Figure 3.6 with some added notation. Use this model to consider the separate effects on the refineries and on society associated with this output restriction.

From the firm's perspective, there is a loss in profit. Notice that as Q falls from 160,000 to 128,000, refineries lose profit, measured as the excess of *MPB* over *MPC* for each unit of output. Aggregating all the $M\pi$ values between $Q_E = 128{,}000$ and $Q_C = 160{,}000$ defines the triangular area *WYZ*, which represents the total loss in profits. However, from the vantage point of society, there is a measurable gain equal to the accumulated *reduction* in *MEC* associated with the output decline. This reduction in external costs represents the decrease in health and ecological damage. Geometrically, this gain is area *WXYZ*. On net, society gains an amount equal to the triangular area *WXY* as efficiency is restored. Notice that, although the movement to the efficient output level achieves a welfare gain, it does not entirely eliminate the exter-

Figure 3.6 ***Assessing the Net Gain to Society of Restoring Efficiency in the Refined Petroleum Market***

To restore efficiency, Q would have to be reduced from 160,000 to 128,000 barrels. As this output restriction occurs, refineries lose profit, represented as the triangular area *WYZ*. At the same time, society gains the associated decline in damages to health and the ecology, shown as area *WXYZ*. On net, society gains by the triangular area *WXY* as a result of the restoration of efficiency.

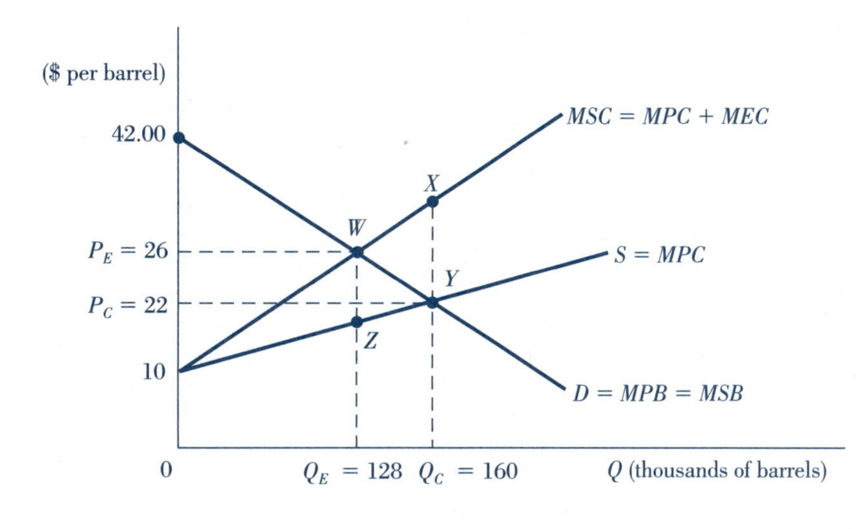

nality. Firms are still generating some amount of pollution at Q_E, but the amount is lower than it would be at Q_C.

In sum, if production of a commodity generates a negative externality, the market will yield an inefficient solution with too many resources allocated to production. If that externality were somehow accounted for within the market, society as a whole would gain. Of course, the operative issue is *how* to account for externalities such that efficiency can be restored.

Market Failure Analysis

It is important to understand the lack of incentive in the natural market process to explicitly account for external costs. Petroleum refineries are motivated by *private* gain, not by *social* gain. Although these firms may be aware of the environmental damage associated with their production, there is no incentive—in fact there is a disincentive—for them to absorb these costs. Doing so would negatively affect profits. It would be as though firms offered to pay for the external costs on society's behalf. However, there is no market incentive for a rational firm to incur higher costs than it has to, even if it is for the good of society.

These assertions should not deter society's efforts to solve the problems of environmental damage. Quite the contrary. Market failure models give us a better understanding of *why* we observe increasing damage to the physical environment as industrial production has intensified throughout the world. The theory also explains the persistence of environmental problems from a market perspective and the need for government intervention where such problems arise. Finding appropriate policy solutions is not easy, but the process is facilitated by an understanding of how and why markets fail. On that point, if we consider both the public goods problem and the externality model we have examined, an important common element leads us to the source of virtually all environmental problems—the **absence of property rights.**

The Absence of Property Rights

In every model presented thus far in this chapter, the market has failed to provide an efficient solution. Using basic elements of economic theory, we have demonstrated *how* the market fails and have even quantified the failure in terms of the overproduction of output. Although in each case the failure has been examined conceptually, we have not yet focused on the underlying root of the problem.

Reconsider the negative externality model, and recall the assertion that an externality is a public good if it affects a broad segment of society. In our model, the external cost is damage to water supplies, which does fit that characterization. Furthermore, clean water possesses the two distinguishing characteristics of a public good. Now recall our public goods model in which the relevant market was defined as air quality. What do these two public goods have in common? The answer is that in each case, the **property rights** of the good are undefined, and, as a result, markets for these goods are virtually nonexistent. **Property rights** are the set of valid claims to a good or resource that permits its use and the transfer of its ownership through sale. These rights are generally limited by law and/or social custom.

property rights
The set of valid claims to a good or resource that permits its use and the transfer of its ownership through sale.

In the context of environmental public goods, it is unclear who "owns" the rights to water supplies, or who "owns" the rights to the air. For example, do swimmers own the right to clean water, or do refineries hold the rights to pollute it? Do individuals own the right to breathe clean air, or do polluting firms own the right to contaminate it? Because the answers to these questions are not clear-cut, there is no built-in market mechanism to solve environmental problems. It turns out that property rights are critically important to the sound functioning of the market system. In fact, as pointed out by Nobel laureate Ronald Coase, the assignment of property rights alone can provide for an efficient solution even in the presence of an externality.[15]

The Coase Theorem

Coase's exposition about property rights and their relevance to externality problems is so significant that it has come to be known as the **Coase Theorem.** This theorem states that proper assignment of property rights to any good, even if externalities are present, will allow bargaining between the affected parties such that an efficient solution can be obtained, no matter which party is assigned those rights. Two important underlying assumptions of this theory are noteworthy:

Coase Theorem
Assignment of property rights, even in the presence of externalities, will allow bargaining such that an efficient solution can be obtained.

- Transactions are costless.
- Damages are accessible and measurable.

To illustrate the Coase Theorem, we revisit our market model of refined petroleum to see how the assignment of property rights and bargaining can restore efficiency. We impose the assumptions given above and test the theory's prediction by examining the outcome under two different assignments of rights to some hypothetical river. One is that private individuals hold the rights to the river for recreational use, and the other is that refineries hold the rights to release toxic chemicals into the river.[16]

[15] In 1991, Ronald Coase received the Nobel prize in economics for his pioneering work in the theory of transactions costs and other concepts that link economic theory and the law (Coase, October 1960). Much of his research is relevant to the economics of environmental problems.

[16] Although it would be more realistic to assume that only a single refinery is involved in a bargaining scenario about the use of a river, it is simpler to approach the problem from an industry perspective so that we can use the same data and graphical models developed in the negative externality discussion.

Figure 3.7 ***Bargaining in the Refined Petroleum Market with the Assignment of Property Rights***

If the refineries own the rights to the river, bargaining begins at Q_C and continues to Q_E because at all output levels between these points, the payment ρ satisfies the condition $MEC > \rho > M\pi$, which is acceptable to both parties. If the recreational users own the rights, bargaining begins at $Q = 0$ and continues up to Q_E. This occurs because at all output levels between these points, the payment ρ satisfies the condition $M\pi > \rho > MEC$, which is acceptable to both parties.

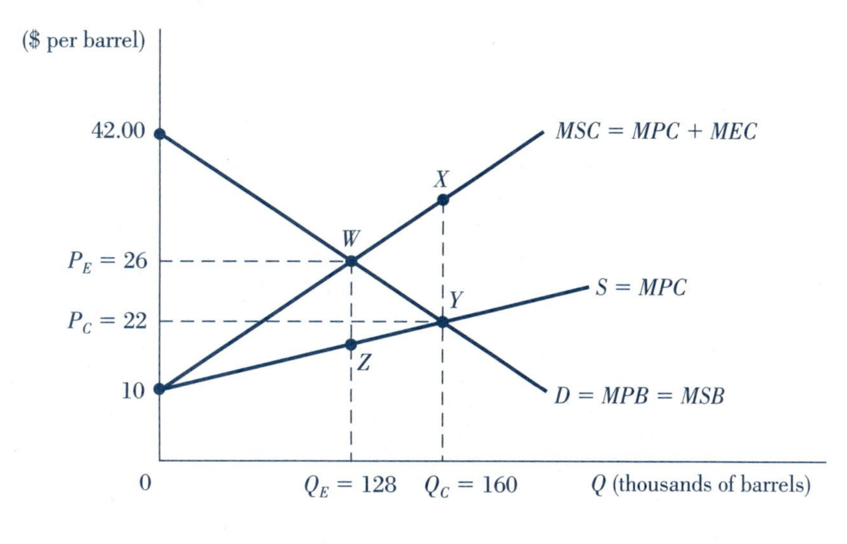

Bargaining When Property Rights Belong to the Refineries

We begin by assigning the property rights to the refineries. This means that the refineries have the right to pollute the river as part of their production processes. Although this assignment of rights may appear to be perverse, recall that according to the Coase Theorem, this should have no effect on the outcome at all.

Remember that the refineries are interested in producing petroleum products up to a level that maximizes profit—by itself, not an objectionable motive in a free enterprise system. The pollution they cause is an unintended by-product of production. Nonetheless, the recreational users of the river are harmed by the associated pollution, or, more formally, their utility is negatively affected. Knowing that the refineries have the right to pollute and given their own motivation to maximize utility, recreational users have an incentive to negotiate. For each unit change in output, they would be willing to pay the refineries *not* to pollute up to an amount equal to the associated negative effect on their utility. The refineries, on the other hand, would be willing to accept payment *not* to pollute as long as that payment is greater than the loss in profits from cutting back production.

The terms of the negotiation can best be understood by examining Figure 3.7, which replicates the negative externality model. As before, Q_C represents the profit-maximizing competitive equilibrium, and Q_E is the efficient equilibrium. We assume that the market is at Q_C when the bargaining begins, because this is the point that the owners of the rights, the refineries, would choose. Each party's bargaining position is given in the context of the cost and benefit curves in the diagram:

Recreational users: Willing to offer a payment (ρ) such that $\rho < (MSC - MPC)$.
Refineries: Willing to accept a payment (ρ) such that $\rho > (MPB - MPC)$.

Note that the value of $(MSC - MPC)$ measured vertically is simply the *MEC*, the value of the marginal damage incurred by the recreational users for each added unit of production. The vertical distance measured by $(MPB - MPC)$ is the $M\pi$ earned by the refineries for each extra unit of output produced. In theory, the bargaining between the two groups should continue as long as the payment is greater than the refineries' loss in profit but less than the recreational users' damage. In the context of the model, payment ρ will be acceptable to both parties as long as the following condition holds:

$$(MSC - MPC) > \rho > (MPB - MPC), \quad \text{or, equivalently,} \quad MEC > \rho > M\pi.$$

Find these vertical distances in Figure 3.7. Note that at the competitive equilibrium (Q_C) the refineries' $M\pi$ equals zero, or $(MPB - MPC) = 0$. On the other hand, the *MEC* at Q_C is some positive value, or $(MSC - MPC) > 0$, represented as distance *XY*. Thus, bargaining between the parties is feasible at the competitive equilibrium, because $MEC > M\pi$ at Q_C. Moreover, because this condition holds for all output levels between Q_C and Q_E, the parties will continue to bargain all the way to Q_E, where negotiation ceases. The reason? Because at Q_E, $(MSC - MPC)$ is exactly equal to $(MPB - MPC)$, or $MEC = M\pi$, each measured as distance *WZ*. Output reductions beyond that point would generate a loss in profits to the refineries greater than what the recreational users would be willing to pay, or $M\pi > MEC$, so bargaining would break down.

This is a profound result. Under the assumptions of Coase's model, assigning the property rights to the refineries will lead to bargaining between the parties, which ultimately generates an efficient outcome without any third-party intervention.

Bargaining When Property Rights Belong to the Recreational Users

Now reconsider the model with the property rights assigned to the recreational users. Remember that according to the Coase Theorem, an efficient outcome can be achieved regardless of which of the affected parties controls the property rights. We should expect to obtain the same result as before. The starting point at which bargaining begins, however, is different. If the recreational users hold the rights to the river, technically the refineries cannot produce at all unless they pay for the rights to pollute. Thus, we assume that the point at which bargaining begins is at $Q = 0$.

The recreational users pursue their rights to a clean river, not to obstruct production of the refineries but to maximize their utility. Nonetheless, the refineries' profit objective is negatively affected. Hence, it is they who have a market-based motivation to bargain with the recreational users in an attempt to get them to accept payment for their rights to the river. The terms of the bargaining are analogous to the previous scenario, except that now the refineries are in the *offering* position and the recreational users are in the *accepting* position.

Refer to Figure 3.7, and consider the two sides' relative positions at $Q = 0$. For each unit change in output, the refineries are willing to pay for the right to pollute (and thus the right to produce output), up to an amount equal to the $M\pi$ received from production, measured as the vertical distance between *MPB* and *MPC*. The recreational users are willing to trade off their rights to clean water only if they receive payment greater than the damage they incur as the river becomes polluted. This damage is of course the *MEC*, measured as the distance between *MSC* and *MPC*. Thus, each party's bargaining stance is as follows:

Refineries: Willing to offer a payment ρ such that $\rho < (MPB - MPC)$.
Recreational users: Willing to accept a payment ρ such that $\rho > (MSC - MPC)$.

Therefore, there is opportunity for bargaining to proceed as long as the following condition holds:

$$(MPB - MPC) > \rho > (MSC - MPC), \quad \text{or, equivalently,} \quad M\pi > \rho > MEC.$$

The model presented in Figure 3.7 confirms that this condition holds at $Q = 0$ and continues to hold for all output levels up to Q_E. At Q_E, bargaining ceases, because at that point $(MPB - MPC)$ is exactly equal to $(MSC - MPC)$, or $M\pi = MEC$, each represented by distance WZ. Note that beyond Q_E, the marginal damage associated with increased production exceeds the addition to profit that the refineries would receive, or $MEC > M\pi$, so bargaining would break down. Once again, the assignment of property rights, this time to the recreational users, leads to an efficient outcome without any government intervention.

Limitations of the Coase Theorem

Coase's model yields a powerful result and one that underscores the importance of property rights to the market process, regardless of who is assigned those rights. However, as was stated at the outset, the model's prediction of an efficient outcome depends on two very limiting assumptions: that transactions are costless and that damages are accessible and measurable. Thus, for the theory to hold in practice, at minimum it must be the case that very few individuals are involved on each side of the market.

The reality of the refinery market, and in fact of most markets, is that there are many affected parties on both sides of the market. Nontrivial costs would be associated with trying to reach a consensus about the bargaining terms within each group even before the negotiation could begin. Undoubtedly, legal counsel would be necessary, adding still more to the transactions costs. Then there is the difficult problem of identifying the sources of the damage and attempting to assign a value to that damage. As more parties get involved, the task becomes increasingly difficult.

Common Property Resources

It turns out that property rights need not be completely missing for a market problem to exist. If property rights exist in some form but are ill defined, the outcome will also be an inefficient one. Such is the case for **common property resources,** which represent another source of externalities. **Common property resources** are those for which property rights are shared by some group of individuals.

common property resources
Those resources for which property rights are shared.

Notice that common property resources fall somewhere on a continuum between the extremes of pure public goods and pure private goods. Unlike pure public goods, common property resources are not accessible to everyone, meaning there is some measure of excludability.[17] However, because the property rights extend to more than one individual, they are not as clearly defined as they are for pure private goods. Some classic examples of common property resources are fisheries and oil pools.

With common property resources, the problem is that public access without any control leads to resource exploitation, which in turn generates a negative externality. The problem arises because each co-owner makes decisions about using the resource based only on private costs and benefits, ignoring how that decision would affect other owners. For example, consider a lake stocked with fish for exclusive use by members of a local community. Unless the catch is limited in some way, each person would fish from the lake based on private motivations, ignoring how the associated decline in the fish population would negatively affect others' ability to share in the benefits. Furthermore, each individual would have no incentive to consider the ultimate cost of restocking the lake, because their personal share of that cost would be very small. The result is overuse or depletion of the scarce resource.

[17] Put another way, one can think of public goods as an extreme case of common property resources, when the group that shares the property rights consists of all individuals.

The Solution: Government Intervention

From an economic perspective, the general solution to externalities, including those affecting the environment, is to **internalize the externality,** that is, to force the market participants to absorb the external costs or benefits. One way this can be accomplished is through the assignment of property rights. In our model, when the refineries owned the rights to pollute, the recreational users internalized the externality through their payment offer. Conversely, when the recreational users owned the rights to clean water, the refineries internalized the external cost by paying for the right to pollute. But how do these rights get assigned in the first place? In practice, the government would have to make this determination as well as enforce limitations on these rights for the good of society.

Other approaches to internalizing environmental externalities are policies that change the effective price of a product by the amount of the associated external cost or benefit. In the petroleum refinery market, for example, the price per barrel of oil could be forced up by the amount of the *MEC,* perhaps by a unit tax. More recent policy prescriptions involve the most direct form of internalizing environmental externalities, which is to establish a market and a price for pollution. These approaches will be investigated in later chapters, all of which are rooted in market failure theory.

Conclusions

From an economic perspective, environmental problems persist because they are market failures. Whether we model pollution as a negative externality or as damage to environmental public goods, we observe conditions that impede natural market forces. At the root of the dilemma is the absence of property rights. Because no one owns the atmosphere or the earth's water bodies, there are no market incentives to pay for the right to protect these resources or for the right to pollute them. The result is a misallocation of economic resources and a decline in society's welfare. Some third-party mediation, typically government, is needed to correct the market failure and reach an efficient equilibrium.

How strong a presence should government have in affected markets, and how should government go about the difficult task of developing effective policy solutions? Although market failure models communicate what the overall approach should be, many practical issues have to be addressed. In theory, we know that the value of an environmental externality must be internalized so that the efficient solution can be identified. But *how* is this accomplished in practice, and is this a practical solution? Are there methods that determine some proxy measure of the costs and benefits of pollution abatement? Or are there better, alternative solutions or more reasonable objectives that should be pursued?

In the next two chapters, we will begin to answer these questions by investigating various policy solutions, ranging from legislated regulations to market-based initiatives. By using the analytical modeling tools developed thus far, we will critically evaluate different policy approaches to reducing environmental pollution.

Summary

- There are two basic explanations for the economic assessment of environmental problems as market failures: Environmental quality is a public good; and pollution-generating products are associated with externalities.
- A pure public good is one that is both nonrival and nonexcludable in consumption.
- Market demand for a public good is found by vertically summing individual demand curves.

- The market failure of public goods exists because demand is not readily identified. The market failure arises because of nonrevelation of preferences, which in turn is due to free-ridership.
- Even if consumers revealed their willingness to pay, the resulting price likely would underestimate the good's true value because of imperfect information.
- Governments respond to the public goods problem through direct provision of public goods, political procedures and voting rules, and education and information.
- An externality is a third-party effect associated with production or consumption. If this effect generates costs, it is a negative externality; if it yields benefits, it is a positive externality.
- In the presence of a negative (positive) externality, the competitive equilibrium is characterized by an overallocation (underallocation) of resources such that too much (too little) of the good is produced.
- In a negative externality model, the competitive price is too low because the marginal external cost (*MEC*) is not captured by the market transaction.
- To identify the efficient equilibrium, the *MEC* must be added to the marginal private cost (*MPC*) to derive the marginal social cost (*MSC*), which must be set equal to the marginal social benefit (*MSB*).
- The source of the public goods problem and of externalities in private markets is that property rights are not defined.
- The Coase Theorem argues that under certain conditions the assignment of property rights will lead to bargaining between the parties such that an efficient solution can be obtained.
- If property rights exist but are ill defined, such as in the case of common property resources, the market solution is inefficient because of the presence of externalities.
- Solutions to market failures typically involve government intervention, which may include regulation, tax policy, or market-based solutions.

Key Concepts

market failure
public good
nonrivalness
nonexcludability
market demand for a public good
nonrevelation of preferences
free-ridership
externality
negative externality
positive externality
marginal social cost (*MSC*)
marginal social benefit (*MSB*)
competitive equilibrium
efficient equilibrium
property rights
Coase Theorem
common property resources

Use the Key Concepts listed above to begin your search for additional articles and information using the InfoTrac College Edition database.

Review Questions

1. Use economic analysis to evaluate the following statement: *The only amount of acceptable pollution is no pollution at all.*
2. Using the theory of public goods, explain the logic of *why* in some resort communities, the ownership of waterfront homes also includes some defined area along the beach.

3. Recall the model of the refined petroleum market given in the text. Use the following graph to answer the questions below.

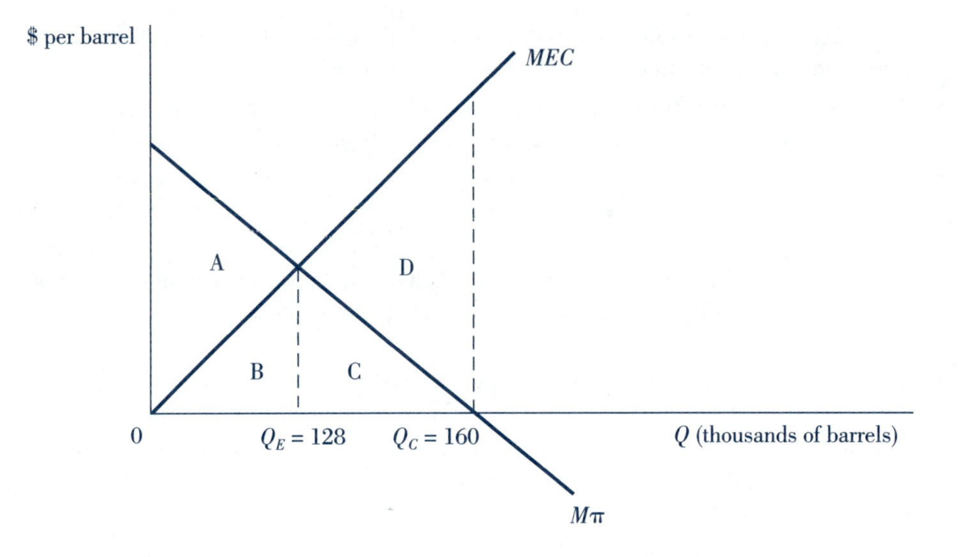

a. Give the economic interpretation of areas A, B, C, and D.
b. Which area represents the loss to the petroleum refineries as a result of the restoration of efficiency? To which area in Figure 3.6 is this area equivalent?
c. Which area represents the net gain to society? Should the reduction in output from Q_C to Q_E take place? Why or why not?

4. Using the graph shown in Question 3, describe the bargaining process between the refineries and the recreational water users, assuming the refineries have the right to pollute.

5. Suppose you serve on an environmental policy planning board for the federal government. Your task is to propose a policy initiative aimed at reducing urban air pollution, using the assignment of property rights according to the Coase Theorem. Assume that the major polluters to be targeted are commuter bus companies and that the major parties affected are the collection of city dwellers—residents and workers. Describe your proposal in detail, and include the following:
a. Whether the rights are to be assigned to city dwellers or to bus companies.
b. How these rights are to be defined.
c. How the rights are to be distributed to the parties.
d. Whether the rights are to be auctioned or sold, and, if so, at what price.

Additional Readings

Bator, F. M. "The Anatomy of Market Failure." *Quarterly Journal of Economics* 72 (1958), pp. 351–79.

Braden, John B., and Charles D. Kolstad, eds. *Measuring the Demand for Environmental Quality.* Amsterdam: North Holland, 1991.

Cornes, Richard, and Todd Sandler. *The Theory of Externalities, Public Goods, and Club Goods.* Cambridge: Cambridge University Press, 1987.

Dasgupta, Partha. "The Environment as a Commodity." In Dieter Helm, ed., *Economic Policy Towards the Environment.* Cambridge: Blackwell Publishers, 1991.

Fullerton, Don, and Robert N. Stavins. "How Economists See the Environment." *Nature* 395 (1998), pp. 433–34.

Hardin, Garrett. "The Tragedy of the Commons." *Science* 162 (December 13, 1968), pp. 1243–48.

Head, John G. "Public Goods and Public Policy." Reprinted in John G. Head, ed., *Public Goods and Public Welfare*, pp. 164–83. Durham, NC: Duke University Press, 1974.

Hyman, David N. *Public Finance: A Contemporary Application of Theory to Policy.* Fort Worth, TX: Dryden Press, 2002.

Popp, David. "Altruism and the Demand for Environmental Quality." *Land Economics* 77 (August 2001), pp. 339–49.

Related Web Sites

http://

Massachusetts Water Resources Authority	**http://www.mwra.state.ma.us**
U.S. Federal Trade Commission, 1992 "Green Guides"	**http://www.ftc.gov/bcp/grnrule/guides92.htm**

PART 2

Modeling Solutions to Environmental Problems

Recognizing the risks of environmental pollution helps us to appreciate what is at stake if the problem goes unchecked. Knowing how the market functions helps us to understand its inability to correct the problem on its own. Solutions to reduce the effects of polluting residuals must come from outside the market, generally in the form of government policies and programs. But how does government go about designing environmental policy? What objectives does it establish to protect society's health and welfare, and what kinds of policy instruments does it use to achieve those objectives? These are tough questions, and in this two-chapter module, we will begin to answer them.

At the outset, we know that government solutions will not eliminate pollution entirely. The laws of science and the materials balance model tell us that much. However, determining exactly where environmental targets should be set is one of the more difficult and even controversial elements of policy development, as is the specific approach undertaken by government to intervene in pollution-generating markets. Governments all over the world use a variety of policy instruments aimed at reducing environmental risks, some more successfully than others. These range from regulations that directly control the activities of polluters to incentive-based plans that use market forces and the price mechanism to achieve a cleaner environment.

In this module we introduce this broad range of policy alternatives by exploring both conventional and economic solutions to environmental degradation. Conventional policy is the subject of Chapter 4, where we investigate the traditional use of standards to define environmental objectives along with the command-and-control approach to implementing these standards. In Chapter 5, we examine the market approach, which has recently been adopted by the United States and other nations as a secondary form of environmental control policy. By using economic modeling and the criteria of allocative efficiency and cost-effectiveness, we evaluate the implications of using these different regulatory strategies to solve the complex problem of environmental pollution.

Chapter 4

Conventional Solutions to Environmental Problems: The Command-and-Control Approach

In Chapter 3, the public goods model and the theory of externalities were used to show how pollution is the result of market failures—failures that arise because of the absence of property rights. Because no one owns the atmosphere or the earth's rivers and streams, there is no market incentive to prevent or correct contamination of these resources. According to the Coase Theorem, assigning property rights would solve the dilemma, but only under certain limiting conditions. The bottom line is that government must act as a third-party mediator in those markets where pollution problems arise.

Recognizing the need for government to correct environmental market failures is an important observation gained through economic modeling, but it is only a first step. We also can use these models to determine *how* government should respond to achieve effective policy solutions. For example, economic theory maintains that government should set objectives to achieve **allocative efficiency,** balancing social benefits and costs at the margin. However, government is generally not motivated by efficiency, and, even when it is, it is unlikely that these benefits and costs can be determined exactly. Nonetheless, the efficiency criterion is useful in assessing whatever policy objectives are set relative to their optimal level. Also, the criterion of **cost-effectiveness** can be used to evaluate how these objectives are being implemented, even an objective set at something other than its efficient level.

Our goal in this chapter is to analyze government's use of conventional policy solutions to respond to environmental market failure. We begin by providing an overview of environmental standards and their role in policy development. We use allocative efficiency to assess the level at which standards are set to define environmental objectives. Next, we provide an overview of the two broadly defined approaches to implementing these standards-based objectives: the **command-and-control approach** and the **market approach.** Finally, we investigate the cost-effectiveness of the command-and-control approach, the more conventional of the two, deferring an analysis of the market approach to Chapter 5.

Use of Standards in Environmental Policy

Standards are the fundamental basis of most environmental policies. In the United States, setting standards follows a lengthy set of procedures involving scientific research and a series of formal reviews. The EPA is charged with oversight of these tasks and for making a formal recommendation about how these standards are to be defined. Ultimately, the standards are legislated by Congress and subsequently monitored for compliance and enforced by the EPA.[1]

ambient standard
A standard that designates the quality of the environment to be achieved, typically expressed as a maximum allowable pollutant concentration.

Types of Environmental Standards

When environmental standards are defined in the law, they can be specified as **ambient standards, technology-based standards,** or **performance-based standards. Ambient standards** designate the desired quality level of some element of the environment, such as

[1] In subsequent chapters, we will elaborate on these processes that define environmental policy development. For access to information on environmental regulations and rulings on the Internet, students can visit the EPA Web site at **http://www.epa.gov/epahome/rules.html.**

the outdoor air or a body of water. These standards typically are expressed as a maximum allowable concentration of some pollutant in the ambient environment. The United States uses ambient standards to define air quality and water quality. In each case, the ambient standard is not directly enforceable but serves as a target to be achieved through a pollution limit, which in turn is implemented through one of the other types of standards.

As its name implies, a **technology-based standard** stipulates the type of abatement control that must be used by all regulated polluting sources. In the United States, the EPA is responsible for researching available technologies and evaluating their relative effectiveness in accordance with certain criteria outlined in the law. It then selects the "best" technology, which subsequently must be adopted by all regulated polluters.[2] The motivation is straightforward—to ensure a specific limit on pollution releases by controlling *how* that limit is to be achieved. For example, to reduce sulfur dioxide emissions, the EPA might require all coal-burning power plants to use a scrubber system, forcing each one to achieve the same level of abatement in precisely the same way.

technology-based standard
A standard that designates the equipment or method to be used to achieve some abatement level.

An alternative type of environmental standard is performance based. A **performance-based standard** specifies an emissions limit to be achieved by every regulated polluter but does not stipulate the technology to be used to achieve that limit. By definition, performance-based standards are more flexible than their technology-based counterparts. They implicitly allow polluting sources to choose *how* they will reduce pollution releases, as long as they meet the statutory emissions limit.

performance-based standard
A standard that specifies a pollution limit to be achieved but does not stipulate the technology.

Economic Implications of Using Standards

Although the use of standards sounds straightforward enough, there are two important economic implications to be considered. The first deals with the *level* at which standards are set. This is an important issue because standards define environmental quality objectives. For example, a standard that limits carbon monoxide emissions defines an "acceptable level" of that pollutant for society. From an economic perspective, the relevant issue is whether that level achieves **allocative efficiency.** If not, there is a welfare loss to society.

A second implication of using standards relates to *how* they are implemented across polluting sources. Policy implementation is concerned with the selection of **control instruments,** such as pollution limits or taxes. The decision determines not only whether the objectives are realized but also whether they are achieved in a **cost-effective** manner. If not, resources are wasted.

We can investigate these implications through a two-part economic evaluation that centers on the following questions:

- Are the standards being used to define environmental objectives set at a level that is **allocatively efficient**? That is, does the marginal social cost of pollution abatement equal the marginal social benefit?
- Given some predetermined environmental objective, is the implementation of that objective conducted in a **cost-effective** manner?

Are Environmental Standards Set at an Allocatively Efficient Level?

allocatively efficient standards
Standards set such that the associated marginal social cost (*MSC*) of abatement equals the marginal social benefit (*MSB*) of abatement.

Because standards define environmental objectives, it is important to determine whether these objectives are set to achieve **allocative efficiency.** This condition holds if resources are allocated such that the associated benefits and costs to society are equal at the margin. Therefore, we need to develop these benefit and cost concepts specifically for the pollution

[2] The meaning of "best" in this context is one that is often the subject of debate, an issue we will investigate in upcoming chapters.

Application

4.1 *Industrial Pollution and Damages to Human Health: Catano, Puerto Rico*

For years, the people of Catano, Puerto Rico, blamed their persistent health problems on the town's severe air pollution. The residents live with a barrage of emissions released by a nearby oil refinery, a sewage sludge incinerator, ships in San Juan Bay, and a parade of 18-wheelers transporting goods from nearby docks. According to the island dwellers, however, most of the problem is caused by two giant power plants operated by the Puerto Rico Electric Power Authority. Together, the two facilities house 10 generating units with a combined capacity of 1,086 megawatts, and they release an average of 100 million pounds of sulfur dioxide emissions into Catano's atmosphere each year. One of the town's residents leading the charge against the utility says that Catano's air can be likened to a "toxic soup."

Acting on the residents' complaints, several agencies conducted health studies. Findings by a Puerto Rico Medical Association study showed that cancer rates in Catano were nearly twice the national average. The report found approximately 362 incidents of cancer for every 100,000 residents in Catano in the 1987–88 study year, a dismal statistic compared to Puerto Rico's average rate of 199 incidents for every 100,000 residents. Still more disturbing evidence came from a U.S. Public Health Service investigation, which found an alarming rate of respiratory disease among the people of Catano. Because it is well known that long-term exposure to sulfur dioxide causes respiratory ailments, this report validated the suspicion that the utility's emissions were the primary cause of health damages in the town.

Responding to these disturbing medical reports, the EPA began its own study of the area, focusing on the power plants. What they found confirmed the accusations of Catano's residents. Pollutant releases from the plants were in violation of air quality regulations. The EPA conducted what are called opacity tests, which measure the amount of light that can pass through emissions. The results showed that the opacity levels of the power plants' emissions were two to four times the allowable limit. These findings were forwarded to the U.S. Justice Department to bring Puerto Rico's utility back into compliance.

Source: Ross (January 7, 1993).

abatement market, expanding on what we presented in Chapter 3. Our objective is to learn precisely what is required to identify an allocatively efficient abatement level so that we can assess the likelihood of the government achieving such an outcome by setting a standards-based environmental objective.

Marginal Social Benefit of Abatement

marginal social benefit (*MSB*) of abatement
A measure of the additional gains accruing to society as pollution abatement increases.

As pollution is abated, the social gains are all the benefits associated with a cleaner environment, such as improvements in health, ecosystems, aesthetics, and property. If we measure how these benefits increase relative to increases in abatement, we arrive at the **marginal social benefit (*MSB*) of abatement.** It is equally correct to think of this *MSB* as a measure of the *reduction in damages or costs* caused by pollution.[3] In theory, if we were to add up all the marginal reductions in environmental external costs across every market where pollution is reduced, we would arrive at the *MSB.* We have actually modeled damage reduction, although in a limited context, in our discussion of bargaining in Chapter 3. By paying refineries to pollute less, recreational users of the river gained the *reduction* in the marginal external cost (*MEC*) associated with refined petroleum production. A real-world case in which damages were shown to be directly attributable to industrial pollution is in the town of Catano, Puerto Rico, the subject of Application 4.1.

[3]Technically, the *MSB* also includes the reduction in social costs from attempting to avoid the effects of pollution, such as the costs of air purification systems, bottled water, or water filtration systems.

4.2 Abatement Costs: The Promise of Remediation Technology

Application

Soil contamination is an environmental problem of major proportions. This seemingly esoteric subject has captured world attention in large part because of the damage caused by major oil spills. The 1989 *Valdez* incident, for example, despoiled Alaskan shorelines with 11 million gallons of oil. Such disasters are international news both because the environmental implications are severe *and* because the abatement costs are so staggering. Adding to these concerns is the well-publicized problem of deliberate hazardous waste dumping. Many landfill sites have become contaminated with cancer-causing substances, ranging from polychlorinated biphenyls (PCBs) to heavy metals. Based on EPA estimates, there were more than 40,000 contaminated sites throughout the United States as of 2000.

Projected abatement costs to clean up the worst 1,400-plus U.S. hazardous waste sites bears an impressive price tag: as high as $750 billion, according to some estimates. In 1992, actual accumulated expenses for abating just 84 sites were reported to be $11 billion. Oil spill abatement expenses are no less disturbing. Exxon spent $3 billion to clean up the beaches damaged by the *Valdez* oil spill. These and other cleanup expenditures prompted new research to find lower cost-abatement technologies.

Recently, scientists have made important strides in contaminated soil abatement using a technology called **bioremediation.** Bioremediation is a relatively low-cost technique that relies on bacteria to consume waste materials. Fertilizer sprays are used to stimulate the feeding of these microorganisms, which accelerates the usually lengthy process of returning soil to its original state. Used for some 50 years to clean wastewater treatment plants, the process is now being employed to abate toxic contaminants. In its most successful application to date, bioremediation helped to clean up the oily mess left by the *Valdez* on Alaskan beaches. Within three weeks, the soil was restored down to a foot below the surface.

Reportedly, contractors are charging in the range of $50 to $100 per ton for this innovative process. The table shows how this cost compares to that of more conventional methods, such as on-site or off-site incineration.

Abatement Method	Per Ton Abatement Cost ($)
On-site soil incineration	300
Off-site soil incineration	1,000
Bioremediation	50–100

Many continue to believe that bioremediation holds great promise for the environment. Such treatment processes provide decision makers with a wider range of alternatives to abate pollution and greater opportunities to find cost-effective solutions.

Sources: U.S. EPA, Office of Communications, Education, and Public Affairs (April 1992) p. 27; Hong and Galen (May 11, 1992); Hof (June 4, 1990); "Updating Previous Big Spills" (January 6, 1993); U.S. EPA, Office of Solid Waste and Emergency Response (October 3, 2000).

From a market perspective, the *MSB* of abatement is society's *demand* for pollution abatement, or, equivalently, its demand for environmental quality.[4] Just as the recreational users were *willing to pay* the refineries for a cleaner river, society is *willing to pay* for a cleaner environment. We expect this willingness to pay to decline with increasing levels of abatement, just as the Law of Demand predicts. Hence, the *MSB* is negatively sloped.

Marginal Social Cost of Abatement

On the supply side, we need to model the costs to society as polluters reduce their releases of contaminating residuals. This relationship is referred to as the **marginal social cost (*MSC*)**

marginal social cost (*MSC*) of abatement
The sum of all polluters' marginal abatement costs plus government's marginal cost of monitoring and enforcing these activities.

[4] Although we generally model demand as the *MPB*, which differs from the *MSB* by the amount of any marginal external benefit (*MEB*), in this context there is no *MEB*, because the demanders are *all* of society. Thus, there are no third parties to which any external benefits could accrue.

Figure 4.1 ***Single Polluter's Marginal Abatement Cost (MAC)***

A typical *MAC* curve is positively sloped and increases at an increasing rate. Notice that as the firm continues to abate from A_1 to A_2, the *MAC* increases by a proportionately greater amount from MAC_1 to MAC_2. This reflects the fact that as the abatement process continues and the environment becomes cleaner, it becomes increasingly difficult and therefore more costly to remove each additional unit of pollution.

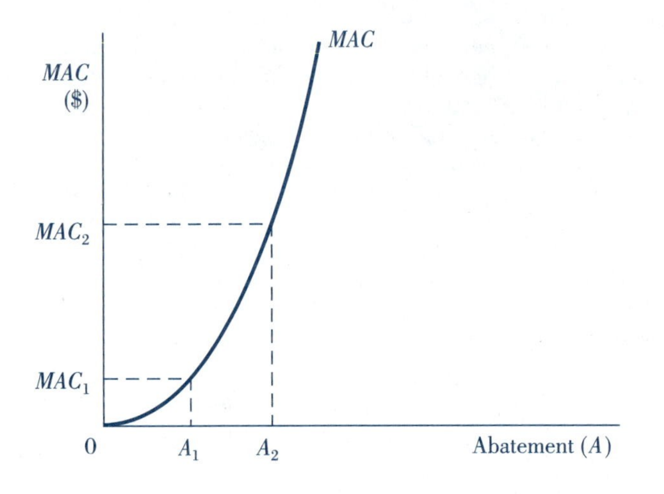

of abatement. To gain a better sense of what the *MSC* represents, it is easier to think of it as containing two parts:

1. an aggregation of the marginal costs of every polluter's abatement activities
2. the marginal costs government incurs to monitor and enforce those activities.

Let's begin with the first part by considering a representative polluter and how it goes about the task of abating its emissions.

Firm-Level Marginal Abatement Cost

Referring once again to our discussion of bargaining under the Coase Theorem, recall that refineries reduced their toxic releases by decreasing output. In that case, the marginal cost to refineries of abating pollution was forgone profit, modeled as a movement from right to left along the marginal profit ($M\pi$) curve. Therefore, if the decision variable is output, each polluting firm faces a marginal abatement cost equal to its forgone $M\pi$. However, such a model implicitly assumes that polluters can meet an environmental standard *only* by reducing output, a very limiting assumption. Because other methods are available, we need to expand on this output-based specification.

As a profit maximizer, the polluting firm is implicitly a cost minimizer. To meet an emissions standard, the firm will consider all available abatement options and select the least-cost method. In addition to reducing the level of production, these options may include installing some type of abatement technology or altering a production process. To allow for the fact that polluters choose from a menu of available abatement methods, we model what is conventionally called a **marginal abatement cost (*MAC*)** function. The *MAC* measures the change in economic costs associated with increasing pollution abatement (*A*), using the least-cost method. Application 4.2 on page 84 discusses the importance of choosing a least-cost abatement technology to clean up oil spills and hazardous wastes.

marginal abatement cost (*MAC*)
The change in costs associated with increasing abatement, using the least-cost method.

Figure 4.2 ***Effect of Cost-Saving Technology on the Polluter's MAC Curve***

Changes in a firm's abatement options change the position of the *MAC* curve. For example, this model shows how the introduction of a new cost-saving abatement technology would pivot the *MAC* downward to *MAC*'.

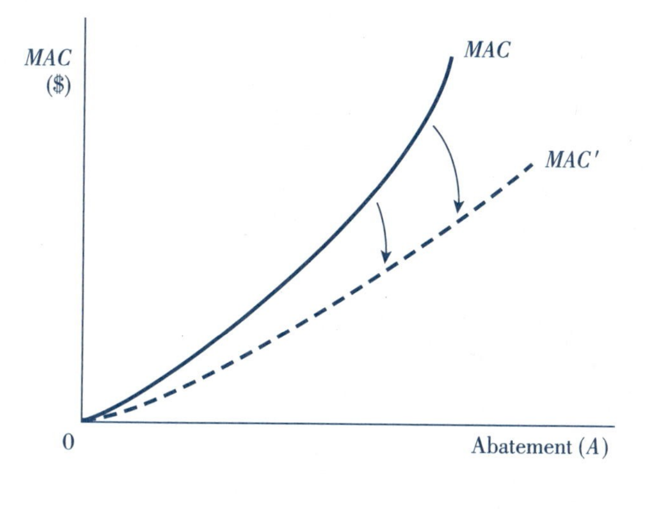

Each polluting source likely faces a unique *MAC* curve. The firm's location, the type of contaminants it releases, the nature of its production, and the availability of technology are among the factors that affect the shape and position of the *MAC* curve. However, a typical *MAC* curve is positively sloped and increases at an increasing rate, as shown in Figure 4.1. Think about what this means. When pollution levels are high, the addition of virtually any type of abatement technology will likely be effective. So although costs are rising, they are doing so at a fairly slow rate relative to the abatement being accomplished. As this process continues and the environment becomes cleaner, however, it becomes difficult to remove still more pollution. Thus, the added costs relative to the abatement achieved increase at a much faster rate.

Should a firm's abatement options change, the position of its *MAC* curve would be affected. For example, the introduction of a cost-saving abatement technology would pivot the *MAC* curve downward, as shown in Figure 4.2.[5] Likewise, a polluter's *MAC* for one pollutant might be lower than that of another. Consequently, an *MAC* function generally is defined for a particular contaminant at a given level of technology.

Market-Level Marginal Abatement Cost

The aggregation of all polluters' *MAC*s represents the **market-level *MAC* (MAC_{mkt})** defined as the horizontal sum of each polluter's *MAC*, or $MAC_{mkt} = \sum MAC_i$ for all *i* firms. This is exactly the same procedure used to derive market supply in Chapter 2. In this context, the horizontal summing ensures that the MAC_{mkt} represents least-cost decisions, because it effectively sets each *MAC* equal at every abatement level.

market-level marginal abatement cost (MAC_{mkt})
The horizontal sum of all polluters' *MAC* functions.

[5] A firm's *total* abatement costs are represented as the area under the *MAC* curve up to the abatement level required by the standard, assuming no fixed costs. Thus, the effect of the cost-saving technology on total abatement costs is implicitly shown as the reduction in the area under the firm's *MAC* curve.

Figure 4.3 ***Deriving the Marginal Social Cost (MSC) of Abatement***

To derive the *MSC* function, the *MCE* is vertically added to the MAC_{mkt}. At any abatement level *A*, the *MCE* is shown as the vertical distance between MAC_{mkt} and *MSC*. Notice that this distance increases with higher abatement levels. As pollution standards become more stringent, polluters have a greater tendency to evade the law, which in turn calls for more sophisticated, and thus more costly, monitoring and enforcement programs.

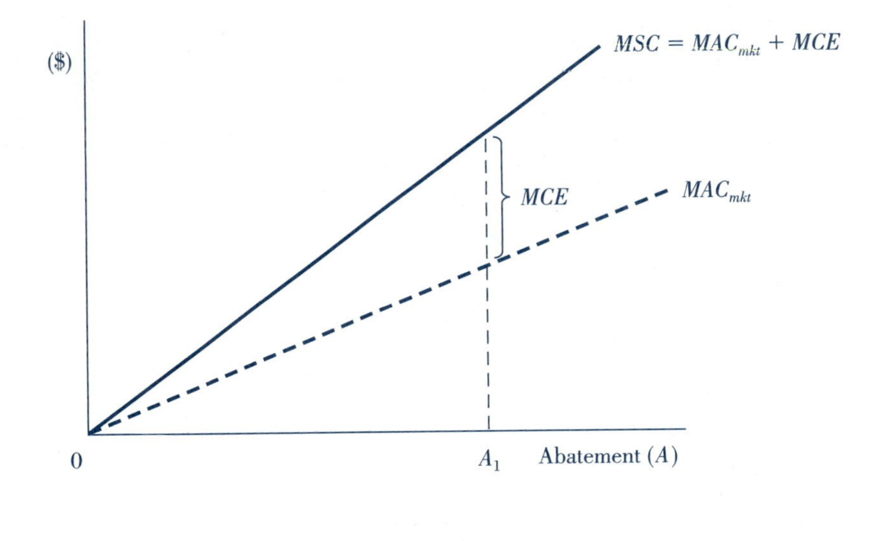

Marginal Cost of Enforcement

marginal cost of enforcement (*MCE*) Added costs incurred by government associated with monitoring and enforcing abatement activities.

Now, consider the second element of the marginal social cost (*MSC*) of abatement. To the MAC_{mkt} function we need to add the marginal costs incurred by government for enforcing and monitoring abatement activities. This component is commonly referred to as the **marginal cost of enforcement (*MCE*).** Figure 4.3 illustrates how the *MCE* is added vertically to the MAC_{mkt} to derive the *MSC* function. (For simplicity, all functions are assumed to be linear.) At any abatement level (*A*), the *MCE* is represented as the vertical distance between MAC_{mkt} and *MSC*. Notice that this distance is increasing with higher abatement levels. As pollution standards become more stringent, polluters have a greater tendency to evade the law, which in turn calls for more sophisticated and thus more costly monitoring and enforcement programs. Figure 4.4 shows U.S. expenditures on abatement, monitoring, and research for 1975–1994.

Are Abatement Standards Set Efficiently?

From our discussion in Chapter 3, we know that the *MSB* and the *MSC* simultaneously determine the efficient level of abatement. This result is illustrated in Figure 4.5 on page 86 where, for simplicity, we assume that both functions are linear. The efficient level of abatement (A_E), occurs at the intersection of the two functions. Whether or not the government sets environmental standards to achieve this level depends on a variety of considerations. Four factors in particular suggest that this outcome is highly unlikely: (1) the existence of legislative constraints, (2) imperfect information, (3) regional differences, and (4) nonuniformity of pollutants.

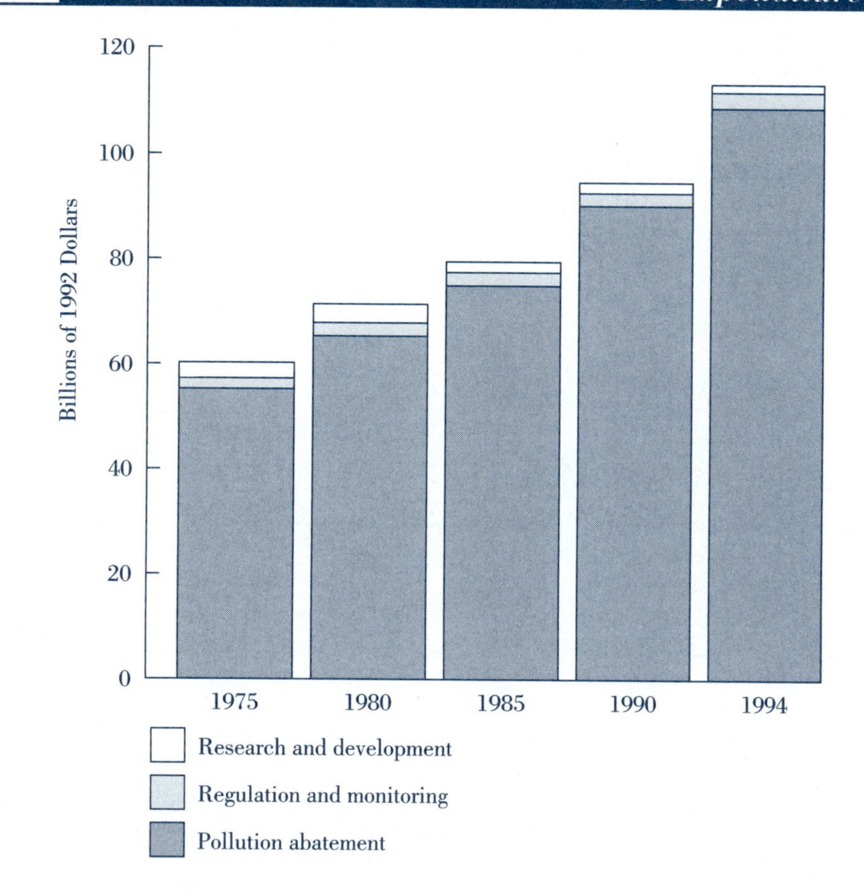

Source: Vogan (September 1996), cited in Council on Environmental Quality (1998), p. 222, table 2.4.

Legislative Constraints

The reality of a standards-based approach is that it does not necessarily set pollution limits to account for the associated benefits and costs. In fact, under U.S. law, many standards are said to be **benefit-based,** meaning that they are set at a level to improve society's well-being with no allowance for balancing the associated costs. For example, under the U.S. Clean Air Act, air quality ambient standards are motivated solely by the anticipated benefits of improved health and welfare. If costs are not accounted for in the standard-setting process, however, resources likely will be overallocated to abatement.

benefit-based standard
A standard set to improve society's well-being with no consideration for the associated costs.

Imperfect Information

Even when a cost-benefit balancing *is* called for by law, the absence of full information would likely prevent the government from identifying the *MSB* and *MSC* of abatement. Let's consider the *MSB* relationship first. Recall from our discussion of market failure in Chapter 3 that pollution abatement is a public good. As such, its demand, which is the *MSB* curve, is not readily identified because of the problem of nonrevelation of preferences. In practice, there are methods to estimate the value society places on the damage reductions associated with

Figure 4.5 ***Allocatively Efficient Amount of Pollution Abatement***

In the market for pollution abatement, the efficient level A_E occurs at the intersection of the *MSB* and *MSC* curves.

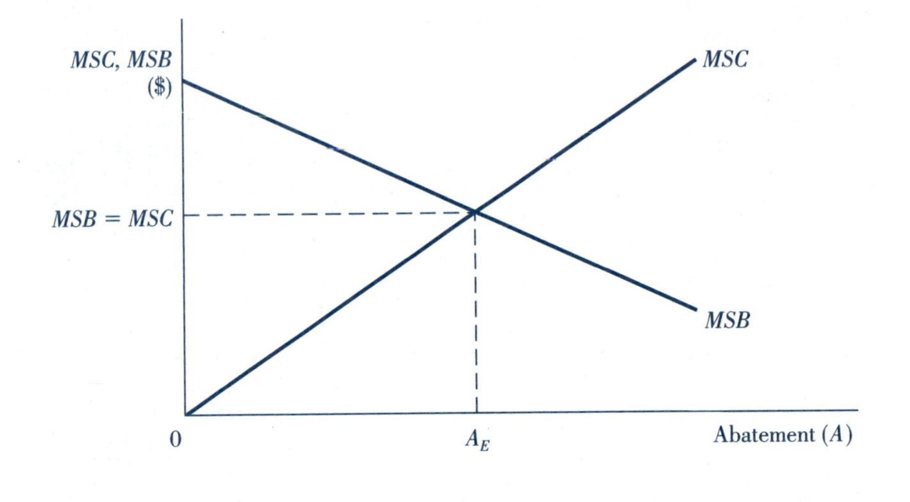

abatement. However, given the inherent difficulty of trying to monetize such intangibles as health improvements and longevity of life, the probability of accurately modeling *MSB* is low.

There are similar problems in identifying the *MSC*. In addition to estimating the *MCE*, the government also would have to know the *MAC* for every polluter. Obtaining this firm-level information would be virtually impossible, given the diversity of production and abatement techniques across polluting sources. Furthermore, the *MAC* must also account for the **implicit costs** of abatement, which are difficult to quantify. In this context, implicit costs would include any unemployment associated with production declines, the potential loss of consumer choice if products were eliminated or altered, and any price and income effects arising from abatement requirements.[6]

In the absence of perfect information, it is highly probable that the government will unknowingly establish the abatement standard at some level other than the allocatively efficient one, even if that was the legislated intent. Figure 4.6 reproduces the market for abatement at equilibrium, comparing the allocatively efficient outcome (A_E) with two other possible abatement levels (A_0 and A_1). If the standard is set at A_0, the *MSB* would be greater than the *MSC*, meaning that society places a higher value on reducing pollution than it must give up in resources to achieve it. Hence, the A_0 standard would be considered too lenient. Conversely, a standard set at A_1 would be considered too restrictive. Only at A_E would society consider the legal limit to be allocatively efficient. However, the information needed to find this optimal level of pollution abatement is immense.

Regional Differences

Even if the law permits a balancing of costs and benefits *and* even if full information were available, there is a qualifier on the use of A_E as a national standard across all polluting sources. Why? Because this optimal level is determined from *MSB* and *MSC*, both of which assume the absence of region-specific abatement benefits and costs. The only way that A_E

[6]In Chapters 7 and 8, we will investigate the methods used in practice to estimate the benefits and costs of improving environmental quality.

Figure 4.6 ***Setting an Environmental Quality Standard: Is It Allocatively Efficient?***

If the government were to set an abatement standard at A_0, the *MSB* would be higher than the *MSC*, meaning that society places a higher value on the gain from reducing pollution than on the resources needed to achieve it. Hence, the A_0 standard would be too lenient. On the other hand, if the standard were set at A_1, it would be considered too restrictive. Only at A_E would society accept the legal limit as allocatively efficient.

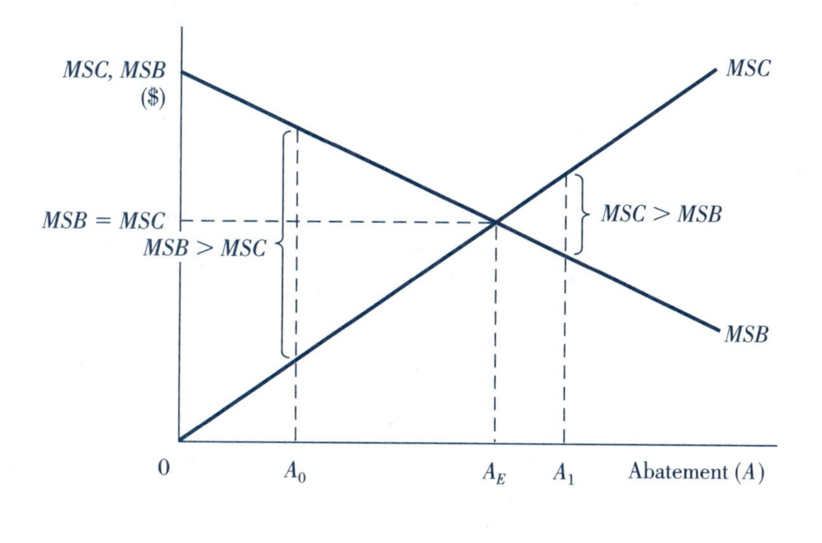

would be allocatively efficient in all regions is if the respective *MSB* and *MSC* functions defined for those locations were identical.

By way of example, consider two hypothetical regions X and Y that have identical *MSC* functions (i.e., $MSC_X = MSC_Y$) but different *MSB* functions, such that MSB_X is *lower* than MSB_Y at all abatement levels. Such a disparity might be due to differences in income, education, or population across the two locations. In any case, at most, only one of the two locations would consider a nationally determined A_E as efficient. Look at Figure 4.7, which superimposes the *MSB* and *MSC* for each region on the same diagram. The allocatively efficient level of abatement in region X (A_X) is much lower than that for region Y (A_Y). So there is no way that a single national standard of abatement—even one that *is* efficient on a national level—would be optimal for both regions.

Nonuniformity of Pollutants

An inefficient outcome also can arise within the same region if changes in releases from polluting sources do not have a uniform impact on the environment. This can occur if the relationship between the change in pollutant releases and the change in exposure is nonlinear, or not directly proportional. The nonuniformity also can arise when polluting sources are located at varying distances from an exposed population or ecosystem, even if their pollution releases are identical.[7] In general, the farther away from a source an affected population is, the lower the associated damage, because there is greater opportunity for dilution of contaminants.[8]

[7] National Acid Precipitation Assessment Program (November 1991), chap. 3, sec. 3.1.
[8] More technically, this relationship can be captured by what is called a transfer coefficient. For detail, see Tietenberg (1996), pp. 339–41.

Figure 4.7 *Effect of Regional Differences on Achieving Allocative Efficiency*

This model shows how region-specific conditions can give rise to different optimal levels of abatement. Although the *MSC* curves are identical (i.e., $MSC_X = MSC_Y$), the *MSB* in region X is *lower* than that in region Y, that is, $MSB_X < MSB_Y$. Therefore, the allocatively efficient abatement level for region X (A_X) is lower than that for region Y (A_Y). This means that a single national standard of abatement cannot be optimal for both regions.

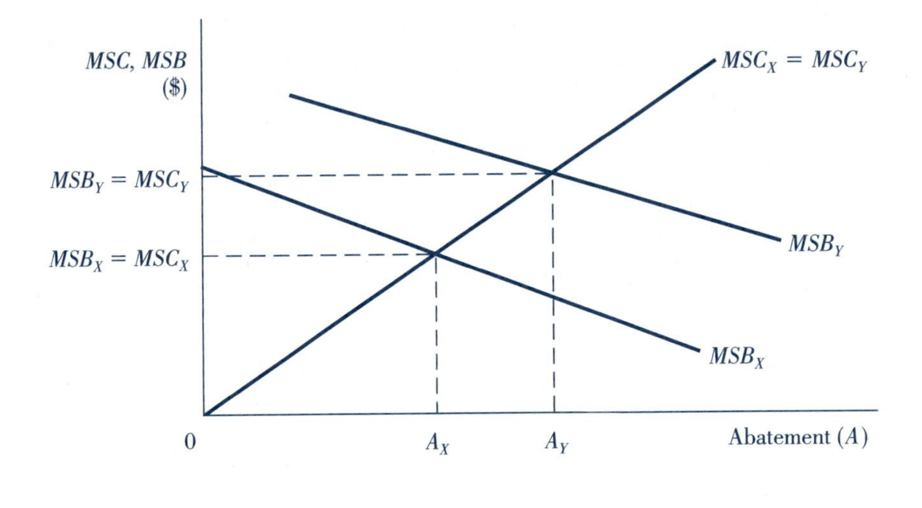

This in turn would mean that the *MSB* of abatement varies inversely with the distance between a source and the affected population or resource. Thus, even with equal *MSC*s of abatement, the efficient level of abatement would not be the same for all polluting sources. Consequently, just as for regional differences, a nationally determined abatement standard would not be optimal for all sources.

What we can surmise from this assessment is that in most real-world applications, at least one of the factors—legislative constraints, imperfect information, regional differences, or nonuniformity of pollutants—will be present. Consequently, there is a low probability that pollution abatement standards will be set at an allocatively efficient level. In accepting this, we must rely on a different criterion to evaluate not where the standards are set but how they are implemented.

General Approaches to Implementing Environmental Policy

cost-effective policy
A policy that meets an objective using the least amount of economic resources.

command-and-control approach
A policy that directly regulates polluters through the use of rules or standards.

Our preceding analysis makes a strong case that a government-established environmental standard will be set at something other than its allocatively efficient level. Yet, it is possible that even a nonoptimal environmental standard can be implemented using the least amount of resources. If so, the policy is said to be **cost-effective.** Whether or not this "second-best" criterion is met depends on the method used to bring about the desired reduction in pollution.

Most governments, including that of the United States, use a number of different policy tools to achieve environmental quality, as Application 4.3 explains. However, it is helpful at the outset to collapse these down into two broad categories. One is the **command-and-control approach,** which uses pollution limits or technology-based restrictions to directly

Application

4.3 Methods Used by Government to Reduce Environmental Pollution

According to the Relative Risk Reduction Strategies Committee of the EPA's Science Advisory Board (**www.epa.gov/history1/topics/risk/01.htm**), there are six categories of risk-reducing policy tools that governments should consider when addressing any environmental problem. Two of these are the more conventional broad-based categories of **conventional regulations** (such as standards, use restrictions, and product design) and **market incentives** (such as pollution charges and permit systems). The other four major groups are as follows:

- **Scientific and Technical Measures**

Research and development to suggest promising solutions and to improve understanding of problems such as the potential for global warming.
Innovations in pollution prevention and pollution control technology, such as the development of ecologically safe, cost-effective methods to manage contaminated sediments.

- **Provision of Information**

Improving communication to producers, consumers, and/or state and local governments about risks of environmental problems and threats to communities. For example, new homebuyers could be provided with the results of radon tests, and state and local governments could be given technical information to support their efforts in addressing indoor air pollution.

- **Enforcement**

Implementation of more vigorous enforcement of existing laws and regulations. Suggested options include (1) the use of statistical techniques in enforcement to ensure that all classes of potential violators are properly inspected and (2) the imposition of penalties that create incentives for compliance with existing environmental laws.

- **Cooperation with Other Government Agencies and Nations**

Promotion of interagency cooperation to reduce environmental risk. (Environmental authorities typically have limited jurisdiction, but other government bodies make decisions that affect the environment. This necessitates a cooperative approach to environmental issues.)
Promotion of cooperation with other nations through such means as international conventions aimed at problems such as global warming or acid rain.

Source: U.S. EPA, Science Advisory Board (September 1990), p. 15.

regulate polluting sources. The second is the **market approach,** which uses incentive-based policy tools to motivate abatement through market forces.

market approach
An incentive-based policy that encourages conservation practices or pollution reduction strategies.

Of the two, the command-and-control approach is the more conventional, and it dominates environmental policy in most countries. This nearly universal reliance on direct regulation seems to have evolved from an attempt to gain immediate control of what was initially an unfamiliar and urgent social dilemma. Although well intentioned, the use of inflexible regulations and pollution limits, often imposed uniformly across all polluters, has not met with consistent success. Hence, over time, policymakers began to look for alternatives.

The United States and other industrialized countries have gradually integrated market-based solutions into their environmental policy programs. Combining incentive-oriented policy instruments with more conventional methods seems to be an evolving trend in environmental policy. However, the *relative* gains of market-based solutions cannot be fully appreciated without assessing the cost-effectiveness of the command-and-control approach.

Is the Command-and-Control Approach Cost-Effective?

The practical way to assess whether the command-and-control approach is cost-effective is to determine whether society is incurring higher costs than necessary to achieve a given level of environmental quality. For discussion purposes, let's consider some abatement standard as the socially desirable (as opposed to efficient) outcome, perhaps motivated to protect human health. In theory, to achieve cost-effectiveness, policymakers must identify the relative costs of all control instruments that can achieve this objective and then select the one that minimizes costs. Given this general premise, we can identify two command-and-control decisions that may violate the cost-effectiveness criterion. The first is the use of a technology-based standard, and the second is the use of uniform standards. We discuss each of these in turn.

Cost-Ineffectiveness of the Technology-Based Standard

Recall from the beginning of this chapter that there are three types of standards: **ambient standards, performance-based standards,** and **technology-based standards.** Take a minute to reread the definitions of each of these control instruments and think about the cost implications of each. What should be apparent is that the technology-based standard potentially prevents the polluter from minimizing the costs of achieving a given abatement level. Remember that the *MAC* curve is defined under the assumption that the polluter selects the *least-cost* available method. If the government forces polluters to use a specific technology to meet an emissions limit, it is impeding the firm's incentive to abate in a cost-effective manner. Unless the mandated technology happens to be the least-cost abatement approach for *all* polluters, at least some will be forced to operate above their respective *MAC* curves. This in turn means that society is incurring costs higher than the *MSC* of abatement. The outcome is a waste of economic resources with no additional benefits accruing to society.

If instead a performance-based standard were used, each polluter could select the means by which it achieves a given objective. Without further guidance, it would follow its self-interest and choose the least-cost abatement method. Society would still gain the benefits of a cleaner environment, but fewer resources would be used to achieve that gain. A word of caution is in order, however. Although using performance-based standards has potential cost advantages over using technology-based standards, this selection does not by itself ensure a cost-effective solution. In fact, regardless of which type of standard is used, resources will be wasted if they are imposed *uniformly* across polluters.

Cost-Ineffectiveness of Uniform Standards

Under a strict command-and-control framework, standards often are imposed *uniformly* across groups of polluting sources.[9] The operative question is whether such a policy approach is cost-effective. The answer? The use of uniform standards across polluting sources will waste economic resources as long as abatement cost conditions differ among those sources. Of course, the reality is that there are many factors that might give rise to such differences. One is the age of the polluter's physical plant. Newer facilities typically are designed and built with the most advanced pollution control equipment, making them capable of meeting an abatement standard at a much lower marginal cost than their less modern counterparts. Another relevant factor is regional differences in input prices. There is no reason to expect firms in different locations to face the same costs of labor, land, and capital. As long as input prices vary, so too will polluters' costs to achieve a given abatement standard.

By accepting that abatement costs will likely differ among polluters, we need to explore why this makes the use of uniform standards cost-ineffective. The problem is that uniform standards force high-cost abaters to reduce pollution as much as low-cost abaters, so more resources than necessary are used to achieve a cleaner environment. Cost savings could be realized by having more of the abatement accomplished by polluters who can do so at a relatively lower cost. A simple analytical model illustrates this important assertion.

To begin, assume there are only two polluting sources in a given region, each of which generates 10 units of pollution for a total of 20 units released into the environment. The govern-

[9]In some contexts, pollution limits vary across major industrial groups but are applied uniformly *within* each of these groups.

ment determines that emissions must be reduced by 10 units across the region to achieve the "socially desirable level of pollution." Each firm faces different abatement cost conditions, modeled as follows:[10]

Polluter 1's Marginal Abatement Cost (MAC_1):	$MAC_1 = 2.5A_1$,
Polluter 1's Total Abatement Costs (TAC_1):	$TAC_1 = 1.25(A_1)^2$,
Polluter 2's Marginal Abatement Cost (MAC_2):	$MAC_2 = 0.625A_2$,
Polluter 2's Total Abatement Costs (TAC_2):	$TAC_2 = 0.3125(A_2)^2$,

where A_1 is the amount of pollution abated by Polluter 1, and A_2 is the amount of pollution abated by Polluter 2.

Now, assume the government implements the 10-unit standard *uniformly*, requiring each polluter to abate by 5 units (i.e., $A_1 = A_2 = 5$). At this level, the *MAC* for Polluter 1 is $12.50 [$MAC_1 = 2.5(5) = \$12.50$], and its *TAC* is $31.25 [$TAC_1 = 1.25(5)^2 = \$31.25$]. For Polluter 2, MAC_2 is $3.13, and TAC_2 is $7.81. Therefore, the total abatement costs for the region (absent the costs of monitoring and enforcement) equal $39.06, which represents the value of resources used by polluters to meet the standard. The question is, could the same standard be achieved at a lower cost?

Notice that Polluter 2 has an abatement cost advantage over Polluter 1. For example, the fifth unit of abatement costs Polluter 2 $9.37 less than Polluter 1. Therefore, it would be cheaper if Polluter 2 were to do more of the abating. Of course, it would have to have an incentive to do this, and the two firms would have to negotiate to arrive at some mutually beneficial agreement.[11] However, no such opportunity is allowed when the government forces every polluter to abate by the same amount. Thus, we conclude that the use of uniform standards under a conventional command-and-control approach does not achieve the cost-effectiveness criterion as long as *MAC* conditions differ across polluters.

Could the government reallocate abatement levels across the two polluters to achieve a cost-effective solution? The answer is yes, and economic theory conveys exactly how this result could be achieved. If each polluter were to abate to the point where the corresponding level of *MAC* is equal across firms, the **cost-effective abatement criterion** would be achieved. This means that the environmental standard would be met at minimum cost. This result is one application of what microeconomic theory calls the **equimarginal principle of optimality.**

cost-effective abatement criterion
Allocation of abatement across polluting sources such that the *MACs* for each source are equal.

To illustrate the cost savings of such an approach, let's return to our two-polluter model. We need to find the abatement levels for each polluter at which their respective *MACs* are equal, holding the combined abatement level at 10 units to meet the standard. Algebraically, the steps in the solution are

Step 1.	Set $MAC_1 = MAC_2$:	$2.5A_1 = 0.625A_2$.
Step 2.	Set $A_1 + A_2$ = abatement standard:	$A_1 + A_2 = 10$.
Step 3.	Solve the equations simultaneously:	$A_1 = 2; A_2 = 8$.

This same result is shown in Figure 4.8. Both firms' *MAC* curves are plotted on the same diagram. For Polluter 1, A_1 is measured horizontally *left to right*, whereas for Polluter 2, A_2 is measured *right to left*. The horizontal axis measured in either direction ranges from 0 units of abatement up to the 10-unit requirement imposed by the regulatory authority. Thus, every point on this axis represents a *combined* abatement level that satisfies the standard. Notice that the intersection of the two *MAC* curves yields the cost-effective solution ($A_1 = 2$ and $A_2 = 8$). At this point, $MAC_1 = MAC_2 = \$5.00$.

[10] Assuming no fixed costs, the total abatement costs (*TAC*) for a given abatement level *A* are found as the aggregation of the *MAC* at each abatement level up to *A*. Graphically, this means that the *TAC* for abatement level *A* is the area under the *MAC* curve up to point *A*.

[11] In a real-world setting, there would also be transactions costs associated with the negotiations between the two firms.

Figure 4.8 **Cost-Effective Solution in a Two-Polluter Model**

The model shows the *MAC* curves for two firms, Polluter 1 and Polluter 2, diagrammed on the same graph. Polluter 1's abatement level (A_1), is measured left to right on the horizontal axis, and Polluter 2's abatement (A_2), right to left. The horizontal axis measures from 0 units of abatement up to the 10-unit limit imposed by the regulatory authority, so that every point represents a *combined* abatement level that satisfies the standard. In accordance with the equimarginal principle of optimality, the intersection of the two *MAC* curves yields the cost-effective solution, where $A_1 = 2$ and $A_2 = 8$. At this point, notice that $MAC_1 = MAC_2 = \$5.00$.

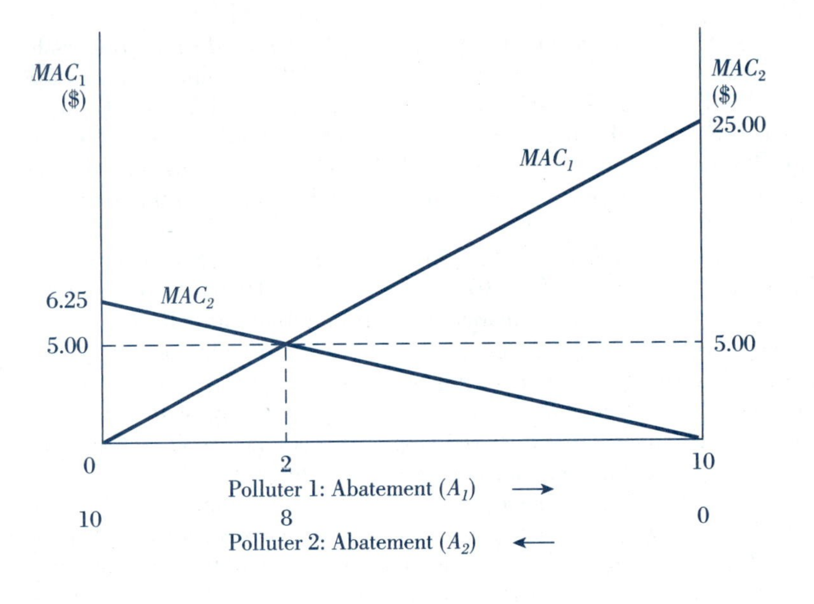

If each polluter were given these firm-specific abatement targets, the total costs of achieving the environmental objective would be minimized. These costs can be calculated from each firm's *TAC* equation, substituting in the cost-effective abatement levels:

$$TAC_1 = 1.25(2)^2 = \$5.00,$$
$$TAC_2 = 0.3125(8)^2 = \$20.00.$$

By following the equimarginal principle of optimality, the total cost to society of achieving the 10-unit abatement standard is \$25.00—a \$14.06 savings over the uniform standards approach. Equivalently, this \$14.06 represents the unnecessary costs incurred by society when a uniform standard is imposed across nonidentical polluters.

As a final point, it is reasonable to ask how in practice the government could arrive at these firm-specific abatement standards within a command-and-control framework. The answer is that it would have to know the abatement cost conditions for every firm it was regulating. Of course, this kind of information would be virtually impossible to determine, particularly when thousands of individual sources are being controlled. But there is a way around the problem, although not within the command-and-control approach. As we will discover in Chapter 5, instruments used in the market approach can arrive at this same cost-effective solution without specific knowledge of polluters' costs. How? By using market incentives and the price mechanism in place of inflexible rules.

Conclusions

In this chapter, we have begun to evaluate solutions to environmental problems by focusing on the more conventional policy tools used in practice—the use of standards to define environmental objectives and the command-and-control approach to implementing those objectives. Even at this introductory level, we were able to reach some important conclusions. It is apparent, for example, that government-mandated environmental standards are not likely to be set at an efficient level. Beyond those instances when the law does not allow for the requisite balancing of benefits and costs, there is still an information problem. Policymakers would need extensive data to accurately measure the marginal costs and benefits of abatement. Hence, in all likelihood, the level at which environmental objectives are set will not be allocatively efficient.

Upon accepting this realization, we find that cost-effectiveness becomes the relevant criterion by which to assess the command-and-control approach and compare it to the market approach. Thus far, our investigation shows that the use of uniform standards under a command-and-control framework likely wastes economic resources. Cost savings can be realized if polluters reduce emissions up to the point where their marginal costs of doing so are equal. But how can such a result be achieved in practice? The answer requires an investigation of alternative control instruments, in particular those characterized as part of the market approach to environmental policy.

In sum, although our findings in this chapter are important, they are nonetheless incomplete. We still must learn how market-based initiatives are designed and how they compare to their command-and-control counterparts—precisely the agenda of our next chapter. This part of our analysis is particularly relevant, given the recent trend in the United States and other nations to integrate market-based instruments into what had been an exclusively command-and-control approach to environmental policy. To appreciate this trend and what it means for society, we must understand how market instruments operate and evaluate their effectiveness in achieving environmental goals.

Summary

- There are three basic types of standards used in environmental control policy: ambient standards, which designate the level of environmental quality as a maximum allowable pollutant concentration; technology-based standards, which indicate the abatement method to be used by polluters; and performance-based standards, which specify an emissions limit to be achieved by polluters.
- An environmental standard achieves allocative efficiency if resources are allocated such that the marginal social benefit (*MSB*) of abatement is equal to the marginal social cost (*MSC*) of abatement.
- The *MSB* measures the additional gains to society associated with the reduction in damages caused by pollution.
- The *MSC* is the horizontal sum of the market-level *MAC* (MAC_{mkt}) plus the government's marginal cost of enforcement (*MCE*).
- Four factors suggest that a government-mandated abatement standard is not likely to meet the allocative efficiency criterion: (1) the existence of legislative constraints, (2) imperfect information, (3) regional differences, and (4) nonuniformity of pollutants.
- Governments generally use one of two approaches to implement environmental policy: the command-and-control approach or the market approach.
- Two aspects of the command-and-control approach may violate the cost-effectiveness criterion: the use of technology-based standards and the use of uniform standards.

- Because technology-based standards dictate a specific abatement method to polluting sources, they prevent the polluter from minimizing costs.
- Uniform standards force high-cost abaters to reduce pollution as much as low-cost abaters, so more resources than necessary are used to achieve the benefits of a cleaner environment.
- To achieve a cost-effective outcome, abatement responsibilities across polluting sources must be allocated such that the *MAC* is equal across polluters.

Key Concepts

ambient standards
technology-based standard
performance-based standard
allocatively efficient standards
marginal social benefit (*MSB*) of abatement
marginal social cost (*MSC*) of abatement
marginal abatement cost (*MAC*)
market-level marginal abatement cost (MAC_{mkt})
marginal cost of enforcement (*MCE*)
benefit-based standard
cost-effective policy
command-and-control approach
market approach
cost-effective abatement criterion

Use the Key Concepts listed above to begin your search for additional articles and information using the InfoTrac College Edition database.

Review Questions

1. One of the major problems in applying the Coase Theorem in practice is the existence of high transactions costs. Propose an approach that a third party could institute that would reduce these costs sufficiently so that bargaining could proceed. How likely is the solution to be efficient, and why?
2. Using a graph of the pollution abatement market, model a situation in which the allocatively efficient level of abatement occurs at 100 percent, or, equivalently, where pollution is zero. Referring to the relative position of the *MSC* and *MSB* curves, explain such an outcome intuitively.
3. a. Under a strict command-and-control framework, suppose abatement standards are set equally across polluters. Assume the total abatement target is 30 units. Show the cost implications using three graphs, each of a different polluter with a unique *MAC* curve drawn to depict a low-cost abater, a moderate-cost abater, and a high-cost abater. On each graph, identify the abatement level corresponding to a uniform standards approach, and show the level of *MAC* at that point and the area corresponding to *TAC*.
   b. Now, refer directly to your model, and summarize what would happen *qualitatively* to the abatement levels of each firm if the equimarginal principle of optimality were used. Explain intuitively why this method would be cost-effective.
4. It is well documented that the carbon monoxide (CO) emissions from combustible engines increase in colder climates. This implies that the associated damages are expected to be less severe in summer months than in winter. Nonetheless, air quality control authorities use a standard for CO that is uniform throughout the year with no allowance for seasonal effects. Use this information and the following model to answer the questions:

MSB of CO abatement in winter	$= 350 - 0.5A$,
MSB of CO abatement in summer	$= 140 - 0.2A$,
MSC of CO abatement	$= 0.2A$,

where A is the level of CO abatement.

a. Graph the *MSB* and *MSC* functions on the same diagram.
b. Assume the government sets a uniform standard for winter and summer at $A = 500$. Support or refute this policy based on the criterion of allocative efficiency, using your model to explain your response.
c. If you were in charge of setting policy for CO emissions, what action would you recommend to ensure an allocatively efficient outcome across the two seasons?

Additional Readings

Buck, Susan. *Understanding Environmental Administration and Law.* Covelo, CA: Island Press, 1996.

Crandall, Robert W. "Is There Progress in Environmental Policy?" *Contemporary Economic Policy* (January 1995), pp. 80–83.

Keohane, Nathaniel O., Richard L. Revesz, and Robert N. Stavins. "The Choice of Regulatory Instruments in Environmental Policy." *Harvard Environmental Law Review* 22 (1998), pp. 313–67.

Kneese, Allen V., and Charles Schultz. *Pollution, Prices, and Public Policy.* Washington, DC: Brookings Institution, 1975.

Portney, Paul R. "EPA and the Evolution of Federal Regulation." In Paul R. Portney, ed., *Public Policies for Environmental Protection*, pp. 7–25. Washington, DC: Resources for the Future, 1990.

Spence, A. M., and M. L. Weitzman. "Regulatory Strategies for Pollution Control." In A. E. Friedlander, ed., *Approaches to Controlling Air Pollution*, pp. 199–219. Boston: MIT Press, 1979.

Related Web Sites

http://

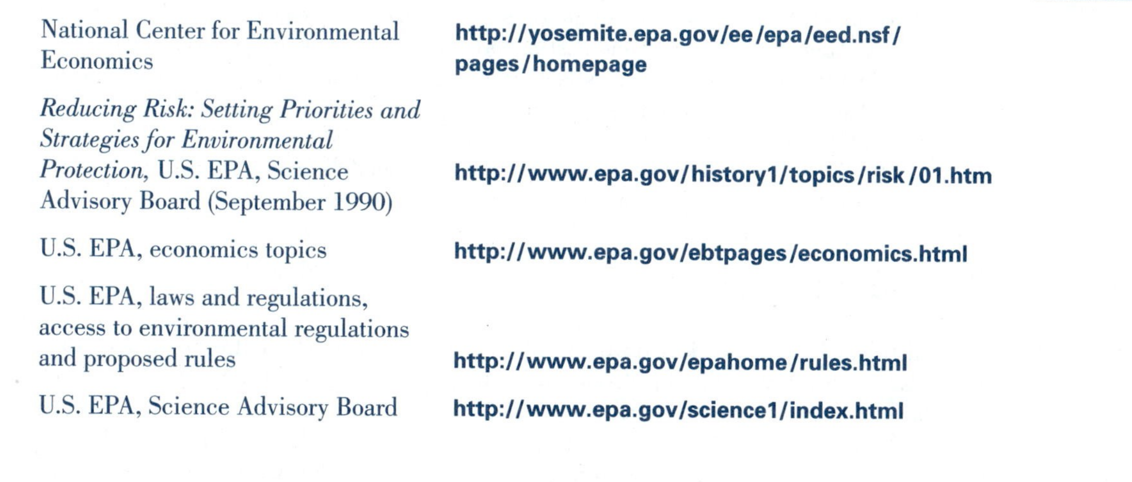

National Center for Environmental Economics	**http://yosemite.epa.gov/ee/epa/eed.nsf/pages/homepage**
Reducing Risk: Setting Priorities and Strategies for Environmental Protection, U.S. EPA, Science Advisory Board (September 1990)	**http://www.epa.gov/history1/topics/risk/01.htm**
U.S. EPA, economics topics	**http://www.epa.gov/ebtpages/economics.html**
U.S. EPA, laws and regulations, access to environmental regulations and proposed rules	**http://www.epa.gov/epahome/rules.html**
U.S. EPA, Science Advisory Board	**http://www.epa.gov/science1/index.html**

Chapter 5

Economic Solutions to Environmental Problems: The Market Approach

Although the market fails to correct environmental problems on its own, the incentives that define the market process can be put to work by policymakers. The **market approach** to environmental policy, recommended for some time by economists, has begun to be adopted by governments as part of their overall response to the risks of pollution. Distinct from the more traditional use of command-and-control instruments, the market approach uses price or other economic variables to provide incentives for polluters to reduce harmful emissions.

Economists are strong proponents of the market approach because it can achieve a cost-effective solution to environmental problems. How? By designing policy initiatives that allow polluters to respond according to their own self-interest. Market instruments are aimed at bringing the external costs of environmental damage back into the decision making of firms and consumers. Taking its cue directly from market failure theory, the market approach attempts to restore economic incentives by assigning a value to environmental quality, or, equivalently, by pricing pollution. Once done, firms and consumers adjust their optimizing behavior to the resulting change in market conditions.

Awareness of the gains associated with market-based instruments is growing in both the private and the public sector. Consequently, it is important to understand this alternative policy approach and its advantages over more traditional forms of regulation. To that end, in this chapter we examine the theory and the practical implications of control instruments based on market incentives. Our primary tool of analysis will be economic modeling, and we will use the criteria of allocative efficiency and cost-effectiveness to make policy assessments.

To provide a framework for our investigation, we begin with a brief overview of the major categories of market-based instruments: pollution charges, subsidies, deposit/refund systems, and pollution permit trading systems. We then use these categories to structure our economic analysis. Taking each category in turn, we develop models of specific instruments, assess the results, and provide examples of how each is used in practice.

Descriptive Overview

market approach
An incentive-based policy that encourages conservative practices or pollution reduction strategies.

What distinguishes the **market approach** from the command-and-control approach is the way in which environmental objectives are implemented, as opposed to the level at which those objectives are set. From a practical perspective, standards-based objectives are set at a socially desirable level rather than at an efficient level. Where the market approach parts company with the command-and-control approach is in *how* it attempts to achieve those objectives, that is, in its design of policy instruments.

Identifying Types of Market Instruments

Because many types of control instruments use market incentives, it is helpful to classify these instruments into major categories: **pollution charges, subsidies, deposit/refund systems,** and **pollution permit trading systems.** A brief description of each of these is provided in Table 5.1.[1]

[1] A good overview of market-based instruments used in the United States is found in U.S. EPA, Office of Policy, Economics, and Innovation (January 2001), which is also available online at **http://yosemite1.epa.gov/ee/epa/eed.nsf/pages/incentives.**

Table 5.1 *Taxonomy of Market-Based Instruments*

Market Instrument	Description
Pollution charge	A fee charged to the polluter that varies with the quantity of pollutants released. It can be implemented through any of the following: **Effluent or emission charge** A fee based on the actual discharge of pollution. **Product charge** An upward adjustment to the price of a pollution-generating product based on its quantity or some characteristic responsible for the pollution. Product charges can be implemented through *tax differentiation,* that is, levying different taxes on goods based on their potential effect on the environment. **User charge** A fee levied on the user of an environmental resource based on the costs for treatment of emissions or effluents that adversely affect that resource. **Administrative charge** A service fee for implementing or monitoring a regulation or for registering a pollutant with an authority.
Subsidy	A payment or tax concession that provides financial assistance for pollution reductions or plans to abate in the future.
Deposit/refund	A system that imposes an up-front charge to pay for potential pollution damages that is returned for positive action, such as returning a product for proper disposal or recycling.
Pollution permit trading system	The establishment of a market for rights to pollute, using either credits or allowances. **Credits** With a **credit system,** polluters earn marketable credits for emitting below an established standard. **Allowances** Under an **allowance system,** permits give polluters the right to release some amount of pollution, which can be increased or decreased through trading.

Sources: U.S. EPA, Office of Policy, Planning, and Evaluation (July 1992); OECD (1989).

Nations all over the world use market-based instruments to help control pollution. In fact, an international survey of 24 nations found that, on average, 11 market instruments were being used in each country as of 1998.[2] Every type of economic instrument was identified by the survey as part of national policy across all environmental media. Although the market approach continues to be a secondary form of control, its use in national policy prescriptions speaks to its importance as part of the full range of available solutions to environmental problems. Therefore, we will discuss each of the four major categories of market instruments, using economics to analyze each one.

[2] OECD (1999a), pp. 100–108, app. 1.

Figure 5.1 ***Implementation of a Pigouvian Tax to Achieve Efficiency***

The model illustrates the use of a Pigouvian tax to achieve efficiency in the market for some pollution-generating product (Q). By setting the per unit tax exactly equal to the *MEC* at Q_E, shown as distance *ab*, the *MPC* curve shifts up to MPC_t. Equilibrium output is then determined by the intersection of MPC_t and *MSB*, which establishes an allocatively efficient production level at Q_E.

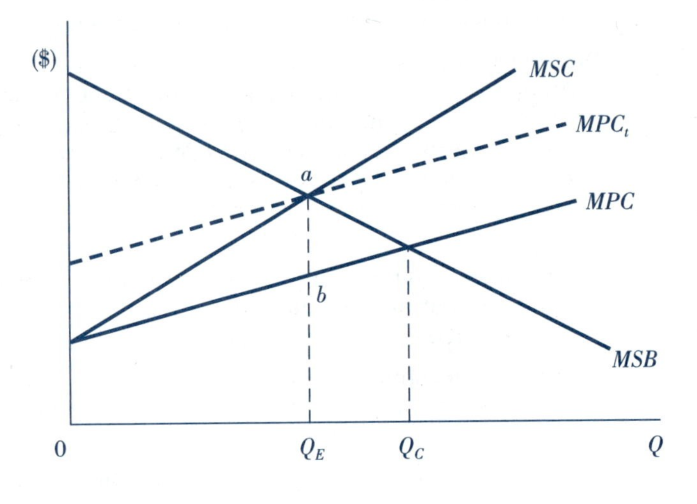

Pollution Charges

pollution charge
A fee that varies with the amount of pollutants released.

product charge
A fee added to the price of a pollution-generating product based on its quantity or some attribute responsible for pollution.

Pigouvian tax
A unit charge on a good whose production generates a negative externality such that the charge equals the *MEC* at Q_E.

The theoretical premise of a **pollution charge** is to internalize the cost of environmental damages by pricing the pollution-generating activity. The motivation follows what is known as the **"polluter-pays principle,"** a position rooted in the belief that the polluter should bear the costs of control measures to maintain an acceptable level of environmental quality.[3] We begin by modeling a **product charge** implemented as a tax—the classical solution to negative externalities.

Modeling a Product Charge as a Per Unit Tax

Consider a good in a competitive market whose production generates a negative environmental externality. Because producers base their decisions solely on the marginal private cost (*MPC*) of production, ignoring the marginal external cost (*MEC*) of the environmental damage, too many resources are allocated to production. As shown in Figure 5.1, firms produce at output level Q_C, where the marginal social benefit (*MSB*) of consuming the good is equal to the marginal private cost (*MPC*) of producing it.[4] Notice that the equilibrium output (Q_C) is higher than the efficient level (Q_E), which corresponds to the point where the *MSB* equals the marginal social cost (*MSC*).

The policy motivation of a **product charge** is to induce firms to internalize the externality by taking account of the *MEC* in their production decisions. One way this can be done is by imposing a unit tax on the pollution-generating product exactly equal to the *MEC* at the efficient output level (Q_E). This type of tax is called a **Pigouvian tax,** named after the English economist A. C. Pigou, who initially formulated the theory. As illustrated in Figure 5.1,

[3] OECD (1989), p. 27.
[4] As we have done in previous discussions, we are implicitly assuming that there is no marginal external benefit (*MEB*) in the market for this good, so $MPB = MSB$.

this policy instrument effectively shifts the *MPC* curve up by distance *ab* to MPC_t, which generates an equilibrium at the efficient output level.

Assessing the Model

In theory, the Pigouvian tax forces firms to lower production to the efficient output level. Although theoretically pleasing, this instrument is difficult to impose in practice and is not commonly used. Why? One problem is the difficulty in identifying the dollar value of *MEC* at Q_E and hence the level of the tax. A second problem is that the model implicitly allows only for an output reduction to abate pollution—an unrealistic restriction. To address both of these reservations in this context, we consider a more practical alternative. The pollution charge can be implemented as an **emission charge,** a tax levied on the pollution, instead of as a product charge. By moving out of the product market, the model does not restrict the polluter's response to an output reduction.

Modeling an Emission Charge: Single-Polluter Case

An **emission or effluent charge** assigns a price to pollution, typically through a tax. Once this price mechanism is in place, the polluter can no longer ignore the effect of its environmental damages on society. The pollution charge forces the polluter to confront those damages, pay for them, and in so doing, consider the damages as part of its production costs. Faced with this added cost, the polluting firm can either continue polluting at the same level and pay the charge or invest in abatement technology to reduce its pollutant releases and lower its tax burden. Based on normal market incentives, the firm will choose whichever action minimizes its costs.

emission or effluent charge
A fee imposed directly on the actual discharge of pollution.

It is fairly simple to model an emission charge that allows polluters to make cost-minimizing decisions. To begin, we assume that the government sets an abatement standard at some "acceptable level" (A_{ST}). Now, we consider a policy that presents the polluter with the following options to be undertaken singularly or in combination:

- The polluter must pay a constant per unit tax (t) on the difference between its existing abatement level (A_0) and the standard (A_{ST}) such that Total Tax $= t(A_{ST} - A_0)$; and/or
- The polluter incurs the cost of abating.

In Figure 5.2, we graph these options from a single firm's perspective using marginal curves. Because the per unit tax is constant at *t*, the marginal tax (*MT*) curve is a horizontal line at that tax level. The cost of abating at the margin is shown as the marginal abatement cost (*MAC*) curve. At each unit of *A*, the cost-minimizing firm will compare *MAC* to *MT* and choose the lower of the two. In our model, the firm will abate up to A_0, because up to that point *MAC* is below *MT.* Assuming no fixed costs, total abatement costs (*TAC*) are represented by the area under the *MAC* curve up to A_0, or area OaA_0. Notice that these costs are lower than what the taxes would be up to A_0, shown as area $OtaA_0$. Beyond A_0 and up to A_{ST}, the firm will opt to pay the tax, because *MT* is lower than *MAC* in that range. The firm's total tax payment for not abating between A_0 and A_{ST} is represented by area A_0abA_{ST}, which is smaller than the cost of abating that amount, which is area A_0acA_{ST}. In sum, the total costs to the polluter of complying with this policy are area $OabA_{ST}$, which consists of the following two elements:

1. Area OaA_0, which is the total cost of abating A_0 units of pollution
2. Area A_0abA_{ST}, which is the tax on pollution not abated up to A_{ST}

Assessing the Model

Be sure to recognize that the emission charge stimulates the natural economic incentives of the polluter. At any point in time, *static* incentives motivate the firm to choose among the available options, given its existing technology. Seeking to satisfy its own self-interest to maximize profit, the polluter makes a least-cost decision between paying the tax or abating. The result is that the externality is internalized, using the least amount of resources.

Figure 5.2 ***Modeling an Emission Charge for a Single Firm***

The model illustrates the decisions of a single polluter, faced with the options of either paying a tax $[t(A_{ST} - A_0)]$ or incurring the cost of abating. The tax burden is represented by the marginal tax (*MT*) curve, and the firm's marginal cost of abating is shown as the *MAC* curve. In this case, the cost-minimizing firm will abate up to level A_0, because up to that point $MAC < MT$. Its total abatement costs (*TAC*) are represented by area $0aA_0$. Beyond A_0 and up to the standard A_{ST}, the firm will opt to pay the taxes, because $MT < MAC$ in that range. Its tax burden is represented by area A_0abA_{ST}. Hence, the total costs to the polluter of complying with this policy are shown as area $0abA_{ST}$.

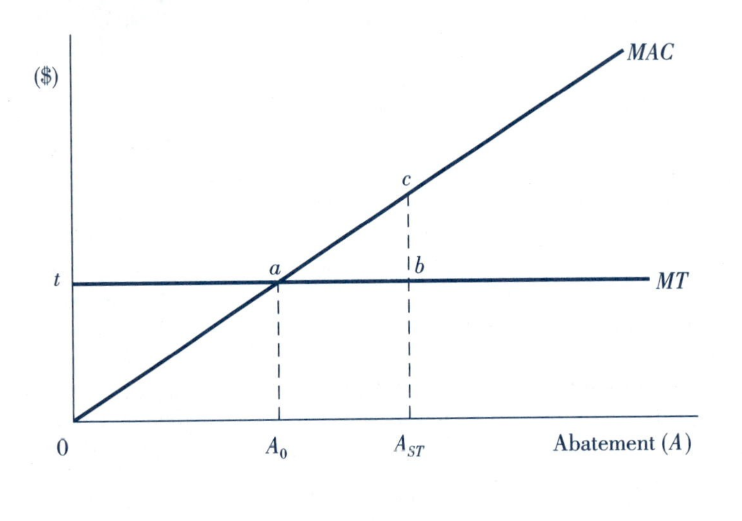

There are also *dynamic* incentives that encourage the firm to advance its abatement technology. More efficient abatement techniques would allow the firm to reduce pollution more cheaply and enjoy the associated cost savings. Furthermore, the lower abatement costs might even allow the firm to avoid paying any emission charges. Consider the effect of a hypothetical technological advance, shown in Figure 5.3 as a downward pivot of *MAC* to *MAC'*. In this case, if the firm were faced with the same set of policy options based on A_{ST}, it would be better off abating all the way up to that standard. The firm would incur total costs of abatement equal to $0bA_{ST}$ and would pay no emission charges. Relative to its expenditures before the technological advance, which were $0abA_{ST}$, it saves an amount equal to $0ab$.

Modeling an Emission Charge: Multiple-Polluter Case

To evaluate the cost-effectiveness of an emission charge across multiple polluters, we return to the two-polluter model used in Chapter 4. As before, we assume that the government imposes a 10-unit abatement standard for the region. We repeat the cost functions from Chapter 4 to facilitate our discussion:

Polluter 1's Marginal Abatement Cost (MAC_1): $MAC_1 = 2.5A_1$,
Polluter 1's Total Abatement Costs (TAC_1): $TAC_1 = 1.25(A_1)^2$,
Polluter 2's Marginal Abatement Cost (MAC_2): $MAC_2 = 0.625A_2$,
Polluter 2's Total Abatement Costs (TAC_2): $TAC_2 = 0.3125(A_2)^2$,

where A_1 is the amount of pollution abated by Polluter 1, and A_2 is the amount of pollution abated by Polluter 2.

Assume that the government imposes the same emission charge as in the single-polluter case, that is, Total Tax $= t(A_{ST} - A_0)$. In this case, because $A_{ST} = 10$, the emission charge be-

Figure 5.3 ***Effect of Technology Improvement on a Firm's Least-Cost Decision Making***

A technological advance causes the *MAC* curve to pivot downward to *MAC'*. As a result, if the firm is faced with the option of abating or paying the unit tax at each abatement level up to A_{ST}, it would be better off abating all the way up to A_{ST}. Hence, the technological change helped the firm to avoid paying any emission charges.

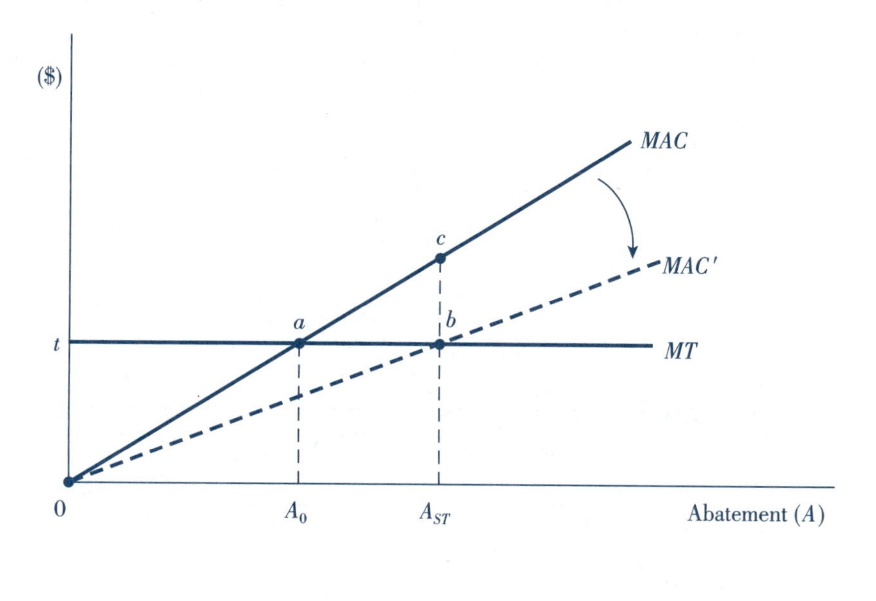

comes $t(10 - A_0)$. We also assume that the tax rate (t) is set at \$5, so Total Tax = \$5$(10 - A_0)$ for each polluter.

Now, consider each firm's response to the tax. As we proceed, refer to Figure 5.4, which reproduces the model used in Chapter 4, adding a horizontal line (*MT*) at \$5 to represent the emission charge. When faced with the \$5 per unit charge, Polluter 1 would compare the relative costs of *MT* and MAC_1 for each incremental unit of abatement, just as in the single-firm case. It would abate as long as $MAC_1 < MT$ and pay the tax when the opposite is true. Thus, Polluter 1 would abate up to the point where $MAC_1 = MT$, which occurs at $A_1 = 2$, and pay the tax on the remaining 8 units. Polluter 2 would proceed in the same way, abating to the point where $MAC_2 = MT$ at $A_2 = 8$ and paying the tax on the remaining 2 units. The corresponding calculations are

Polluter 1

Abates up to the point where $MAC_1 = MT$: $2.5(A_1) = \$5$, or $A_1 = 2$.
Incurs Total Abatement Costs of: $TAC_1 = 1.25(2)^2 = \$5$.
Incurs Total Tax payment of: Total Tax $= 5(10 - 2) = \$40$.

Polluter 2

Abates up to the point where $MAC_2 = MT$: $0.625(A_2) = \$5$, or $A_2 = 8$.
Incurs Total Abatement Costs of: $TAC_2 = 0.3125(8)^2 = \$20$.
Incurs Total Tax payment of: Total Tax $= 5(10 - 8) = \$10$.

Figure 5.4 Effect of an Emission Charge in a Two-Polluter Model

The model shows the *MAC* curves for Polluter 1 and Polluter 2 and the emission charge imposed as a unit tax of $5, represented by the *MT* curve. The horizontal axis measures the 10-unit abatement standard imposed by the government. Each firm abates as long as its $MAC < MT$, and each firm pays the emission charge on all units of pollution not abated. In this case, Polluter 1, the high-cost abater, abates 2 units and pays $40 in taxes on the remaining 8 units. Polluter 2, the low-cost abater, abates 8 units and pays taxes of $10 on the remaining 2 units. At this point, $MAC_1 = MAC_2 = \$5$, which indicates the least-cost allocation of abatement responsibilities across the two firms.

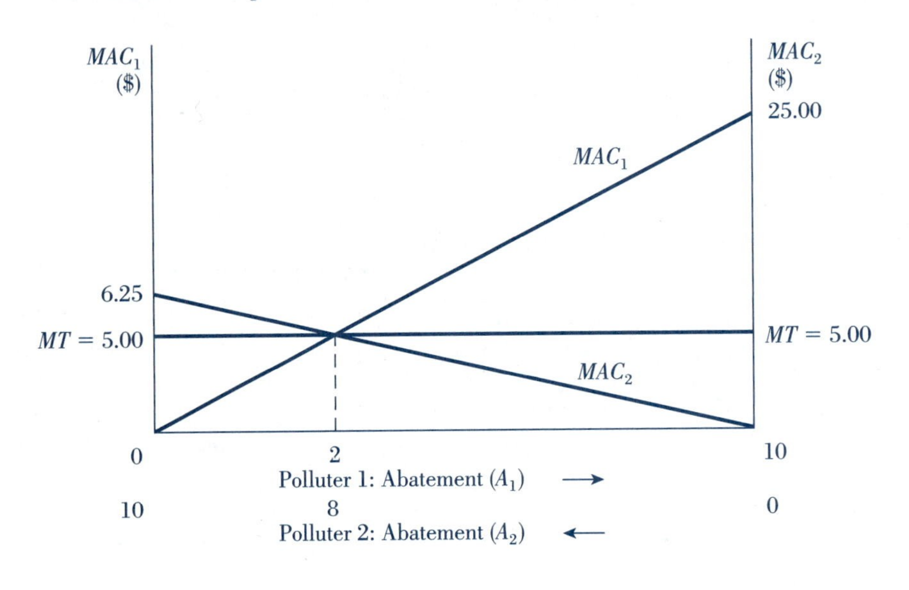

Combined Values for the Region

Total level of abatement $= 10 = A_{ST}$.
Total Abatement Costs $= \$25$.
Total Tax Payments $= \$50$.

There are two important observations to make about these results. First, the $5 unit tax achieves the 10-unit abatement level. Second, this objective has been achieved using the cost-effective allocation of abatement resources across polluters.[5] Because each polluter abates to the point where its *MAC* equals *MT*, all *MAC*s are equal, precisely in accordance with the cost-effective abatement criterion. Notice that most of the abating is done by the low-cost abater, Polluter 2. The high-cost polluter, Polluter 1, abates less but pays much higher emission charges in the form of taxes.

Assessing the Model

The emission charge exploits each polluter's natural incentive to pursue a least-cost strategy. As a result, the low-cost abaters do most of the cleaning up, and the high-cost abaters pay more in taxes to cover the greater damages they cause. The benefits to society are not affected

[5] Recall from Chapter 4 that using a uniform standard under a command-and-control approach generates total abatement costs of $39.06.

by *who* does the abating, but the costs are. In this case, costs are minimized because Polluter 2 does most of the abating. It does so, not because it is motivated by society's objectives, but because doing so is in its own best interest. An added advantage of this approach is that the tax generates revenues to the government, and these revenues can be used to help finance the costs of enforcement and monitoring.

Having underscored the advantages of the emission charge, there are nonetheless several caveats that deserve mention. The first concerns the setting of the emission charge itself. Realistically, the government will not know the tax rate at which polluters' abatement levels collectively meet the standard and therefore will have to adjust the tax until the environmental objective is achieved.[6] Such an adjustment process can be time intensive. Another consideration is the potential increase in monitoring costs. Monitoring is likely to be more complex when each polluting source responds to a policy based on its own internal operations. Distributional implications also have to be considered. Because polluting firms pay higher taxes, part of the tax burden is shared with consumers in the form of higher prices. Job losses may also occur as polluters adjust to the tax burden or change their technologies to increase abatement. Finally, firms may try to evade a tax by illegally disposing of pollutants. To minimize the potential of such activity, the government may have to strengthen its monitoring programs, which adds to administrative costs.

Pollution Charges in Practice

Internationally, the pollution charge is the most commonly used market-based instrument. The Organisation for Economic Co-operation and Development (OECD) surveyed 24 nations and found that 131 pollution charges were in use as of 1998, out of 270 identified market-based instruments. Several countries, including Australia, Italy, Japan, Switzerland, and Turkey, use **effluent charges** to control the noise pollution generated by aircraft. France and Germany are among the nations using effluent charges to protect water resources.[7]

A real-world application of a **product charge** is one levied on lubricant oils, which is done by Finland, Hungary, and Italy. Another example is a fee charged on automobile tires. Denmark and Finland use this particular product charge to help cover the costs of collecting and recycling used tires. Other targeted products used by various countries include motor vehicles, packaging, pesticides, fertilizers, batteries, and fuels.[8] Application 5.1 gives an overview of the international experience with gasoline taxes aimed at internalizing the external costs of consumption.

Environmental Subsidies

An alternative market approach to reducing environmental damage is to pay polluters *not* to pollute through an **environmental subsidy.** There are two major types of subsidies—**abatement equipment subsidies** and **pollution reduction subsidies.** We discuss each of these in turn.

Modeling an Abatement Equipment Subsidy

abatement equipment subsidy A payment aimed at lowering the cost of abatement technology.

Abatement equipment subsidies are aimed at reducing the costs of abatement technology. Because subsidies are "negative taxes," they have a similar incentive mechanism to pollution charges except that they reward for *not* polluting, as opposed to penalizing for engaging in polluting activities. In practice, abatement equipment subsidies are implemented through

[6] For example, if the government had set the tax at too low a level, say, $4, each polluter would have abated to the point where its *MAC* equals $4. The result would have been $A_1 = 1.6$ and $A_2 = 6.4$ for a combined abatement level of 8.0 units—too low to satisfy the objective. So the government would have had to raise the tax until the combined total reached the 10-unit abatement standard.
[7] OECD (1999a), pp. 22–24, 26, 27, and app. 1, pp. 100–108.
[8] OECD (1999a), pp. 29–35.

Application

5.1 *Taxing Gasoline Consumption: An International Comparison*

According to Kazuo Aichi, former chief of Japan's environmental agency, "Gasoline is too cheap in the United States and should be taxed more to cut energy use." Aichi's argument is motivated in part by ongoing disagreements between the United States and some of its industrialized counterparts about the appropriate response to such problems as natural resource depletion and global air pollution. Aichi's criticism of U.S. policy on fuel taxes underscores the importance of the price mechanism in encouraging conservation.

To better understand the objection to American policy, consider how the U.S. tax on gasoline compares to what is imposed by other industrialized countries. The data are given in the accompanying table.

The data validate Aichi's claim that the price of gasoline in the United States *is* cheap in a relative sense and that much of the difference is due to a lower tax rate. Notice that the tax rate is between 65 percent and 81 percent of the unit price in European nations, 50 percent in Canada, and 56 percent in Japan. The comparatively low tax rate of 35 percent in the United States reflects its fuel tax policy.

Beyond this cursory assessment of Aichi's commentary, there is the more important issue of the economic intent of a gasoline tax. In addition to boosting government revenues, taxing a commodity such as gasoline is designed to internalize the negative externalities of consumption. External to the market for gasoline are air pollution problems caused by operating gasoline-powered motor vehicles. Other negative externalities are highway congestion and the increased risk of traffic accidents. These adverse effects have one thing in common — they extend to parties beyond those engaged in the market transaction. Consequently, the costs of these damages are not reflected in the price, and too much gasoline is brought to market. As other nations have apparently learned, raising the tax on gasoline can reduce consumption and bring the associated social costs and benefits closer together.

Gasoline Prices and Taxes by Country (December 2001)

Country	Total Gasoline Price per Liter (U.S. dollars)	Tax Rate % of Price
United States	0.286	35
United Kingdom	0.999	81
France	0.847	77
Germany	0.851	76
Italy	0.886	71
Japan	0.827	56
Spain	0.654	65
Canada	0.365	50

Source: International Energy Agency (December 2001).

Sources: "Japanese Environmentalist Says" (February 2, 1992); Reifenberg and Sullivan (May 1, 1996); Tanner (June 9, 1992).

grants, low-interest loans, or investment tax credits, all of which give polluters an incentive to invest in abatement technology.

Pigouvian subsidy
A per unit payment on a good whose consumption generates a positive externality such that the payment equals the *MEB* at Q_E.

From a theoretical perspective, we argue that these subsidies attempt to internalize the *positive* externality associated with the consumption of abatement activities. If a subsidy were offered for the installation of specific abatement equipment, such as scrubbers, the effect would be to increase quantity demanded of this equipment by lowering the effective price. To achieve an efficient equilibrium, the subsidy would have to equal the marginal external benefit (*MEB*) of scrubber consumption measured at the efficient output level. Notice that this is analogous to a Pigouvian tax, and in fact this type of subsidy is known as a **Pigouvian subsidy.**

Figure 5.5 ***A Pigouvian Subsidy in the Market for Scrubbers***

In the market for scrubbers, there is a positive externality associated with their consumption. Therefore, the true measure of benefits to society is given by the *MSB,* which is the vertical sum of the *MPB* and the *MEB.* If a Pigouvian subsidy equal to the *MEB* at the efficient output level were provided to purchasers of scrubber systems, 210 scrubbers would be traded in the market instead of the competitive output level of 200. In this case, the Pigouvian subsidy is \$14 million, labeled as distance *KL.* Notice that the effective price to polluters would be the efficient market price *less* the amount of the subsidy paid by the government, which in this case is (\$175 million − \$14 million), or \$161 million.

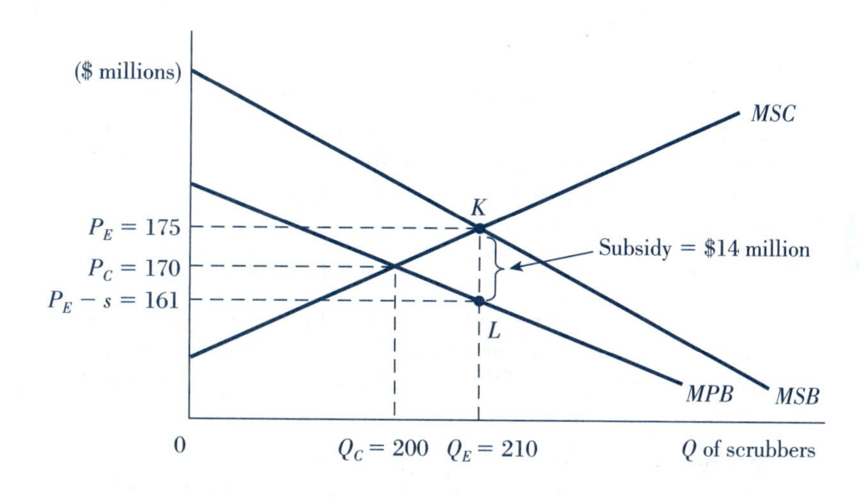

A model of a hypothetical competitive market for scrubbers would be

$$MSC = 70.0 + 0.5Q, \qquad \begin{array}{rl} & MPB = 350.0 - 0.9Q \\ + & MEB = \ \ 56.0 - 0.2Q, \\ \hline & MSB = 406.0 - 1.1Q \end{array}$$

where Q is the number of scrubber systems produced in a year, and *MSC, MPB,* and *MEB* are denominated in millions of dollars.[9] The corresponding graph is given in Figure 5.5. The competitive equilibrium is found where $MPB = MSC$, or at $Q_C = 200$ and $P_C = \$170$ million. However, the efficient equilibrium occurs where $MSB = MSC$, or at $Q_E = 210$ and $P_E = \$175$ million. In an unregulated competitive market, too few scrubbers are exchanged at too low a price because the external benefits of a cleaner environment are not recognized by the market participants. If a subsidy (s) equal to the *MEB* at the efficient output level was provided to demanders (i.e., polluters), more scrubbers would be traded. In this case, the Pigouvian subsidy would equal $MEB = 56.0 - 0.2(210) = \$14$ million, shown as distance *KL* in Figure 5.5. The effective price to polluters would be the efficient market price *less* the subsidy, or $(P_E - s)$, which in this case is (\$175 million − \$14 million) or \$161 million.

Assessing the Model

Just as in the case of a Pigouvian tax, one problem with implementing a Pigouvian subsidy is measuring the *MEB.* Monetizing the marginal external benefits of such intangibles as better health and more stable ecosystems is difficult at best. Hence, it is not likely that a subsidy of

[9]For simplicity, we assume there are no marginal external costs (*MEC*), so that $MPC = MSC$.

abatement equipment will achieve allocative efficiency. However, its associated effect of encouraging greater consumption because of an effectively lower price should still occur.

Even setting aside the difficulty in achieving an efficient outcome, equipment subsidies may have other drawbacks. One commonly cited criticism is that this type of control instrument biases polluters' decisions about how best to abate. Subsidies affect relative prices, making other alternatives less attractive to polluters from a financial perspective. However, some of these abatement alternatives might be more effective in reducing pollution. For the same reason, innovation of a potentially superior abatement system could be discouraged as long as the government is subsidizing existing equipment. Finally, subsidies must be financed through taxes or government borrowing. Thus, they effectively redistribute income from society to polluters, an outcome some view as unacceptable despite the associated gains of a cleaner environment.

Modeling a Per Unit Subsidy on Pollution Reduction

per unit subsidy on pollution reduction
A payment for every unit of pollution removed below some predetermined level.

An alternative type of subsidy is one based on emission or effluent reductions, called a **per unit subsidy on pollution reduction.** In this case, the government agrees to pay the polluter a subsidy (s) for every unit of pollution removed below some predetermined level or standard (Z_{ST}). This is modeled as

$$\text{Total Subsidy} = s(Z_{ST} - Z_0),$$

where Z_0 is the actual level of pollution. Suppose, for example, that Z_{ST} is set at 200 tons of emissions per month and the subsidy (s) is set at \$100 per ton per month. Then, if a polluter reduces its emissions to 180 tons per month, it would receive a subsidy of \$100(200 − 180), or \$2,000.

Assessing the Model

On the plus side, a per unit pollution reduction subsidy *might* be less disruptive than an equipment subsidy, because it is established independent of the abatement method used and thus avoids any technological bias. On the other hand, these subsidies can have the perverse effect of elevating pollution levels in the aggregate. How does such a paradox arise? Because a per unit subsidy effectively lowers a polluter's unit costs, which in turn raises its profits. If the industry has limited entry barriers, these profits would signal entrepreneurs to enter the industry. In the long run, although each individual polluter reduces its emissions, the subsidy may cause the market to expand such that *aggregate* emissions end up higher than they were originally.[10] The dilemma could be solved if entry were prohibited or at least limited in some way. Whether or not this is feasible or even desirable depends on the industry structure, the extent of environmental damage, and the associated costs.

Environmental Subsidies in Practice

There are many instances where environmental subsidies are used in practice. Internationally, a common application of environmental subsidies is in the form of grants or low-interest loans, which are being used in many countries, including Austria, Denmark, Finland, Japan, the Netherlands, and Turkey.[11] In the United States, the most common use of subsidies is federal funding for such projects as publicly owned treatment works. Federal subsidies also are used to promote the use of pollution control equipment and to encourage the use and development of cleaner fuels and low-emitting vehicles. Federal subsidies are implemented in a variety of ways, such as through grants, rebates, tax exemptions, and tax credits. At the state level, a common application of environmental subsidies is in the form of tax incentives to encourage recycling activities.[12] An overview of some of these programs is given in Application 5.2.

---

[10] For further details, see Baumol and Oates (1975), chap. 12.
[11] OECD (1999a), pp. 49–55.
[12] U.S. EPA, Office of Policy, Planning, and Evaluation (July 1992), p. 6–1; U.S. EPA, Office of Policy, Economics, and Innovation (January 2001), chap. 7.

5.2 *State Subsidies for Recycling Programs*

Application

Some state governments have implemented subsidies as a way to encourage recycling. Many subsidies are imposed as one-time tax incentives. Among the more common subsidy instruments are investment tax credits for recycling facilities and sales tax exemptions on the purchase of recycling equipment. The premise for these applications is that once the incentives establish the infrastructure to support recycling, the activity should become self-directed.

The accompanying table itemizes some of the approaches being used by state governments throughout the United States.

State	Description of Tax Incentive
California	Tax credits for the cost of equipment used to manufacture recycled products.
	Development bonds for manufacturing products with recycled materials.
Colorado	Income tax credits for investment in plastics recycling technology.
Florida	Sales tax exemptions on recycling machinery purchased after July 1, 1988.
	Tax incentives to encourage affordable transport of recycled materials from collection centers to processing sites.
Illinois	Sales tax exemptions for recycling equipment.
Indiana	Property tax exemptions for buildings, equipment, and land used in recycling wastes into new products.
Iowa	Sales tax exemptions.
Kentucky	Property tax exemptions to recycling industries.
Maine	Tax credits to businesses of 30 percent of the cost of recycling equipment.
	Subsidies to municipalities for transport costs of scrap metal.
Maryland	Income tax deduction of 100 percent of expenses for furnace conversion to burn used oil and for the purchase and installation of equipment for recycling used Freon.
New Jersey	Fifty percent investment tax credit to businesses for recycling vehicles and equipment.
	Six percent sales tax exemption on recycling equipment purchases.
North Carolina	Corporate income tax credits and exemptions for recycling equipment and facilities.
Oregon	Income tax credits for recycling equipment and facilities.
	Special tax credits for property or machinery used to collect, transport, or process reclaimed plastics.
Texas	Franchise tax exemptions for sludge recycling corporations.
Utah	Payment of $21 per ton to tire recyclers for tires recycled or incinerated for energy recovery.
Virginia	Tax credit of 10 percent of the cost of machinery or equipment used for processing recyclable materials.
Washington	Exemption from motor vehicle rate regulation for vehicles used to transport recovered materials.
Wisconsin	Sales tax exemptions for waste reduction and recycling equipment and facilities. Property tax exemptions for certain equipment.

Source: Drawn from National Solid Waste Management Association (October 1990) with permission of the National Solid Waste Management Association.

Deposit/Refund Systems

deposit/refund system
A market instrument that imposes an up-front charge to pay for potential damages and refunds it for returning a product for proper disposal or recycling.

The potentially perverse consequences of abatement subsidies suggest that pollution charges might be a better alternative. In some contexts, however, pollution charges can be costly to administer because of the associated expense of monitoring and enforcement. Recall that one of the drawbacks of pollution charges is that they may encourage illegal disposal of contaminants. This potential problem is an important motivation for using a **deposit/refund system.** Operationally, deposit/refund systems attach a front-end charge (the deposit) for the *potential* occurrence of a damaging activity and guarantee a return of that charge (the refund) upon assurance that the activity has not been undertaken. This market instrument combines the incentive characteristic of a pollution charge with a built-in mechanism for controlling monitoring costs. In general, its intent is to capture the difference between the private and the social costs of improper waste disposal, with its most common targets being beverage containers and lead-acid batteries.

Economics of Deposit/Refund Systems

Improper or illegal waste disposal gives rise to a negative externality. The external costs include health damages, such as lead contamination from discarded lead-acid batteries, and aesthetic impairment from litter and trash accumulation. Deposit/refund systems are intended to force the potential polluter to account for both the marginal private cost (MPC) and the marginal external cost (MEC) of improper waste disposal, should that activity be undertaken.

As with the pollution charge, the deposit component of the system is intended to capture the MEC of improper waste disposal. The deposit forces the polluter to internalize the cost of any damage it may cause by making it absorb this cost *in advance.* Unique to the deposit/refund system is the refund component, which introduces an incentive to properly dispose of wastes and prevent environmental damage from taking place at all. Taken together, the deposit/refund system targets the *potential* polluter instead of penalizing the actual polluter, using the refund to reward appropriate behavior.

Modeling a Deposit/Refund System

A model of a deposit/refund system is shown in Figure 5.6. From left to right, the horizontal axis measures *improper* waste disposal (IW) as a percentage of all waste disposal activity. Implicitly, then, the percentage of *proper* waste disposal (PW) is measured right to left. Thus, if 25 percent of all wastes is improperly disposed of, then by default 75 percent is disposed of appropriately and safely.

The MPC_{IW} includes expenses for collecting and illegally dumping wastes plus the costs of improperly disposing of recyclable wastes, such as the expense of trash receptacles, collection fees paid to refuse companies, and the opportunity costs of forgone revenue associated with recycling. The MSC_{IW} includes the MPC_{IW} plus the MEC_{IW}, represented implicitly as the vertical distance between MSC_{IW} and MPC_{IW}. The MPB_{IW} is the demand for improper waste disposal. It is motivated by the avoidance of time and resources to collect wastes, bring nonrecyclables to a landfill, and haul recyclables to a collection center.[13] Because we assume no external benefits in this case, $MPB_{IW} = MSB_{IW}$. In the absence of environmental controls, equilibrium is determined by the intersection of MSB_{IW} and MPC_{IW}, or Q_{IW}. The efficient equilibrium occurs where MSC_{IW} equals MSB_{IW}, or at Q_E, which is smaller than Q_{IW}. Once again, we observe that in the presence of a negative externality, too much improper waste disposal is produced because market participants do not consider the full impact of their actions.[14]

[13] The MPB_{IW} measured left to right is equivalent to the MPC_{PW} measured right to left.

[14] This analysis could equivalently be modeled by specifying the market as proper waste disposal. In such a specification, there would be a *positive* external benefit from *proper* disposal of wastes, which would be equivalent to the reduction in external costs from reducing *improper* waste disposal.

Figure 5.6 *Modeling a Deposit/Refund System in the Market for Waste Disposal*

Improper waste disposal (IW) is measured left to right on the horizontal axis as a proportion of all waste disposal activity. Hence, the percentage of proper waste disposal (PW) is measured right to left. To correct the negative externality associated with improper waste disposal, a deposit/refund system is instituted with the deposit set equal to the MEC_{IW} measured at Q_E. The deposit, labeled as distance ab, elevates the MPC_{IW} to MSC_{IW}, forcing the market participants to establish a new equilibrium at the socially optimal level. The result is an increase in the proportion of wastes properly disposed of from Q_{IW} to Q_E

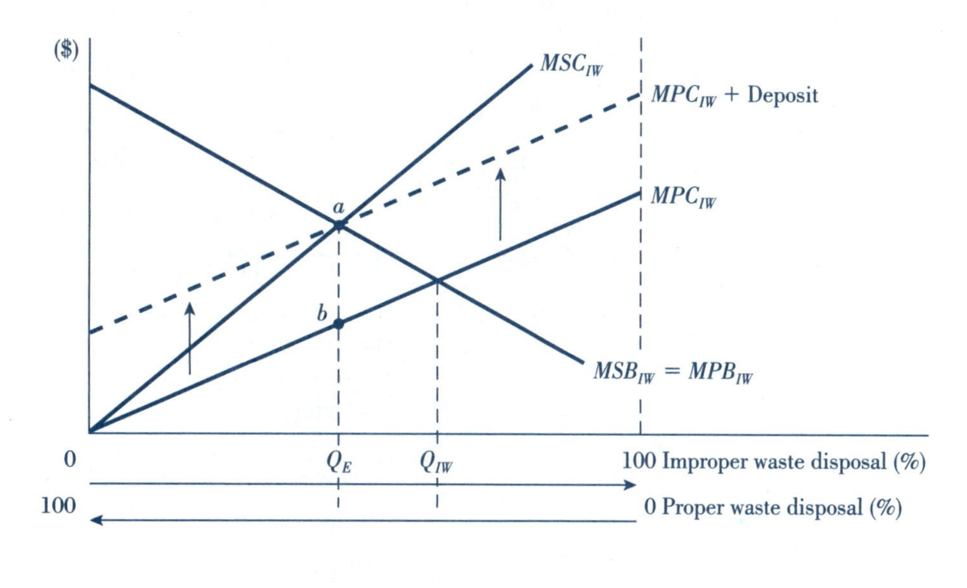

To correct the negative externality, assume that a deposit/refund system is instituted whereby the deposit equals the MEC_{IW} at Q_E. This is labeled distance ab in Figure 5.6. Once imposed, the deposit effectively elevates MPC_{IW} by distance ab, forcing the market participants to a new equilibrium at Q_E. In so doing, a percentage of waste disposal is converted from improper methods to appropriate ones, measured by the distance $(Q_{IW} - Q_E)$. Notice that the deposit serves the same function as a pollution charge. The critical difference is that the refund helps to deter improper waste disposal. The potential polluter has an explicit incentive to properly dispose of wastes because doing so allows it to reclaim the deposit. Should disposers choose instead to illegally discard their wastes, at least they will have paid for the external costs in advance. Authorities also have the flexibility to adjust the deposit or refund amounts to enhance the built-in incentives.

Assessing the Model

The value added of the deposit/refund system is that the refund encourages environmentally responsible behavior without adding significantly to government's monitoring and compliance costs. What makes this instrument unique is that once established, the incentives operate with limited supervision.

Another advantage of the deposit/refund instrument is that it can be used to encourage more efficient use of raw materials. An inordinate amount of used products and materials ends up in landfills or burned in incinerators, when they could be recycled. The availability of

Application

5.3 *Mechanics of a Deposit/Refund System: Bottle Bills*

Currently, 10 states have passed bottle bills: California, Connecticut, Delaware, Iowa, Maine, Massachusetts, Michigan, New York, Oregon, and Vermont. Although each state's law is unique, there are similarities that characterize how a deposit/refund system is designed for beverage containers. The following list is a brief description of the phases of such a system.

Phase 1: A deposit is paid by the retailer to the bottler or wholesaler for each beverage container received. If the product is a soft drink, retailers pay the deposit to the bottler; for beer, the retailer pays the wholesaler.

Phase 2: Consumers pay the same deposit to the retailer as part of the product's purchase price.

Phase 3: After the beverage is consumed, the consumer returns the used container to the retailer, who refunds the consumer for the amount of the initial deposit.

Phase 4: Retailers reclaim the deposit from either the bottler or the wholesaler when they return the empty container. In addition, bottlers and wholesalers typically pay a per unit handling fee to retailers to cover their costs to collect and return the containers.

Notice from this four-step process that bottle bills are self-implementing. Once the deposit system is in place, market forces take over. There is a natural incentive for consumers and retailers to return used containers so that they can recover their deposits.

Why haven't more states passed bottle bills? Voter opposition to these bills exists for a number of reasons. One is the perverse outcome caused by the availability of substitute containers not covered by a bottle bill. For example, if a bottle bill requires a deposit only on aluminum containers, some consumers may purchase the product in plastic just to avoid the deposit. Such a response occurred in New York, where consumption of soft drinks in plastic containers rose from 39 percent to 52 percent following the passage of the state's bottle bill. The response is problematic because plastic is not as recyclable as aluminum, and disposal of plastic containers adds more to the volume of the solid waste stream than other types of containers.

Another reason for voter opposition is the opportunity cost of participating in a deposit/refund program — the value of time to collect, clean, store, and transport containers to retailers. Some consumers view these costs as outweighing the value of refunds. As cost minimizers, these individuals would make the rational decision *not* to participate in a deposit/refund program. Finally, there is some controversy about what proportion of collected containers are recycled versus the proportion that ends up in landfills. Other skeptics believe that bottle bills have a relatively small effect on the size of the solid waste stream.

How valid are these concerns? Benefit-cost analysis of deposit/refund systems in states that have bottle bills could help answer this question. The appraisal should use information about the associated energy savings, reduction in littering, recovery rates, and changes in consumption patterns. As these data are examined, consumers and firms will be better able to make informed decisions about proposals in their own states and to tailor a deposit/refund system to better meet the needs of their regions.

Sources: U.S. EPA, Office of Policy, Planning, and Evaluation (March 1991), chap. 2; U.S. EPA, Office of Policy, Planning, and Evaluation (July 1992), chap. 4; U.S. EPA, Office of Policy, Economics, and Innovation, Office of the Administrator (January 2001), chap. 5.

recycled products and wastes can help slow the depletion of such virgin raw materials as aluminum and timber and may result in associated price declines as well. Charging firms a deposit on raw materials acts as a tax, encouraging more efficient use of resources *during* the production process. The refund encourages proper disposal or recycling of raw material waste at the *end* of the production phase. Firms that elect to ignore this incentive face not only conventional disposal costs but also the opportunity cost of the forgone refund.

Deposit/Refund Systems in Practice

Perhaps the best-known applications of deposit/refund systems are those used to encourage proper disposal of beverage containers. In the United States, such programs typically are initiated through state bottle bills. As of 2002, 10 states had passed legislation requiring deposits on beer and soft drink containers. Deposits range from 2.5 cents to 15 cents per container. Application 5.3 discusses the mechanics of these state-level initiatives. Oregon, the first state to enact a bottle bill, reported that roadside litter was reduced by 75–85 percent just two years after its bill became law.[15] Michigan reported a 95 percent return rate in the first year after it instituted its program.[16] Similar results also have been observed in other countries. Examples include Australia, the Czech Republic, Denmark, and Norway, which report return rates on beverage containers ranging from 62 percent to 99 percent after instituting deposit/refund systems.[17]

Other deposit/refund initiatives are aimed at encouraging responsible disposal of such products as used tires, car hulks, and lead-acid batteries. Proper discard of lead-acid batteries is of particular concern because of the health risks of lead exposure. According to the EPA, lead-acid batteries represent approximately 65 percent of the lead found in municipal solid wastes.[18] Consequently, some states have imposed mandatory deposit/refund systems on lead-acid batteries. Typically, the deposit is $5 or $10 per battery, and the consumer can obtain a refund by returning a used battery within a specified period along with proof that the deposit was paid.[19] Denmark has achieved considerable success with its deposit/refund systems, meeting its annual objectives for return rates of both batteries and lead accumulators. Norway and Sweden are using a deposit/refund system to reduce the improper disposal of junk cars. Interestingly, both charge a deposit that is actually less than the refund, presumably to further enhance the incentive to return unwanted vehicles for proper disposal or recycling.[20]

Pollution Permit Trading Systems

Thus far, we have illustrated that market instruments can be used to set prices for polluting and abatement activities. It is also possible for government to use the price-quantity relationship in the opposite direction — by establishing the quantity of pollution or abatement to be achieved and letting the market determine the price. With perfect information, either approach is viable, and both will lead to the same outcome. However, pricing instruments can be problematic in that the government does not know in advance what price will achieve a quantity-based environmental objective. Consequently, it has to monitor the quantity response to some initially established price and continually make adjustments until the proper pollution level is achieved, essentially a trial-and-error process.

It may be more efficient to use a policy instrument that operates from the *known* variable, that is, the socially desirable quantity of pollution or abatement, and let the market establish the price. Such is the underlying premise of a **pollution permit trading system,** which can be implemented through the use of credits or allowances. Under a **pollution credit** system, a polluter earns marketable credits only if it emits below an established standard. If, instead, the trading system uses **pollution allowances,** each permit gives the bearer the right to release some amount of pollution. These too are marketable, so that polluters can buy and sell allowances as needed, based on their access to abatement technologies and their cost conditions.

pollution permit trading system
A market instrument that establishes a market for rights to pollute by issuing tradeable pollution credits or allowances.

pollution credits
Tradeable permits issued for emitting below an established standard.

pollution allowances
Tradeable permits that indicate the maximum level of pollution that may be released.

[15] U.S. EPA, Office of Policy, Economics, and Innovation (January 2001), chap. 5, pp. 58–59, 63–64.
[16] Porter (1983).
[17] OECD (1999a), pp. 39–42.
[18] U.S. EPA, Office of Policy, Planning, and Evaluation (March 1991), pp. 2–18.
[19] U.S. EPA, Office of Policy, Economics, and Innovation (January 2001), p. 64.
[20] OECD (1999a), pp. 42–43.

Structure of a Pollution Permit Trading System

A system of marketable pollution permits has two components:

- the issuance of some **fixed number of permits** in a region
- a provision for **trading** these permits among polluting sources within that region

The total number of permits issued is bound by whatever level of pollution is mandated by law as "acceptable." For example, if the level were set at 200 units of emissions, a maximum of 200 one-unit permits could be issued. Any polluter releasing emissions not authorized by permits would be in violation of the law. Once the permits are distributed, polluters are allowed to trade them with one another. A bargaining process should develop, which gives rise to a **market for pollution rights.** Following their own self-interest, polluters either purchase these rights to pollute or abate, whichever is the cheaper alternative. High-cost abaters will have an incentive to bid for available permits, whereas low-cost abaters will have an incentive to abate and sell their permits on the open market. The result is a cost-effective allocation of abatement responsibilities.[21]

The tradeable permit system accommodates environmental objectives, defined at an aggregate level. For example, in the United States, air pollution policies are designed to achieve national ambient air quality standards within well-defined regions. Within any region, however, some polluters might perform above the standard and others below it, which is acceptable as long as in the aggregate the region is in compliance. This is exactly how the permit system operates — controlling the total amount of emissions in a region but not the releases for each source within that region.[22] The trading component of the permit system capitalizes on differences in polluters' abatement technologies and opportunities. Sources that can abate efficiently are given the incentive to do so because they can sell their unused permits to their less efficient counterparts. As long as the environmental goal is achieved in the aggregate, the benefit to society is the same whether the task is undertaken by a select few or by all firms doing an equal amount of abating. However, the costs will be markedly lower if abatement is done by more efficient polluters.

Modeling a Pollution Permit System for Multiple Polluters

To illustrate the operation of a permit system, we return to our two-polluter model, where each firm faces distinct abatement costs:

$$\begin{aligned} \text{Polluter 1:} \quad & TAC_1 = 1.25(A_1)^2, \\ & MAC_1 = 2.5A_1, \\ \text{Polluter 2:} \quad & TAC_2 = 0.3125(A_2)^2, \\ & MAC_2 = 0.625A_2. \end{aligned}$$

Before any government intervention, we assume that each firm releases 10 units of pollution for an aggregate amount of 20 units in their region. The government has determined that the "acceptable level" of pollution for this region is 10 units, and it decides to reach this objective by using a tradeable permit system. It therefore issues 10 permits, each of which allows the bearer to emit 1 unit of pollution. For simplicity, assume that the government allocates 5 permits to each polluter.[23] Under the rules of the permit system, each firm is required

[21] Trading is critical to the cost-effective outcome. For example, if the permits were allocated equally across all polluters and no trading were allowed, the result would be no different from a command-and-control system of uniform standards.

[22] Contrast this scenario with a command-and-control instrument that forces every polluting source to meet identical standards of emission or effluent levels. Such instruments equalize the *level of control* across polluters rather than the *marginal costs of control.*

[23] The government could have introduced the permits through a direct sale, assigning a price to each permit, or through an auction. Either method has the advantage of generating revenue to the government, and this revenue could be used to absorb some of the administrative costs of the permit system.

to hold a permit for each unit of pollution released and to undertake abatement on all remaining units.

Based on the *initial* allocation of permits, each polluter must abate 5 units of pollution. This initial condition, termed Round 1 of the permit system, is summarized as follows:

Round 1: Government Issues Five Permits to Each Polluter

Polluter 1: Current pollution level: 10 units.
Number of permits held: 5.
Abatement required: 5 units.

$$MAC_1 = 2.5A_1 = 2.5(5) = \$12.50.$$
$$TAC_1 = 1.25(A_1)^2 = 1.25(5)^2 = \$31.25.$$

Polluter 2: Current pollution level: 10 units.
Number of permits held: 5.
Abatement required: 5 units.

$$MAC_2 = 0.625A_2 = 0.625(5) = \$3.125.$$
$$TAC_2 = 0.3125(A_2)^2 = 0.3125(5)^2 = \$7.81.$$

If the permit system did not allow for trading, each firm would have no choice but to abate 5 units each. Although the environmental objective would be met, it would not be achieved in a cost-effective manner. The combined abatement cost for both sources without trading is \$39.06.[24]

Now, consider how the result changes when permit trading is allowed. Because the two firms face different *MAC* levels at the end of Round 1, there will be an incentive for trade. Polluter 1 has an incentive to buy permits from Polluter 2 as long as the purchase price of each permit is less than its MAC_1. Likewise, Polluter 2 has an incentive to sell permits to Polluter 1 as long as it can obtain a price greater than its MAC_2. Suppose that in Round 2 of the trading process, the two firms agree on the purchase and sale of one permit at a price of \$8.00.[25] Polluter 1 purchases one permit from Polluter 2, giving Polluter 1 the right to pollute 6 units and the obligation to abate 4 units. Polluter 2 now possesses the right to release 4 units of pollution, which means it must abate 6 units. Round 2 is summarized as follows:

Round 2: Polluter 1 Purchases One Permit from Polluter 2

Polluter 1: Current pollution level: 10 units.
Number of permits held: 6.
Abatement required: 4 units.

$$MAC_1 = 2.5A_1 = 2.5(4) = \$10.00.$$
$$TAC_1 = 1.25(A_1)^2 = 1.25(4)^2 = \$20.00.$$
$$\text{Cost of one permit purchased} = \$\ 8.00.$$

Polluter 2: Current pollution level: 10 units.
Number of permits held: 4.
Abatement required: 6 units.

$$MAC_2 = 0.625A_2 = 0.625(6) = \$\ 3.75.$$
$$TAC_2 = 0.3125(A_2)^2 = 0.3125(6)^2 = \$11.25.$$
$$\text{Revenue from one permit sold} = \$\ 8.00.$$

[24] Recall from Chapter 4 that this is precisely the same expenditure incurred by the two polluters if a uniform standard is used under a command-and-control approach.

[25] Any negotiated price between MAC_1 and MAC_2 would be acceptable. The ultimate selling price within that range would be determined by the two firms' relative bargaining strengths.

This outcome can be analyzed from two perspectives—that of society and that of the firm. From the vantage point of society, the total costs of abating 10 units of pollution are now \$31.25, which is \$7.81 *less* than the costs without permit trading. Qualitatively, this is exactly what should result, because trading has brought the two firms' *MAC* values closer together. Polluter 1 now faces a lower MAC_1 of \$10 (compared to \$12.50 in Round 1), and Polluter 2 now has a higher MAC_2 of \$3.75 (compared to \$3.125 in Round 1).

Next, consider the gains that accrue to each firm as a result of the trade. Polluter 1 is better off because its total expenditures have decreased. Its outlay for abating plus the cost of the added permit is \$28.00 (i.e., TAC_1 of \$20.00 plus the cost of the additional permit, \$8.00), which is \$3.25 less than its TAC_1 in Round 1. Likewise, Polluter 2 is better off because its net expenditures associated with abating and trading are \$3.25 (i.e., TAC_2 of \$11.25 minus the revenue received from selling one permit, \$8.00), which is \$4.56 less than its TAC_2 at the end of Round 1.

Because there is an incentive for trade as long as the two firms face different *MAC* levels, it should be apparent that Round 2 does not represent a cost-effective solution. After the exchange of one permit, Polluter 1 still faces a higher *MAC* level than Polluter 2 (i.e., \$10.00 for Polluter 1 versus \$3.25 for Polluter 2). Thus, it is in each polluter's best interest to continue to trade. The rule of thumb is that in the presence of differing *MAC* levels among polluting sources, high-cost abaters have an incentive to purchase permits from low-cost abaters, and low-cost abaters have an incentive to sell them. The result? Low-cost abaters will do what they do best—clean up the environment—and high-cost abaters will pay for the right to pollute by buying more permits. Trading will continue until the incentive to do so no longer exists, that is, when the *MAC* levels across both firms are equal. At precisely this point, the cost-effective solution is obtained.

By applying this equimarginal principle to our model, we come to the final round, or equilibrium, as follows:

Final Round:* *Polluter 1 Purchases Three Permits from Polluter 2: Equalization of MAC Levels Across Polluters Is Achieved

Polluter 1: Current pollution level: 10 units.
Number of permits held: 8.
Abatement required: 2 units.

$MAC_1 = 2.5A_1 = 2.5(2) = \$ 5.00.$
$TAC_1 = 1.25(A_1)^2 = 1.25(2)^2 = \$ 5.00.$
Cost of three permits purchased[26] = \$20.00.

Polluter 2: Current pollution level: 10 units.
Number of permits held: 2.
Abatement required: 8 units.

$MAC_2 = 0.625A_2 = 0.625(8) = \$ 5.00.$
$TAC_2 = 0.3125(A_2)^2 = 0.3125(8)^2 = \$20.00.$
Revenue from three permits sold = \$20.00.

At this point, each polluter faces a *MAC* of \$5, and society's total cost to achieve the environmental objective is \$25. (Notice that the \$20 payment for permits should not be included in society's abatement costs, because this amount is just a transfer from one firm to another.) As predicted, the low-cost abater, Polluter 2, is doing most of the abating at 8 units, whereas the high-cost abater, Polluter 1, abates only 2 units.

[26] We assume that the second permit was sold for \$7 and the third for \$5. These values plus the \$8 price for the first permit results in a total payment of \$20 for permits.

Assessing the Model

It is no coincidence that the final abatement allocation for these two firms is identical to what results if a $5 pollution charge is imposed. Logically, the outcome is the same because both instruments use incentives that operate through the polluter's *MAC*. There are, however, three important differences. First, with a pollution charge, the government has to search for the price that will bring about the requisite amount of abatement. In the permit system, trading establishes the price of a right to pollute without outside intervention. Second, the pollution charge generates tax revenues on all units of pollution not abated, whereas no tax revenues are generated from the permit system. This distinction may be critical in jurisdictions facing tight fiscal budgets. However, a trading system can be designed to generate revenues if the government sells or auctions off the initial allocation of permits. Third, the trading system is more flexible; the number of permits can be adjusted to change the environmental objective. If the objective is too stringent, more permits can be introduced. If it is too lenient, the government, environmental groups, or concerned citizens can buy up permits, effectively reducing the amount of pollution allowed in the affected region.

Some measure of controversy surrounds the use of trading systems. Economists typically tout the advantages of a system that so explicitly uses the market process. On the other side of the coin, opponents argue that trading systems can create pollution hot spots, that is, localized areas facing high concentrations of pollutants where most of the permit buying takes place. Another objection is the potential for elevated administrative costs to keep records of trades and the emissions of buyers and sellers.[27] Hypotheticals aside, the true test will be in observing how these permit systems perform in practice.

Pollution Permit Trading Systems in Practice

Of all the available control instruments, tradeable permit systems are by far the most market-oriented form of environmental policy. Proposed for years by economists,[28] these systems are still in the early stages of development, and most have been used in air quality contexts. Although some trading systems are in use internationally, they are few in number and limited in scope. Examples include a program in Canada for ozone-depleting substances and one in Denmark for carbon dioxide emissions. Much of the evolution in tradeable permit systems is taking place in the United States at both the state and the federal level.[29] In attempting to combat the adverse effects of acid rain, the Clean Air Act Amendments of 1990 establish an allowance-based trading program to control sulfur dioxide emissions.[30] Application 5.4 describes the major events that characterized the first day of official trading on March 31, 1993.

Conclusions

No environmental policy instrument is without flaws. In truth, this less-than-perfect outcome should be expected. The market process works as well as it does because it operates autonomously, without external guidance. Therefore, it should not be surprising that any attempt to correct a market problem by imposing third-party controls is likely to have its share of pitfalls.

This line of defense is precisely the motivation of market-based policy approaches. The aim is not to add more restraint but to restore the market forces that broke down in the first place. In one form or another, market-based instruments effectively assign a price to environmental goods, such as clean air and clean water. Once this signaling mechanism is in place, polluters are forced to internalize the costs of pollution damage and adjust their decisions accordingly.

[27] Mandel (May 22, 1989).
[28] See, for example, Ruff (spring 1970).
[29] OECD (1999a), pp. 36–39.
[30] These important amendments and the legislation dealing with emissions trading will be discussed at length in Chapter 12.

Application

5.4 *Fighting Acid Rain with Pollution Rights: The First Annual Auction*

In March 1993, the first annual auction of rights to release sulfur dioxide (SO_2) emissions was held. The event was administered by the largest commodity exchange in the world, the Chicago Board of Trade (CBOT). Although most of the available permits are allocated by the EPA directly to the nation's largest polluting sources, a relatively small number (150,000 one-ton permits per year) are set aside in what is called an auction subaccount for direct sale. The CBOT was selected by the EPA to conduct the annual one-day auction, and it receives no remuneration from the EPA and may not charge for its services.

Most of the bids in the 1993 auction came from the nation's major utilities, who are the largest SO_2 polluters. A case in point is Illinois Power, a utility responsible for releasing about 240,000 tons of SO_2 emissions each year. This facility was unable to operate on the 171,000 permits issued by the EPA, so it bought some of the 75,000 to 125,000 more it needed from other utilities, at about $225 each. It subsequently submitted bids for another 5,000 permits at the 1993 auction. For Illinois Power and others like it, the costs to abate were apparently greater than the expected outlay to purchase permits.

Despite the predominance of utilities in the bidding, there was at least one important exception—a nonprofit environmental group called National Healthy Air License Exchange. According to the group's president, any permits bought in the auction would be retired and kept off the market. That private citizens can exercise such a tangible influence over environmental policy is one of the advantages of an emissions trading program. National Healthy Air License Exchange submitted bids for 1,100 permits but came away with only one, for which it paid $350. Nonetheless, the organization was able to participate in the auction and eliminate some emissions from the atmosphere, all without the help of political lobbyists or public officials.

In an apparently philanthropic move, Northeast Utilities of Connecticut donated 10,000 of its permits to the American Lung Association just before the auction. The association retired the permits to keep them out of the bidding process. The utility, which did not need the permits to operate, could have sold them for an estimated value of $3 million on the open market. However, the gesture was not totally without financial incentive. Northeast Utilities was expected to enjoy tax deductions for the contribution that would offset any sacrifice of pollution rights' revenues. Hoping to encourage other such donations, Northeast Utilities and the American Lung Association established a repository for permits donated by other utilities.

How did the bidders fare in the first pollution rights auction? Rights to emit the 150,000 tons of SO_2 were purchased by utilities, brokers, and environmentalists for a total of $21 million. Permit prices ranged from $122 to $450. The largest single purchaser was Carolina Power and Light Company, a utility that bid for and won over 85,000 permits. In accordance with the law, all auction proceeds went to the EPA, which then allocated the funds to those utilities from which the permits were originally obtained.

Sources: Taylor (March 31, 1992); Taylor and Kansas (March 26, 1992); Allen (March 20, 1993); Taylor and Gutfeld (September 25, 1992).

Of course, not all market-based instruments are well suited to all environmental problems. Both the nature of the problem and the market context must be understood before any policy can be implemented with success. Environmental problems are complex, both in origin and in implication. Likewise, today's markets are sophisticated and dynamic. However, evidence is beginning to accumulate that the link between the two is a fundamental step toward finding solutions.

Summary

- The market approach to pollution control uses economic incentives and the price mechanism to achieve a government-mandated environmental standard.
- The major categories of these market-based instruments are pollution charges, subsidies, deposit/refund systems, and pollution permit trading systems.
- A pollution charge is a fee that varies with the quantity of pollutants released. It can be implemented as an effluent or emission fee, a product charge, a user charge, or an administrative charge.
- A Pigouvian tax is a unit charge on the pollution-generating product equal to the *MEC* at the efficient level of output.
- An emission charge is a fee levied directly on the actual release of pollutants. Given a choice of abating, paying the fee, or employing a mix of abatement technology and tax payments, the profit-maximizing polluter will choose the most cost-effective strategy.
- When multiple polluters face a given abatement standard, the emission charge yields a cost-effective allocation of abatement responsibilities where the *MAC* levels for all sources are equal.
- Abatement equipment subsidies are aimed at reducing the costs of abatement.
- If the subsidy equals the *MEB* at the efficient level of production, it is called a Pigouvian subsidy.
- Per unit pollution reduction subsidies pay the polluter for abating beyond some predetermined level.
- A deposit/refund system imposes an up-front charge to pay for potential pollution damage and later refunds the charge for returning a product for proper disposal or recycling. The deposit is intended to capture the *MEC* of improper disposal, and the refund provides an incentive to properly dispose of or recycle wastes.
- A pollution permit trading system involves the issuance of tradeable rights to pollute based on a given environmental objective. Following natural incentives, polluters will either purchase these rights or abate, whichever is the cheaper alternative.

Key Concepts

market approach
pollution charge
product charge
Pigouvian tax
emission or effluent fee
abatement equipment subsidy
Pigouvian subsidy
per unit subsidy on pollution reduction
deposit/refund system
pollution permit trading system
pollution credits
pollution allowances

Use the Key Concepts listed above to begin your search for additional articles and information using the InfoTrac College Edition database.

Review Questions

1. In an article titled, "Environmentalist Predicts Marked Change in U.S. Policy," which appeared in the December 7, 1992, issue of the *Boston Globe,* Jessica Mathews, then vice president of the World Resources Institute, is quoted as saying that U.S. environmental policy will be undergoing major changes. She argues that subsidies encourag-

ing the use of scarce resources will be abandoned and that the prices of goods using these resources will rise to reflect their true environmental cost.

a. Illustrate Mathews's prediction by using appropriate market diagrams, both before and after a subsidy is in use.
b. Mathews also asserts that future policy will include tax differentials, whereby taxes will be reduced or removed on productive activity, such as savings, and imposed on non-productive negative externalities, such as pollution.
   (i) Give the economic intuition of this tax proposal.
   (ii) Give an example of a *current* polluting activity that might be taxed in the future. Graphically illustrate your example, and explain its significance to environmental quality control.

2. Despite economists' support of a market approach to environmental policy, the command-and-control approach continues to dominate the policy of most nations. Explain why this is the case. In your response, cite and then comment on some of the common criticisms of market-based initiatives.

3. Assume that there are two firms, each emitting 20 units of pollutants into the environment, for a total of 40 units in their region. The government sets an aggregate abatement standard of 20 units. The polluters' cost functions are as follows:

Polluter 1: $TAC_1 = 10 + 0.75(A_1)^2$, $MAC_1 = 1.5A_1$,

Polluter 2: $TAC_2 = 5 + 0.5(A_2)^2$, $MAC_2 = A_2$.

a. What information does the government need to support an assertion that the 20-unit abatement standard is allocatively efficient?
b. Suppose that the government allocates the abatement responsibility equally such that each polluter must abate 10 units of pollution. Graphically illustrate this allocation, and analytically assess the cost implications.
c. Now, assume that the government institutes an emission fee of $16 per unit of pollution. How many units of pollution would each polluter abate? Is the $16 fee a cost-effective strategy for meeting the standard? Explain.
d. If instead the government used a pollution permit system, what permit price would achieve a cost-effective allocation of abatement? Compare this allocation to the equal allocation standard described in part (b).

Additional Readings

Bohringer, C. "Industry-Level Emission Trading Between Power Producers in the EU." *Applied Economics* 34 (2002), pp. 523–33.

Fullerton, Don. "A Framework to Compare Environmental Policy." *Southern Economic Journal* 68 (October 2001), pp. 224–48.

Hanneman, W. Michael. "Improving Environmental Policy: Are Markets the Solution?" *Contemporary Economic Policy* 13 (January 1995), pp. 74–79.

McCann, Richard J. "Environmental Commodities Markets: 'Messy' versus 'Ideal' Worlds." *Contemporary Policy Issues* 14 (July 1996), pp. 85–97.

Nichols, Albert L. *Targeting Economic Incentives for Environmental Protection.* Cambridge, MA: MIT Press, 1984.

Nugent, Rachel. "Teaching Tools: A Pollution Rights Trading Game." *Economic Inquiry* 35 (July 1997), pp. 679–85.

Oates, Wallace. "Green Taxes: Can We Protect the Environment and Improve the Tax System at the Same Time?" *Southern Economic Journal* 61 (April 1995), pp. 915–22.

Repetto, Robert, Roger C. Dower, Robin Jenkins, Jacqueline Geoghegan. *Green Fees: How a Tax Shift Can Work for the Environment and the Economy.* Washington, DC: World Resources, November 1992.

Stavins, Robert N., "Market-Based Environmental Policies." In Paul R. Portney and Robert N. Stavins, eds., *Public Policies for Environmental Protection,* pp. 31–76. Washington, DC: Resources for the Future, 2000.

Stavins, Robert N., and Bradley W. Whitehead. "Dealing with Pollution: Market-Based Incentives for Environmental Protection." *Environment* 34 (September 1992), pp. 6–11, 29–42.

Tietenberg, T. H., ed. *Emissions Trading Programs,* vols. 1 and 2. Burlington, VT: Ashgate, 2001.

Walbert, Mark S., and Thomas J. Bierma. "The Permits Game: Conveying the Logic of Marketable Pollution Permits." *Journal of Economic Education* 19 (fall 1988), pp. 383–89.

Wirth, Timothy E., and John Heinz. *Project 88: Harnessing Market Forces to Protect Our Environment: Initiatives for the New President.* Washington, DC, December 1988.

Related Web Sites

Energy Information Administration	**http://www.eia.doe.gov/index.html**
International Energy Agency, "Selected Statistics"	**http://www.iea.org/statist/index.htm**
National Solid Wastes Management Association (NSWMA)	**http://www.nswma.org**
Organisation for Economic Co-operation and Development	**http://www.oecd.org**
"Pollution Rights Trading Game"	**http://www.marietta.edu/~delemeeg/expernom/f93.html#nugent1**
Resources for the Future	**http://www.rff.org**
The United States Experience with Economic Incentives for Protecting the Environment. U.S. EPA, National Center for Environmental Economics (January 2001)	**http://yosemite1.epa.gov/ee/epalib/incent2.nsf**
U.S. EPA, economics topics	**http://www.epa.gov/ebtpages/economics.html**
U.S. EPA, National Center for Environmental Economics	**http://yosemite.epa.gov/ee/epa/eed.nsf/pages/homepage**

PART 3

Analytical Tools for Environmental Planning

Economics uses powerful models to explain environmental market failures and the policy solutions used to address them. With these models as a foundation, we can now move to the practical implications of environmental planning—the process through which government identifies environmental risks, prioritizes them, and responds with a policy plan. The planning process involves difficult decisions: determining which hazards pose the greatest threat to society, where to set policy objectives, and which control instruments to use. These decisions are guided by analytical tools designed to evaluate environmental risks and assess the costs and benefits of minimizing them. In this module we conduct an in-depth investigation of two key analytical tools used in environmental planning and decision making: risk analysis and benefit-cost analysis.

In Chapter 6, we study the two components of risk analysis: risk assessment, which is the identification of risk; and risk management, which is the formulation of a risk response. In our study of risk assessment, we will explore how many scientific fields contribute to policymakers' understanding of environmental hazards. Even more disciplines are involved in risk management, including economics and the law. Among several risk management strategies available to policymakers is benefit-cost analysis, an economic tool used in guiding environmental policy decisions. Because of its potential significance, the final three chapters of the module are devoted to an investigation of this important analytical tool. Chapter 7 explains how economists measure and monetize environmental benefits. Chapter 8 presents the analogous discussion for costs, and Chapter 9 shows how both sets of results are used in a comparative evaluation to guide policy decisions.

Chapter 6

Environmental Risk Analysis

As we learned in the last module, there are many kinds of policy instruments government can use to control pollution; ranging from uniform standards to tradeable pollution permits. Now, we are prepared to investigate *how* such policies are designed and implemented in practice. It turns out that the means by which government becomes involved in any market failure and how it ultimately responds through policy are elements of a fairly complex process. First, scientists must identify and evaluate the risks of an environmental hazard and communicate their findings to policymakers. Once done, government must use this information to formally assess the magnitude of the environmental problem and the risk it poses to society—a procedure referred to as **risk assessment.** Second, public officials must decide on an appropriate course of action or policy response to reduce that risk—part of the process known as **risk management.** In this chapter, we will elaborate on these important risk-based procedures, which are the key elements of risk analysis.

Risk assessment and risk management involve difficult and sometimes controversial decisions. The decision making is difficult because there is uncertainty about environmental hazards and the implications for the ecology and human health—particularly over the long term. The controversy arises because there is no clear consensus about how government should respond to what *is* known about a given hazard. Accepting that all environmental risk cannot be eliminated, policymakers have to determine how much risk society can tolerate—a decision about which there is usually much debate. They also must decide what policy to use to achieve whatever risk level is deemed acceptable. How do public officials justify choosing one policy instrument over another? What criteria guide this decision-making process, and are they appropriate?

From an economic perspective, managing environmental risks should be guided by the costs and benefits associated with abatement. Unfortunately, the data to fully assess these benefits and costs are often insufficient. So policymakers have to rely on best available estimates or use an alternative risk management strategy. What are these alternative strategies, and what criteria are used to motivate them? These and other issues in risk analysis comprise our agenda in this chapter.

Concept of Risk

risk
The chance of something bad happening.

Risk is a somewhat obscure notion, yet all of us have some intuitive sense of what it is. After all, while most of us choose not to dwell on it, risk—or the chance of something bad happening—is part of life. Accepting that risk is a pervasive phenomenon, we also know that some risks can be minimized or even avoided, provided they have been recognized. This realization suggests that dealing with risk involves two tasks: (1) *identifying* the degree of risk and (2) *responding* to it. Individuals participate in both activities, although usually not in any systematic fashion. As a matter of course, they usually formulate a perception of risk based not on scientific data but on a subjective or even instinctive level. These perceptions, however unscientific, are what determine how an individual responds to risk. A person might choose to accept the risk as is, find ways to reduce it, or try to avoid it entirely.

All of this seems to imply that dealing with risk is strictly a private exercise, but it turns out that risk analysis is also an important part of public policy development. Why? Because at its core, policy is a formal response to risks faced by society. Because the government is devising a response for society as a whole, it cannot rely on individual perceptions of risk that are highly subjective, often uninformed, and possibly off the mark. Instead, the policymaker must use a systematic assessment of risk before devising a policy response.

Classifying Risk: Voluntary and Involuntary Risk

Attempting to impose structure on obscure concepts is always difficult. Yet it is precisely in these instances that order is needed. That researchers and other analysts classify risk in a number of ways seems to bear out this observation. One of the more common approaches is to consider two broad risk categories: **voluntary risk** and **involuntary risk.**

Voluntary Risk

As the label suggests, **voluntary risks** are those that are deliberately assumed at an individual level. That is, they are the result of a conscious decision. Every day, we make private decisions to engage in activities that implicitly add or subtract some amount of risk that we as individuals elect to accept.

voluntary risk
A risk that is deliberately assumed at an individual level.

Most voluntary risks arise from personal activities, such as driving a car, flying in an airplane, or drinking coffee. Because these risks are self-imposed, individuals can and do make decisions to respond to them. In particular, they adjust their personal exposure level to the underlying hazards. For example, skydiving is a personal activity associated with the risks of serious injury or premature death. An individual might choose to respond to these risks by avoiding the activity entirely. In so doing, exposure to the hazard and the risks to that individual are reduced to zero.

Since voluntary risks are self-imposed and the potential outcome is confined to a single person, the public sector typically is not involved or at most plays a limited role. When government does intervene, it is usually confined to identifying potentially dangerous conditions or products and communicating that information to society. In so doing, the government is helping people with the tasks of identifying and assessing risk, but it is not imposing a response decision. Providing information helps individuals perceive risks more accurately, so that they can make more informed private decisions about how to minimize or avoid them. This communication might be accomplished through product labeling—a sort of one-on-one information stream directly to the consumer—or through public service announcements that inform the general population about certain hazards.[1]

Involuntary Risk

People also are exposed to hazards that are beyond their control. Here, the risks are **involuntary,** since they do not arise from a willful decision. A classic example of involuntary risk is the likelihood of property damage and personal injury caused by a natural disaster. The risk of being harmed by a hurricane or an earthquake is not self-imposed, yet the chance of such an event, while relatively small, exists nonetheless. Environmental hazards, such as air pollution or toxic waste sites, are another source of involuntary risk. Here, the hazard is often chemical exposure that arises as an externality of production. In such cases, the risk is considerable because the effects are pervasive and because the potential harm extends to human health and the ecology.

involuntary risk
A risk beyond one's control and not the result of a willful decision.

Characteristic of involuntary risk is that individual responses are limited. Personal exposure to the hazard can be adjusted, but, absent extreme behavior, the risk cannot be reduced to zero. For example, to reduce personal risk of harm from an earthquake, an individual can avoid living in places prone to such occurrences, such as the San Francisco area. Such a

[1] Recently, the EPA has undertaken what it calls its Consumer Labeling Initiative to improve information on and understanding of household product labels. To learn more about this project, visit the EPA Web site at **http://www.epa.gov/opptintr/labeling/**.

Application

6.1 *EPA Declares Secondhand Smoke a Carcinogen*

In January 1993, the EPA released a long-awaited report, presenting the agency's conclusion that environmental tobacco smoke, also known as passive or secondhand smoke, is a human carcinogen responsible for the lung cancer death of 3,000 nonsmokers each year. Other reported risk estimates include the increased incidence of asthma in children and the higher risk of bronchitis and pneumonia in infants younger than 18 months. The EPA's report was based on the findings of a scientific advisory panel, which reviewed evidence from a collection of U.S. studies.

The EPA's announcement alarmed the general public, but it also stimulated debate about the agency's treatment of the matter. Harsh criticism was mounted about the two-year delay between the EPA's initial draft report on environmental smoke and its formal announcement. In 1990, the EPA had concluded its review of the risks of passive smoke and prepared a first draft report. However, the final report was delayed while the tobacco industry disputed the validity of the EPA's findings. At the same time, the EPA's indoor air research program terminated its work on tobacco smoke. According to the agency's critics, this halt in tobacco research *and* the delay in announcing its conclusions were the result of the EPA's succumbing to pressure from the tobacco industry. The EPA denied the allegation, asserting that its research on tobacco ended because the work had been completed, and it wished to move on to study other pollutants. EPA officials further argued that the 1993 announcement was connected to their risk assessment division, whose research was not affected by any changes within the agency's indoor air program.

Other questions were raised about the methods the EPA used to arrive at the risk estimates. In fact, soon after the report was issued, tobacco growers and cigarette manufacturers jointly filed suit against the EPA. The objective was to obtain a declaration that the agency's report was null and void because it was based on faulty scientific evidence, invalid procedures, and manipulated data. Notwithstanding the motivation of the plaintiffs, it has been reported that the agency *did* adjust the statistical confidence level of the scientific studies downward, from 95 to 90 percent. This adjustment allowed one of the studies to show a statistically significant increased risk of lung cancer in nonsmokers from exposure to passive smoke. Without the adjustment, no single study arrives at such a conclusion.

Responding to the allegations, then EPA administrator Carol Browner stated that she stood firmly behind the report. In July 1993, the Justice Department filed a motion to dismiss the lawsuit but to no avail. In 1997, a North Carolina district court ruled to vacate several chapters in the EPA's report, including its classification of secondhand smoke as a human carcinogen. The judge argued that the methods used in the study were questionable and that its conclusion was motivated by the EPA's desired outcome rather than by science and standard statistical tests. The following year the Clinton administration countered with an appeal. Although the legal challenge of the 1993 report is still in dispute, a more recent government study confirms the link between secondhand smoke and lung cancer incidence and also finds that it causes higher rates of heart disease, sudden infant death syndrome, and asthma.

Sources: National Research Council, Committee on Passive Smoking, Board on Environmental Studies and Toxicology (1986); Noah (January 6, 1993); Shalal-Esa (June 23, 1993); "EPA Warns of Exposure to Smoke" (July 23, 1993); "Quietly, EPA Drops Some Tobacco Research" (January 7, 1993); Raeburn and DeGeorge (September 15, 1997) U.S. EPA, Office of Communications, Education, and Public Affairs (September 15, 1998); Raul and Smith (January 8, 1999).

response would decrease that individual's risk of harm from earthquakes, but the risk still exists. Similarly, to decrease one's risk of the effects of dirty air, one could avoid living in such cities as Los Angeles or Mexico City. Again, the risk of harm is reduced but not eliminated.

Because the sources of involuntary risk are beyond the control of private individuals, the associated threat is a *public* problem. Consequently, government tries to control society's exposure to some involuntary risks. For chemical contaminants, government assumes most of the responsibility by using legislated controls. Occasionally, regulations are passed that forbid both production and consumption of an environmental contaminant. By banning a dangerous chemical, government reduces society's risk of exposure to zero. In the more conventional and less extreme case, government limits the use of the chemical or controls its release into the environment.

Distinguishing Between Voluntary and Involuntary Risks

To illustrate the difference between voluntary and involuntary risks, think about the voluntary risk of smoking cigarettes. As researchers learned more about the associated health risks, such as lung cancer and heart disease, the government began to disseminate this information through public service announcements and by placing warnings on cigarette packages. However, the individual decision to smoke, or, in this context, the response to the voluntary risk, remained a private decision because the risk was believed to extend *only* to the smoker. As research progressed, it became apparent that passive or secondhand smoke presents a risk to persons *other than* the smoker. This discovery meant that what was originally perceived as a voluntary risk posed a threat to others that was very much beyond their control—an involuntary risk. As such, passive smoke was considered a public problem, and government began to assume a more aggressive position on smoking. For more on the risks of passive smoke, see Application 6.1.

Defining Environmental Risk

One of the more important concerns of environmental decision makers is determining the involuntary risk of exposure to hazards such as pollutant emissions and toxic substances, known as **environmental risk.** Notice that two elements determine the extent of environmental risk: the hazard itself and exposure to that hazard. The **hazard** is the source of the damage or the negative externality, such as poisonous emissions from factories or toxic chemicals dumped into a river. **Exposure** refers to the pathways between the source of the damage and the affected population or resource. Although both hazard and exposure define environmental risk, each can independently affect the outcome. That is, some hazards are relatively minor but affect a large part of the population; others, like certain chemicals, are dangerous, but exposure to them is limited.

environmental risk
The probability that damage will occur due to exposure to an environmental hazard.

hazard
The source of the environmental damage.

exposure
The pathways between the source of the damage and the affected population or resource.

Because risk analysis is central to environmental decision making, policymakers have devised methods to assess, characterize, and respond to environmental risk. These interdependent methods are referred to as **risk assessment** and **risk management,** each of which will be investigated in some detail.

Risk Assessment in Environmental Decision Making[2]

Risk assessment refers to the qualitative and quantitative evaluation of the risk posed to health or the ecology by an environmental hazard. In practice, environmental risk assessment is conducted by scientists who gather, analyze, and interpret data about a given contaminant. In the United States, the EPA uses a paradigm of this process first presented by the National Academy of Sciences in 1983.[3] This model characterizes risk assessment as a series of four

risk assessment
Qualitative and quantitative evaluation of the risk posed to health or the ecology by an environmental hazard.

[2] Much of the following discussion is drawn from Patton (January–March 1993).
[3] National Academy of Sciences (1983).

Figure 6.1 *Risk Assessment Process*

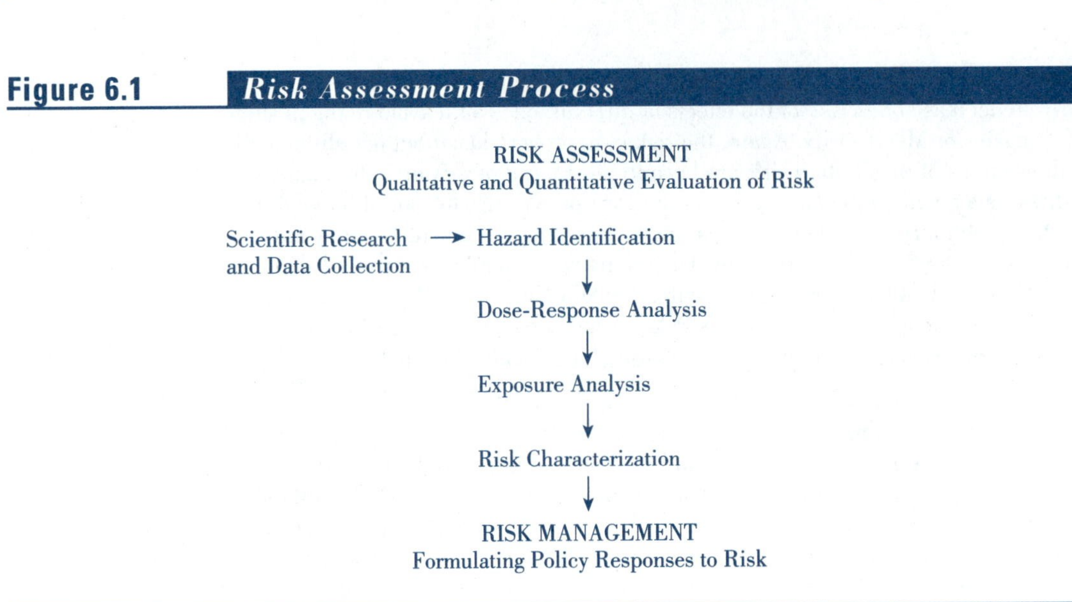

Source: National Academy of Sciences (1983).

steps, or fields of analysis, as they are known. These are **hazard identification, dose-response analysis, exposure analysis,** and **risk characterization.** Figure 6.1 shows the flow of information through these four steps as well as between risk assessment and risk management.

Hazard Identification

hazard identification Scientific analysis to determine whether a causal relationship exists between a pollutant and any adverse effects.

The first step of risk assessment is known as **hazard identification.** It is in this stage that scientists analyze data to determine whether a causal relationship exists between a pollutant and adverse effects on the ecology or human health. Causality in this context refers to a linkage between an environmental agent and the observed effect that is *believed* to exist based on a consensus within the scientific community.[4]

Ecological effects are any changes in the natural environment, such as crop damage, soil contamination, or fish kills. Under most public policies, these are viewed as secondary to human health effects. However, ecological effects and human health effects are not independent. Over time, human health is adversely affected if ecological health deteriorates. For example, damage to soil and crops may negatively affect economic productivity, human fitness, and the quality of life. In fact, a report on environmental risk conducted by the EPA's Science Advisory Board recommends that the EPA devote more attention to reducing ecological risks and to recognizing the link between ecological health and human health.[5]

Identifying the health consequences of exposure to an environmental contaminant is a high priority. Several scientific methods are used to gather the evidence needed to identify an environmental health hazard, some more reliable than others. Three common methods are **case clusters, bioassays,** and **epidemiology,** which are described in Table 6.1.[6]

[4]Kuhn (1970), cited in Lave (1982a), pp. 36–37. The interested reader should refer to Lave (1982b) for more detail on the problems of establishing causality in the context of hazard identification.
[5]U.S. EPA, Science Advisory Board (September 1990). In response, the EPA developed guidelines aimed specifically at ecological risk assessment. These guidelines can be accessed at **http://www.epa.gov/ncea/ecorsk.htm.**
[6]For more detail on these and other hazard identification methods, see Lave (1982a), which also provides an analysis of the strengths and weaknesses of each approach.

Table 6.1 ***Scientific Methods to Identify an Environmental Health Hazard***

Scientific Method	Definition
Case Cluster	A study based on the observation of an abnormal pattern of health effects within some population group.
Animal Bioassay	A study based on the comparative results of laboratory experiments on living organisms both before and after exposure to a given hazard.
Epidemiology	The study of the causes and distribution of disease in human populations based on characteristics such as age, gender, occupation, and economic status.

Source: Lave (1982a).

Dose-Response Analysis

Once a chemical substance has been identified as a hazard, scientists must investigate its potency by quantifying the ecological or human response to various doses. This element of risk assessment determines the **dose-response relationship.** Using data collected in the hazard identification stage, dose-response analysis attempts to develop a complete profile of the effects of an environmental pollutant. An important aspect of this analysis is determining whether some level of exposure to the hazard is safe. More formally, scientists call this a **threshold** level of exposure, which is the point up to which no response exists based on scientific evidence.

dose-response relationship
A quantitative relationship between doses of a contaminant and the corresponding reactions.

threshold
The level of exposure to a hazard up to which no response exists.

To determine the dose-response relationship, researchers first conduct two types of extrapolations from the data obtained through hazard identification:

1. **High-to-low dose extrapolation** adjusts for the high exposure levels used in laboratory or other test conditions.
2. **Laboratory-to-natural extrapolation** infers how the effects observed in the laboratory or test environment would differ under conditions existing in nature. (This includes the adjustment to infer a human response from the results of laboratory animal studies.)

Then, the researcher assigns a general functional form to the expected relationship between exposure and response and uses a statistical model to quantitatively estimate it.

Consider, for example, estimating the dose-response relationship associated with exposure to carbon monoxide (CO), a gas released from the incomplete combustion of carbon-based fuels like gasoline. At relatively small doses of CO, the individual may experience drowsiness. At higher exposure levels, visual perception and learning ability may be impaired. As the exposure level increases, death results. The purpose of the dose-response analysis is to quantitatively estimate at what exposure levels each of these effects occurs and all the intermediate dose-response relationships as well.

As in any statistical study, estimating a dose-response relationship requires the researcher to make initial assumptions. These include which factors are being controlled when defining the relationship and what the underlying relationship looks like.[7] Figure 6.2 (a), (b), and (c) illustrate three hypothetical dose-response functional forms. Common to each graph is a positive relationship between dose level and response. What differs among them is the *rate* at

[7]See Lave (1982a), pp. 43–47, for further information.

Figure 6.2 ***Hypothetical Dose-Response Relationships***

Panel (a) shows a linear dose-response function, meaning that the rate of increase between dose and response is constant. Because the function starts at the origin, no threshold level is observed. The function in panel (b) shows a constant rate of change between dose levels and response beyond dose level D_t. Up to and including that point, there is no response, meaning that there is a threshold at D_t. Panel (c) depicts a cubic relationship drawn from the origin, showing that the response level increases at an increasing rate up to dose level D_0 and then increases at a decreasing rate thereafter.

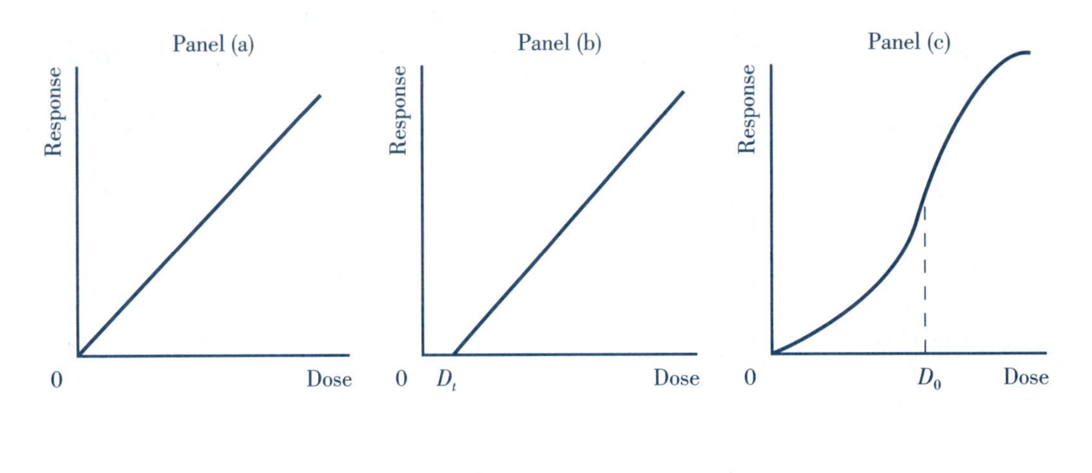

which the response increases with the dosage. For example, in Figure 6.2a, the dose-response relationship is linear, meaning that the rate of increase between dose and response is constant. Notice also that this curve begins at the origin. This means that a response is expected no matter how small the dose, or, equivalently, that no threshold level has been observed. Contrast this with the relationship shown in Figure 6.2b. In this case, there is a constant rate of change between dose levels and response but only beyond dose level D_t. Up to and including that point, there is no response at all, meaning that the hazard represented by this relationship has an identified threshold at D_t. Finally, Figure 6.2c shows a cubic (or S-shaped), relationship drawn from the origin. Here, the response level increases at an increasing rate up to dose level D_0 and then increases at a decreasing rate beyond that point.

The scientific results derived from hazard identification and dose-response analysis provide general information about the risks of an environmental hazard based on some known population defined by the laboratory or test conditions. This general information can then be used as a basis for assessing the risk to a potentially exposed population in a specific context. The EPA has established a database of identified environmental hazards and their estimated dose-response relationships for use by researchers and the general public. Referred to as the **Integrated Risk Information System (IRIS),** this database is designed to improve risk assessment by lending consistency and efficiency to what is a difficult and time-intensive process. See Application 6.2 for more on what IRIS offers and how it is used in practice to facilitate the remaining steps of the risk assessment process.

exposure analysis Characterizes the sources of an environmental hazard, concentration levels at that point, pathways, and any sensitivities.

Exposure Analysis

The process through which a generalized dose-response relationship is applied to specific conditions for an affected population is called **exposure analysis.** Exposure analysis characterizes the following:

6.2 *EPA's IRIS: A Health Effects Database*

Application

Risk assessment of environmental health hazards has become a well-structured process. In large part, this is a consequence of the paradigm presented by the National Academy of Sciences (NAS) in its 1983 publication *Risk Assessment in the Federal Government: Managing the Process.* Of the four steps outlined in this model—hazard identification, dose-response analysis, exposure analysis, and risk characterization—the first two provide scientific findings about the risk of a contaminant based on laboratory studies. By design, these findings then can be used to assess the risk posed by that same contaminant under actual conditions, using the latter two steps of the NAS model.

Because of the applicability of hazard identification and dose-response analysis, the EPA added structure to the risk assessment process by creating a repository of EPA consensus views on the health risks of environmental contaminants. Established in 1986, this database of consensus opinions comprises the Integrated Risk Information System, referred to by its acronym IRIS. Originally accessible only to EPA officials and staff, IRIS was made available to the general public in 1988. Today, the public can access IRIS through the Internet at **http://www.epa.gov/iris/index.html.**

There are three sections in the database: carcinogen assessment data for oral and inhalation exposure, noncancer health effects from oral exposure, and noncancer health effects from inhalation exposure. The database was developed and is maintained by two work groups, each of which comprises a panel of scientists from a multitude of disciplines and several EPA program areas.

As each group arrives at a consensus on a given pollutant, a summary is prepared and added to IRIS. Each summary includes a risk assessment table, which gives a risk number or relative risk value [i.e., a probability value for carcinogenic effects, a reference dose (RfD) for oral exposure to noncarcinogenic effects, or a reference concentration (RfC) for inhalation exposure to noncarcinogenic effects]. The risk value is supported by a synthesis and discussion of the relevant preliminary scientific data used to form the consensus. A full reference list also is provided to direct the reader to the studies upon which the consensus is based. Drinking water health advisories and any EPA regulatory actions relevant to the pollutant are itemized along with supplemental data, which may include acute health hazard information, chemical and physical properties of the substance, and major uses of the pollutant. These summaries ensure consistency and efficiency in conducting exposure analysis and risk characterization.

Sources: Tuxen (January–March 1993); U.S. EPA, Environmental Criteria and Assessment Office (1993).

- the sources of the environmental hazard
- the concentration levels at the source point
- the pathways from the source to the affected population
- any sensitivities within the population group

To illustrate, consider conducting an exposure analysis for lead. The potentially affected target group is the general population. The sources of lead are many, including painted surfaces, factory emissions, improperly fired ceramic cookware, lead-acid batteries, lead water pipes, and lead-soldered cans. Since lead is ubiquitous, this contaminant can reach the population through many pathways. Lead exposure can occur through inhalation of contaminated air, ingestion from drinking contaminated water or eating contaminated food, and direct ingestion of lead particles. Existing research also shows that sensitivity to lead exposure is greater for developing fetuses, infants, and young children.

Application

6.3 *Dynamics of Risk Assessment: The Case of Dioxin*

A family of chemical compounds called dibenzo-*p*-dioxins are considered by some scientists to be the single, most deadly of all synthetic compounds. Named as one of the contaminants responsible for the evacuation of residents in Love Canal, New York, in 1980 and Times Beach, Missouri, in 1983, dioxins were classified by the United States in 1985 as probable, highly potent human carcinogens. Over time, new evidence has called this initial risk assessment into question, suggesting the need for further scientific inquiry. In a broad sense, what this reevaluation implies is that risk assessment is a dynamic process.

The chronicle of events surrounding the reassessment of dioxins is a good case study for observing this dynamic process in action. A summary is given in the following list:

1980–1985	Following the disaster in Love Canal, the EPA begins to assess the risks of exposure to dioxins. In 1985, the EPA concludes that dioxin is a probable human carcinogen.
1988	Based mainly on scientific judgment, the EPA writes a draft document to revise its initial risk assessment of dioxin. The draft suggests that dioxin may not be as potent as originally stated in the 1985 assessment.
1990	A conference of 30 scientific experts agrees that the health effects of dioxin in humans can be predicted from the measured effects obtained through animal studies. They further agree that a new risk assessment model needs to be developed to accommodate current understanding of how dioxin affects the human body.
1991	In January, the National Institute for Occupational Safety and Health releases new data relating to the cancer mortality of dioxin-exposed workers. In April, the EPA begins work on a reassessment of *all* risks of dioxin. As part of its tasks, the agency is to develop a dose-response model with the help of outside prominent scientists. The EPA holds meetings to inform the public of its progress and to ask for comment.
1992	In August, the EPA issues drafts of human health and exposure assessment documents. As part of the review process, the EPA schedules a series of public meetings.
1994	The EPA releases its draft report, reaffirming its earlier findings that human exposure to dioxin at high levels may cause cancer and at low levels may lead to serious health consequences. Public meetings are held to receive comments.
1995	The EPA Science Advisory Board (SAB) reviews the documents and requests that two sections of the reassessment be reviewed again.
1997	Peer review and SAB re-review of the revised sections are scheduled.
2000–2001	The EPA prepares preliminary draft reassessment documents and provides them to its SAB for review. A final reassessment report is to follow. Detailed information on the reassessment is available at **http://cfpub.epa.gov/ncea/cfm/dioxreass.cfm.**

Sources: Preuss and Farland (January–March 1993); "Study: Dioxin Health Threat Much Worse Than Suspected" (September 12, 1994); U.S. EPA, Office of Research and Development, National Center for Environmental Assessment (September 19, 2001).

Risk Characterization

The final phase of risk assessment is called **risk characterization,** which is the objective of the entire process. **Risk characterization** is a complete description of the form and dimension of the expected risk based upon the assessment of its two components: the identified hazard and the exposure to that hazard. More than just an assimilation of the previous steps, the description includes both a **quantitative** and a **qualitative** risk evaluation.

risk characterization
Description of risk based upon an assessment of a hazard and exposure to that hazard.

The **quantitative** component identifies the magnitude of the risk and provides a way to compare one risk with another. Risk can be measured as a probability that an event will occur, using a numerical value that quantifies the likelihood of occurrence in some time period. Some probabilities are based on what are called **actuarial risks,** those determined from factual data. Actuarial risk measures are found by calculating the number of victims of a given hazard relative to the total number exposed. For example, the actuarial risk of death per year from driving an automobile has been estimated at 24 in 100,000, or 0.024 percent. The likelihood of premature death from being struck by lightning is 0.00005 percent, or 5 people for every 10 million. Other probability measures, such as the carcinogenic risks associated with chemical exposure, are based not on actual data but on inferences derived from animal bioassays or epidemiological studies. For example, the risk of getting cancer in a year from drinking chlorinated water has been estimated to be 0.0008 percent or 8 in 1 million persons exposed.[8]

Other environmental risks, such as noncarcinogenic health risks, are quantified as the exposure level to a hazard that can be tolerated over a lifetime without harm. This is communicated as a **reference dose (RfD),** expressed as

$$\text{RfD} = \text{milligrams of a pollutant per body weight (in kilograms) per day.}$$

Thus, an RfD for pollutant X of 0.005 milligrams/kilogram/day means that exposure to 0.005 milligrams of pollutant X per kilogram of body weight each day over a lifetime should cause no harm.

The **qualitative** element of the characterization gives context to the numerical risk value. It gives a description of the hazard, an assessment of exposure that notes any susceptible population groups, an identification of the data used, the scientific and statistical methods employed, and all underlying assumptions. Any scientific uncertainties, data gaps, or measurement errors that distinguish the findings are pointed out as well. All this information characterizes the reliability of the results and facilitates further research.

It is important to realize that risk assessment is not a fixed evaluation but rather part of a dynamic process. The assessment changes as new information and better analytical methods become available—exactly what has occurred in assessing the risk of exposure to dioxin, the subject of Application 6.3.

Risk Management in Environmental Decision Making: Responding to Risk

While the objective of risk assessment is to *identify* risk, it is the goal of **risk management** to *respond* to it. More to the point, risk management is concerned with evaluating and selecting from alternative policy instruments to reduce society's risk of a given hazard. To evaluate various policy options, the decision maker must consider not only the information given by the risk characterization but also such factors as technological feasibility, implementation costs, and other economic implications. Notice in Figure 6.3 that, although risk assessment is dominated by the work of scientists, risk management relies on many fields.

risk management
The decision-making process of evaluating and choosing from alternative responses to environmental risk.

[8]The estimates of risk presented in this section are given in Scheuplein (January–March 1993).

Figure 6.3 ***Disciplines in Risk Analysis***

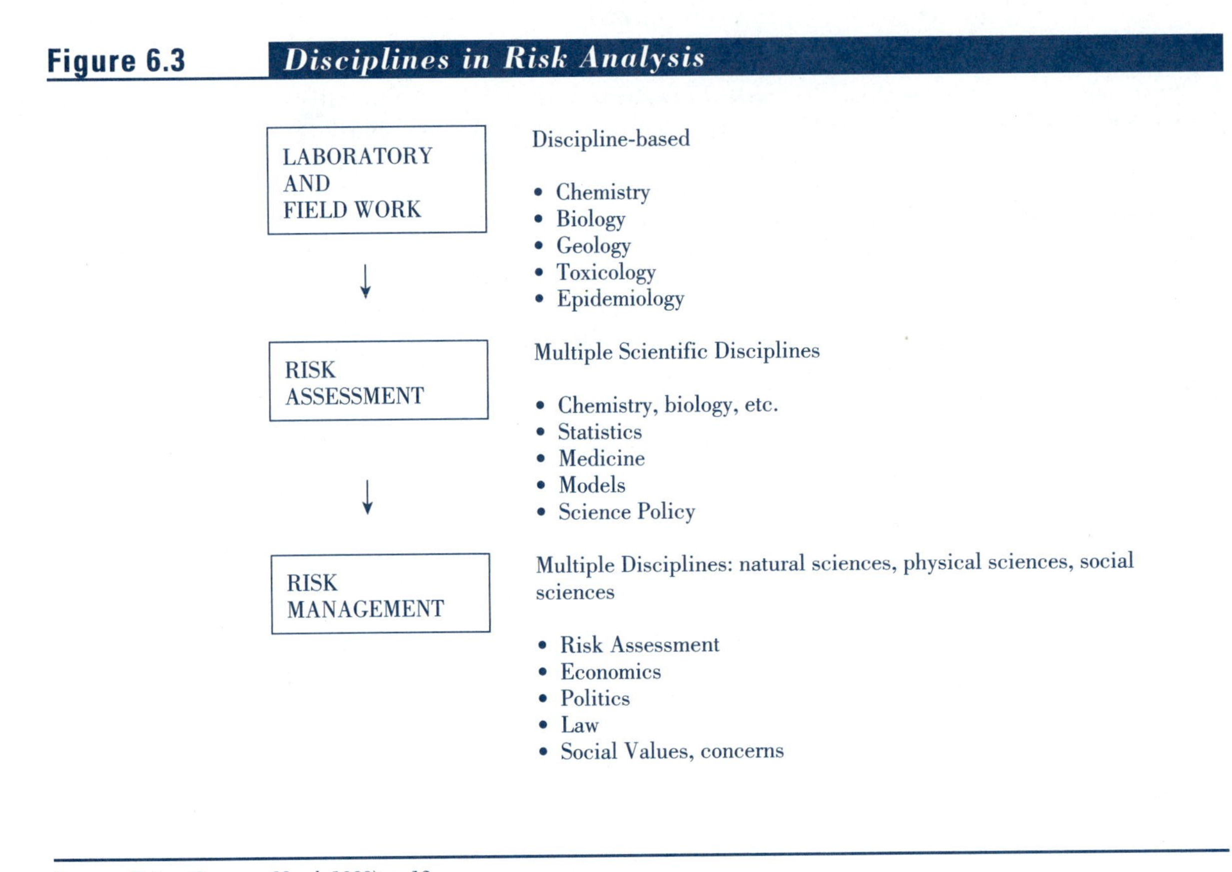

Source: Patton (January–March 1993), p. 12.

Implementation of the risk management process involves a series of decisions aimed at two major tasks:

1. determining what level of risk is "acceptable" to society
2. evaluating and selecting the "best" policy instrument to achieve that risk level

None of the underlying decisions are unidimensional, and realistically none can be made with complete objectivity. However, there are strategic approaches used to guide the decision making, some of which are mandated by law. We begin with a brief discussion of these two major tasks of risk management, laying the groundwork for an analysis of risk management strategies.

Tasks of Risk Management

Determining "Acceptable" Risk

"acceptable" risk The amount of risk determined to be tolerable for society.

The universal objective of all risk management strategies is to reduce risk. However, for each policy proposal, the public official must decide how much of a reduction is appropriate. Although risk is a function of both hazard and exposure, only one of these—exposure—can be controlled. So when the risk manager decides the amount of risk reduction to be achieved, the exposure level is implicitly determined. This in turn dictates how stringent public policy must be. In setting the level of **"acceptable" risk,** the policymaker confronts a difficult but necessary question: Should the policy eliminate the risk by reducing exposure to zero, or should some compromise be struck at some positive risk level, and, if so, where?[9]

If the acceptable risk level is set at zero, the policy must reduce society's exposure to zero (assuming no threshold level). Although this action eliminates the associated health and ecological damage, such a stringent policy is likely to cause economic problems. For example, if the hazard is a chemical used in manufacturing, reducing exposure to zero means that its use must be prohibited. Such a ban may force plant closings and create job losses.

Conversely, if some positive level of risk is deemed "acceptable," then the decision maker is setting policy that allows exposure and therefore some amount of damage. To help guide this difficult decision, policymakers sometimes use the concept of ***de minimis* risk.** ***De minimis* risk** refers to a negligible level of risk such that reducing it further would not justify the costs of doing so.[10] This concept is sometimes equated to the risk of a natural hazard, such as a lightning strike or an earthquake.[11] Once the baseline is established, the decision maker might use **comparative risk analysis** to evaluate how the positive risk level compares to other risks currently faced and accepted by society. This sort of analysis has been used to communicate the relatively unfamiliar risks associated with exposure to radon, a naturally occurring gas that can be harmful when trapped indoors. See Application 6.4 for a discussion of how the risks of radon exposure are expressed in terms of the risks from more familiar activities such as X-rays and cigarette smoking.

***de minimis* risk**
A negligible level of risk such that reducing it further would not justify the costs of doing so.

Evaluating and Selecting a Policy Instrument

Once the degree of risk and hence the stringency of policy have been determined, the second task of risk management is to decide what type of policy instrument to use. Here, the decision maker has to first evaluate alternative policies that can achieve the "acceptable" risk level and then select the "best" option from among them. Recall from Chapters 4 and 5 that the options likely will include some that are command-and-control and others that are market based. In making the decision, the risk manager considers the magnitude of risks, benefits, and/or costs associated with each available control instrument. In fact, there are risk management strategies that define exactly how this evaluation is to be done.

Risk Management Strategies

Executing the two tasks of risk management—determining the "acceptable" risk level and choosing the appropriate policy instrument—requires a systematic evaluation of available options. From an economic perspective, the most important considerations are

- the level of risk established
- the benefits that accrue to society from adopting the policy
- the associated costs of implementing the policy

Several risk management strategies have been developed over time, each of which outlines *how* these factors are to be evaluated. The most prevalent of these are **comparative risk analysis, risk-benefit analysis,** and **benefit-cost analysis.**

comparative risk analysis
An evaluation of relative risk.

Comparative Risk Analysis

Just as comparative risk analysis helps the risk manager select an "acceptable" risk level, it also can be used to help officials identify which risks are most in need of an official response. The EPA's Science Advisory Board has prepared a ranking of environmental problems by

[9] For an excellent discussion of the political and legal processes involved in environmental standard setting as it applies to "acceptable" risk, see Marchant and Danzeisen (1989).
[10] For example, the United States generally does not regulate cancer risks measured at less than 1 additional case per 1 million people. This does not mean that they will never intervene if the risk is smaller—only that as a general rule they choose not to.
[11] Krimsky and Golding (1991), p. 105.

Application

6.4 Using Comparative Risks to Communicate the Dangers of Radon

Over the last 10 years, public awareness of radon has improved, but many people still are unaware of the potential risks. The result? Government is faced with a public policy problem. Although the risk of exposure to radon is involuntary, government cannot directly intervene because radon pollutes a part of the environment that is beyond its jurisdiction—the *indoor* air.

Exposure to radon is an involuntary risk because it is a natural hazard. It is a radioactive gas caused by the decaying of uranium found in soil deposits and rocks. Outdoors, as radon is released, it dissipates quickly and poses no adverse health effects. It becomes a potential problem only if it enters homes and buildings through cracks in basements or foundations. If radon is trapped indoors, it can accumulate to dangerous levels. It is odorless and colorless, so people are unaware of its presence and are exposed to it quite involuntarily. Research studies suggest that long-term exposure to radon causes lung cancer. The EPA estimates that chronic exposure may be responsible for 7,000 to 30,000 lung cancer deaths per year, ranking it second only to smoking as a cause of lung cancer.

The EPA has joined forces with the surgeon general and various national health associations to educate society about the little-known but potentially dangerous problem of radon. People have been encouraged to measure the radon level in their homes using readily available test kits. Brochures have been prepared by government and other associations to disseminate facts about radon to homeowners. Yet all of this well-intentioned information is useful only if it is communicated in a way that is meaningful to the general public. To accomplish this, the government uses **comparative risks.**

Radon concentration levels are measured in units of picocuries per liter (pCi/L). Risk estimates for lifetime exposure to various levels are given as the number of people who get lung cancer in a given population. To communicate these estimates more effectively, comparable lifetime risks of more familiar activities are provided, as shown in the accompanying table.

If the indoor radon level is 4 pCi/L or higher, the EPA recommends that some action be taken to correct the problem. But the first step is proper risk communication. Using comparative risks to explain an otherwise unfamiliar hazard can be an effective means to help society recognize and respond to the health risk of radon exposure.

Sources: U.S. EPA (August 1988), pp. 35–37; U.S. EPA, Office of Communications, Education, and Public Affairs (April 1992), p. 15; U.S. EPA, U.S. Department of Health and Human Services, and U.S. Public Health Service (September 1992).

Radon Level (pCi/L)	Risk of Getting Lung Cancer (If You've Never Smoked) (number of persons per 1,000)	Comparative Risk
20	8	Risk of being killed in a violent crime
10	4	
8	3	10 times the risk of dying in an airplane crash
4	2	Risk of drowning
2	1	Risk of dying in a home fire
1.3	<1	Average indoor radon level

Source: U.S. EPA, U.S. Department of Health and Human Services, and U.S. Public Health Service. (September 1992). (Online version of this document can be accessed at **http://www.epa.gov/iaq/radon/pubs/citguide.html.**)

Table 6.2 ***Scientific Rankings of Environmental Problems***

Relative Risk Ranking	Environmental Problem
High risk to human health	Ambient air pollutant Worker exposure to chemicals in industry and agriculture Indoor pollution Contamination of drinking water
High risk to natural ecology and human welfare	Habitat alteration and destruction Species extinction and loss of biological diversity Stratospheric ozone depletion Global climate change
Medium risk to natural ecology and human welfare	Herbicides/pesticides Contamination of surface waters Acid deposition Airborne toxics
Low risk to natural ecology and human welfare	Oil spills Groundwater contamination Radionuclides Thermal pollution Acid runoff to surface waters

Sources: U.S. EPA, Science Advisory Board (September 1990); U.S. EPA, Office of Communications, Education, and Public Affairs (April 1992), p. 9.

degree of risk, shown in Table 6.2.[12] The ranking was based on the best available data and scientific information and considered such factors as severity of effects and the number of people exposed. The SAB specifically advises that EPA programs should be guided by the principle of **relative risk reduction,** meaning the agency should order its policy decisions to reduce the most severe environmental risks first. The board's view on using comparative risk analysis is clear. In its report, the SAB asserts:

> If priorities are established based on the greatest opportunities to reduce risk, total risk will be reduced in a more efficient way, lessening threats to both public health and local and global ecosystems.[13]

One difficulty associated with setting risk-based priorities is that the rankings set by government are often different from how society *perceives* environmental risks. Look at Table 6.3 to see the ranking of environmental problems based on general public perception, and compare it to the SAB's ranking given in Table 6.2. This dichotomy of views presents a dilemma to officials attempting to gain support for policy proposals. Therefore, it is important that government communicate scientific findings to the public to improve its understanding of environmental risk.[14]

Comparative risk analysis also can be used to select from among alternative control instruments. Used in this context, the approach is often called **risk-risk analysis.** This risk management strategy involves a comparison of the estimated risk probabilities or risk-ranking scores from two or more policy options. For example, the decision maker might compare the relative risks of two different mandates for hazardous waste treatment, such as land disposal

[12] U.S. EPA, Science Advisory Board (September 1990); "Two Faces of Risk" (January–March 1993).
[13] U.S. EPA, Science Advisory Board (September 1990), p. 2.
[14] U.S. EPA, Science Advisory Board (September 1990), p. 12.

Table 6.3 ***Public Perceptions of Environmental Problems***

Environmental Problem	Percentage Responding That Problem Is Very or Extremely Serious
Hazardous waste	89
Oil spills	84
Air pollution	80
Solid waste disposal	79
Atmospheric damage	79
Nuclear waste	78
Contaminated water	77
Forest destruction	76
Ocean pollution	75
Endangered species	67
Threats to wildlife	65
Pesticide use	60
World population	57
Poor energy use	56
Global warming	56
Reliance on coal/oil	53
Wetland development	50
Radon gas	35
Indoor air pollution	27
Electromagnetic fields	19

Source: U.S. EPA, Office of Communications, Education, and Public Affairs (April 1992), p. 9.

versus incineration, and propose whichever approach is more effective in reducing risk.[15] Implicitly, the objective of a risk-risk strategy is risk minimization, with no explicit consideration allowed for associated costs.

Risk-Benefit Analysis

risk-benefit analysis An assessment of risks of a hazard along with the benefits to society of not regulating that hazard.

An alternative risk management strategy, called **risk-benefit analysis,** simultaneously considers the benefits to society of *not* regulating an environmental hazard along with the level of associated risk. Here, the objective is to simultaneously maximize the expected benefits *and* minimize the risk. Although it may seem perverse, it is true that the source of some environmental hazards offers benefits to society. Think about gasoline, which during combustion gives off emissions that present a health risk. Nonetheless, gasoline *does* benefit society by fueling motor vehicles. Hence, if the risk manager were to assess only the risks in this case, the solution might be to ban gasoline. Yet, if the risk-reduction strategy were balanced with a consideration of benefits, the risk manager would have to consider how a reduction in gasoline usage would diminish society's well-being.

The use of a risk-benefit strategy is commonly mandated in environmental law. An example is found in the Toxic Substances Control Act (TSCA). This law requires the EPA to simultaneously consider the health and environmental effects, the degree of exposure to the substance, *and* the benefits the substance provides to society in use. A similar mandate is called for in the Federal Insecticide, Fungicide, and Rodenticide Act (FIFRA).[16]

[15] U.S. Congress, Office of Technology Assessment (1983), pp. 226–27.
[16] "Some Statutory Mandates on Risk" (January–March 1993). We will have much more to say about TSCA and FIFRA in Chapter 19.

Benefit-Cost Analysis

President Reagan's Executive Order 12291 is largely responsible for the more intensified use of **benefit-cost analysis** in formulating environmental policy. President Clinton's Executive Order 12866 continued to support this risk management strategy, and the current Bush administration is adhering to this Executive Order until the president issues a new or modified version.[17] Benefit-cost analysis can identify an "acceptable" risk level based on the criterion of **allocative efficiency.** For incremental risk reductions, the decision maker would compare the monetized value of social benefits with the associated costs to find the efficient risk level, where the marginal social benefit (*MSB*) and the marginal social cost (*MSC*) of risk reduction are equal. Equivalently, this corresponds to the risk level that maximizes the difference between the total social benefits (*TSB*) and the total social costs (*TSC*).

benefit-cost analysis
A strategy that compares the *MSB* of a risk reduction policy to the associated *MSC*.

Often, environmental law establishes the risk reduction to be achieved, which means that the level of associated benefits has been predetermined. In such cases, the risk manager still can use benefit-cost analysis but with a different objective in mind. Here, the goal would be to select a policy instrument that meets the legislated risk objective at least cost. If the selection is made properly, the initiative will achieve the economic goal of **cost-effectiveness** rather than allocative efficiency.

Finally, benefit-cost analysis can be used in the policy appraisal stage much in the same way it is used to find an "acceptable" risk level. At this stage, allocative efficiency is used to evaluate the effectiveness of an ongoing initiative. In practice, estimates of the *MSB* and *MSC* at the risk level achieved by policy are compared to see if they are equivalent. If not, the risk manager knows that the policy needs to be amended to correct the resource misallocation.

Conclusions

While most policy decisions are difficult and even controversial, those made in the context of environmental issues are particularly so. Indeed, much of what environmental policymakers struggle with is how to deal objectively and fairly with the risks posed by environmental hazards. Such is the purpose of risk analysis and its two components, risk assessment and risk management.

Scientists provide the data and analysis needed for risk assessment. As a result of their research, public officials gain valuable information about the nature of environmental hazards and the risks of exposure. Armed with a characterization of the risks involved, government can make better and more informed policy decisions. Through risk management strategies, such as comparative risk analysis and benefit-cost analysis, "acceptable" risk levels can be determined and alternative policy instruments can be evaluated objectively using well-defined criteria. Independent of how the "acceptable" risk level is determined, estimates of the social benefits and costs of a policy are useful in evaluating the effectiveness of policy initiatives after they have been adopted into law. These data can guide proposals for legislative amendments that characterize the dynamic process of environmental policy development.

The use of benefit-cost analysis as a decision rule is becoming more prevalent in public policy decision making. Monetizing environmental costs and benefits is an attempt to provide an impartial guideline to the risk manager, but the task can be difficult to execute in practice. It also has been the source of some controversy. Think about the dilemma of assigning a dollar value to saving a life or restoring the quality of coastal waters. Consequently, economists are continuing to research better methods and more comprehensive data to improve their estimates of benefits and costs. In the next three chapters, we will explore the fundamental steps of benefit-cost analysis and assess the contribution of this risk management strategy to environmental decision making.

[17] To access Executive Order 12866 on the Internet, visit **http://www.epa.gov/fedrgstr/eo/eo12866.htm.**

Summary

- Voluntary risks are deliberately assumed at an individual level. Involuntary risks arise from exposure to hazards beyond the control of individuals.
- Environmental risk measures the likelihood that damage will occur due to exposure to an environmental hazard. The hazard is the source of the damage, and exposure refers to the pathways between this source and the affected population or resource.
- Risk assessment is the qualitative and quantitative evaluation of the health or ecological risk posed by an environmental hazard. This can be modeled as a series of four steps: hazard identification, dose-response analysis, exposure analysis, and risk characterization.
- Hazard identification uses scientific data to determine whether a causal relationship exists between an environmental agent and adverse health or ecological effects. Several methods are used, including case clusters, bioassays, and epidemiology.
- A dose-response relationship quantitatively shows how a biological organism responds to a toxic substance as exposure changes. An important objective is to identify whether there is a threshold level of exposure, the point up to which no response is observed.
- Exposure analysis characterizes the conditions faced by the potentially affected population.
- Risk characterization is a quantitative and qualitative description of expected risk.
- The quantitative component of risk characterization provides a means to gauge the relative magnitude of the risk. Risk might be measured as a probability or as a reference dose (RfD).
- The qualitative component of risk characterization gives context to the numerical measure of risk and includes a description of the hazard, an assessment of exposure, an identification of the data, the scientific and statistical methods used, and any uncertainties in the findings.
- Risk management is concerned with evaluating and selecting from alternative policy instruments to reduce society's risk of a given hazard. Several risk management strategies are used in practice, including comparative risk analysis, risk-benefit analysis, and benefit-cost analysis.
- Comparative risk analysis, known in some contexts as risk-risk analysis, involves an evaluation of relative risk. This can be used to help officials identify which risks are most in need of an official response. It also can be used to select among alternative control instruments.
- Risk-benefit analysis is aimed at simultaneously maximizing expected benefits *and* minimizing risk.
- Benefit-cost analysis evaluates alternative risk levels by comparing the value of the expected gains with the associated costs. If the "acceptable" risk level maximizes the difference between the total social benefit (*TSB*) and the total social costs (*TSC*), the outcome will be allocatively efficient. If the law establishes the risk level to be achieved, a cost-effective solution can be realized by selecting the least-cost policy instrument that achieves the risk objective.

Key Concepts

risk
voluntary risk
involuntary risk
environmental risk
hazard
exposure
risk assessment
hazard identification

dose-response relationship
threshold
exposure analysis
risk characterization
risk management
"acceptable" risk
de minimis risk
comparative risk analysis
risk-benefit analysis
benefit-cost analysis

Use the Key Concepts listed above to begin your search for additional articles and information using the InfoTrac College Edition database.

Review Questions

1. Refer to Tables 6.2 and 6.3, and investigate how the public's perception of the risk of indoor pollution compares with the ranking given by the EPA's Science Advisory Board. How is this comparison relevant to the use of comparative risk analysis in communicating the hazards of radon, as discussed in Application 6.4? (Or access the radon report directly at **http://www.epa.gov/iaq/radon/pubs/citguide.html#riskcharts.**)

2. Comment on the following statement: "*Without exposure, there is no risk.*"
3. Other than those mentioned in the chapter, give several real-world examples of how government has provided public information to enhance the identification of a voluntary risk.
4. a. Interpret the shape of the dose-response function in Figure 6.2c.
   b. Does this dose-response relationship suggest the presence of a threshold level? If so, where is it? If not, why not?
5. a. Verbally describe what an RfD of 0.002 for some pollutant Z means.
   b. Graphically sketch a dose-response function for pollutant Z, assuming that the dose-response relationship increases at a decreasing rate throughout. Label the RfD on your diagram.
6. Suppose you are using risk-benefit analysis to evaluate a policy aimed at limiting the use of a pesticide applied to grain crops. Describe the risks and benefits that would have to be estimated to conduct this analysis properly.

Additional Readings

Ahearne, John F. "Integrating Risk Analysis into Public Policymaking." *Environment* 35 (March 1993), pp. 16–20, 37–39.

Carnegie Commission on Science, Technology, and Government. *Risk and the Environment: Improving Regulatory Decision Making.* New York: Carnegie Commission, June 1993.

Chess, Caron, and Billie Jo Hance. "Opening Doors: Making Risk Communication Agency Reality." *Environment* 31 (June 1989), pp. 11–15, 38–39.

Davies, Terry. "Congress Discovers Risk Analysis." *Resources* (winter 1995), pp. 5–8.

Gerrard, Simon, R. Kerry Turner, and Ian J. Bateman, eds. *Environmental Risk Planning and Management.* Northampton, MA: Elgar, 2001.

Hattis, Dale. "Drawing the Line: Quantitative Criteria for Risk Management." *Environment* 38 (July/August 1996), pp. 10–15, 35–39.

Johnson, F. Reed, Ann Fisher, V. Kerry Smith, and William H. Desvouges. "Informed Choice or Regulated Risk? Lessons from a Study in Radon Risk Communication." *Environment* 30 (May 1988), pp. 13–15, 30–35.

Manes, Christopher. *Green Rage.* Boston: Little Brown, 1990.

Smith, Kirk R. "Air Pollution: Assessing Total Exposure in the United States." *Environment* 30 (October 1988), pp. 10–15, 33–38.

U.S. Environmental Protection Agency. *Setting the Record Straight: Secondhand Smoke Is a Preventable Health Risk.* EPA 402-F-94-005. Washington, DC, June 1994.

U.S. Environmental Protection Agency, Office of Health and Environmental Assessment, Office of Research and Development. *Respiratory Health Effects of Passive Smoking: Lung Cancer and Other Disorders.* EPA/600/6-90/006F. Washington, DC, December 1992.

Viscusi, Kip. "The Value of Risks to Life and Health." *Journal of Economic Literature* 31 (December 1993), pp. 1912–46.

Related Web Sites

A Citizen's Guide to Radon: The Guide to Protecting Yourself and Your Family from Radon, U.S. EPA, Office of Air and Radiation (September 1992)	**http://www.epa.gov/iaq/radon/pubs/citguide.html**
Draft Dioxin Reassessment	**http://cfpub.epa.gov/ncea/cfm/dioxreass.cfm**
Guidelines for Ecological Risk Assessment	**http://www.epa.gov/ncea/ecorsk.htm**
Integrated Risk Information System (IRIS)	**http://www.epa.gov/iris/index.html**
National Center for Environmental Assessment	**http://cfpub.epa.gov/ncea/cfm/nceahome.cfm**
President Clinton's Executive Order 12866	**http://www.epa.gov/fedrgstr/eo/eo/12866.htm**
Risk Assessments for Toxic Air Pollutants: A Citizen's Guide, U.S. EPA, Office of Air and Radiation (March 1991)	**http://www.epa.gov/oar/oaqps/air_risc/3_90_024.html**
U.S. EPA, Consumer Labeling Initiative (CLI)	**http://www.epa.gov/opptintr/labeling/**

Chapter 7

Assessing Benefits for Environmental Decision Making

Risk assessment and risk management are central to the decision-making process that guides environmental policy. Once the degree of environmental risk has been identified, public officials begin the critical task of formulating policy. Ultimately, the objective is to minimize risk, which is a benefit to society. Meeting this objective is not an unconstrained decision, however. There are opportunity costs. Public officials must consider that resources used to reduce smog are no longer available to clean the Great Lakes or to save the California condor or to improve public education. How do policymakers come to grips with such tough decisions? There is no simple answer, and, in fact, most would argue that the public sector wrestles with this problem on an ongoing basis. Yet there are decision-making strategies that can be effective in environmental policy development, among them, **benefit-cost analysis.**

Benefit-cost analysis underlies much of economic theory. For example, the balancing of revenues and costs at the margin to maximize profit is an application of benefit-cost analysis. In the broader context of policy decisions, benefit-cost analysis is used to evaluate the associated gains and losses to society as a whole. In every case, the basic principle is the same: An efficient solution results if benefits and costs are balanced at the margin.

Having said this, it should be apparent that the theoretical footing of benefit-cost analysis is sound. However, applying this theory in practice is not so clear-cut. In order to use benefit-cost analysis to guide environmental decisions, policymakers must quantify the associated social benefits and costs. Yet many intangibles that are difficult to measure in monetary terms are involved, for example, longevity of human life, improved aesthetics, and the preservation of ecosystems. Although the process is difficult, it is critically important. Governments everywhere are spending huge sums to develop and implement environmental policy. In the United States alone, the annual expenditure is about $120 billion. Such large appropriations cannot be made without understanding the economic implications on both sides of the ledger. Hence, there is no debate that policymakers need reliable measures of social benefits and costs to help guide these important decisions.

In this chapter, we explore the motivation for valuing **environmental benefits** and the methods used to measure them. We start by presenting the conceptual issues of how and why society values natural resources and environmental quality. This theoretical framework supports our subsequent investigation of various benefit estimation methods. Once this side of the analysis is complete, we conduct an analogous study of costs in Chapter 8. All of this lays the groundwork for studying how the two elements come together in a benefit-cost analysis, which is the subject of Chapter 9.

Identifying and Valuing Environmental Benefits: Conceptual Issues

As a starting point, we need to establish the appropriate level of analysis for assessing policy-induced environmental benefits. From previous chapters, we know that these health and ecological gains can be assessed as damage reductions. The key is to recognize that the relevant measure is the *change* in damage reductions brought about by policy. In practice, these changes are called **incremental benefits.**

Defining Incremental Benefits

To assess the social benefits attributable to environmental policy, decision makers must find out how health, ecological, and property damages change as a consequence of that policy. This focus on the *change* in damages instead of on their absolute level is not new. Economic theory is concerned with effects that occur *at the margin.* These too are changes, although they are infinitesimal—measured at a point. When the relevant change is over a discrete range, it is referred to as **incremental** rather than as **marginal.** Because policy evaluation is concerned with identifying damage reductions over some discrete time period, the appropriate measure of benefits is incremental.

incremental benefits
The reduction in health, ecological, and property damages associated with an environmental policy initiative.

To identify incremental benefits, analysts must compare the actual or expected benefits to society after some policy is implemented to a baseline measure of current conditions. Environmental benefits are commonly separated into categories, such as improvements in human health, aesthetics, the economy, recreation, property, and the ecology. Application 7.1 discusses how incremental benefits were estimated to support an important revision in U.S. air quality standards for particulate matter.

Defining Primary and Secondary Environmental Benefits

Within the broad category of incremental benefits are two types of damage-reducing effects: **primary environmental benefits** and **secondary environmental benefits.** A **primary environmental benefit** is a damage-reducing effect that is the *direct* consequence of implementing policy. Compare this with the concept of a **secondary environmental benefit,** which is an *indirect* gain to society associated with the implementation of policy.

primary environmental benefit
A damage-reducing effect that is a direct consequence of implementing environmental policy.

secondary environmental benefit
An indirect gain to society that may arise from a stimulative effect of primary benefits or from a demand-induced effect to implement policy.

Primary Environmental Benefits

Most environmental policy actions are aimed at increasing primary benefits, particularly those associated with human health. Health benefits include both decreased mortality (e.g., a reduction in the risk of cancer deaths) and reduced morbidity (e.g., a lower incidence of respiratory ailments). Other primary benefits include more stable ecosystems and improved aesthetics. Still other primary benefits are economic, such as a more prosperous fishing industry that results from the enactment of clean water regulations. What these benefits have in common is that they are a *direct* outcome of environmental policy.

Secondary Environmental Benefits

Secondary environmental benefits arise *indirectly* from a policy change. One source might be the stimulative effect of a primary benefit, such as higher worker productivity that results from the primary benefit of improved health. Higher productivity increases the availability of goods and services, which may lead to a decline in prices. Because these gains are stimulated by a primary benefit and arise indirectly, they are considered secondary benefits. An alternative source is a demand-induced change, such as the increased demand for labor to implement a new policy. In this case, the economic gains of an improved labor market are a secondary benefit.[1]

Conceptually Valuing Environmental Benefits

What is the value to society of cleaner air or cleaner water? What value does society place on cleaning up a hazardous waste site? As we have discussed in previous chapters, both questions could be answered directly if the commodity in each case were a private good traded in the open market. Then demand prices would convey the marginal benefit of each additional unit of the good. The problem is that environmental quality is a public, nonmarketed good. The absence of prices and the dilemma of nonrevelation of preferences cloud a determination of how society values a cleaner environment. In theory, if we could infer society's demand for

[1]There is some debate about whether secondary benefits should be considered when assessing public policy proposals. It is argued that these indirect gains in one market or region are offset by losses in another. There is also the practical problem of trying to measure these types of benefits. Hence, many researchers exclude them from their analyses. See, for example, Haveman and Weisbrod (1975).

Application

7.1 *Incremental Benefit Estimates for Revising U.S. Particulate Matter Standards*

In 1983, the EPA prepared a Regulatory Impact Analysis (RIA) for a proposal to tighten the air quality standard for particulate matter. (The RIA was required under President Reagan's Executive Order 12291.) Particulate matter (PM) refers to a broad class of contaminants that are emitted into the air as small particles. As part of the requirements of the RIA, the EPA had to estimate the incremental benefits of this proposal.

Relying on the findings of scientific studies, the EPA determined that exposure to PM is linked to such health problems as respiratory and cardiovascular disease. The agency further found that the associated welfare effects include (1) soiling of buildings and materials, (2) increased acidic deposition through releases of sulfate particles, and (3) visibility impairment. Using this qualitative assessment as a basis, the EPA had to estimate how much these damages would be reduced by the proposed change in the PM standard and then monetize its findings.

To illustrate the inherent variability in quantifying environmental benefits, the accompanying table presents the series of benefit estimates actually used in the EPA's evaluation. For each benefit class, the letters A through F indicate the various procedures used to derive the estimates.

Procedure A is the most conservative, since its total of $1.24 billion includes only the benefits of reduced mortality and chronic illness. At the other end of the spectrum, Procedure F is the most comprehensive. It includes *all* expected health and welfare benefits, that is, the reductions in mortality and morbidity plus the reduced soiling and material damages. According to this latter approach, the incremental benefits of tightening the PM standard are valued at $52.36 billion.

Beyond the differences in aggregation across the six procedures, there is also variability in how the individual benefit categories are valued. For example, according to Procedures A through C, mortality benefits are monetized at $1.12 billion, but Procedures D and E place a value on this same category of $12.72 billion, and Procedure F an even higher value of $13.84 billion.

An important inference to be made from these data is that benefit assessment is not an exact science. Rather, it relies on approximations based on a consensus of experts and supported by available scientific evidence. Often, as is the case here, a *range* of benefit estimates is considered, with each estimate based on certain underlying assumptions.

Sources: Mathtech Inc. (March 1983), pp. 1–52; U.S. EPA, Office of Policy Analysis, Office of Policy, Planning, and Evaluation (August 1987).

Incremental Benefit Estimates, by Aggregation Procedure (billions of 1980 dollars)

Benefit Class	A	B	C	D	E	F
Mortality	1.12	1.12	1.12	12.72	12.72	13.84
Acute morbidity	0.00	1.32	10.65	10.65	10.65	11.97
Chronic morbidity	0.12	0.12	0.12	0.12	11.40	11.40
Soiling and materials (household sector)	0.00	0.00	0.73	0.73	3.14	13.85
Soiling and materials (manufacturing sector)	0.00	0.00	0.00	0.00	1.30	1.30
Total incremental benefits	1.24	2.56	12.63	24.24	39.22	52.36

Note: Some totals may not agree due to independent rounding.

Figure 7.1 **Marginal Social Benefit (MSB) and Total Social Benefits (TSB) of Air Quality (% SO_2 Abatement)**

The market demand curve for SO_2 abatement represents the marginal social benefit (*MSB*) of air quality. At the hypothetical level of abatement A_1, the corresponding *MSB* is the vertical distance from the horizontal axis at that point up to the curve. The total social benefits (*TSB*) associated with A_1 are shown as the shaded area under the *MSB* up to that point.

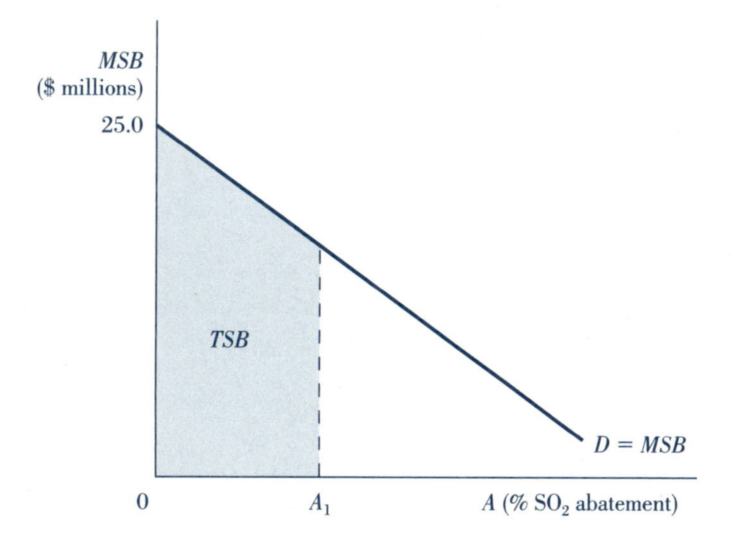

environmental quality, we could then measure the incremental benefits associated with any environmental policy.

To illustrate this assertion, we return to our model of air quality based on SO_2 pollution abatement, which we introduced in Chapter 3. Recall that since demand for this public good represents society's decisions, it is both the marginal private benefit (*MPB*) and the marginal social benefit (*MSB*) of air quality. Thus, we can refer to the demand for SO_2 abatement as $MSB=25-0.3A$, where *MSB* is measured in millions of dollars and *A* is the percentage of SO_2 abated.

The graph of this relationship is shown in Figure 7.1. At each level of abatement, *MSB* is measured as the vertical distance from the horizontal axis up to the demand curve. The total social benefits (*TSB*) for any abatement level are measured as the aggregation of these vertical distances, or the area under the *MSB* curve up to that point.[2] In Figure 7.1, the *TSB* for some hypothetical abatement level A_1 is shown as the shaded area under the demand curve up to A_1. Using this model, we can measure the incremental benefits from a policy-induced increase in SO_2 abatement in three steps:

1. Find the baseline level of *TSB* *before* the policy is undertaken
2. Find the new level of *TSB* that would arise *after* the policy is implemented
3. Subtract the baseline *TSB* from the post-policy *TSB* to determine incremental benefits.

[2]Recall from Chapter 2 that the area beneath the demand curve and above the market price is consumer surplus, or the *net* benefit enjoyed by consumers. Notice that this net benefit is exactly equal to the total benefits received minus the total dollar value paid for the good. In fact, if a commodity has a zero market price, the total benefits would be exactly equal to consumer surplus.

Figure 7.2 **Modeling Incremental Social Benefits for Air Quality (% SO_2 Abatement) Using the MSB Function**

The *MSB* at the baseline abatement level of 20 percent is $19 million, and the *TSB* are shown as the area under the *MSB* curve up to that point, or $440 million. If a policy were proposed to increase SO_2 abatement to 25 percent, *MSB* would be $17.5 million and *TSB* would rise to $531.25 million. Thus, the incremental benefits are measured as the difference between the two *TSB* values, or $91.25 million. In this model, incremental benefits are shown as the shaded area under the *MSB* curve between the two abatement levels.

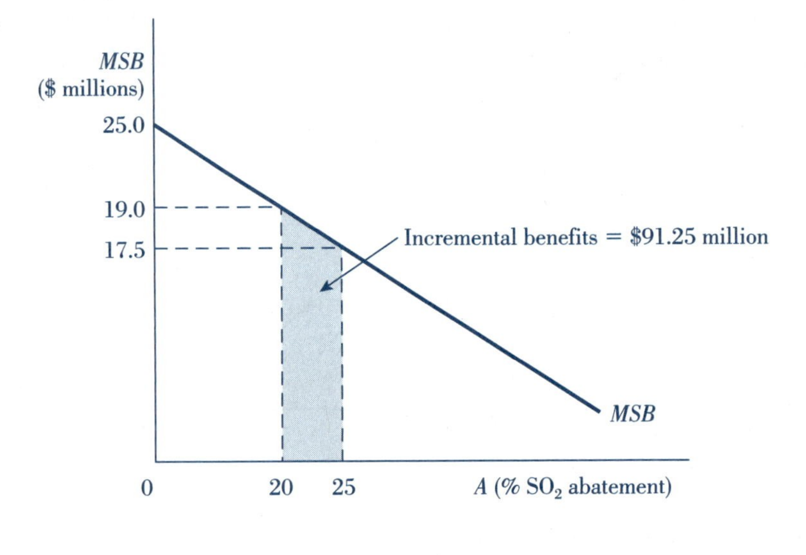

Suppose that the current level of SO_2 abatement is 20 percent and the objective is to find the incremental benefits of a policy that increases abatement to 25 percent. First, find the baseline *TSB*. Referring to Figure 7.2, notice that *MSB* at the 20 percent abatement level is $19 million. *TSB* at this level are shown as the area under the *MSB* curve up to that point, or $440 million. This dollar value represents society's **willingness to pay (WTP)** for the benefits achieved when 20 percent of SO_2 emissions are abated. Next, do the analogous calculations for the proposed abatement increase to 25 percent. At this post-policy abatement level, *MSB* is $17.5 million and *TSB* equal $531.25 million. Finally, the incremental benefits are found as the difference between the two *TSB* values, or $91.25 million. This is shown as the shaded area in Figure 7.2.

An alternative model of the same result is shown in Figure 7.3, where the *TSB* are graphed directly with SO_2 abatement. In this model, the *TSB* associated with a given abatement level are measured simply as the vertical distance up to the curve and not as an area. Identify on the graph the pre- and post-policy levels of *TSB* corresponding to the 20 percent and 25 percent abatement levels. Notice that the incremental benefits of $91.25 million are measured as the vertical distance between these two *TSB* levels.

Both models assume that society's valuation of environmental quality, or, equivalently, its demand for pollution abatement, can be identified. However, because there is no explicit market for this commodity, the valuation cannot be obtained from observing market-determined prices. Instead, inferences must be made about *how* society derives value or utility from various levels of environmental quality.

Figure 7.3 ***Modeling Incremental Social Benefits for Air Quality (% SO_2 Abatement) Using the TSB Function***

An alternative way to model the incremental benefits of improving air quality is to graph the relationship between the *TSB* and SO_2 abatement. At each abatement level, the *TSB* are shown as the vertical distance from the horizontal axis up to the curve. The model shows the *TSB* for the 20 percent baseline abatement level and for the 25 percent post-policy abatement level as $440 million and $531.25 million, respectively. The incremental benefits are shown as the vertical distance between the two points on the *TSB* curve, or $91.25 million.

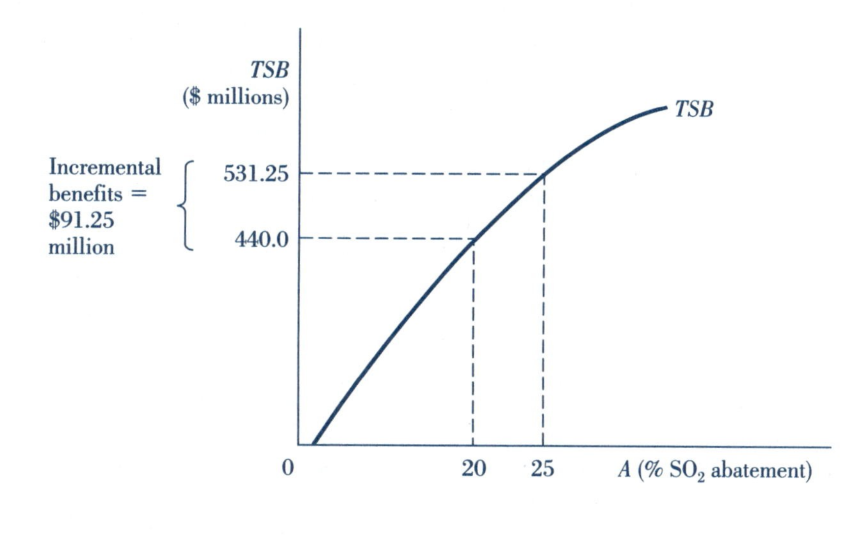

User Versus Existence Value[3]

Discovering how society values a good is difficult in the absence of market prices. Even if the dilemma of nonrevelation of preferences could be overcome, the value of a good like clean air or water is difficult to quantify because of the many intangibles involved. Although economists recognize that some of these intangibles are immeasurable, they still need some sense of how the benefits of environmental quality are *perceived* by society. Fortunately, theories have been advanced that can help.

From a purely conceptual vantage point, it is generally recognized that society derives utility from environmental quality through two sources of value: **user value** and **existence value. User value** refers to the utility or benefit received from usage of or access to an environmental good. In contrast, **existence value** is the utility or benefit received from an environmental good simply through its continuance as a good or service. Collectively, these components measure society's total valuation of an environmental good, expressed as follows:

user value
Benefit derived from physical use or access to an environmental good.

existence value
Benefit received from the continuance of an environmental good.

Total value of environmental quality = User value + Existence value.

User Value

To better understand user value, consider the benefits of using a lake. If an individual swims in a lake, that person is deriving utility from physically using the natural resource. Likewise, a commercial fishing fleet derives user benefit from catching fish from the lake. In both cases,

[3]Much of the following discussion is drawn from Mitchell and Carson (1989), chap. 3.

direct user value Benefit derived from directly consuming services provided by an environmental good.

indirect user value Benefit derived from indirect consumption of an environmental good.

benefits are derived from directly consuming services provided by the resource. These activities, both recreational and commercial, generate benefits that yield **direct user value.** This valuation helps to determine what the individual or the fishing fleet would be willing to pay to maintain or improve the lake's quality.

Continuing with the same example, another individual might receive utility from simply looking at a view of the lake. Here, the utility is derived from the lake's aesthetic qualities. This type of activity involves using the lake in a less immediate way than swimming in it or fishing from it. Hence, the lake is said to yield **indirect user value.** Both direct and indirect user value are elements of society's total valuation of environmental quality.[4]

Existence Value

Society also receives benefits from environmental goods beyond the utility associated with direct or indirect use. Think about how people value such natural resources as the great rain forests, the Grand Canyon, or the bald eagle. Consumption does not explain how or why society values these resources. Yet we know that society as a whole is willing to pay to preserve them. In such circumstances, benefits accrue to society from simply knowing that these resources exist and are being preserved. This component of total valuation is referred to as **existence value.**

While seemingly abstract, existence value is an important motivation for privately funded conservation efforts and for many environmental policy initiatives. A case in point is the Endangered Species Act, which provides for the protection and preservation of certain animals, birds, fish, and plants threatened with extinction.[5] As discussed in Application 7.2, this act is an example of the U.S. government's recognition of existence value. Other tangible evidence is the willingness of society to support the work of environmental groups, such as the National Wildlife Federation, the Sierra Club, and the National Audubon Society—groups whose agendas focus on preserving resources that many of their benefactors never expect to use or even see firsthand.

One of the earlier discussions of existence value is presented by Krutilla (1967), who asserts: "When the existence of a grand scenic wonder or a unique and fragile ecosystem is involved, its preservation and continued availability are a significant part of the real income of many individuals."[6] Since this early work, economists have been studying various theories about the motivations for existence value. In a text about valuing public goods, Mitchell and Carson (1989) classify the motives for existence value as **vicarious consumption** and **stewardship,** among others.

vicarious consumption The utility associated with knowing that others derive benefits from an environmental good.

stewardship The sense of obligation to preserve the environment for future generations.

Vicarious consumption refers to the notion that individuals value a public good for the benefit it provides to others whether or not these others are known personally. This suggests that the utility derived is *interdependent,* that is, that an individual can and does receive benefit from the knowledge that others are enjoying the public good. **Stewardship** arises both from a sense of obligation to preserve the environment for future generations *and* from the recognition of the intrinsic value of natural resources. In sum, we can express the total valuation of environmental quality as:

Total value = **User value** + **Existence value.**
(direct and indirect user value) (vicarious consumption and stewardship value)

[4]Although these examples involve present period consumption, economists also have begun to examine how society expects to benefit from consumption in some future period. This valuation concept, which adds uncertainty to benefit assessment, is referred to as option value. For detail on this concept, see Mitchell and Carson (1989), chap. 3, and Johansson (1991).

[5]To view the text of the Endangered Species Act and to research information on endangered species, visit **http://www.nmfs.noaa.gov/prot_res/laws/ESA/esatext/esacont.html.**

[6]Krutilla (1967), p. 779. Krutilla attributes the birth of conservation economics to A. C. Pigou, an English economist who wrote: "It is the clear duty of government, which is the trustee for unborn generations as well as for its present citizens, to watch over, and if need be, by legislative enactment, to defend, the exhaustible natural resources of the country from rash and reckless spoliation" (Krutilla 1967, p. 777, citing Pigou 1952).

Application

7.2 The Endangered Species Act

In 1973, Congress passed the Endangered Species Act to "provide a means whereby the ecosystems upon which endangered species and threatened species depend may be conserved" (sec. 2b). Officially, the act protects the biodiversity of the earth: the diversity of genes, species, ecosystems, and the interaction among them. An important outcome of this act was the creation of a formal list of biological organisms in danger of extinction, *regardless of their direct or indirect use to humans.* Originally numbering 109 in 1973, the list of endangered and threatened species in the United States has grown to include over 1,250 species as of 2002.

When the Endangered Species Act was originally proposed and when it went through its subsequent reauthorizations, arguments were made to justify its passage on economic grounds. One argument was that species were to be protected because they may serve a direct benefit to mankind that scientific study has not yet discovered. Some justification for this argument can be found in the recent discovery of taxol. This compound, found in the bark of Pacific yew trees, holds promise for the treatment of certain cancers. Approximately 60 pounds of yew bark are needed to produce enough taxol to treat one cancer patient, and the Pacific yew is a slow-growing species being harvested by timber companies at a rapid rate. Notice that this discourse focuses on the **user value** of environmental resources.

Just as strong in their position were those who asserted that environmental resources offer benefits that span a much broader range than their direct or indirect value in human consumption. Here, the premise is that species should be protected based on a presumed right of survival—evidence of the role of **existence value** in benefit assessment.

There is not now, nor is there likely to be, a consensus about whether user value is relevant to the benefit assessment of biodiversity. What *is* clear is that society recognizes and appreciates existence value—an issue that continues to emerge in the ongoing deliberation about biodiversity and economic development.

Sources: *Endangered Species Act of 1973,* 16 U.S.C. § 153–44; Adler and Hager (April/May 1992); Council on Environmental Quality (January 1993), pp. 17–28.

Recognizing how society values an environmental resource is important for identifying the social benefits of a policy proposal. It also helps economists decide which estimation method might be most effective in quantifying those benefits. Yet the question remains: How do economists assign dollar values to nonmarketed environmental goods such as clean water, human health, and the spotted owl? Remember that the objective is to *monetize* or find the WTP for changes in user and existence value arising from a policy-driven increase in environmental quality.

Approaches to Measuring Environmental Benefits: Overview[7]

Economists have made great strides in developing methods to estimate the benefit of environmental quality improvements. For the most part, these methods are aimed at estimating primary benefits, implicitly assuming that secondary benefits are insignificant and likely offset by secondary costs. A review of the extensive literature on the subject shows that several methods are used in practice. Some are better than others at quantifying the more intangible benefits of improved environmental quality, including the somewhat elusive concept of existence value.

[7]This section is drawn mainly from Mitchell and Carson (1989), particularly pp. 74–78, and from Cropper and Oates (June 1992).

Table 7.1 *Synopsis of Benefit Estimation Methods*

Approach and Method	Description
Physical linkage approach	
Damage function method	Uses a model of the relationship between levels of a contaminant and observed (or statistically inferred) environmental damage to estimate the damage reduction arising from a policy-induced decline in the contaminant.
Behavioral linkage approach	
Direct methods	
Political referendum method	Uses the *actual* market of a public good by monitoring voting results from political referenda on proposed changes in environmental quality.
Contingent valuation method (CVM)	Uses surveys to inquire about individuals' willingness to pay (WTP) for environmental improvements based on *hypothetical* market conditions.
Indirect methods	
Averting expenditure method (AEM)	Assesses changes in an individual's spending on goods and services that are *substitutes* for personal environmental quality to assign value to changes in the overall environment.
Travel cost method (TCM)	Values a change in the quality of an environmental resource by assessing the effect of that change on the demand for a *complementary* good.
Hedonic price method (HPM)	Uses the theory that a good is valued for the attributes it possesses to estimate the implicit or hedonic price of an environmental attribute and identify its demand as a means to assign value to policy-driven improvements in quality.

To organize our discussion of benefit measurement methods, we rely upon a general classification introduced by Smith and Krutilla (1982), which places the various measurement techniques into two broad categories: the **physical linkage approach** and the **behavioral linkage approach.** A summary of selected benefit valuation methods within each approach is presented in Table 7.1.

Physical Linkage Approach to Environmental Benefit Valuation

physical linkage approach
Estimates benefits based on a technical relationship between an environmental resource and the user of that resource.

Methods in the **physical linkage approach** use some tangible attribute of the environment to make a connection to an individual through which benefits can be observed or inferred and subsequently valued. More formally, the **physical linkage approach** measures benefits based on a technical relationship between an environmental resource and the user of that resource. A common estimation procedure that uses this approach is the **damage function method.** This method uses a functional relationship to capture the link between a contaminant and any associated damages. Based on this function, incremental benefits are measured as the reduction in damages arising from a policy-induced decrease in the contaminant. This damage reduction is then monetized to obtain a dollar value of the benefits brought about by the policy.

Behavioral Linkage Approach to Environmental Benefit Valuation

behavioral linkage approach
Estimates benefits using observations of behavior in actual markets or survey responses about hypothetical markets.

In general, the **behavioral linkage approach** to quantifying benefits is based upon observations of behavior in *actual* markets or survey responses about *hypothetical* markets for environmental goods. Another element of behavioral linkage methods deals with how closely the behavior or responses are linked to the environmental good. Techniques that assess responses immediately related to environmental changes are broadly termed **direct methods.** As shown in Table 7.1, two types of direct methods are the **political referendum method,** which relies on *actual* market information, and the **contingent valuation method (CVM),** which uses *hypothetical* market data. **Indirect methods** are those that examine responses not about the environmental good itself but about some set of market conditions related to it. Three examples of indirect benefit estimation methods are the **averting expenditure method (AEM),** the **travel cost method (TCM),** and the **hedonic price method (HPM).**

The published literature on benefit valuation methods is extensive. Many research papers present specific empirical findings, and numerous texts focus on the methodology itself. We offer only an overview of the operational issues associated with the more common methods used by researchers.[8]

Estimation Under the Physical Linkage Approach

Damage Function Method

When using the **damage function method,** the researcher specifies a model of the relationship between an environmental contaminant and some type of observed damage.[9] A generalized damage function is shown in Figure 7.4. The levels of some environmental contaminant (C) are measured horizontally, and the total damages (TD) resulting from exposure to that contaminant are measured vertically.[10] Once the function is specified, the analyst uses the model to estimate the damage reduction from any policy-induced decline in the contaminant. At this point, the reduction is measured in nonmonetary units, for example, the number of acidified lakes, acres of damaged forests, or number of premature deaths. Ultimately, this reduction must be assigned a monetary value, either by using market prices (if available) or by employing some estimation technique.

damage function method
Models the relationship between a contaminant and its observed effects as a way to estimate damage reductions arising from policy.

To illustrate the procedure, look at Figure 7.4, and assume that a policy initiative is expected to reduce the contaminant from C_0 to C_1. Based on the damage function model, this proposal would reduce damage (or, equivalently, increase benefits) by the vertical distance between TD_0 and TD_1. If, for example, the damage reduction was diminished injury to wheat crops, this vertical distance might be measured as thousands of bushels of wheat. Having quantified the incremental benefit, a simple way to monetize it would be to multiply the number of bushels by the market-determined price.

Assessing the Damage Function Method

Although the damage function method is useful, it has limitations. First, by construction, it estimates only one aspect of incremental benefits. In our example, the measured benefits are increased wheat crops. In most cases, a contaminant reduction would give rise to other types of gains, perhaps increases in other crops or improvements in human health. Hence, a full assessment of benefits using the damage function approach would require that the same estimating procedure be performed for every type of damage reduction. Second, the procedure is only a first-step approach in that it is not capable of simultaneously monetizing the benefits it quantifies.

Applications of the Damage Function Method

Recognizing the limitations of the damage function method, analysts typically use it for measuring a specific type of incremental benefit as opposed to performing a comprehensive benefit assessment. Furthermore, the context is often one in which market-determined prices can be used to monetize the gain.

---

[8]To study any of these methods in detail, consult the references cited along the way and the list of additional readings given at the end of the chapter.

[9]If the benefit assessment is the reduction in adverse effects to a *biological organism,* the analyst would use a **dose-response function,** which is a particular type of damage function. Recall from Chapter 6 that scientists use dose-response functions in the risk assessment process.

[10]The model in Figure 7.4 is a cubic relationship, but this is just one possible functional form for a damage function. For more on this issue, see, for example, Halvorsen and Ruby (1981), p. 106.

Figure 7.4 Measuring Incremental Benefits Using the Damage Function Method: A Physical Linkage Approach

A damage function shows the relationship between an environmental contaminant (*C*) and the total damages (*TD*) due to exposure to that contaminant. Once the function is specified, the analyst can use it to estimate the damage reduction from any policy-induced decline in the contaminant. This damage reduction represents incremental benefits measured in non-monetary units.

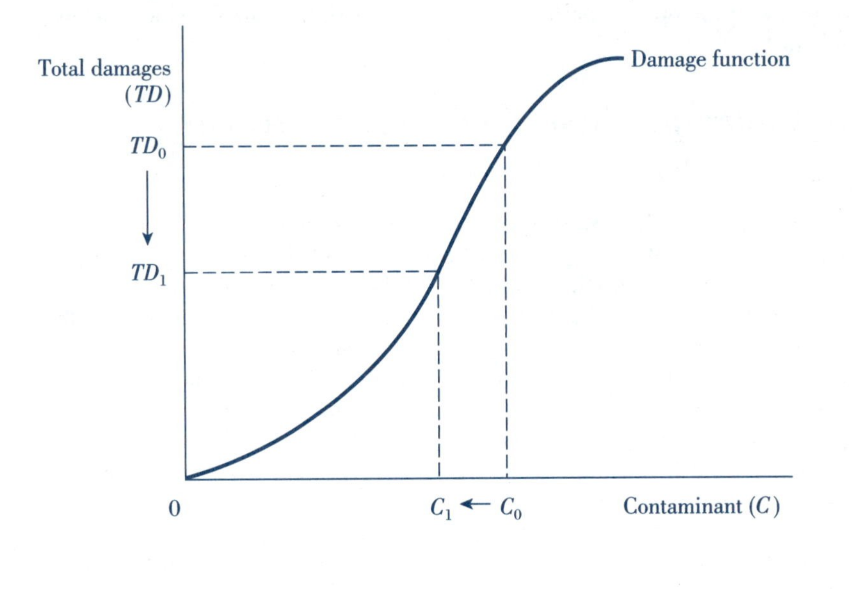

To illustrate, consider a benefit assessment of the Clean Air Act provisions aimed at reducing ozone in the lower atmosphere or troposphere. According to scientific evidence, one type of benefit associated with reducing tropospheric ozone is an increase in crop yields.[11] Conceptually, these agricultural benefits could be modeled by measuring the change in consumer and producer surpluses associated with an increase in crop yields. Figure 7.5 models this policy-induced effect as an increase in crop supply from S_0 to S_1, which in turn causes a price decline from P_0 to P_1.

Consider first the size of the surpluses before any policy change, using supply curve S_0. Consumer surplus is the area below the demand curve and above the market price, or area P_0ab. Producer surplus is the area above the supply curve and below the market price, or area P_0be. Thus, the total surplus *before* a policy change is area *eab*. After the ozone-reducing policy is implemented and supply shifts to S_1, consumer surplus becomes area P_1ac, and producer surplus becomes area P_1ce. Thus, the total surplus *after* the policy change is area *eac*. The operative question is whether society is better off as a result. Since the total surplus increased from *eab* to *eac*, the answer is yes, and the incremental benefit can be quantified as area *ebc* (i.e., $eac - eab$).

Notice, however, that the *distribution* of benefits is not as easily determined. Consumers definitely gain, because consumer surplus rises from P_0ab to P_1ac, but the same assertion cannot be made about producers. Some of the original producer surplus, area P_0bfP_1, has

[11] The interested reader may wish to consult Kopp and Krupnick (December 1987) for results of an empirical investigation that estimates the agricultural benefits of ozone reduction.

Figure 7.5 Modeling Incremental Benefits of an Ozone-Reducing Policy

The increase in crop yields associated with a hypothetical ozone-reducing policy can be modeled as a shift in crop supply from S_0 to S_1. Before the policy is implemented, consumer surplus is area P_0ab and producer surplus is area P_0be, for a total of area *eab*. After the policy is implemented and supply shifts to S_1, consumer surplus becomes area P_1ac and producer surplus becomes area P_1ce, for a total of area *eac*. Thus, the incremental benefit is area *ebc* (i.e., *eac*−*eab*).

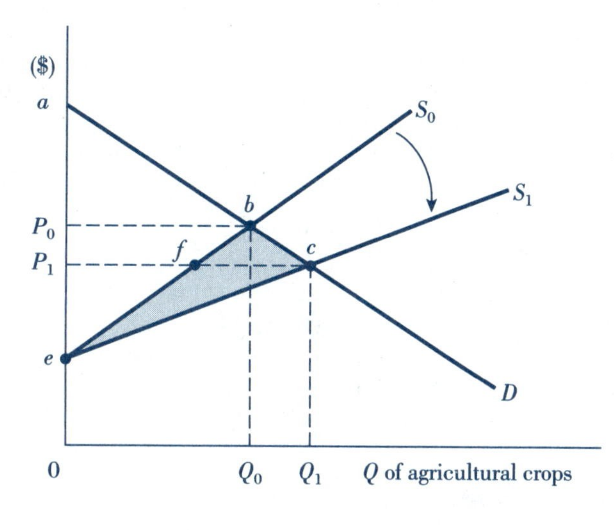

been transferred to consumers, but there is also a gain to producers of area *efc*. Whether or not the gain of *efc* exceeds the transfer of P_0bfP_1 depends on the shapes of the supply and demand curves and the magnitude of the shift in supply. Consequently, to measure the incremental benefits of policy *and* to determine the distribution of these benefits, sophisticated models must be used. Application 7.3 presents the actual results of such a benefit estimation conducted as part of a U.S. air quality assessment.

Direct Estimation Methods Under the Behavioral Linkage Approach

The so-called **direct methods** under the behavioral linkage approach estimate environmental benefits according to responses or observed behaviors directly tied to environmental quality. Although a number of direct methods are available to researchers, we consider only one representative example: the **contingent valuation method (CVM).**

Contingent Valuation Method (CVM)

contingent valuation method (CVM) Uses surveys to elicit responses about WTP for environmental quality based on hypothetical market conditions.

When market data are unavailable or unreliable, economists can use alternative estimation methods that rely on *hypothetical* market conditions. Such methods typically use surveys to inquire about individuals' willingness to pay (WTP) for some environmental initiative. This survey approach to benefit estimation is known as the **contingent valuation method**

Application

7.3 Valuing Agricultural Benefits: The Case of Tropospheric Ozone Reductions

The National Acid Precipitation Assessment Program (NAPAP) was launched as an investigative plan to gather information for formulating policy on acidic deposition. According to NAPAP's 1990 report, scientific research has been unable to find a consistent adverse effect on crop yields caused by acidic deposition. However, there *is* well-documented evidence that elevated levels of tropospheric ozone can diminish crop yields. In fact, estimates of this damage range from 2 to 56 percent, depending on such factors as crop species, location, and exposure levels. Thus, a key part of NAPAP's research dealt with quantifying the incremental benefits that would result from an ozone-reducing policy initiative.

Modeling techniques to measure environmental benefits were used in the NAPAP study of ozone's effects on agricultural yields. The final report presents estimates of changes in consumer and producer surplus for alternative ozone policies (see accompanying table). Under the first scenario, a 10 percent reduction in ozone concentrations is estimated to increase total surplus by approximately $739 million. Of this amount, consumer surplus would increase by $785 million, and producer surplus would decline by $46 million. A more restrictive ozone policy, such as a 25 percent reduction, would increase the benefits to both producers and consumers and elevate the change in total surplus to $1,732 million.

These findings have important implications about the distribution of these incremental benefits across buyers and sellers. Notice that consumers and producers do not respond symmetrically to increases in ozone levels. In fact, if ozone levels were to *increase* by either 10 or 25 percent, society as a whole would lose but producers would actually benefit. According to the NAPAP report, this gain to producers arises because of farmers' ability to pass on the higher costs of depleted supplies to consumers in the form of higher prices.

Source: NAPAP (November 1991), pp. 55, 154–56, 398–401.

	Change in Surplus (billions of 1989 dollars)		
Changes in Ozone (%)	**Consumer Surplus**	**Producer Surplus**	**Total Surplus**
−10	0.785	−0.046	0.739
−25	1.637	0.095	1.732
+10	−1.044	0.215	−0.829
+25	−2.659	0.453	−2.206

Source: NAPAP (November 1991), p. 401, table 4.9–6.

(CVM) because the results are dependent or *contingent* on the devised hypothetical market. This market serves as the context for a series of survey questions. The critical assumption is that properly designed surveys can elicit responses comparable to those arising in actual situations. In some sense, the survey instrument helps to finesse the problem of nonrevelation of preferences that characterizes public goods.

Implementing this survey approach involves the following three tasks:[12]

- Constructing a detailed model of the hypothetical market, including the characteristics of the good and any conditions that affect the market
- Designing a survey instrument to obtain an unbiased estimate of individuals' WTP.
- Evaluating the truthfulness of survey respondents' answers

[12] For more detail on the specific elements of a CVM analysis, see Mitchell and Carson (1989), Cummings, Brookshire, and Schulze (1986), and Smith and Desvousges (1986).

Assessing the CVM

The CVM is favored by researchers because it can be applied to a variety of environmental goods and because it can assess existence value as well as user value. However, because this approach makes inferences about actual markets from a hypothetical model, it is subject to the biases that typically plague a survey-dependent study, such as an individual's unwillingness to reveal a WTP because of the free-ridership problem. See Exhibit A.1 in the Appendix for common sources of bias associated with the CVM.

Responding to the potential biases, economists continue to make improvements to the CVM. For instance, some researchers add more detail to their hypothetical models. Others improve the design of the survey instrument. Some surveys include maps to illustrate the location of the good or photographs of the commodity and the area affected by its provision.[13] Whatever the form, the objective is the same: to make the hypothetical market situation as factual and as close to actual conditions as possible.

Applications of the CVM

Researchers have used the CVM in a variety of contexts to estimate environmental benefits. An important application is estimating the value of a statistical human life.[14] In a recent review of these findings, estimates were reported as falling within a range of $1.6 million to $4.0 million ($1986).[15]

Another common focus of CVM studies is to measure society's WTP for water quality improvements. Two examples are a study by Smith and Desvousges (1986), which examines a specific water body (the Monongahela River in Pennsylvania), and an analysis by Carson and Mitchell (1988), which estimates a generalized measure across all U.S. water sites. The study by Smith and Desvousges (1986) finds that the average household in five western Pennsylvania counties is willing to pay $25 ($1981) per year to improve the Monongahela River from boatable to fishable quality. Carson and Mitchell's (1988) nationwide survey shows that the average respondent is willing to pay $80 ($1983) per year for water quality improvements. How can the difference be explained?

Since the valuation in the national survey is higher than the more localized finding of Smith and Desvousges (1986), the difference may be attributable to existence value. Why? Because the respondents in the more general survey are willing to pay for water quality improvements throughout the United States, even though they do not expect to use these water bodies themselves.[16]

Incremental benefits from air quality improvements also have been estimated using the CVM. In fact, some argue that the CVM is particularly useful for valuing visibility improvements at national parks, where existence value is likely to be significant. One study by Schulze and Brookshire (1983) seems to support this hypothesis. These researchers find that the *user value* of improving visibility at the Grand Canyon from 70 to 100 miles is under $2.00 ($1988) per visitor per day. In contrast, they estimate the comparable *existence value* at $95 ($1988) per household per year to prevent diminished visibility at the Grand Canyon.[17]

Because the CVM is capable of capturing existence value, it has been used to value ecological benefits, such as preserving an endangered species. For example, one study estimates that individuals would be willing to pay $22 ($1983) per year to save the whooping crane.[18]

---

[13] Brookshire and Crocker (1981).

[14] A statistical life saved is related to the concept of environmental risk introduced in Chapter 6. For example, if an environmental policy lowers the risk of death from 2 in 100,000 persons exposed to 1 in 100,000 exposed, the incremental benefit of that policy is one human life saved. To give this context, in a recent EPA study, a statistical life saved is valued at $4.8 million in 1990 dollars (U.S. EPA, Office of Air and Radiation, November 1999).

[15] Cropper and Oates (June 1992), p. 713. Note that the notation, ($1986), means that the values are expressed in 1986 dollars.

[16] Cropper and Oates (June 1992), pp. 716–17.

[17] Notice that for user value the WTP is appropriately given on a *per visitor* basis, whereas the existence value is quoted for *households* who may never visit the site.

[18] Bowker and Stoll (May 1988).

Another finds that people would pay $11 per year to preserve the bald eagle.[19] There is, however, an important caveat. Although the value of preserving an entire ecosystem is typically what is of interest in policy analysis, many CVM studies focus on a single species. Yet the sum of species-specific valuations is generally higher than the valuation of an entire area.[20]

Indirect Estimation Methods Under the Behavioral Linkage Approach

For some environmental proposals, direct estimation procedures such as the CVM might not be viable. In these cases, economists use **indirect methods,** which make inferences about markets or conditions that are linked to the environmental good under investigation. Three such methods dominate the literature: the **averting expenditure method (AEM),** the **travel cost method (TCM),** and the **hedonic price method (HPM).**

Averting Expenditure Method (AEM): An Indirect Approach Using Substitutes

To indirectly estimate the WTP for such nonmarketed commodities as cleaner air or water, the **averting expenditure method (AEM)** uses changes in spending on goods that are *substitutes* for environmental quality. The motivation for this approach is quite intuitive. Exposure to pollution causes damages that negatively affect an individual's utility. Consequently, people undertake averting action by purchasing goods and services that improve their *personal* environmental quality, such as the indoor air or a private drinking water supply.[21]

averting expenditure method (AEM)
Estimates benefits as the change in spending on goods that are *substitutes* for a cleaner environment.

Table 7.2 gives some common examples of what individuals do to reduce the effects of pollution on their personal environment. Notice that in each case the averting action involves an expenditure on a substitute good or service. Therefore, if the general environment is improved by some policy initiative, the individual can spend *less* on these substitute goods. It is pre-

Table 7.2 *Averting Actions to Reduce Risks of Exposure to Pollution*

Pollution	Effect	Averting Action
Air pollution	Material soiling	Clean or repaint material surfaces; use protective covers; move to new location.
	Health problems	Install air purifiers or air conditioners; schedule more frequent visits for medical examinations; purchase medications to alleviate respiratory symptoms; move to new location.
Water pollution	Material soiling	Install water filtration system; purchase cleaning products and rust removers; move to new location.
	Health problems	Install water filtration system; purchase bottled water; move to new location.
Hazardous waste site	Aesthetic degradation	Install fencing or shrubbery; move to new location.
	Health problems	Test water supply for contamination; install air filtration or air conditioner; move to new location.
Noise pollution	Health problems	Install sound-deadening insulation; purchase medication to aid sleeping; move to new location.

Source: Adapted from Bartik (1988), pp. 111–126, table 1.

[19] Boyle and Bishop (1987).
[20] Cropper and Oates (June 1992), pp. 719–20.
[21] The reference to an individual's "personal" environment was first used by Bartik (1988) to describe the use of the AEM to value nonmarginal improvements in the environment.

Figure 7.6 ***Measuring Incremental Benefits Using the Averting Expenditure Method (AEM): A Behavioral Linkage Approach***

When overall environmental quality is E_0, personal environmental quality is X_0, and averting expenditures are area $0abX_0$. After a policy increases overall environmental quality to E_1, personal environmental quality increases to X_1, and averting expenditures change to area $0acX_1$. To achieve X_1 in the absence of the policy change, the individual would be willing to spend an amount equal to area $0abcX_1$. Thus, the individual's WTP for the incremental benefits is the difference between $0abcX_1$ and $0acX_1$, or triangular area *abc*.

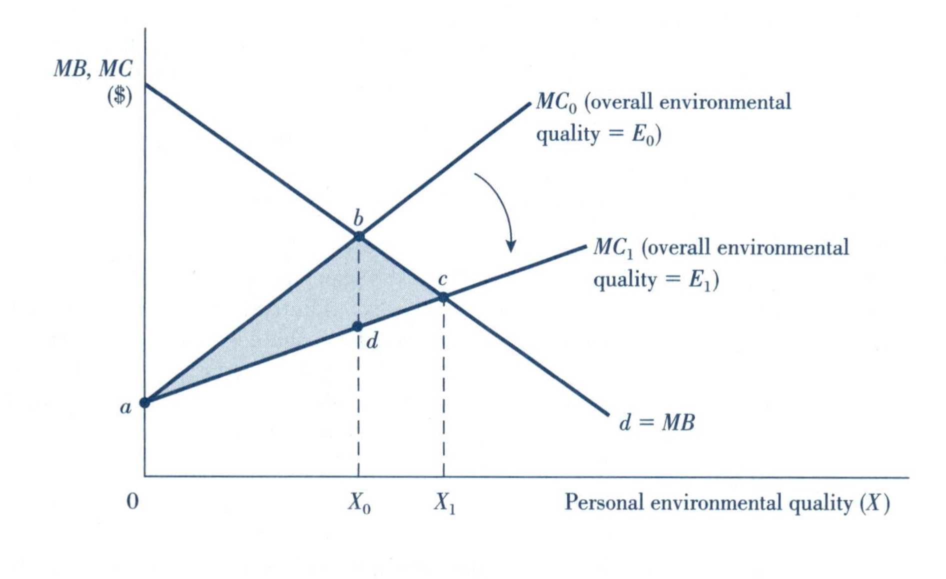

cisely this decline in averting expenditures that gives an *indirect* estimate of the individual's WTP for the associated incremental benefits. For instance, faced with a contaminated drinking water supply, an individual might purchase bottled water or install a water-filtering system. If a government policy improves the public drinking water supply, the individual can spend less on these substitute commodities. This reduction in spending identifies the incremental benefits provided by the drinking water policy.

We model the AEM in Figure 7.6, where the relevant market is defined as *personal* environmental quality (*X*). The demand curve (*d*) is also the marginal benefit (*MB*) function, and each supply (*s*) relationship is modeled as a marginal cost (*MC*) curve.[22] The critical assumption is that each *MC* curve represents the averting expenditures on environmental quality substitutes to achieve various levels of *personal* environmental quality (*X*), given some level of *overall* environmental quality (*E*). In our diagram, MC_0 represents the marginal cost of averting expenditures at the existing level of overall environmental quality, E_0. As overall environmental quality improves to E_1, the individual spends less (or incurs lower costs) to achieve each level of personal environmental quality, and the *MC* curve moves downward to the right, becoming MC_1.

At the initial equilibrium when overall environmental quality is E_0, the individual's personal environmental quality is X_0, where *MB* and MC_0 intersect. At this point, total averting expenditures are the area under the MC_0 curve up to X_0, or area $0abX_0$. After the policy improves overall environmental quality to E_1, the individual's marginal cost curve shifts to

[22] Because we are modeling *personal* environmental quality, the marginal benefits and costs accrue to a single individual. To avoid confusion, we use the simple labels of *MB* and *MC* so as not to infer a distinction between private and social decisions, which is irrelevant in this unique situation.

MC_1. At the new equilibrium, where MB and MC_1 intersect, personal environmental quality increases to X_1, and averting expenditures change to area $OacX_1$. Now, we can use this information to monetize the incremental benefits of the improvement in overall environmental quality from E_0 to E_1.

The key is to compare averting expenditure levels before and after the policy change *for the same level of personal environmental quality.* As we determined, post-policy averting expenditures for X_1 are represented by area $OacX_1$. Let's compare this to what the individual would be willing to spend to achieve X_1 without the influence of the policy. This is found by calculating the area under the original MC_0 curve up to X_1 and bounded by MB, or area $OabcX_1$. Thus, the individual's WTP for the incremental benefit is the difference between the two areas ($OabcX_1$ and $OacX_1$), or the triangular area *abc.*

An alternative valuation could be based on a constraint that holds personal environmental quality level at its original level (X_0). This approach might be preferred because the calculation is simpler, requiring information on only the two *MC* curves rather than on both the *MC* curves and the *MB* curve. Under this scenario, the incremental benefits would be the difference between areas $OabX_0$ and $OadX_0$, or the area *abd.* This smaller area can be interpreted as a lower bound for the WTP valuation.[23]

Assessing the AEM

A drawback of the AEM arises from the phenomenon known as *jointness of production.* This refers to the fact that some averting expenditures yield benefits beyond those associated with a cleaner environment. Consider, for example, the averting expenditures on an air-conditioning system. While the system does reduce certain health risks of air pollution, it also provides comfort. Hence, the savings in expenditures arising from a clean air policy initiative cannot be attributed solely to the incremental benefits of that policy.

Applications of the AEM

A number of studies have used the AEM to value a statistical life. Blomquist (1979) focuses on the averting activity of wearing seat belts in automobiles to reduce mortality risk. This study estimates the incremental benefit of a life saved between $380,000 and $1.4 million ($1986). Dardis (1980) conducts the same sort of analysis using expenditures on smoke detectors and monetizes the value of a statistical life at $460,000 ($1986).[24] Notice how these estimates are lower than those derived using the CVM. One explanation is that the averting behavior of using seat belts and smoke detectors is not a decision of degree but rather one of use versus nonuse. Consequently, individuals answer affirmatively as long as the marginal benefit exceeds the marginal cost. In such instances, the risk-reduction estimate is based upon the marginal individual who just finds that the averting behavior is worthwhile, causing the resulting estimate to be generally undervalued.[25]

Travel Cost Method (TCM): An Indirect Approach Using Complements

travel cost method (TCM)
Values benefits by using the *complementary* relationship between the quality of a natural resource and its recreational use value.

An alternative approach to valuing environmental benefits is the **travel cost method (TCM),** which uses the *complementary* relationship between the quality of a natural resource and its recreational use value. Simple observation suggests that the demand for the recreational use of an environmental resource, such as a lake or a national forest, increases as its quality improves. Therefore, as this demand function shifts with a change in environmental quality, the resulting change in consumer surplus can be used to assess the associated incremental benefits.

We model the travel cost method in Figure 7.7, assuming that recreational demand has been properly identified.[26] Two demand curves for the recreational use of a lake are shown in

---

[23] Bartik (1988).

[24] These estimates are reported by Fisher, Violette, and Chestnut (1989).

[25] Cropper and Oates (June 1992), pp. 713–14.

[26] One technique used to identify a recreational demand curve is called the Clawson-Knetch method (Clawson and Knetch, 1966). This method uses travel costs to a recreational site, visitation rates, and other socioeconomic data to estimate recreational demand. Smith and Desvousges (1986) provide an excellent review of how this approach has been revised over time.

Figure 7.7 ***Measuring Incremental Benefits Using the Travel Cost Method (TCM): A Behavioral Linkage Approach***

D_0 is the recreational demand for a lake at some preexisting environmental quality level, E_0. D_1 is the new demand curve after a policy improves the lake's quality to E_1. The price line at P_0 is the admission fee. Before the policy is implemented, the number of visits to the site is V_0, and consumer surplus is area abP_0. After the policy is put into effect, the number of visits increases to V_1, and consumer surplus increases to area cdP_0. The *change* in consumer surplus, area *acdb*, represents the incremental benefits of improving the lake's quality from E_0 to E_1.

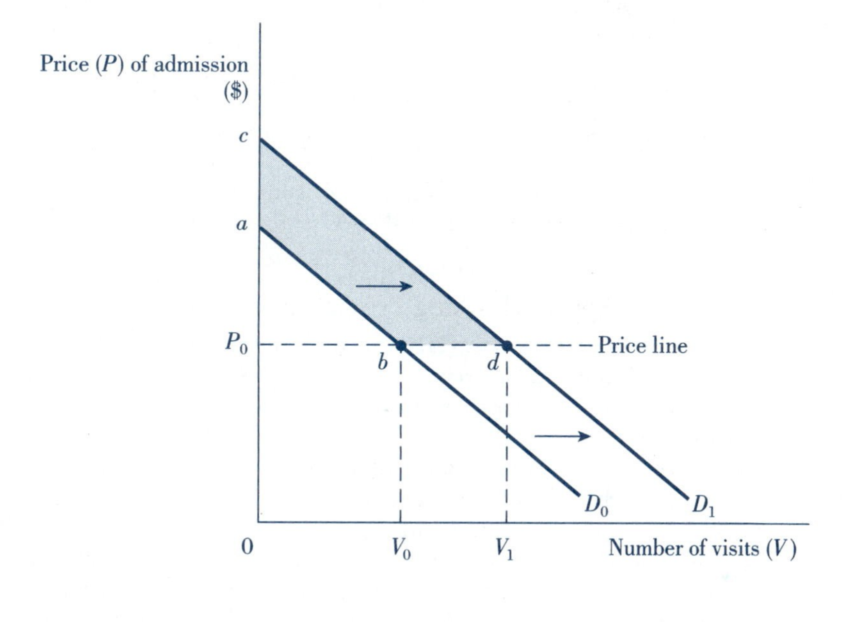

the diagram, D_0 and D_1. D_0 is the relevant demand at some preexisting environmental quality level E_0. D_1 is the new demand curve after a policy has been implemented to improve the lake's quality to E_1. A price line is drawn at P_0 to represent the admission fee to use the lake. Before the policy is implemented, the number of visits to the site is V_0, where visitors enjoy a consumer surplus equal to area abP_0. After the policy is put into effect, the total number of visits increases to V_1, and consumer surplus increases to area cdP_0. The resulting *change* in consumer surplus, shown as area *acdb* (i.e., area cdP_0 minus area abP_0), estimates the incremental benefits to visitors associated with improving the lake's quality.[27]

Assessing the TCM

One disadvantage of the TCM is that it estimates only user value and not existence value—an omission likely to create bias. Another is that it focuses on recreational use, making it ineffective for estimating benefits that might accrue to commercial users of a resource. Finally, the TCM has been found to generate estimates that are biased downward if access to a site is deterred by congestion.[28]

[27] Notice that even if there were no entry fee to use the lake, the model would be the same, except that consumer surplus would be the entire area under each demand curve. In such an instance, the resulting change in consumer surplus would extend to the horizontal axis.
[28] Smith and Desvousges (1986), p. 220.

Applications of the TCM

Because of its limitations, the TCM is commonly used to value improvements to water bodies used mainly for recreation. For example, a study by Mullen and Menz (1985) uses the TCM to value the effect of acid rain damages to the Adirondack, New York, lake region. Other studies have used the procedure to value the benefits of improving water quality from boatable to fishable conditions. These findings tend to vary considerably. Consider the following results from three independent analyses, all measured as WTP per person per day in 1982 dollars:

- Vaughan and Russell (1982): Between $4.68 and $9.37.
- Smith and Desvousges (1985): Between $0.06 and $29.92.
- Smith, Desvousges, and McGivney (1983): Between $1.04 and $2.15.[29]

Part of the reason for the inconsistent valuations is that TCM estimates tend to be sensitive to the site under study. Although demographic variables across regions can be controlled for, other site differences, such as aesthetics, access to major highways, and substitute recreational opportunities, are difficult to quantify and control. Consequently, it is unlikely that the TCM can determine a *generalized* value of improved water quality.[30]

Hedonic Price Method (HPM): An Indirect Approach Using Product Attributes

hedonic price method (HPM)
Uses the estimated hedonic price of an environmental attribute to value a policy-driven improvement.

The **hedonic price method (HPM)** is based on the theory that a good or service is valued for the attributes or characteristics it possesses.[31] This perception of value suggests that *implicit* or *hedonic prices* exist for product attributes and that these prices can be determined from the explicit price of the product. In environmental economics, researchers use this technique to value the environmental attributes of certain commodities.

Housing markets have been a classic context for hedonic environmental pricing studies. Such analyses assume that the market price of a home is determined by the implicit value of its many characteristics, such as location, number of baths, lot size, and the environmental quality of the community. Therefore, changes in any of these characteristics are capitalized into the price of the property. The conventional model specifies the market price of a house, P, as a function of its attributes. A simplified version of such a model is:

$$P = f(X_1, X_2, \ldots, X_n, E),$$

where each X variable represents some housing attribute, such as lot size or number of baths, and E signifies the associated environmental quality. As any one of these characteristics increases in magnitude, the price of the property P increases.[32] It is this *marginal price* that is the implicit value of that attribute. Thus, as environmental quality improves, the resulting increase in property value can be used to estimate the associated incremental benefits.

Another common context for the HPM is wage analysis. Here, the method is used to explain wage differentials that might be associated with occupational risks, including those of environmental origin. Similar to hedonic studies of housing prices, hedonic wage (w) models generally are specified as:

$$w = g(Z_1, Z_2, \ldots, Z_n, E).$$

the Z variables would include such attributes as the age, gender, education, and experience of the worker. The E variable would be some measure of occupation-specific environmental risk, such as the dangers faced by hazardous waste handlers or by engineers in a nuclear

[29] These comparisons are discussed by Smith and Desvousges (October 1985).
[30] For more detail on the TCM, readers should consult Cropper and Oates (June 1992) or Mitchell and Carson (1989), chap. 3.
[31] Lancaster (1966).
[32] This assumes that each attribute is specified as a favorable property characteristic.

power plant. These workers are expected to receive a higher wage than those not so exposed to compensate them for the added risk exposure. This wage differential is used to value the incremental benefits to workers of reducing occupational risk.

The HPM uses a statistical procedure known as regression analysis to determine the implicit price of any environmental variable. In simple terms, the idea is to decompose an *explicit* price, such as the price of a house or the wage rate, into its *implicit* price components, one of which is the price of environmental quality (E). Once the implicit price of E is determined, the demand for environmental quality can be estimated. This in turn can be used to measure changes in consumer surplus arising from policy-driven improvements in environmental quality.

Assessing the HPM

The appeal of the HPM is that it is highly intuitive. It approaches the problem of monetizing incremental benefits in a logical way, directly using market prices. Its major disadvantage is that it requires a fairly complicated empirical model. Furthermore, the method calls for extensive data on product characteristics, which often are unavailable or incomplete. If an important product attribute is missing, it will not be possible to determine the connection between a change in an explicit price and a change in environmental quality.

Applications of the HPM

One application of the HPM is in measuring how the siting of hazardous waste facilities affects prices of nearby properties. An example is a study by Kohlhase (1991), who finds that housing prices in the Houston area are positively affected by distance from a declared Superfund site up to 6.2 miles. According to this research, an additional mile in distance from a site adds $2,364 ($1985) to a property's value. In a similar study of single-family homes in Woburn, Massachusetts, Kiel (1995) estimates the analogous marginal benefit to be $1,377 for the period when the waste facilities were declared Superfund sites.[33] Hedonic wage differential studies have been used to estimate the value of a statistical life saved as a result of reducing occupational risk. For example, Thaler and Rosen (1976) estimate that society's WTP for saving one additional life is between $440,000 and $840,000 ($1986). Another finding is presented by Moore and Viscusi (1988), who estimate the value at approximately $5.4 million.[34] The complex specification and data requirements of the HPM are partly responsible for the varying results across studies.

Conclusions

As a risk management strategy, benefit-cost analysis can be used to set policy objectives and to select the best available control instrument to achieve those objectives. However, its success in guiding these important decisions hinges on the accuracy of benefit and cost measurements. In this chapter, we have focused on the conceptual issues of benefit valuation and some of the estimation methods used in practice. Of all the available benefit valuation techniques, there is no clear consensus about which is consistently superior. The diversity of approaches reflects both the complexity of the task *and* a recognition of its importance in public policy decision making.

For each estimation method, we examined the underlying theory and the measurement technique, assessed the effectiveness of the approach, and surveyed a selection of empirical results. From a general perspective, our investigation uncovered some of the difficulties inherent in any social benefit estimation—difficulties that often are magnified in an environmental context. The primary challenge is in monetizing gains that involve many intangibles

[33] Examples of hedonic price models that attempt to value air quality include a study by Harrison and Rubinfeld (March 1978) and an analysis done by Freeman (1979). Freeman also provides an excellent survey of the major issues surrounding the HPM.

[34] These results are reported by Fisher, Violette, and Chestnut (1989).

not traded in the marketplace. Hence, economists have had to devise methods to quantify these intangibles using something other than explicit prices.

On balance, research efforts in measuring social benefits have been fruitful. Progress has been made in fine-tuning estimation procedures, in recognizing which methods are most useful for which contexts, and in interpreting the results. The efforts are important to support the use of benefit-cost analysis as a risk management strategy and ultimately to devise better policy solutions. Of course, all of this necessitates that comparable progress be made on the cost side of the analysis, which is the focus of the next chapter.

Summary

- To assess incremental benefits attributable to environmental policy, policymakers must determine how health, ecological, and property damages change as a result of that policy initiative.
- A primary environmental benefit arises as a *direct* consequence of implementing policy, whereas a secondary environmental benefit is an *indirect* gain arising either from the primary benefit or from some demand-induced effect.
- If we could infer society's demand for environmental quality, we could measure incremental benefits, because this demand represents the marginal social benefit of abatement.
- It is generally recognized that society derives utility from environmental quality based on its user value and its existence value.
- User value refers to the benefit received from physical utilization or access to an environmental resource. Existence value is the benefit received from the continuance of the resource based on motives of vicarious consumption and stewardship.
- There are two major types of benefit measurement techniques: the physical linkage approach and the behavioral linkage approach.
- A common procedure that uses the physical linkage approach is the damage function method. This method is based on a model of the relationship between levels of a contaminant and the associated damages. Incremental benefits are estimated as the damage reduction achieved from any policy-induced decline in the contaminant.
- Two direct methods that use the behavioral linkage approach are the political referendum method and the contingent valuation method (CVM). Examples of indirect methods using this approach are the averting expenditure method (AEM), the travel cost method (TCM), and the hedonic price method (HPM).
- The contingent valuation method (CVM) is a survey approach that determines individuals' willingness to pay (WTP) for some environmental improvement based on hypothetical market conditions.
- The averting expenditure method (AEM) uses expenditures on goods that are substitutes for environmental quality to indirectly determine the willingness to pay for a cleaner environment.
- The travel cost method (TCM) relies on identifying the recreational demand for an environmental resource, which is a complementary good to environmental quality. As environmental quality improves, recreational demand increases, and the associated benefits can be estimated as the change in consumer surplus.
- The hedonic price method (HPM) is based on the theory that implicit or hedonic prices exist for individual product attributes, including those related to environmental quality.

Key Concepts

incremental benefits
primary environmental benefits
secondary environmental benefits
user value
existence value
direct user value
indirect user value
vicarious consumption
stewardship
physical linkage approach
behavioral linkage approach
damage function method
contingent valuation method (CVM)
averting expenditure method (AEM)
travel cost method (TCM)
hedonic price method (HPM)

Use the Key Concepts listed above to begin your search for additional articles and information using the InfoTrac College Edition database.

Review Questions

1. Is it possible for an *individual's* valuation of an environmental commodity to include both user value and existence value? Explain briefly.
2. Contrast the averting expenditure method (AEM) with the travel cost method (TCM) and discuss the relative strengths and weaknesses of each.
3. Refer to Application 7.3 and the estimated changes in consumer and producer surplus reported by the NAPAP for various ozone policies. Graphically model the result of a 25 percent *increase* in tropospheric ozone to *qualitatively* show the distribution of benefits reported by the NAPAP. (Do not attempt to arrive at the numerical values for the changes in surplus values. Show only how the consumer, producer, and total surpluses *change* in accordance with the reported findings.)
4. One of the strengths of the contingent valuation method (CVM) is its ability to capture existence value. How can the researcher take advantage of this, yet avoid some of the biases of such a survey-based approach?
5. a. Suppose you are part of a research team evaluating a proposal to clean up a hazardous waste site. You are in charge of assessing the incremental benefits. Which method would you choose to derive the estimation? Explain briefly.
   b. Based on your selection, outline your research plan for this specific estimation problem. Be sure to identify the following in your outline: a general description of your model, the relevant market for your model, the primary variables of interest, the data requirements, and any potential bias in your results.

Additional Readings

Abdalla, Charles W., Brian Roach, and Donald J. Epp. "Valuing Environmental Quality Changes Using Averting Expenditures: An Application to Groundwater Contamination." *Land Economics* 68 (May 1992), pp. 163–69.

Boyle, Melissa A., and Katherine A. Kiel. "A Survey of House Price Hedonic Studies of the Impact of Environmental Externalities." *Journal of Real Estate Literature* 9, no. 2 (2001), pp. 117–44.

Bresnahan, Brian W., Mark Dickie, and Shelby Gerking. "Averting Behavior and Urban Air Pollution." *Land Economics* 73 (August 1997), pp. 340–57.

Brookshire, D. S., M. A. Thayer, W. D. Schulze, and R. C. d'Arge. "Valuing Public Goods: A Comparison of Survey and Hedonic Approaches." *American Economic Review* 72 (1982), pp. 165–76.

Brown, Gardner M., Jr., and Jason F. Shogren. "Economics of the Endangered Species Act." *Journal of Economic Perspectives* 12 (summer 1998), pp. 3–20.

Carson, Richard. *Contingent Valuation: A Comprehensive Bibliography and History.* Northampton, MA: Elgar, 2002.

Clawson, Marion. *Methods for Measuring the Demand for and Value of Outdoor Recreation.* Washington, DC: Resources for the Future, 1959.

Common, M., I. Reid, and R. Blamey. "Do Existence Values for Cost Benefit Analysis Exist?" *Environmental and Resource Economics* 9 (March 1997), pp. 225–38.

Giraud, K. L., J. B. Loomis, and J. C. Cooper. "A Comparison of Willingness to Pay Estimation Techniques from Referendum Questions." *Environmental and Resource Economics* 20, no. 4 (2001), pp. 331–46.

Hanneman, W. Michael. "Valuing the Environment Through Contingent Valuation." *Journal of Economic Perspectives* 8 (fall 1994), pp. 19–43.

Kneese, Allen V. *Measuring the Benefits of Clean Air and Water.* Washington, DC: Resources for the Future, 1984.

Mendelsohn, Robert. "Modeling the Demand for Outdoor Recreation." *Water Resources Research* 23 (May 1987), pp. 961–67.

Portney, Paul R. "The Contingent Valuation Debate: Why Economists Should Care." *Journal of Economic Perspectives* 8 (fall 1994), pp. 3–18.

Smith, V. Kerry. "Can We Measure the Economic Value of Environmental Amenities?" *Southern Economic Journal* 56 (April 1990), pp. 865–78.

Smith, V. Kerry, and William H. Desvousges. "Averting Behavior: Does It Exist?" *Economics Letters* 20, no. 3 (1986), pp. 291–96.

Wilhelmsson, Mats. "The Impact of Traffic Noise on the Values of Single-Family Houses." *Journal of Environmental Planning and Management* 43 (November 2000), pp. 799–815.

Related Web Sites

Economics and the Endangered Species Act (at Resources for the Future)	**http://www.rff.org/proj_summaries/files/ando_econ_esa.htm**
Endangered Species Act (full text)	**http://www.nmfs.noaa.gov/prot_res/laws/ESA/esatext/esacont.html**
Environmental Valuation Reference Inventory (EVRI)	**http://www.evri.ec.gc.ca/evri/english/default.htm**

Chapter 8

Assessing Costs for Environmental Decision Making

In Chapter 7, we began our formal study of benefit-cost analysis with the theory and measurement of environmental benefits. Equally critical is the analysis of **environmental costs,** that is, the costs of environmental improvement. In this phase of risk management, the policymaker must consider the value of all economic resources allocated to reducing environmental risk. Unlike the benefit side, the challenge here is not assigning a monetary value to costs, since most are already expressed in money terms. Rather, the more critical issue is identifying all the resources used to design, implement, and execute the policy prescription.

Few would debate that identifying environmental costs is a major undertaking. Think about the amount of government spending necessary to support the scientific research, the network of administrative agencies, and the labor force needed to implement a major environmental initiative. Add to that the billions of dollars spent by private businesses on abatement equipment and labor to comply with environmental regulations. Identifying these expenses on such a massive scale is by itself a tremendous task. But there is another element of cost analysis that adds to the difficulty—the premise that *economic* costs, and not simply *accounting* costs, are to be determined.

In this chapter, we discuss all these issues, beginning with an overview of the fundamentals, much as we did on the benefit side. In particular, we start by defining incremental costs, the distinction between explicit and implicit costs, and the concept of valuing environmental costs. Once done, we address actual cost estimation methods used in practice today. Finally, we discuss the more prevalent ways in which environmental costs are classified and reported.

Identifying and Valuing Environmental Costs: Conceptual Issues

Just as is the case for benefits, the appropriate level of analysis for evaluating the cost of an environmental initiative is **incremental costs.** The rationale is to allow a comparison between post-policy expenditures and their pre-policy level, which we call the baseline.

Defining Incremental Costs

Starting with the basics, we know that environmental costs must be defined in incremental terms. The motivation for using incremental variables is to capture *changes* brought about by policy. In this case, the relevant change is the increase in costs associated with policy-induced improvements in environmental quality. **Incremental costs** are calculated by first identifying the existing level of environmental expenditures, then estimating the costs after the policy is implemented, and finally finding the difference between the two. By way of example, the United States incurred millions of dollars in incremental costs to address the environmental damage associated with the terrorist attacks of September 11, 2001. An overview of these costs is provided in Application 8.1.

incremental costs
The change in costs arising from an environmental policy initiative.

From an economic perspective, incremental costs should reflect changes in *economic* costs. As discussed in Chapter 2, economic costs are a more accurate measure of resource utilization than are accounting costs because they include both **explicit** (i.e., out-of-pocket) **costs**

Application

8.1 Incremental Environmental Costs of the September 11, 2001, Terrorist Attacks

Characterized as the worst terrorist attack on U.S. soil in history, the events of September 11, 2001, imposed grave harm to virtually every aspect of American living. Beyond the horrific human loss in New York, Washington, DC, and Pennsylvania, there was enormous damage to the Pentagon and to the entire infrastructure of Lower Manhattan. The explosions and structural ruin placed environmental quality in the two metropolitan areas at risk. Because of the collapse of the World Trade Center and the ensuing fires at ground zero, the environmental risks were of greater concern in the New York City area.

When the massive twin towers fell, the debris field was enormous. As a consequence, the air was filled with smoke, dust, and such substances as asbestos, chemical gases, and fiberglass. Former Mayor Giuliani and other officials donned dust masks as they made their way through the third largest business district in America. The image was nothing less than extraordinary. The EPA and other government agencies were summoned to the scene to assist with cleanup efforts and to test the air and water supplies. Of major concern was the potential exposure of rescue workers and others on the scene to elevated levels of hazardous contaminants.

The EPA's work in the wake of the disaster was immense. Working with the Occupational Safety and Health Administration (OSHA), the Centers for Disease Control and Prevention (CDC), and other government agencies, the EPA provided expertise on the abatement of hazardous materials, sampled drinking water supplies, tested runoff, and established a monitoring network to test air and water quality in New York City and in the area adjacent to the Pentagon. About 20 fixed air monitors were placed in and around the World Trade Center site, with others located in the Bronx, Brooklyn, Queens, and Staten Island. Portable equipment also was used to gather data from other locations. The results were assessed relative to standards and benchmarks established by various federal regulations. Daily updates of the environmental monitoring results were made available by the EPA at **http://www.epa.gov/wtc/data_summary.htm.**

http://

These environmental initiatives necessitated additional spending by the EPA, which by definition represents the incremental environmental costs associated with the agency's response to the disaster. Initially, these expenditures were financed by emergency funding of $23.7 million. However, on September 18, 2001, the EPA administrator, Christine Todd Whitman, announced that the Federal Emergency Management Agency (FEMA) had given the EPA up to $83 million to cover the incremental costs of their abatement, advisory, and monitoring efforts in New York City and Washington, DC. Because of the difficulty in identifying implicit costs, the dollar values referenced by these agencies were necessarily confined to explicit cost estimates. Clearly, if the full economic costs were considered, including all implicit costs, the dollar values cited by these agencies would be considerably higher.

Sources: U.S. EPA (September 18, 2001); U.S. EPA (September 24, 2002a); U.S. EPA (September 24, 2002b).

and **implicit costs.** However, since the latter are not readily identifiable, analysts often derive incremental cost values based solely on explicit expenditures.

Explicit Environmental Costs

The **explicit costs** of implementing an environmental policy include the administrative, monitoring, and enforcement expenses paid by the public sector as well as the compliance costs incurred by virtually all sectors of the economy. Explicit costs and their components are easier to identify than their benefit counterparts, since most of the resources used to implement pollution control policies are traded in private markets. Those associated with the use of economic resources—land, labor, and capital—are rents, wages, and interest, respectively. Since these resources are traded on the open market, expenditures are based on

explicit costs
Administrative, monitoring, and enforcement expenses paid by the public sector plus compliance costs incurred by all sectors.

market-determined input prices. However, these expenditures are not made simultaneously, and some are less controllable than others in the short run.

Recognizing these distinctions, economists classify costs into two components: (1) **fixed costs,** which are not controllable in the short run and do not depend on production levels, and (2) **variable costs,** which have the opposite characteristics. In the context of environmental policy implementation and compliance, the accounting equivalents of these categories are **capital costs** and **operating costs,** respectively.

Capital and Operating Costs

According to U.S. government guidelines provided by the Department of Commerce and used by the EPA, capital and operating costs have specific meanings. **Capital costs** are expenditures for plant, equipment, construction in progress, and the costs of changes in production processes that reduce or eliminate pollution generation. **Operating costs** are those incurred in the operation and maintenance of pollution abatement processes, including spending on materials, parts and supplies, direct labor, fuel, and research and development.[1]

capital costs
Fixed expenditures for plant, equipment, construction in progress, and production process changes associated with abatement.

operating costs
Variable expenditures incurred in the operation and maintenance of abatement processes.

A critical distinction between these two categories is in how costs are related to the level of abatement. A capital cost is incurred regardless of the amount of pollution abated, analogous to the economic definition of fixed cost. Examples include the installed price of a scrubber system to control air pollution, construction costs of a wastewater facility, and the purchase of acreage to use as a landfill. Notice that these capital costs are incurred before making any of the investments operational and are independent of whether the capital is ever used to abate pollution. Conversely, operating costs are directly related to the quantity of abatement, comparable to the economic definition of variable costs. Examples include the costs of monitoring emissions and the costs of labor to run an abatement facility.

Implicit Environmental Costs

Although the explicit costs of abatement appear to be comprehensive, it turns out that these costs convey only part of the story. There are also **implicit costs**—those concerned with any nonmonetary effects that negatively affect society's well-being. These include the value of diminished product variety arising from a ban on certain inputs, the time costs of searching for substitutes, and the reduced convenience that environmental control policies might impose. Although implicit costs are arguably an important factor, identifying and measuring them in practice is another issue entirely. In fact, most analyses fail to fully capture these values, ignoring effects that represent real costs to society. The result? Many environmental cost assessments are seriously understated, a problem that has yet to be fully resolved.

implicit costs
The value of any nonmonetary effects that negatively influence society's well-being.

Conceptually Valuing Environmental Costs

What economic resources are used to achieve cleaner air or water? And what are the costs of these resources? Ideally, one should determine the *social costs* of environmental policy to answer these questions.

In theory, the **social costs** of any policy initiative are the expenditures needed to compensate society for the resources used so that its utility level is maintained. This compensation would have to account for all the price, output, and income effects that arise from a given regulation.[2] One example of such an effect is the increased price of automobiles caused by the cost of abatement equipment like catalytic converters. Another is the higher price of household products arising from the use of recyclable packaging and the costs of more highly regulated labeling. There are also real income changes, such as those caused by higher taxes needed to support regulatory expenses. Beyond these effects are implicit or nonmonetary costs, which also have to be identified and monetized. Examples include the inconvenience

social costs
Expenditures needed to compensate society for resources used so that its utility level is maintained.

[1] U.S. EPA, Office of Policy, Planning, and Evaluation (December 1990), pp. 1-2–1-3, in accordance with U.S. Department of Commerce, Bureau of the Census, *Government Finances* (various years), and U.S. Department of Commerce, Bureau of Economic Analysis, "Pollution Abatement and Control Expenditures," published periodically in *Survey of Current Business.*

[2] In earlier chapters, we argued that one of these effects might be the forgone profits caused by firms reducing production levels to lower pollution releases.

Figure 8.1 ***Marginal Social Cost (MSC) and Total Social Costs (TSC) of Air Quality (% SO_2 Abatement)***

The market supply for sulfur dioxide (SO_2) abatement represents the marginal social cost (*MSC*) of air quality. For each abatement level *A*, the associated *MSC* is shown as the vertical distance from the horizontal axis up to the supply curve. Total social costs (*TSC*) for any abatement level are measured as the area under the *MSC* curve up to that point. Thus, the shaded area represents the *TSC* to achieve abatement level A_1.

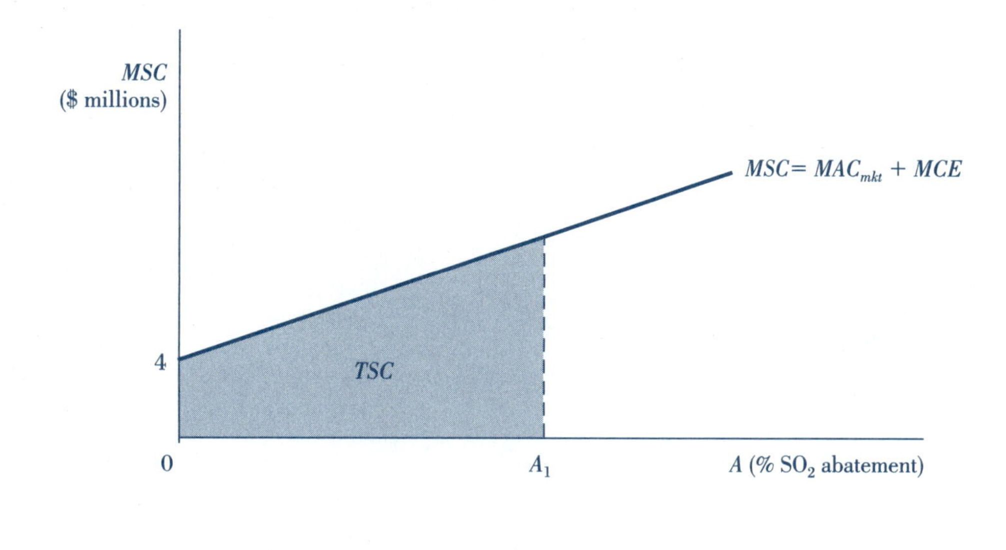

of using public transportation or car pools in response to new urban air quality policies or the search costs to find nontoxic substitute products.

From a modeling perspective, we capture marginal social costs (*MSC*) by the supply of the public good, environmental quality. Although the demand for a public good can only be inferred, the supply function can be identified directly in the same manner as that used for a private good. Let's reconsider the market supply model presented in Chapter 3 that represents the marginal social cost (*MSC*) of sulfur dioxide (SO_2) abatement. The function is specified as $MSC = 4+0.75A$, where *MSC* is measured in millions of dollars, and *A* is a percentage of SO_2 abatement. In Chapter 4, we argued that the *MSC* is the vertical sum of the market-level marginal abatement costs (MAC_{mkt}) plus the government's marginal cost of enforcement (*MCE*). The graph is shown in Figure 8.1.

Analogous to the benefit side, notice that for each abatement level, *MSC* is shown as the vertical distance from the horizontal axis up to the supply curve. Total social costs (*TSC*) for any abatement level are measured as the area under the *MSC* curve up to that point.[3] The shaded area in Figure 8.1 represents the *TSC* to achieve an abatement level of A_1.

Just as on the benefit side, the incremental cost assessment follows a logical three-step process:

1. Find the baseline level of *TSC* *before* the policy is undertaken
2. Find the new level of *TSC* that would arise *after* the policy is implemented
3. Subtract the baseline *TSC* from the post-policy *TSC* to determine incremental costs.

[3] Technically, the area under any marginal cost curve represents only total *variable* costs. Hence, to argue that the area under the *MSC* curve represents the *TSC* implicitly assumes there are no fixed costs.

Figure 8.2

Modeling Incremental Costs for Air Quality (% SO_2 Abatement) Using the MSC Function

At the baseline abatement level of 20 percent, the *MSC* equals $19 million, and the *TSC* are shown as the area under the *MSC* curve up to that point, or $230 million. At the post-policy abatement level of 25 percent, the *MSC* is $22.75 million, and the corresponding *TSC* are $334.375 million. Therefore, the difference between the two *TSC* values, or $104.375 million, represents the incremental costs to achieve the additional 5 percent abatement of SO_2 emissions. These incremental costs are shown as the shaded area.

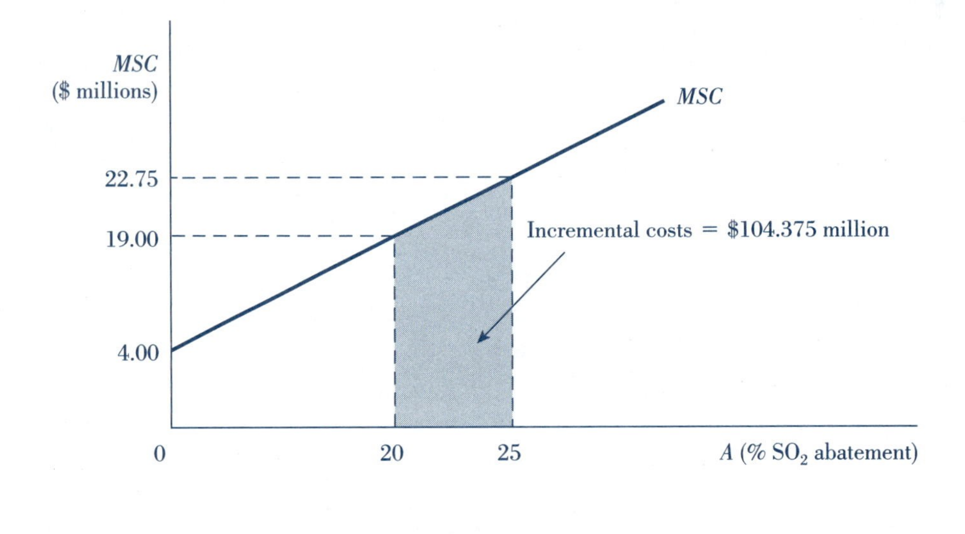

To illustrate this procedure, consider the incremental costs of a policy-induced increase in SO_2 abatement from 20 to 25 percent. First, identify the baseline level of *TSC* when *A* equals 20 percent. Referring to Figure 8.2, notice that the *MSC* at this abatement level is $19 million and the corresponding *TSC* are shown as the area under the *MSC* curve up to that point, or $230 million. Second, we determine the post-policy *TSC* to increase abatement to 25 percent. Following the same steps, the *MSC* when *A* equals 25 percent is $22.75 million, and the *TSC* are $334.375 million. Hence, the difference between the two *TSC* values, or $104.375 million, represents the incremental costs to achieve the additional 5 percent abatement of SO_2 emissions.

Figure 8.3 presents an alternative model that graphs the *TSC* directly with the percentage of SO_2 abatement. In this case, the *TSC* associated with any abatement level can be found simply as the vertical distance from the horizontal axis up to the *TSC* curve. To determine incremental costs from this model, begin by identifying the pre- and post-policy abatement levels (20 and 25 percent, respectively) and the corresponding *TSC* levels ($230 million and $334.375 million, respectively). The difference, or $104.375 million, is shown as the vertical distance between the two cost values.

In practice, to properly assess the social costs of an environmental initiative, the analyst would have to measure changes in social welfare, perhaps by estimating changes in consumer and producer surplus. This is a complex undertaking, but there is some evidence that doing so improves the accuracy of cost assessments. A study by Hazilla and Kopp (1990) estimates the social costs of U.S. air and water quality control policies. According to this research, there is a significant difference between social costs and the explicit cost estimates typically used

Figure 8.3 ***Modeling Incremental Costs for Air Quality (% SO_2 Abatement) Using the TSC Function***

An alternative way to model the incremental costs of SO_2 abatement is to graph the *TSC* directly as a function of various abatement levels. Then, the *TSC* associated with any abatement level are found simply as the vertical distance from the horizontal axis up to the *TSC* curve. Notice that at the pre- and post-policy abatement levels, 20 and 25 percent, respectively, the corresponding *TSC* levels are \$230 million and \$334.375 million, respectively. Thus, the difference, or \$104.375 million, is simply the vertical distance between the two cost values.

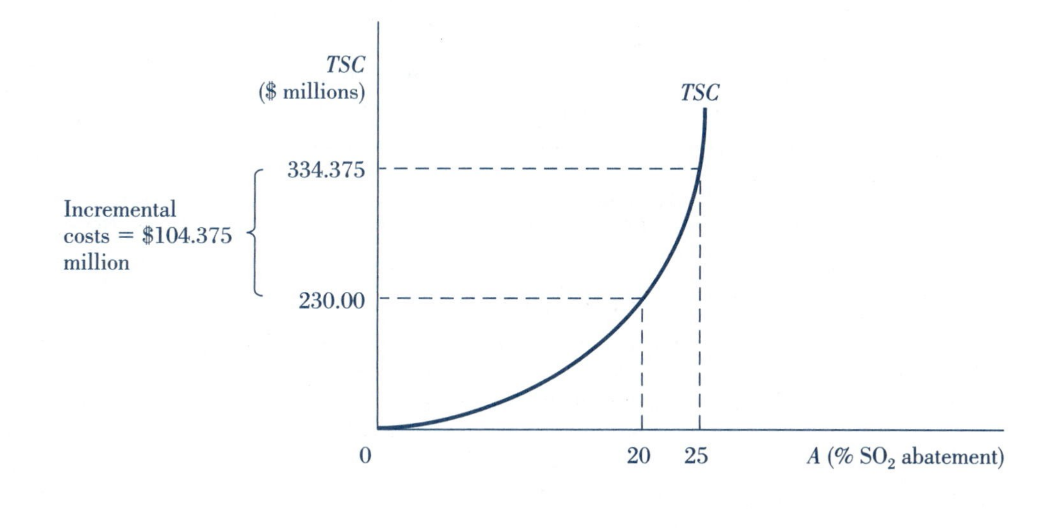

in practice. In fact, these researchers recommend that policymakers adjust their procedures to consider more of a general equilibrium approach to cost analysis than is currently in place. However, available methods to execute this more comprehensive cost assessment are too cumbersome and unreliable to be used as a regular practice. So at least for the present, cost analysis continues to be based solely upon explicit expenditures, although it is generally acknowledged that such an approach understates the true social costs.

Estimation Methods for Measuring Explicit Costs

Economists use two major approaches to estimate incremental environmental costs: the **engineering approach** and the **survey approach.** The **engineering approach** estimates abatement expenditures based on the least-cost available technology needed to achieve some level of pollution abatement. The **survey approach** derives estimated abatement expenditures directly from polluting sources. Although the two approaches are distinct, a common practice is to use a combination approach that draws from both. First, the survey approach is used to solicit information about available technologies and existing market conditions. Then, the engineering approach is employed to estimate dollar values based on the collected data. What follows is a brief discussion of each approach, with some mention of their respective shortcomings.

engineering approach
Estimates abatement expenditures based on least-cost available technology.

survey approach
Polls a sample of firms and public facilities to obtain estimated abatement expenditures.

Table 8.1 *Survey Approach to Pollution Abatement and Control Expenditures for 1994*

Title of Survey	Source	Proportion of Cost Data Determined by the Survey
Pollution Abatement Costs and Expenditure (MA-200)	Bureau of Census	21
Government Finances	Bureau of Census	22
Value of New Construction Put in Place	Bureau of Census	10
Federal Funding for Pollution Control	Bureau of Economic Analysis	5
Structures and Equipment Survey	Bureau of Census	5
Steam-Electric Plant Operation and Design	Department of Energy	2

Source: Drawn from Vogan (September 1996), p. 54, table 10.

Notes: The Government Finances line item is not the title of one particular survey. Rather, several sources make up the survey of government finance data. For example, the Bureau of Economic Analysis gathers information from Energy Information Agency reports and from the EPA.

Engineering Approach

The **engineering approach** to expenditure estimation relies upon the knowledge of experts in abatement technology. Based on the state of the art in abatement, engineers and scientists are called upon to identify combinations of equipment, labor, and materials needed by polluters to comply with a policy mandate. Then, capital and operating costs for all feasible abatement designs are estimated. Finally, the analyst selects the least-cost model from among these technology-based designs and uses the result to estimate the aggregate incremental cost for all affected polluting sources.

While theoretically reasonable, the engineering approach is not without problems. For one thing, the procedure is difficult to implement for *proposed* environmental controls, since there is uncertainty about price movements, availability of raw materials, and future energy costs. Even absent these forecasting problems, any aggregation procedure of this magnitude has inherent difficulties. Most comprehensive policy initiatives affect a wide range of industries and public facilities. To properly account for these, the engineering approach would have to be customized to suit each type of production setting. If ignored, the results likely would suffer from the averaging process. Even if industry-specific values were derived, the heterogeneity of firms and the unique market conditions that each faces could not be captured in such a generalized estimate. Finally, because the approach is based on the least-cost abatement design, it explicitly assumes that all firms are cost-effective or technically efficient entities—an assumption that likely understates the true costs incurred.

Survey Approach

In contrast to the engineering approach, the **survey approach** relies directly upon polluting sources instead of external experts to provide data for the estimation. Similar to the contingent valuation method for estimating benefits, the survey approach polls a sample of firms and public facilities to inquire about existing or projected environmental expenditures. To illustrate the applicability of this approach, in Table 8.1 we list the surveys used to gather U.S. cost data in 1994. Also given are the government sources responsible for administering each survey. Approximately 65 percent of the information used for 1994 cost reporting was obtained from some type of survey. Notice that these surveys target both private and public sector costs of abatement and control.[4]

[4]For the remaining 35 percent of the cost information needed, the Bureau of Economic Analysis uses various *indirect* costing methods, most of which also use some type of survey information. See Vogan (September 1996), pp. 54–55.

On the plus side, the survey method is a more direct means to obtain abatement cost data than the engineering approach. However, there are also disadvantages. First, it implicitly assumes that polluting sources are sufficiently well informed to provide reasonable estimates. Second, there is an inherent bias. Polluting sources have an incentive to offer inflated values to government officials, because they recognize that higher costs will increase the probability that the proposed regulation will be rejected.

Just as in benefit assessment, no cost estimation method is perfect. In fact, this realization is what has motivated researchers to use a combination of the engineering and survey methods. The objective is to try to have the best of both worlds and minimize potential bias. Given the growing emphasis on benefits and costs in environmental policy decisions, efforts such as these are ongoing to improve the accuracy of environmental cost estimation.

Cost Classifications in Practice

As part of cost assessment, analysts are interested in determining the composition of environmental control costs. The motivation may be that some component of costs, such as spending on water quality control or the abatement costs incurred by private industry, has particular relevance to a policy decision. More important, information about the composition of control costs gives policymakers a clearer sense of how resources are being allocated to achieve environmental objectives—critical information for evaluating policy on the basis of allocative efficiency. Similarly, adjustments to correct for environmental inequities require knowledge about how the cost burden is distributed across economic sectors.

Because of the diversity of objectives in cost assessment, different cost classifications are used to obtain different kinds of information. An overview of two of the more commonly used classifications follows.

Cost Classifications by Economic Sector

Policy analysts often are concerned with how environmental costs are distributed across the public and private sectors of the economy. For instance, the costs of policy revisions that strengthen monitoring and enforcement programs shift a higher cost burden to the public sector, possibly at all levels of government. This in turn would shift up the *MCE* curve and hence the *MSC* of abatement. Conversely, tighter abatement requirements will change the cost distribution more toward private industry. To illustrate, consider the U.S. trend data for 1975–1994 in Table 8.2. The expenditures are broken down by major function: abatement, regulation and monitoring, and research and development.

Table 8.2 ***U.S. Pollution Abatement and Control Expenditures by Function, 1975–1994 (in billions of 1992 dollars)***

	Pollution Abatement			Regulation and	Research and	
Year	Personal	Business	Government	Monitoring	Development	Total
1975	7.72	38.99	16.37	1.54	2.61	67.24
1980	11.02	49.71	18.42	2.09	2.90	84.14
1985	15.48	53.53	18.52	1.59	1.76	90.87
1990	9.97	62.29	25.63	1.91	1.52	101.30
1994	9.29	72.92	29.72	2.09	1.89	115.91

Source: Vogan (September 1996).

Notes: Some totals may not agree because of independent rounding. All of personal spending is used to purchase and operate motor vehicle emission abatement devices. Real 1992 dollars are calculated using implicit price deflators (1992=100).

Figure 8.4 Decomposition of U.S. Pollution Abatement and Control Expenditures for 1994

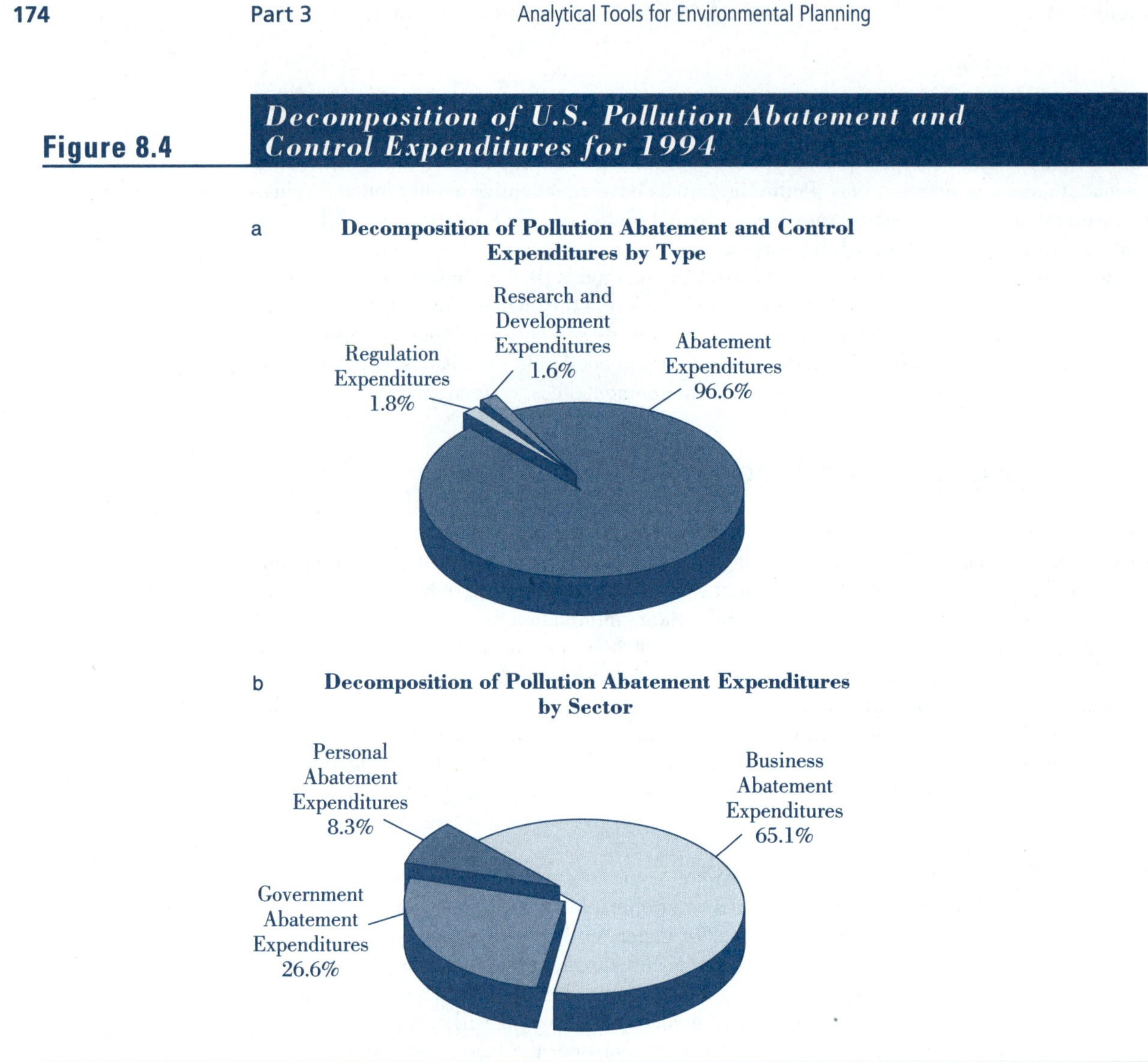

Source: Vogan (September 1996).

As the data in Table 8.2 indicate, the aggregate level of real spending on environmental issues has grown fairly steadily in the United States, up to $115.91 billion ($1992) in 1994.[5] Furthermore, spending on abatement consistently represents the vast majority of total environmental expenditures over time. Within the abatement category, spending by each economic sector has shown a similar growth pattern, so much so that the relative proportions borne by each have remained fairly constant. However, the cost burden is not shared equally across the economy. Over time, private business spending on pollution abatement represents about 63 percent of the total on average, public sector spending accounts for about 24 percent, with the personal private sector responsible for the remainder.[6]

To see this distribution more clearly, look at Figure 8.4, which presents just the 1994 data but in percentage terms. Notice from Figure 8.4a that aggregate spending on abatement was

[5] Internet access to the report from which these data are drawn is available at **http://www.bea.doc.gov/bea/an/0996eed/maintext.htm.** Further updates to these data (beyond 1994) were discontinued for budgetary reasons.
[6] Within the private business sector, certain industries spend proportionately more than others on abatement. For an overview of abatement expenditure growth for selected industries, see Exhibit A.2 in the Appendix.

Table 8.3 ***Abatement and Control Expenditures Disaggregated by Environmental Media for 1975–1994 (billions of 1992 dollars)***

Year	GDP	Air	Water	Solid Waste
1975	3,873.9	28.32	28.68	10.74
1980	4,615.0	37.05	34.21	14.12
1985	5,323.5	39.07	33.32	19.33
1990	6,136.3	30.27	39.67	32.74
1994	6,610.7	35.78	40.33	39.72

Source: U.S. Department of Commerce, Bureau of Economic Analysis (August 1998), as cited in Council on Environmental Quality (1998), p. 219, table 2.1; Vogan (September 1996).

Note: Real 1992 dollars are calculated using implicit price deflators (1992=100).

96.6 percent of the total, with spending on regulation and on research and development representing less than 2 percent each. Taking this one step further, we can assess the relative cost burden within the abatement category, which is shown in Figure 8.4b. Notice that most of these expenditures were borne by the business sector. In 1994, 65.1 percent of all abatement spending was attributable to private business outlays, government accounted for 26.6 percent, and the personal private sector accounted for 8.3 percent. According to Vogan's (1996) report, the entire $9.29 billion ($1992) of personal private spending is for durable goods, specifically, the purchase and operation of abatement equipment for motor vehicles. Application 8.2 takes a closer look at how this category of spending has been affected by tougher regulations on emissions from motor vehicles.

Cost Classifications by Environmental Media

Another important policy consideration is the distribution of expenditures by environmental media (i.e., air, water, and solid waste). Table 8.3 shows U.S. abatement and control expenditures for 1975–1994 disaggregated by environmental media. Such information helps policymakers examine how resources are allocated across environmental problems and how this allocation changes from period to period.

Notice, for example, that a rough estimate of the incremental costs of air quality regulations during the 1990–1994 period is $5.51 billion ($1992). Of course, a more accurate assessment would require a more complex series of calculations. For one thing, analysts are usually interested in assessing the costs of a policy once it has been fully implemented. So the appropriate analysis would extend over the relevant period of years, and the expenditure data would have to be adjusted to account for overlap with other regulations.

Finally, to give context to all these cost values, real gross domestic product (GDP) (also in 1992 dollars) data are provided in Table 8.3. This helps to illustrate the *relative* magnitude of environmental spending in the United States. Note that in 1994, for example, environmental spending on water resources was $40.33 billion. However, this is a relatively small percentage of GDP, approximately 0.61%.

Conclusions

Environmental cost assessment and estimation is a critical element of risk management. Although more tangible than benefit assessment, the valuation of environmental costs has its share of complexities. Chief among these is the divergence between the social costs of an environmental initiative and the explicit costs estimated in practice. In the absence of better

Application

8.2 Abatement Control Costs on Motor Vehicles

Virtually all cars on the road today are equipped with pollution abatement equipment. In response to tougher legislation, automakers have shifted from using so-called low-tech equipment, such as engine timing devices, to more sophisticated high-tech equipment, such as catalytic converters and computerized emission control systems. As this technological transition has evolved, analysts have examined how this trend has affected private spending on abatement control and how it has altered the composition of that spending.

The data in the accompanying table are expenditures by the personal private sector on motor vehicle abatement devices over the 1975–1994 period. Durable goods expenditures refer to capital costs, and nondurable goods expenditures refer to operating and maintenance costs. These data show how expenditures have changed with strengthening U.S. air quality policy initiatives. Total personal consumption grew by over 100 percent between 1975 and 1985, and the 1975 value is nearly twice what it was in 1972 when these data were first recorded. Notice also that in 1994, spending was $9.3 billion, reflecting a 40 percent decline from the 1985 level. However, much of this decline is attributable to a fall in unit sales of automobiles over the period.

Another important observation is how the trend toward more capital-intensive regulations influenced the *composition* of abatement costs. For example, spending on catalytic converters began with the 1975 model year, which explains why the $3.3 billion spent on durable goods in 1975 is an 80 percent increase over the 1974 level of $1.8 billion. Similarly, rising spending throughout much of the 1980s reflects the costs of more sophisticated high-tech equipment introduced during that period.

The nondurable goods component of expenditures shows relatively little movement through the 1970s and 1980s. Yet major shifts were taking place in the *composition* of these costs as well. For example, in the pre-1975 model years, most operating expenses were attributable to fuel consumption penalties (i.e., higher fuel usage due to abatement control equipment) and increased maintenance costs. In later-model cars, fuel price penalties (i.e., higher prices for unleaded gasoline) became the more important component of nondurable expenditures. Why? Because devices like catalytic converters and computer controls improve fuel economy, decreasing the fuel consumption penalty, and extend the life of exhaust systems, reducing maintenance costs. Since the 1980s, however, fuel price penalties also have shown signs of decline. Again, the primary explanation is technology. Improvements in the production of nonlead octane boosters as well as scale economies caused a reduction in the price differential between leaded and unleaded fuel. As a result, nondurable goods expenditures fell to zero in 1994.

Source: Vogan (September 1996), p. 57–63, table 11.

Personal Expenditures on Air Pollution Abatement and Control 1975–1994 (billions of 1992 dollars)

Year	Durable Goods	Nondurable Goods	Total Personal Consumption
1975	3.3	4.4	7.7
1980	5.9	5.1	11.0
1985	10.6	4.9	15.5
1990	9.6	0.4	10.0
1994	9.3	0	9.3

Source: Vogan (September 1996), pp. 57–63, table 11.

techniques that can capture implicit costs, incremental cost estimates are generally assumed to be biased downward.

As economists search for a practical solution to this problem, they also look for ways to improve their estimates of explicit costs. The combined use of the survey and engineering approaches appears to be one way that greater accuracy can be achieved. Related to these efforts are attempts to classify cost data in ways that are meaningful to policy evaluation procedures and to the formulation of new programs and initiatives. The motivation is simple. Policymakers have to make important decisions based in part on the results of environmental cost estimation. In the next chapter, we will explain precisely how environmental policy is influenced by both benefit and cost estimates when they are systematically considered in a formal benefit-cost analysis.

Summary

- Environmental cost analysis is concerned with incremental costs, or the change in explicit and implicit costs to society, incurred as a result of government policy. Because implicit costs are not readily identifiable, analysts generally derive incremental costs based solely on explicit expenditures.
- Explicit costs of implementing environmental policy include the administrative, monitoring, and enforcement expenses paid by the public sector as well as the compliance costs incurred by virtually all sectors of the economy.
- Capital costs are expenditures for plant and equipment used to reduce or eliminate pollution.
- Operating costs are those incurred in the operation and maintenance of abatement processes.
- Implicit costs are those concerned with any nonmonetary effects that negatively affect society's well-being.
- The social costs of any policy initiative are the expenditures needed to compensate society for the resources used so that its utility level is maintained.
- Conceptually, the supply of environmental quality can be used to model the marginal social cost of abatement (MSC), which is the vertical sum of the market-level marginal abatement cost function (MAC_{mkt}) and the marginal cost of enforcement (MCE).
- The engineering approach to cost estimation is based on the least-cost available technology needed to achieve a given level of abatement.
- The survey approach to cost estimation relies on estimated abatement expenditures obtained directly from polluting sources.
- Because of the diversity of objectives in environmental cost assessment, costs are commonly communicated through various classifications, such as by environmental media or by economic sector.

Key Concepts

incremental costs
explicit costs
capital costs
operating costs
implicit costs
social costs
engineering approach
survey approach

Use the Key Concepts listed above to begin your search for additional articles and information using the InfoTrac College Edition database.

Review Questions

1. Consider a policy proposal to impose more stringent controls on automobile tailpipe exhausts. Distinguish between the explicit and implicit costs of this proposal, and support your discussion with several specific examples of each.

2. Of the two approaches to cost estimation, which in your view likely produces the most reliable estimates? Explain.

3. Suppose the *MSC* of cleaning Puget Sound is modeled as $MSC = 10 + 1.4A$, where A is the percentage of phosphorus abated and *MSC* is measured in millions of dollars.

   a. Find the incremental costs of a policy initiative that increases the phosphorus abatement level from its baseline of 30 to 45 percent.

   b. Graphically illustrate using the *MSC* function, labeling clearly where incremental costs are shown.

   c. Repeat part (b) using the *TSC* function directly.

Additional Readings

Cairncross, Frances. *Costing the Earth.* Boston: Harvard Business School Press, 1991.

Harrington, Winston, Richard D. Morgenstern, and Peter Nelson. "Predicting the Costs of Environmental Protection." *Environment* 41 (September 1999), pp. 10–19.

Hartman, Raymond S., David Wheeler, and Manjula Singh. "The Cost of Air Pollution Abatement." *Applied Economics* 29 (June 1997), pp. 759–74.

Lee, Dwight R. "Economics and the Social Costs of Smoking. *Contemporary Policy Issues* 9 (January 1991), pp. 83–92.

Morgenstern, Richard D., William A. Pizer, and Jhih-Shyang Shih. "Are We Overstating the Real Economic Costs of Environmental Protection?" Discussion Paper 97-36-REV. Washington, DC: Resources for the Future, June 1997.

Palmer, Karen, Hilary Sigman, and Margaret Walls. "The Cost of Reducing Municipal Solid Waste." *Journal of Environmental Economics and Management* 33 (June 1997), pp. 128–50.

U.S. Environmental Protection Agency, Office of Solid Waste and Emergency Response. *Full Cost Accounting for MSW Management: A Handbook.* Washington, DC, September 1997.

Weyant, John P. Costs of Reducing Global Carbon Emissions. *Journal of Economic Perspectives* 7 (fall 1993), pp. 27–46.

Related Web Sites

http://

Environmental Management Accounting International	**http://www.emawebsite.org/about_ema.htm**
Research on the Cost of Environmental Regulation	**http://www.rff.org/proj_summaries/99files/harrington_99clean_enviro.htm**
U.S. Census Bureau, Survey of Pollution Abatement Costs and Expenditures	**http://www.census.gov/ftp/pub/econ/www/mu1100.html**
U.S. EPA, Daily Environmental Monitoring Summary of Post 9/11 New York City	**http://www.epa.gov/wtc/data_summary.htm**
Vogan, Christine R. "Pollution Abatement and Control Expenditures, 1972–94" (September 1996)	**http://www.bea.doc.gov/bea/an/0996eed/maintext.htm**

Chapter 9

Benefit-Cost Analysis in Environmental Decision Making

Should the national limit on sulfur dioxide emissions be tightened? Would society be better off if a tradeable permit system were used instead of technology-based standards to control water pollution? These are the kinds of questions that environmental decision makers deal with. Analytical tools such as **benefit-cost analysis** can help them find answers. Benefit-cost analysis begins with identifying and monetizing environmental benefits and costs, the subject of the previous two chapters. From this point, these preliminary estimates must be adjusted and then systematically compared to arrive at a decision. These critical steps that link benefit and cost estimates to a decision rule complete the strategic process of benefit-cost analysis.

Adjustments to estimated values are necessary because incremental costs and benefits from a given policy initiative are not realized immediately. Instead, they accrue over a period of years. Think about the implementation of a major policy initiative such as the Clean Air Act or the Superfund Amendments. Such legislation takes years to put into effect. Although some policy costs are incurred right away, others accrue at various points in time in the future. Similarly, the benefits are not realized in a single period. It takes time before the gains from an environmental policy are achieved. In the first place, there is a lag between risk reduction and realized improvements in human health and welfare. Furthermore, even if the effects were immediate, government-mandated reductions in contamination typically are phased in over a period of years.

How do these time differences affect the result? There are two considerations. One is that costs and benefits realized in some future period are not valued as highly as those achieved immediately. Thus, future costs and benefits have to be adjusted downward to be comparable to those incurred in the present. A second adjustment is needed to account for expected changes in the price level over time. For example, the value of costs and benefits measured in today's dollars will be much higher in future dollars during periods of inflation. Without adjusting for these time-oriented differences, the benefit-cost analysis would yield biased results, and any policy decision based upon them would be misguided.

Once benefit and cost estimates are adjusted for time differences, they then must be compared to one another. But on what basis is the comparison made? Is it sufficient for the benefits to outweigh the costs, or must the numerical difference between the two be at a particular level? What if the law precludes the analysis by predetermining the benefit level to be achieved? The answers to these questions are found in understanding the decision rules that guide benefit-cost analysis. Of particular interest are the implications of those rules for meeting economic criteria—an important issue in policy evaluation.

The growing prominence of benefit-cost analysis as a risk management strategy is an important reason for understanding how it is undertaken, what it implies from an economic perspective, and how it has come to be required by the U.S. government in major policy decisions. To that end, this chapter presents and analyzes these aspects of benefit-cost analysis. We begin by giving a thorough exposition of the procedures used to adjust preliminary benefit and cost estimates for time differences. Next, we explore the policy evaluation process and the decision rules used in practice to guide that process. Once done, we consider how the role of benefit-cost analysis in U.S. policy making has evolved into a major tool for guiding legislative decisions. We conclude with an inside look at how benefit-cost analysis was used in the U.S. decision to tighten the lead standard for gasoline.

Adjusting for the Time Dimension of Environmental Benefits and Costs

Critical to any environmental benefit-cost analysis is reconciling the timing of benefits and costs. First, benefits and costs do not necessarily accrue at the same time. Furthermore, even if they did, they are not realized immediately. Although an environmental policy evaluation is made in the current period, the effects of that policy typically extend well into the future. Consequently, decision makers must be forward thinking in their evaluation of a policy proposal and must make projections about its future implications.

To support these forecasts, benefit and cost estimates must be adjusted to account for the fact that the value of a dollar is not constant over time. Two types of time-oriented modifications are necessary. One is **present value determination,** which accounts for the opportunity cost of money. The second is **inflation correction,** which adjusts for changes in the general price level. We examine each of these procedures in turn.

Present Value Determination

Opportunity cost is one of the most pervasive concepts in economic thinking. It means that the highest valued alternative of any decision represents the full cost of that decision, whether it arises in production, consumption, or even purely financial transactions. To understand the concept in this latter context, consider the following simple scenario.

Assume a friend asks to borrow \$200 today, promising to pay back the loan one year from today. What amount of money would the payback have to be to maintain your well-being or utility level? (To keep things simple, assume there is no inflation.) Economically, the answer depends upon the opportunity cost of money, that is, its highest valued alternative use. If the \$200 could be invested to yield a 5 percent return, then the opportunity cost of the loan is \$10. Therefore, the appropriate payback should be \$210/[\$200 + (\$200)(0.05)]. Technically, this calculation represents the conversion of the loan value in the *present period* into its value in the *future period.* Mathematically, the conversion for a one-year period is achieved using the simple formula:

$$FV = PV + r(PV) = PV(1 + r),$$

where

$$FV = \text{future value}$$
$$PV = \text{present value}$$
$$r = \text{rate of return.}$$

This equation shows that the future value of a dollar is equal to its present value plus the opportunity cost of not using the dollar in the present period $[r(PV)]$. If the future valuation is to account for more than one time period (t), the above formula becomes

$$FV = PV(1 + r)^t,$$

where $t = 0, 1, 2, \ldots, T$ is the number of periods.

Inverting the problem, we can argue that \$210 received one year from now is equivalent to having \$200 today, or [\$210/(1 + 0.05)]. This conversion procedure is called **present value determination** because it involves **discounting** a future value (FV) into its present value (PV). Mathematically, the calculation is performed using a rearrangement of the first formula:

$$PV = FV[1/(1 + r)].$$

In this form of the equation, r is called the **discount rate,** and the term $1/(1 + r)$ is called the **discount factor.** If the discounting involves more than one time period (t), the formula is written as

$$PV = FV[1/(1 + r)^t],$$

present value determination
A procedure that discounts a future value (FV) into its present value (PV) by accounting for the opportunity cost of money.

discount factor
The term, $1/(1 + r)^t$, where r is the discount rate, and t is the number of periods.

where

$$[1/(1 + r)^t] \text{ is the discount factor.}$$

Discounting is the procedure economists use to adjust the value of environmental benefits and costs accruing in the future.

Notice that the discount rate (r) is the one variable element in present value determination. As this rate is elevated, the PV is decreased, and of course the converse is true. Because the magnitude of the discount rate affects the conversion, its selection is critically important to benefit-cost analysis. In the context of environmental policy development, this rate can directly affect which proposals meet a given criterion and which are rejected out of hand. Because of the implications, discount rate selection is one of the more commonly debated points in the literature on present value analysis.

How *is* the discount rate selected for public policy decision making? There are many considerations, but one important position is that this rate should reflect the **social opportunity cost** of funds allocated to the provision of a public good. Why? Because monies used to support public policy initiatives are perceived as a transfer from the private sector. Therefore, it is argued that the discount rate used for public policy—called the **social discount rate**—should reflect the rate of return that *could* be realized through private spending on consumption and investment, assuming the same level of risk.[1]

social discount rate
Discount rate used for public policy initiatives based on the social opportunity cost of funds.

Since 1992, the U.S. government has maintained a policy that its agencies use a social discount rate of 7 percent in their benefit-cost analyses.[2] The operative issue is whether this 7 percent rate reflects the social opportunity cost of public expenditures. Because this rate is stated in **real** terms (i.e., net of inflationary effects), the **nominal** rate is roughly equivalent to the sum of the inflation rate and the 7 percent rate. Inflation over the last several years has been about 3 percent, which means that the nominal social discount rate is about 10 percent. If this rate is meant to capture the social opportunity cost of funds, it must be the case that society could earn a 10 percent return through private investment. Beyond this issue, there are also conceptual arguments about what *should* determine this rate, a question about which there is no clear consensus.[3]

Inflation Correction

Preliminary cost and benefit estimates also must be modified to account for movements in the general price level over time. This adjustment is referred to as **inflation correction.** To adjust a dollar amount in the present period for expected inflation in the next future period, the value must be converted to its **nominal value** for that period. The conversion uses a measure of price, such as the consumer price index (CPI), as shown by

inflation correction
Adjusts for movements in the general price level over time.

nominal value
A magnitude stated in terms of the current period.

$$\text{Nominal value}_{\text{period } x+1} = \text{Real value}_{\text{period } x}(\text{CPI}_{\text{period } x+1}/\text{CPI}_{\text{period } x}), \text{ or}$$
$$\text{Nominal value}_{\text{period } x+1} = \text{Real value}_{\text{period } x}(1 + p),$$

where p is the rate of inflation between period x and period $x+1$. The more generalized formula for any number of (t) periods is

$$\text{Nominal value}_{\text{period } x+t} = \text{Real value}_{\text{period } x}(1 + p)^t.$$

Inverting the formula allows for the conversion of a nominal value to its **real value:**

real value
A magnitude adjusted for the effects of inflation.

$$\text{Real value}_{\text{period } x} = \text{Nominal value}_{\text{period } x+t}/(1+p)^t.$$

---

[1] For an excellent discussion of the social discount rate and the opportunity cost of public funding, see Sassone and Schaffer (1978), chap. 6.

[2] U.S. Office of Management and Budget (October 29, 1992).

[3] For further reading on this important public policy matter, see a series of articles in the March 1990 issue of the *Journal of Environmental Economics and Management*, and Quirk and Terasawa (January 1991).

This conversion procedure, sometimes called **deflating,** is used to assess changes over a period during which there has been inflation.[4] To illustrate, look back at the cost data presented in Table 8.2. Notice that all the annual dollar values are expressed in 1992 dollars. This is done to avoid confounding changes in control costs over time with changes in the general price level.

deflating
Converts a nominal value into its real value.

By way of example, the *Survey of Current Business* reports that the *nominal* value of abatement and control costs rose from \$104.8 billion in 1992 to \$110.1 billion in 1993.[5] Although these values imply a \$5.3 billion increase in expenditures, part of the increase is due to a 2.6 percent inflation rate over the period. To correct for the price effect, the \$110.1 billion measured in 1993 dollars must be deflated to 1992 dollars. Using the appropriate formula, we obtain

$$\$110.1 \text{ billion}/[1 + (0.026)^1] = \$107.3 \text{ billion}.$$

Thus, the *real* increase in abatement and control costs over the 1992–1993 period is (\$107.3–\$104.8) billion, or \$2.5 billion measured in 1992 dollars.

Summary of Deriving Time-Adjusted Benefits and Costs[6]

The time-adjusted magnitudes for incremental benefits and costs are referred to as the **present value of benefits (*PVB*)** and the **present value of costs (*PVC*),** respectively. The steps to measuring the **present value of benefits (*PVB*)** in *real* terms are:

present value of benefits (*PVB*)
The time-adjusted magnitude of incremental benefits associated with an environmental policy change.

1. Monetize all present and future incremental benefits in nominal terms (B_t), where t refers to the appropriate time period (t=0, 1, 2, . . ., T).
2. Deflate each of the B_t, converting to real dollars (b_t):

$$b_t = B_t/(1 + p)^t.$$

3. Select the appropriate real social discount rate (r_s).
4. Discount the b_t for each period as $b_t/(1 + r_s)^t$.
5. Sum the discounted b_t values over all t periods to find the **present value of benefits** in real dollars as $PVB = \sum[b_t/(1 + r_s)^t]$.

Analogous steps yield a measure for the **present value of costs (*PVC*)** in *real* terms, as follows:

present value of costs (*PVC*)
The time-adjusted magnitude of incremental costs associated with an environmental policy change.

1. Monetize all present and future incremental costs in nominal terms (C_t), where t refers to the appropriate time period (t = 0, 1, 2, . . ., T).
2. Deflate each of the C_t, converting to real dollars (c_t):

$$c_t = C_t/(1 + p)^t.$$

3. Maintain the same real social discount rate (r_s) used in discounting future benefits.
4. Discount the c_t for each period as $c_t/(1 + r_s)^t$.
5. Sum the discounted c_t values over all t periods to find the **present value of costs** in real dollars as $PVC = \sum[c_t/(1 + r_s)^t]$.

Example: Time-Adjusted Incremental Benefits

To illustrate how these steps might be carried out in practice, let's consider a hypothetical benefit assessment. Suppose an analyst is using the averting expenditure method (AEM) to monetize the incremental benefits of improved water quality. As part of the assessment, the

[4] Notice that the deflating formula is identical in form to the present value formula, except that the inflation rate (p) is substituted for the discount rate (r).
[5] Vogan (September 1996).
[6] The following presentation derives the time-adjusted values in *real* terms. The procedure is analogous for finding *nominal* values, as long as all magnitudes, including the discount rate, are properly adjusted.

Table 9.1 Assessing the Value of Future Operating Costs of a Water-Filtering System

	Adjustments for Inflation (assuming an annual inflation rate of 5%)		Adjustments for Opportunity Cost (assuming a real annual discount rate (r) of 10%)		
	[Real value$_{period\ x}$ = Nominal value$_{period\ x+t}$/ $(1 + p)^t$]		[$PV = FV[1/(1 + r)^t]$]		
Year	**Expenditure in Nominal Terms**	**Future Expenditure in Real Dollars**	**Future Value (FV)**	**Discount Factor [$1/(1 + r)^t$]**	**Present Value (PV)**
0	–	–	$200.00	$1.0000 = 1/(1.10)^0$	$200.00
1	$30.00	$\$28.57 = \$30/(1.05)^1$	$ 28.57	$0.9090 = 1/(1.10)^1$	$ 25.97
2	$30.00	$\$27.21 = \$30/(1.05)^2$	$ 27.21	$0.8264 = 1/(1.10)^2$	$ 22.49
3	$30.00	$\$25.92 = \$30/(1.05)^3$	$ 25.92	$0.7513 = 1/(1.10)^3$	$ 19.47
4	$30.00	$\$24.68 = \$30/(1.05)^4$	$ 24.68	$0.6830 = 1/(1.10)^4$	$ 16.86
5	$30.00	$\$23.51 = \$30/(1.05)^5$	$ 23.51	$0.6209 = 1/(1.10)^5$	$ 14.60
Total Present Value in Real Terms					$299.39

analyst must estimate an individual's averting expenditures on a water-filtering system.[7] Assume the system has an initial capital cost (i.e., purchase and installation) of $200 and a useful economic life of 5 years. Further assume that annual operating costs for general maintenance and filters are quoted as $30 in nominal terms, due at the end of each year.[8] Part of the objective of the AEM is to determine the overall cost to the individual of purchasing, installing, and maintaining this filtration system, accounting for both the opportunity cost of money and expected inflation.

Assume for simplicity that prices are expected to rise on average, by 5 percent each year. Using the conversion formula defined earlier, the real value of the operating costs at the end of the first year will be $\$30.00/(1.05)^1 = \$28.57$. At the end of the second year, the real costs will be $\$30.00/(1.05)^2 = \$27.21$, and so forth. Table 9.1 presents these calculations for each year of the system's economic life.

A further adjustment is necessary to find the present value of these future expenditures based on the opportunity cost of money. Assuming that the proper discount rate (r) is 10 percent in real terms,[9] the present value (PV) of each year's inflation-adjusted expenditures is found using the discounting formula. For example, the PV of the first year's operating costs is $\$25.97 = \$28.57[1/(1.10)^1]$. For the second year, the PV of the $27.21 real expenditure is $\$22.49 = \$27.21[1/(1.10)^2]$, and so forth. These calculations are given along with the price adjustments in Table 9.1.

To complete the analysis, all the time-adjusted annual operating costs are added to the initial capital outlay of $200. Notice that the $200 is unaffected by the conversion procedures, since it is incurred in the present period. Overall, the PV of the filtering system expressed in real dollars is $299.39.

[7] For an actual study of this type of analysis, see Abdalla (June 1990).

[8] The timing of the annual expenditures is a critical element of the discounting process. For illustrative purposes, assume that all annual costs are incurred at the *end* of each year. For more detail on this or any other discounting issue, see any public finance text, such as Rosen (2002).

[9] Because the annual costs have been converted to their real values, the discount rate must also be expressed in real terms to adjust for the effect of inflation on the rate of return. Alternatively, one can express all the cost values in nominal terms and use the nominal discount rate.

The Final Analysis: Comparing Environmental Benefits and Costs

The final phase of any benefit-cost analysis involves comparing the time-adjusted incremental benefits and costs and arriving at a decision based on their relative values. In the context of environmental policy, this phase would be used either to set a policy objective or to select a control instrument. Such an analysis would be carried out over a series of possible options and would be used to identify the "best" solution among them.[10]

The first step of this comparative evaluation is to determine whether or not an option is *feasible* from a benefit-cost perspective. This step is essentially a zero-one decision made for every option under consideration: to accept or to reject. All the acceptable options are then evaluated in a second step, where they are assessed relative to one another on the basis of a decision rule. At this point, a single "best" solution can be identified.

Step One: Determining Feasibility

To distinguish feasible options from infeasible ones, the analyst must compare the time-adjusted values of incremental benefits and incremental costs for each option under study. A common way to do this is to form a **benefit-cost ratio** of (*PVB/PVC*) and compare the result to 1. If the ratio for a policy option exceeds 1, it is counted among the feasible solutions; if not, it is rejected:

benefit-cost ratio
The ratio of *PVB* to *PVC* used to determine the feasibility of a policy option if its magnitude exceeds unity.

If $(PVB / PVC) > 1$ for a given option, the option is considered feasible.

An equivalent test for feasibility is to find the **present value of net benefits (*PVNB*)**, which is (*PVB*−*PVC*), and compare the result to zero. If this differential is greater than 0, the policy option is feasible; if not, it is rejected:

present value of net benefits (*PVNB*)
The differential of (*PVB*−*PVC*) used to determine the feasibility of a policy option if its magnitude exceeds zero.

If $(PVB - PVC) > 0$ for a given option, the option is considered feasible.

What both rules communicate is that feasibility is implied if the benefits associated with a policy proposal *outweigh* the costs incurred.

Although any policy option satisfying the $(PVB / PVC) > 1$ condition necessarily meets the requirement that $(PVB - PVC) > 0$, the equivalency ends there. First, the magnitudes of the two expressions have different interpretations. The numerical value of the *PVB/PVC* ratio conveys the benefits of a policy option *per dollar of costs incurred.* For example, a ratio of 4.2 means that for every dollar of costs imposed on society, there are $4.20 in realized benefits. The value of (*PVB* − *PVC*) measures the dollar value of *excess benefits,* so it directly communicates the net gain to society. Second, although it might be tempting to think that either measure could be used to rank feasible projects and arrive at the same "best" solution, such a hypothesis is incorrect. In fact, of the two measures, only the benefit-cost differential can be used to establish such a ranking. Why? Because the benefit-cost ratio is not reliable for comparing feasible policy proposals.

Attempts to use the benefit-cost ratio for any kind of ranking among options will lead to ambiguous results. The problem arises because of the inherent uncertainty about whether to consider an event as an increase in costs or a reduction in benefits. Suppose, for example, that a policy option is expected to create some amount of unemployment. Should the value of that unemployment be counted as a negative benefit or a positive cost? Intuitively, it should be apparent that either approach is correct and should not affect the outcome. However, although the choice would not affect the value of the *PVNB,* it would change the benefit-cost ratio. A simple example will illustrate.

Assume that this hypothetical policy option is expected to cause unemployment valued at $2 million after discounting and adjusting for inflation. Excluding this effect, suppose also that the *PVB* is $18 million and that the *PVC* is $10 million. If the $2 million is counted as a

[10] Benefit-cost analysis is also used to evaluate an *existing* policy initiative.

negative benefit, the benefit-cost ratio would have a value of 1.6. If instead the \$2 million is counted as a cost, the ratio would be 1.5. Try the same experiment with the *PVNB* to see that in either case the result is consistent at \$6 million. What this example implies is that in the second step of evaluation, only the *PVNB* is useful in guiding the decision to select among feasible projects.

Step Two: Decision Rules to Select Among Feasible Options

In practice, the decision-making process used to evaluate feasible options is guided by one of the following economic criteria: **allocative efficiency** or **cost-effectiveness.** Adding the dimension of time to these criteria, we arrive at the following decision rules:

For **allocative efficiency:** Maximize the present value of net benefits (*PVNB*).
For **cost-effectiveness:** Minimize the present value of costs (*PVC*) based on a preestablished benefit objective.

In the United States, the determination about which decision rule to use generally is dictated by the legislation or by presidential executive order—an issue we will explore later in the chapter. For now, notice that both rules involve some type of optimization to guide the selection of the "best" available option.

Decision Rule: Maximize the Present Value of Net Benefits (*PVNB*)

maximize the present value of net benefits (*PVNB*) A decision rule to achieve allocative efficiency by selecting the policy option that yields greatest excess benefits after adjusting for time effects.

As noted, the **present value of net benefits (*PVNB*)** is a benefit-cost differential found as the difference between the **present value of benefits (*PVB*)** and the **present value of costs (*PVC*).** Thus, this decision rule calls for choosing the option that yields to society the highest amount of excess benefits after adjusting for time effects. According to basic microeconomic theory, the point at which benefits exceed costs by the greatest amount corresponds to where $MB=MC$, or where resources are efficiently allocated. Hence, we can summarize this decision rule in the following way:

To achieve allocative efficiency, maximize $PVNB = (PVB - PVC) = \sum [b_t/(1 + r_s)^t] - \sum [c_t/(1 + r_s)^t] = \sum [(b_t - c_t)/(1 + r_s)^t]$ for all t periods $(t = 0, 1, 2, \ldots, T)$ among all feasible alternatives.[11]

Decision Rule: Minimize the Present Value of Costs (*PVC*)

minimize the present value of costs (*PVC*) A decision rule to achieve cost-effectiveness by selecting the least-cost policy option that achieves a preestablished objective.

The decision rule to **minimize the present value of costs (*PVC*)** guides the policymaker to select the least-cost option among those capable of achieving some preestablished objective. This is an explicit directive to set policy based on the criterion of cost-effectiveness. Of the two economically based decision rules, this one is the more common. The reason is that the law often predetermines the level of environmental benefits to be achieved through its definition of an environmental quality objective. When this is the case, the *PVB* is essentially fixed, leaving only *PVC* as a decision variable.

Before the cost comparison is done, the analyst has to first eliminate any options that do not achieve the requisite benefit level. Once done, the *PVC* is calculated for the remaining possibilities, and the option that incurs the lowest costs is selected. This decision rule can be summarized as follows:

To achieve cost-effectiveness, minimize $PVC = \sum [c_t/(1 + r_s)^t]$
for all t periods $(t = 0, 1, 2, \ldots, T)$
among all feasible alternatives that achieve a predetermined benefit level.[12]

---

[11] An important caveat is that the efficiency criterion will be ensured only if *all* feasible options are considered. In practice, only a subset of possibilities is ever evaluated. If the efficient solution does not happen to be in this subset, the decision rule will identify the most efficient option of those considered, but the efficiency criterion will not be satisfied in a strict sense.

[12] Just as with the efficiency-based decision rule, the cost-effectiveness solution will be obtained only if it is among the feasible options being considered. Because it is not practical for decision makers to consider more than a small number of possibilities, it is likely that the selected option will not be the true cost-minimizing solution, although it will be the least-cost alternative among those evaluated.

Reservations About the Use of Benefit-Cost Analysis

Although benefit-cost analysis is a viable risk management strategy, it is not without flaws. In fact, as a tool of environmental decision making, it has been the object of a fair amount of scrutiny. Critics of benefit-cost analysis generally point to two sources of concern. The first is the inherent problem of accurately measuring and monetizing environmental costs and benefits. The second refers to potential equity problems not addressed by this approach.

Measurement Problems

Attempting to measure and assign a dollar value to environmental benefits and costs is unquestionably a major challenge of using benefit-cost analysis. As we indicated in Chapter 7, estimation is particularly problematic on the benefit side, where many intangibles are involved. Not only is it difficult to identify all the health and ecological benefits of a policy proposal, but also assigning a dollar value to these gains is difficult at best. In fact, this measurement problem is often used to support the use of cost-effectiveness as an alternative criterion to allocative efficiency. On the opposite side of the analysis, capturing implicit costs is the primary source of difficulty. Solving this problem continues to be the objective of ongoing research. Finally, the selection of the social discount rate, which affects the present value of both benefit and cost estimates, is the subject of much debate, even among proponents of the benefit-cost approach.

Equity Issues

In an objective assessment of benefit-cost analysis, equity concerns are not unfounded. The decision rules in benefit-cost analysis do not consider how the benefits and costs are distributed across various segments of society. Yet it is possible that the distribution of incremental benefits might be highly skewed, so that some group of consumers or some industrial sector receives less than its fair share. Similarly, the distribution of costs might be such that some economic sector bears an inequitable share of the burden. Of course, either problem is less troublesome if the same inequity occurs on both the benefit side and the cost side, such as lower benefits matched by lower costs. However, there is no assurance that this will be the case.

Even proponents of benefit-cost analysis do not dismiss the measurement problems or potential inequities associated with benefit-cost analysis. Therefore, it is important that policymakers understand the implications and make adjustments for any noted shortcomings. In the interim, economists are continuing to research better estimation techniques, and many researchers and government officials are working to analyze and correct inequities associated with environmental policy decisions. Finding ways to improve the results is worth the effort. As a risk management strategy, benefit-cost analysis adds vital information to a very difficult undertaking. Furthermore, in the United States, benefit-cost analysis is required of all major policy proposals—a decision that has a long history behind it.

U.S. Government Support of Benefit-Cost Analysis[13]

Most policy analysts point to the Flood Control Act of 1936 as the first U.S. federal legislation that explicitly acknowledged benefits and costs in a public policy initiative. According to this mandate, federal funds could be allocated to water projects only if the associated benefits were found to exceed the costs. Notice that this is precisely the feasibility rule based on $(PVB - PVC) > 0$ that we discussed earlier. Since this historic first, the use of economic

[13] Part of this discussion is drawn from the following: Andrews (1984), pp. 43–85; and U.S. EPA, Office of Policy Analysis, Office of Policy, Planning, and Evaluation (August 1987), pp. 2-1–2-6, which is also available online at **http://yosemite.epa.gov/ee/epalib/ee222.nsf.**

Application

9.1 *Benefits, Costs, and Risk Analysis in Environmental Rule Making*

During the 103d Congress, over a dozen bills and amendments were introduced that dealt with risk analysis. Among these was Senate Bill 110, proposed as the Environmental Risk Reduction Act of 1993 and introduced by Senator Daniel Patrick Moynihan (D–N.Y.). One of the bill's objectives was to make the major findings of the EPA's 1990 report called *Reducing Risk* part of federal law. This report summarizes an EPA Science Advisory Board study of how the United States could improve its efforts to reduce environmental risk.

According to the proposal, the Environmental Risk Reduction Act would have advanced the integration of risk assessment and benefit-cost analysis in the environmental decision-making process. As Senator Moynihan stated in a recent article: "I am convinced that risk ranking and cost-benefit analysis are valuable tools for making environmental decisions. They are not our only tools, but they do offer the means to set priorities and to measure success." (Moynihan, January–March 1993, p. 46). As part of its proposed requirements, the Senate bill called for quantitative estimates of risk exposure, an understanding of the alternatives available to reduce this exposure, and accurate estimates of the benefits and costs associated with these alternatives.

Senator Moynihan was aggressive about the bill's intent. Referring to U.S. annual spending of 2.2 percent of its gross national product on environmental protection, he argued: "While this may not be too much money to spend on environmental protection, it is too much to spend unwisely" (p. 46). Despite Moynihan's efforts, the proposed legislation was not passed by the 103d Congress.

Similar bills have been proposed by other congressional officials, most notably two amendments proposed by Senator Johnston of Louisiana. The original amendment would have required the EPA to consider relative risk reduction, costs, and benefits for all proposed and final regulations. A revised version was introduced in Congress's second session. Intended to provide statutory strength to the intent of Clinton's Executive Order 12866, the Johnston amendments were added to many pending environmental bills. Interestingly, according to Davies (1995), "The lack of action on environmental legislation during the 103d Congress was due, to a great extent, to an inability to reach an acceptable compromise on the amendment's language" (p. 5). Ultimately, the amendments were attached to a Department of Agriculture bill that was enacted, but the amendments were applicable only to regulations initiated by that department.

A more recent Congressional effort dealing with environmental risk is a bill titled the "Regulatory Improvement Act of 1999." Introduced by Senator Levin of Michigan and Senator Thompson of Tennessee to the 106th Congress, this bill closely follows Clinton's executive order. One of its major provisions deals explicitly with benefit-cost analysis, requiring this decision rule for all regulations imposing costs over $100 million or otherwise significantly affecting the economy. At the initial and final rule-making stages, a benefit-cost analysis would include an evaluation of the anticipated benefits and costs, an analysis of regulatory alternatives, an assessment of whether the benefits justify the costs, and a determination of whether the rule-making objectives can be achieved in a more cost-effective manner or with greater net benefits than other alternatives.

Sources: Lee (January 3, 1995); Moynihan (January–March, 1993); Heath (July 1993); U.S. EPA, Science Advisory Board (September 1990); Davies (winter 1995); U.S. Congress (106th), "Regulatory Improvement Act of 1999," Senate Bill S746.

theory as a decision-making tool has been promoted in many public policy settings, a practice that continues into present-day rule making. See Application 9.1 for recent examples of congressional efforts to legislate the use of costs, benefits, and risk analysis in environmental rule making.

Over the last 30 years, U.S. presidents have used the power of the office to require economic analysis in federal regulatory decision making. Table 9.2 lists some of the presidential exec-

Table 9.2 ***Recent History of U.S. Regulation Using Economic Analysis***

Act/Executive Order	Year	Title of Analysis	Type of Analysis
OMB, October 1971 Memo	1971	Quality of Life (QOL) review	Costs, benefits
Executive Order 11821	1974	Inflation Impact Statement (IIS)	Cost, benefits, inflationary impacts
Executive Order 11949	1976	Economic Impact Statement (EIS)	Costs, benefits, economic impacts
Executive Order 12044	1978	Regulatory Analysis	Costs, economic consequences
Regulatory Flexibility Act	1980	Regulatory Flexibility Analysis	Impacts on small businesses
Executive Order 12291	1981	Regulatory Impact Analysis (RIA)	Costs, benefits, net benefits
Executive Order 12866	1993	Economic Analysis (EA)	Costs, benefits, net benefits

Source: U.S. EPA, Office of Policy Analysis, Office of Policy, Planning, and Evaluation (August 1987), p. 2–2, table 2–1.

utive orders and other directives that specifically call for the consideration of benefits and/or costs in policy decisions. To understand the economic thinking in policy decisions, we consider the instruments called for in the most recent of these directives, namely, the **Regulatory Impact Analysis (RIA)** and the **Economic Analysis (EA).**

Executive Order 12291: Regulatory Impact Analysis (RIA)[14]

Although the framework for considering the benefits and costs of regulatory proposals had existed under previous administrations, it was not until President Reagan's first term in office that the efficiency criterion was explicitly made a part of the regulatory review process. In 1981, Reagan signed Executive Order 12291, which outlined requirements that federal agencies had to follow in proposing or reviewing any major rule or regulation. Under Section 1 of this Executive Order, a "major rule" referred to any regulation expected to have an annual effect of at least $100 million or otherwise adversely affect the economy.

Section 2 outlined the requirements that all agencies had to follow in developing new regulatory actions or assessing existing ones, as shown in Table 9.3. Unlike prior references to economic considerations, this set of requirements explicitly called for meeting economic criteria in adopting regulatory actions. Notice that **allocative efficiency** is referenced in paragraphs (c) and (e), which called for maximizing net benefits. The criterion of **cost-effectiveness** is referenced in paragraph (d), which required that the selection from among alternatives fulfilling a given objective be made on the basis of least net cost to society.

Taken out of context, paragraphs (c) and (d) suggest that these economic criteria were to have been met by every new policy proposal. However, there is a caveat in the first sentence of Section 2, given in the phrase "to the extent permitted by law." Because of this wording, many major environmental laws precluded the EPA from using either allocative efficiency or cost-effectiveness in establishing and implementing policy. Some rulings, such as air quality standards established by the Clean Air Act, do not consider economic costs at all. Hence, the level of environmental quality under these provisions is set to achieve some objective other than maximizing net benefits or cost-effectiveness.[15]

[14] For more detail concerning this executive order, see *U.S. Federal Register* 46 (February 17, 1981), pp. 13193–98.

[15] To access information on how various U.S. regulations require the consideration of certain benefit and cost categories, visit **http://yosemite.epa.gov/ee/epalib/riaepa.nsf.** At this Web site, select the "table of contents view," and then select "Statutory Authorities for Economic Analysis."

Table 9.3 ***Key Sections of President Reagan's Executive Order 12291***

Section 2 [emphasis added]

In promulgating new regulations, reviewing existing regulations, and developing legislative proposals concerning regulations, all agencies, to the extent permitted by law, shall adhere to the following requirements:

(a) Administrative decisions shall be based on adequate information concerning the need for and consequences of proposed government action;

(b) Regulatory action shall not be undertaken unless the *potential benefits to society for the regulation outweigh the potential costs to society;*

(c) Regulatory objectives shall be chosen to *maximize the net benefits to society;*

(d) Among alternative approaches to any regulatory objective, the alternative involving the *least net cost to society shall be chosen;* and

(e) Agencies shall set regulatory priorities with the aim of *maximizing the aggregate net benefits to society,* taking into account the condition of the particular industries affected by regulations, the conditions of the national economy, and other regulatory actions contemplated for the future.

Section 3(d) [emphasis added]

To permit each proposed major rule to be analyzed in light of the requirements stated in Section 2 of this Order, each preliminary and final *Regulatory Impact Analysis* shall contain the following information:

(1) A description of the *potential benefits* of the rule, including any beneficial effects that cannot be quantified in monetary terms, and the identification of those likely to receive the benefits;

(2) A description of the *potential costs* of the rule, including any adverse effects that cannot be quantified in monetary terms, and the identification of those likely to bear the costs;

(3) A determination of the *potential net benefits* of the rule, including an evaluation of effects that cannot be quantified in monetary terms;

(4) A description of alternative approaches that could substantially achieve the same regulatory goal at lower cost, together with an analysis of the *potential benefits and costs* and a brief explanation of the legal reasons why such alternatives, if proposed, could not be adopted; and

(5) Unless covered by the description required under paragraph (4) of this subsection, an explanation of any legal reasons why the rule cannot be based on the requirements set forth in Section 2 of this Order.

Source: *U.S. Federal Register* 46 (February 17, 1981), pp. 13193–98.

Regulatory Impact Analysis (RIA)
A requirement under Executive Order 12291 that called for information about the potential benefits and costs of a major federal regulation.

Although the law was to prevail over the requirements of the executive order, all major regulations still had to be accompanied by what is called a **Regulatory Impact Analysis (RIA),** described in Section 3(d) of the executive order. As shown in Table 9.3, the guidelines of Section 3(d) left little doubt about the importance of identifying and monetizing the potential costs and benefits of any policy proposal or regulatory review process through an RIA.

Executive Order 12866: Economic Analysis (EA)

In September 1993, President Clinton signed Executive Order 12866, which replaced Reagan's Executive Order 12291.[16] Nonetheless, it continued Reagan's commitment to economic fundamentals in policy formulation and evaluation. Essentially, the Clinton directive requires that all costs and benefits of available regulatory alternatives be considered in deciding whether to regulate and how to regulate. Among the executive order's "Principles of Regulations" that pertain directly to economic criteria are the following [emphasis added]:

> When an agency determines that a regulation is the best available method of achieving the regulatory objective, it shall design its regulations in the most *cost-effective* manner to achieve the regulatory objective.
>
> Each agency shall assess both the costs and the benefits of the intended regulation and . . . propose or adopt regulation only upon a reasoned determination that the *benefits of the intended regulation justify its costs.*

Similar to Reagan's executive order, Clinton's order applies to all "significant regulatory actions," which includes those expected to have an annual impact on the economy of at least $100 million or some adverse effect. For each such action, the agency involved is required to prepare an **Economic Analysis (EA),** which acts as the successor to the RIA.

Economic Analysis (EA)
A requirement under Executive Order 12866 that called for information on the benefits and costs of a "significant regulatory action."

The Office of Management and Budget formed an interagency group to determine the "best practices" for preparing an EA under the new executive order. After two years of effort, the group released its guidance on implementation of the president's executive order in 1996. The report states that any EA should include information that allows decision makers to determine the following [emphasis added]:

> There is adequate information indicating the need for and consequences of the proposed action;
>
> The *potential benefits to society justify the potential costs,* . . . , unless a statute requires another regulatory approach;
>
> The proposed action will *maximize net benefits to society* . . . , unless a statute requires another regulatory approach;
>
> Where a statute requires a specific regulatory approach, the proposed action will be the *most cost-effective,* including reliance on performance objectives to the extent feasible;
>
> Agency decisions are based on the best reasonably obtainable scientific, technical, economic, and other information.[17]

Notice how the EA follows the same logic and economic criteria outlined in an RIA. It too explicitly calls for a maximization of net benefits, or, where otherwise required under the law, that a proposed action be the most cost-effective.

These fundamental guidelines that support the use of economic criteria continue to be followed by President George W. Bush. In fact, the Bush administration specifically announced that it would adhere to Executive Order 12866 until the president issues a new or modified version.[18]

---

[16] To access the complete text of Clinton's executive order on the Internet, visit **http://www.epa.gov/fedrgstr/eo/eo12866.htm**

[17] The text of this report can be accessed at **http://www.whitehouse.gov/omb/inforeg/riaguide.html.**

[18] Introduction to "Economic Analysis of Federal Regulations Under Executive Order 12866" (January 11, 1996). Visit **http://www.whitehouse.gov/omb/inforeg/oira_review-process.html** to read a memo from the President's Management Council about exactly this issue, with an attachment citing the principles and procedures to be used in implementing Executive Order 12866.

Implementing either an EA or an RIA to satisfy the requirements to properly assess environmental costs and benefits is a major undertaking. To better illustrate exactly how economic criteria are used in a regulatory context, it is helpful to study a real-world example. So we next investigate the EPA's use of the net benefits decision rule in its RIA for the phasedown of lead in gasoline.

Reducing Lead in Gasoline: A Regulatory Impact Analysis[19]

Petroleum refiners began using lead additives as an inexpensive source of octane in the 1920s. By adding lead to gasoline, refiners were able to reduce engine knock and increase engine performance. However, these gains were not without consequence. Coincident with the increased use of lead additives, researchers observed a link between lead exposure and adverse health effects, such as mental and cardiovascular disorders. Acting on the growing body of scientific evidence, the EPA was charged with the responsibility of achieving a reduction in the lead content of gasoline.

In 1985, the existing U.S. lead standard was 1.1 grams per leaded gallon (gplg). As part of its plan to reduce society's exposure to lead, the EPA proposed tightening this standard to 0.1 gplg, effective January 1, 1986. Since such a policy change came under the legal definition of a major rule, the EPA was required to conduct an RIA. This in turn meant that the agency had to examine the costs, benefits, and net benefits of approaches that could be used to meet the more stringent lead standard.

Estimating the Incremental Benefits of the Lead Standard Proposal

Health Benefits

An important element of the benefit assessment for the RIA was an evaluation of the predicted health improvements associated with the lead standard proposal. Based on the findings of scientific studies, the EPA conducted tests to determine the relationship between human blood lead levels and leaded gasoline usage. Figure 9.1 shows the correlation between these two variables.

As part of the risk assessment process, the EPA also analyzed scientific findings to find out whether any population groups were more susceptible to the health risks of lead exposure. Even then, there was already an abundance of scientific evidence about the effects of lead exposure on children. According to the final RIA (cited in note 19), "these effects range from relatively subtle changes in biochemical measurements at low doses to severe retardation and even death at very high levels (p. E-6)." Based on the evidence, it was clear that an important incremental benefit of tightening the lead standard would be improvements in the health of children.

To monetize children's incremental health benefits, the EPA used two measures: (1) savings in medical care expenditures and (2) savings in compensatory educational expenditures. These attempted to capture the benefits of improving physiological health and cognitive development. Using these predefined measures, the EPA estimated that the value of incremental health benefits for children would be approximately $600 million in 1986, the first full year of the revised lead standard. (This value and all others given in the RIA report are expressed in 1983 dollars.)

Another element of the EPA's study of lead-related health effects was an examination of the evidence linking lead exposure to blood pressure levels in adult males. Much of the EPA's analysis was based on the findings of two recent scientific studies. Because of this, the EPA

[19]The following discussion is drawn from U.S. EPA, Office of Policy Analysis (February 1985). The executive summary of this document is available online by visiting the EPA's National Center for Environmental Economics at **http://yosemite.epa.gov/ee/epalib/ee222.nsf/vwl** and clicking on "Lead in gasoline."

Figure 9.1 ***Correlation Between Lead in Gasoline and Average Blood Lead Levels***

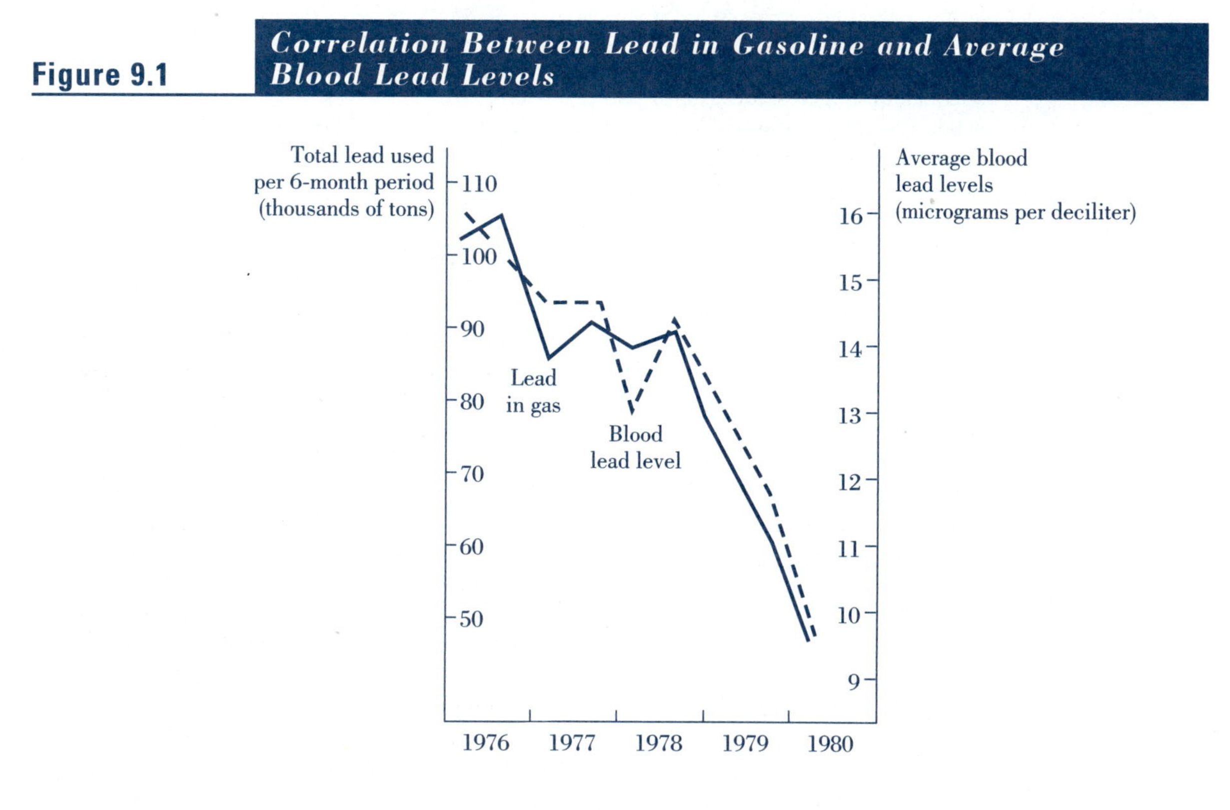

Source: Whiteman (May/June 1992), p. 38.

stated that their findings on this health issue were to be viewed as preliminary and *not* to be considered for the subject rule making. Hence, results were presented in the RIA for informational use only.

Using statistical models, the EPA estimated the relationship between blood lead levels in adult males and blood pressure levels. Based on the results, models were constructed to estimate the associated incremental benefits of reductions in cardiovascular-related death and various disorders. To monetize the benefits of reducing *nonfatal* health effects, the EPA used the dollar value of medical care costs, expenditures on medications, and lost wages. For *fatal* cases, the EPA used existing estimates of a statistical life and selected a magnitude of $1 million per life saved. Taken together, the monetized values of the incremental benefits per case were reported in the RIA as follows: $220 per case of hypertension, $60,000 per heart attack, $44,000 per stroke, and $1 million per life saved. In the aggregate, the incremental benefits of the lead ruling associated with adult male health were reported over a six-year period and ranged from $5.897 billion in 1986 to $4.692 billion in 1992.

Nonhealth Benefits

Three types of nonhealth benefits were analyzed in the RIA:

- reduction in harmful emissions caused by misfueling (i.e., the use of leaded gasoline in vehicles requiring unleaded gasoline)
- lower maintenance costs
- increased fuel economy

Misfueling causes harmful emissions because leaded gasoline damages a car's catalytic converter, which is designed to abate such pollutants as hydrocarbons (HC), carbon monoxide (CO), and nitrogen oxides (NO_X). Thus, the lower lead content was expected to slow the rate at which catalytic converters are damaged, which in turn would reduce these emissions. Furthermore, the new standard was expected to reduce misfueling because it would increase the production costs of leaded gasoline, thus reducing the price differential between leaded and unleaded fuel.

To estimate the incremental benefits of reducing misfueling damage, the EPA assumed that the new lead standard would eliminate about 80 percent of the misfueling. Accounting for all the associated damage reductions, the EPA estimated that the resulting decrease in emissions of HC, CO, and NO_X would yield incremental benefits of approximately $222 million in 1986. In addition, the values for reduced maintenance expenditures and for increased fuel economy for 1986 were estimated to be $914 million and $187 million, respectively.

Estimating the Incremental Costs of the Lead Standard Proposal

On the cost side of the analysis, the EPA had to define, quantify, and monetize the social costs of the proposed lead ruling. The RIA defines real social costs as "the costs of real resources that are used to comply with the rule (i.e., the extra energy, capital, labor, etc. that are needed to meet the tighter lead standard)" (p. II-2). To approximate these costs, the EPA used estimates of the change in manufacturing costs of gasoline and other petroleum products arising from the new lead ruling. The agency used an **engineering cost model** of the refinery industry, which was originally developed for the Department of Energy. First, the model was specified under the existing or baseline lead content standard. Then, it was reestimated, using the new standard. The difference between the two yielded the incremental cost estimate.

In general, real manufacturing costs were found to decrease over the seven-year period under study, mainly because of the expected decline in the demand for leaded gasoline. Year-to-year cost estimates suggested a relatively small impact on refiners. For example, tightening the standard from 1.1 gplg to the proposed rule of 0.1 gplg yielded an incremental cost estimate of $608 million in 1986, the first full year of the proposed rule, and $441 million in 1992.

Putting It All Together: Benefit-Cost Analysis

Table 9.4 summarizes the EPA's estimates of the incremental benefits and costs (in real terms) associated with tightening the lead standard from 1.1 gplg to 0.1 gplg. Notice that the net incremental benefits of the proposed rule are positive and substantial. Including the benefits of reducing adverse effects on adult blood pressure, the values range from $7,213 million in 1986 to $5,770 million in 1992. Because the evidence for the adult blood pressure benefits was not well established, the EPA chose to rely on the net benefit estimates *excluding* these benefits to defend the economic feasibility of the proposed rule change. Using this more conservative estimate, net benefits still were found to be positive for each forecasted year. In fact, the estimates are above $1 billion for each of the seven full years affected by the proposed ruling.

Finally, to adjust for time differences over the period, the EPA conducted a present value analysis of their net benefit estimates. Using the conservative estimate that excludes blood pressure effects and selecting a *real* social discount rate of 10 percent, the present value of net benefits (*PVNB*) over the period was estimated to be $5.9 billion ($1983).

The Final Decision

Following the estimation of incremental benefits and costs required in the RIA, the EPA officially announced a low-lead standard of 0.1 gplg, effective January 1, 1986. The decision was supported by the economic evidence that the more stringent lead standard would be beneficial to society.

Table 9.4 ***Regulatory Impact Analysis (RIA) for Reducing Lead in Gasoline: EPA Estimates of Incremental Benefits and Costs***

	Monetized Benefits and Costs of Reducing Lead in Gasoline (millions of 1983 dollars)							
Benefits or Costs	**1985**	**1986**	**1987**	**1988**	**1989**	**1990**	**1991**	**1992**
Monetized benefits								
Children health effects	223	600	547	502	453	414	369	358
Adult blood pressure	1,724	5,897	5,675	5,447	5,187	4,966	4,682	4,692
Conventional pollutants	0	222	222	224	226	230	239	248
Vehicle maintenance	102	914	859	818	788	767	754	749
Fuel economy	35	187	170	113	134	139	172	164
Total monetized benefits	2,084	7,821	7,474	7,105	6,788	6,517	6,216	6,211
Monetized costs								
Total refining costs	96	608	558	532	504	471	444	441
Net benefits	1,988	7,213	6,916	6,573	6,284	6,045	5,772	5,770
Net benefits excluding adult blood pressure	264	1,316	1,241	1,125	1,096	1,079	1,090	1,079

Source: U.S. EPA, Office of Policy Analysis (February 1985), p. E-12.

Notes: The estimates are reported under the assumption of partial misfueling. Sums may not equal totals shown because of rounding.

Conclusions

In its most fundamental form, environmental policy is aimed at minimizing society's risk of exposure to environmental hazards. Although all of society is involved in public policy development, most of the detail of formulating, implementing, and monitoring regulatory provisions falls in the hands of public officials. By itself, this fact supports the need for analytical tools such as benefit-cost analysis to guide the decisions that define environmental policy.

Benefit-cost analysis, although not without flaws, is a useful strategic approach to environmental decision making. Its objectivity is its primary strength in helping officials evaluate the social gains and the opportunity costs of their decisions. Choosing between a policy that reduces mortality risk to save one more life in a million lives exposed and a regulation that will reduce the risk of birth defects by 10 times that amount is a tough decision by any measure. Yet these kinds of decisions are precisely what environmental policy development is all about. Every resource allocated to save a national forest is one less resource available to clean rivers and streams. Every dollar spent to clean a hazardous waste site is one less dollar available to save an endangered species. Such is the dilemma that confronts every society as it makes decisions about how to allocate scarce resources, and such is the fundamental premise of economic thought.

Guiding decision making is the purpose of benefit-cost analysis. Explaining what this analytical tool can accomplish, how it is implemented, and what it fails to achieve have been the primary objectives of this chapter. Because benefit-cost analysis is becoming a more dominant force in public policy decisions, it is critical that we comprehend how it influences environmental regulations and how, in so doing, it affects the quality of our lives.

Summary

- In benefit-cost analysis, two types of time-oriented adjustments are necessary: present value determination and inflation correction.
- To discount a future value (FV) into its present value (PV), use the conversion formula: $PV = FV\,[1/(1 + r)^t]$, where t is the number of periods and $1/(1 + r)^t$ is the discount factor.
- To adjust a value in the present period for expected inflation in the future period, it must be converted to its nominal value for that period, using the formula: Nominal value$_{\text{period } x+t}$ = Real value$_{\text{period } x}$ $(1 + p)^t$, where p is the rate of inflation.
- Time-adjusted incremental benefits and costs are referred to as the present value of benefits, $PVB = \sum b_t/(1 + r_s)^t$, and the present value of costs, $PVC = \sum c_t/(1 + r_s)^t$, respectively.
- The first step of benefit-cost analysis identifies feasible options. A policy option is feasible if $(PVB/PVC) > 1$ for that option or if $(PVB - PVC) > 0$.
- The second step of benefit-cost analysis evaluates all acceptable options *relative* to one another on the basis of a decision rule.
- One decision rule is to maximize the present value of net benefits ($PVNB$), which equals $\sum [b_t/(1 + r_s)^t] - \sum [c_t/(1 + r_s)^t]$, among all feasible alternatives to achieve allocative efficiency. Another decision rule is to minimize the present value of costs (PVC), which equals $\sum c_t/(1 + r_s)^t$, among all feasible alternatives that attain a predetermined benefit level to achieve cost-effectiveness.
- Measuring and monetizing intangibles, the selection of the social discount rate, and capturing implicit costs are among the concerns of using benefit-cost analysis. Another problem is the potential for an inequitable distribution of costs and benefits.
- Over the last 30 years, U.S. presidents have required economic analysis in federal regulatory decision making. Two instruments that call for some type of benefit and/or cost considerations in policy decisions are the Regulatory Impact Analysis (RIA) required by Executive Order 12291 and the Economic Analysis (EA) called for in Executive Order 12866.
- Section 2 of President Reagan's Executive Order 12291 and more recently President Clinton's Executive Order 12866 explicitly call for the achievement of allocative efficiency and cost-effectiveness in adopting regulatory actions.
- In the RIA for evaluating a proposal to tighten the lead standard, the EPA estimated that the present value of the net benefits ($PVNB$) was \$5.9 billion (\$1983). The EPA officially announced a low-lead standard of 0.1 gplg effective January 1, 1986.

Key Concepts

present value determination
discount factor
social discount rate
inflation correction
nominal value
real value
deflating
present value of benefits (*PVB*)
present value of costs (*PVC*)
benefit-cost ratio
present value of net benefits (*PVNB*)
maximize the present value of net benefits (*PVNB*)
minimize the present value of costs (*PVC*)
Regulatory Impact Analysis (RIA)
Economic Analysis (EA)

Use the Key Concepts listed above to begin your search for additional articles and information using the InfoTrac College Edition database.

Review Questions

1. Suppose industry abatement costs rise from $850 million in 1998 to $1,000 million in 1999 in nominal terms and that the CPI is 100 in 1998 and 106 in 1999.
   a. Evaluate the change in costs over the period in real terms, first in 1998 dollars and then in 1999 dollars.
   b. Are your answers the same? Explain why or why not.
2. To examine the implications of selecting various discount rates, reconsider the water filtration system example in the text but change the discount rate from 10 to 5 percent.
   a. Find the present value of the system.
   b. Now compare this present value with the one calculated under the assumption of a 10 percent rate. Explain the difference intuitively.
3. Refer to the RIA for the revised lead ruling discussed in the chapter.
   a. Mathematically confirm that the estimate of the present value of net benefits (*PVNB*) is approximately $5.9 billion, as stated.
   b. Discuss how the *PVNB* would have changed if the EPA had used a social discount rate of 8 percent.
4. In 1997, the EPA proposed an increase in the national air quality standards for particulate matter (PM). Because the change was considered a major rule, the agency was required to conduct an RIA and assess the potential costs and benefits of the proposed standard. The completed RIA can be accessed at **http://www.epa.gov/ttn/oarpg/naaqsfin/ria.html.** As part of this RIA, benefit and cost data for the *existing* standard were given and are replicated here:

Control Region	Annual Control Cost ($1990 millions)
Midwest/Northeast	380
Southeast	2
South Central	230
Rocky Mountain	210
Northwest	140
West	130
National	1,100

Source: Appendix C, table C1: "Costs and Benefits of Achieving the Current PM10 and Ozone Standard," available at the EPA Web site provided in this question.

Benefit Category	Monetized Benefits ($1990 billions)
Mortality	
Short-term exposure	$2.950
Long-term exposure	2.860
Chronic bronchitis	2.010
Hospital admissions	
Total respiratory	0.022
Congestive heart failure	0.003
Ischemic heart disease	0.005
Upper respiratory systems	
Asthma attacks	0.001
Work loss days	0.015
Minor restricted activity days	0.057
Consumer cleaning cost savings	0.039
Visibility	0.320
Total Benefits	
Using short-term mortality	$5.400
Using long-term mortality	$5.300

Source: Appendix C, table C1: "Costs and Benefits of Achieving the Current PM10 and Ozone Standard," available at the EPA Web site provided on p. 197.

a. Examine the control cost estimates, and provide a plausible explanation of why there are incremental cost differences across regions for meeting the PM standard.
b. Economically, can you justify the PM standard? Explain.
c. List some of the limitations associated with benefit-cost analysis. Which of these are applicable to this particular economic analysis? Explain.

Additional Readings

Baumol, William J. "On the Social Rate of Discount." *American Economic Review* 58 (September 1968), pp. 788–802.

Crandall, Robert W. "Is There Progress in Environmental Policy?" *Contemporary Economic Policy* 13 (January 1995), pp. 80–83.

Cropper, Maureen L. "Has Economic Research Answered the Needs of Environmental Policy?" *Journal of Environmental Economics and Management* 39 (May 2000), pp. 328–50.

Cropper, Maureen L., and Wallace E. Oates. "Environmental Economics: A Survey." *Journal of Economic Literature* 30 (June 1992), pp. 675–740.

Cropper, Maureen L., and Paul R. Portney. "Discounting Human Lives." *Resources* 108 (summer 1992), pp. 1–4.

Dorfman, Robert. "An Introduction to Benefit-Cost Analysis." In Robert Dorfman and Nancy Dorfman, eds., *Economics of the Environment: Selected Readings,* pp. 297–322. New York: Norton, 1993.

Farrow, Scott, and Michael Toman. "Using Benefit-Cost Analysis to Improve Environmental Regulations." *Environment* 41 (March 1999).

Halvorsen, Robert, and Michael G. Ruby. *Benefit-Cost Analysis of Air-Pollution Control.* Lexington, MA: Heath, 1981.

Kopp, Raymond J., Alan J. Krupnick, and Michael Toman. "Cost-Benefit Analysis and Regulatory Reform: An Assessment of the Science and the Art." Discussion Paper 97–19. Washington, DC: Resources for the Future, January 1997.

Krupnick, Alan J., and Paul R. Portney. "Controlling Urban Air Pollution: A Benefit-Cost Assessment." *Science* 252 (April 26, 1991), pp. 522–28.

Merrifield, John. "Sensitivity Analysis in Benefit-Cost Analysis: A Key to Increased Use and Acceptance." *Contemporary Economic Policy* 15 (July 1997), pp. 82–92.

Morgenstern, Richard D. *Economic Analyses at EPA: Assessing Regulatory Impact.* Baltimore, MD: World Resources Institute, 1997.

Peskin, Henry M., and Eugene P. Seskin. *Cost Benefit Analysis and Water Pollution Policy.* Washington, DC: Urban Institute, 1973.

Related Web Sites

http://

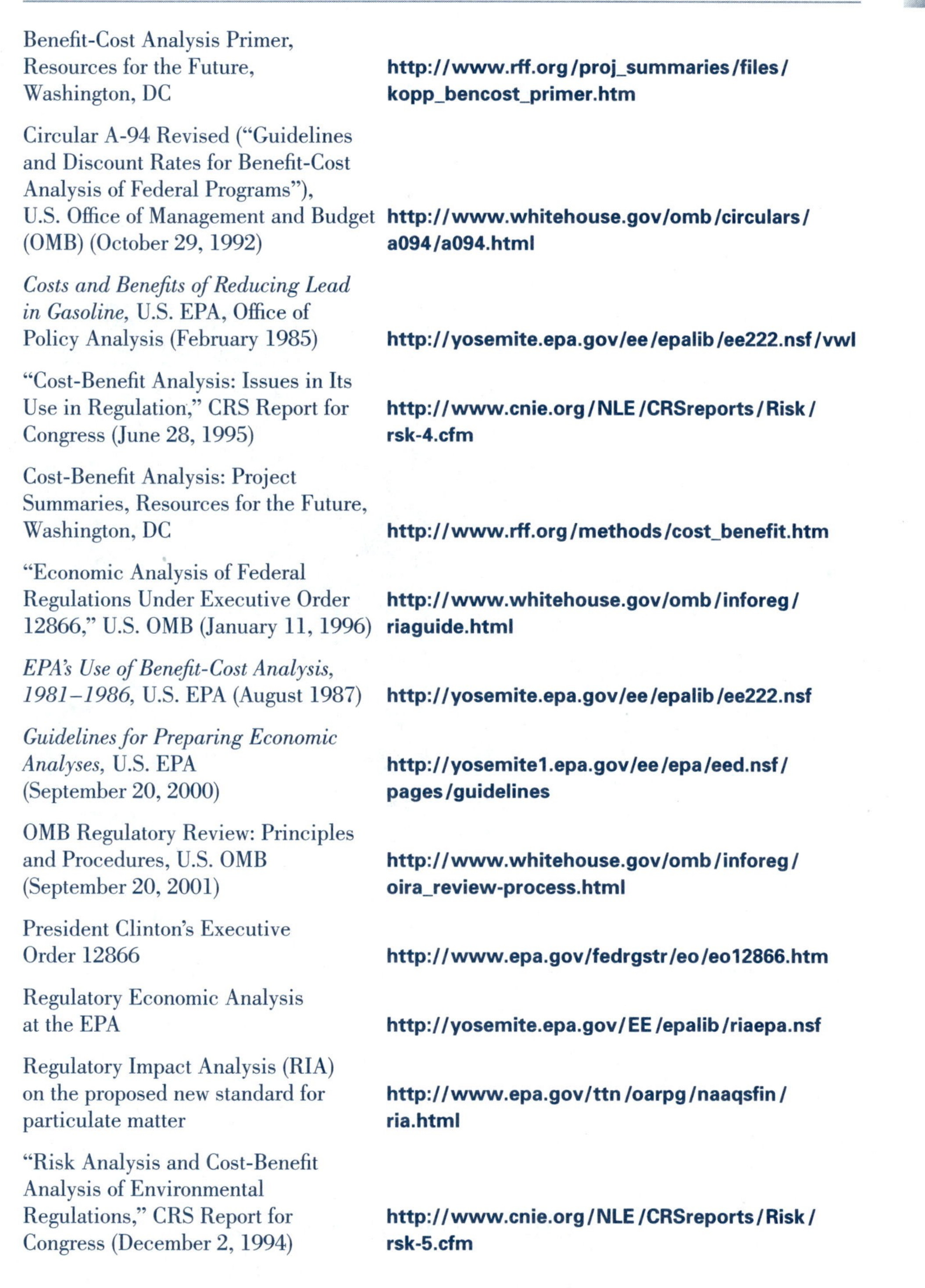

Benefit-Cost Analysis Primer, Resources for the Future, Washington, DC	**http://www.rff.org/proj_summaries/files/kopp_bencost_primer.htm**
Circular A-94 Revised ("Guidelines and Discount Rates for Benefit-Cost Analysis of Federal Programs"), U.S. Office of Management and Budget (OMB) (October 29, 1992)	**http://www.whitehouse.gov/omb/circulars/a094/a094.html**
Costs and Benefits of Reducing Lead in Gasoline, U.S. EPA, Office of Policy Analysis (February 1985)	**http://yosemite.epa.gov/ee/epalib/ee222.nsf/vwl**
"Cost-Benefit Analysis: Issues in Its Use in Regulation," CRS Report for Congress (June 28, 1995)	**http://www.cnie.org/NLE/CRSreports/Risk/rsk-4.cfm**
Cost-Benefit Analysis: Project Summaries, Resources for the Future, Washington, DC	**http://www.rff.org/methods/cost_benefit.htm**
"Economic Analysis of Federal Regulations Under Executive Order 12866," U.S. OMB (January 11, 1996)	**http://www.whitehouse.gov/omb/inforeg/riaguide.html**
EPA's Use of Benefit-Cost Analysis, 1981–1986, U.S. EPA (August 1987)	**http://yosemite.epa.gov/ee/epalib/ee222.nsf**
Guidelines for Preparing Economic Analyses, U.S. EPA (September 20, 2000)	**http://yosemite1.epa.gov/ee/epa/eed.nsf/pages/guidelines**
OMB Regulatory Review: Principles and Procedures, U.S. OMB (September 20, 2001)	**http://www.whitehouse.gov/omb/inforeg/oira_review-process.html**
President Clinton's Executive Order 12866	**http://www.epa.gov/fedrgstr/eo/eo12866.htm**
Regulatory Economic Analysis at the EPA	**http://yosemite.epa.gov/EE/epalib/riaepa.nsf**
Regulatory Impact Analysis (RIA) on the proposed new standard for particulate matter	**http://www.epa.gov/ttn/oarpg/naaqsfin/ria.html**
"Risk Analysis and Cost-Benefit Analysis of Environmental Regulations," CRS Report for Congress (December 2, 1994)	**http://www.cnie.org/NLE/CRSreports/Risk/rsk-5.cfm**

PART 4

The Case of Air

Technically, there has been some form of air pollution as long as the earth has supported life. Yet air pollution did not begin to be a pervasive problem until the Industrial Revolution. The combination of population growth, motorized transportation, and manufacturing activity that characterized the nineteenth century also marks the period when air quality became a real concern. As industrial growth proceeded on its course, so too did air pollution, though few realized the severity of the problem in those early days of industrial development.

In the United States, as urban smog became more apparent in cities like Los Angeles and Philadelphia, pressure began to mount for public officials to take action. Local communities and some state governments began to enact air quality laws. But it was literally decades before the federal government took an active role in what was fast becoming a worldwide problem.

Over the last four decades, U.S. air quality control policy has evolved into a comprehensive body of laws, with the most recent revisions embodied in the 1990 Clean Air Act Amendments. These extensive provisions are supported by an equally sophisticated infrastructure to implement controls and monitor compliance. In addition, international conferences have begun to address worldwide air pollution problems, such as global warming and ozone depletion.

Overall, there has been a good deal of progress. We have a much better sense of the implications of air pollution, thanks in part to medical and scientific research. As knowledge advanced, we have become better able to find solutions, and many nations have begun substituting economic incentives for command-and-control approaches to regulating air quality. Understanding the extent of this progress and analyzing what has been accomplished require a fairly thorough examination of the facts and some sense of the evolution of policy.

To that end, this module provides a multistep approach to learning about air quality problems and policy solutions, using economics as an analytical tool. We begin in Chapter 10 with a broad-based analysis of U.S. air quality control policy and the standard-setting process that defines clean air for the nation. In Chapter 11, we investigate the problem of urban smog and how U.S. standards are implemented through controls aimed at mobile sources. Chapter 12 follows with an analogous presentation for stationary sources. A particular focus in this chapter is how well these initiatives reduce acidic deposition. Finally, in Chapter 13, global air pollution is explored with an in-depth discussion of international policy and various proposals aimed at ozone depletion and global warming.

Chapter 10

Defining Air Quality: The Standard-Setting Process

One of the most important elements of clean air policy development is determining the target level of air quality that legislative initiatives are to achieve. In simpler terms, policymakers have to decide how clean is clean air. In the United States, standards or emissions limits are used to define a national level of air quality that will protect human health and the ecology. Many factors influence the standard-setting process that ultimately guides U.S. policy. Among them are such determinants as technological feasibility, energy requirements, and economic considerations. But more fundamentally, a determination has to be made about which substances in the atmosphere present a risk to society and the extent to which these substances can and should be controlled.

This important determination proceeds from the basic premise that the earth's atmosphere is composed of a specific combination of various gases. Nitrogen and oxygen account for 99 percent, and other gases such as carbon dioxide and helium make up the remainder—a composition critical to supporting life on earth. Any substances that contaminate or disrupt this natural composition pose an environmental risk and are potential targets for clean air policy.

Identifying these contaminants is important, but it's only part of the story. Ultimately, a decision must be made about how much of any contaminant society can tolerate—an issue of some debate. What *is* clear is that we cannot entirely avoid the risks of air pollution without incurring unreasonable opportunity costs.

To understand this assertion, consider the following. Some contaminants in the atmosphere are **natural pollutants,** such as pollen, dust particles from volcanic disturbances, gases from decaying animals and plants, even salt spray from the oceans. Because these pollutants occur naturally, they are virtually beyond human control.

natural pollutants
Contaminants that come about through nonartificial processes in nature.

anthropogenic pollutants
Contaminants associated with human activity, including polluting residuals from consumption and production.

Other contaminants are **anthropogenic,** meaning they are caused by human activity. These include such substances as carbon monoxide from automobile tailpipe exhausts, sulfur dioxide emissions from electricity generation, and toxic chemicals released from manufacturing plants. Although these types of pollutants are controllable and generally present a greater environmental risk than natural contaminants, they cannot be avoided completely without incurring the unrealistic opportunity cost of the absence of industrial activity. Consequently, we must accept the reality that our air quality will not be synonymous with a zero pollution level. But just what level of pollution is "acceptable," and how should it be determined?

Based on economic theory, we know that an **efficient** level of air quality is identified where the marginal social benefits of cleaner air are balanced with the marginal social costs incurred to achieve it. Furthermore, we recognize that market-based instruments can be used to reduce the negative externalities from pollution-generating activities and to encourage the positive externalities associated with abatement. These theories can and sometimes do guide policy decisions. However, as is often the case, the real-world complexities of government procedures, scientific uncertainty, and political pressures tend to delay or even prohibit the realization of this economic approach.

Our goal in this chapter is to explore the realities of clean air policy development by studying how the process has evolved in the United States. Of particular interest is to examine how the government uses standards to define air quality for the nation and to critically evaluate the implications of using a standards-based approach. To organize the presentation, we begin by discussing the following aspects of U.S. air quality policy:

- its evolution and development
- its primary objectives
- its standards-based definition of air quality
- the infrastructure it establishes for implementation.

Once done, we conduct a two-part benefit-cost analysis—the first as an overall appraisal of the Clean Air Act and its amendments, and the second as an evaluation of the standard-setting process. At the end of the chapter is a helpful reference list of the acronyms used in our discussion.

Overview of Air Quality Legislation in the United States

On November 15, 1990, President George H. W. Bush signed into law extensive changes in air quality policy in the form of the 1990 Clean Air Act Amendments. This legislative landmark received strong congressional support. Its hallmark is the integration of market-based policy instruments. The amendments came only after years of political battles, false starts, and a chronicle of events that underscores the importance of environmental issues and the complexity of drafting policy to deal with them.

In the Beginning

Prior to the 1950s, all air quality legislation in the United States had been enacted by state and local governments, most of which have a rather extensive history of laws aimed at cleaner air. Logically, the earliest of these were passed during the Industrial Revolution.[1] Even in contemporary times, state and local governments typically have taken the lead. The most notable example is California, whose often path-breaking legislation has been in direct response to the smog problem in Los Angeles. In 1961, California passed the first state-level air pollution law to control motor vehicle emissions. This legislation followed many years of struggle between California officials and the American automobile industry, the subject of Application 10.1. In fact, a decade had passed since a 1951 study by A. J. Haagen-Smit, of the California Institute of Technology, scientifically confirmed that automobile emissions were a major contributor to smog formation.

At the federal level, the legislative history of air pollution initiatives is much shorter. In fact, there were no *national* air quality laws until the Air Pollution Control Act of 1955 was passed, and there was no truly comprehensive legislation until the Clean Air Act of 1963 was enacted, nearly a century after state and local governments had begun to take action. From that point on, a series of revisions and new initiatives helped to form U.S. policy as it is now defined. Look at Table 10.1 for a synopsis of the major federal laws that map out this legislative evolution.

Current U.S. Policy

With this chronicle of policy evolution behind it, Congress passed into law some of the most comprehensive legislation in its history—the 1990 Clean Air Act Amendments. These map out national directives for reducing the risks of air pollution. The 1990 Amendments are ex-

[1] See Stern (January 1982) for a thorough historical perspective on U.S. air quality laws.

Application

10.1 *California Smog and the Automobile Industry*

California's severe smog problem motivated its early attempts in the 1950s to gain the support of the automobile industry in solving its dilemma. Despite the evidence connecting motor vehicle emissions to smog production revealed in A. J. Haagen-Smit's 1951 study, the automobile industry denied that auto emissions were a major contributor to the problem and argued that the source of urban smog required further investigation. The result was a volley between the industry and the state government that accomplished little throughout the 1950s.

In 1959, manufacturers acknowledged that the engine's design was an important source of the emissions problem. In 1961, automakers began to install engine ventilation devices on all new cars sold in California to control these emissions, a technology change that was eventually mandated under California law. However, this technology was not new. In fact, it had been developed in the 1930s. Apparently weary of the lack of progress, the California government attempted to force the issue. Legislation was passed calling for the installation of equipment to control vehicle exhausts as soon as it was developed either by the major automakers or by any three independent manufacturers. The automobile producers asserted that such technology simply did not exist. Yet, in June 1964, the state confirmed that independent parts manufacturers could provide the needed add-on equipment at a reasonable cost and mandated its installation on all new vehicles starting with the 1966 model year. Predictably, the automakers announced in August 1964 that they had the capability to install their own devices on 1966 models, despite the fact that in March they had claimed that 1967 was the earliest model year by which they could accommodate the change.

Not only were the major auto producers using delay tactics, but the measure of their innovative effort was unimpressive. Ford and General Motors developed only a simple air pump, while Chrysler's solution, marketed as a clean air package, involved only minor alterations to its fuel and carburetion system. It was later discovered that Chrysler's "solution" contributed significantly to nitrogen oxide emissions.

Why were there such significant delays? Part of the answer is based on economic theory. If there had been a demand for cleaner running cars, competitive firms would have recognized the advantage of being the first to satisfy that demand as well as the threat to their survival of being left behind in what is often a race to innovate. However, it is not at all clear that such a demand existed, and, even if it had, market incentives were dampened by the market power possessed by the Big Three—General Motors, Ford, and Chrysler. The giant automakers enjoyed a position of market strength that was protected by entry barriers from the threat of any innovative entering firms.

In 1953, the three firms formally pooled their efforts by forming a committee to jointly investigate the pollution problem. Later, in 1955, they executed a cross-licensing agreement to share access to any patents on emission controls. In so doing, they effectively removed the incentive for any one of them to find an innovative solution. (It turns out that in 1969 the Department of Justice filed suit against the Big Three, charging them with collusive attempts to hinder the advancement of pollution control technology. The suit was settled by a consent decree that ended the cross-licensing agreement.)

Sources: White (1982), chap. 3; U.S. Senate, Staff of the Subcommittee on Air and Water Pollution of the Committee on Public Works (October 1973); Seskin (1978).

Table 10.1 *Brief Retrospective of U.S. Air Quality Legislation*

Legislation	Brief Summary
Air Pollution Control Act of 1955	This was the first federal legislation on air pollution. Its focus was limited, aimed mainly at providing federal appropriations to state governments in support of research and training on air quality.
Clean Air Act of 1963	This was the first comprehensive federal air quality legislation, though it continued to place the onus of air quality control on states. Emissions regulations for stationary sources were established, and a liaison committee was to be formed with the auto industry to study the effects of motor vehicle emissions.
Motor Vehicle Air Pollution Control Act of 1965	The Department of Health, Education, and Welfare (HEW) was authorized to set emissions standards on new motor vehicles, but no statutory deadline for doing so was set.
1965 Clean Air Act Amendments	HEW was authorized to establish the first federally mandated mobile emissions standards. Maximum emissions of carbon monoxide and hydrocarbons were set for new motor vehicles, starting with the 1968 model year.
Air Quality Control Act of 1967	Air quality control regions (AQCRs) were to be established. HEW was to determine air quality criteria for common air pollutants. States were to set and implement ambient air quality standards based on the results of federal research on air pollution.
1970 Clean Air Act Amendments	National Ambient Air Quality Standards (NAAQS) were established for stationary sources, and emissions limits were set for mobile sources. Both were to be implemented through State Implementation Plans (SIPs). New source performance standards (NSPS) were established at more stringent levels than for existing sources.
1977 Clean Air Act Amendments	Previously set deadlines for meeting air quality objectives were modified. AQCRs were reclassified into attainment and nonattainment regions to protect regions that were cleaner than what was required by the NAAQS. These regions were termed PSD areas to indicate prevention of significant deterioration.

Sources: Stern (January 1982); Mills (1978), pp. 189–94; Portney (1990a); Wolf (1988), chap. 2.

tensive. As shown in Table 10.2, there are 11 major sections that comprise the new and revised statutes. Exhibit A.3 in the Appendix provides a summary of each section.

Certain of the titled sections within the 1990 Amendments use market-based approaches, such as Title IV and Title VI, but the underlying structure continues to be command-and-control oriented. In particular, the objectives of U.S. policy are to be met using national air quality standards, which effectively define a common air quality level for the nation. These standards in turn are implemented through an extensive infrastructure that facilitates federal oversight. The uniformity of such an approach impedes the achievement of allocative efficiency or cost-effectiveness. To verify this assertion, we begin by examining the statutory objectives of the Clean Air Act.

Table 10.2 *Titled Sections of the Clean Air Act Amendments of 1990*

Section	Amendment Purpose
Title I. Provisions for Attainment and Maintenance of National Ambient Air Quality Standards (NAAQS)	Amends and extends the legislation to achieve the NAAQS.
Title II. Provisions Relating to Mobile Sources	Controls vehicle emissions and regulates the use of clean fuels and clean vehicles.
Title III. Hazardous Air Pollutants	Identifies toxic pollutants for setting emissions controls.
Title IV. Acid Deposition Control	Controls sulfur dioxide using market-based allowances and mandates a reduction in nitrogen oxide emissions.
Title V. Permits	Requires states to establish permit programs and requires major air polluters to have federal permits.
Title VI. Stratospheric Ozone Protection	Establishes allowance trading to reduce the production and use of ozone depleters.
Title VII. Provisions Relating to Enforcement	Strengthens enforcement through penalties and by establishing new authorities.
Title VIII. Miscellaneous Provisions	Includes statutes on various issues, including energy conservation and visibility.
Title IX. Clean Air Research	Calls for research on air pollution and reauthorizes the National Acid Precipitation Assessment Program (NAPAP).
Title X. Disadvantaged Business Concerns	Requires the EPA to allocate a proportion of research funding to disadvantaged firms.
Title XI. Clean Air Employment Transition Assistance	Provides employment services if job loss is related to compliance with the 1990 Amendments.

Source: U.S. EPA, Office of Air and Radiation (November 30, 1990).

Note: For an online overview of the 1990 Clean Air Act Amendments, visit **http://www.epa.gov/oar/caa/overview.txt.**

http://

Defining the Objectives of Air Quality Control

The current objectives of U.S. air quality policy were originally defined in the first comprehensive federal body of law on air quality control: the Clean Air Act of 1963. Chief among these is:

> to protect and enhance the quality of the Nation's air resources so as to promote the public health and welfare and the productive capacity of its population.[2]

To achieve the nation's objectives, the government must understand the risks of air pollution and the abatement necessary to reduce these risks to an acceptable level. The first step is to identify the primary causes of air pollution and to isolate those contaminants deemed most harmful.

[2]42 U.S.C. §740(b)(1), July 14, 1955, as amended 1963.

Identifying Major Air Pollutants

Consider the extraordinary responsibility of determining which pollutants are most responsible for air pollution and setting the proper level at which to control them. Such a decision-making process is complex, given the wide range of human sensitivity to pollutants, the uncertainty about health and welfare effects—particularly over the long term—and the enormous task of assessing the effects of various combinations of pollutants.

In the United States, official reports called **criteria documents** present available scientific evidence on the properties and effects of any known or suspected pollutant.[3] This evidence is used to identify common air pollutants known to present a risk to health and the environment. Officially, these **criteria pollutants** are identified substances known to be hazardous to health and welfare. As of 2002, there were six identified criteria pollutants in the United States:

criteria documents Reports that present scientific evidence on the properties and effects of known or suspected pollutants.

criteria pollutants Substances known to be hazardous to health and welfare, characterized as harmful by criteria documents.

- Particulate matter (PM-10 and PM-2.5),[4]
- Sulfur dioxide (SO_2),
- Carbon monoxide (CO),
- Nitrogen dioxide (NO_2),
- Tropospheric ozone (O_3), and
- Lead (Pb).

Table 10.3 summarizes the health and welfare effects of exposure to these pollutants, which are critical to the identification process.

A second group of contaminants identified by U.S. legislation is **hazardous air pollutants,** or **air toxics.** These are noncriteria pollutants that may contribute to irreversible illness or increased mortality.[5] What distinguishes these substances from the criteria pollutants is that the associated risk is much greater, although typically a much smaller segment of society is affected. The 1990 Clean Air Act Amendments include a list of 189 identified hazardous air pollutants, which is to be periodically revised, as needed. As of 2002, the number of identified hazardous air pollutants under the EPA's control was 188.[6] Prior to 1990, only eight such substances had been identified and placed under the EPA's control.[7] It was precisely because of this lack of regulatory action that a more aggressive policy was enacted in the 1990 Amendments.

hazardous air pollutants Noncriteria pollutants that may cause or contribute to irreversible illness or increased mortality.

Setting Standards as a National Definition of Air Quality

Once the EPA identifies the major air pollutants, it then establishes **national standards** for them. These standards set maximum allowable levels for each pollutant to be met by all pol-

[3]The documents are so named because the assessment process is based on descriptive factors called **criteria.** These criteria are characteristics of the pollutants and their potential health and welfare effects.

[4]PM-10 refers to particles less than 10 micrograms in diameter, and PM-2.5 has an analogous meaning. Standards for PM-2.5 were added by the EPA in 1997 and were subsequently challenged in the courts by some states and business groups, who claimed that the standards were unconstitutional. After a lengthy legal battle, a federal appeals court ruled in March 2002 that the tougher standards could be issued. However, because of the time delay, the standards had expired and therefore need to be reevaluated. Information on the status of this controversial series of actions is available at **http://www.epa.gov/ttn/oarpg/naaqsfin/** or at **http://www.epa.gov/airlinks/airlinks4.html.**

[5]Visit **http://www.epa.gov/ttn/atw/hapindex.html** to access the EPA's Health Effects Notebook for Hazardous Air Pollutants.

[6]For current information on hazardous substances, visit the EPA's Air Toxics Web site at **http://www.epa.gov/ttn/atw/pollsour.html.**

[7]The eight substances are asbestos, beryllium, mercury, vinyl chloride, benzene, radionuclides, inorganic arsenic, and coke oven emissions.

Table 10.3 *Selected Health and Welfare Effects of the Six Criteria Pollutants*

Criteria Pollutant	Major Effects
Particulate matter (PM)	Health: aggravation of existing respiratory and cardiovascular disease (coarse particles); decreased lung functioning, increased respiratory symptoms, premature death (fine particles). Welfare: visibility impairment; interference with plant metabolism; damage to materials; soiling.
Sulfur dioxide (SO_2)	Health: temporary breathing impairment; respiratory illness; changes in the lungs' defenses. Welfare: contributes to acidic deposition; increased foliar injury; reduced plant growth and yield.
Carbon monoxide (CO)	Health: at elevated levels causes impairment of cognitive skills, vision, and work capacity.
Nitrogen dioxide (NO_2)	Health: increased changes in airway responsiveness and pulmonary function in people with preexisting respiratory illness; long-term exposure may increase susceptibility to respiratory infection. Welfare: contributes to global warming, stratospheric ozone depletion, and acidic deposition.
Ozone (O_3)	Health: lung inflammation; aggravation of preexisting respiratory diseases; reduced lung functioning; increased respiratory symptoms. Welfare: reduction in crop and forest yields; increased plant susceptibility to disease and pests; foliar damage to plants and trees.
Lead (Pb)	Health: damage to kidneys, liver, nervous system, and other organs; excessive exposure can cause neurological impairments and/or behavioral disorders; at low doses can cause damage to the central nervous system of fetuses and children. Welfare: ingestion by animals of airborne lead deposits on plant or soil surfaces; can inhibit plant growth.

Source: U.S. EPA, Office of Air Quality Planning and Standards (March 2001).

luting sources. The two major categories of potentially controllable sources are stationary and mobile

stationary source
A fixed-site producer of pollution, such as a building or manufacturing plant.

mobile source
Any nonstationary polluting source, including all transport vehicles.

- A **stationary source** is any building or structure that emits pollution, such as a coal-burning power plant.
- A **mobile source** refers to any transport vehicle that generates pollution, such as an automobile or truck.

By establishing national standards, the federal government implicitly defines air quality for the entire country—a practice that is not exclusive to the United States. For example, Japan and some European countries also use a standard-setting approach to communicate an ac-

10.2 *Is Standard Setting Under the Clean Air Act Unconstitutional?*

Application

A recent revision of the NAAQS for particulate matter (PM) and ozone (O_3) became the subject of a lengthy legal battle that sparked controversy and debate in Washington, DC, and across the nation. Indeed, the constitutionality of standard setting under the Clean Air Act was called into question in the high profile case that ensued. Richard A. Epstein, of the University of Chicago Law School, dubbed *American Tracking Associations, Inc. v. EPA* "the most closed watched administrative law case of the past decade" (Epstein, March 1, 2001).

It all began in 1996, when the EPA proposed tighter standards for both PM and tropospheric ozone. For the first time, standards for PM were to include smaller particles than had been previously controlled: PM-2.5, or particles less than 2.5 micrograms in diameter. According to the Regulatory Impact Analysis for the proposal, annual health benefits of the new standard were estimated at between $19 and $104 billion ($1990) and the comparable costs at $8.6 billion ($1990). The 8-hour ozone standard was to be tightened from a concentration of 0.12 parts per million to 0.08 parts per million. For this part of the revision, annual benefits were estimated at between $0.4 and $2.1 billion ($1990), with annual costs at $1.1 billion ($1990). For a variety of reasons, there was much debate about the accuracy of these estimates on both sides of the ledger.

Challenging the new standards on a number of fronts, the American Trucking Associations, a consortium composed of three states and a number of business groups, filed suit against the EPA. Among the claims filed by the petitioners were that the science used to support the standards was faulty, that the EPA should have considered the associated costs when setting the air quality standards, and that the EPA had overstepped its authority in issuing the standards.

On May 14, 1999, the D.C. District Circuit Court of Appeals ruled on the case. The court did not question the science on which the EPA had relied, and it rejected the claim that the EPA should have considered costs in setting the air quality standards. However, the court did rule that the Clean Air Act "effects an unconstitutional delegation of legislative power." The essence of the court's argument was as follows. Because the law allows only for **benefit-based** standards with a margin of safety, the EPA must justify anything other than a zero standard for a nonthreshold pollutant, yet it "failed to state intelligibly how much is too much." So the court ruled that the standard-setting process is an unconstitutional delegation of power granted to the EPA by Congress. The court further held that the new standards could not be enforced. Both the EPA and the American Trucking Associations appealed the decision to the U.S. Supreme Court.

In February 2001, the Supreme Court announced its ruling. It upheld the constitutionality of the EPA's actions, and it confirmed that the Clean Air Act called for standards to be set based on public health benefits and *not* on cost considerations. It further held that states and the EPA could continue to consider costs in implementing these standards. That said, the court did identify problems with the implementation of the ozone standard and ordered the EPA to reconsider that issue. Other matters were sent back to the U.S. Circuit Court, including the EPA's setting of a nonzero standard. In March 2002, a federal appeals court upheld the tougher ozone and PM standards initially issued by the EPA. However, because the standards had been issued in 1997, they actually had expired, which meant they had to be reevaluated.

On net, the EPA claimed victory, though it faces many obstacles before the new standards can be implemented. Moreover, although the constitutionality issue was resolved, the court's ruling about cost considerations reinforces an interesting irony that characterizes standard setting under the Clean Air Act: The EPA must conduct a benefit-cost analysis as part of the requisite Regulatory Impact Analysis, but it must consider only benefits when setting the standards.

Sources: U.S. EPA, Office of Air Quality Planning and Standards (July 17, 1997); U.S. EPA, Office of Air and Radiation (June 28, 1999); U.S. EPA, Office of Communications, Education, and Media Relations (February 27, 2001); Freeman (winter 2002); Wald (March 27, 2002); Epstein (March 1, 2001).

ceptable level of air quality for society. Although the specifics vary across countries, in almost every case the list of identified air pollutants coincides with those named as criteria pollutants in the United States. For an overview of other nations' environmental policies as well as links to related sites, visit the Web site of the Organisation for Economic Co-operation and Development (OECD), Environment Directorate, at **http://www.oecd.org/env** or the Website of the United Nations Environment Programme (UNEP) at **http://www.unep.org.**

Standards for Criteria Air Pollutants

National Ambient Air Quality Standards (NAAQS) Maximum allowable concentrations of criteria air pollutants.

primary NAAQS Set to protect public health from air pollution, with some margin of safety.

secondary NAAQS Set to protect public welfare from any adverse, nonhealth effects of air pollution.

In the United States, the standards for the six criteria pollutants are called **National Ambient Air Quality Standards (NAAQS).** Within this group are two subcategories: primary and secondary NAAQS.

- **Primary NAAQS** are set to protect public health, with some margin of safety.
- **Secondary NAAQS** are intended to protect public welfare.

Originally established under the 1970 Clean Air Act Amendments, these standards have been revised from time to time. In fact, the law requires that the criteria and the NAAQS be reviewed by the EPA every five years. An interesting legal case arose over a recent revision to the NAAQS proposed in 1997, a matter discussed in Application 10.2 on p. 209.

The primary and secondary NAAQS in effect as of 2002 are given in Table 10.4. Notice that none of the standards are set at a zero concentration level. Thus, if we accept these standards as the nation's definition of air quality, then we must conclude that acceptable air quality does not mean the absence of criteria pollutants.

Standards for Hazardous Air Pollutants

National Emission Standards for Hazardous Air Pollutants (NESHAP) Set to protect public health and the environment, applicable to every major source of any identified hazardous air pollutant.

maximum achievable control technology (MACT) The control technology that achieves the degree of reduction to be accomplished by the NESHAP.

U.S. law also calls for the establishment of **National Emission Standards for Hazardous Air Pollutants (NESHAP)** for every major source of the listed hazardous air pollutants. They are intended to protect public health and the environment, taking into account the costs to attain the standards, any non–air quality health and environmental impacts, and energy requirements.

The NESHAP are to attain the maximum degree of reduction for each air toxic achievable, referred to as **maximum achievable control technology (MACT).** Where possible, this reduction should achieve a complete ban on the substance. Polluters can use various methods to meet these standards, such as substituting less harmful inputs or enclosing production processes to eliminate hazardous emissions.

Establishing an Infrastructure to Implement the Standards

To implement national standards, the United States has established an infrastructure that involves both federal and state governments. It is defined through two components:

- State Implementation Plans (SIPs)
- Air Quality Control Regions (AQCRs)

State Implementation Plans

State Implementation Plan (SIP) A procedure outlining how a state intends to implement, monitor, and enforce the NAAQS and the NESHAP.

Coordination between the two major levels of government is achieved through State Implementation Plans (SIPs). A **State Implementation Plan (SIP)** is an EPA-approved procedure of how a state intends to implement, monitor, and enforce the NAAQS and the NESHAP.[8]

[8]The NAAQS represent the *minimum* requirements to be attained by every state. However, at its discretion, a state can submit a plan to achieve more stringent standards. Internet access to Section 110 in the Clean Air Act that refers to SIPs is available at **http://www.epa.gov/oar/caa/caa110.txt.**

Table 10.4 *National Ambient Air Quality Standards in Effect in 2002*

Pollutant	Standard Value		Standard Type
Carbon monoxide (CO)			
8-hour average	9 ppm	(10 mg/m³)	Primary
1-hour average	35 ppm	(40 mg/m³)	Primary
Nitrogen dioxide (NO_2)			
Annual arithmetic mean	0.053 ppm	(100 μg/m³)	Primary and secondary
Ozone (O_3)			
8-hour average	0.08 ppm	(157 μg/m³)	Primary and secondary
1-hour average	0.12 ppm	(235 μg/m³)	Primary and secondary
Lead (Pb)			
Quarterly average	1.5 μg/m³		Primary and secondary
Particulate (PM-10)[a]			
Annual arithmetic mean	50 μg/m³		Primary and secondary
24-hour average	150 μg/m³		Primary and secondary
Particulate (PM-2.5)[b]			
Annual arithmetic mean	15 μg/m³		Primary and secondary
24-hour average	65 μg/m³		Primary and secondary
Sulfur dioxide (SO_2)			
Annual arithmetic mean	0.03 ppm	(80 μg/m³)	Primary
24-hour average	0.14 ppm	(365 μg/m³)	Primary
3-hour average	0.50 ppm	(1300 μg/m³)	Secondary

Source: U.S. EPA, Office of Air Quality Planning and Standards (March 29, 2002).

Notes: Parenthetical value is an approximately equivalent concentration; ppm=parts per million; μg/m³=micrograms per cubic meter; mg/m³=milligrams per cubic meter.

The ozone 8-hour standard and the PM-2.5 standards were originally proposed in 1997 and were subsequently challenged in the courts. In 2002, the courts upheld the new standards, but by that time the standards had expired and hence had to be reevaluated. Further information is available at **http://www.epa.gov/ttn/oarpg/naaqsfin/**.

[a] Particles with diameters of 10 micrometers or less.

[b] Particles with diameters of 2.5 micrometers or less.

The SIP system follows a federalist format by delegating certain tasks to different jurisdictions. The standard setting is assigned mainly to the federal level to standardize air quality across the country. State governments are responsible for implementing the standards and monitoring polluters within their jurisdictions, since their knowledge of the immediate region gives them an advantage in doing so.

Air Quality Control Regions

To coordinate states' responsibilities, **Air Quality Control Regions (AQCRs)** are defined within each state's jurisdiction. These are geographic areas designated by the federal government within which common air pollution problems are shared by several communities.

Air Quality Control Region (AQCR)
A geographic area designated by the federal government within which common air pollution problems are shared by several communities.

Currently, 247 AQCRs have been designated across the United States. These well-defined geographic areas are monitored to determine whether or not the region is in compliance with the national standards. Today, a number of AQCRs still have not met the current NAAQS for one or more of the six criteria pollutants. Table 10.5 provides information on these nonattainment areas as of 2001. As these data indicate, regions have the greatest difficulty

Table 10.5 ***Number of Nonattainment Areas for the NAAQS Pollutants as of 2001***

Pollutant	Number of Nonattainment Areas
Particulate matter (PM-10)	61
Sulfur dioxide (SO_2)	24
Carbon monoxide (CO)	11
Nitrogen dioxide (NO_2)	0
Ozone (O_3)	36
Lead (Pb)	4

Source: U.S. EPA, Office of Air Quality Planning and Standards (June 8, 2001).

Note: Unclassified areas are not included in the totals.

achieving the particulate matter and ozone standards, with 61 and 36 regions, respectively, still classified as nonattainment for these pollutants.

Reclassification of AQCRs to Protect "Clean Air Areas"

In 1972, the Sierra Club filed suit against the EPA for failing to protect areas that were cleaner than what was required by law.[9] The environmental group argued that the existing NAAQS were aimed solely at *improving* air quality in areas that did not meet the standards. The concern was that nondegradation areas, as they are called, would be allowed to deteriorate to the existing national standards. The Sierra Club claimed that the lack of protection for nondegradation areas was a violation of the law, because one of the statutory purposes of the Clean Air Act is to *protect* as well as *enhance* the quality of the nation's air resources. Ultimately, the group won the suit, and in 1974 a new program was established to protect these already clean areas.

The new "clean air areas" program redefined the structure of federal air pollution control and played a significant role in directing future policy decisions. States had to reassess their AQCRs so that they could identify the following three groups:

- regions that met or exceeded the standards
- regions that did not meet the standards
- regions with insufficient data to confirm a classification

prevention of significant deterioration (PSD) area
An AQCR meeting or exceeding the NAAQS.

nonattainment area
An AQCR not in compliance with the NAAQS.

Regions meeting or exceeding national standards were redesignated as areas targeted for **prevention of significant deterioration (PSD) areas.** Those that did not were designated as **nonattainment areas.**

Once determined, every PSD area was to be designated as Class I, II, or III—a progressive classification based on the maximum concentration of criteria pollutants allowed. Class I areas, the most stringently controlled, include wilderness areas and national parks. The law specifically prohibits a redesignation of these areas.[10]

1990 Reclassification of Nonattainment Areas by Criteria Pollutant

In response to the nation's persistent problem of urban air pollution, the 1990 Amendments reclassify all nonattainment areas for the pollutants most responsible—ozone, carbon monoxide, and particulate matter—into new progressive categories based upon existing pol-

[9] See *Sierra Club* v. *Ruckelshaus*, 344 F. Supp. 253 (D.D.C. 1972).

[10] Other than these initially classified Class I areas and certain other exceptions, states can redesignate classifications for PSD areas in accordance with very specific rules.

Figure 10.1 Summary of Changes in U.S. Emissions of Criteria Pollutants, 1970–2000

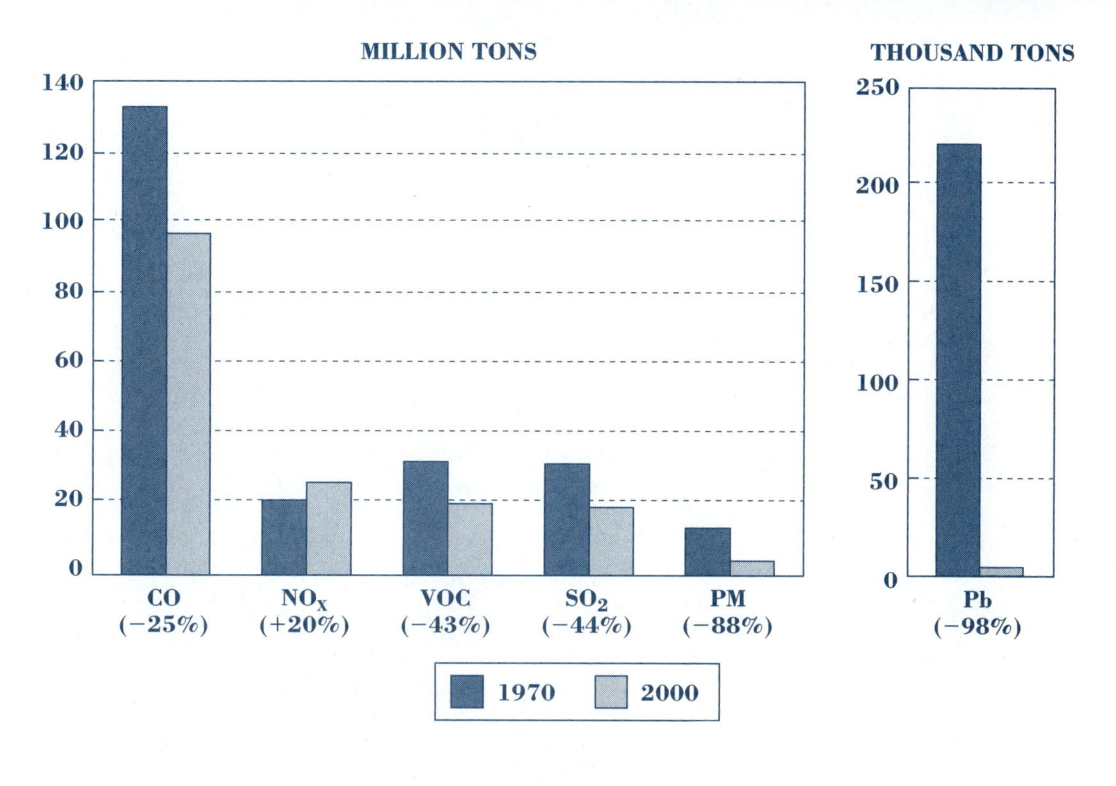

Source: U.S. EPA, Office of Air Quality Planning and Standards (September 2001), p. 3.

Note: For Internet access to EPA's *Latest Findings on National Air Quality: 2000 Status and Trends* (September 2001), visit **http://www.epa.gov/air/aqtrnd00/index.html.**

http://

lutant concentrations. For example, ozone nonattainment areas are reclassified into five groups: marginal, moderate, serious, severe, and extreme. These new categories not only identify the severity of pollution but also provide some justification to set more stringent regulations in those areas with higher pollution levels.

Monitoring Air Quality Across Regions

Determining the compliance status of each region and the assessment of air quality for the nation depends upon a systematic measurement of the six criteria pollutants. In general, this is accomplished either by estimating the emissions level of each pollutant or by measuring the ambient concentration of each in some volume of air.

Estimating Pollutant Emissions Levels

It is not feasible to measure actual pollutant emissions levels on a national scale, so instead they are estimated. In the United States, best available **engineering methods** are used to derive annual emissions estimates for over 450 source categories, which include almost all anthropogenic sources. These are then aggregated to determine regional and national emissions trends.

Figure 10.1 shows emissions estimates for the six criteria pollutants in 2000 compared to those for 1970. According to these estimates, there have been reductions in most emissions (the exception being NO_X) over the long run. Some of these reductions are the result of policy

Application

10.3 Market Incentives to Phase Out Lead Emissions

Even skeptics of market solutions to environmental problems are hard-pressed to deny the success story of lead banking and trading. Touted by economists and environmentalists alike, the use of market instruments in reducing lead emissions was undeniably a victory.

Lead is a toxic heavy metal that poses a serious health hazard if ingested or inhaled. Unfortunately, it is ubiquitous, showing up in paints, plastics, lead-acid batteries, plumbing compounds, and gasoline. Its use in gasoline began in the 1920s, when refineries recognized it as the cheapest available source of octane. By the 1960s and 1970s, however, evidence had begun to accumulate that linked lead exposure to mental disorders and cardiovascular disease. Research also showed that lead levels in the bloodstream were directly associated with changes in the lead content of gasoline. In addition to the health threat, lead was also found to adversely affect the performance of catalytic converters being installed in new cars to meet tougher standards on tailpipe emissions.

In 1973, the EPA responded by requiring that unleaded gasoline be made available by 1974. The agency also established a limit on the average lead concentration of *all* gasoline (both leaded and unleaded) sold by a given refiner. Nonetheless, lead concentrations remained high, averaging about 2.0 grams per gallon in 1975, in part because of misfueling (i.e., using leaded fuels in vehicles designed to use only unleaded gasoline). During the 1980s, new evidence revealed that the health threat was even more serious than originally believed. On every front, the consumption of leaded gasoline generated negative externalities, and existing policy was not tough enough or effective enough to solve the problem.

The EPA's ultimate policy response was to tighten the lead standard and use market incentives to achieve it. In 1982, the EPA established a new standard based only on leaded gasoline: 1.1 grams per leaded gallon (gplg) effective July 1, 1985. Subsequently, the standard was made progressively more stringent through a series of steps, calling for a concentration of 0.5 gplg after July 1, 1985, 0.1 gplg after January 1, 1986, and a complete ban by 1996. To accomplish this phasedown, the agency introduced a program of **credits** issued to refineries for reductions that surpassed those required by law. These credits could then be banked for use during a subsequent stage of the phasedown or traded with older refineries unable to meet the reductions by the statutory deadlines.

The end result was that the refineries best able to meet the required standards did so, compensating for their less efficient counterparts, who were effectively given more time to retrofit their production processes. According to EPA reports, 73 percent of all refiners participated in the trading program in the second quarter of 1984. Ultimately, the plan cost the nation $220 million less than what it would have under a strict command-and-control approach of standard setting, and this savings came about with no change in the benefits of lead reduction. Since 1970, airborne lead emissions have fallen by a remarkable 98 percent. The difference is only that the method used to achieve the reductions minimized costs.

Sources: Whiteman (May/June 1992); U.S. EPA, Office of Policy, Planning, and Evaluation (July 1992), pp. 5-7–5-8; U.S. EPA, Office of Policy Analysis (February 1985).

initiatives. The most substantial decline is for lead, which is the result of the EPA's lead phase-out program, a market-based plan discussed in Application 10.3.

Measuring Pollutant Concentrations

Pollutant concentration levels are actually measured at air-monitoring station sites located throughout the country. Most of these sites are in urban regions characterized by relatively high pollutant concentrations and population exposure. All sites report their data to the EPA by means of an air-monitoring network. A wealth of information on air pollution monitoring in the United States is available at **http://www.epa.gov/oar/oaqps/montring.html**.

Economic Analysis of U.S. Air Quality Policy

If we consider the evolution of U.S. air quality legislation that began in 1955, it is clear that the nation's control policy has become more comprehensive over time. A vast infrastructure now exists to implement and monitor national air quality objectives. But is it reasonable to assess U.S. accomplishments by the number of provisions that now define air quality legislation or by the more elaborate implementation and monitoring systems that have been established? These may be indicators that the United States is attempting to strengthen policy, but do they suggest that policy is any more effective?

It seems simple enough to argue that the bottom line in assessing regulatory policy is whether or not any headway has been made in achieving cleaner air. However, such an argument is a gross oversimplification of what is in fact a much more complex problem. Look back at the trend data shown in Figure 10.1. These data suggest that U.S. air quality has improved over time. Although this is likely true to a point, many factors must be considered before drawing conclusions. Among these are the accuracy of the emissions estimates and the determination of what part of any observed improvement is actually attributable to effective policy.

Even if these factors are accounted for, there is still the challenge of converting any policy-driven air improvements to monetized values of social benefits, such as reduced health risks, increased visibility, and the preservation of the ecology. Finally, one must consider the full extent of the costs borne by society to achieve whatever progress has been made. These costs should include not only explicit expenditures for compliance, monitoring, and enforcement but also implicit costs resulting from shifts in production technology, substitution of fuels and other inputs, and changes in available goods and services.

Collectively, these are the tasks that comprise an economic analysis of environmental policy. By systematically estimating and evaluating the costs and benefits associated with the Clean Air Act, one can determine whether U.S. air control policy is **allocatively efficient.** Recall from previous chapters that the decision rule ensuring an efficient outcome is the **maximization of net benefits**—the point where the associated marginal social costs and marginal social benefits are equal. To illustrate this approach, we present a two-part benefit-cost investigation of U.S. air policy. The first part assesses the overall efficiency of the Clean Air Act. The second considers the efficiency of the standard-setting process itself.

Benefit-Cost Analysis of the Clean Air Act

Benefit-Cost Analysis of the Pre-1990 Clean Air Act

There is a fair amount of research that evaluates the efficiency of specific air quality control provisions. Far less prevalent are comprehensive analyses assessing the Clean Air Act in its entirety. Drawn from the work of other researchers, one such investigation was conducted by Paul Portney (1990a), an economist and researcher at Resources for the Future in Washington, DC. With careful reservations, Portney offers a very telling comparison of the social benefits and costs associated with U.S. policy prior to the enactment of the 1990 Clean Air Act Amendments. Beyond the implications of the results, the study illustrates the practical value of benefit-cost analysis and the difficulties of using this strategy to evaluate policy on a comprehensive scale.

Total Social Costs

Any cost analysis of environmental policy involves the challenge of estimating the total costs to society from having to adapt to legislated provisions. These include both explicit and implicit costs. In the United States, explicit or out-of-pocket expenditure data are collected mainly by the Department of Commerce. These are the actual compliance and operating costs

Table 10.6 ***Annualized Control Costs for Air (in millions of 1986 dollars)***

Target	1972	1980	1987	1995	2000
Stationary sources	6,230	13,298	18,960	25,188	29,725
Mobile sources	1,345	4,010	7,469	11,097	14,140
Undesignated sources	341	327	250	207	184
Air pollution total	7,916	17,635	26,679	36,493	44,049

Source: U.S. EPA, Office of Policy, Planning, and Evaluation (December 1990), p. 3-2.

Notes: Data for 1995 and 2000 are projections. Sums may not equal totals shown due to rounding.

associated with air quality controls. Cost data for selected years are shown in Table 10.6.[11] Although important, these explicit cost data are insufficient for an economic benefit-cost analysis. Why? Because they do not account for implicit costs to society, such as higher priced products, loss of employment from production and technology conversions, and restrictions on consumers' choice sets.

Portney's analysis derives social cost estimates of the Clean Air Act primarily from the work of two other researchers, Michael Hazilla and Raymond J. Kopp (1990). These researchers use a model of the U.S. economy to estimate how prices and income change with the implementation of policy aimed at both air and water quality. Ultimately, their measure of social costs is based on an assessment of how social welfare changed as a result of the environmental legislation. Portney then approximates what proportion of these social costs were attributable to *just* the air quality regulations. Three annual social cost estimates are derived, each stated in 1984 dollars, which are listed below:

- $4.5 billion for 1975
- $13.7 billion for 1981
- $33.0 billion for 1985

Each of these values is an estimate of the explicit *and* implicit costs of the pre-1990 U.S. clean air laws at different points in time, and each can be interpreted as the value of resources given up by society to implement air quality control legislation.

Total Social Benefits

Estimating the total social benefits of the Clean Air Act requires a series of steps:

- Analyze the data that measure the trend in air quality over time
- Control for external influences on air quality to determine the amount of improvement due solely to public policy initiatives
- Determine the benefits gained by society from the air quality improvement and monetize them so they can be compared to the associated costs.

Given the availability of air quality trend data, it might seem that the only difficulty in executing the first step is in choosing which of these data to employ. While there is some truth

[11] These costs are deflated to remove any inflationary effects and are broken down by year (to present annualized data), by sector of the economy (i.e., EPA, non-EPA federal, state government, local government, and private sector), and by type (i.e., capital versus operating costs).

to this supposition, it is nonetheless shortsighted. It turns out that most of the available air quality data are not totally reliable or at least are limited in some way. This is not to say that these data are not useful; rather, they have limitations that must be acknowledged for the analysis to have validity.

A case in point are U.S. emissions data. Recall that these data are estimates determined by the EPA. It is important to understand how these estimates are calculated, to know exactly what information the data are conveying, and to be aware of the limitations imposed by the estimation procedure. Ambient concentration data are also imperfect measures of U.S. air quality. Although these data are based on actual measurements, the monitoring system in the United States, while improving, is currently less than adequate.

Another important consideration is that all national statistics, by construction, disguise regional and local differences. Even when national data suggest that the ambient air is improving, some parts of the country, such as urban areas, continue to face serious air pollution problems. Such localized variations are hidden by the averaging process. Finally, it is necessary to recognize that not all observed changes in air quality are attributable to policy. In truth, many factors other than regulatory controls can influence air pollution, and many of them are simply beyond our control.

Acknowledging these issues, Portney's estimate of social benefits for the pre-1990 clean air laws relies on a comprehensive survey and synthesis report conducted by another noted economist, A. Myrick Freeman (1982). Using ambient air concentration data from 1970 to 1978, Freeman translates improvements in pollution levels for the period into benefit categories, using links established in existing research studies. These categories are improvements in human health, decrease in damage to vegetation, decrease in materials damage, lower cleaning costs, and enhanced aesthetics and visibility. Monetized values for each are derived, resulting in the following estimates, expressed in 1984 dollars:

- Health: $27.2 billion
- Vegetation: 0.5 billion
- Materials: 1.1 billion
- Cleaning: 4.8 billion
- Property values: 3.7 billion

Aggregating these values, Portney estimates the total value of social benefits as of 1978 at $37.3 billion. The estimate, although only an approximation, suggests that society ought to be willing to pay $37.3 billion for the improvements to health, property, and the environment brought about by the clean air laws in effect at that time.[12]

Benefit-Cost Comparison

The final step in the analysis is to compare the estimates of social costs and social benefits and interpret the result. Notice, however, that in the context of the Portney study, the two independently estimated components are valued for different periods. The relevant social cost estimate is as of 1981, but the comparable estimate of social benefits is as of 1978.

Making one final concession, Portney suggests that the social benefits in 1981 are likely to be higher than their estimated value in 1978. He justifies this noting observed improvements in air quality through the period and increases in the general population, meaning that more people would gain from the improvements. As a result, he treats the $37.3 billion magnitude as a viable, albeit conservative, benefit estimate for 1981. Comparing this value with the 1981 social cost estimate of $13.7 billion, Portney offers a qualified conclusion that the total social

[12] For a complete analysis of Freeman's work, see Portney (1990a), pp. 54–60, or consult the original source, Freeman (1982).

benefits (*TSB*) of pre-1990 clean air legislation outweigh the total social costs (*TSC*), at least for 1981. We can summarize these findings as:

- Estimated *TSB* of pre-1990 clean air policy for 1981: **$37.3 billion** ($1984)
- Estimated *TSC* of pre-1990 clean air policy for 1981: **$13.7 billion** ($1984)
- Estimated net benefits of pre-1990 clean air policy for 1981: **$23.6 billion** ($1984)

While Portney suggests that this assessment is likely reasonable for 1981, he argues that the conclusion cannot be made without reservations. For example, Freeman attributes all the observed changes in air quality to the Clean Air Act, an assumption that biases the benefit estimates upward. Further, Portney cautions that even if the estimates are reasonable, they cannot be generalized beyond 1981. For instance, he points out that estimated social costs rise sharply in 1985 to $33.0 billion, an increase that is not likely to be matched by higher social benefits. Moreover, the benefit and cost magnitudes are estimated as *total* rather than as *marginal* values, which does not convey sufficient information to assess allocative efficiency.

Even though the *TSB* of clean air policy exceed the *TSC*, there is no reason to assume that abatement controls are at their efficient level. Such an evaluation can be made only by finding out whether the marginal social benefit (*MSB*) equals the marginal social cost (*MSC*) at that point. In fact, only through a marginal analysis can the implications of subsequent legislative changes be assessed.

To illustrate this point, consider three hypothetical depictions of the 1981 social benefit and cost estimates shown in Figure 10.2. Costs and benefits in billions of 1984 dollars are measured on the vertical axis, and pollution abatement (*A*) associated with air control policy is measured on the horizontal axis. Notice that in all three cases, at the 1981 level of abatement, A_{1981}, *TSB* of $37.3 billion are higher than *TSC* of $13.7 billion. What differs across the three diagrams is the location of A_{1981} relative to the efficient level of abatement A_E.

In Figure 10.2a, A_{1981} is shown as equivalent to the efficient level, A_E, the point at which the vertical distance between *TSB* and *TSC* is maximized or where the slopes of the two functions are equal. This corresponds to the abatement level that maximizes net benefits to society, or the point where *MSC* and *MSB* of abatement are equal. However, it is quite possible that A_{1981} is to the *left* of A_E, as is shown in Figure 10.2b. Such a result would indicate that *too little* abatement has been legislated. Just the opposite case is also possible where *too much* abatement is imposed by law. This alternative is shown in Figure 10.2c. The point is, without specific *marginal* benefit and cost data, there is no way to know whether the regulatory controls in place in 1981 correspond to an efficient level of abatement.[13]

Benefit-Cost Analysis of the 1990 Amendments

In an excellent and very readable article, Portney (1990b) extends and updates his benefit-cost analysis to assess the 1990 Amendments. In particular, he assigns estimated benefits and costs to each of the three major sets of provisions of the 1990 law: those for acidic deposition (Title IV), those for urban air pollution (Titles II and V), and those for hazardous air pollutants (Title III). There are two major differences between this analysis and the one Portney conducted for the earlier period. First, in this study *marginal* costs and benefits are examined as opposed to *total* costs and benefits. This use of marginal values allows for a determination of whether the additional controls imposed by the new amendments are allocatively efficient.

[13] In 1997, the EPA presented a retrospective analysis of the Clean Air Act over the 1970–1990 period, as directed by law. The comprehensive study includes benefit and cost estimates of U.S. clean air policy on an annualized basis for selected years (U.S. EPA, Office of Air and Radiation, October 1997). For 1980 (close to Portney's study year), the EPA's estimate of annual benefits is $930 billion ($1990) and its estimate of annual costs is $20.8 billion ($1990). In a recent article, Freeman (2002) offers an assessment of these and other EPA benefit-cost findings.

Figure 10.2 ***Benefit-Cost Analysis of U.S. Air Quality Controls as of 1981***

In all three depictions of U.S. policy, the *TSB* of $37.3 billion at the 1981 abatement level, A_{1981}, are higher than the *TSC* of $13.7 billion. What differs is the location of A_{1981} relative to the efficient level of abatement, A_E. In Figure 10.2a, A_{1981} is equal to A_E, since at this point, the slope of *TSB* is equal to the slope of *TSC*, meaning that the vertical distance between *TSB* and *TSC* is maximized. In Figure 10.2b, A_{1981} is to the *left* of A_E, indicating that *too little* abatement has been legislated. Just the opposite case is depicted in Figure 10.2c, where *too much* abatement is imposed by law. Without specific *marginal* benefit and cost data, there is no way to know whether the 1981 controls corresponded to an efficient abatement level.

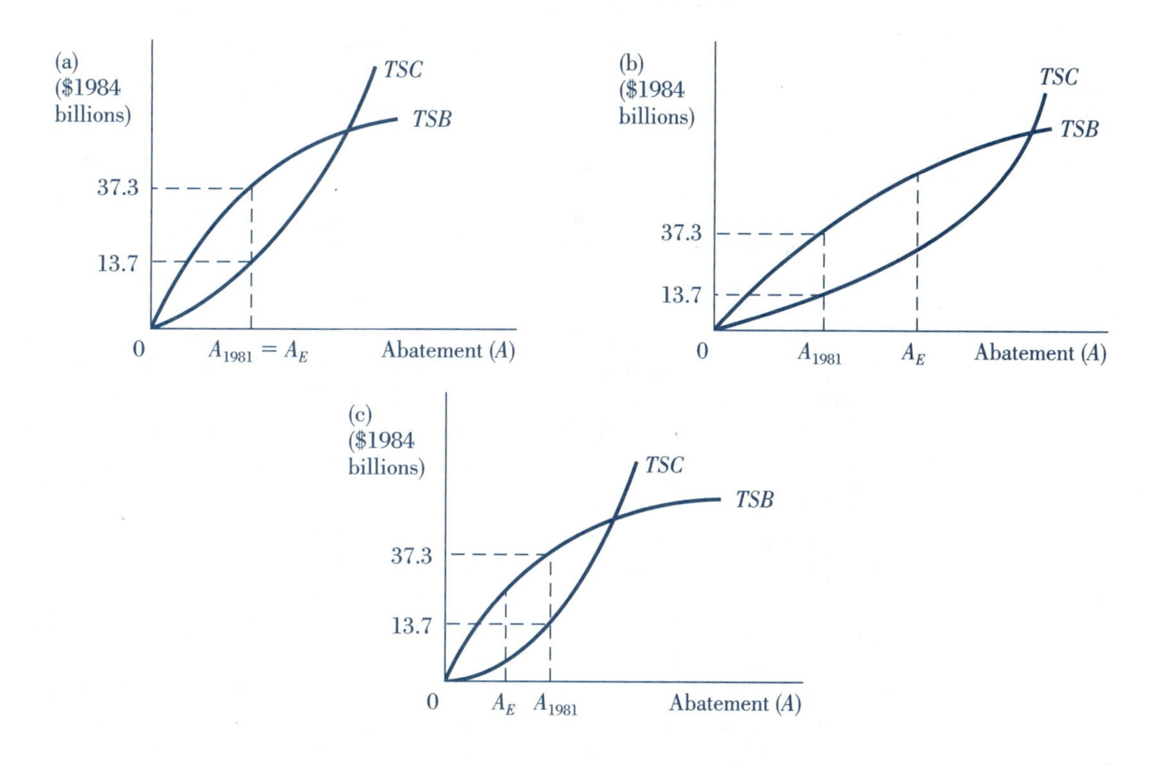

Sources: Values drawn from Portney (1990a); Hazilla and Kopp (1990); Freeman (1982).

Second, this study is a "seat-of-the-pants" evaluation, as Portney put it, because only *explicit* private costs are considered.

Marginal Costs and Benefits

To facilitate the discussion, we present Portney's (1990b) estimates of marginal benefits and costs in Table 10.7, which we assume are points lying on the true *MSB* and *MSC* curves.[14] The table lists the reported values in 1990 dollars for the three sets of provisions as well as for the aggregate of all three. For ease of comparison to the values Portney derived for the pre-1990 period, the magnitudes also are stated in 1984 dollars in parentheses. Notice that most of the estimates are reported as a range of values, since there is some amount of guesswork involved.

[14] Recall from Chapters 7 and 8 that technically these values are *incremental* as opposed to *marginal*, since the magnitudes represent a discrete change from one policy initiative to another.

Table 10.7 ***Estimated Annual Marginal Costs and Benefits of the 1990 Clean Air Act Amendments***

	Stated in billions of 1990 dollars (1984 dollars given in parentheses)				
	Acidic Deposition	**Urban Air Quality**	**Hazardous Air Pollutants**	**Aggregate Range**	**Aggregate Point Estimate**
Marginal costs	\$4 (\$3.2)	\$19–\$22 (\$15.1–\$17.5)	\$6–\$10 (\$4.8–\$7.9)	\$29–\$36 (\$23.1–\$28.6)	\$32 (\$25.4)
Marginal benefits	\$2–\$9 (\$1.6–\$7.2)	\$4–\$12 (\$3.2–\$9.5)	\$0–\$4 (\$0–\$3.2)	\$6–\$25 (\$4.8–\$19.9)	\$14 (\$11.1)

Source: Drawn from Portney (1990b).

Overall, Portney suggests that abatement linked to the 1990 Amendments should yield an *MSB* in the range of \$6–\$25 billion annually, with most of the benefit resulting from urban air quality improvements. Portney also offers a rough point estimate for *MSB* of \$14 billion, which is \$11.1 billion in 1984 dollars. The comparable estimate for the *MSC* is \$29–\$36 billion annually, or \$23.1–\$28.6 billion in 1984 dollars. Again, most of the costs are attributable to urban air quality controls. A point estimate for *MSC* to correspond to the benefit side is \$32 billion, or \$25.4 billion in 1984 dollars. From a relative perspective, the \$25.4 billion cost estimate suggests that the 1990 Amendments are expected to be very costly, particularly when the *total* social costs of U.S. policy as of 1981 were estimated to be only \$13.7 billion.

Benefit-Cost Comparison

Based on these estimates, one possible conclusion is that Titles II through V of the 1990 Amendments might be overregulating society, since *MSC* far outweighs *MSB*.[15] This relationship holds even if we conservatively compare the low end of the range for *MSC* (\$29 billion) with the upper end of the range for *MSB* (\$25 billion), giving an excess of \$4 billion in nominal terms. On the other hand, the excess could be as much as \$30 billion if we use the high end of the range for *MSC* (\$36 billion) with the low end of the range for *MSB* (\$6 billion).

To further illustrate this important result, Figure 10.3 presents two models of these estimates expressed in 1984 dollars. Figure 10.3a depicts the relationship between *MSC* and *MSB*. Notice that the hypothetical abatement level corresponding to the 1990 Amendments, A_{1990}, is located to the *right* of the efficient abatement level, A_E. It follows, therefore, that society would be better off with less regulation, which translates to a movement on the graph from A_{1990} to A_E.

The model in Figure 10.3b illustrates these same results using an approximation of the *TSC* and *TSB* as of 1990. To obtain the estimate for *TSC*, we add the estimated *MSC* for the 1990 law in 1984 dollars, \$25.4 billion, to Portney's (1990a) earlier estimate of *TSC* for 1981 measured in 1984 dollars, \$13.7 billion. This yields an estimate for the 1990 *TSC* of \$39.1 billion:

TSC effective 1981:	\$13.7 billion
+ *MSC* for 1990 Amendments:	+ \$25.4 billion
***TSC* effective 1990:**	**\$39.1 billion**

[15]There is, however, an important caveat. If, for example, Portney's (1990b) marginal cost estimates do not reflect least-cost decisions, meaning that they lie *above* the true *MSC* curve, then the appropriate conclusion is that the 1990 Amendments are not being implemented in a cost-effective manner. If this is the case, the efficiency of the abatement level being achieved by the 1990 law is indeterminate.

Figure 10.3 Analyzing the Marginal and Total Social Costs and Social Benefits Associated with the 1990 Clean Air Act Amendments

According to Portney's estimates, the *MSC* associated with the 1990 Amendments outweighs the *MSB*. Thus, the *MSC* is above the *MSB* at A_{1990}. Notice that A_{1990} is higher than the efficient abatement level, A_E, where the *MSB* curve intersects the *MSC* curve. This implies that society would be better off with less regulation, which translates to a movement from A_{1990} to A_E.

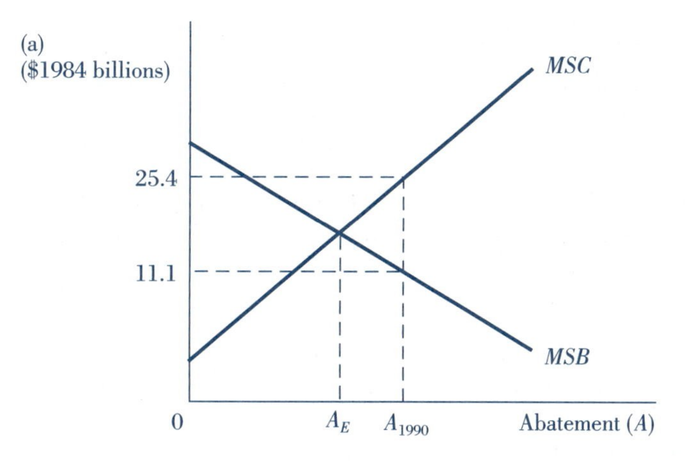

This model shows the estimated values of *TSB* and *TSC* associated with the hypothetical abatement level established by the 1990 Amendments, A_{1990}. Although *TSB* exceeds *TSC* at this point, A_{1990} is higher than the efficient abatement level, A_E. Just as in part (a), this model implies that the 1990 Amendments overregulate the private sector to achieve cleaner air.

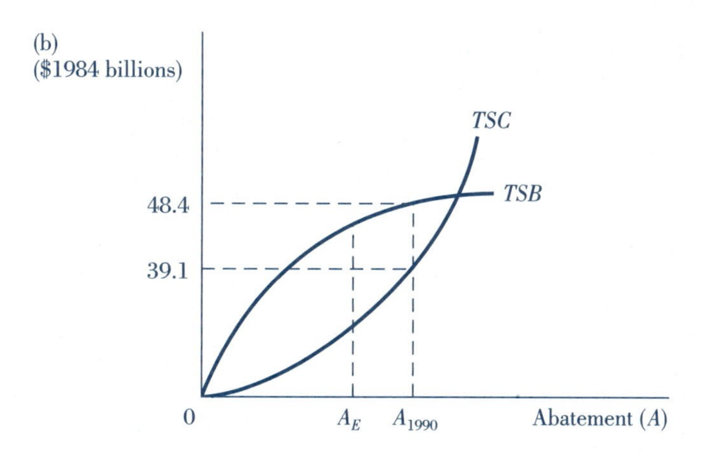

Sources: Values derived from Portney (1990b); Portney (1990a).

Similarly, to derive an approximation of *TSB* from all air quality controls in effect as of 1990, the estimated *MSB* for the 1990 Amendments, $11.1 billion, is added to Freeman's (1982) estimate of *TSB*, $37.3 billion, to get an approximation for the *TSB* as of 1990 of $48.4 billion:

	TSB effective 1978:	$37.3 billion
+	*MSB* for 1990 Amendments:	+ $11.1 billion
	***TSB* effective 1990:**	**$48.4 billion**

These values for *TSB* and *TSC* are shown in Figure 10.3b, corresponding to A_{1990}. Notice that while *TSB* exceeds *TSC*, the associated abatement level is nonetheless in excess of the efficient level, A_E.

The implication of these findings is that by implementing the 1990 Amendments, the U.S. government has overregulated the private sector to achieve cleaner air. Overallocation of resources to implement this legislation means that resources are being underallocated toward other uses. Many would agree with this conclusion, arguing, for example, that the allocation of government spending among environmental problems is disproportionate to the associated risks.

A common outcry is aimed at the relatively small amount of federal monies devoted to reducing the risks of such hazards as radon and tobacco smoke, which are known health hazards. For example, an article in *Fortune*, citing EPA data, points out that, although radon is responsible for 5,000 to 20,000 deaths per year, only about $100 million is spent annually to fight the problem. These statistics are troubling when compared to the $1.2 billion being spent on global warming, to which zero deaths are attributed.[16] But as we stated at the outset, benefit-cost analysis of environmental issues is not an exact science. These values are just estimates, and the scientific evidence on most if not all environmental problems is incomplete.

Taken in this context, Portney's research should be recognized for what it is *and* for what it is not. Although the results have been validated by some existing data and by the work of other researchers, they are nonetheless a series of estimates and guesses, albeit educated ones. There are no hard-and-fast conclusions that can be drawn from such preliminary results. Rather, more advanced research, using new data as they become available, is necessary to make a more concrete determination about the efficiency implications of the 1990 Amendments.

EPA's Benefit-Cost Analysis of the 1990 Amendments

Under Section 812 of the 1990 Clean Air Act, the EPA must prepare periodic reports to Congress on the benefits and costs of this major legislation. The first of these reports, a retrospective analysis, was presented in 1997 for the 20-year prior period from 1970 to 1990. The second report, the first in a series of prospective analyses, assesses the 1990–2010 period and was completed in 1999.[17] In this latter report, the EPA estimates that the present value of net benefits associated with the 1990 Amendments (Titles I through V) for the 20-year period is $510 billion ($1990). More specific estimates also are derived for individual titled sections and for specific annual periods.[18]

Because the EPA's approach and methodology are vastly different from Portney's, it is not surprising that the resulting estimates likewise are markedly dissimilar, and, in truth, the two sets of estimates are not directly comparable. For one thing, Portney's analysis is much less rigorous than the study conducted by the EPA. That said, however, the EPA's quantitative results, though recognized as based on sound methods and data, are considered to be controversial on a number of fronts.

---

[16] See Main (May 20, 1991).
[17] U.S. EPA, Office of Air and Radiation (October 1997, November 1999).
[18] Selected benefit and cost data from the 1999 report are shown in Exhibit A.4 in the Appendix. The actual reports should be consulted for detail on the methodology, underlying assumptions, modeling techniques, and uncertainties. Both reports can be accessed on the Internet at **http://www.epa.gov/air/sect812/index.html.**

As discussed in a recent article by Freeman (2002), the controversies arise on both the benefit and cost sides. For example, in the benefit assessment, the EPA study assumes that the statistical value of a human life is $4.8 million ($1990). However, this value is based on labor market studies for people younger than those most at risk for premature death due to poor air quality. On the cost side, the study assesses only direct costs, which means that changes in prices and employment are not considered. Hence, until greater clarity is achieved and further research is done, the efficiency implications of U.S. air quality policy remain uncertain.

Benefit-Cost Analysis of the Air Quality Standards

One of the more important contributions of the EPA's reports to Congress and studies such as Portney's is that they motivate the need for further investigation of the Clean Air Act. To that end, we conduct a benefit-cost analysis of the standards established through this legislation, focusing on the *efficiency* implications of two specific issues:

- the **absence of cost considerations** in setting the standards
- the **uniformity of the standards**

The *equity* implications of the Clean Air Act are discussed in Application 10.4. Although equity considerations are not the mainstay of economic analysis, they have taken on increasing importance in U.S. environmental policy making. In 1994, President Clinton signed Executive Order 12898, which directs all federal agencies to make environmental justice part of their missions.[19] Under the current Bush administration, there is an ongoing commitment to environmental justice, which was formally reaffirmed by EPA Administrator Christine Todd Whitman in August 2001.

Absence of Cost Considerations in the Standard-Setting Process

Setting the National Ambient Air Quality Standards (NAAQS)

From an efficiency perspective, a common criticism of both the primary and secondary NAAQS is that they are motivated solely by the anticipated benefits from protecting public health and welfare with no mention of the economic feasibility of doing so. Effectively, this means that costs, including implicit costs, are not to be explicitly considered. Such an omission is particularly problematic for the primary standards for health, which are to include a "margin of safety." This wording suggests that there is some level of pollution that will not cause any harm to public health. But is there a pollution level that does not harm at least one individual? Probably not. So, without the balance of cost considerations, the law seems to suggest setting the primary standards for criteria pollutants at zero. Although such an extreme was not Congress's intent, the law does not provide appropriate guidelines to set the NAAQS at an efficient level.[20]

Setting the National Emission Standards for Hazardous Air Pollutants (NESHAP)

An analogous argument has been made about the NESHAP, as originally defined in the Clean Air Act. Recall that the United States had made little progress in controlling air toxics prior to the changes made in the 1990 Amendments. To a great extent, this lack of progress was the result of using a **benefit-based decision rule** to control these substances. Specifically, for any identified air toxic, the EPA was required to set a standard that "provides an ample margin of safety to protect the public health."[21]

benefit-based decision rule
A guideline to improve society's well-being with no allowance for a balancing of the associated costs.

It is difficult to conceptualize what an "ample margin of safety" might be for such substances as carcinogens, which pose a threat even at low emission levels. Again, since the law

[19]This executive order can be accessed at **http://es.epa.gov/program/exec/eo-12898.html.**
[20]This discussion is elaborated by Burtraw and Portney (1991) and Portney (1990a).
[21]See Clean Air Act, sec. 112(b)(1)(B).

Application

10.4 *The Inequities of Air Pollution—Who Suffers More?*

While efficiency criteria are the mainstay of economic analyses, an evaluation of any public policy also must consider issues of equity. Does the policy correct for any preexisting inequities across population groups? Does its implementation affect all segments of society in the same way? These sorts of questions are particularly relevant to an assessment of national air quality policy. Consider the difference in air quality between rural and urban communities or between attainment and nonattainment areas. Although these differences seem to be strictly regional in orientation, it turns out that they translate to inequities across income, ethnic, and racial population groups.

The linkage is simple. Much of the racial and ethnic-based differences in air pollution exposure arise because a high proportion of minorities live in urban centers, where the ambient air is dirtier than in suburban and rural areas. In fact, some 63 percent of the nation's air-polluting facilities are located in urban areas. Given these residential differences, it is not surprising that certain minority groups are at relatively greater risk for the health hazards associated with exposure to urban air pollution.

The accompanying figure presents results of a study that examined the proportion of Hispanic, Asian/Pacific, African American, White, and Native American people living in areas of reduced air quality. Other analyses also find evidence of these environmental inequities. For example, Brajer and Hall (1992) examine the distribution of income, ethnic, and racial groups affected by exposure to ozone and fine particulate matter in the South Coast Air Basin of California. The results show a positive correlation between particulate matter exposure and the percentage of lower income families and between exposure and the percentage of African Americans and Hispanics. Similar results are found for ozone exposure, although some of the correlations are statistically weaker.

So what is being done about the problem? In the United States, the EPA formed an Environmental Equity Workgroup in 1990 to assess the evidence about risk to minority and low-income groups and to consider appropriate responses to any inequities it identifies. (Its findings are given in a formal report titled *Environmental Equity: Reducing Risk for All Communities,* issued in June 1992.) In 1994, the EPA also established its Office of Environmental Justice. In that same year, President Clinton issued Executive Order 12898, *Federal Actions to Address Environmental Justice in Minority Populations and Low-Income Populations.* This directive required agencies to develop strategies that incorporate environmental justice into their operations. Under the new Bush administration, the EPA has reaffirmed its commitment to environmental justice and continues to have an agenda in support of this issue. More information is available online at **http://www.epa.gov/compliance/environmentaljustice/.**

http://

Sources: U.S. EPA, Office of Policy, Planning, and Evaluation (June 1992); Council on Environmental Quality (1997); Wernette and Nieves (March/April 1992); Brajer and Hall (April 1992).

U.S. Populations Exposed to Poor Air Quality, 1993

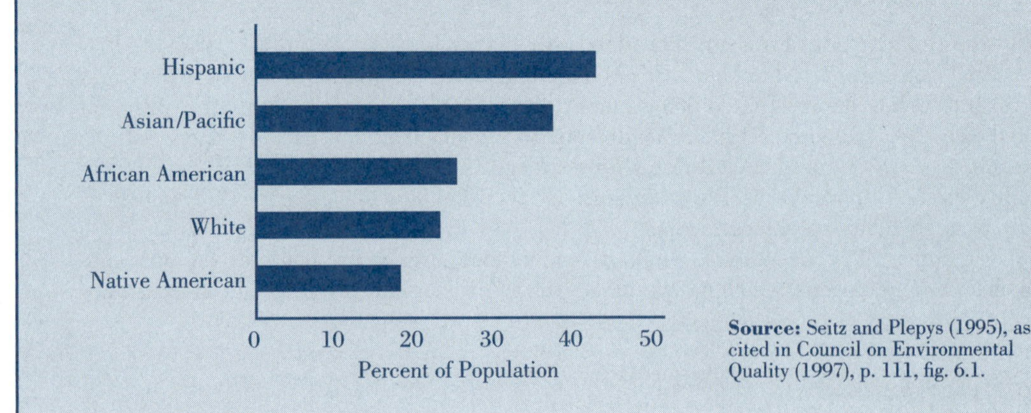

Source: Seitz and Plepys (1995), as cited in Council on Environmental Quality (1997), p. 111, fig. 6.1.

did not allow for any consideration of costs, taken literally it seemed to be calling for the setting of the NESHAP at zero emission levels. Such a radical decision would have had serious consequences for American industry, because many of these substances are critical to manufacturing processes. Recognizing the problem, the EPA was reticent to identify a substance as a hazardous air pollutant, and little was done to control these dangerous toxics for many years.[22]

Through the 1980s, there was some controversy about the identification and subsequent standard setting for air toxics, particularly for carcinogens. One suggestion was to allow for economic feasibility in the standard-setting process, a solution supported by then EPA administrator William Ruckelshaus. Opponents argued that such an objective approach was inappropriate for dealing with matters as delicate as human health.

Today, U.S. regulation of hazardous air pollutants is better defined, thanks to the 1990 Amendments. All air toxics subject to federal controls are identified explicitly in the law, and the EPA has been charged with setting emissions standards for these contaminants within a 10-year time frame. Furthermore, although the Clean Air Act continues to require that the standards protect public health and the environment, it also allows for the costs of achieving these standards to be considered—an important step toward efficiency in the setting of the NESHAP.

Uniformity of the National Ambient Air Quality Standards (NAAQS)

A critical observation to make about the NAAQS is that, since they are nationally based, they ignore any region-specific cost or benefit differences associated with meeting them. All nonattainment regions must meet the same uniform standards regardless of such differences as existing pollution levels, access to technology, demographics, and traffic patterns. Yet the marginal costs and benefits of reducing pollution to the same level across these highly diverse regions likely will be quite dissimilar. This might explain why some regions met the legislated deadlines with relative ease, whereas others continue to struggle with compliance. Thus, while the shift to federally mandated standards was intended to strengthen U.S. policy, it nonetheless contributed to the inefficiency that typically arises with a command-and-control approach.

One legislative change that allows for some region-specific differences is the recognition of prevention of significant deterioration (PSD) areas and the use of existing air quality in those regions as the relevant standard. This policy reform effectively elevates the standard for PSD areas above the NAAQS applicable to nonattainment areas. The relevant issue is whether this particular use of differentiated standards represents an efficient allocation of resources across both types of regions.[23] Some economists question the wisdom of assigning more resources to already clean areas in order to achieve a higher standard of air quality.[24] To investigate this issue, we use benefit-cost analysis in a simple economic model.

Benefit-Cost Analysis of Higher PSD Standards

To economically justify the relatively higher standards in PSD areas, the associated *MSC* and *MSB* curves of abatement for these regions must intersect at a higher level of abatement than the comparable intersection for nonattainment areas. For this to occur, there must be certain differences in either or both of the *MSC* and *MSB* across the two types of regions. In Figure 10.4, we analyze three possible scenarios under which the use of different standards across PSD and nonattainment areas would be efficient. The marginal cost and benefit curves for a representative PSD area are labeled as MSC_{PSD} and MSB_{PSD}, and those for a nonattainment area as MSC_{NON} and MSB_{NON}.

[22] National Commission on Air Quality (March 1981), pp. 76–77.
[23] In order for the differentiated abatement levels to be efficient, the efficient level of abatement in a PSD area must be higher than that in a nonattainment area.
[24] See Portney (1990a), pp. 78–79.

Figure 10.4 ***Economic Modeling of Higher Air Quality Standards in PSD Areas Relative to Nonattainment Areas***

In panel (a), $MSB_{PSD}=MSB_{NON}$. In this case, the efficient abatement level in the PSD area A_{PSD}, would be higher than the efficient level in a nonattainment area A_{NON} only if MSC_{PSD} is *lower* than MSC_{NON}. In panel (b), $MSC_{NON}=MSC_{PSD}$. Here, efficiency holds only if MSB_{PSD} is *above* MSB_{NON}. Finally, in panel (c), both the *MSC* and *MSB* curves are unique to each area. Under this assumption, efficiency holds only under certain sets of conditions.

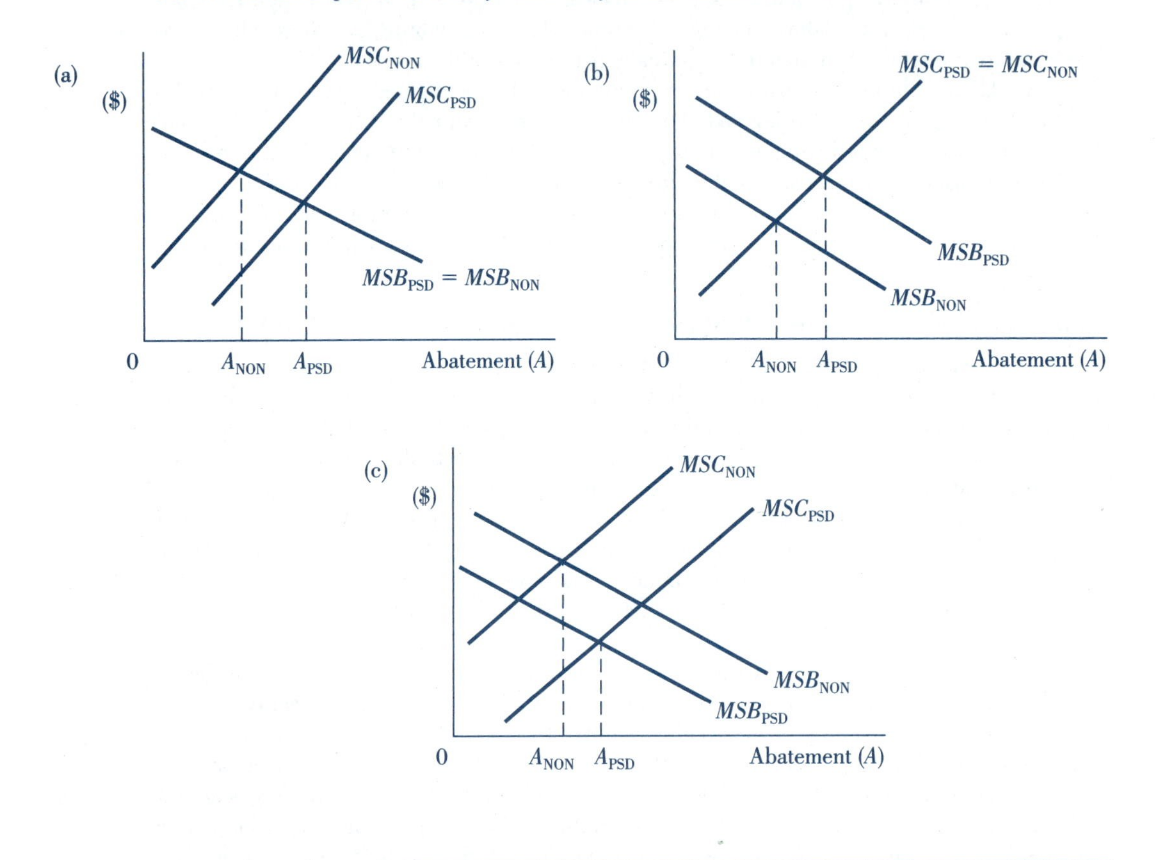

In Figure 10.4a, the *MSB* functions are identical across the two types of areas, that is, $MSB_{PSD} = MSB_{NON}$ at all abatement levels (A). If this relationship holds, the efficient abatement level in the PSD area (A_{PSD},) would be higher than the efficient level in a nonattainment area, A_{NON}, only if MSC_{PSD} is *lower* than MSC_{NON}. In Figure 10.4b the opposite set of conditions is modeled, with the *MSC* functions drawn as identical curves. Here, efficiency holds only if MSB_{PSD} is *above* MSB_{NON}. Finally, in Figure 10.4c, both the *MSC* curve and the *MSB* curve are allowed to be unique to each area. Under this latter assumption, one way that efficiency results is if the following three conditions hold:

1. MSB_{PSD} is *below* MSB_{NON},
2. MSC_{PSD} is *below* MSC_{NON}, and
3. the vertical distance between the *MSC* curves is sufficiently *greater* than that between the *MSB* functions to support the relative position of A_{PSD} and A_{NON}.[25]

---

[25] Other scenarios are possible under which both the *MSC* and the *MSB* curves are allowed to be unique and the A_{PSD} is higher than A_{NON}. To see this, try modeling a case different from the one shown in Figure 10.4c.

So far, what we have illustrated is that the use of relatively higher standards for PSD areas can be efficient, but only under certain conditions. Now, we need to consider the logic of each scenario based on the expected relationship between the respective marginal costs and benefits for the two areas. For context, think of the differences between nonattainment areas and the most stringently controlled PSD areas, which are Class I. Recall that these PSD areas include national parks and wilderness regions. One reason these areas have cleaner air is that they are less populated and less industrialized than their nonattainment counterparts. How might these conditions affect the *MSB* and *MSC?*

On the cost side, it is reasonable to expect that MSC_{PSD} is lower than MSC_{NON}, since there are fewer polluting sources to control in PSD regions. By itself, this assertion suggests that the model in Figure 10.4b is not a likely representation of actual conditions.

Now consider the benefit side. The primary benefits of maintaining higher air quality in PSD areas are nonhealth gains, such as enhanced recreational uses, aesthetic improvements, and protection of the ecology. In contrast, the major benefits of cleaning up the more populated nonattainment areas are improvements in human health. Since health improvements are generally valued more at the margin than nonhealth gains, MSB_{NON} is likely higher than MSB_{PSD} at all abatement levels. Based on this hypothesis, the model shown in Figure 10.4a appears to be inaccurate.

Taking these assumed benefit differences into account along with the expected cost differences, the model in Figure 10.4c appears to be more accurate than the other two models in Figure 10.4. However, recall that for this model to support the efficiency criterion, several conditions have to hold. Hence, all we can conclude is that the higher standards in the cleaner PSD areas *may* be justifiable on efficiency grounds under certain economic circumstances.

Another reform that acknowledges region-specific differences was the reclassification of nonattainment areas for ozone, carbon monoxide, and particulate matter enacted within the 1990 Amendments. By imposing progressively more stringent controls on the dirtiest of these areas, these revisions may have improved efficiency. Again, this issue can be examined through benefit-cost analysis.

Benefit-Cost Analysis of Progressively Regulated Nonattainment Areas

To give our analysis context, we consider two of the new classifications for nonattainment ozone areas: marginal areas at the cleaner end of the continuum and extreme areas at the other end. Now, we can apply benefit-cost analysis to examine the efficiency of strengthening regulatory controls in extreme ozone areas.

Since the United States is imposing more stringent regulations on extreme ozone regions, the *MSC* in extreme regions will necessarily be higher than the *MSC* in marginal ozone regions at all levels of abatement. Hence, in order for the use of different control methods to make sense from an economic perspective, the *MSB* at the existing level of abatement in extreme regions must be greater than the *MSB* at the current abatement level in marginal regions.[26] As long as this is correct, we can argue that the higher incremental costs of more stringent regulations are at least to some extent justified by higher incremental benefits.

Conclusions

U.S. air quality policy has evolved considerably over the last several decades. Legislative initiatives and the infrastructure to implement them have become more extensive and more complex. Indeed, the 1990 Clean Air Act Amendments are far more comprehensive than any U.S. environmental legislation enacted to date. Today, there is a massive network of agencies and interagencies working toward the objective of improving national air quality. In some sense,

[26] If the *MSB* curves are the same for both regions, this relative difference will necessarily hold, because the abatement level in extreme regions is by definition lower than that in marginal regions. If the two curves are distinct, then the *MSB* in extreme regions lies everywhere above the *MSB* in marginal regions.

such an evolution is not surprising. Air pollution is a difficult problem, and the sheer size of the United States means that air quality controls must accommodate a variety of meteorological, geographic, and economic conditions.

Despite the numerous revisions and new initiatives, U.S. legislation continues to be based primarily on a command-and-control approach, evidenced largely by the use of uniform standards to define national air quality. The problem is that such an approach is not likely to achieve an efficient solution to the nation's air pollution problems. In fact, benefit-cost analysis suggests that clean air policy initiatives have been more costly to society than necessary. Similar conclusions follow from a qualitative evaluation of the standard-setting process. From an economic perspective, uniform standards are a troubling element of U.S. air legislation. Likewise, the absence of cost considerations in establishing these standards defies time-tested economic theory.

Of course, there are no simple answers, nor can there be any sweeping conclusions without careful reservations. Air quality issues involve many intangibles that are difficult to quantify. Moreover, many of the nation's new clean air policies have not been fully implemented, so part of the assessment must rely on estimated projections. Nonetheless, the analytical process guided by economic theory provides a way to better understand the motivation and economic implications of the Clean Air Act and to set more realistic expectations about achieving the nation's air quality objectives.

Summary

- There were no federal air pollution laws until 1955, when the Air Pollution Control Act was passed, and no comprehensive legislation until the Clean Air Act of 1963. The 1970 Clean Air Act Amendments marked dramatic policy changes in the United States, followed by still more amendments in 1977.
- On November 15, 1990, President George H. W. Bush signed into law extensive changes in U.S. air quality control policy in the form of the 1990 Clean Air Act Amendments.
- Two groups of air pollutants have been identified as causing the greatest damage to outdoor air quality: criteria pollutants and hazardous air pollutants.
- The six criteria pollutants are particulate matter, sulfur dioxide, carbon monoxide, nitrogen dioxide, tropospheric ozone, and lead. An initial list of 189 hazardous air pollutants is identified in the 1990 Clean Air Act Amendments.
- Established by the EPA, the National Ambient Air Quality Standards (NAAQS) state the maximum allowable concentrations of criteria air pollutants that may be emitted from stationary or mobile sources into the outside air.
- Primary NAAQS are standards set to protect public health, with some margin of safety. Secondary NAAQS are intended to protect public welfare.
- For the hazardous air pollutants, National Emission Standards for Hazardous Air Pollutants (NESHAP) are established for every major source of one or more of the listed hazardous air pollutants.
- Coordination of air quality control policy between federal and state governments is formalized through a State Implementation Plan (SIP). An SIP is an EPA-approved procedure outlining how a state intends to implement, monitor, and enforce the NAAQS and the NESHAP.
- There are currently 247 air quality control regions (AQCRs) defined in the United States. These are classified into nonattainment areas and prevention of significant deterioration (PSD) areas. In the 1990 Clean Air Act Amendments, new classifications are established for certain of the nonattainment areas.

- The research of Paul Portney suggests that the total social benefits of U.S. clean air legislation as of 1981 outweigh the total social costs. In a more current study, Portney finds that the 1990 Amendments may abate pollution beyond the efficient level.
- According to the EPA's second report to Congress on the benefits and costs of the Clean Air Act, the present value of net benefits associated with the 1990 Amendments (Titles I through V) for the 20-year period (1990–2010) is $510 billion in 1990 dollars.
- The NAAQS ignore region-specific cost or benefit differences associated with meeting them. This uniformity suggests an inefficient allocation of resources across the AQCRs.
- The primary and secondary NAAQS are motivated solely by anticipated benefits from protecting public health and welfare, meaning that costs are not explicitly considered.
- The historical lack of progress in setting the NESHAP was the result of using a benefit-based decision rule. The 1990 Amendments have improved air toxics regulation by identifying substances to be controlled in the law and by revising the standard setting to allow for costs to be considered.
- The higher standards in the cleaner PSD areas *may* be justifiable on efficiency grounds under certain economic conditions.
- The use of progressively more stringent controls on the dirtiest nonattainment regions may have improved resource allocation if the marginal benefit of abatement in these regions is higher than that in cleaner nonattainment areas.

Key Concepts

natural pollutants
anthropogenic pollutants
criteria documents
criteria pollutants
hazardous air pollutants
stationary source
mobile source
National Ambient Air Quality Standards (NAAQS)
primary NAAQS
secondary NAAQS
National Emission Standards for Hazardous Air Pollutants (NESHAP)
maximum achievable control technology (MACT)
State Implementation Plan (SIP)
Air Quality Control Region (AQCR)
prevention of significant deterioration (PSD) areas
nonattainment areas
benefit-based decision rule

Use the Key Concepts listed above to begin your search for additional articles and information using the InfoTrac College Edition database.

Review Questions

1. If you were responsible for setting the NAAQS for lead, what key determinants would you consider if the standard were established to meet the efficiency criterion? Be sure to itemize separately the benefits and costs associated with your decision.
2. Using the efficiency criterion, carefully analyze the problem faced by the EPA in identifying hazardous air pollutants prior to the 1990 Amendments.
3. a. Briefly explain the significance of prevention of significant deterioration (PSD) areas to the setting of air quality standards.
   b. From an economic perspective, explain the paradox associated with setting higher standards in PSD areas relative to nonattainment areas. Show how an inefficient result may arise using an *MSB-MSC* model.
4. Briefly summarize Portney's overall assessment of the 1990 Clean Air Act Amendments, and discuss the major implications of these findings for society.
5. Refer back to Table 10.7 of the marginal benefit and cost estimates for the 1990 Clean Air Act Amendments. Conduct an individual benefit-cost analysis for each of the three major components of the amendments, namely, acidic deposition, urban air quality, and hazardous air pollutants. What do you conclude?

6. Visit the site of EPA's Green Book: Nonattainment Areas for Criteria Pollutants, which can be accessed at **http://www.epa.gov/oar/oaqps/greenbk/**, and use the data to investigate which AQCRs in your state have nonattainment status for each of the criteria pollutants. Summarize your findings.

Additional Readings

Aeppel, Timothy. "Clean Air Act Triggers Backlash as Its Focus Shifts to Driving Habits." *Wall Street Journal,* January 25, 1995, pp. A1, A10.

Anderson, J. W. "Revising the Air Quality Standards: A Briefing Paper on the Proposed NAAQS for PM and O_3." Washington, DC: Resources for the Future, February 1997.

Brannigan, Martha. "CAT Scan May Soon 'Map' Air Pollution." *Wall Street Journal,* November 10, 1994, p. B7.

Clement, Douglas. "Cost v. Benefit: Clearing the Air?" *The Region* (Minneapolis: Federal Reserve Bank of Minneapolis, December 2001), pp. 19–21, 48–57.

"Environmental Protection—Has It Been Fair?" *EPA Journal* 18 (March/April 1992).

Freeman, A. Myrick, III. "Air and Water Pollution Policy." In Paul R. Portney, ed., *Current Issues in U.S. Environmental Policy.* Baltimore: Johns Hopkins University Press, 1978.

Hall, Jane V. "Air Quality in Developing Countries." *Contemporary Economic Policy* 13 (April 1995), pp. 77–85.

Hall, Jane V., and Amy L. Walton. "A Case Study in Pollution Markets: Dismal Science vs. Dismal Reality." *Contemporary Economic Policy* 14 (April 1996), pp. 67–78.

Koch, Gayle S., and Paul R. Ammann. "Current Trends in Federal and State Regulation of Hazardous Air Pollutants." *Journal of Environmental Regulation* 4 (Autumn 1994), pp. 25–41.

Krupnick, Alan J., and J. W. Anderson. "Revising the Ozone Standard." *Resources* 125 (fall 1996), pp. 6–9.

Palmer, Karen, Wallace E. Oates, and Paul R. Portney. "Tightening Environmental Standards: The Benefit-Cost or the No-Cost Paradigm?" *Journal of Economic Perspectives* 9 (winter 1995), pp. 129–32.

Solomon, Caleb, and Oscar Suris. "Hot Weather Hurts Efforts to Clean Up Air." *Wall Street Journal,* July 7, 1994, pp. B1, B6.

http:// Related Web Sites

Air Pollution Monitoring	http://www.epa.gov/oar/oaqps/montring.html
The Benefits and Costs of the Clean Air Act: 1970 to 1990, U.S. EPA, Office of Air and Radiation (October 1997)	http://www.epa.gov/air/sect812/index.html
The Benefits and Costs of the Clean Air Act: 1990 to 2010, U.S. EPA, Office of Air and Radiation (November 1999)	http://www.epa.gov/air/sect812/index.html
Clean Air Act, Section 110 on SIPs	http://www.epa.gov/oar/caa/caa110.txt
Green Book: Nonattainment Areas for Criteria Pollutants	http://www.epa.gov/oar/oaqps/greenbk/
Health Effects Notebook for Hazardous Air Pollutants	http://www.epa.gov/ttn/atw/hapindex.html
Latest Findings on National Air Quality: 2000 Status and Trends, U.S. EPA, Office of Air Quality Planning and Standards (September 2001)	http://www.epa.gov/air/aqtrnd00/index.html
Organisation for Economic Co-operation and Development, Environment Directorate	http://www.oecd.org/env
Overview of the 1990 Clean Air Act Amendments	http://www.epa.gov/oar/caa/overview.txt
The Plain English Guide to the Clean Air Act, U.S. EPA, Office of Air Quality Planning and Standards (April 1993)	http://www.epa.gov/oar/oaqps/peg_caa/pegcaain.html
Presidential Executive Order 12898, Environmental Justice	http://es.epa.gov/program/exec/eo-12898.html
United Nations Environment Programme (UNEP)	http://www.unep.org
Updated NAAQS for ozone and for PM	http://www.epa.gov/ttn/oarpg/naaqsfin/ http://www.epa.gov/airlinks/airlinks4.html
U.S. Air Quality Nonattainment Areas	http://www.epa.gov/airs/nonattn.html
U.S. EPA Air Toxics site	http://www.epa.gov/ttn/atw/pollsour.html
U.S. EPA AIRTrends Reports	http://www.epa.gov/airtrends/
U.S. EPA Environmental Justice	http://www.epa.gov/compliance/environmentaljustice/
U.S. NAAQS	http://www.epa.gov/airs/criteria.html

A Reference to Acronyms and Terms Used in Air Quality Control Policy

Environmental Economics Acronyms

TSB	Total social benefits
TSC	Total social costs
MSB	Marginal social benefit
MSC	Marginal social cost
MSC_{PSD}	Marginal social cost of abatement in a PSD area
MSB_{PSD}	Marginal social benefit of abatement in a PSD area
MSC_{NON}	Marginal social cost of abatement in a nonattainment area
MSB_{NON}	Marginal social benefit of abatement in a nonattainment area

Environmental Science Terms

CO	Carbon monoxide
CO_2	Carbon dioxide
mg/m^3	Micrograms per cubic meter
mg/m^3	Milligrams per cubic meter
NO_x	Nitrogen oxides
NO_2	Nitrogen dioxide
O_3	Ozone
Pb	Lead
PM	Particulate matter
PM-2.5	Particulate matter of less than 2.5 micrograms in diameter
PM-10	Particulate matter of less than 10 micrograms in diameter
ppm	Parts per million
SO_2	Sulfur dioxide
SO_x	Sulfur oxides

Environmental Policy Acronyms

AQCR	Air Quality Control Region
BAT	Best Available Technology
MACT	Maximum Achievable Control Technology
NAAQS	National Ambient Air Quality Standards
NAPAP	National Acidic Precipitation Assessment Program
NESHAP	National Emission Standards for Hazardous Air Pollutants
PSD	Prevention of Significant Deterioration
SIP	State Implementation Plan

Chapter 11

Improving Air Quality: Controlling Mobile Sources

Of the two major classes of air pollutants identified by the Clean Air Act, the criteria pollutants are more common. Because they are pervasive, these substances are responsible for most of the air pollution in the world, even though they are less dangerous than hazardous air pollutants. The relevant point is that everyone is exposed to the risks of criteria pollutants, which explains why the Clean Air Act places such a strong emphasis on them and why Congress elected to legislate strategies to implement the National Ambient Air Quality Standards (NAAQS). Although the motivation for this decision may be well placed, it turns out that this top-down policy approach has had some adverse consequences, an assertion that we will support with economic analysis in this chapter.

Starting from a general perspective, we know that the Clean Air Act establishes stringent regulations on both mobile and stationary sources, because both contribute to the release of all six criteria pollutants. In fact, controlling these sources of criteria pollutants is the means by which the NAAQS are implemented. In this chapter, we focus on mobile sources, deferring our discussion of stationary sources to Chapter 12.

Automobiles and other mobile sources are controlled by provisions in Title II of the Clean Air Act. Some of the policy strategies outlined in the law are tailored to these types of polluting sources and to the air quality problems with which they are commonly associated. To illustrate this point, we present some interesting data in Figure 11.1 based on 1999 emissions in the United States. The pie charts in this figure show clearly how mobile sources are more responsible for some of the criteria pollutants than other polluting sources. Notice, for example, that transportation sources released 77 percent of the carbon monoxide (CO) emissions, 56 percent of the nitrogen oxide (NO_X) emissions, and 47 percent of the volatile organic compound (VOC) emissions.

In fact, because motor vehicles are the major source of emissions that cause urban smog, we begin this chapter with a discussion of urban air pollution. Once done, we begin our investigation of mobile source controls. This includes a brief retrospective of what has transpired over time between the government and the American automobile industry and an overview of U.S. mobile source policy based on the 1990 Clean Air Act Amendments. An economic analysis follows, which examines the efficiency and cost-effectiveness of mobile source initiatives. A list of commonly used acronyms and terms is provided for reference at the end of the chapter.

Urban Air Pollution: An Important Policy Motivation

Environmentalists as well as policymakers are concerned about the air pollution that characterizes urban centers. In these metropolitan areas, the high concentration of human population, traffic, and industrial activity intensifies the concentration of criteria pollutants and hence increases the environmental risks of exposure. Because a large proportion of the population is exposed to the associated health hazards, increased abatement efforts in cities should yield a higher level of marginal benefit than in rural communities.

Figure 11.1 *Sources of Criteria Pollutant Emissions for 1999*

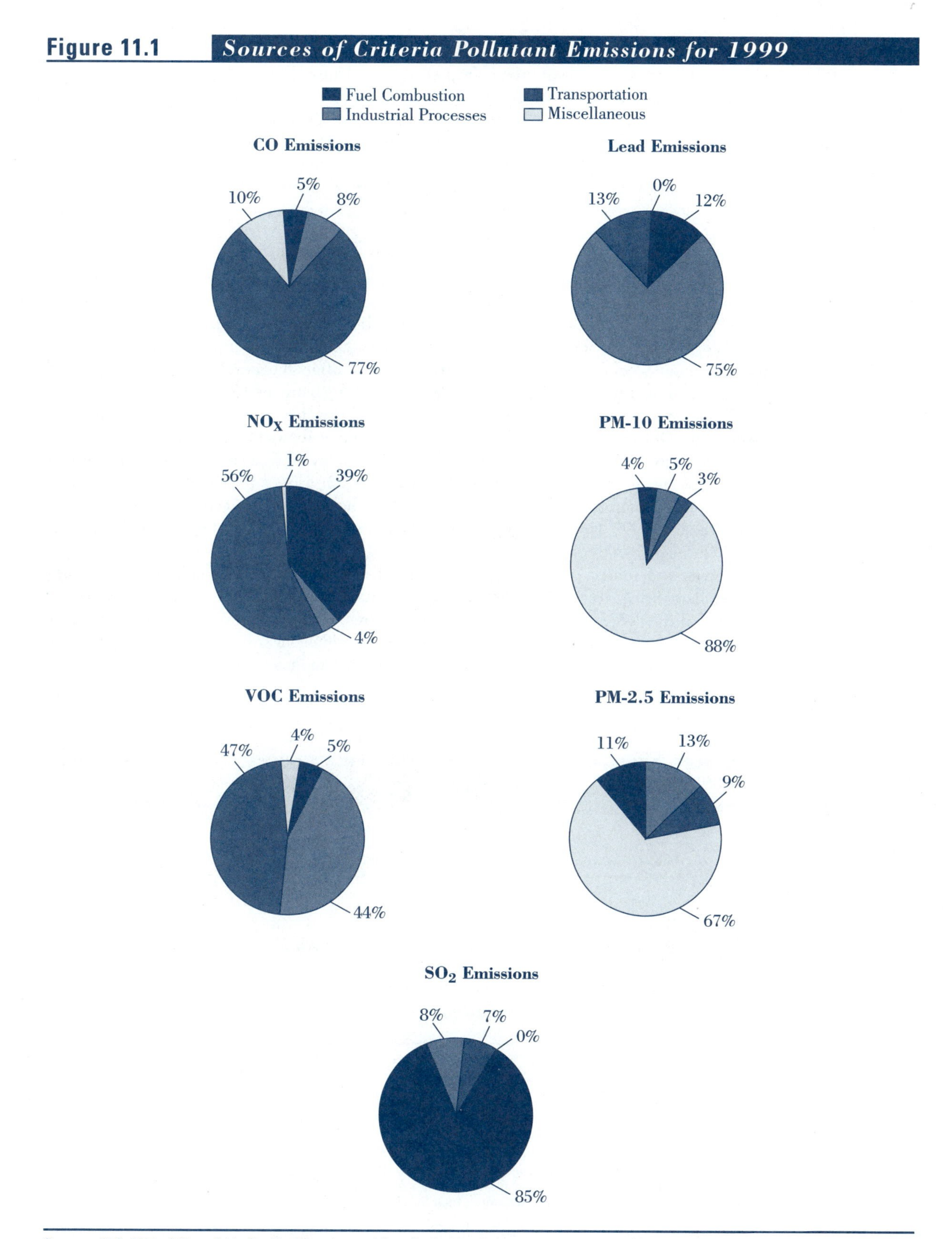

Source: U.S. EPA, Office of Air Quality Planning and Standards (March 2001), pp. 132–59, tables A2–A9.

Measuring U.S. Urban Air Quality[1]

Air Quality Index (AQI)
An index that signifies the worst daily air quality in an urban area over some time period.

To develop a better sense of urban air quality in the United States, the EPA monitors the air in those metropolitan statistical areas (MSAs) with populations greater than 350,000. Starting in 1976, the agency began reporting part of its findings using the Pollutant Standards Index (PSI), which was updated in June 2000 and renamed the **Air Quality Index (AQI).**[2] The AQI value uses data obtained from a monitoring network that measures five of the criteria pollutants—particulate matter (PM), sulfur dioxide (SO_2), carbon monoxide (CO), ozone (O_3), and nitrogen dioxide (NO_2). From these values, a single index number is formed based on the short-term national standards for each substance.[3]

The resulting AQI value, ranging from 0 to 500, signifies the worst daily air quality in an urban area over a given time period. An AQI of 100 is considered the standard set by the Clean Air Act. The descriptor words and color codes that identify the health effects associated with various ranges of the AQI are shown in Table 11.1. The AQI, its set of descriptors, and corresponding color codes are used by various media in the United States to communicate air quality conditions to the general public. In addition, this new index has been adopted by other nations around the world, such as Mexico, Singapore, and Taiwan.[4]

To gain a sense of urban air quality in the United States, look at Table 11.2, which presents selected AQI trend data over a 10-year period. The table shows the number of days in a year that each city experienced an air quality level signified by an AQI greater than 100, that is, a value not meeting the standard set by the Clean Air Act. Notice that in all the cities shown, with the exception of Seattle, the number of days that the AQI exceeded 100 was lower in 2000 than in 1995. However, only seven of the cities show a consistent downward trend from 1990 through 2000.

The data in Table 11.2 also indicate that air quality in Los Angeles is far worse than any other major urban area in the sample, a distinction that has persisted for several decades. As early as the 1940s, Los Angeles residents noticed frequent episodes of a brownish haze on the horizon—an awareness that eventually led to nationwide concern about urban air pollution. Today, we know that the brownish haze in Los Angeles and many other metropolitan areas is **urban smog.**

Table 11.1 *Air Quality Index (AQI) Categories, Color Codes, and Ranges*

AQI Values	Level of Health Concern	Color
0–50	Good	Green
51–100	Moderate	Yellow
101–150	Unhealthy for sensitive groups	Orange
151–200	Unhealthy	Red
201–300	Very unhealthy	Purple
301–500	Hazardous	Maroon

Source: U.S. EPA, Office of Air and Radiation (June 2000).

[1] Much of the following discussion is obtained from U.S. EPA, Office of Air Quality Planning and Standards (March 2001), chap. 3, and U.S. EPA, Office of Air and Radiation (June 2000).

[2] For more information on the AQI, consult the EPA brochure *Air Quality Index: A Guide to Air Quality and Your Health* (U.S. EPA, Office of Air and Radiation, June 2000) which is accessible online at **http://www.epa.gov/airnow/aqibroch/.**

[3] The criteria pollutant omitted from the index is lead, because there are no short-term NAAQS for this substance.

[4] U.S. EPA, Office of Air Quality Planning and Standards (March 2001), p. 71.

Urban Smog

Coined from the words *smoke* and *fog*, the word **smog** originally referred to the overall air quality first observed in London at the turn of the century. Used in this context, smog refers mainly to the presence of particulate matter and other emissions, such as sulfur oxides, in the air.[5] Because of Los Angeles's history and its dubious distinction as America's "king of smog," many people erroneously think of this city as the only one facing a severe smog problem. It is true that the dense traffic there and the meteorological conditions make this region particularly vulnerable to smog formation. However, most major cities suffer from the effects of smog. The problem is more pervasive in highly industrialized and heavily populated parts of the world, such as Europe and Japan.

In urban areas, another type of smog can form from a chemical reaction involving several of the criteria pollutants. Scientists refer to this type of smog as **photochemical smog** to distinguish it from the more conventional "smoke and fog" variety. Photochemical smog is formed from certain air pollutants that chemically react in the presence of sunlight (thus the prefix *photo*) to form entirely new substances. Its principal component is tropospheric ozone

photochemical smog
A type of smog caused by pollutants that chemically react in sunlight to form new substances.

Table 11.2 *Trend Data for Major Urban Areas: Number of Days with AQI Greater Than 100*

Metropolitan Statistical Area	1990	1995	2000
Atlanta	42	35	26
Baltimore	29	36	16
Boston	7	11	1
Chicago	4	23	0
Cleveland	10	27	5
Dallas	24	36	20
Denver	9	3	2
Detroit	11	14	3
El Paso	19	8	3
Houston	51	65	42
Kansas City	2	23	10
Los Angeles	173	113	48
Miami	1	2	0
Minneapolis–St. Paul	4	5	0
New York	36	19	12
Philadelphia	39	38	18
Phoenix	12	22	10
Pittsburgh	19	25	4
St. Louis	23	36	14
San Diego	96	48	14
San Francisco	0	2	0
Seattle	9	0	1
Washington, DC	25	32	11

Source: U.S. EPA, Office of Air Quality Planning and Standards (March 2001), table A-17 (for 1990 data); U.S. EPA, Office of Air and Radiation, (January 16, 2002) (for 1995 and 2000 data).

Note: AQI is the Air Quality Index, a dimensionless number ranging from 0 to 500 converted from daily monitoring data for PM, SO_2, CO, O_3, and NO_2 based on their short-term NAAQS, federal episode criteria, and significant harm levels. (Lead is excluded because it has none of these criteria.) Data are reported from trend sites, that is, those sites for which there is sufficient historical data to be used for trends.

[5] Pryde (1973), p. 147.

Application

11.1 *Mexico City's Serious Smog Problem*

Mexico City faces an ongoing battle to reduce its air pollution problem, said to be the worst in the world. At points in time, its ozone and carbon monoxide levels have been measured at almost four times the acceptable level by U.S. standards. The smog in this famous capital city is so severe that the future of many of its historic buildings, statues, and monuments is seriously threatened. Workers are unable to keep up with the continual degradation of the city's buildings, some of them centuries old. The city's historian, Jorge Hernandez, laments the lost detail on Mexico City's treasured statues, which can never be replaced.

Officials are taking steps to reverse some of the damage. A primary target is the automobile, allegedly responsible for two-thirds of the smog problem. Such a statistic is not surprising, given that some 3 million drivers operate vehicles that are, on average, 10 years of age. In an attempt to rid the city of the aging automobile population, new car buyers must turn in an old one, a requirement started in 1993. To mitigate the problem in the future, automobile manufacturers in Mexico made a commitment to begin installing catalytic converters on their new vehicles, but residents demanded more immediate action.

In response, Mexico City officials ordered the installation of pollution abatement devices on all taxicabs and small buses. This mandate may extend to private citizens as part of the license renewal process. This official stance sparked the entrepreneurial spirit in literally hundreds of firms, who submitted prototypes of pollution devices for the city's approval.

Another directive begun in 1990 requires that every car in the city be prohibited from the streets one day each week. Under this plan, each automobile is assigned a colored sticker corresponding to the particular day of the week on which driving is banned. The city's billboards herald which color corresponds to which day, and teams of enforcers work the intersections to make certain the mandate is followed. Violators are fined 30 days of minimum wages, or approximately $120.

In 1997, forest fires so worsened the city's already poor air quality that officials had to ban half of the 3 million automobiles from roadways and urge the city's 60,000 factories to cut back production by 33 percent. According to one official, these stringent measures have had a substantial economic impact, estimated to be $8 million per day for lost production and reduced labor hours. Added to these economic costs were associated health costs of some $3.5 million per day.

Sources: Cormier (November 23, 1992); Baker (June 25, 1990) Marshall (February 25, 1992) "Mexico's Bad Air Exacts Financial Toll, Official Says." (May 30, 1998).

(O_3), also referred to as ground-level ozone. Along with more than 100 different compounds, ground-level ozone is produced by a chemical reaction of nitrogen oxides (NO_X), (VOCs) (primarily hydrocarbons), and sunlight.[6]

Because the chemical reaction is stimulated by sun and temperature, peak ozone levels are observed in the warmer seasons of the year. In most parts of the United States, the ozone season is from May to October. In warmer regions such as the South and Southwest, ozone levels might be elevated for the entire year.

Mobile sources are major contributors to photochemical smog formation, since they are responsible for the largest proportion of NO_X and VOC emissions, the precursors of smog, across the major source categories. (Refer to Figure 11.1 on p. 235.) Among all transportation sources, those emitting the largest amounts of these substances are ordinary highway vehicles, particularly those fueled by gasoline. It is important to recognize that in urban centers, officials are more concerned with emissions from transportation sources because of traffic

[6] Visit **http://www.epa.gov/airnow/health/index.html** to view the EPA document *Smog—Who Does It Hurt?* (U.S. EPA, Office of Air Quality Planning and Standards, July 1999).

congestion. In Mexico City, for example, local officials are implementing drastic controls on mobile sources, including a program that prohibits the use of every automobile for one day per week. Application 11.1 discusses the air quality conditions faced by Mexico's smoggy capital city and officials' attempts to find a solution.

Controlling Mobile Sources

Of all the transportation vehicles, the primary focus of U.S. policy is highway vehicles, such as passenger cars and trucks. As Figure 11.2 shows, these polluting sources, particularly those that are gasoline powered, emit the largest quantity among mobile sources of nearly every criteria pollutant.[7] These data explain the policy focus on the manufacture of these

Figure 11.2 *National Emissions Estimates for Mobile Sources by Major Category for 1999*

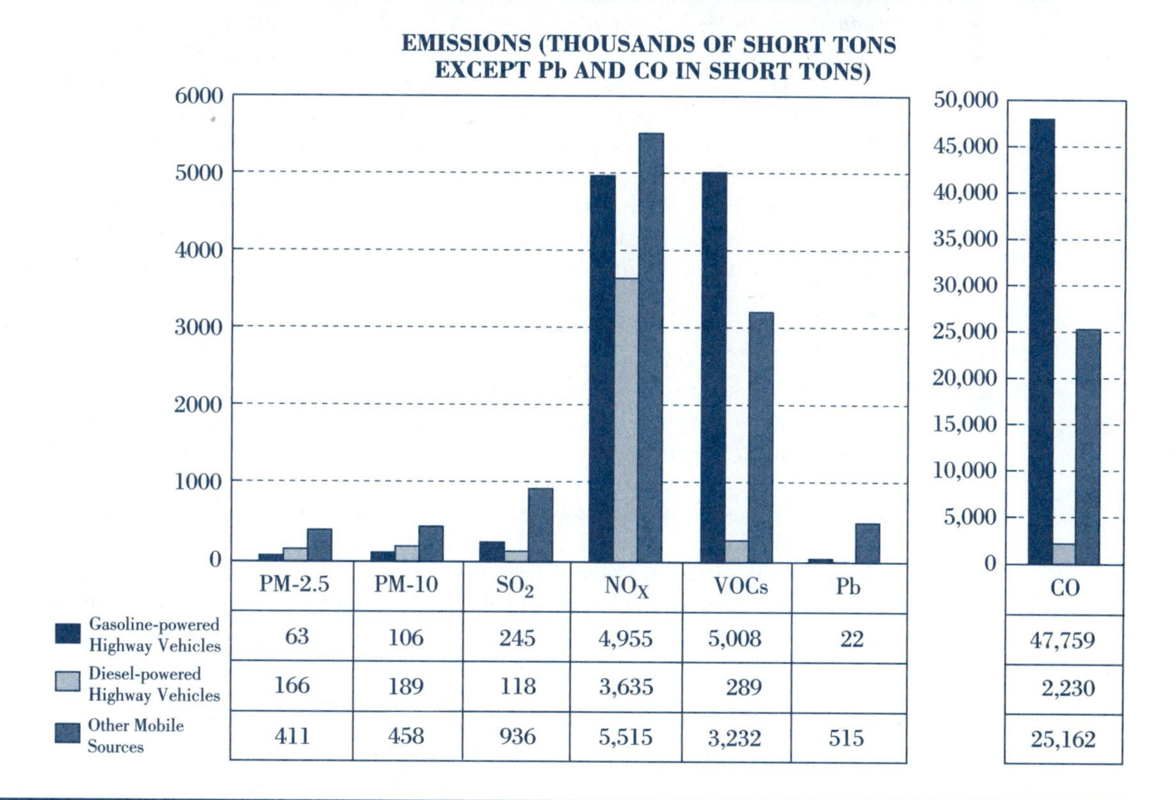

	PM-2.5	PM-10	SO_2	NO_X	VOCs	Pb	CO
Gasoline-powered Highway Vehicles	63	106	245	4,955	5,008	22	47,759
Diesel-powered Highway Vehicles	166	189	118	3,635	289		2,230
Other Mobile Sources	411	458	936	5,515	3,232	515	25,162

Source: U.S. EPA Office of Air Quality Planning and Standards (March 2001), pp. 132–59, tables A2–A9.

Note: CO is shown separately because of the difference in its vertical scale; 1 short ton = 2,000 pounds.

VOC emissions are used to measure the formation of ozone (O_3).

Data shown for Pb are for all on-road vehicles.

Other mobile sources are nonroad gasoline, nonroad diesel, aircraft, railroads, and marine vessels.

Data shown for Pb are for all nonroad engines and vehicles.

[7] As noted in the figure, the emissions of VOCs are used as a proxy measure of ground-level ozone formation. More information on automobile emissions can be found at the Web site of the EPA's Office of Transportation and Air Quality at **http://www.epa.gov/otaq/ld-hwy.htm.**

vehicles and the increasing number of initiatives aimed at developing alternative fuels. To research the emissions, fuel efficiency, and environmental ranking of any make and model vehicle on the Internet, go to **http://www.epa.gov/greenvehicles.**

For its part, the American automobile industry has begun to take a more aggressive position toward reducing motor vehicle pollution, not only to meet government controls but also to accommodate environmentally conscious consumers. In 1992, the Big Three automakers (General Motors, Ford, and Chrysler) formed a research consortium to develop cleaner-running automobiles, a cooperative effort discussed in Application 11.2.

Such public and private initiatives were unheard of 25 years ago. The evolution of U.S. controls on motor vehicles underscores the inherent problems associated with regulation of a major industrial sector—in this case, the U.S. auto industry. Scientific evidence tying the automobile to the formation of smog had existed since the early 1950s, but no federal mandate to control auto emissions was legislated until 1965, and it was not until 1970 that emission limits appeared within the law itself.

Brief Retrospective on Motor Vehicle Emission Controls[8]

By most accounts, national legislation on motor vehicle emissions was slow to start in the United States. In 1963, Congress passed into law the Clean Air Act, the first extensive set of air quality controls. Notably absent from its provisions was any direct regulation of automobile emissions. This was a startling omission, given that California had passed new laws that same year to do exactly that. In contrast to California's activism, the Clean Air Act of 1963 took a much more cautious approach, giving virtually no power to federal officials to directly control mobile sources.

It was not until 1965, when the Clean Air Act was first amended, that federal controls on mobile sources were strengthened, calling for uniform emission standards. Somewhat ironically, this transition was motivated in large part by the historically uncooperative automobile industry. The reason was an economic one. Automakers had an incentive to lobby for *national* controls, because they wanted to avoid the costly alternative of having to meet state-specific standards. The 1965 law set maximum emissions for new automobiles, starting with the 1968 model year, but for only two of the major pollutants: carbon monoxide and hydrocarbons.

Mobile source controls in the 1970 Amendments were precedent setting in that new car emission standards were explicitly stated in the statutes themselves, and they were extremely stringent. In retrospect, most argue that achieving them was not feasible with available technology. Emissions of carbon monoxide and hydrocarbons were to be reduced by 90 percent by the 1975 model year from their 1970 levels, with comparable reductions for nitrogen oxide emissions by the 1976 model year from their 1971 level. These standards were to be applicable over the entire useful life of the vehicle, a requirement that placed even more pressure on the automobile industry. Manufacturers had to guarantee their emission control systems for this entire period, not only to the initial buyer but to every subsequent purchaser as well. Predictably, Detroit's automakers sought extensions to the legislated deadlines and adjustments to the targeted emission levels. Ultimately, the compliance dates originally set for 1975 and 1976 were postponed for five years.

All told, the early years of mobile source controls were marked by a series of extensions and a chronicle of delays. As the power struggle between Washington and Detroit continued, society grew increasingly impatient with the lack of progress. Washington officials knew that federal policy on motor vehicle emissions was in need of major revision.

Current U.S. Controls on Motor Vehicles and Fuels

Attempting to correct errors of the past, Congress passed the 1990 Clean Air Act Amendments, which strengthened U.S. controls on motor vehicle emissions and fuels. In addition to tougher command-and-control regulations, the new laws also incorporate incentives to en-

[8]To read more about this history, see "Milestones in Auto Emissions Control" at **http://www.epa.gov/otaq/12-miles.htm** (U.S. EPA, Office of Transportation and Air Quality, August 1994b).

11.2 *The Big Three Form a Research Consortium in the 1990s*

Application

Stiffer requirements for auto emissions in the 1990 Clean Air Act Amendments and under current California law are partly responsible for the formation of an emissions control research consortium of the three largest U.S. automakers: General Motors, Ford, and Chrysler. A similar arrangement formed in 1953 and a related cross-licensing agreement ultimately were dissolved by a 1969 antitrust suit settled through a consent decree. Interestingly, the provisions of that consent decree, which prohibited cooperation for any technological research in emissions controls, expired in 1987.

Of course, times have changed since the 1950s, when General Motors, Ford, and Chrysler enjoyed unquestionable dominance of the automobile industry. Their control of the market in that era was in large part responsible for the legal action against the former research consortium. Government officials had reason to believe that this earlier research effort was another delay tactic made possible through implicit collusion.

Today, the Big Three face pressure from foreign producers and, as a result, control a much less commanding market share. Imports currently account for about 17 percent of all U.S. new car sales, up from a 0.4 percent share in the post–World War II period. Foreign-built transplants represent another 24 percent. These facts, coupled with a probusiness political atmosphere, make the new consortium an unlikely target for antitrust officials. In fact, the U.S. government is now supporting collaborative research efforts, validating that support through legislation passed in 1984 by Congress. For their part, GM, Ford, and Chrysler argue that their cooperative effort not only will help achieve environmental policy objectives but also will strengthen their position against foreign competition.

The research effort, named USCAR (United States Council for Automobile Research) is one of eight such agreements among the automakers. It is aimed at investigating a variety of technologies to develop cleaner-running automobiles as well as cleaner fuels. For an up-to-date perspective on the outcome of this research effort, visit **http://www.uscar.org/**.

http://

Sources: Miller (June 9, 1992); Brock (2001).

courage technological development of cleaner-running vehicles and cleaner alternative fuels. Among the more salient features of this law are:

- emission reductions for cars and trucks
- onboard pollution control systems for conventional motor vehicles
- fuel quality controls
- initiatives to develop clean fuel vehicles

A brief summary of each of these initiatives follows.[9]

Emission Reductions for Motor Vehicles

The 1990 Amendments impose tougher regulations on tailpipe emissions from all types of highway vehicles. More refined than under existing law, the new standards have two tiers. The first refers to the initial 5 years or 50,000 miles of use, and the second establishes less stringent standards for the remainder of the vehicle's useful life redefined at 10 years or 100,000 miles. Beyond this extension of a car's economic life, the new emissions limits are implemented much as they have been in the past: uniformly applicable to every new model car.

[9]Title II of the Clean Air Act includes the provisions relating to mobile sources. To access this titled section on the Internet, visit **http://www.epa.gov/oar/caa/contents.html**.

Onboard Pollution Control Systems for Light-Duty Vehicles

To control refueling emissions, the 1990 law requires onboard vapor recovery systems to be installed on new light-duty vehicles. Other provisions call for emission control diagnostic systems on all light-duty vehicles and trucks starting with the 1994 model year. At minimum, these systems must be able to accurately identify a failure in the vehicle's catalytic converter and oxygen sensor over its useful life.

Fuel Quality Controls

Just as vehicles are more stringently regulated under the 1990 law, so too are fuels and fuel additives. For example, after December 31, 1995, any fuel containing lead or lead additives was prohibited for highway use. Other provisions require the use of newly developed cleaner fuels in certain of the nonattainment areas. For instance, **reformulated gasoline** must be used in some ozone nonattainment regions, and **oxygenated fuel** must be used in designated carbon monoxide nonattainment areas.

reformulated gasoline
Newly developed fuels that emit less hydrocarbons, carbon monoxide, and toxics than conventional gasoline.

oxygenated fuel
Formulations with enhanced oxygen content to allow for more complete combustion and hence a reduction in CO emissions.

- **Reformulated gasoline** refers to formulations that emit less hydrocarbons, carbon monoxide, and toxics than conventional gasoline.[10]
- **Oxygenated fuel** contains more oxygen to allow for more complete combustion and a reduction in carbon monoxide emissions.

To encourage compliance with these new fuel laws, market-based incentives are used. Marketable credits are issued for fuels that exceed legal requirements. These credits can be used by the recipient or transferred to another individual in the same nonattainment area.

Recently, there has been increasing discussion about using ethanol as an oxygen-boosting additive to gasoline. Why? Because the once common use of methyl tertiary-butyl ether (MTBE) as an oxygenated additive has come under scrutiny for posing an environmental risk. The substance, which had been touted as a relatively inexpensive way to meet oxygenated requirements, was detected in drinking water supplies around the nation, apparently leaking from underground storage tanks. Beyond the concerns that the additive is known to add an offensive taste and odor to drinking water, research also suggests that MTBE is a possible human carcinogen at high doses. As a consequence, the Clinton-Gore administration recommended congressional action to amend the Clean Air Act to reduce or eliminate the use of MTBE, and some states decided to phase out use of the substance entirely.[11] These actions prompted the need for substitute additives to comply with regulations on clean fuels, which led to renewed support for ethanol.

In addition to its environmental advantages, production of the corn-based ethanol would be a boon to the ailing agricultural industry, according to its proponents. However, use of ethanol is not without debate. Some argue that it is a costly solution, with an estimated production cost of $1.75 per gallon. Furthermore, a recent and controversial analysis by a Cornell University professor argues that more energy is required to produce the fuel than the energy it generates.[12]

Clean Fuel Vehicles

clean fuel vehicle
A vehicle certified to meet stringent emission standards.

Responding further to urban air quality pollution, the 1990 legislation establishes a **clean fuel vehicles** program. A **clean fuel vehicle** is one that has been certified to meet stringent emission standards for substances such as CO, NO_X, PM, and formaldehyde over prescribed time periods. For designated ozone and carbon monoxide nonattainment areas, states must set up programs for the adoption of clean fuel vehicles by owners or operators of fleets of more than 10 vehicles. These programs were to be phased in starting with the 1998 model year. At

[10] For more information on reformulated gasoline, visit **http://www.epa.gov/otaq/rfgvehpf.htm.**
[11] Alvarez and Barboza (July 23, 2001) and U.S. EPA (March 20, 2000). For information and updates about MTBE, visit the EPA's MTBE home page at **http://www.epa.gov/mtbe.**
[12] Hall (September 25, 2001) cites the analysis by David Pimentel, which was published in September 2001 in the *Encyclopedia of Physical Sciences and Technology.*

this point, a fixed percentage of all new fleet vehicles bought in each area had to be clean fuel vehicles and use **clean alternative fuels.** These are any fuels, such as methanol, ethanol, or other alcohols, or any power sources, including electricity, used in a clean fuel vehicle. Application 11.3 presents the pros and cons of using clean fuels to reduce air pollution.

clean alternative fuels
Fuels such as methanol, ethanol, or other alcohols, or power sources, such as electricity, used in a clean fuel vehicle.

Again, market incentives are integrated into the new provisions. Credits are to be issued to fleet operators or owners who surpass the requirements in the law. They might do this by purchasing clean fuel vehicles in advance of the deadline or by buying more of these vehicles than the law requires. The awarded credits can be held, banked for future use, or traded. They are to be weighted by the emission reduction achieved by each vehicle, which gives the fleet owner a further incentive to go beyond what the law requires.

Economic Analysis of Mobile Source Controls

Historically, U.S. attempts to control emissions from mobile sources have been frustrated by a litany of problems and delays. In a very real sense, many of the difficulties were predictable, even avoidable. In retrospect, much of the problem was that the law failed to consider the dynamics of the marketplace and the importance of benefit-cost analysis. To provide structure to our evaluation of these issues, we focus our analysis on the following characteristics of mobile source emissions control policy:

- absence of benefit-cost analysis in setting standards for tailpipe emissions
- uniformity of auto emissions standards
- bias against new vehicles
- implications of clean fuel alternatives

Absence of Benefit-Cost Analysis: An Inefficient Decision Rule

It is an interesting irony that after years of delay in establishing federal limits on auto emissions, the 1970 Clean Air Act Amendments imposed controls that by most accounts were extraordinarily stringent. Even at that point in legislative history, there was virtually no argument that the stated objectives for emission reductions were ambitious. In part, the stringent controls were a response to accumulating evidence that motor vehicles were a major source of pollution. However, there was also reason to believe that Washington was attempting to take a strong position against the American automobile industry.

Beyond the motivations for tougher standards, the more critical issue is whether the emissions targets were technologically feasible and, if not, whether they were intentionally set beyond what was possible given the existing knowledge base. Most argue that the standards were not attainable and in fact were meant to be **technology forcing,** by design—that is, specifically set to compel the auto industry to find solutions. Such a strategy may have been in order, given the industry's track record to that point. However, most have questioned the government's wisdom in setting emission reductions so far out of reach that manufacturers had a strong case to seek adjustments and postponements.

Just as important is the fact that the implied decision rule used to establish the emissions controls was not supported by benefit-cost analysis. Rather, the standards were set to protect public health and welfare, a purely **benefit-based** objective. As shown in Figure 11.3 on page 245, this is linked to abatement activity that maximizes total social benefits, which corresponds to the point where the marginal social benefit (*MSB*) curve crosses the horizontal axis (i.e., where $MSB = 0$), at A_0. Notice that A_0 is *higher* than the efficient level A_E where the *MSB* curve intersects the marginal social cost (*MSC*) curve (i.e., where $MSB = MSC$), which implies overregulation.

As it turned out, the 1970 standards were never implemented in their original form. In fact, they were modified considerably. Worse, the haggling over the unrealistic standards between

Application

11.3 Changing Fuels for America's Automobiles: Pros and Cons

Two characteristics of petroleum-based fuels are linked to air pollution: their burning efficiency and their volatility. When fuels burn inefficiently, carbon monoxide (CO) emissions are produced as a by-product of combustion. This problem is more prevalent in cold weather and is exacerbated when fuels are burned in poorly maintained engines. As efficiency improves, carbon dioxide (CO_2) emissions are formed instead. A fuel's volatility, or its capacity to evaporate, is also important to air pollution because evaporation releases toxic vapors into the air. Fuel volatility arises from the presence of volatile organic compounds (VOCs), which are added to enhance a fuel's octane rating or antiknock properties. The higher the volatility of gasoline, the more VOC emissions are released into the atmosphere.

The 1990 Clean Air Act Amendments explicitly call for the use of several types of cleaner fuels under certain well-defined conditions. These are broadly classified into three groups: **oxygenated fuels, reformulated gasoline,** and **clean alternative fuels.**

http://

- **Oxygenated Fuels:** The U.S. oxygenated fuels program is aimed at reducing CO emissions and therefore is used in CO nonattainment areas. Fuel additives provide the oxygen content needed to enhance fuel efficiency and allow for more complete combustion. Fuel prices increased by a few cents per gallon as a result.
- **Reformulated Gasoline:** Launched in 1995, the reformulated gasoline program is applicable to regions facing severe ozone problems. Aimed specifically at fuel volatility and toxicity, reformulated fuel must contain at minimum 2 percent oxygen and at maximum 1 percent benzene. Standards for this fuel require specific reductions of hydrocarbon and toxic emissions.
- **Clean Alternative Fuels:** Clean alternative fuels may include reformulated gasoline or other alternatives in place of the more conventional petroleum-based products. These alternatives include ethanol, methanol, compressed natural gas, propane, and electricity.

The accompanying table summarizes the major advantages and disadvantages of these fuels based on information provided by the EPA. For further information, there is a comprehensive Web site on fuels established by the EPA's Office of Transportation and Air Quality at **http://www.epa.gov/otaq/fuels.htm.**

Sources: U.S. EPA, Office of Transportation and Air Quality (August 1994a); U.S. EPA, Office of Transportation and Air Quality (August 1994c); Kinsman, (April 1992); Younger (November/December 1992).

Fuel	Advantages	Disadvantages
Ethanol	Very low emissions of ozone-forming hydrocarbons and toxics; made from renewable sources	High fuel cost; somewhat lower vehicle range
Methanol	Very low emissions of ozone-forming hydrocarbons and toxics; can be made from renewables	Fuel initially could be imported; somewhat lower vehicle range
Natural gas (methane)	Very low emissions of ozone-forming hydrocarbons, toxics, and CO; can be made from renewables	Higher vehicle cost; lower vehicle range; less convenient refueling
Propane	Cheaper than gasoline today; widely available; somewhat lower emissions of ozone-forming hydrocarbons and toxics	Cost will rise with demand; limited supply; no energy security or trade balance benefits
Reformulated gasoline	Can be used without changing fuel distribution system; somewhat lower emissions of ozone-forming hydrocarbons, NO_X, and toxics	Somewhat higher fuel cost; few energy security or trade balance benefits
Electricity	Potential for zero vehicle emissions; can recharge at night when power demand is low	Higher vehicle cost; lower vehicle performance; lower vehicle range; less convenient refueling

Figure 11.3 ***Inefficiency of a Benefit-Based Decision Rule on Motor Vehicle Emissions***

If mobile source standards are set to protect public health and welfare, they represent a purely benefit-based objective. Such an objective calls for abatement activity that maximizes total social benefits, which corresponds to the point where the *MSB* curve crosses the horizontal axis (i.e., where *MSB*=0), at A_0. Notice that A_0 is *higher* than the efficient level A_E (i.e., where *MSB*=*MSC*), which implies overregulation.

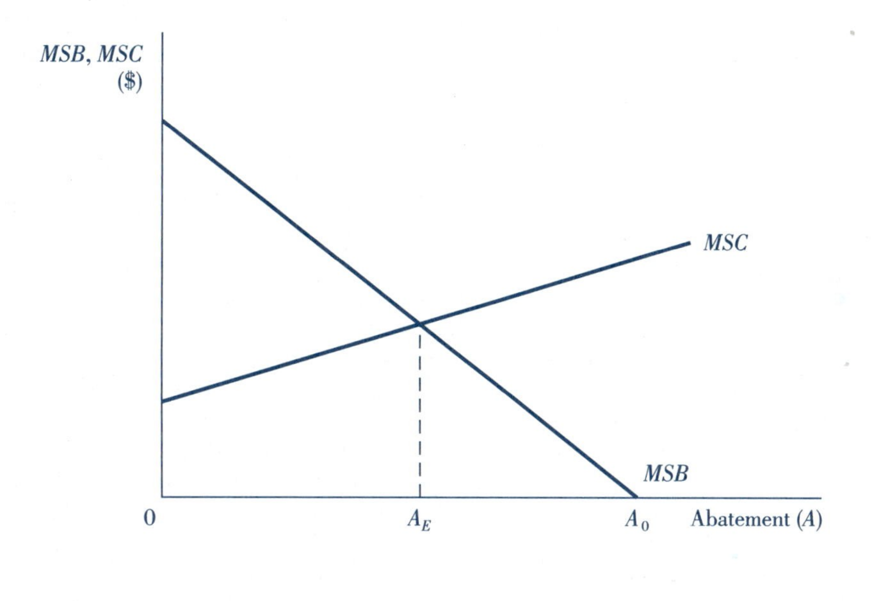

the government and the automobile industry added further to the already long delays. Had the standards been set at levels that accounted for the full extent of the associated costs, it is quite possible that the automobile industry would not have had sufficient support for its lobbying efforts to extend the statutory deadlines. Instead, the lack of efficient decision making left policymakers unarmed for the assault mounted by the automakers. The result? The ultimate burden was borne by society, and the ill-defined decisions of that era accomplished little more than to postpone the provision of cleaner-running vehicles to the marketplace.

Uniformity of Auto Emissions Standards

An ongoing problem with U.S. mobile source controls is that with few exceptions the emission standards are applicable across the board on every model produced and without regard to where the vehicle will be driven. Another uniform requirement calls for all cars and light-duty trucks starting with 1984 models to meet high-altitude standards, despite the fact that only about 3.5 percent of cars sold in the United States are purchased in high-altitude areas. Ultimately, such an approach needlessly increases the costs of reducing pollution with no offsetting benefit to society.[13]

Disregard for the location where a vehicle is operated overregulates vehicles sold in relatively clean areas and undercontrols those sold in more polluted areas. This is illustrated in

[13] Council of Economic Advisers (1982), pp. 145–46; Seskin (1978), p. 91; and White (1982), pp. 67–68. White reports that, according to General Motors, the per unit cost in current dollars of this high-altitude standard is approximately $65 for a gas-powered automobile and $230 for a diesel-powered car.

Figure 11.4, where we logically assume that the *MSB* of abatement in dirty areas is higher than in clean areas, *ceteris paribus.*[14] Notice that if the standard, A_{ST}, is the same for both regions, the dirty area is undercontrolled, because A_{ST} is below the efficient level, A_E. The opposite holds in relatively clean areas, where A_{ST} is above the efficient level A_E.

An alternative would be to have two sets of standards to be administered at the state level based on the degree of pollution in that area. Proponents of this more federalistic approach believe this would yield important cost savings. A 1974 study conducted by the National Academy of Sciences estimates the costs of a **uniform standard** of 0.4 grams per mile (gpm) for NO_X relative to the costs of a **two-tiered standard,** calling for a 0.4 gpm standard in seriously polluted areas (about 37 percent of motor vehicles) and a 3.1 gpm standard in all other areas.[15] The study predicts that a $23 billion savings over the 1975–1985 period would have resulted if the two-tiered system had been used.

Similarly, it is argued that the emission controls need not be applicable to every model produced. Substantial savings could be achieved by allowing each producer to meet legislated targets based on *average* emissions of all new cars produced. The benefits based on overall emission reductions would be unchanged, but the costs of compliance would be much lower.[16] The forgone savings help to explain the growth rate of mobile source control costs since the 1970 Amendments from $1.3 billion ($1986) in 1972 to $7.5 billion ($1986) in 1987.[17]

Current law does allow some exceptions to these uniform standards. Perhaps the most notable is the use of more stringent standards for vehicles sold in California. Other states have followed suit. California law also calls for 2 percent of vehicle sales in the state to be zero-emission vehicles (i.e., electric cars) by 2003.[18] Massachusetts and New York have enacted similar requirements.

These regional variances from national requirements have set off a round of debates. The auto industry points to the increased costs of complying with several sets of emissions standards. Others question the wisdom of moving to electric vehicles that have to be fueled by power plants. In some regions, such as the Northeast, electric power plants are more important contributors to smog than are mobile sources, so the net gain in air quality may not be worth the effort. Furthermore, the technology of electric car engineering is still in its infancy, particularly in addressing the harsher conditions of cold-weather climates. An alternative to electric cars that seems to hold greater promise is the so-called hybrid vehicle, which is discussed in Application 11.4 on page 248.

Inherent Bias Against New Versus Used Automobiles

One of the more troubling dilemmas in environmental law is the inherent bias caused by more stringent controls imposed on *new* polluting sources. This partiality is particularly problematic for motor vehicles, because engine performance and sticker price are at issue. As long as there are differences between the regulation of new and used vehicles, there will be price and performance impacts that can create market distortions.

To fully understand how the law can influence market decisions, consider the following scenario, which is modeled in Figure 11.5 on page 249. The automobile is a durable good, meaning its economic life extends beyond a single period. Because of this, new and used automobiles are substitutes for one another with the usual relationship between their relative prices. Regulating new car emissions adds to the production costs of those vehicles, shifting

[14]It is not unreasonable in this context to assume that the *MSC* is the same for both regions, because it is referring to abatement of mobile source emissions, and abatement of a mobile source is by definition not specific to a geographic location.
[15]See National Academy of Sciences and National Academy of Engineering (1974).
[16]This is discussed in the economic analysis of mobile source emissions by the Council of Economic Advisers (1982), p. 146, which asserts that such an averaging plan could save millions of dollars.
[17]U.S. EPA, Office of Policy, Planning, and Evaluation (December 1990).
[18]Originally, California regulations called for 10 percent of cars sold in 2002 to be zero-emission vehicles, but the plan was scaled down in 2001 (Pollack, January 27, 2001).

Figure 11.4 **Uniform Abatement Standards Across Regions for Mobile Source Emission Controls**

Disregard for the location where a vehicle is operated overregulates vehicles sold in relatively clean areas and undercontrols those sold in more polluted areas. Notice that if the standard, A_{ST}, is the same for both regions, the dirty area is undercontrolled, because A_{ST} is below the efficient level A_E. The opposite holds in relatively clean areas, where A_{ST} is above the efficient level A_E.

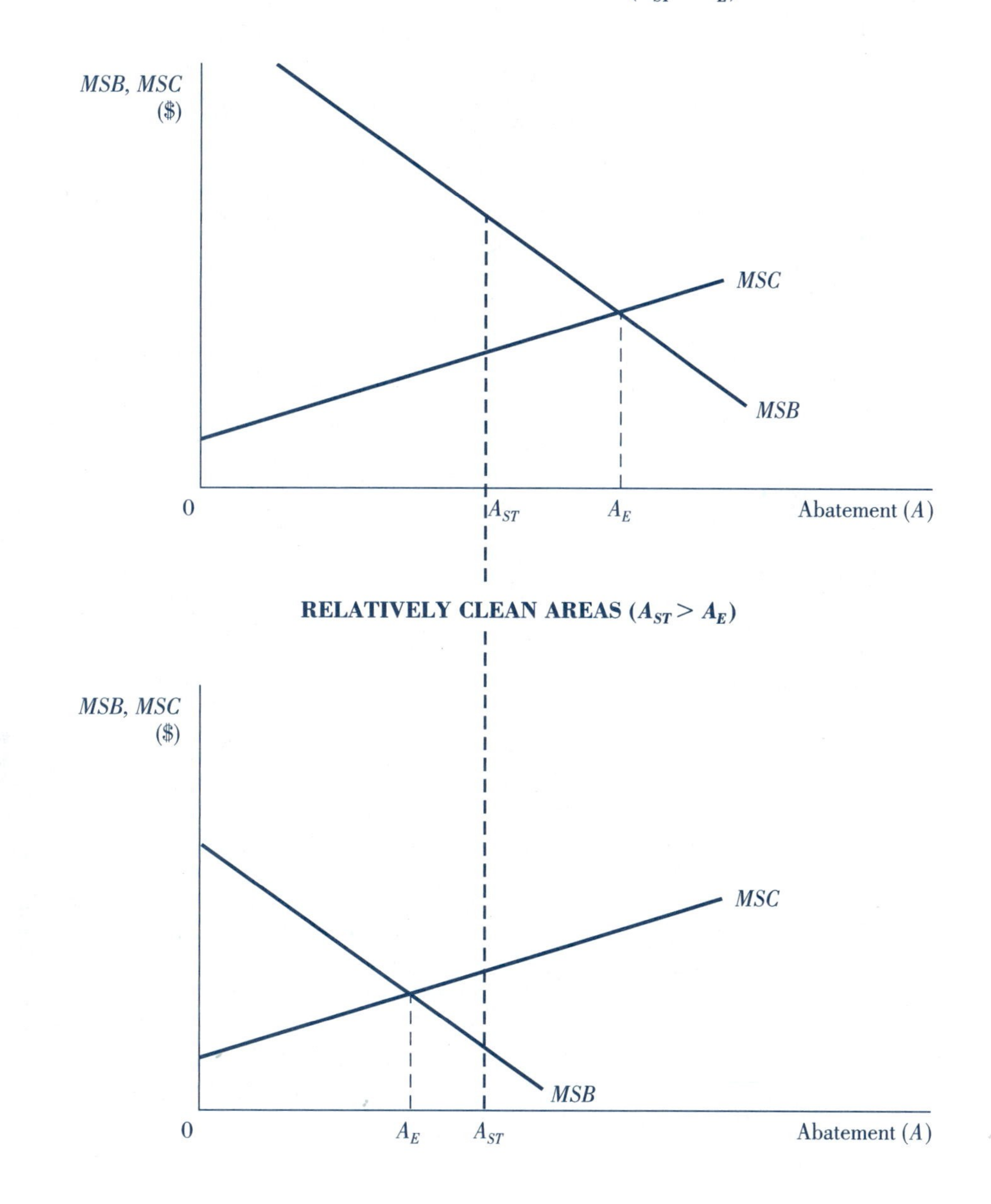

Application

11.4 *The United States Must Play Catch-Up in Hybrid Vehicles*

As their name implies, hybrid vehicles are powered by a combination of two energy sources: electricity and conventional fuels. These vehicles, first introduced to the mass market by Honda and Toyota, utilize an electric motor for start-up and low speeds and rely on a small conventional engine for cruising at normal speeds. A major advantage over electric cars is that the battery is recharged not from an external socket but rather from the vehicle's gas-powered generator and from kinetic energy generated when the brakes are applied. An onboard computer controls the "switching" process between the two power sources based on the most fuel-efficient allocation.

These high-tech, low-emission vehicles can get over 65 miles per gallon (mpg), offering drivers significant savings on fuel costs and resulting in reduced emissions. The fuel cost savings rather than environmental concerns may be the strongest incentive to buy a hybrid car. This was particularly evident in 2000 when hybrid demand increased as gasoline prices rose sharply. Nonetheless, society still gains the environmental benefit of lower fuel emissions, no matter what incentive triggered the demand.

Just as Japanese automakers led the industry in producing small cars during the energy crisis of the 1970s, so too were these companies the first to develop hybrid vehicles and market them on a mass scale. Honda's 3-cylinder Insight achieves an impressive 68 mpg, selling for about $20,000. Toyota's hybrid is the Prius, a 4-cylinder, five-passenger sedan that gets 58 mpg and sells for about the same price. Both companies admit that their hybrids sell at a loss in the United States, and estimates suggest that Honda's losses are $8,000 per vehicle. Nonetheless, American automakers are jumping on the hybrid wagon. Why? The major driving force is a regulatory one. The Detroit companies are in danger of not meeting Corporate Average Fuel Economy (CAFE) rules due in large part to robust sales of their popular but gas-guzzling SUVs and pickups. Despite this, or perhaps because of it, the SUV market is precisely where the Big Three plan to launch their own hybrids.

By 2003, Ford plans to offer a four-cylinder hybrid version of its small SUV, the Escape, that will get 40 miles to the gallon. General Motors will introduce a hybrid version of its full-size pickup truck in the 2004 model year that will increase fuel efficiency by 15 percent. DaimlerChrysler plans to launch a hybrid version of its Durango in 2003, which should achieve a 20 percent improvement in fuel economy, but that translates to just 18 mpg, compared to 15 mpg, for its conventional gas-powered model. There are also plans for so-called mybrids, or mild hybrids, which have electric motors that offload certain features from the engine, such as the air conditioner and power steering. Savings at the gas pump are lower: 15 percent compared to 30 percent for true hybrids. However, the incremental price to consumers is lower as well, estimated to be between $500 and $1,000 relative to a $1,500–$3,000 increase for true hybrids.

As further evidence of the marketability of this technology, Toyota and Ford have been discussing the prospect of jointly offering a new hybrid. If successful, Ford may achieve its objective to improve the fuel economy of its SUVs by 25 percent by 2005. Toyota stands to gain the savings of scale economies in hybrids that it cannot achieve on its own. Adding still more incentives to boost this market is the U.S. government. In May 2001, the Bush administration announced a new energy plan that includes a proposal for a $2,000 tax credit for buyers of hybrid vehicles.

Sources: Armstrong (July 9, 2001); Ball (October 24, 2000); Bradsher (February 20, 2001); Shirouzu (November 30, 2001); U.S. Department of Energy and U.S. EPA (2002); U.S. Government, White House (2001); Welch and Woellert (August 14, 2000).

Figure 11.5 *Modeling the Bias Against New Automobiles*

Regulating new car emissions adds to the production costs of those vehicles, shifting supply in the new car market to the left from S_1 to S_2. The result is that the higher costs are passed on to consumers, elevating price from P_1 to P_2. Faced with higher relative prices for new cars, some consumers will purchase a used car or keep a deteriorating one, either of which is likely to be a relatively high emitter of pollutants. This would cause a shift rightward in the demand for used cars from D_1 to D_2. Because the regulation encourages consumers to substitute in favor of used cars, it effectively extends the economic life of cars already on the road.

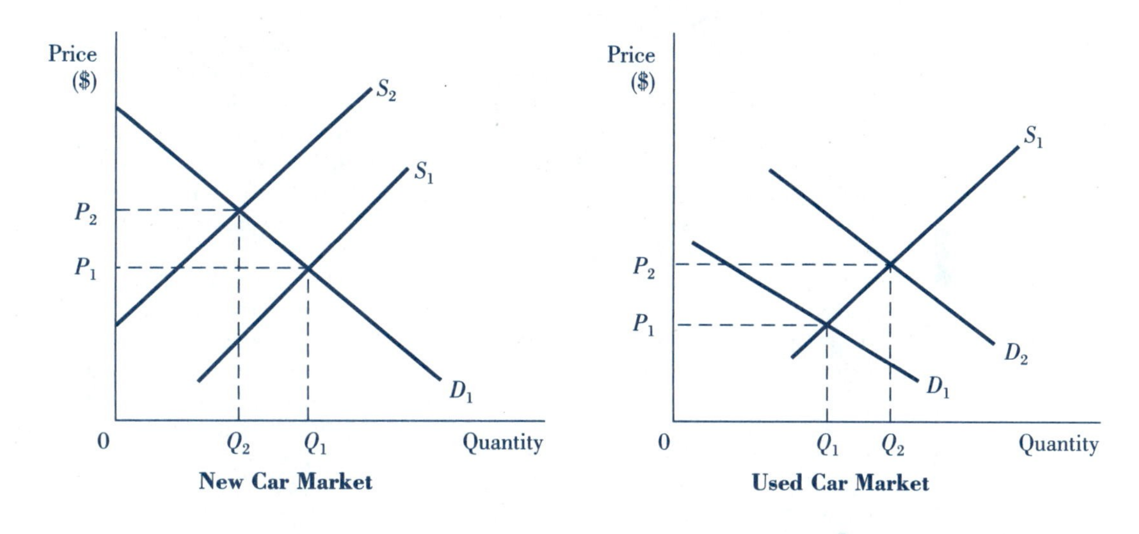

the supply curve to the left from S_1 to S_2. The result is that the higher costs are passed on, at least in part, to consumers, elevating price from P_1 to P_2.

Faced with higher relative prices for new cars, some consumers will purchase a used car or opt to keep a deteriorating one, either of which is likely to be a relatively high emitter of pollutants. As illustrated in Figure 11.5, we then would observe a shift rightward in the demand for used cars from D_1 to D_2. Because the regulation encourages consumers to substitute in favor of used cars, it effectively extends the economic life of cars already on the road. There is also the issue that emission control devices negatively affect a car's acceleration and gas mileage. Thus, the legislation confounds the signaling mechanism of price by diminishing the performance of a relatively higher priced vehicle.

Some policy changes address this dilemma. For example, the 1990 Amendments extend the economic life of the automobile over which emission standards are mandated by increasing the regulatory period from its former 5 years or 50,000 miles to 10 years or 100,000 miles. Although the standards for the second half of the period are less stringent, the extension to a longer time period helps to reduce the bias against future new vehicles, because most cars on the road ultimately will be subject to at least some degree of emissions control.

An alternative idea, still in the experimental stage, is a program called Cash for Clunkers. This incentive-based voluntary program, discussed in Application 11.5, allows polluters to earn emissions credits in return for buying older vintage cars and scrapping them to get these high emitters off the road.

Application

11.5 *The Cash for Clunkers Program*

In March 1992, the United States announced an innovative proposal aimed at encouraging the scrapping of older model, high-emitting automobiles. This so-called Cash for Clunkers program promotes the issuance of emission reduction credits to firms that purchase and scrap old cars. As in other emissions trading programs, these credits can be applied against legislated reductions imposed on the firm as a stationary source. This voluntary program, now called the Accelerated Retirement of Vehicles Program, ties together controls on stationary sources with those imposed on mobile sources. In fact, President George H. W. Bush called it "mobile-stationary source trading and emissions control." A fact sheet on the program with contact information is available online at **http://www.epa.gov/otaq/consumer/42097031.pdf.**

http://

To some extent, the national program is based on a prototype experiment put in place in Los Angeles. In 1990, an oil company called Unocal Corporation purchased 8,400 pre-1971 cars for $700 each over a one-year period and scrapped them. The oil firm estimates that the plan reduced hydrocarbon emissions at an average cost of $7,000 per ton removed. Based on these estimates, it has been suggested that Unocal's approach is cost-effective in a severely polluted area like Los Angeles.

In 1993, the EPA released a document titled *Guidance for the Implementation of Accelerated Retirement of Vehicles Programs.* (This document is available on the Internet at **http://www.epa.gov/otaq/transp/trancont/scrapcrd.pdf.**) Through this guidance document, the agency attempts to assist state and local officials with the design and implementation of their own programs.

Several issues need to be resolved in implementing any such plan. For one thing, some have argued that $700 is not likely to be a sufficient inducement for an individual to give up their used automobile and face the expense of buying a newer car. Responding to this objection, a Unocal official cited a survey it conducted that showed that 360 of the 800 individuals surveyed did in fact use the $700 to purchase newer and presumably cleaner-running cars. Another potential problem is in determining the price to be paid for these old cars. Related to this issue is controlling the migration of older vehicles from one region to another to seek out the best available price. Finally, there is the matter of how to calculate credits for the purchased vehicles. Should the credits be uniform, or should they reflect the expected reduction in pollution per mile over the remaining economic life of the car?

Although still in its infancy, the program is capable of providing an incentive to pull high-emitting vehicles off American highways. In fact, according to former acting assistant EPA director Richard Morgenstern, 38 percent of all cars on the road in 1992 were built before 1980. These vehicles generated 86 percent of the hydrocarbons and carbon dioxide released by mobile sources into the atmosphere. Appealing to the effectiveness of contrasts, Morgenstern said: "The dirtiest 6 percent of the cars on the road today emit 50 percent of the hydrocarbons. The cleanest 50 percent of cars emit only 3 percent of these hydrocarbons."

Sources: " 'Cash for Clunkers' to Cut Pollution" (May/June 1992); U.S. EPA, Office of Policy, Planning, and Evaluation (July 1992); U.S. EPA, Office of Mobile Sources (December 1997).

Implications of Clean Fuel Alternatives

Clean fuel provisions in the 1990 Amendments are an extension of technology-based standards, which were formerly directed almost entirely toward engines and emission control systems. Furthermore, the use of more advanced fuel compositions is required only in the dirtier regions of the country. Targeting these areas makes good sense from an economic perspective. The development and use of alternative fuels generates relatively high marginal costs, which aligns with the higher expected marginal benefits that will accrue in the designated nonattainment regions. Hence, this selective legislation follows the fundamental decision rule of benefit-cost analysis, which may approach an efficient solution.

A separate issue is whether the fuel-switching program is a cost-effective way to reduce mobile source emissions. ARCO, the oil company that pioneered the development of reformulated fuel, estimates that it can produce a reformulated gasoline to satisfy the 1990 laws that will cost consumers 15 cents more per gallon.[19]

Conclusions

After decades of delays and false starts, U.S. air quality policy aimed at motor vehicles has evolved into a fairly comprehensive set of provisions. The 1990 Clean Air Act Amendments strengthen federal regulations on tailpipe emissions for cars and trucks and establish a two-tiered standard that effectively extends the economic life of a motor vehicle. There are also requirements for onboard vapor recovery systems, emission control diagnostic systems, and initiatives to develop clean fuel vehicles. Beyond the controls aimed directly at vehicles, the 1990 Amendments also set policy on fuels, including provisions to mandate the use of cleaner fuels, such as requirements to use reformulated gasoline in certain ozone nonattainment areas.

Most mobile source initiatives are command-and-control in orientation, though a few provisions are market based, such as credits for purchasing more clean fuel vehicles than required by law. These incentives should enhance compliance and help the nation achieve its overall air quality objectives. That said, economic analysis of mobile source controls suggests the need for further revision. One example is the absence of benefit-cost analysis in setting standards for tailpipe emissions. Another is the regulatory bias against new cars inadvertently caused by the nation's policy on mobile source emissions.

Taken together, these observations remind us that legislating air quality is an evolving process and that there are no perfect solutions. On net, there has been measurable progress in controlling the emissions of motor vehicles in the United States, particularly given the delays in launching policy initiatives at the federal level. Yet there is more to be done, not only in policy development but also in technological innovation, to develop more fuel-efficient engines, better emissions diagnostics, and cleaner fuels.

Summary

- In metropolitan areas, the concentration of human population, traffic patterns, and industrial activity intensifies the concentration of criteria pollutants and hence increases the environmental risks of exposure. In the United States, the EPA gathers data in major urban centers and reports part of its findings using an Air Quality Index (AQI).
- In urban areas, photochemical smog can form from a chemical reaction involving several of the criteria pollutants. The principal component of photochemical smog is ground-level or tropospheric ozone. Mobile sources contribute to photochemical smog formation, because they emit NO_X and VOCs, the precursors of smog. Those transportation sources emitting the largest amounts of these substances are ordinary highway vehicles, particularly those fueled by gasoline.
- The Clean Air Act of 1963 gave virtually no power to federal officials to control mobile source emissions. It was not until the 1970 Amendments that emission standards for new automobiles were explicitly stated in the law.
- The 1990 Amendments strengthen federal control over motor vehicle emissions and fuels. The new law also incorporates market-based incentives to encourage development of cleaner-running vehicles and alternative fuels.
- The stringent controls on mobile sources initially established by the 1970 Clean Air Act Amendments may have been purposefully unrealistic to force technology development by

[19] Dickerson (January/February 1991).

the automobile industry. The implied decision rule used to establish the controls was benefit based, set to protect public health and welfare.

- The uniformity of the national emission standards for automobiles inflates the costs of reducing pollution with no added benefit to society.
- Different controls on new versus used motor vehicles affect relative prices and engine performance, which creates market distortions. Because new and used cars are substitutes, tougher regulations on new vehicles can bias buying decisions in favor of higher emitting used cars.
- The use of reformulated fuels and oxygenated fuels is being called for in the nation's dirtiest regions, where they can yield the most benefit to society.

Key Concepts

Air Quality Index (AQI)
photochemical smog
reformulated gasoline
oxygenated fuel
clean fuel vehicle
clean alternative fuels

Use the Key Concepts listed above to begin your search for additional articles and information using the InfoTrac College Edition database.

Review Questions

1. In the 1990 Clean Air Act Amendments, Congress and the EPA rely on the automobile industry to develop a cleaner automobile. At the same time, the government imposes a relatively minor federal tax on gasoline.

a. Do you see any problem with the implicit signals the federal government is sending to American auto manufacturers and to American car drivers through these policies? Briefly discuss.

b. Formulate a hypothetical economic policy to motivate automobile manufacturers to advance the technology of cleaner motor vehicles.

2. Refer to the discussion of a two-tiered system of automobile emissions standards studied by the National Academy of Sciences. Use a graph to model this system and illustrate how the cost savings are achieved.

3. New source bias may exist for mobile sources. Briefly discuss why this bias leads to a solution that is *not* cost-effective. What policies would you implement to eliminate this bias?

Additional Readings

Ball, Jeffrey. "Detroit Again Attempts to Dodge Pressures for a 'Greener' Fleet." *Wall Street Journal*, online edition, January 28, 2002.

Banerjee, Neela, and Danny Hakim. "U.S. Ends Car Plan on Gas Efficiency; Looks to Fuel Cells." *New York Times*, January 9, 2002, p. A1.

Blumenstein, Rebecca. "Auto Makers, New York Collide on Electric Cars." *Wall Street Journal*, December 16, 1997, pp. B1, B7.

Carey, John. "We Can Fight Smog Without Breaking the Bank." *Business Week*, October 3, 1994, pp. 128–29.

Fullerton, D., and S. E. West. "Can Taxes on Cars and on Gasoline Mimic an Unavailable Tax on Emissions?" *Journal of Environmental Economics and Management* 43 (January 2002), pp. 135–57.

Hubbard, Thomas N. "Using Inspection and Maintenance Programs to Regulate Vehicle Emissions." *Contemporary Economic Policy* 15 (April 1997), pp. 52–62.

Krupnick, Alan J., and Paul R. Portney. "Controlling Urban Air Pollution: A Benefit-Cost Assessment." *Science* 252 (April 26, 1991), pp. 522–28.

Lareau, Thomas J. "The Economics of Alternative Fuel Use: Substituting Methanol for Gasoline." *Contemporary Policy Issues* 8 (October 1990), pp. 138–52.

Rusco, Frank W., and W. David Walls. "Vehicular Emissions and Control Policies in Hong Kong." *Contemporary Economic Policy* 13 (January 1995), pp. 50–61.

Sevigny, Maureen. *Taxing Automobile Emissions for Pollution Control.* Northampton, MA: Elgar, 1998.

United Nations Environment Programme and the World Health Organization. "Air Pollution in the World's Megacities." *Environment* 36 (March 1994), pp. 4–13, 25–37.

Welch, David, Kathleen Kerwin, Laura Cohn, Lorraine Woellert, and Larry Armstrong. "Detroit Rides Alone." *Business Week,* July 9, 2001, pp. 32–34.

Zesiger, Sue. "Reinventing the Wheel." *Fortune,* October 25, 1999, pp. 184–90.

Related Web Sites

Accelerated Vehicle Retirement Programs	**http://www.epa.gov/otaq/consumer/42097031.pdf**
Air Quality Index: A Guide to Air Quality and Your Health, U.S. EPA, Office of Air and Radiation (June 2000)	**http://www.epa.gov/airnow/aqibroch/**
Alternative Fuels	**http://www.epa.gov/otaq/consumer/fuels/altfuels/altfuels.htm**
Clean Air Act, Title II, provisions on mobile sources	**http://www.epa.gov/oar/caa/contents.html**
Consumer information on motor vehicles, fuels, and emissions	**http://www.epa.gov/otaq/consumer.htm**
Fuels	**http://www.epa.gov/otaq/fuels.htm**
Guidance for the Implementation of Accelerated Retirement of Vehicles Programs, U.S. EPA, Office of Mobile Sources (February 1993)	**http://www.epa.gov/otaq/transp/trancont/scrapcrd.pdf**
"Milestones in Auto Emissions Control," U.S. EPA, Office of Transportation and Air Quality (August 1994b)	**http://www.epa.gov/otaq/12-miles.htm**
Model Year 2002 Fuel Economy Guide, U.S. Department of Energy and U.S. EPA (2002)	**http://www.fueleconomy.gov**
Motor Vehicle Emissions	**http://www.epa.gov/otaq/ld-hwy.htm**
Reformulated Gasoline and Vehicle Performance	**http://www.epa.gov/otaq/rfgvehpf.htm**
Smog—Who Does It Hurt? U.S. EPA, Office of Air Quality Planning and Standards (July 1999)	**http://www.epa.gov/airnow/health/index.html**

United States Council for Automobile Research (USCAR)	**http://www.uscar.org**
U.S. EPA, Green Vehicle Guide	**http://www.epa.gov/greenvehicles/**
U.S. EPA, Market Incentives Resource Center	**http://www.epa.gov/orcdizux/transp/traqmkt.htm**
U.S. EPA, MTBE home page	**http://www.epa.gov/mtbe**
U.S. EPA, Office of Transportation and Air Quality	**http://www.epa.gov/otaq/**

A Reference to Acronyms and Terms Used in Mobile Source Control Policy

Environmental Economics Acronyms

MSB	Marginal Social Benefit
MSC	Marginal Social Cost

Environmental Science Terms

CO	Carbon monoxide
CO_2	Carbon dioxide
gpm	Grams per mile
MTBE	Methyl tertiary-butyl ether
NO_X	Nitrogen oxides
NO_2	Nitrogen dioxide
O_3	Ozone
Pb	Lead
PM	Particulate matter
PM-2.5	Particulate matter less than 2.5 micrograms in diameter
PM-10	Particulate matter less than 10 micrograms in diameter
SO_2	Sulfur dioxide
SOx	Sulfur oxides
VOC	Volatile organic compound

Environmental Policy Acronyms

AQI	Air Quality Index
NAAQS	National Ambient Air Quality Standards

Chapter 12

Improving Air Quality: Controlling Stationary Sources

In this chapter, we continue our discussion of national air quality policy aimed at criteria pollutants, but here we focus our attention on policy to control stationary sources. The motivation for these controls is the same as it is for mobile sources. Because criteria pollutants affect all of society, the Clean Air Act imposes stringent regulations on *all* major sources of these emissions. As we noted in the previous chapter, although the motivation is common to both source groups, there are important differences in policy design and implementation.

Some of these distinctions are a direct function of the fixed location of stationary sources, which makes certain policy instruments easier to implement and monitor. Others are related to the pollutant being controlled. For example, mobile sources contribute relatively more to the level of carbon monoxide (CO) emissions, whereas stationary sources are bigger emitters of sulfur oxides (SO_X). Consequently, some control instruments are uniquely motivated by specific emissions and their associated environmental risks.

To this latter point, an important context for our analysis in this chapter will be an environmental problem tied primarily to stationary sources—acidic deposition, more commonly known as acid rain. Of particular interest is an economic evaluation of the market-based program aimed specifically at this problem, which involves emissions trading.

To organize our discussion, we begin with an overview of how stationary source controls are defined under the 1990 Clean Air Act Amendments. In particular, we examine both location-specific and age-specific distinctions in U.S. air quality policy. Once these fundamentals are established, we move to a discussion of acidic deposition and the specific policies and programs that address this important pollution problem. Lastly, we conduct an economic analysis of several aspects of federal policy to control stationary sources. A list of the primary acronyms and terms used in this context is provided for reference at the end of the chapter.

Controlling Stationary Sources

Stationary sources contribute to the emissions levels of all the criteria pollutants. This source group covers a lot of ground. Included are electric power plants, chemical plants, steel mills, and even residential furnaces. Obviously, certain categories are more important to U.S. control policy than others, as the data in Figure 12.1 attest. These data show the 1999 emissions of each criteria pollutant released by the two major stationary source categories: fuel combustion and industrial processes.

Notice in Figure 12.1 how fuel combustion sources, which include fossil-fueled electric power plants, are major contributors of sulfur dioxide (SO_2) and nitrogen oxide (NO_X) emissions, the primary causes of acid rain. Furthermore, because NO_X is a precursor of photochemical smog, as discussed in Chapter 11, this means that fuel combustion sources also contribute significantly to urban smog. In fact, this observation motivated the NO_X Budget Program developed by the Ozone Transport Commission (OTC). This innovative program, which involves NO_X emissions trading, is discussed in Application 12.1. Notice also in Figure 12.1 that industrial sources (e.g., petroleum refineries and chemical manufacturers) are responsible for most of the volatile organic compound (VOC) emissions, so these stationary sources also are important targets for policies aimed at reducing urban smog.

Figure 12.1 ***National Emissions Estimates for Stationary Sources by Major Category, 1999***

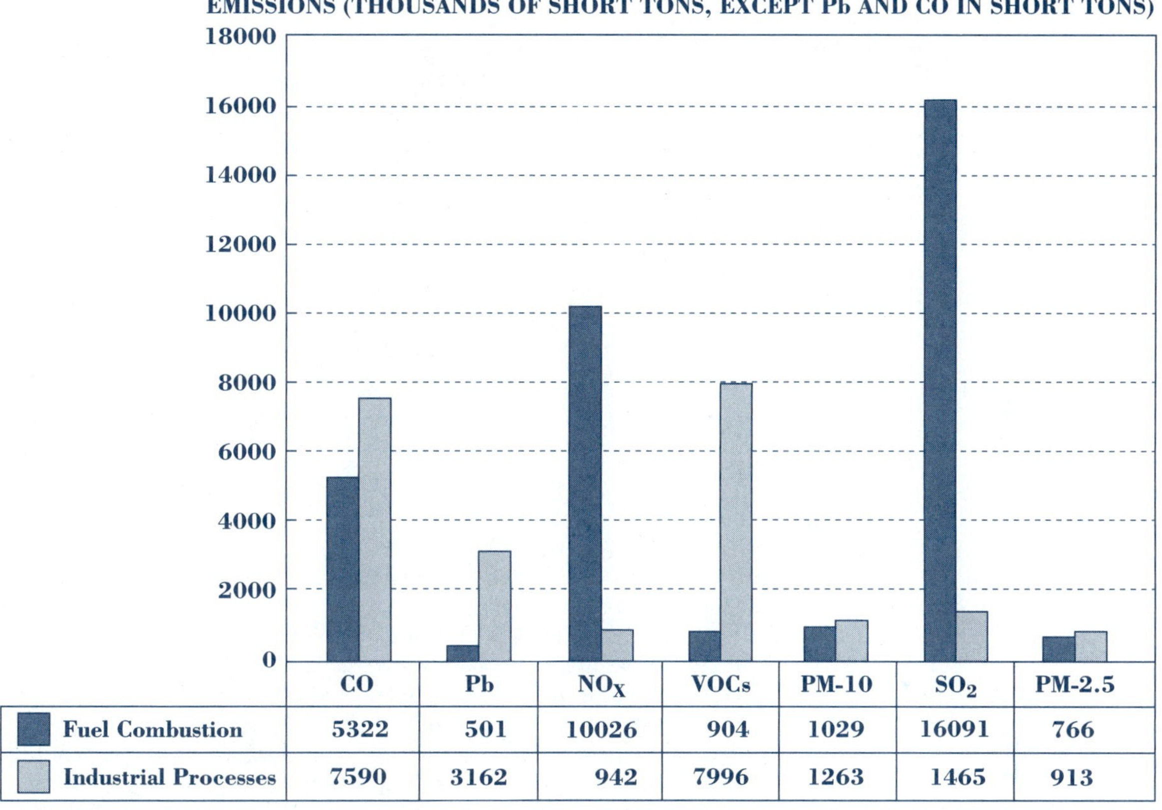

	CO	Pb	NO_X	VOCs	PM-10	SO_2	PM-2.5
Fuel Combustion	5322	501	10026	904	1029	16091	766
Industrial Processes	7590	3162	942	7996	1263	1465	913

Source: U.S. EPA, Office of Air Quality Planning and Standards, (March 2001), pp. 132–59, tables A2–A9.

Emissions of all stationary sources are regulated primarily by uniform **technology-based standards,** which indicates use of a command-and-control policy approach. In the late 1970s, the EPA did introduce some market-based programs, although these were intended to be subordinate to the ongoing use of standards. For the most part, both the technology-based standards and the market programs are defined and implemented in accordance with two features of stationary sources:

- facility age (new or existing)
- facility location (in a prevention of significant deterioration [PSD] area or a nonattainment area).

These characteristics form the structural basis for our discussion. To serve as a guideline for what is to follow, refer to Table 12.1 for a summary of the various standards and market-based programs that have evolved based on these factors. More detail is available in Exhibit A.5 in the Appendix.

Application

12.1 *Developing a Regional NO_X "Cap and Trade" Plan*

In addition to the NO_X reductions called for under Title IV of the 1990 Clean Air Act Amendments, Section 184 of this legislation requires additional cuts in NO_X to combat the ground-level ozone problem in the Northeastern and Mid-Atlantic states. To accomplish this task, the Ozone Transport Commission (OTC) was created, which includes Connecticut, the District of Columbia, Delaware, Maine, Maryland, Massachusetts, New Hampshire, New Jersey, New York, Pennsylvania, Rhode Island, Vermont, and the northern counties of Virginia.

The OTC's objective is to assist each member with their State Implementation Plan to meet the NAAQS for ground-level or tropospheric ozone. Following through on this goal, members of the OTC (with the exception of Virginia) signed a 1994 agreement with the EPA to design and implement a multiphase program aimed at reducing regional NO_X emissions. The program's initial phase required all identified major stationary sources (e.g., electric utilities and large industrial boilers) to install reasonable available control technologies (RACT). What followed next was the more innovative aspect of the OTC's plan.

Working with the EPA and a myriad of special interest groups, such as utilities and environmentalists, the OTC designed what is known as the NO_X Budget Program. It is through this program that a regional cap and trade NO_X plan was established. The annual cap is set at 219,000 tons of NO_X emissions for the 1999–2002 period and at 143,000 tons beginning in 2003. Relative to a 1990 baseline, these caps represent more than a 50 percent reduction of regional NO_X emissions.

As for the trading component of the program, the approach is analogous to the national SO_2 emissions trading program aimed at the problem of acid rain. In this case, NO_X emissions are not permitted without allowances, and the number of allowances issued is limited to the specified cap amounts. Each NO_X allowance permits the release of one ton of emissions. Allowances can be traded or banked (with some exceptions) for future use. The EPA is responsible for tracking the NO_X allowances and for monitoring stationary source emissions, but individual states are in charge of compliance and enforcement. From an economic perspective, the development of this allowance market should provide a cost-effective approach to achieving the OTC's regional air quality targets.

Although trading has been active for only three years, results to date suggest that the NO_X Budget Program is working quite well. During the ozone season of May to September in 2001, regional NO_X emissions by targeted sources were more than 60 percent lower than in the 1990 baseline period. Furthermore, some major sources were able to reduce emissions below their budgeted amount. So these polluters have the option of either selling their excess allowances to other sources or banking them to meet the more stringent standards called for in 2003.

Currently, the EPA is expanding the use of the NO_X Budget Program as a cost-effective means of assisting non-OTC member states in attaining their own ground-level ozone objectives. To learn more about the OTC's regional program and the EPA's federal programs using the NO_X Budget Program, visit **http://www.epa.gov/airmarkets/progsregs/noxview.html.**

http://

Sources: U.S. EPA, Office of Policy, Economics, and Innovation, Office of the Administrator (January 2001), pp. 82–84; U.S. EPA, Office of Air and Radiation, Clean Air Markets Division (May 9, 2001); Ozone Transport Commission (April 4, 2002).

Age-Specific Control Differences: New Versus Existing Sources

Defining Technology-Based Emissions Limits

Age-specific control differences were established in the 1970 Clean Air Act Amendments. In general, the rulings require that facilities constructed or significantly modified after 1970 are to meet more stringent emissions limits than those already in existence. The reasoning is that new or modified stationary sources presumably can integrate the technology needed to meet stringent standards more easily than their older counterparts. Existing facilities would likely

Table 12.1 ***Technology-Based Standards and Market-Based Programs Under the Clean Air Act***

	New or Modified Sources	Existing Sources
PSD Areas		
Standards	Best Available Control Technology (BACT)	Best Available Retrofit Technology (BART)
Market-based program	Netting	Bubble policy
Nonattainment Areas		
Standards	Lowest Achievable Emission Rate (LAER)	Reasonably Available Control Technology (RACT)
Market-based program	Offset plan	Bubble policy

Note: For more detail, see Exhibit A.5 in the Appendix.

face high retrofitting costs, which could hurt local economies. Furthermore, they eventually would deteriorate to the point of needing to be modified or replaced. Hence, over time, the tougher standards would have to be met by *all* stationary sources.

Emissions limits applicable to new and modified stationary sources are called **New Source Performance Standards (NSPS).** According to the Clean Air Act, these are to be technology based in accordance with the "best system of emission reduction." Such a phrase suggests that the NSPS were meant to be stringent, but in this case the stringency is tempered somewhat by allowing for costs and energy requirements to be considered as well.[1]

Because technological factors are the basis for these limits, the NSPS are allowed to vary across industry categories designated by the EPA. However, each set of industry standards is to be applied *uniformly* across all firms within a given category. Because of the long-term importance of these more stringent controls, Congress established a **dual-control approach,** placing the responsibility for controlling **new or modified stationary sources** with the EPA and assigning the comparable duties for **existing stationary sources** to state governments.

New Source Performance Standards (NSPS)
Technology-based emissions limits established for new stationary sources.

new or modified stationary source
A source for which construction or modification follows the publication of regulations.

existing stationary source
A source already present when regulations are published.

The Bubble Policy

In December 1979, the EPA launched its **bubble policy,** originally conceived as an option for states to use in implementing the national standards for **existing** sources. The intent of this program was to allow a plant to measure emissions of a pollutant, not from each of the release points within the plant but as an *average* of all such points, as if the emissions were captured within a bubble. The aim was to offer firms flexibility in deciding *how* they could achieve the requisite emissions limits at least cost.

In 1980, the EPA introduced another market-based program, called **emissions banking.** Designed to complement the bubble concept, this plan allows a source to accumulate **emission reduction credits** if it abates more than required by law. These credits are then deposited through the banking program for future use, establishing the foundation for an emissions market. Emissions banking has been used only in a limited way, with only 24 emission banks identified in a 1994 report. The number has changed little since that time.[2]

bubble policy
Allows a plant to measure its emissions as an average of all emission points emanating from that plant.

emissions banking
Allows a source to accumulate emission reduction credits if it reduces emissions more than required by law and to deposit these through a banking program.

[1] However, in practice, this cost consideration is meant to imply only that the technology be affordable and *not* that it be used as part of a formal benefit-cost analysis in setting the standards (Portney 1990a, p. 38).
[2] Crookshank (1994), as cited in U.S. EPA, Office of Policy, Economics, and Innovation (January 2001), p. 73. For an interesting and comprehensive discussion on emissions banking and the bubble concept, see Liroff (1986).

Location-Specific Control Differences: PSD Versus Nonattainment Areas

Defining Emissions Limits in PSD Areas

With the official introduction of PSD areas in 1977, stationary source controls had to be revised to accommodate the change. Because the air quality standard in PSD regions is higher, controls for sources located in these areas had to be made more stringent. However, new facilities in PSD areas would face tougher standards than existing ones. According to the new policy, which is implemented through a permit procedure, any new or modified source in a PSD area has to meet emissions limits based on what is called the **best available control technology (BACT).** What this means is that a new source has to meet a standard that aligns with the *maximum* degree of pollution reduction available.

Defining Emissions Limits in Nonattainment Areas

Standards for nonattainment areas are relatively less stringent, but just like their PSD counterparts, these sources face different standards based on their age. Existing sources must use **reasonably available control technology (RACT),** which is the least stringent of all the technology-based standards. New or modified sources in these areas must comply with the **lowest achievable emission rate (LAER),** roughly defined as the most stringent emission limit achieved in practice by the same type of source.

Emissions Trading in PSD Areas

netting
Developed for PSD areas to allow emissions trading among points within a source for the same type of pollutant, such that any emissions increase due to a modification is matched by a reduction from another point within that same source.

To facilitate control in PSD areas, a trading program known as **netting** was developed for use by modified sources.[3] A type of bubble policy, netting allows emissions trading to take place among release points within a facility for the same type of pollutant. Its distinguishing feature is that any added emissions associated with a plant modification must be *exactly matched* by a reduction from somewhere else within that same plant. Hence, netting ensures that any construction project will have no detrimental effect on the air quality in a PSD area.

Emissions Trading in Nonattainment Areas

In some sense, implementing pollution controls in nonattainment areas must be more strategic. Obviously, there is concern about the environmental impact of allowing new or modified facilities into an area already not meeting the national ambient standards. Yet preventing new construction could potentially worsen air quality, because new plants likely operate more efficiently and cleanly than older ones.

offset plan
Developed for nonattainment areas to allow emissions trading between new or modified sources and existing facilities such that releases from the new or modified source are more than countered by reductions achieved by existing sources.

In an attempt to achieve the best of both worlds, the 1977 Amendments introduced the **offset plan.** Implemented through a permit procedure, an offset plan ensures that emissions from a new or modified source be *more than countered* by the reductions achieved by existing sources. Unlike the bubble policy or netting, the offset plan involves trades between existing and new sources rather than within the same facility or plant complex. To facilitate the plan, an **emissions bank** of accumulated **emission reduction credits** was set up where facilities could deposit and access offsets as needed.

Sorting It Out

Although the collection of technology-based standards can be confusing, the key points to remember about these standards are:

- Emissions limits in **PSD areas** are more stringent than those in **nonattainment areas.**
- Limits for **new sources** are more stringent than those for **existing sources** within both types of areas.[4]

---

[3] Although netting was developed for use in PSD areas, it technically can be used in all areas.
[4] Furthermore, although the new source standards for PSD and nonattainment areas (i.e., BACT and LAER, respectively) are supposed to be more stringent than the NSPS, in practice, the NSPS are generally accepted for either (Portney 1990a, pp. 37–38).

Another important observation is the way in which market incentives are used in the bubble policy, emissions banking, offsets, and netting. Although these programs were intended to be subordinate to the standards, they nonetheless represented a shift toward integrating some economic-based mechanisms into what had been almost exclusively a command-and-control approach. Title IV of the 1990 Clean Air Act Amendments continues this trend by establishing another type of trading program for stationary sources specifically aimed at the problem of acid rain.

Controlling Acidic Deposition

Acid rain, or more accurately **acidic deposition,** has been the subject of political and media attention in recent years. Research on the problem has intensified, and both public officials and private citizens have grown increasingly cognizant of the implications. As scientific knowledge has grown, so too has the motivation to enact stronger legislation. Effectiveness of policy initiatives is limited by the fact that acid rain is a *regional* air pollution problem, meaning that the source of the contamination is often hundreds of miles from where the detrimental effects are felt. Consequently, some form of federal intervention generally is necessary to control polluters.

The Problem of Acidic Deposition

Acidic deposition occurs when sulfuric and nitric acids mix with other airborne particles and fall to the earth either as dry deposition or as fog, snow, or rain, thus the descriptive phrase "acid rain."[5] Table 12.2 gives an overview of the causes, sources, and effects of this important regional problem. These acidic compounds result from the chemical reaction of sulfur dioxide (SO_2) and nitrogen oxide (NO_X) emissions with water vapor and oxidants in the earth's atmosphere.[6]

acidic deposition Arises when sulfuric and nitric acids mix with other airborne particles and fall to the earth as dry or wet deposits.

Of the two responsible pollutants, the more significant is SO_2. It is formed when the sulfur found naturally in coal or oil is released during combustion and reacts with the oxygen in the atmosphere. Primary generators of SO_2 emissions are fossil-fueled electric power plants,

Table 12.2 ***The Problem of Acidic Deposition: A Summary***

Causes	Anthropogenic Sources	Major Effects
SO_2	Fossil fuel burning, primarily by electric power plants	Ecological and forestry effects: Acidification of surface waters; localized forest damage.
NO_X	Fossil fuel burning, primarily by electric power plants and motor vehicles	Health effects: Direct effects include respiratory and cardiovascular problems; some indirect effects possible from increased lead or methyl mercury exposure under extreme conditions.
		Aesthetic effects: Visibility impairment.
		Property effects: Damage to buildings, statues, and monuments.

Note: For more complete information on the effects of acid rain, visit **www.epa.gov/airmarkets/acidrain/effects/index.html.**

[5] For online information about acid rain, visit **http://www.epa.gov/airmarkets/acidrain/index.html.**

[6] An oxidant is any oxidizing agent, such as oxygen (O_2) or ozone (O_3), capable of chemically reacting in the atmosphere to form a new substance.

refineries, pulp and paper mills, and any sources that burn sulfur-containing coal or oil. According to recent estimates, about two-thirds of SO_2 emissions and 25 percent of the NO_X emissions in the United States are generated by coal-burning electric power plants.

The Policy Response: Title IV of the 1990 Amendments

Title IV of the 1990 Clean Air Act Amendments is dedicated to a reduction plan for NO_X emissions and an innovative "cap and trade" allowance program to decrease SO_2 emissions. The reduction plan for NO_X is to be achieved through various performance standards set by the EPA. Ultimately, emissions are to be reduced by 2 million tons from their 1980 level.

For SO_2, the law establishes a national cap of 8.95 million tons of annual emissions for electric power plants (the major sources of SO_2) and a cap of 5.6 million tons for nonutility industrial sources. Both caps were to be met by 2000. This is the first time any national limit had been imposed on a single pollutant. These limits were designed to achieve a 10 million ton reduction in SO_2 emissions below their 1980 level.

SO_2 Emissions Allowance Program[7]

tradeable SO_2 emission allowances Permits issued to stationary sources, each allowing the release of one ton of SO_2, which can be either held or sold through a transfer program.

To achieve the SO_2 emissions cap, the 1990 law sets up an emissions market that operates through an allocation and transfer system. At the start of the program, the EPA issues **tradeable SO_2 emission allowances,** or rights to pollute, to stationary sources; each allowance permits the release of one ton of SO_2. Because no emissions are permitted unless authorized by an allowance, the aggregate number of allowances issued by the EPA effectively sets the national limit.[8]

Once the emissions allowances are distributed, they can be used by the recipient or sold to other sources through a transfer program. This program sets up a market through which allowances can be bought and sold. In May 1992, the first allowance trade was announced between the Tennessee Valley Authority (TVA), a utility operating some 59 coal-fired units, and Wisconsin Power and Light Company, one of the cleanest utilities in the country. The transaction involved TVA's purchase of 10,000 allowances at a value of \$2.5–\$3 million. The market incentive for this precedent-setting trade is clear, once the facts are known. To meet the new federal standards *without* trading, the TVA would have had to invest in abatement technology, estimated to cost between \$750 million and \$850 million.

An extension of this emissions market is provided through a Special Allowance Reserve, a bank of allowances available for direct sale by the EPA. Because the success of this program depends greatly on external trades, the EPA placed the Chicago Board of Trade (CBOT) in charge of running the auctions. Because of its experience, the CBOT can facilitate the exchange of allowances and may reduce the transactions costs of matching up buyers and sellers. According to the law, any individual, corporation, or government can enter the market for SO_2 allowances.[9]

As it turns out, a number of private firms have anticipated the profit opportunities of the newly created pollution market and are competing with the CBOT as allowance brokers. In fact, one of these, Clean Air Capital Markets, arranged the first \$3 million trade between the TVA and Wisconsin Power and Light. Thus far, Clean Air Capital Markets has arranged transactions totaling \$72 million. A brokerage firm, Cantor Fitzgerald Inc., has launched a computer-assisted plan to match up long-term exchanges. These entrepreneurial entities are capturing part of the CBOT's market, making some question the longevity of the CBOT's role in SO_2 allowance trading.[10]

[7] A comprehensive fact sheet on the SO_2 Allowance Trading Program with links to related sites is available online at **http://www.epa.gov/airmarkets/arp/allfact.html.** Internet access to annual compliance reports for the U.S. Acid Rain Program is available at **http://www.epa.gov/airmarkets/cmprpt/.**

[8] In Phase I of the plan for the 1995–2000 period, each source received allowances based on the following formula: average fuel usage over the 1985–1987 period multiplied by an emission rate of 2.5 pounds of SO_2 per million Btu, all divided by 2,000. In Phase II, covering the period from 2000 to 2009, the allowance formula is similar except that the emission rate is reduced to 1.2 pounds.

[9] Current data on the trading activity of SO_2 emissions allowances can be accessed on the Internet at **http://www.epa.gov/airmarkets/trading/index.html** or at **http://www.epa.gov/airmarkets/tracking/index.html.**

[10] Taylor (August 24, 1993) and Hong (July 20, 1992).

Economic Analysis of Stationary Source Controls

Whether or not the gains of stationary source controls justify the costs can be determined through a **benefit-cost analysis.** To structure such an investigation, we focus on four aspects of national policy:

- Cost of command-and-control methods relative to market-based incentives
- Uniform technology-based NSPS
- Dual-control approach for new versus existing sources
- Emissions trading policy

Relative Cost of Using Command-and-Control Instruments

Because of the absence of benefit-cost balancing in the standard-setting process, it would only be by chance that the resulting stationary source regulations would correspond to those that maximize the net benefits to society. Despite the failure of the standards to meet the efficiency criterion, it is still possible to minimize the cost of meeting them, efficient or not. Unfortunately, the lack of flexibility in a standards-based approach adds significantly to society's costs and offers no incentive to low-cost abaters to clean up beyond the statutory level.

There is a growing body of research that attempts to measure the extent of cost inefficiency associated with command-and-control policy instruments. These studies use computer simulations to model existing conditions in a given region and compare the cost of implementing command-and-control instruments with that of the least-cost alternative. Table 12.3 summarizes a sample of those studies that specifically examine stationary control costs for meeting the National Ambient Air Quality Standards (NAAQS). Each study quantifies the cost of using a command-and-control instrument *relative* to the least-cost method. This comparison is expressed as a ratio, shown in the last column of the table.

Notice the range of results. At the high end, Spofford (1984) found that the costs of using a uniform percentage reduction to meet the particulate matter (PM) standard in the Lower Delaware Valley exceeded the least-cost alternative by a factor of 22. This implies that the costs of using a command-and-control approach are 2,200 percent of what they would be if the policy allowed for equal marginal abatement cost (*MAC*) levels across all polluters in that region. At the other end of the spectrum are the findings of Hahn and Noll (1982). Their ratio of 1.07 suggests that the command-and-control approach is not that far away from the least-cost method. In this case, however, the researchers suggest that their unusual finding may be the result of stringent controls imposed in Los Angeles by the California control authority with the specific intent of achieving cost-effective standards.

Despite the wide range of values for the cost ratio, in nearly every case the magnitude is significantly greater than 1. It is difficult to dispute such a consistent finding that points to the excess costs of using command-and-control instruments to implement the NAAQS. These results suggest that in most instances society should realize important cost savings from a shift to more flexible policy instruments with no reduction in air quality benefits.

Uniform Technology-Based NSPS

Implicitly, there are two potential problems with the new source performance standards (NSPS). First, they are implemented **uniformly** across all firms in a given category. Second, because the standards are **technology based,** firms have no flexibility in selecting *how* to meet the national emission limits. Both of these characteristics suggest that cost-effectiveness is not being achieved.

The intent of using uniform emission standards was to prevent regional differences in regulations from affecting new firms' decisions about where to locate. Nonetheless, just as is the case for the uniform emissions limits for new cars, this particular command-and-control approach generally disallows the achievement of a cost-effective outcome. Why? Because the

Table 12.3 ***Cost-Effectiveness and Air Quality Regulation: Some Empirical Studies***

Study and Year	Command-and-Control Approach	Pollutant Controlled	Geographic Area	Ratio of Command-and-Control Cost to Least Cost
Atkinson and Lewis (1974)	SIP regulations	Particulate matter	St. Louis	6.0
Roach et al. (1981)	SIP regulations	Sulfur dioxide	Four Corners: Utah, Colorado, Arizona, and New Mexico	4.25
Hahn and Noll (1982)	California emissions standards	Sulfates	Los Angeles	1.07
McGartland (1984)	SIP regulations	Particulate matter	Baltimore	4.18
Spofford (1984)	Uniform percentage reduction	Sulfur dioxide Particulate matter	Lower Delaware Valley	1.78 22.0
Maloney and Yandle (1984)	Uniform percentage reduction	Hydrocarbons	All domestic Du Pont plants	4.15
Krupnick (1986)	Proposed RACT	Nitrogen dioxide	Baltimore	5.9
Oates et al. (1989)	Equal proportional treatment	Particulate matter	Baltimore	4.0
ICF Resources International (1989)	Uniform emission limit	Sulfur dioxide	United States	5.0
SCAQMD (1992)	Best Available Control Technology	Reactive organic gases and NO_2	Southern California	1.5 in 1994 1.3 in 1997
Krupnick et al. (2000)	SIP call provisions	Nitrogen oxides	Eastern United States	1.83 (utilities) 2.00 (all sources)

Sources: U.S. EPA, Office of Policy, Economics, and Innovation, Office of the Administrator (January 2001) p. 25, table 3-2; Tietenberg (1985); and the original sources cited in the table.

Note: SCAQMD is South Coast Air Quality Management District.

cost-effective solution calls for pollution abatement up to the point where all firms face the same marginal abatement cost (*MAC*), *not* to the point where all firms abate at the same level.

In Figure 12.2, a model of two hypothetical stationary sources, A and B, is shown with the cost-effectiveness solution indicated where MAC_A intersects MAC_B at A_0. (Source A's allocation is measured left to right up to A_0, and source *B*'s allocation is measured right to left.) The total costs of abatement, assuming no fixed costs, are shown as the shaded area under the two *MAC* curves up to the cost-effective abatement allocation. If instead a uniform standard were used, the abatement allocation across the two firms would force each to abate equally to the point labeled A_1.[11] The total costs again would be measured as the area under the two *MAC* curves up to the solution point. However, notice that the total costs under the uniform standard approach are higher by the triangular area labeled *XYZ*.

[11] A_1 represents half the total abatement standard.

Figure 12.2 **Cost-Ineffectiveness of the Uniform New Source Performance Standard**

The model shows two hypothetical stationary sources, A and B, with the cost-effectiveness solution at the point where the MAC_A curve intersects the MAC_B curve at A_0. The total costs of abatement, assuming no fixed costs, are shown as the shaded area under the two *MAC* curves up to the cost-effective abatement allocation. If instead a uniform standard is used, the abatement allocation across the two firms would force each to abate equally to the point labeled A_1. Notice that the total costs under the uniform standard approach is higher by the triangular area labeled *XYZ*.

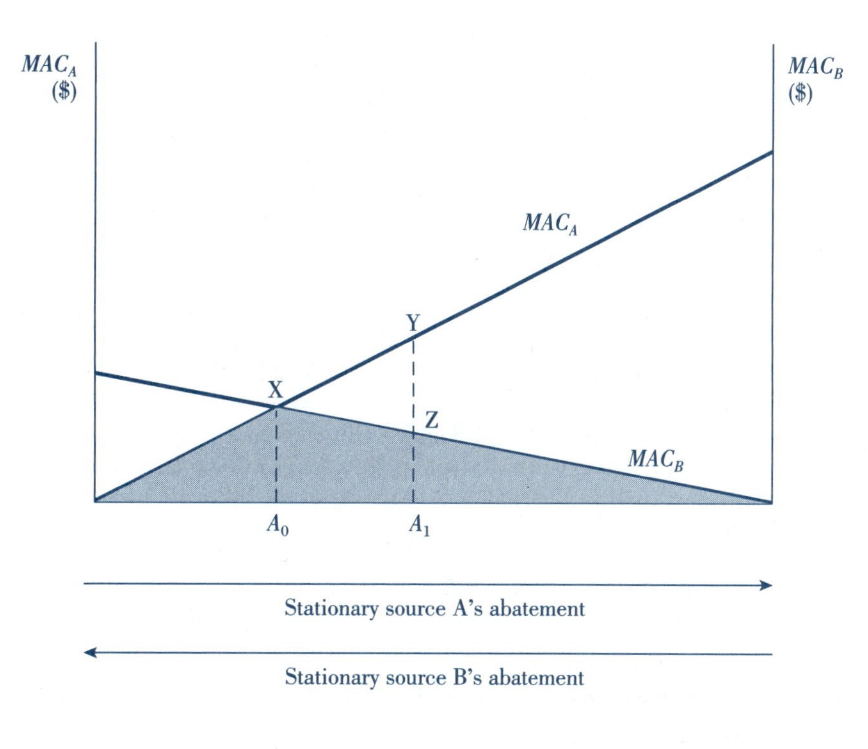

Using technology-based NSPS instead of performance-based standards can also produce a result that is not cost-effective. By dictating the type of abatement technology to be used across the board, firms are prevented from finding and using cheaper alternative methods. The result is a waste of scarce economic resources.

To illustrate the potential problem with using the less flexible technology-based standards, we consider one set of NSPS for coal-burning electric power plants established through the 1977 Amendments to control SO_2 emissions, perhaps the most scrutinized set of standards in all U.S. air quality legislation.

Revisions in the 1977 Amendments included a seemingly innocuous change in the wording of the NSPS (emphasis added):

> a standard of performance shall reflect the degree of emission limitation and the *percentage reduction achievable* through application of the best technological system of continuous emissions reduction.[12]

[12] 42 U.S.C. § 7411, as amended August 7, 1977 (Pub. L. 95-95, sec. 109-c-1-A).

The change to refer to a "*percentage reduction achievable*" coupled with the specific reference to "*the best technological system of continuous emissions reduction*" forced coal-burning power plants to install scrubbers to meet the SO_2 emissions limit rather than switch to cleaner, low-sulfur coal.[13] The result? The new NSPS created an artificial incentive for coal-burning plants to use high-sulfur fuel with scrubbers, even though substituting low-sulfur fuel would have been a more cost-effective approach.

There is some evidence to support the hypothesis that this technology-forcing revision was costly. A 1986 Congressional Budget Office report presents estimates that the implicit restriction on fuel switching would increase the cost of an 8 million ton reduction in SO_2 emissions by \$2.7 billion (in discounted 1985 dollars) and a 10 million ton reduction by \$16.3 billion.[14]

Similar findings were obtained in an independent study conducted by two economists, Lewis J. Perl and Frederick C. Dunbar (1982).[15] Their analysis begins with a baseline estimation of the costs and benefits of SO_2 emission reductions based on the former definition of the NSPS under the 1970 Amendments. At that time, power plants could choose either scrubbers or fuel switching to comply with the law. To achieve the reduction of 20.38 million tons called for in the 1970 law, total annual costs would have been \$3.3 billion and total benefits would have been \$5.41 billion. Hence, had the former NSPS remained in effect, benefits would have exceeded costs by \$2.1 billion.

Perl and Dunbar (1982) then repeat the exercise based on the 1977 NSPS, which called for a reduction of 23.91 million tons of SO_2 emissions to be achieved through the installation of scrubbers. Under this scenario, they estimated that annual costs would rise to \$7.05 billion and benefits would rise to \$6.22 billion, so that net benefits would become −\$0.83 billion.

On an incremental basis, notice that to achieve the additional reduction in emissions of 3.53 million tons, costs would rise *by* \$3.75 billion (a cost of \$1,062 per additional ton of SO_2 removed) and benefits would rise *by* \$0.81 billion (a benefit of \$229 per additional ton removed):

Additional SO_2 reduction = 3.53 million tons
(23.91 million tons_{1977} − 20.38 million tons_{1970}).

MSC = \$3.75 billion (\$7.05 billion_{1977} − \$3.30 billion_{1970}).
MSC per additional ton removed = \$1,062 (\$3.75 billion/3.53 million tons).

MSB = \$0.81 billion (\$6.22 billion_{1977} − \$5.41 billion_{1970}).
MSB per additional ton removed = \$229 (\$0.81 billion/3.53 million tons).

This scenario is illustrated in Figure 12.3. Using benefit-cost analysis, the findings show that the 1977 revision in the NSPS was simply not worth it.

Based on these estimates, it is natural to ask why Congress would force such costly controls on firms. The answer is a complex one, involving political battles, obscure legal interpretations, and special-interest groups. Yet, even a brief review of the facts leaves little doubt that the major force behind these legislative changes was successful lobbying by the United Mine Workers Union, a labor group consisting primarily of high-sulfur coal miners. It was in their best interest to limit fuel switching toward low-sulfur coal, leaving high-intensity coal users,

[13] The revision elevated the former 1971 standard on SO_2 emissions from 1.2 pounds per million Btu of coal consumed to a tougher standard that varied with the sulfur content of the coal. Specifically, for high-sulfur coal the new standard became 1.2 pounds per million Btu of fuel consumed plus a 90 percent emissions reduction, and for low-sulfur coal the limit was 0.6 pounds per million Btu of fuel consumed plus a 70 percent emissions reduction.

[14] U.S. Congress, Congressional Budget Office (June 1986), chap. 2.

[15] All of Perl and Dunbar's estimates are expressed in 1980 dollars for the year 1990, when the mandated reductions were to have been achieved.

Figure 12.3 ***Analysis of the NSPS in the 1977 Clean Air Act Amendments According to Perl and Dunbar (1982)***

The 1977 NSPS called for a reduction of 23.91 million tons of SO_2 emissions to be achieved through the installation of scrubbers. On an incremental basis, notice that, to achieve the additional reduction in emissions of 3.53 million tons, costs would rise *by* \$3.75 billion (a unit cost of \$1,062 per ton of removed SO_2) and benefits would rise *by* \$0.81 billion (a per unit benefit of \$229 per ton).

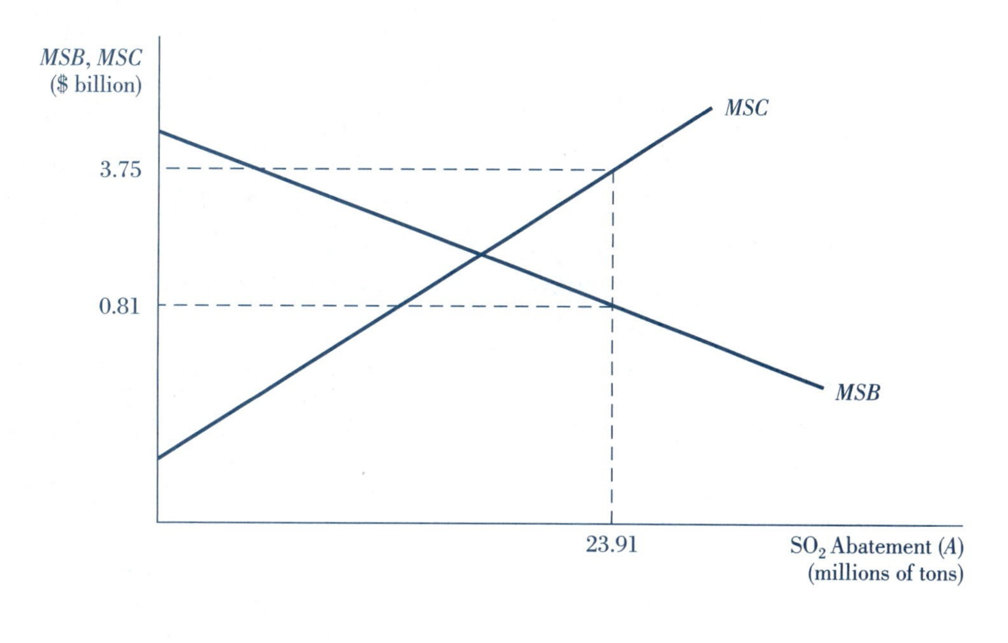

such as electric power plants, no choice but to install expensive scrubbers.[16] Although the miners' success in saving their jobs did not adversely affect air quality, it *did* elevate abatement costs with no comparable increase in benefits—a clear case of inefficient policy making. (As an interesting postscript, the reference in the law to a "*percentage reduction*" was omitted in the 1990 Amendments.)

The Dual-Control Approach and the New Source Bias

The dual-control approach gives states more direct supervision over those firms on which their economies have come to depend. Yet, despite its apparent logic, a dual-control system can generate market distortions. There is no reason to expect that state-established controls on existing stationary sources will be the same as those set at the federal level for new or modified sources. As long as there are two sets of standards, firms have both the incentive and the opportunity to avoid the more stringent, and therefore the more costly, of the two. This in turn means that firms' decisions about whether to build a new facility or modify an existing one will be influenced by the legislation. In fact, the logical expectation is that state-determined emissions limits will be relatively more lenient, which suggests that firm decision making will be biased *against* new construction—a true market distortion.

[16]For more detail, see Ackerman and Hassler (1981). Further evidence of the mining industry's influence on U.S. law is found in sec. 125 of the Clean Air Act. These provisions specifically allow government intervention to force the use of locally or regionally available coal to avoid economic disruption in the area.

Application

12.2 *Should the New Source Review for Stationary Sources Be Revoked?*

In keeping with an aggressive enforcement policy begun in the mid-1990s, the Clinton administration proceeded against 13 utilities in 1999 for violating the Clean Air Act at more than 51 of their plants across 12 states. The Justice Department and several states filed suits against the firms for what amounted to billions of dollars in penalties. The claim was that these utilities had failed to obtain a New Source Review (NSR) permit to undertake "significant modifications" to their facilities. According to the Clean Air Act, new or significantly modified stationary sources are subject to more stringent standards than their existing counterparts and are required by law to install more advanced technology for pollution control.

Predictably, there was an intense, negative response from the utilities. Among their objections was that the effective litmus test, "significant modification," was not well defined in the law but, in any case, should not be interpreted to include ordinary maintenance. Perhaps even more important was their assertion that the NSR program is destructive. As written, it discouraged power companies from updating or improving their facilities, because doing so would be met with tougher and more costly regulations. Such an outcome is environmentally perverse, since newer power plants generally emit less pollution, and is economically damaging if it interferes with meeting rising energy demands.

A lengthy debate ensued, involving the EPA, government officials, private industry, and environmental groups, and it was not at all clear how the gridlock would be resolved. Some utilities settled or reached tentative settlements with the EPA, but others sought a decision at a higher level. Joined by refineries and other targets of analogous enforcement action, the utilities lobbied strongly for a review of the NSR requirements, even before the November 2000 election. For their part, environmentalists lobbied in support of the legal action, concerned that the power plants were disguising major plant modifications as routine maintenance to evade the law. Before a resolve was reached, President Clinton's term of office ended, leaving the dilemma and the ultimate decision to the incoming Bush administration.

In May 2001, the new administration's energy task force, headed by Vice President Cheney, recommended that the EPA review the NSR requirements to ascertain how these provisions affected energy efficiency and environmental quality. In response, the EPA, in consultation with the Department of Energy, the Council on Environmental Quality, and other federal entities, conducted a 90-day review of the disputed provisions. The White House also asked the Justice Department to review the string of lawsuits filed by the Clinton administration.

After the EPA's review of the NSR requirements was completed, the agency concluded that the determination of which plant changes constituted a significant modification had been too inflexible. The EPA also announced that it would recommend legislative or administrative changes in the NSR program to the White House. At the same time, the EPA administrator, Christine Todd Whitman, stated that the Bush administration wants Congress to consider consolidating certain air pollution programs that have become contentious into a more comprehensive, less intrusive plan.

In February 2002, President Bush proposed his so-called Clear Skies Initiative, a market-based, multipollutant program addressing SO_2, NO_X, and mercury emissions released by power plants. (This initiative is discussed in Application 12.3.) This new initiative, if passed by Congress, would be applicable to both new and existing sources, making the NSR procedures redundant. In fact, Whitman announced in March 2002 that the Bush administration will propose an end to the NSR as it applies to power plants if Congress approves the Clear Skies Initiative. As for the lawsuits, the Justice Department ruled in January 2002 that it would pursue the cases filed against the power plants.

Sources: Cohn, Carey, and Palmer (June 11, 2001); U.S. EPA, Office of Air Quality Planning and Standards (June 22, 2001); Pianin and Mintz (August 8, 2001); "Bush Won't Drop Utility Pollution Lawsuits" (January 15, 2002); "EPA's Whitman Sees Overhaul of Pollution Control Program" (March 20, 2002).

To understand this latter hypothesis, consider that states have an incentive to attract business as a way to create jobs for their constituents and to maintain a healthy tax base to support fiscal spending. States also compete against one another to attract industry. It is generally in a state's best interest to avoid policies that increase firms' operating costs and reduce profits. In this context, states have a political incentive to avoid setting more stringent, and thus more costly, emissions limits that may encourage firms to exit from their jurisdictions and relocate where regulations are more lenient.[17]

What this means for firms is that they can avoid meeting the federally established standards for new or modified sources simply by maintaining their existing facilities, which would be subject to less stringent state limits. The irony is that the law sets up a disincentive for businesses to build new facilities that would likely operate more efficiently and more cleanly. Hence, the controls on existing stationary sources are to some extent self-defeating because the disincentive to modernize or replace deteriorating facilities can cause short-term pollution to increase.[18] Interestingly, this process was the subject of litigation and considerable debate that began under the Clinton administration and continued into the Bush presidency, an issue discussed in Application 12.2.

From a benefit-cost perspective, the variability in controls for new versus existing stationary sources may cause marginal abatement costs (MAC) to be higher per ton of pollutant removed in new facilities than in existing ones, despite the fact that the associated benefits are identical. This is modeled in Figure 12.4. The figure shows two MAC curves: that of an existing source (MAC_{EX}) and that of a new stationary source (MAC_N). Notice that the MAC_N curve is lower than the MAC_{EX} curve at every level of abatement to show the more advanced control technology available to a new, more modern facility. Now, consider the cost implications of two distinct abatement levels: A_{EX1}, a more lenient level to represent the state-determined standard on existing sources, and A_{N1}, a level that represents the more stringent controls imposed by the NSPS. Notice that the MAC_{N1} corresponding to A_{N1} is higher than the MAC_{EX1} corresponding to A_{EX1}. Thus, the inconsistency in controls on stationary sources has generated a solution that is *not* cost-effective.

Theoretically, by altering the emission standard for either existing sources or new sources, the control authority can achieve a cost-effective abatement solution. For example, if the state official increases the required abatement level for existing sources to A_{EX2}, then MAC_{EX2} would equal MAC_{N1}. (Alternatively, the federal control authority could reduce the abatement requirement for new sources until the corresponding marginal abatement cost level equals MAC_{EX1}.) In so doing, the bias against new construction could be eliminated. The outcome would be difficult to realize, however, because the control authority would need detailed information about the control costs for each and every polluting source in its jurisdiction.[19]

Economics of Market-Based Trading Programs

Although the legislative effort in the United States up through the 1977 Amendments was primarily a command-and-control approach, national policy was beginning to use more market-based instruments to control stationary sources. Perhaps the best example of this trend was the initiation of the EPA's emissions trading program in the mid-1970s implemented through offsets, netting, bubbles, and banking.

Analysis of Emissions Trading

To illustrate how the market process operates through these instruments, reconsider the motivations inherent in the offset plan. If Firm A wishes to construct a new plant in a nonattainment area, it might choose to provide the offsetting emissions through one of its own existing

[17] In practice, there are limits as to how permissive state laws can be. Recall that under the State Implementation Plan (SIP) each state's standards must be approved by the EPA, so they must conform to the EPA's range of reasonableness.

[18] For more on this argument, see Gruenspecht and Stavins (January 26, 2002).

[19] For a more detailed description of the implications concerning alternative abatement levels for new versus existing sources, refer to Crandall (1984).

Figure 12.4

Analyzing the Dual-Control Approach and the New Source Bias: Marginal Abatement Cost Levels for New Versus Existing Sources

The marginal abatement cost of an existing source is MAC_{EX}, and the marginal abatement cost of a new source is MAC_N. A_{EX1} represents the more lenient abatement level of a state-determined standard on existing sources, and A_{N1} represents the more stringent controls imposed by the NSPS. The result is not cost-effective because the MAC_{N1} corresponding to A_{N1} is higher than the MAC_{EX1} corresponding to A_{EX1}. To achieve cost-effectiveness, the state could increase the required abatement for existing sources to A_{EX2} so that MAC_{EX2} equals MAC_{N1}. Alternatively, the federal authority could reduce the abatement requirement for new sources until the marginal abatement cost equals MAC_{EX1}.

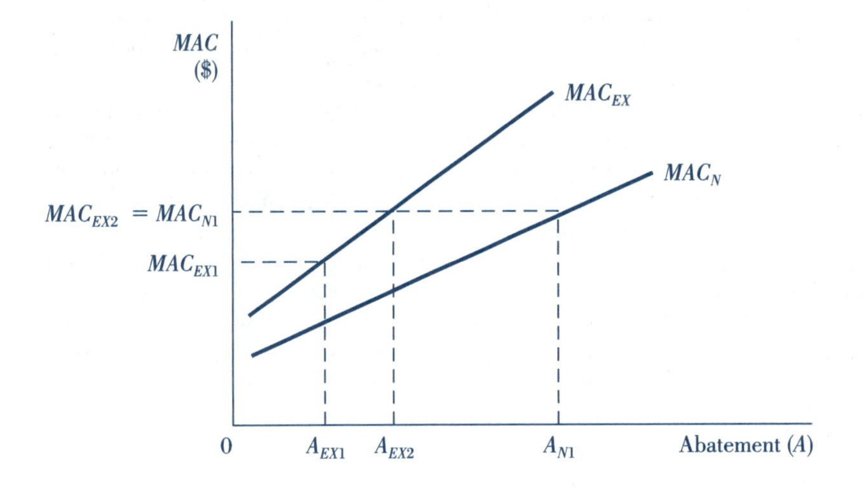

plants in the region. Alternatively, it could negotiate the cost of having some other firm in the region, Firm B, provide the emission reductions in its place. Firm B has an incentive to provide this service if its costs of doing so are less than what Firm A is willing to pay for it. From an economic perspective, Firm A will contract out the task of emission reductions if it is cheaper than performing the task itself. In the end, those firms controlling most of the pollution in a given region will be the ones that can do so more cheaply. Ultimately, all sources will control pollution to the point where the marginal abatement cost of doing so is equal across firms—a cost-effective solution.

Attempts to quantify the gains of emissions trading have been done by estimating the associated control cost savings. The major findings of two such studies that examine trading activity in the pre-1986 period are presented in Table 12.4. Notice that the activity is delineated according to the four major instruments: offsets, bubbles, netting, and banking. Within each of these categories, the data are further refined to distinguish between trades that are *internal* (i.e., between sources within the same plant) and those that are *external* (i.e., between sources in different facilities).

Overall, Table 12.4 shows that trading can be lucrative, with cost savings ranging from $135 million for state-approved bubbles to between $525 million and $12,000 million for netting transactions. The overwhelming majority of trades are internal, which is explained by the inherently higher transactions costs associated with interfirm trades. In theory, banking

Table 12.4 *Emissions Trading Activity*

Activity	Cost Savings Through 1985 (millions)	Number of Trades	
		Internal	External
Offsets[a]	—	1,800	200
Bubbles, EPA approved	$300	40	2
Bubbles, state approved	$135	89	0
Netting[b]	$525–$12,000	5,000–12,000	None
Banking	Very small	<100	<20

Sources: Drawn from Hahn and Hester (1989); Hahn (spring 1989).

[a]Cost savings are not applicable for offsets, because firms do not avoid any emission reduction under the offset program.

[b]According to the EPA, 1984 is the only year for which detailed data exist for netting transactions. Thus, some of the estimates are based on an extrapolation done by Hahn and Hester (1989) from the year 1974. For more detail, see U.S. EPA, Office of Policy, Economics, and Innovation, Office of the Administrator (January 2001), chap. 6.

should help to reduce these costs and foster external trades, but the data indicate otherwise. It seems that some other mechanism is needed to bring firms together so that the potential savings can be realized. Such an alternative is evident in the 1990 trading program aimed at acid rain—to date the most ambitious of its kind in U.S. or international legislative history.

Analysis of the 1990 SO_2 Allowance Trading Plan

Let's consider the cost-effectiveness of the allowance trading established by Title IV of the 1990 Amendments. Polluting sources that can reduce SO_2 emissions relatively cheaply will do so and sell off their excess allowances. They have an incentive to do so as long as their respective *MAC*s are below the market price of an allowance (P), as shown in Figure 12.5. In the model shown, the polluter would abate A_0 units of SO_2. Excess allowances are an asset to these firms, who effectively become **suppliers** in the market for SO_2 allowances. The excess can be used either internally to offset emissions at another source within the firm, banked for future use, or sold on the open market.

On the other side of the market are buyers of allowances, or **demanders.** These are primarily firms unable to match their emissions level to the number of allowances held. For these polluters, abatement is relatively costly, so they will pay for the right to release excess emissions as long as the market price of an allowance (P) is below their respective *MAC* levels. Other potential buyers include brokers and environmental organizations, each of which have different motivations. Brokers are motivated by profit, expecting to resell the allowances for a lucrative rate of return. Environmentalists want to buy allowances and remove them from the market.

On net, buyers and sellers will trade for allowances, and emissions should reach the level where *MAC*s are equal across firms. Creating an emissions market and allowing the usual incentives to operate freely can accomplish abatement in a cost-effective manner with no compromise to social benefits. Notice that the national emissions limit is achieved, because the number of allowances issued determines the maximum amount of SO_2 emissions that may be released. Hence, all the health and ecological benefits of pollution abatement still accrue to society, but the associated costs are lower. These advantages of allowance trading are important motivations for President Bush's Clear Skies Initiative, which is aimed at achiev-

Figure 12.5 *Cost-Effectiveness of the SO_2 Allowance Trading Plan*

Polluters that can reduce SO_2 emissions relatively cheaply will do so and sell their excess allowances. They have an incentive to do so as long as their respective *MAC*s are below the market price of an allowance (*P*). In the graph shown, the polluter would abate A_0 units of SO_2. Buyers of allowances are primarily firms unable to match their emissions to the number of allowances held. For these polluters, abatement is relatively costly, so they will pay for the right to release excess emissions as long as *P* is below their respective *MAC* levels. On net, buyers and sellers will trade for allowances, and emissions should reach the level where *MAC*s are equal across firms.

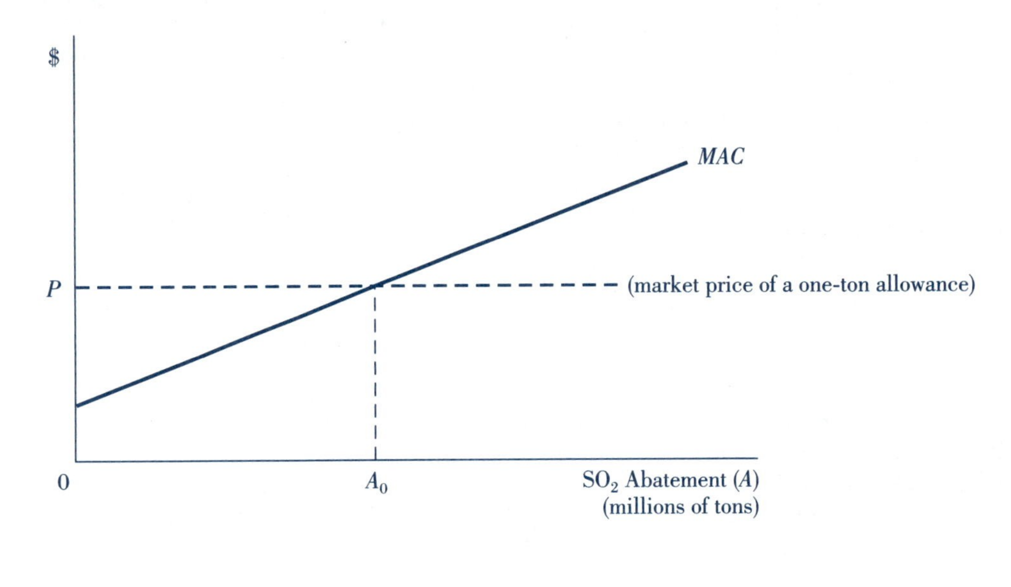

ing further reductions of SO_2 and other emissions from power plants, as discussed in Application 12.3.

Using the results of the second annual spot auction for SO_2 allowances held in March 1994, we can gain a better perspective of how this market operates. On the supply side, the EPA offered 50,000 allowances to be sold to the highest bidders.[20] Private utilities supplied another 58,001 allowances. These were to be sold at various offer prices, ranging from $165 to $550. On the demand side, there were 103 bids submitted for 294,354 allowances. The highest per unit bid was $400, and the lowest was $24. At the close of the auction, all 50,000 EPA-issued allowances had been sold to 37 individual demanders, who bid between $150 and $400, for an average of $159. Once the EPA-offered allowances were sold to the highest bidders, the remaining 58,001 allowances were made available for sale. However, none were sold because the remaining bidders were not willing to pay the asking prices.[21]

According to most analysts, the Acid Rain Program has been deemed a success because firms have achieved emission reduction targets at less than 50 percent of the projected cost. Trading, however, has been less active than anticipated, attributable in part to the existence of transaction costs and the fact that many transactions are intrautility trades (i.e., within the

[20]The 50,000 allowances are sold in the spot auction for use in the same year unless banked for future use. There are also 100,000 allowances sold in an advance auction for use in the seventh year after the year of sale, unless banked for future use.

[21]U.S. EPA, Office of Communications, Education, and Public Affairs (March 29, 1994).

12.3 *Debating the Merits of President Bush's Clear Skies Initiative*

Application

In February 2002, President Bush announced a new proposal to improve air quality called the Clear Skies Initiative. The plan is aimed at reducing power plant emissions of three important air pollutants: sulfur dioxide (SO_2), nitrogen oxides (NO_X), and mercury. The specific objectives for each of these are:

- To reduce SO_2 emissions by 73 percent from the current level of 11 million tons to a cap of 4.5 million tons in 2010 and 3.0 million tons in 2018.
- To reduce NO_X emissions by 67 percent from the current level of 5 million tons to a cap of 2.1 million tons in 2008 and 1.7 million tons in 2018.
- To reduce mercury emissions by 69 percent from current levels of 48 tons to a cap of 26 tons in 2010 and 15 tons in 2018. This would be the first ever national cap established for mercury.

Why are power plants targeted for the president's new policy? Because these stationary sources are responsible for 67 percent of the nation's SO_2 emissions, 25 percent of the NO_X emissions, and 37 percent of mercury emissions.

Beyond the explicit emissions reductions enumerated in the initiative, the plan has drawn attention for the market-based approach it has outlined to achieve its objectives. The Bush proposal is based squarely on the cap and trade approach used successfully in the nation's Acid Rain Program. If passed by Congress, the new initiative would extend the gains of SO_2 allowance trading already established by Title IV of the Clean Air Act and would add two new pollutants to the nation's trading program.

According to the plan, the number of 1-ton allowances issued for each pollutant would be controlled by the government in keeping with the established caps and would be reduced in number over time. Allowance trading among power plants would allow these caps or emissions limits to be reached in a cost-effective manner. Natural economic incentives will promote abatement activity by those firms able to abate more cheaply than the going price of an allowance and will incite those that cannot to purchase needed allowances. The White House claims that the new program will save the nation up to $1 billion annually in compliance costs. Such cost savings achieved by electricity generators should help keep energy prices lower for consumers.

Proponents of the president's plan also point to the advantages of a multipollutant approach. By addressing three important air pollutants simultaneously, the initiative is expected to be more effective and efficient than the current uncoordinated single-pollutant regulations. Expected benefits of the Clear Skies Initiative include health gains (e.g., reduced risk of respiratory infection and lung damage) and greater environmental protection from the effects of acid rain caused by SO_2 and NO_X, (e.g., reduced visibility and acidification of lakes and streams).

Not surprisingly, the new initiative is not without critics. Some environmental groups, including the Natural Resources Defense Council, have expressed concern that the president's plan actually eases controls on power plants, giving them more time to comply than they would under existing law. They further argue that emissions would actually be cut further under the Clean Air Act as currently written. Citing a recent EPA briefing, environmentalists point to the agency's assertion that NO_X emissions would be cut to 1.25 million over the next 10 years, when the Clean Air Act is fully implemented, which is lower than the Clear Skies Initiative would achieve by 2018. They further claim that the EPA said that SO_2 could be reduced to 2 million tons by 2012 under existing law and mercury to 7.5 tons under regulations soon to be issued. The agency responded that the numbers given at the briefing were never meant to be a projection about the effectiveness of existing policy and that the new plan will bring about much greater reductions in air pollution than under current law.

Such debates are not uncommon when new environmental policy is proposed. Ultimately, Congress will decide the fate of the president's proposal and with it the future of more comprehensive allowance trading in U.S. air pollution policy.

Sources: U.S. Government, White House (February 14, 2002); White House (February 20, 2002).

Figure 12.6 ***Comparison of Intrautility and Interutility SO_2 Allowance Trades Through 2001***

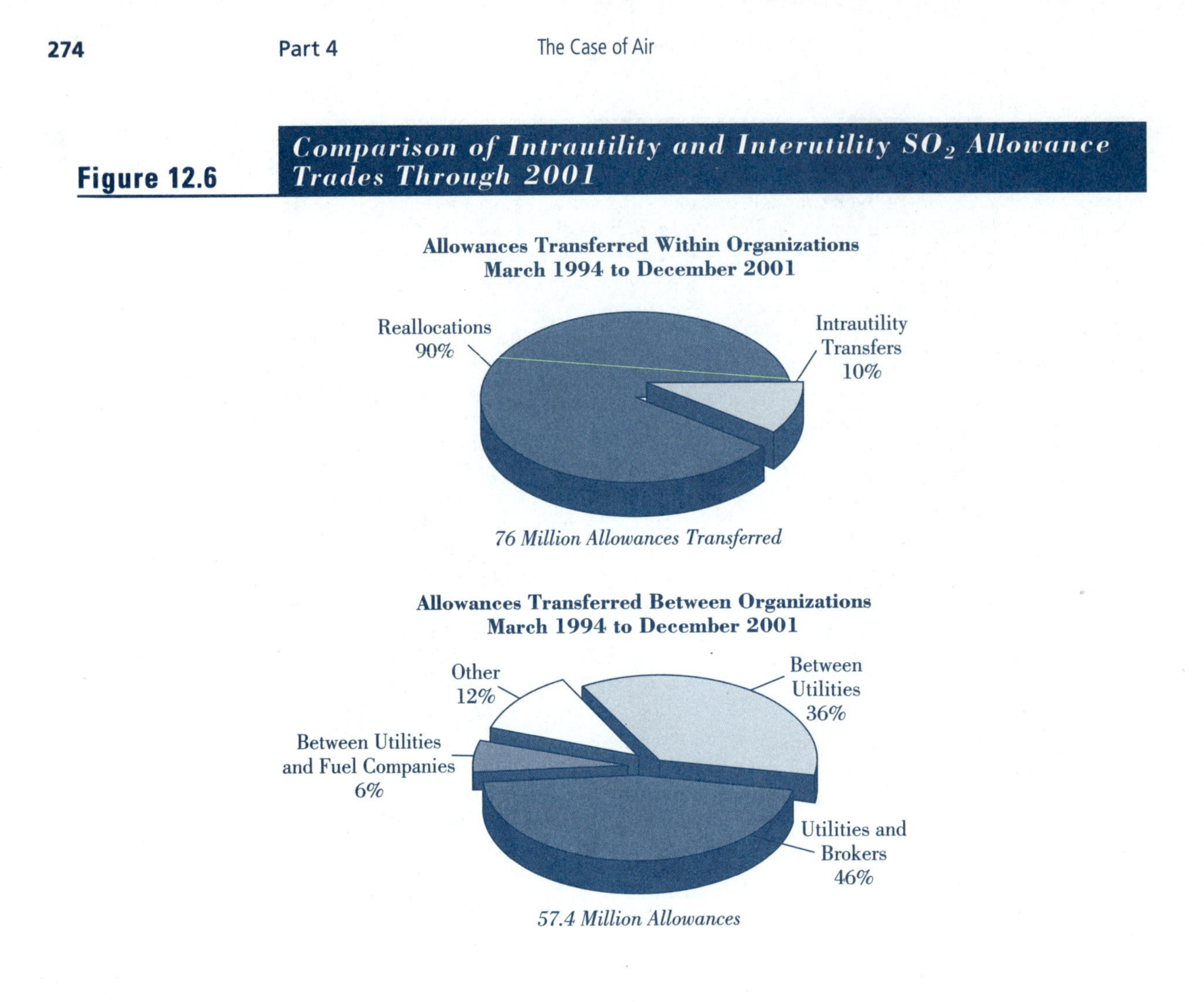

Source: U.S. EPA, Office of Air and Radiation, Clean Air Markets Division (January 29, 2002b).

same firm), which are not reported in conventional data. Figure 12.6 shows a comparison of interutility and intrautility trades. Through December 2001, over 17,800 transfers involving some 133.4 million SO_2 allowances had been reported through the national Allowance Tracking System. Of this total, 76 million allowances (about 57 percent) were transferred *within* utilities, leaving the remaining 57.4 million traded *between* firms.[22]

Allowance prices are much lower than originally anticipated. The EPA had predicted that prices would be about $750 per ton, but these forecasts proved to be inaccurate. Since the launch of trading activity, allowance prices have ranged from about $66 per ton to about $300 per ton.[23] Figure 12.7 shows the change in allowance prices from August 1994 to August 2001. The highest price over this period was $217. By December 2001, the price had fallen to $173. Allowance price data are accessible on the Internet at **http://www.epa.gov/airmarkets/trading/so2market/prices.html.**[24]

http://

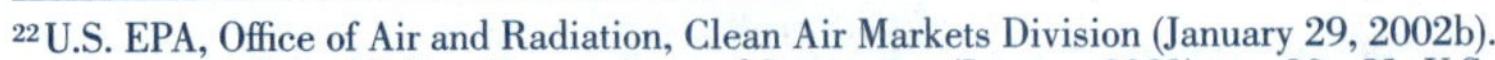

[22] U.S. EPA, Office of Air and Radiation, Clean Air Markets Division (January 29, 2002b).

[23] U.S. EPA, Office of Policy, Economics, and Innovation (January 2001), pp. 80–81; U.S. EPA, Office of Air and Radiation, Clean Air Markets Division (January 29, 2002a).

[24] A very readable article that analyzes the relatively low allowance prices and other aspects of the SO_2 allowance trading program is Schmalensee et al. (summer 1998). Related issues are discussed in the accompanying article by Stavins (summer 1998).

Figure 12.7 *Monthly Average Price of SO_2 Allowances for 1994–2001*

Sources: Data are presented with permission of Cantor Fitzgerald Brokerage, LP, Environmental Brokerage Services (800-228-2955), available at **www.emissionstrading.com,** and cited by the U.S. EPA, Office of Air and Radiation, Clean Air Markets Division, (July 1, 2002).

Note: Prices are rounded to the nearest whole dollar.

Part of the impetus for developing a national market for SO_2 allowances was the expectation of significant cost savings linked to trading activity. Initially, the EPA estimated these savings to be in the range of $0.7–$1 billion per year.[25] As the program has evolved, a number of researchers have addressed this important issue. For example, Ellerman et al. (2000) estimates that the cost of compliance with Title IV of the Clean Air Act is $2.5 billion less annually because of allowance trading. In addition, there may be further gains as this national program serves as a prototype for similar cost-saving strategies being developed by state environmental officials.

One example is a Massachusetts plan called MassIMPACT that allows emissions trading for several types of pollutants. Launched in 1994, this program is touted as the first state program of its kind in the United States. On the West Coast, California's South Coast Air Quality Management District developed what it calls the Regional Clean Air Incentives Market (RECLAIM) program. Also started in 1994, this program uses a bubble-type approach whereby polluters are issued tradeable permits for releases of sulfur oxides, nitrogen oxides, and reactive organic gases. It's the focus of Application 12.4.[26] Other states with active emissions trading programs include Illinois, Michigan, New Jersey, Pennsylvania, and Texas.[27]

[25] *U.S. Federal Register* 56 (December 3, 1991), p. 63097.

[26] Allen (January 8, 1993); Allen (September 29, 1993); U.S. EPA, Office of Policy, Planning, and Evaluation (July 1992), pp. 7-1–7-2.

[27] For information on any state market-based program, visit **http://yosemite.epa.gov/aa/programs.nsf/,** find the "View Programs" menu, and click on "By Program Location."

Application

12.4 Trading Programs at the State Level: California's RECLAIM

Compared to other U.S. cities, Los Angeles is typically cited as having the poorest air quality in the nation. As part of an ongoing effort to address this problem, California officials implemented a regional trading market for sulfur oxide (SO_X) and nitrogen oxide (NO_X) emissions in 1994. The targeted region is the South Coast Air Quality Management District (SCAQMD), which includes four counties around the Los Angeles basin. The trading program is officially called the Regional Clean Air Incentives Market and commonly referred to by its acronym, RECLAIM.

The basic premise of the RECLAIM program is the same as the national SO_2 allowance trading system: pricing pollution. However, instead of using allowances, SCAQMD polluters are required to hold RECLAIM trading credits (RTC), where each RTC is equivalent to one pound of emissions. To be in compliance, each polluting source must possess the requisite number of credits for each pound of pollution released. The RTCs are tradeable within a current year or for a future year, but each credit is valid only for the year it is issued.

Approximately 400 stationary sources are included in the RTC market: 370 for the NO_X market and 40 for the SO_X market. To reduce overall emissions, the program calls for an annual reduction in the number of RTCs issued to each participating source between 1994 and 2003, at which point the allocation remains level. On net, the plan is expected to reduce NO_X emissions by 8.3 percent per year on average, or from 105 tons per day in 1994 to 27 tons per day by 2003. For SO_X emissions, the anticipated decline is 6.8 percent per year, or a cumulative reduction of 15 tons per day by 2003.

Beyond the expected environmental gains, RECLAIM's tradeable credit approach also makes economic sense. Compared to a conventional command-and-control approach, the RTC market allows polluters to achieve air quality objectives in a cost-effective manner. Indeed, compliance cost savings associated with the California trading program have been estimated at approximately $58 million per year.

Just as in any other trading program, polluting sources in the RECLAIM plan have an incentive to trade as long as their respective marginal abatement costs (*MACs*) are unequal. Such a situation would arise if, for example, a stationary source installs new pollution control technology, reducing its emissions and lowering its *MAC*. With lower emissions, the firm would have more RTCs than it needs for the year. And as long as the market price of an RTC is greater than the *MAC*, it would have an incentive to sell the excess credits. Conversely, a polluter with poor abatement technology would incur relatively high abatement costs and face a shortage of RTCs. That firm would have an incentive to buy RTCs, as long as the going price is lower than its *MAC*.

Trading activity in the RECLAIM program has increased markedly over time since the 1994 launch. The most active buyers have been utilities and refineries. Another observed trend is the growth rate of RTC prices for NO_X, which has accelerated in recent years. For example, NO_X credit prices were about $1,500 per ton in 1999, but the price increased tenfold to approximately $15,000 in 2001. This enormous price rise is attributable to increased demand for electricity during the recent California energy crisis. Such unanticipated price levels are now being addressed by California officials. In contrast, SO_X prices have remained fairly stable at about $1,500 per ton, although prices of vintage SO_X credits for future years show an upward trend.

The RECLAIM program is not without critics, but it has been commended for achieving higher air quality, for creating new jobs, and for encouraging technological innovation. Furthermore, it has become a prototype for emissions trading programs in other regions. In fact, several states, including Illinois, Michigan, and Texas, have implemented similar trading plans to meet their respective air quality goals.

To learn more about California's RECLAIM program or other state initiatives, visit the EPA's Directory of Air Quality Economic Incentive Programs at **http://yosemite.epa.gov/aa/programs/nsf**.

Sources: U.S. EPA, Office of Policy, Economics, and Innovation, Office of the Administrator (January 2001) pp. 93–96; "California's Smog Market: Right to Pollute." (October 30, 1993); "SCAQMD Moves to Relieve Power Plants" (January 26, 2001).

Conclusions

Current U.S. air quality policy is ambitious and complex, implemented in part through a comprehensive set of controls on stationary sources. There is no doubt that the trend toward integrating more market-based initiatives is a landmark in environmental legislative history. Evidence is accumulating that market instruments are a more cost-effective means to achieve air quality than command-and-control policies. Other aspects of existing air control policy are less satisfying and continue to be debated. The dual-control approach to setting standards for stationary sources is one example of how government policy can create market distortions.

All of this means that, as a nation, we still have much to learn—about air quality and about policy instruments that work effectively to protect it. As we have come to discover, policy development and appraisal are ongoing processes. In the context of air quality controls, the simple truth is that it is extremely difficult to assess exactly what has been accomplished through all the revisions of the last several decades. Even more difficult is to project the success of the 1990 Amendments, which will not be fully implemented until 2005. In the interim, the United States must continue to make policy adjustments and to respond to the challenge of achieving both its air quality objectives *and* economic prosperity.

Summary

- Standards applicable to new stationary sources are called New Source Performance Standards (NSPS). The EPA is in charge of controlling new or modified sources, whereas states set standards for existing facilities.
- In 1979, the EPA initiated the bubble policy for stationary sources and launched its emissions banking program in 1980.
- In PSD areas, standards for any proposed new or modified source are to be based on the best available control technology (BACT). Existing facilities releasing emissions that might impair visibility in PSD areas are required to install best available retrofit technology (BART).
- In nonattainment areas, existing sources must use, at minimum, reasonably available control technology (RACT). New or modified sources must comply with the lowest achievable emission rate (LAER).
- To facilitate control in PSD areas, netting was developed for use by modified sources. For nonattainment areas, the permit program includes trading for new or modified sources through an offset plan.
- Acidic deposition is caused by the reaction of SO_2 and NO_X emissions with water vapor and oxidants in the earth's atmosphere. These chemical reactions form sulfuric acid and nitric acid, respectively, which mix with other airborne particles and fall to the earth as dry or wet deposition.
- To control acidic deposition, Title IV of the 1990 Amendments is dedicated to a reduction plan for NO_X emissions and an allowance program to reduce SO_2 emissions.
- Once emission allowances for SO_2 are issued, they may be exchanged through an allowance market. Small banks of allowances are available for direct sale by the EPA or through annual auctions supervised by the Chicago Board of Trade.
- The NSPS are implemented uniformly across all firms in a given category and are technology based. As a result, firms have no flexibility to find the least-cost method of achieving them.
- Under the dual-control system, state-determined limits will likely be more lenient. Thus, firms can avoid meeting the NSPS by maintaining existing plants instead of building new ones.
- The EPA's Emissions Trading Program can yield a cost-effective solution, because sources abate to the point where the *MAC* of doing so is equal across firms.

Key Concepts

New Source Performance Standards (NSPS)
new or modified stationary source
existing stationary source
bubble policy
emissions banking
netting
offset plan
acidic deposition
tradeable SO_2 emission allowances

Use the Key Concepts listed above to begin your search for additional articles and information using the InfoTrac College Edition database.

Review Questions

1. New source bias can exist for stationary sources. Discuss why this bias leads to a solution that is *not* cost-effective. What policies might eliminate this bias?
2. Distinguish between the technology-based emission standards RACT and LAER used to control stationary sources.
3. a. Carefully explain how economic theory supports (i) the bubble policy and (ii) the emission allowance program for sulfur dioxide.
   b. Other than these two programs, briefly summarize any two U.S. policies that use market incentives to control air pollution.
4. a. Using the results presented in Table 12.4, discuss why transactions costs are so important in explaining the success or failure of the EPA's bubble program.
   b. How would you devise a program that minimizes the transactions costs of bringing polluters together so that they could effectively equalize the level of their individual *MAC*s?
5. In July 1997, the EPA announced new air quality standards for small (2.5 micrometers in diameter) particulate matter, referred to as PM-2.5. Steel mills are a major source of these smaller particles and therefore must find ways to abate. To analyze the implications, consider the following hypothetical model of two steel plants, one owned by Bethlehem Steel (B) and one owned by National Steel (N), both located in Pittsburgh:

   Bethlehem: $MAC_B = 1.2A_B$
   $TAC_B = 0.6A_B^2$

   National: $MAC_N = 0.3A_N$
   $TAC_N = 0.15A_N^2$

   Assume that each plant emits 40 units of PM-2.5, for a total of 80 units. In order for the Pittsburgh area to meet the new standard, the EPA determines that the combined abatement for both plants must total 30 units.

   a. If the new abatement standard is implemented *uniformly* across the two firms, find the total cost of abatement.
   b. Find the *cost-effective* solution and illustrate graphically, labeling all curves, intercepts, and relevant intersections. Calculate the associated cost savings.

Additional Readings

Burtraw, Dallas. "Markets for Clean Air: The U.S. Acid Rain Program." *Regional Science and Urban Economics* 32 (January 2002), pp. 139–44.

Carlson, Curtis, Dallas Burtraw, Maureen Cropper, and Karen L. Palmer. "Sulfur-Dioxide Control by Electric Utilities: What Are the Gains from Trade?" *Journal of Political Economy* 108 (December 2000), pp. 1,292–1,326.

Chestnut, Lauraine G., and Robin Dennis. "Economic Benefits of Improvements in Visibility: Acid Rain Provisions of the 1990 Clean Air Act Amendments." *Journal of Air and Waste Management Association* 47 (March 1997), pp. 395–402.

Feldman, Stephen L., and Robert K. Raufer. *Emissions Trading and Acid Rain: Implementing a Market Approach to Pollution Control.* Totowa, NJ: Rowman and Littlefield, 1987.

Hahn, Robert W., and Gordon L. Hester. "Where Did All the Markets Go? An Analysis of EPA's Emissions Trading Program." *Yale Journal on Regulation* 6 (1989), pp. 109–53.

Hall, Jane V., and Amy L. Walton. "A Case Study in Pollution Markets: Dismal Science vs. Dismal Reality." *Contemporary Economic Policy* 14 (April 1996), pp. 67–78.

Klaassen, Ger, and Andries Nentjes. "Creating Markets for Air Pollution in Europe and the USA." *Environmental and Resource Economics* 10 (September 1997), pp. 125–46.

Kruger, Joseph, and Melanie Dean. "Looking Back on SO_2 Trading: What's Good for the Environment Is Good for the Market." *Public Utilities Fortnightly* 135 (August 1997), pp. 30–37.

National Science and Technology Council. *NAPAP Biennial Report to Congress: An Integrated Assessment.* Silver Spring, MD, May 1998.

Related Web Sites

http://

Acid rain effects	**http://www.epa.gov/airmarkets/acidrain/effects/index.html**
Acid rain overview	**http://www.epa.gov/airmarkets/acidrain/index.html**
Acid Rain Program SO_2 Allowances Fact Sheet	**http://www.epa.gov/airmarkets/arp/allfact.html**
Allowance Price Data	**http://www.epa.gov/airmarkets/trading/so2market/prices.html**
Allowance Trading	**http://www.epa.gov/airmarkets/trading/index.html**
Allowance Tracking	**http://www.epa.gov/airmarkets/tracking/index.html**
Annual compliance reports for the U.S. Acid Rain Program	**http://www.epa.gov/airmarkets/cmprpt/**
Clean Air Market Programs: NO_X Trading Programs	**http://www.epa.gov/airmarkets/progsregs/noxview.html**
"The Costs and Benefits of Reducing Acid Rain," Burtraw et al. (July 1997)	**http://www.rff.org/disc_papers/abstracts/9731.htm**
Ozone Transport Commission (OTC), NO_X Budget Program	**http://www.epa.gov/airmarkt/otc/index.html**
Ozone Transport Commission (OTC)	**http://www.sso.org/otc/**

Regional Clean Air Incentives Market (RECLAIM)	http://www.aqmd.gov/reclaim/reclaim.html
SO_2 Allowance Market Analysis	http://www.epa.gov/airmarkets/trading/so2market/
U.S. EPA, Directory of Air Quality Economic Incentive Programs	http://yosemite.epa.gov/aa/programs/nsf
U.S. EPA, Market Incentives Resource Center	http://www.epa.gov/orcdizux/transp/traqmkti.htm

A Reference to Acronyms and Terms Used in Stationary Source Control Policy

Environmental Economics Acronyms

MAC	Marginal abatement cost
$\boldsymbol{MAC_{EX}}$	Marginal abatement cost for an existing stationary source
$\boldsymbol{MAC_{N}}$	Marginal abatement cost for a new stationary source
MSB	Marginal social benefit
MSC	Marginal social cost
TAC	Total abatement cost

Environmental Science Terms

Btu	British thermal unit
CO	Carbon monoxide
$\mathbf{NO_X}$	Nitrogen oxides
$\mathbf{NO_2}$	Nitrogen dioxide
$\mathbf{O_2}$	Oxygen
$\mathbf{O_3}$	Ozone
Pb	Lead
PM	Particulate matter
PM-2.5	Particulate matter less than 2.5 micrograms in diameter
PM-10	Particulate matter less than 10 micrograms in diameter
$\mathbf{SO_2}$	Sulfur dioxide
$\mathbf{SO_X}$	Sulfur oxides
VOC	Volatile organic compound

Environmental Policy Acronyms

BACT	Best available control technology
BART	Best available retrofit technology
LAER	Lowest achievable emission rate
NAAQS	National Ambient Air Quality Standards
NSPS	New Source Performance Standards
OTC	Ozone Transport Commission
PSD	Prevention of significant deterioration
RACT	Reasonably available control technology
RECLAIM	Regional Clean Air Incentives Market
SCAQMD	South Coast Air Quality Management District
SIP	State Implementation Plan

Chapter 13

Global Air Quality: Policies for Ozone Depletion and Global Warming

While most air pollutants produce localized effects, others have more far-reaching implications. Such is the case for contaminants that alter atmospheric conditions, posing a risk that is geographically without bound and generating a free-ridership problem that crosses national boundaries. Of course, the effects can vary by degree across different locations. In any case, since the associated damage is widespread and since the source cannot be linked to a specific site or region, this air quality problem is termed **global air pollution.** Controlling global air pollution is a unique policy challenge, because solutions must be developed not only through domestic initiatives but also through international treaties and programs.

In this chapter, we investigate the principal issues associated with global air pollution by studying **ozone depletion** and **global warming.** In each case, we consider theories about the causes and sources of the respective atmospheric disturbance and the available evidence to support these theories. Using this as a foundation, we then explore policy responses that have been set in motion in the United States and other nations along with proposals for alternatives. Ultimately, our objective is to evaluate the effectiveness of existing and proposed policy responses economically, given what we know about the origin of the problem and the associated risks. As in the previous three chapters, there is a reference list of acronyms and terms at the end of the chapter.

The Problem of Ozone Depletion[1]

ozone layer
Ozone present in the stratosphere that protects the earth from ultraviolet radiation.

Starting in the 1950s, scientists began measuring the earth's **ozone layer**—the ozone present in the stratosphere, which is the layer of the atmosphere lying between 7 and 25 miles above the earth's surface. The effort was motivated by more than scientific curiosity. Stratospheric ozone protects the earth from ultraviolet radiation.

Some variability in the depth of the ozone layer was assumed normal, including an observed thinning above Antarctica during the Southern Hemisphere spring. This would generally fill back in by November each year. However, in the early 1980s, scientists became concerned when this thinning was found to be increasing in size and persisting into December. In 1985, an "ozone hole" the size of North America was discovered over Antarctica. It was then that world attention was drawn in earnest to the problem of **ozone depletion** and the pollutants responsible for the damage.[2]

ozone depletion
Thinning of the ozone layer, originally observed as an ozone hole over Antarctica.

All the implications are not known with certainty, but there are some consequences of increased ultraviolet radiation about which there seems to be some agreement. Scientists tell us that rising levels of ultraviolet radiation can alter delicate ecosystems, diminish human immune systems, and increase the risk of skin cancer. The National Academy of Sciences estimates that 10,000 more cases of skin cancer per year would result for every 1 percent decline in stratospheric ozone.[3]

[1] A good resource on the science of ozone depletion is available at **http://www.epa.gov/ozone/science/science.html.**
[2] For the published report, see Farman, Gardiner, and Shanklin (1985).
[3] More information on the health effects associated with ozone depletion is available online at **http://www.epa.gov/sunwise/uvandhealth.html.**

Searching for the Causes of Ozone Depletion

Research scientists debate about the principal cause of the ozone hole, which extends approximately 9 million square miles over the Antarctic.[4] Although no one theory has been able to fully explain the extent of ozone depletion, scientists agree that the presence of **chlorofluorocarbons (CFCs)** in the atmosphere is the most likely explanation—a theory originally advanced in 1974 by F. Sherwood Rowland and Mario Molina, two University of California researchers. The pair won the Nobel Prize in Chemistry for this theory in 1995.

chlorofluorocarbons (CFCs)
A family of chemicals that scientists believe contributes to ozone depletion.

CFCs are a family of chemicals that were commonly used in refrigeration, air conditioning, packaging, and insulation and as aerosol propellants. They are sometimes referred to by their trade names, Freon and Styrofoam. In fact, the energy crisis in the 1970s was responsible for an even greater use of CFCs as foaming agents in the production of home insulation. The rise of the fast-food industry was also a contributing factor to intensified CFC use, because polymer foams were utilized to produce disposable cups and food containers.[5] These long-lived compounds are not destroyed in the lower atmosphere and therefore are able to drift up into the stratosphere, where their chlorine components destroy ozone. In addition, because of their long atmospheric lifetimes, CFCs released today affect the ozone layer for decades to come.

Another major group of ozone depleters are **halons,** which also have long atmospheric lifetimes. Before government controls, these substances were becoming increasingly important in the production of fire extinguishants. Their use is not as widespread as CFCs, but halons are known to have a higher potency for ozone depletion than their chlorine-containing counterparts.

Despite the lack of hard evidence at the time, the United States opted to ban the use of CFCs in most aerosol sprays in 1978, and other countries followed suit. However, other uses of these ozone depleters were not controlled, and little effort was aimed at finding substitutes. As a result, domestic and international CFC use continued to grow. Data on cumulative CFC production worldwide are given in Figure 13.1. As these data suggest, there was little question that a stronger policy position was needed to control ozone-depleting substances.

Controlling Ozone Depletion

As a global air pollution problem, ozone depletion cannot be controlled without an integrated international effort. More formally, think of this environmental problem as an externality with transboundary implications. The 1990 Clean Air Act Amendments specifically call for the president to enter into international agreements that encourage joint research on ozone depletion and to establish regulations consistent with those in the United States. Although not without political implications, a number of international agreements and multilateral treaties have been executed or are on the negotiating table. A brief summary of the most significant of these follows.

International Agreements to Control Ozone Depletion[6]

Montreal Protocol

In 1987, 24 countries as well as the European Community Commission signed the Montreal Protocol on Substances That Deplete the Ozone Layer. Among the signatories were the major producers of CFCs. This landmark agreement called for a 50 percent reduction in CFC consumption and production, a target that was to be achieved gradually by 2000. To achieve this

[4] Research indicates that ozone depletion is truly a global issue, occurring over latitudes that include Asia, parts of Africa, Australia, Europe, North America, and South America (U.S. EPA, Office of Air and Radiation, February 5, 2002).
[5] Kemp (1990), pp. 127–28.
[6] General information in the subsequent discussion is drawn from Council on Environmental Quality (March 1992; January 1993; 1997).

Figure 13.1 *Cumulative Global Production of CFCs: 1960–1998*

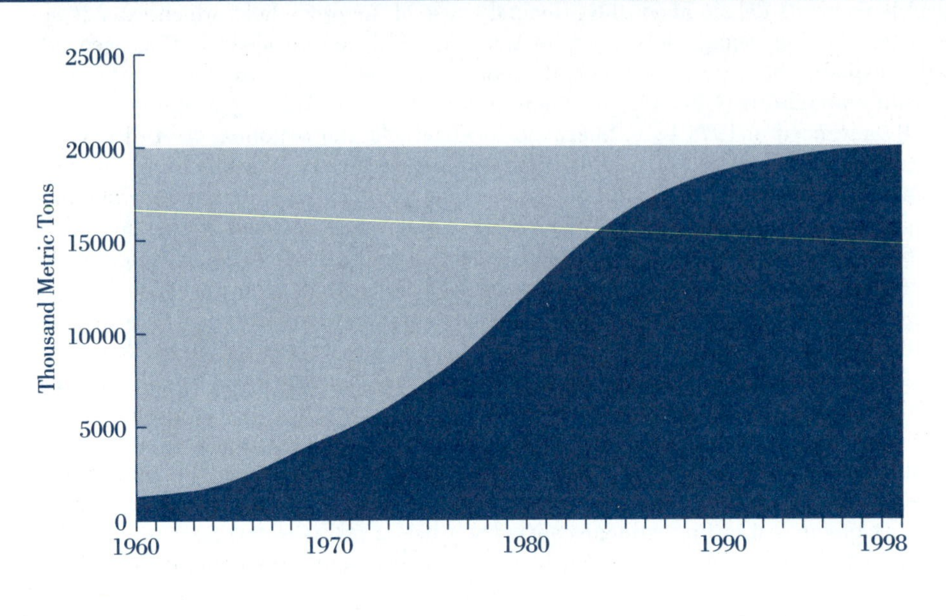

Source: Alternative Fluorocarbons Environmental Acceptability Study, as cited in Council on Environmental Quality (1998), p. 349, table 11.3.

Note: Production data are voluntarily reported by the chemical industry through a survey conducted by an independent account, Grant Thornton LLP.

objective, each party to the protocol was responsible for designing and implementing an effective control program in accordance with the agreed-upon deadlines. As of 2002, over 180 nations have ratified the Montreal Protocol.[7]

Amendments to the Protocol

In 1990, 59 countries executed the London Amendments to the protocol. These amendments, which strengthened the worldwide commitment to protecting the ozone layer, were in direct response to reports that ozone depletion might be more severe than originally believed. The new agreement outlined a full phase-out plan for CFCs, halons, and other ozone-depleting substances, such as methyl chloroform, carbon tetrachloride, and hydrochlorofluorocarbons (HCFCs). At subsequent conferences, the phase-out deadlines were advanced (in the Montreal Amendments in 1997 and in the Beijing Amendments in 1999). Table 13.1 offers a summary of how these phase-out agreements have evolved over time.

International Allowance Trading

A market approach was also integrated as part of the international effort to protect the ozone layer. Specifically, production and consumption allowances were issued to the protocol participants, and transfers were permitted under certain guidelines. To ensure that the phase-outs were achieved, trading was conditioned upon revision of each country's aggregate production limits to levels lower than what would have occurred without the transfers.

[7] The current ratification status of the Montreal Protocol and subsequent amendments is available at **http://www.unep.ch/ozone/ratif.shtml.**

Table 13.1 ***International Agreements to Phase Out Ozone Depleters***

	1987 Montreal Protocol	1990 London Amendments	1992–1999 up through Beijing Amendments
CFCs	50% cut by 2000 (5 types covered)	Phase out by 2000 (15 types covered)	Phase out by 1996 (15 types covered)
Carbon tetrachloride	No controls	Phase out by 2000	Phase out by 1996
Halons	Freeze at 1986 levels in 1992	Phase out by 2000	Phase out in 1994
HCFCs	No controls	Phase out by 2040	Freeze in 2004 at 1989 levels; phase out by 2020
Methyl bromide	No controls	No controls	Freeze in 1995 at 1991 levels; phase out by 2005
Methyl chloroform	No controls	Phase out by 2005	Phase out by 1996

Sources: UN Environment Programme, cited in Dumanoski (November 26, 1992); Council on Environmental Quality (January 1993); Parker (July 12, 2000).

Notes: For more detail on the Montreal Protocol or any of the subsequent amendments, visit the Ozone Secretariat at the United Nations Web site at **http://www.unep.ch/ozone/index.shtml.** Deadlines shown are for developed nations; developing nations generally were given grace periods that extended the phase-out deadlines. After the London Amendments, there was the 1992 Copenhagen Conference, the 1997 Montreal Amendments, and the 1999 Beijing Amendments. Carbon tetrachloride is a solvent used in CFC production; HCFCs are CFC substitutes; Methyl bromide is a pesticide; Methyl chloroform is a solvent.

Multilateral Fund

Ongoing negotiations are aimed at encouraging more nations to ratify the protocol's amendments. Some countries have been hesitant to participate because of the high costs of converting production technology to eliminate the use of ozone-depleting substances. This is particularly problematic for developing nations. In response to these real concerns, a three-year Interim Multilateral Fund of $160 million was established in 1990 by the protocol participants to assist developing countries in transitioning toward the requisite CFC replacement technologies.

Support for developing nations, such as China and India, is crucial to the success of the global initiative. As these nations evolve economically, their increased need for refrigeration (and the associated use of ozone-depleting substances) coupled with their large populations will have serious implications for the earth's ozone layer. Accordingly, the parties to the protocol agreed to increase the Multilateral Fund to $240 million if China and India signed the agreement, committing the incremental amount directly to these two countries. In June 1991, China did agree to sign, and in June 1992, India did as well.[8] To better understand the importance of gaining the cooperation of developing nations, consider the following. Given India's expected growth rate, refrigerator consumption in that nation is predicted to rise to nearly 80 million units by the year 2010, a dramatic increase from its 1989 level of 6 million.[9]

For its part, the United States has charged the EPA with formulating and implementing domestic efforts to meet the protocol's phase-out schedule. We next investigate U.S. policy on ozone depletion with an emphasis on market-based instruments that are integral to achieving national and international objectives.

[8]In 1992, the fund became permanent, and it has continued to grow. As of 2001, 32 industrialized nations contributed $1.3 billion to the Multilateral Fund. More information is available at the Web site of the United Nations Environment Programme, **http://www.unmfs.org/general.htm.**

[9]World Resources Institute (1992b), pp. 152–53.

Table 13.2 Phaseout of U.S. Production of Ozone-Depleting Substances

Substance	End of U.S. Production
CFCs	January 1, 1996
Halons	January 1, 1994
Carbon tetrachloride	January 1, 1996
Methyl chloroform	January 1, 1996
HCFCs[a]	January 1, 2020
Methyl Bromide	January 1, 2005

Sources: U.S. EPA, Office of Atmospheric Programs, Global Programs Division. (November 1, 2001); U.S. EPA, Office of Atmospheric Programs, Global Programs Division (June 17, 1998).

Note:
[a]Production of HCFCs with the most severe ozone-destroying effects will end by January 1, 2020; production of the rest of the HCFCs will end by January 1, 2030.

U.S. Policy to Control Ozone Depletion

The 1990 Clean Air Act Amendments significantly strengthened U.S. policy on ozone-depleting substances. Title VI of these amendments is dedicated to protecting the ozone layer. Chief among its rulings are those that define how ozone-depleting substances are to be identified and those that establish a plan to phase them out of usage. These provisions must comply with the nation's commitment to the protocol.[10]

ozone depletion potential (ODP)
A numerical score that signifies a substance's potential for destroying stratospheric ozone relative to CFC-11.

Congress charged the EPA with the responsibility of publishing a complete list of all ozone-depleting substances, including those already identified in the 1990 Amendments. This list distinguishes between Class I and Class II substances, where Class I refers to those having a greater potential for damage. Each listed substance is assigned a numerical value signifying its **ozone depletion potential (ODP)** relative to chlorofluorocarbon-11 (CFC-11). (ODP values for every listed substance can be accessed on the Internet at the EPA at **http://www.epa.gov/ozone/ods.html.**) For each substance class, specific phase-out schedules are outlined, which state the maximum amount that can be produced or consumed each year through the phase-out period. Table 13.2 gives the phase-out dates for U.S. production of all ozone-depleting substances.

Recognizing the industry's dependence on CFCs, the 1990 Amendments include provisions to establish a mandatory national recycling program for Class I and II substances, with the intent that recycled refrigerants could be used as substitutes for virgin materials. A related set of provisions calls for federal programs and research aimed at finding safe alternatives to identified ozone depleters.[11] There are also two legislated instruments that explicitly use market incentives to eliminate ozone-depleting substances: an **excise tax** and a **marketable allowance system.**

excise tax on ozone depleters
An escalating tax on the production of ozone-depleting substances.

Excise Tax on Ozone Depleters

One market-based instrument used to control ozone depletion and achieve the phase-out deadlines was an escalating **excise tax** on the production of ozone-depleting substances. This tax was enacted by Congress in 1990. The tax rate per pound was a base dollar amount

[10]Other provisions deal with safety requirements, such as the proper handling of refrigerants in servicing motor vehicle air conditioners and the safe use and disposal of listed substances during the servicing or disposal of appliances and refrigeration equipment. Strict labeling regulations are also mandated.

[11]See sec. 608 and sec. 612 of the 1990 Amendments as well as Lee (May/June 1992) for a discussion of these and other provisions in Title VI of the Clean Air Act. For direct information about the legislation itself, visit **http://www.epa.gov/ozone/title6/index.html#stitlevi.**

multiplied by the chemical's ODP, where the base amount was higher for each successive year in the phase-out schedule. The tax was initially set at $1.37 per pound, and by 1995 it had increased to $5.35 per pound. Starting in 1996, the tax increased by $0.45 per pound each year, bringing it to $8.50 per pound in 2002. Although the phase-out deadline has passed, the tax is still applicable to imported recycled CFCs.

From an economic perspective, notice that such a tax elevated the effective price of ozone depleters and thus motivated a reduction in quantity demanded. According to Cook (1996), consumption of ozone-depleting substances (expressed in CFC-11 equivalents) decreased from 318,000 metric tons in 1989 to 200,000 metric tons in 1990, the year the tax was put into effect.

Allowance Market for Ozone-Depleting Chemicals

A proposal is currently under review to establish an **allowance market** in the United States to facilitate the phaseout of HCFCs. If approved, the implementation will generally follow the allowance program put in place prior to 1996 to control CFCs and certain other ozone-depleting substances. Under this program, firms were allowed to produce or import these substances only if they held an appropriate number of allowances. Each allowance authorized a one-time release of some amount of the substance based on its ODP.[12] In the aggregate, the number of baseline allowances was set to freeze domestic production of ozone-depleting substances at their 1986 levels. Given the large number of potential buyers for these allowances, the EPA chose to allocate them only to the largest domestic consumers and producers. Over time, the number of available allowances was gradually reduced and eventually brought to zero.

allowance market for ozone-depleting chemicals
A system that allows firms to produce or import ozone depleters only if they hold an appropriate number of tradeable allowances.

In the interim, market exchanges of allowances were permitted under strict guidelines—even with firms in other signatory nations of the protocol. Transfers were of two types: **trades with other parties** and **interpollutant transfers.** Interpollutant transfers are exchanges of production of one substance in a given year for production of another in that same year, using a weighting scheme based on the ozone depletion values. The proviso for allowance transfers was that trading had to yield greater reductions in annual production or consumption than would have occurred without the transfers.

Regulatory Impact Analysis (RIA) of the Phaseout

As part of its Regulatory Impact Analysis (RIA), the EPA conducted a benefit-cost study of the formal phase-out plan.[13] Given the long life of ozone depleters, the agency considered the regulatory implications over a long time period, out to 2075. The agency's benefit assessment assigned a value to the damages that would be prevented by controlling these substances. These included health effects associated with increased exposure to ultraviolet radiation and nonhealth effects, such as reduced crop yields, rising sea levels, and property damage.[14] In total, the EPA estimated that accumulated damages would be approximately $6.5 trillion by 2075.[15]

On the cost side, a value had to be assigned to all anticipated market disruptions that would arise from a proposed phase-out plan. Some 84 distinct use categories for CFCs were analyzed—the two largest being mobile air conditioning and refrigeration.[16] All told, the EPA's

[12] For example, CFC-11 and CFC-12 each have an ODP of 1.0, whereas CFC-113 has an ODP of 0.8. Using these ODPs as weights, 100 allowances could be used to produce either 100 tons of CFC-11 or CFC-12 or 125 tons of CFC-113. Notice how this weighting system gives firms more flexibility in terms of how they comply with the law.
[13] For more detail, see U.S. EPA, Stratospheric Protection Program, Office of Air and Radiation (December 1987). Also see U.S. EPA, Office of Air and Radiation (November 1999) for a recent benefit-cost analysis of Title VI.
[14] In addition to destroying stratospheric ozone, CFCs have been identified as greenhouse gases, an issue we address later in this chapter. Hence, some of the effects identified in the RIA are associated with the contribution of CFCs to global warming.
[15] Cogan (1988), p. 88.
[16] That CFCs were widely used at the time is evidenced by the following. In 1985, 90 million cars and light trucks used 120 million pounds of CFCs in their air-conditioning units. Another 95 million pounds were used in over 100 million refrigerators, 30 million freezers, 180,000 refrigerated trucks, and hundreds of thousands of food service businesses. In all, approximately 660 million pounds of CFCs were consumed during 1985 (Cogan 1988, p. 15).

Table 13.3 ***Value of CFCs to American Industry***

Usage	Value of Products and Services
Refrigeration	$ 6.0 billion
Air conditioning	$ 10.9 billion
Mobile air conditioning	$ 2.0 billion
Plastic foams	$ 2.0 billion
Cleaning agents	Billions of dollars
Food freezants	$400 million
Sterilants	$100 million

Source: Drawn from Alliance for Responsible CFC Policy (December 1986).

estimate of control costs associated with a phase-out plan was $27 billion through 2075. Clearly, certain CFC-dependent industries would be more affected than others, as shown in Table 13.3.

In any case, although the costs of the phase-out plan are significant, they pale in comparison to the dollar value of damages that would result if the United States took no action at all. Consequently, U.S. regulations to control ozone depleters were announced in August 1988, less than one year after the signing of the Montreal Protocol.[17]

Economic Analysis of U.S. Policy on Ozone Depletion

Critical to U.S. policy on ozone-depleting substances was the market-based allowance system. Because the number of allowances was controlled by the government and declined over time, the objective of gradually eliminating the substances was achieved. Furthermore, fundamental economics tells us that this trading program should have approached a more cost-effective solution than command-and-control instruments. Those firms best able to find substitutes for the controlled chemicals would have done so and would have sold their allowances to less efficient producers. But is there any evidence to support this presumption?

Assessing Cost-Effectiveness

Following the 1978 ban of CFCs in all so-called nonessential aerosols, the EPA began investigating the feasibility of further controls. In an EPA-commissioned study conducted by the Rand Corporation, three alternative control approaches were analyzed: a technology-based command-and-control approach, a fixed emission charge, and a tradeable emission permit system.[18] Each approach was modeled to achieve a given level of reductions over a 10-year period so that the accumulated costs of each plan could be compared. The study showed that the estimated costs for each approach examined were as follows:

- Technology-based command-and-control approach: **$185.3 million**
- Fixed emission charges: **$107.8 million**
- Tradeable emissions permit system: **$ 94.7 million**

These estimates support the expectation that allowance trading would approach a cost-effective solution. At the same time, trading should act as an incentive for the development of

[17] See *U.S. Federal Register* 53, p. 30598.
[18] Palmer et al. (1980).

Figure 13.2 ***Price Adjustments of CFCs and CFC Substitutes***

As the phase-out plan advanced, there was reduced availability of CFCs. In addition, there was an excise tax levied on production. These events are illustrated in panel (a) by the shift leftward of the supply of CFCs, which elevated their price. As this occurred, the demand for CFC substitutes rose, as shown in panel (b). Prices of substitutes were relatively high at the outset, but this price differential diminished as technological advance and government support of that innovation brought about cost declines for substitutes, which in turn shifted the supply curve to the right, as shown in panel (b).

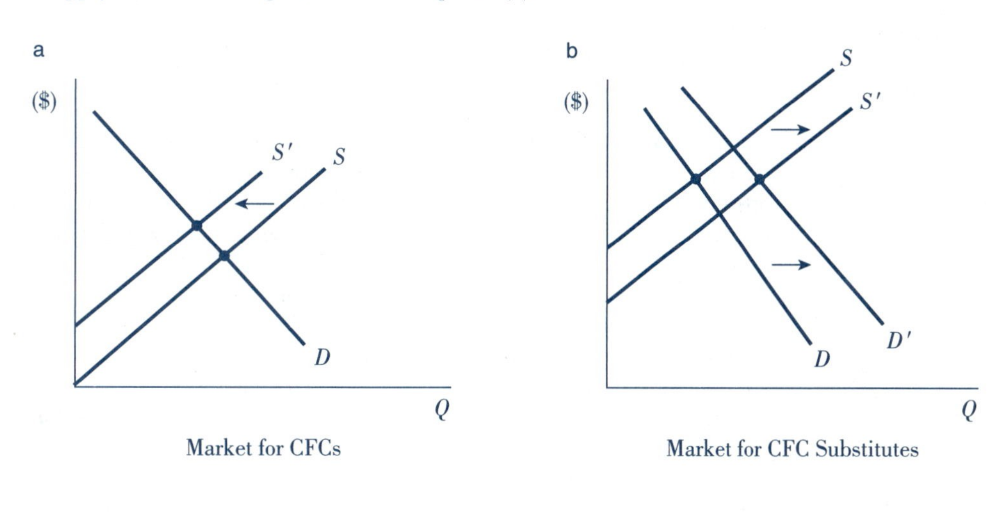

substitutes by those firms that could do so at least cost. Hence, the phaseout in the CFC market had implications for other related markets, in particular, the market for CFC substitutes, which can be analyzed through examining supply and demand and relevant movements in price.[19]

Price Changes

Because U.S. policy was implemented through market-oriented instruments, its progress could be observed through the price mechanism. Prices of CFCs and other ozone-depleting chemicals signaled the impact of the phaseout and the underlying market adjustments. As the phase-out plan advanced, there was reduced availability of CFCs. In addition, there was an excise tax levied on production, as discussed earlier. These events are illustrated in Figure 13.2a by the shift leftward of the supply of CFCs, which in turn elevated their price. Manufacturers of CFC-dependent products faced higher production costs as a result and passed on at least some of this cost increase to consumers. Thus, buyers of commodities such as refrigerators and auto air-conditioning units paid higher prices over time.[20]

Also, as CFC prices increased, the demand for CFC substitutes rose, as shown in Figure 13.2b. One consequence of these events was the evolution of a black market for CFCs. This

[19] Of interest is the EPA's report on the supply and demand of CFC-12 in 1999 (U.S. EPA, Stratospheric Protection Division, Office of Air and Radiation, June 9, 1999). This report can be accessed on the Internet at **http://www.epa.gov/ozone/geninfo/sdreport99.html.**

[20] Of course, the ultimate change in price depended on the elasticity of demand for these products, which in turn depended on the availability of substitutes. The more substitutes there are for a product, the more elastic is demand for that product. Because there are few good substitutes for refrigerators, we would expect demand to be price inelastic. Hence, much of the cost increase from the phase-out plan likely was passed on to refrigerator buyers.

developed in large part because the phase-out dates established by industrialized nations preceded those set by developing nations.[21] The excise tax also may have contributed to this problem.[22]

Incentives and Disincentives to Develop CFC Substitutes

Because costs and prices were allowed to move naturally, the usual incentives encouraged a market adjustment to the observed industry declines and price changes. Theoretically, two opposing reactions were possible.

On the one hand, firms may have perceived a profit advantage in developing ozone-friendly substitute products. Recall that such a market reaction was explicitly supported by U.S. policy on developing safe alternatives. Prices of CFC substitutes were relatively high at the outset. For example, the 1987 price of CFC-12, which was commonly used in automobile air-conditioning units, was approximately $0.50 per pound, whereas a substitute, HFC-134a, was estimated to be $3 per pound.[23] However, this price differential diminished as technological advance and government support of that innovation brought about cost declines for substitutes, which in turn shifted the supply curve to the right. This is shown in Figure 13.2b. Application 13.1 illustrates how industry responded to the market incentives brought about by these changes.

A second possibility is that the relatively small number of firms possessing allowances gained some measure of market power and price control. These firms could have enjoyed above-normal profits at society's expense and made little if any movement toward providing alternative solutions to the marketplace. This explains why some opponents of the allowance plan argued that the EPA compromised its intended cost-effective plan by issuing the valuable rights only to a select few. One possible solution would have been to transfer any excess profit to the government, which in turn could redistribute the windfall in a more effective and equitable manner. Such a safeguard actually was implemented in 1990 when Congress approved the escalating excise tax on ozone-depleting chemicals. Notice how a redistribution of income was achieved when fiscal spending was funded by the tax revenues collected from producers of CFCs and other such substances.

In another step to lower the costs associated with the phaseout, the 1990 Amendments called for a national recycling program for CFCs used in refrigeration and air conditioners. Consider the economics of this approach. By making recycled substances available in the market, firms reduce their demand for virgin compounds to produce more ozone-depleting products. Furthermore, firms that depend on these substances can use recycled materials beyond the phase-out deadlines, thus avoiding costly retrofitting until substitutes are developed and made available for sale.[24]

Overall, the use of the tradeable allowance plan along with the excise tax, the recycling program, and the safe alternatives policy achieved the phase-out objectives in a more cost-effective manner. Such a control program was less disruptive than an immediate ban on production, which would have affected virtually every segment of society with no time to make proper adjustments.

The Problem of Greenhouse Gases and Global Warming

A source of controversy is the predicted climate response to the increasing production of what are termed **greenhouse gases (GHGs).** On the agenda of the 1992 Rio Summit, the issue of accumulating greenhouse gases and the associated predictions of global warming is one that

greenhouse gases (GHGs)
Gases collectively responsible for the absorption process that naturally warms the earth.

[21] For further discussion on this issue, see Arnst (September 29, 1997); Arnst and McWilliams (July 7, 1997); and Council on Environmental Quality (July 7, 1997), p. 202.
[22] See Daley and Barry (August 1, 2001) and U.S. EPA, Stratospheric Protection Division, Office of Air and Radiation (June 9, 1999), p. 6.
[23] Putnam, Hayes, and Bartlett Inc. (1987).
[24] Lee (May/June 1992).

13.1 *Searching for Alternatives to CFCs: The Corporate Response*

Application

Private efforts to respond to the phase-out plan on CFC production and consumption were the subject of much attention from Capitol Hill to Wall Street. Some of the nation's largest corporate entities had to make major adjustments in the way they did business or suffer the consequences. For example, in 1988, Du Pont announced plans to voluntarily halt all production of CFCs by 2000, a segment of its business that generated revenues of $750 million per year. In 1990, the conglomerate had allocated $170 million to research aimed at developing substitutes for ozone depleters and reportedly committed up to $1 billion to continue that effort.

Notwithstanding the importance of these ventures, only the naive would view these actions as corporate altruism. There is no question that firms recognize the value of an environmentally conscious image and the financial retribution that befalls a careless decision. In a speech delivered in London titled "Corporate Environmentalism," Du Pont's CEO Edgar Woolard argued, "Avoiding environmental incidents remains the single greatest imperative facing industry today."

Beyond aggressive efforts on the production side, industrial consumers of CFCs also exerted a measure of influence. Most made dramatic cuts in consumption of CFCs and pushed hard for the development of viable alternatives to make up the difference. Over a two-year period, IBM reduced its CFC consumption by approximately 31 percent and announced a complete halt in usage by 1993. The automobile industry as a whole, already important to air pollution issues, needed to find alternatives to CFCs. Of the 355–425 million pounds of CFCs produced in 1991, approximately 20 percent was used as a refrigerant in motor vehicle air-conditioning units.

Fortunately, the market responded. The statutory deadline for phasing out most ozone-depleting substances was 1996. Spurred by entrepreneurial motives, the effort of research teams paid big dividends. What came to market were usable and less environmentally harmful alternatives. For example, many 1993 cars were equipped with a redesigned air-conditioning unit that used a hydrofluorocarbon by the trade name of R-134a. This was the successor to the ozone-depleting refrigerant R-12, also known as Freon. The new compound contains no chlorine, greatly diminishing its adverse effect on the earth's protective ozone layer. Still more innovative advances have been brought to market, some aimed at replacing aerosol propellants. Recent innovations use new packaging with flexible pouches to house the product, surrounded by a mass of compressed air.

To facilitate substitute development and use, the EPA has established the Significant New Alternatives Policy (SNAP) program. The purpose of the SNAP program is to identify substitutes and publish lists of those that are acceptable and unacceptable. These lists are available in hard copy or online at **http://www.epa.gov/ozone/snap/lists/index.html.** Any producer of a potentially viable substitute must apply for a listing with the EPA. The application must include health and safety studies, which are carefully reviewed by the agency. Applications are granted or denied within a 90-day review period. The EPA also requires a 90-day notice before any new or existing alternative substance is introduced to interstate commerce.

http://

Sources: Kirkpatrick (February 12, 1990); Smith et al. (April 23, 1990); Smith (June 12, 1989); "'Greener' Cooling" (March 1993); "CFC-Free A/C" (March/April 1993); U.S. EPA, Office of Atmospheric Programs, Global Programs Division (February 4, 1998).

continues to be debated. In fact, during 1997, the international community formulated the Kyoto Protocol, which continues the climate change initiative first discussed in Rio. The scientific community is not at all in agreement about this complex phenomenon. Because of the uncertainty, national and international policy responses to this hypothesized global air pollution problem have been tentative and complex at best.

Understanding the Potential Problem

global warming
Caused by sunlight hitting the earth's surface and radiating back into the atmosphere where its absorption by GHGs heats the atmosphere and warms the earth's surface.

The premise of **global warming** is based on the following accepted scientific facts. Sunlight, passing through the atmosphere, hits the surface of the earth and is radiated back into the atmosphere, where it is absorbed by naturally present gases such as carbon dioxide. This absorption process heats the atmosphere and warms the earth's surface. This is somewhat like a greenhouse that allows sunlight through the glass exterior but prevents the heated air from escaping back outside, thus the phrase "**greenhouse effect.**" This natural phenomenon is responsible for the existence of life on earth as we know it. Without the so-called greenhouse gases, the earth's temperature would be some 30° to 40° Celsius cooler.[25]

There are about 20 such gases collectively responsible for this warming phenomenon. The primary ones are carbon dioxide (CO_2), methane (CH_4), chlorofluorocarbons (CFCs), nitrous oxide (N_2O), and tropospheric ozone (O_3), all of which arise from a number of different sources. Look at Figure 13.3, which itemizes the primary anthropogenic sources of each gas and gives estimates of their proportionate contribution to global warming.

global warming potential (GWP)
Measures the global warming effect of a unit of any GHG relative to a unit of CO_2 over some time period.

The ability of a GHG to trap heat in the atmosphere is typically measured relative to carbon dioxide over some time period by something called its **global warming potential (GWP).** The GWP values used in U.S. policy making and reporting are based on a 100-year time frame and are listed online at **http://www.epa.gov/globalwarming/emissions/national/gwp.** Because the presence of GHGs affects the earth's temperature, any significant disruption to their natural levels would have climatological impacts.

A landmark study conducted by the National Academy of Sciences in 1979 predicted that a doubling of CO_2 would generate a rise in the earth's temperature of 1.5° to 4.5° Celsius (or 2° to 8° Fahrenheit).[26] Similar estimates have been found by current analyses, such as those conducted by the Intergovernmental Panel on Climate Change (IPCC).[27] Climate changes in turn would alter the distribution and productivity of agricultural regions, weather conditions, and the level of the earth's seas. Although there is a consensus about the linkage between rising levels of CO_2 and temperature change, there is great uncertainty about the timing and magnitude of the outcome.

Although scientists estimate that there have been a number of cooling and warming cycles throughout the earth's history, current concern about a warming trend stems from the rising anthropogenic production of the most prevalent GHG, which is CO_2. CO_2 is a by-product of fossil fuel combustion, which is basic to the production of energy that supports most industrial activities. CO_2 accumulation is exacerbated by widespread deforestation, such as the burning of tropical rain forests. As plant life is destroyed, less photosynthesis (a process that naturally absorbs CO_2 from the atmosphere) takes place. On these basic facts, most scientists seem to agree. Many also agree that the magnitude of increased levels of atmospheric CO_2 is likely to be substantial—estimated to grow to twice the level of preindustrial times by 2050. Table 13.4 on page 294 gives an overview of the growth and projected trend of atmospheric concentrations of CO_2 and the other primary GHGs. However, predicting the *effect* of this trend on the environment is difficult.

---

[25] For more information on the science of global warming, visit the climate page on EPA's global warming Web site at **http://www.epa.gov/globalwarming/climate/index.html.**

[26] This estimate was subsequently supported by a number of other organizations in 1985, namely, the World Meteorological Organization, the United Nations Environment Programme, and the International Council of Scientific Unions. See National Research Council (1979), cited in U.S. EPA, Office of Policy, Planning, and Evaluation, Office of Research and Development (December 1989).

[27] See, for example, IPCC (2001).

Figure 13.3 *Greenhouse Gases and Their Contribution to Global Warming*

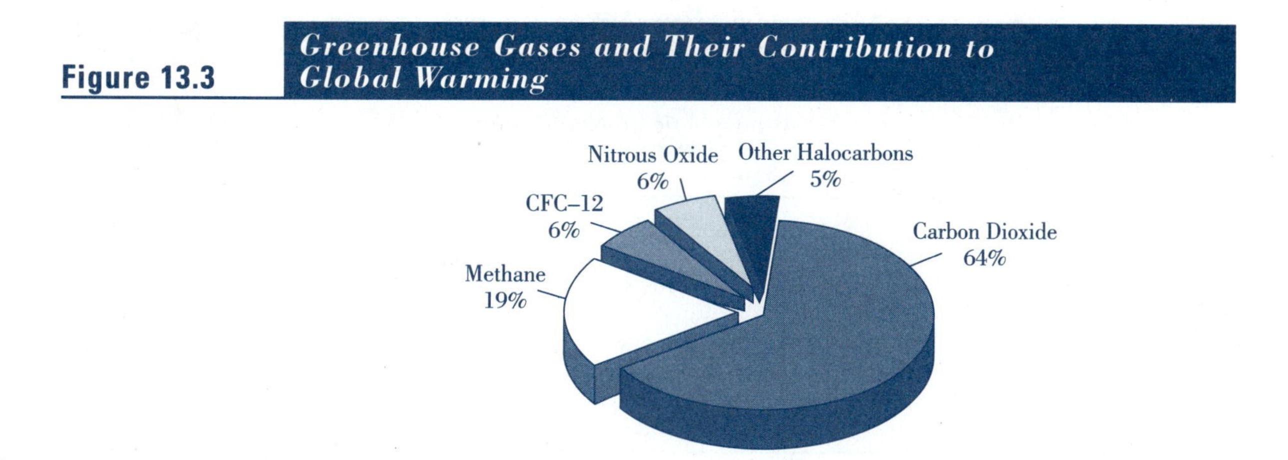

Source: Houghton et al. (1996).

Anthropogenic Sources of Greenhouse Gases

Greenhouse Gas	Major Sources
Carbon dioxide (CO_2)	Burning of solid waste, fossil fuels, and wood; deforestation.
Methane (CH_4)	Production and transport of coal, natural gas, oil; decomposition of organic wastes in landfills; raising of livestock.
Chlorofluorocarbons (CFCs)	Refrigeration, air conditioning, packaging, insulation, and aerosol propellants.
Nitrous oxide (N_2O)	Agricultural and industrial activities; combustion of solid waste and fossil fuels.
Other Halocarbons	Various industrial applications, including fire extinguishers.

Source: U.S. EPA, Office of Air and Radiation, Global Warming Site (March 7, 2002).

Scientific Uncertainty

Although there is some consensus that rising amounts of CO_2 will change the earth's climate, no one knows with certainty the timing or the extent of the outcome, in part because there are many factors to consider. Not the least of these is the influence of other GHGs that play a role in this complex process. Increasing amounts of methane, nitrous oxides, and CFCs are cause for concern, because they are believed to be more damaging than CO_2, even though they are present in lower amounts.[28]

Even more problematic to scientific predictions are the so-called feedback effects that can either lessen or intensify the warming phenomenon. For example, volcanic dust acts to filter the sun's warming rays, which would counter some of the influence of accumulating GHGs. Similarly, scientists argue that sulfur particles in the air can have a cooling effect on the earth. On the other side of the ledger is the role of the oceans and forests, which act as **carbon sinks,** or major absorbers of CO_2. However, higher temperatures likely diminish this capacity, thus intensifying the warming effect. Even cloud cover is a factor, although the qualitative

carbon sinks
Natural absorbers of CO_2, such as forests and oceans.

[28] Kemp (1990), pp. 154–55, and Kerr (May 16, 1997).

Table 13.4 *Growth of Greenhouse Gas Concentrations*

Gas	Atmospheric Concentration Pre-1850	Atmospheric Concentration 1987	Projected Concentration, Mid-21st Century
CO_2	275.00 ppmv	348.00 ppmv	400.00–550.00 ppmv
CH_4	0.70 ppmv	1.70 ppmv	1.80–3.20 ppmv
N_2O	0.29 ppmv	0.34 ppmv	0.35–0.40 ppmv
CFC-11	0	0.22 ppbv[a]	0.20–0.60 ppbv
CFC-12	0	0.39 ppbv[a]	0.50–1.10 ppbv
O_3	0–25% lower than present day	10.00–100.00 ppbv[b]	15–50% higher than present day

Sources: U.S. EPA, Office of Policy, Planning, and Evaluation, Office of Research and Development (December 1989) p. 13, table 2-1, citing Ramanathan (1988) and Lashof and Tirpak (1989).

Notes: ppmv = parts per million by volume; 1 ppmv=0.0001% of the atmosphere; ppbv=parts per billion by volume; 1 ppbv = 0.001 ppmv.

[a]Estimated value given is for 1986.

[b]Estimated value given is for 1985.

impact is uncertain. Clouds contribute directly to the greenhouse effect, but they may have an even stronger cooling influence because of their ability to reflect sunlight back into space.[29] Because so little is known with certainty, scientific research is ongoing in an attempt to settle at least some of the controversy.

Predicting the Potential Effects of Global Warming

Because the science of global warming is itself the subject of debate, it is not surprising that there is a symmetric lack of substantive information on what the eventual outcome of accumulated GHGs might be. Nonetheless, through the use of computer simulation models and laboratory experimentation, researchers have been able to assimilate some information about the presumed implications.

By way of example, Exhibit A.6 in the Appendix presents a summary of expert predictions assembled by the National Academy of Sciences about the estimated climate responses to increased GHGs in the atmosphere.[30] Notice that the climate responses deal with such issues as precipitation and sea level and that predictions span a range of expected outcomes.

Another comprehensive study was conducted as part of the U.S. Global Change Research Program. The findings were published in a 2001 report.[31] Estimates include predictions about forest ranges, biodiversity, the earth's sea levels, agricultural productivity, water and air quality, and health risks. There are also region-specific effects estimated in the study.

Rising sea levels are considered to be one of the more probable outcomes of global warming, arising from thermal expansion of the earth's waters and melting of glaciers. Estimated U.S. changes cited in the report range from an increase of 13 centimeters to as much as 95 centimeters (or 5–37 inches) by 2100. Other changes include substantial losses of U.S. coastal wetlands, regional flooding, and beach erosion.[32]

Agricultural changes are particularly difficult to predict, in large part because of the uncertainty of the agricultural system. Even absent the influence of global warming, crop yields are sensitive to weather, insect damage, and soil conditions, none of which can be predicted

[29]Ramanathan et al. (1989).

[30]A similar set of predictions is presented in IPCC (2001) or in the accompanying *Summary Report for Policymakers* (2001), available online at **http://www.ipcc.ch/pub/spm22-01.pdf**.

[31]National Assessment Synthesis Team (2001), which is also available online at **http://www.usgcrp.gov/usgcrp/Library/nationalassessment/foundation.htm**.

[32]National Assessment Synthesis Team (2001), chap. 16.

with certainty. Nonetheless, scientific predictions attempt to identify both gains and losses to the world's agricultural production.

On the plus side, there is some evidence to support the possibility of a beneficial fertilization effect from increased levels of CO_2, since it is a necessary component of photosynthesis. Furthermore, certain parts of the world would profit from the northward shift of viable agricultural land predicted to occur with a warming trend. For example, most models predict that Canada, Russia, and Northern Europe would likely gain valuable agricultural land, and some existing regions would enjoy extended growing seasons and enhanced productivity. Of course, some areas, such as the southern areas of the United States, may suffer losses should this shift occur. Hence, most global warming models predict that, because some regions would gain while others lose, the net economic effect on agriculture would be relatively minor.[33]

Other effects, such as changes in ecosystems, are also the subject of research. For example, there is concern about the ability of plants and animals to adjust at the same rate as a relatively fast-changing climate. Scientists are concerned that ecosystems may be detrimentally affected and that some species may even die off in the process.[34]

Scientific predictions about the effects of global warming are not at all conclusive. There is disagreement about which of the conjectured events may occur, the degree of impact, and the timing of any associated outcome. Even with sophisticated modeling techniques, forecasts are based on many assumptions, the accuracy of which directly affects the precision of the prediction. So the research continues. In the interim, policymakers must decide how to respond to rising CO_2 levels when there is much uncertainty about the implications.

Policy Response to Global Warming

Setting policy in response to accumulating GHGs is a difficult problem for two reasons. First, as discussed, the concept of global warming is complex, and little is known with certainty. Second, because both the source of the problem and the predicted effects are global in scope, any effective policy solution relies on international agreement. Many nations contribute significantly to the aggregate level of GHG emissions. The top 10 national emitters of the primary GHG, CO_2, are shown in Figure 13.4.

International Response

U.N. Framework Convention on Climate Change (UNFCCC)

At the 1992 Rio Summit, global climate change was an important agenda item for the many national representatives who gathered at the 12-day worldwide conference. Among the major agreements produced at the summit was the U.N. Framework Convention on Climate Change (UNFCCC), which deals with global warming and other air quality issues. Among its major provisions are:

- Countries must implement national strategies to limit GHG emissions with the objective of reducing emissions to their 1990 levels by 2000.
- Differences in political and economic conditions among nations are accommodated by avoiding uniform emission targets and timetables for only one GHG.
- Signatories are encouraged to recognize climate change in the formulation of economic, social, and environmental policies.
- Industrialized nations will assist developing countries in obtaining data and in limiting emissions.
- Countries will elevate public awareness of climate change through education and training.
- Nations will participate in a continuing international research effort.

[33] National Assessment Synthesis Team (2001), chap. 13.
[34] Read more about the expected effects of global warming at **http://www.epa.gov/globalwarming/impacts/index.html.**

Figure 13.4 *Top 10 National Emitters of Carbon Dioxide, 1999*

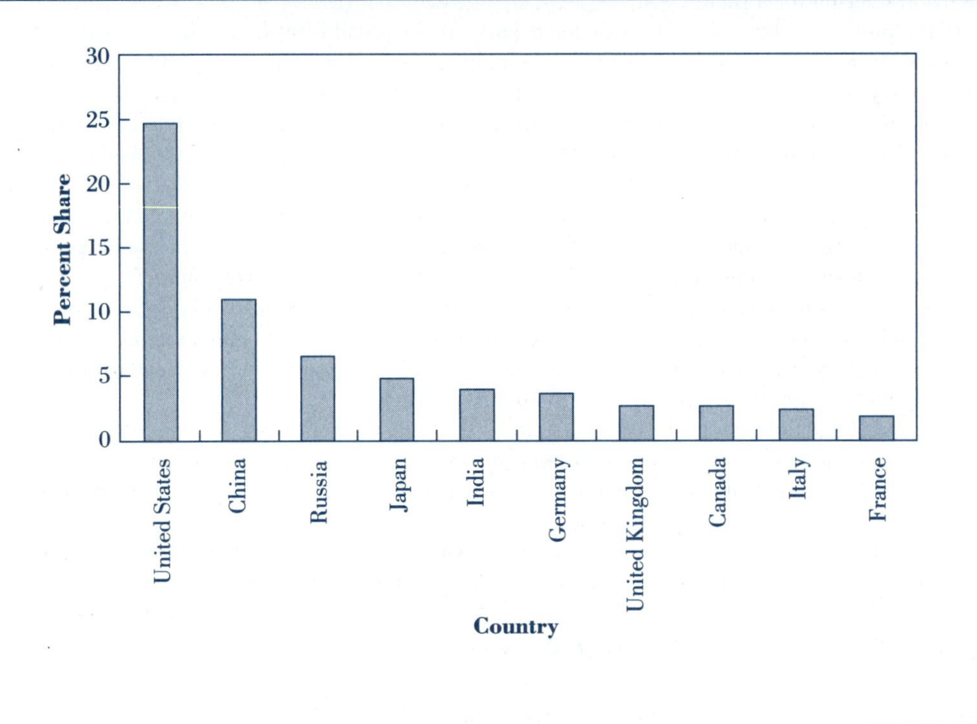

Source: U.S. Department of Energy (1999), Energy Information Administration.

The UNFCCC was to become effective following ratification by 50 nations. In October 1992, the United States became the first industrialized country, and the fourth overall, to do so, following a unanimous vote by the U.S. Senate.[35] By the close of 1993, the requisite number of nations had ratified the treaty, and it became legally binding in March 1994.

National Action Plan (NAP)
A nation's policy to control GHGs and a statement of its target emissions level for the future.

Following ratification, certain of the signatories were required to develop a **National Action Plan (NAP)** for their respective countries. A NAP includes a specific description of a nation's policy prescription to control GHGs and its estimated target level of emissions for the future. NAPs are to be reviewed by the Conference of the Parties (COP), the ruling body of the UNFCCC, and the participants are then bound by specific requirements of the treaty.[36]

Kyoto Protocol

In December 1997, a COP was held in Kyoto, Japan. The goal of this meeting was to reach an agreement or protocol that would address the issue of GHG emissions after 2000. A key outcome of the Kyoto meeting was the establishment of emissions targets for developed nations. According to the preliminary agreement, the European Union was to achieve a limit 8 percent below its 1990 level by 2008–2012, the U.S. limit was 7 percent below, and Japan's was 6 percent below. These limits were to become effective once 55 nations ratified the protocol, including developed countries accounting for at least 55 percent of CO_2 emissions. Achieving these limits was to be accomplished through emissions trading within two blocs: one

[35] Council on Environmental Quality (January 1993), pp. 142–43. In keeping with its commitment to the UNFCCC, the United States completed its third formal climate change report in 2002, which is available online at **http://www.epa.gov/globalwarming/publications/car/**.
[36] World Resources Institute (1994), p. 202.

among the European nations, called the European Bubble, and another, called the Umbrella Group, which includes Australia, Canada, New Zealand, Russia, and the United States.[37]

Further details of the international accord, including its market-based approach, were discussed during a November 1998 meeting in Buenos Aires. After two weeks of negotiations, the participants agreed to a two-year deadline by which rules for cutting GHGs would be reached. Also noteworthy was Argentina's agreement to meet a formal limit on emissions, the first such commitment among developing nations. Beyond these events, most agreed that little had been accomplished.

While the Buenos Aires meeting was in progress, the United States signed the protocol, joining about 60 other countries at the time. Though the signing was intended to indicate U.S. commitment to climate change policy, the action was more symbolic than substantive. The president did not submit the protocol to the Senate for ratification, nor did he plan to in the near future. Why? Because there was strong opposition to the treaty among senators for environmental and economic reasons. In addition to apprehensions about the lack of scientific evidence about global warming, the Senate also was concerned that the stated emissions limits would negatively affect industry and hence the American economy. Furthermore, the United States had adopted the position that it would not ratify the treaty until developing countries such as China and India made a commitment to meet binding emissions targets along with developed nations.

Discussions, often contentious, continued at a series of COPs (in Bonn, Germany, in 1999, then in The Hague, Netherlands, in 2000, and in Marrakech, Morocco, in 2001), generally preceded by preliminary rounds of talks. In addition to deliberations about the role of developing nations and the extent to which emissions trading should be allowed, there was also considerable debate about whether to allow credits for activities that generate carbon sinks, such as forest preservation or certain agricultural practices. Sometimes the meetings ended at an impasse.

Before the Marrakech conference, President Bush announced his opposition to the Kyoto Protocol and took the United States out of the agreement in March 2001, a decision discussed in Application 13.2. At the same time, the Bush administration promised to deliver a domestic initiative to combat global warming, which signaled the nation's intent to address climate change. Nonetheless, the absence of the United States in the international accord might prove to be a critical issue to the remaining Kyoto participants.

At long last, in a round of talks held in Bonn, Germany, in July 2001, 178 nations reached an agreement—without the United States. The accord calls for 38 industrialized countries to cut their GHG emissions to 5.2 percent below 1990 levels by 2012, leaving developing countries without any emissions requirements. Credits for carbon-absorbing forestry practices will be allowed, and emissions trading will be permitted on a global scale with greater flexibility than originally planned. Final language was decided the following November in Marrakech at the seventh COP.

Despite the accomplishments at Bonn and Marrakech, some say that ratification may be at risk without the United States, given the requirement that ratifying nations must represent at least 55 percent of carbon emissions. Moreover, some nations are hesitant to proceed without the United States. In fact, in mid-2002, Australia decided to pull out of the protocol as well. As of June 2002, 44 nations had ratified the Kyoto accord, including the European Union and Japan.[38] Ratification status of the Kyoto Protocol can be viewed online at **http://unfccc.int/resource/kpstats.pdf.**

[37]Fialka (December 11, 1997). For an excellent overview of the Kyoto Protocol, see Fletcher (March 6, 2000), which is available online at **http://cnie.org/NLE/CRSreports/climate/clim-3.cfm.**
[38]Revkin (July 24, 2001; June 6, 2002); Winestock (July 24, 2001); Fialka (November 12, 2001).

Application

13.2 *Why the U.S. Withdrew from the Kyoto Protocol*

In November 2001, the parties to the United Nations Framework Convention on Climate Change (UNFCCC) met in Marrakech, Morocco, to finalize the language and operational elements of the Kyoto Protocol. Through negotiations among the participating nations, provisions were added to allow emissions trading and credits for protection of forests, which act as carbon sinks. In attendance at the Morocco meetings were 171 national governments and 4,500 participants. The protocol becomes legally binding once it is ratified by at least 55 parties to the convention that account for over 55 percent of 1990 GHG emissions from industrialized nations. Interestingly, the United States will not be one of them. Why? Because in March 2001, President Bush declared the Kyoto Protocol to be "fatally flawed in fundamental ways" and proceeded to take the U.S. out of the accord.

The Bush administration's position is based on the initial findings of a cabinet-level climate-change working group established by the president. In its analysis of the Kyoto Protocol, the group expressed concern that the agreement does not impose emissions cuts on developing countries, even though the net emissions from those nations exceed those of industrialized countries. Another reservation cited is that the treaty's emission targets are based on political negotiations rather than on science. There are also economic concerns. The protocol's emissions reduction target for the United States is 7 percent from 1990 levels for each year in the 2008–2012 period. However, this target ignores emissions growth between 1990 and 2012, which, if considered, translates to more than a 30 percent reduction for the period. Achieving such an objective would have serious economic consequences, according to the analysis. In fact, most models predict that U.S. GDP would decline by 1–2 percent as a result.

Beyond the White House report, independent economists have been using models to assess and quantify the protocol's implications for the U.S. economy. For example, Richard Schmalensee, an economist and dean at MIT's Sloan School of Management, forecasts that the United States would have to shut down all its coal-fired facilities by 2012 just to meet the Kyoto targets halfway. Furthermore, analysts at MIT's Joint Program on the Science and Policy of Global Change estimated implementation costs of the Kyoto treaty on a per household basis. Their predictions range from a high of $1,000 to a low of $140, with the latter assuming efficient implementation of emissions trading.

In lieu of participating in the Kyoto Protocol, the United States is planning independent action to research climate change and to implement strategies aimed at reducing GHG emissions. On the research front, a $120 million National Aeronautics and Space Administration (NASA) project is planned to study climate modeling. As for strategic initiatives, plans include developing government-business partnerships to promote source reduction and recycling, promoting energy efficiency in government buildings, developing cleaner technologies for electricity generation and transmission, and supporting the development of fuel-efficient vehicles and renewable resources, such as solar energy and hydropower.

Source: Fialka and Winestock (July 16, 2001); Hilsenrath (August 7, 2001); Revkin (July 24, 2001); U.S. Government, White House (June 2001).

U.S. Response

Clinton Administration[39]

In preparation for the Kyoto meeting in December 1997, President Clinton announced his climate change proposal that October. Its key components included:

[39]The following discussion is drawn from U.S. Government, White House (October 22, 1997); Fialka and Calmes (October 23, 1997); and Rowen and Weyant (December 2, 1997). The Clinton proposal is available in its entirety on the Internet at **http://www.epa.gov/globalwarming/publications/actions/Clinton/index.html**.

- Binding emissions targets to reach 1990 levels by 2008–2012 and further reductions in the next five years;
- A $5 billion program of tax cuts and research and development for new technologies;
- Industry-by-industry consultations on how to reduce emissions and rewards for early actions;
- A requirement that developing countries must participate in addressing climate change; and
- The use of domestic and international emissions trading systems after 10 years of experience with other incentives and federal efforts.

This proposal was submitted to U.S. negotiators who were in Bonn, Germany, working on the preliminary draft of what would become the Kyoto Protocol.

Even at this early stage, the U.S. plan was met with mixed reviews. One United Nations official characterized it as a step, but generally insufficient to meet the objectives of the convention, and the European Union viewed the proposed emissions targets as meager. Of course, the requirement for participation by developing nations was in large part responsible for much of the contentious discussion that followed over the next several years. Also noteworthy is that the proposal for emissions trading did survive the lengthy process, although ultimately without U.S. participation.

Bush Administration

After removing the United States from the Kyoto agreement, President Bush called for a cabinet-level review of the nation's policy on climate change, and a climate change working group was formed. The group in turn requested a report from the National Academy of Sciences (NAS) on what is known about the science of climate change. Initial findings of the working group are presented in a report that includes proposals to advance needed technologies, an analysis of the Kyoto Protocol, proposals to establish international partnerships, and an assessment of existing domestic strategies, including those promoting voluntarism through incentives. An example is the Energy Star program discussed in Application 13.3, which the working group recommends expanding in scope to improve energy efficiency. Both the working group's report and the NAS report were released in June 2001.[40]

President Bush took the reports under advisement and then formulated his climate change plan to act as the U.S. alternative to the Kyoto Protocol. Presented as the *Global Climate Change Policy Book*, the new program was released by the White House in February 2002.[41] According to the plan, the overall objective for the nation is to reduce **GHG intensity** by 18 percent over the next decade, where GHG intensity refers to the ratio of GHG emissions to economic output. This translates to a reduction in emissions from 183 metric tons per million dollars of GDP in 2002 to 151 metric tons per million dollars of GDP in 2012, which the president argues is equivalent to the average outcome for Kyoto Protocol participants. Notice that the use of GHG intensity explicitly considers economic growth in the nation's assessment of progress.

greenhouse gas (GHG) intensity
The ratio of GHG emissions to economic output.

The Bush plan includes both domestic and international initiatives, for example:

- Improve the existing registry program for voluntary GHG emissions reductions for which transferable credits are provided;

[40]The working group's report is called *Climate Change Review—Initial Report,* and the NAS study is titled *Climate Change Science: An Analysis of Some Key Questions,* written by the Committee on the Science of Climate Change and the National Research Council. Internet access to these and other U.S. position papers on climate change is available at **http://www.epa.gov/globalwarming/publications/actions/us_position/index.html.**
[41]U.S. Government, White House (February 2002).

Application

13.3 *The Energy Star Program*

To promote energy efficiency through collaboration between the public and private sectors, the EPA launched the Energy Star Program in 1992. The initiative is a voluntary labeling program that identifies energy-efficient products and allows them to bear the Energy Star® logo. This logo acts as an explicit market signal that communicates to prospective consumers the potential cost savings and the emissions reduction associated with more efficient energy use. In 1996, Energy Star teamed up with the U.S. Department of Energy, with each entity assuming responsibility for certain product categories.

The now-familiar logo first appeared on qualifying computers and monitors. By 1995, it had been made available to other types of office equipment, such as fax machines and copiers, as well as to heating and cooling products for private residences. Today, the label appears on lighting, major appliances, windows, consumer electronics, even new homes and commercial buildings. For new homes to qualify, they must be 30 percent more efficient than the energy code requires, and only office buildings performing in the top 25 percent of the market qualify.

To communicate information about energy-efficient products, Energy Star has set up partnerships with more than 7,000 private and public organizations. These alliances are also used to provide technical assistance. The program touts that these efforts have generated significant cost savings to participants along with important air quality gains. In 2001, the estimated cost savings were $5 billion, and the associated pollution reduction was reportedly equivalent to taking 10 million cars off America's highways.

As new products are added to the Energy Star list, even more savings can be achieved. A case in point is the recent addition of cordless telephones and answering machines to the Energy Star product family. If all such products sold in the next 10 years were Energy Star approved, the nation would save $4 billion in electricity costs and would reduce GHG emissions by 66 billion pounds. All told, Energy Star officials estimate that total energy costs in the United States could be reduced by $200 billion if all consumers, firms, and organizations used its program to make consumption and investment decisions over the next decade.

In addition to the explicit cost savings and the environmental gains of partnering with Energy Star, participating businesses might observe increased demand for their energy-saving products, which translates directly to revenue gains. Increased demand can arise directly from the positive image linked to the logo and from national Energy Star performance awards that receive some measure of media attention. In 2002, 36 Energy Star winners were announced in 5 categories. Major companies such as Maytag Corporation and GE Lighting were among the award recipients for efficient products, and Verizon Communications won for excellence in corporate commitment. Top-performing buildings in the United States also are recognized by the EPA. According to EPA administrator Christine Todd Whitman, the 2001 winners—729 school and office buildings—have collectively saved $134 million in energy costs since 1999 and have released 1.9 billion fewer pounds of CO_2 than average structures.

Programs such as this one operate on the premise that the private sector will cooperate to help improve the environment as long as their market-based objectives are not undermined. The EPA is winning support for its Energy Star Program by illustrating to potential participants that they can contribute to a cleaner environment and at the same time reduce operating costs and enhance profitability.

Sources: U.S. EPA and U.S. Department of Energy (April 24, 2002a; April 24, 2002b; April 24, 2002c); U.S. EPA (March 21, 2002); U.S. EPA (March 26, 2002).

- Increase the federal budget for climate change to $4.5 billion in 2003 and dedicate funding for climate change and energy research;
- Provide $4.6 billion over the next 5 years for energy tax credits to encourage investment in such technologies as renewable energy, hybrid cars, and fuel cell vehicles;
- Promote the development of fuel-efficient vehicles and cleaner fuels through research partnerships with industry;
- Allocate $155 million for the U.S. Agency for International Development, which helps transfer technology to developing countries to help them reduce GHG emission growth; and
- Develop joint research projects with other nations on forestry conservation, energy efficiency, and other climate change issues.

Investigating Market-Based Policy Options

An economic policy response to global warming would use market-based instruments to reduce the accumulation of GHGs, primarily CO_2. Many such initiatives have been proposed in the economic literature and in practice. These are designed to correct the market failure aspect of global warming. Anthropogenic emissions of GHGs are a **negative externality.** The effects of these gases are not captured within the market transaction and therefore are borne by society. To correct the problem, policy instruments must internalize the externality so that the market participants absorb the cost of the damages. Two types of market-based controls that have been proposed in both domestic and international policy forums are a **pollution charge** and a **tradeable permit system.**[42]

pollution charge
A fee that varies with the amount of pollutants released.

gasoline tax
A per unit tax levied on each gallon of gasoline consumed.

Btu tax
A per unit charge based on the energy content of fuel, measured in British thermal units (Btu).

carbon tax
A per unit charge based on the carbon content of fuel.

corrective tax
A tax aimed at rectifying a market failure and improving resource allocation.

tradeable permit system for GHG emissions
Based on the issuance of marketable permits, where each allows the release of some amount of GHGs.

Pollution Charge

The use of some type of **pollution charge** to reduce CO_2 emissions has received a fair amount of attention in both domestic and international policy discussions. However, the specific form of the charge has been the subject of debate. In general, three types of product charges have been proposed as possible candidates: a gasoline tax, a Btu tax, and a carbon tax.

- A **gasoline tax** is a per unit tax levied on each gallon of gasoline consumed.
- A **Btu tax** is a per unit charge based on the energy or heat content of fuel measured in British thermal units (Btu).
- A **carbon tax** is a per unit charge based on the carbon content of fuel.

Unlike taxes on income or consumption that generate market distortions, these charges are referred to as **corrective taxes,** because they are aimed at internalizing a negative externality and hence at correcting a market failure. By design, these taxes should *reduce* market inefficiency. In addition, all three taxes are revenue generating, an attribute that is sometimes used to promote this type of policy, particularly if there are national budget deficits.[43] Beyond these common characteristics, these taxes differ in terms of their applicability, ease of implementation, and overall effectiveness in achieving environmental objectives.

Tradeable Permit System

An alternative market instrument to control global warming is a **tradeable permit system for GHG emissions,** precisely the approach recommended by President Clinton and ultimately incorporated into the Kyoto Protocol. The United States has experience with such in-

[42]For an excellent discussion of a variety of proposals aimed at reducing GHGs, see Wirth and Heinz (May 1991), chap. 2.

[43]Interestingly, however, some believe that environmental taxes should be revenue neutral, so that the motivation and the outcome are exclusively oriented in favor of environmental quality. See, for example, OECD (1989), pp. 13–14.

struments, using domestic trading of sulfur dioxide permits to mitigate acid rain and national and international trading of CFC permits to combat ozone depletion. Conceptually, the design of a GHG emissions market would follow those already in place. However, the implementation of this permit program on an international scale would be far more difficult to accomplish.

Although market-based proposals have appeal, there are also drawbacks that must be considered. Evaluating the pros and cons is facilitated through economic analysis, using the criterion of efficiency and the decision rule of maximizing net benefits.

Economic Analysis of Global Warming Control Policies

Ideally, environmental policy should achieve an efficient allocation of resources, where net benefits are maximized, or equivalently, where the marginal social benefit from implementing policy is exactly offset by the marginal social cost. Although assessing benefits is always the more difficult process, it is particularly difficult for global warming because of the gray areas that weaken scientific predictions. Hence, before we can analyze specific policy proposals, we need to first consider the dilemma of estimating the potential benefits from *any* initiative designed to control climate change.

Estimating the Benefits of Controlling Global Warming: Two Opposing Views

To illustrate how benefit assessment is at the root of the global warming policy dilemma, we consider the findings of two research efforts, each of which arrived at a very different conclusion. The first effort estimates short- and long-term expected benefits, as reported in a recent publication by the Organisation of Economic Co-operation and Development (OECD) in Paris. The second effort is a published commentary on the *relative* benefits of climate change control policy, written by economist Wilfred Beckerman.

Short- and Long-Term Expected Benefits: A Report from the OECD

A report prepared as part of the activity of the OECD Environment Committee includes some new estimates of the expected benefits from controlling global warming. These are presented as estimated damages associated with climate change.[44] Some of the projections for the U.S. economy are given in Table 13.5. Notice that two columns of dollar values are given. The first gives damage estimates from the more conventional prediction of a temperature rise of 2.5° Celsius, and the second presents estimates based on a temperature rise of 10° Celsius over a very long term. According to this report, the benefits of controlling global warming based on conventional climate change predictions would be $61.6 billion (or approximately 1.1 percent of GDP) and as high as $338.6 billion (or at least 6 percent of GDP) over the very long run of 250 to 300 years.

Assessment of Relative Benefits: A Study by Beckerman

In a published article, economist Wilfred Beckerman argues that not only are the damages from global warming difficult to assess, but even if the most dire predictions are correct, most are not sufficient to warrant the high costs of avoidance.[45] Beckerman (1990) cites a 1988 EPA report that estimates the net effect of global warming on U.S. agriculture to be within a range between a net gain of $10 billion and a net loss of $10 billion. Although the media often focus on the potential for financial loss, a net gain is actually possible, because some parts of the nation will enjoy longer growing seasons, enhanced by increased precipitation and the fertilization of increased CO_2 concentrations. At the other extreme, even if the maximum estimated loss of $10 billion is incurred, Beckerman points out that this is only about 0.2 percent

[44] Cline (1992).
[45] Beckerman (1990).

Table 13.5 ***Estimated Effects of Global Warming on the U.S. Economy***

	Annual Damage to U.S. Economy (billions of dollars at 1990 prices)	
Major Damage	**2.5°C**	**10°C, Very Long Term Warming**
Agriculture	$17.5	$95.0
Forest loss	3.3	7.0
Species loss	4.0	16.0
Sea level rise		35.0
Dykes, levees	1.2	
Wetlands losses	4.1	
Drylands losses	1.7	
Electricity requirements	11.7	67.0
Nonelectric heating	−1.3	−4.0
Human morbidity	5.8	33.0
Migration	0.5	2.8
Hurricanes	0.8	6.4
Leisure activities	1.7	4.0
Water supply	7.0	56.0
Urban infrastructure	0.1	0.6
Tropospheric ozone	3.5	19.8
Total	**$61.6**	**$338.6**

Source: Cline (1992), p. 55, table 5. Reprinted with permission.

of U.S. GDP. Furthermore, these estimates ignore the gains from the inevitable progression of agricultural technology advances, genetic engineering, and the like, all of which would likely counter any losses from global warming.

In sum, Beckerman argues that we should be cautious about proceeding with any policy initiative to significantly reduce CO_2 emissions. He argues strongly that the costs associated with a 50 percent reduction in CO_2 emissions called for by most environmentalists would far outweigh the expected benefits. Such a massive reduction would likely increase energy prices by 400–500 percent. To burden society with such a high cost should be undertaken only with a full understanding of what the reduction will ultimately gain. In this case, it is not at all clear that the gain, measured in terms of reduced damages, would offset the expected costs.

The implications of the OECD report and of Beckerman's analysis are quite different, even though both share the same perspective—that benefit assessment is critical to developing policy aimed at global warming. Beckerman argues that the associated benefits do not appear to outweigh the social costs. However, the research reported by the OECD suggests that *long-term* benefits might be more relevant to this particular environmental problem. If this time element is considered, policy development might take a very different direction. Taken together, the two analyses suggest the need for more research. In the interim, studies such as these explain the challenge policymakers face in deciding how to respond to this environmental issue.

Recognizing this challenge, economists strongly promote market-based policies designed to consider the benefits and costs of government controls. Although there are many types of instruments that use market forces to operate, all of them can be motivated by modeling the cause of global warming, rising GHG emissions, as a market failure.

Figure 13.5 **Modeling the Negative Externality of GHG Emissions Associated with Electricity Generation**

In the absence of CO_2 emission controls, competitive equilibrium is determined by market demand (D) and *private* market supply (S_P), yielding a quantity and price of Q_C and P_C. At P_C, fossil fuel utilities supply Q_{F1}, and alternative fuel users supply less at Q_{A1}. The efficient equilibrium is determined by D and the *social* market supply (S_S). The resulting efficient price (P_E) is higher than P_C because demanders now pay the full cost of their consumption activity. At P_E, the efficient quantity is Q_E, representing the correct mix of production using fossil and alternative fuels. Fossil fuel–based production has declined from Q_{F1} to Q_{F2}, reducing CO_2 emissions, and alternative fuel–based production has risen from Q_{A1} to Q_{A2}.

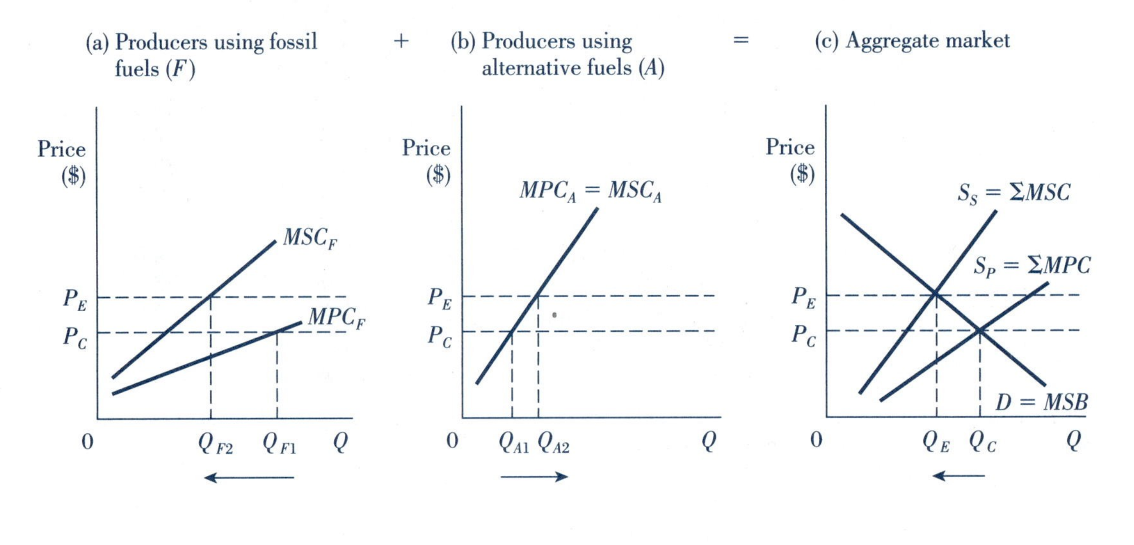

Economic Model of the Market Failure

To simplify our analysis and give it context, we focus on the release of CO_2—the most prevalent GHG, arising mainly from the production of electricity. Figure 13.5 illustrates a hypothetical market for electricity generation, which consists of two groups of electric power plants: one using fossil fuels that generate CO_2 emissions and the other using alternative fuel technologies such as solar, wind, or nuclear power.

Figure 13.5a depicts the marginal private cost of fossil fuel users (MPC_F) and the associated marginal social cost (MSC_F). MSC_F, by definition, is the vertical sum of the MPC_F and the marginal external cost (MEC_F) of electricity production. Thus, MEC_F is implicitly shown as the vertical distance between MSC_F and MPC_F and represents the cost of health and property damages associated with CO_2 emissions. Figure 13.5b shows the marginal social cost of alternative fuel users (MSC_A). For simplicity, it is assumed that this segment of the market does not generate any negative externalities. Thus, MSC_A is exactly equal to MPC_A. Figure 13.5c presents the aggregate market for electricity, shown with a hypothetical demand curve (D) measuring the marginal social benefit of electricity usage (MSB) and two distinct supply curves (S_P and S_S). The S_P is the private market supply of electricity, found as the horizontal sum of MPC_F and MPC_A, and the S_S is the social market supply, found as the horizontal sum of MSC_F and MSC_A.

In the absence of CO_2 emission controls, equilibrium is determined by the intersection of market demand (D) and *private* market supply (S_P). Thus, the competitive equilibrium quantity and price are Q_C and P_C, respectively. At P_C, fossil fuel utilities are willing to supply Q_{F1}, whereas the alternative fuel users supply a much smaller amount, Q_{A1}.[46] Notice that fossil fuel users can supply most of the electricity to the market at price P_C because their MPC_F is relatively low. Without policy controls, these utilities do not consider the external costs of their CO_2 emissions, so private market incentives allocate too many resources to their production processes and too few resources to alternative fuel users. Consequently, P_C is sending a false signal about how to efficiently allocate productive inputs, and the market fails.

To correct the market failure, the external cost of fossil fuel emissions must be brought into the market transaction. As shown in Figure 13.5c, this means that market price must be determined by market demand (D) and the *social* market supply (S_S), which includes both external and private costs. Notice that the efficient equilibrium price (P_E) is higher than what the private market determines because electricity demanders are now paying the full cost of their consumption activity. That is, the per unit price (P_E) paid for electricity is exactly equal to the marginal social cost incurred to produce it.[47] At this higher price level, the quantity of electricity has been reduced to its efficient level (Q_E). This new price level corrects the mix of fossil and alternative fuels used to produce electricity. At P_E, fossil fuel–based production appropriately has declined from Q_{F1} to Q_{F2}, effectively reducing the emissions of CO_2. At the same time, the higher price provides an incentive to alternative fuel users to supply more electricity, shown as the increase from Q_{A1} to Q_{A2}.

Evaluating Market-Based Policies[48]

Analysis of Pollution Charges

As discussed, a pollution charge can be implemented in the form of a **gasoline tax,** a **Btu tax,** or a **carbon tax.** All these instruments attempt to attach a price to the negative externality of anthropogenic GHG emissions, but each uses a different basis for the tax. This in turn suggests that the implications vary as well.

Because gasoline is a carbon-based fuel, its combustion produces CO_2 as a by-product. By levying a tax on gasoline, its effective market price will increase, discouraging consumption and encouraging the use of cleaner alternative fuels. As fewer gallons of gasoline are burned, less CO_2 is emitted into the atmosphere.[49]

In the United States, gasoline taxes already are being collected by state and federal governments, so an increase in the tax rate to bring about a reduction in carbon emissions would be relatively easy to implement. The major drawback from an environmental perspective is that the tax has limited applicability. It targets only polluting sources using gasoline, which are relatively minor emitters of CO_2, and ignores more significant sources that burn other fossil fuels such as oil and coal. Furthermore, the gasoline tax imposes a disproportionate burden on some segments of the economy, such as rural communities that lack good public transportation systems and certain industries, such as interstate trucking firms. Hence, the broader based carbon tax or the Btu tax is often proposed as a superior alternative. Application 13.4 presents an analysis of the Clinton version of the Btu tax originally proposed in 1993.

Although the Btu tax and the carbon tax each use a slightly different tax base, the general purpose of each is the same—to encourage fuel switching and conservation by elevating fuel

[46] Although the market segments modeled in Figure 13.5 are hypothetical, it is interesting to note that as of mid-2001, approximately 71 percent of electricity generated in the United States was produced by burning carbon-based fuels (U.S. Department of Energy, Energy Information Administration, 2001, table 2).

[47] It is important to note that the socially desirable level of electricity production (Q_E) still includes some fossil fuel–based production, meaning that CO_2 emissions are not totally eliminated. However, they are reduced to the level that society believes is acceptable, based on the trade-off between the marginal social benefit of electricity usage and the marginal social cost.

[48] For an interesting overview of market-based policies on global warming, see Parker (March 12, 2002), which is a Congressional Research Service report accessible online at **http://www.cnie.org/nle/crsreports/climate/clim-5.pdf.**

[49] To the extent that the gasoline tax reduces the consumption of gasoline, other harmful emissions such as nitrogen oxides (NO_X) and volatile organic compounds (VOCs) also will be reduced, helping to mitigate other air quality problems, such as urban smog.

Application

13.4 *The Btu Tax: A Market-Based Energy Proposal*

One economic instrument that can be used to reduce GHGs is the Btu tax. A type of **product charge,** the Btu tax is so named because it taxes the energy content of fuels measured in British thermal units (Btu). If levied on all types of fuels, the Btu tax encourages energy conservation across the board by elevating the prices of all fuels. If levied only on fossil fuels, the Btu tax acts much like a carbon tax, raising the relative price of fossil fuels to discourage their use and to mitigate the effects of global warming caused by CO_2 emissions.

In 1993, the Btu tax became national news when President Clinton introduced it as part of his economic package. The Clinton version of this tax was initially proposed as a levy of 25.7 cents per million Btu on natural gas, coal, nuclear power, and crude oil for the first year of implementation, with an *additional* charge on crude oil of 34.2 cents per million Btu to be phased in thereafter. The motivation of the energy tax was obvious: to encourage conservation and to generate revenues to reduce the deficit. On a relative basis, the price differential against crude oil was meant to lessen U.S. dependence on foreign oil supplies.

To see the price effects more clearly, the estimated change in various fuel prices associated with the Clinton Btu tax proposal are provided in the accompanying table.

However well intended, Clinton's energy tax proposal received mixed reviews. Not surprisingly, the sharpest criticism came from utilities and major organizations, such as the American Petroleum Institute, which objected to the price implications and to the government's interference in the marketplace. Beyond this general resistance, there were also objections from certain parts of the country, such as New England, a region heavily dependent on oil for home heating. Congressional representatives from that region opposed the elevated tax on crude oil from an equity perspective, asserting that their constituents would bear a higher than average burden from the tax.

Responding to some of these arguments and other lobbying efforts, the Clinton administration revised the proposal to eliminate the additional levy on oil. Nonetheless, although the general motivation of the Clinton Btu tax was to elevate energy prices to encourage conservation by Americans, it met stiff political opposition and never materialized as a policy instrument.

Sources: Kranish (February 19, 1993); Wartzman (April 2, 1993).

Energy Source	Tax-Induced Price Increase
Gasoline	7.5 cents per gallon
Natural gas	26.3 cents per 1,000 cubic feet
Home heating oil	3.6 cents per gallon
Coal	$5.57 per ton
Nuclear and hydro power	$2.66 per 1,000 kilowatts

Source: U.S. Treasury Department, as cited in Wartzman (April 2, 1993).

prices. Of the two, the carbon tax is more specific, because it targets only carbon-based fuels. In fact, it is considered the more relevant form of taxation to mitigate CO_2 emissions, because the carbon content of fuel and carbon emissions are generally proportional to one another. Effectively, the carbon tax changes *relative* fuel prices, and theoretically could elevate the price of fossil fuel by the marginal cost of the environmental damage caused by its combustion. In the context of Figure 13.5a, the tax amount should equal the vertical distance between the MPC_F and the MSC_F for fossil fuel users, measured at the efficient output level, Q_{F2}. Such a per unit charge would successfully internalize the external costs associated with the burning of fossil fuels.

In practice, the carbon tax is being used in Denmark, Finland, the Netherlands, Norway, and Sweden.[50] In the United States, the EPA has studied this tax as part of its investigation of alternative control options. In a 1991 report, the EPA stated that a tax of $5 per ton of carbon would reduce CO_2 emissions by approximately 1–4 percent by 2000 with estimated revenues of $7–$10 billion per year. By increasing the tax to $25 per ton, emissions would decline by 8–17 percent and revenues would be between $38 billion and $50 billion per year.[51] In addition, a carbon tax would increase the effective price of crude oil, leading to a reduction in oil imports and less dependence on foreign oil sources.

Evaluating a Tradeable Permit System

One of the more critical aspects of the Kyoto Protocol is a GHG tradeable permit system for developed nations. Although the details of this arrangement have not yet been decided, any such agreement specifies a worldwide emissions limit and allocates a predetermined number of permits to each national participant. Each country likely will be responsible for the initial distribution of permits to sources within its own borders and the oversight of any trading among those sources. Countries able to reduce emissions below the amount initially allowed by agreement could sell their excess permits to the highest bidding country.

Just as in the trading of SO_2 allowances and CFC permits, the trading of GHG permits can lead to a cost-effective solution. Nations and individual sources best able to reduce emissions would do so, whereas those that could not achieve the needed reductions would buy the permits. If the system operates efficiently, the price of the permit should be the dollar value of the external cost associated with the emissions. Again, as shown in Figure 13.5a, this price should equal the vertical distance between MSC_F and MPC_F.

The tradeable permit system, while conceptually appealing, does have operational problems. For one thing, arriving at an international agreement is complex, because global warming is expected to have varying impacts on different parts of the world. In fact, some nations stand to gain economically. To persuade these countries to agree to an emissions limit is difficult.

Another problem is setting the initial allowance allocations for each nation. Matters of population, expected economic growth, and existing emission levels are just some of the relevant considerations. For example, if existing emission levels were used as the cap, this would severely limit the growth of developing nations. Still another dilemma is how to monitor emissions of the national participants, particularly if reforestation were allowed as a means to earn emission credits.[52] This type of globally based program unfortunately lends itself to the public goods problem of free-ridership. So, despite the potential for a cost-effective solution, many tough issues still need to be resolved.

Conclusions

Formulating sound policy in response to global air pollution is a major undertaking. Scientific knowledge about atmospheric disturbances and the associated implications is still limited, particularly for global warming. As long as the extent of environmental risk is unknown, the benefits of corrective policy initiatives are likewise indeterminate. Consequently, public officials are unable to justify the social costs of policy controls with reliable information about the comparable benefits.

[50] Parker (March 12, 2002). For more information on various international pollution charges aimed at GHGs, consult Anderson, Lohof, and Carlin (August 1997) or OECD (2001).
[51] These estimates are provided by U.S. EPA, Office of Policy, Planning, and Evaluation (March 1991), p. 3-5.
[52] Satellite monitoring likely would have to be implemented if reforestation were part of any such program. See Wirth and Heinz (May 1991), pp. 18–22; and U.S. EPA, Office of Policy, Planning, and Evaluation (March 1991), pp. 3-7–3-10.

Even when the knowledge base is stronger, such as for ozone depletion, policy development is still complicated by the global nature of the problem. A successful resolution depends critically on international commitment, supported in turn by domestic initiatives. On this front, there have been several achievements. Examples include the Montreal Protocol and its subsequent amendments and the ratification of the U.N. Framework Convention on Climate Change. Nonetheless, there are still important issues to be worked out, not the least of which is how to gain the cooperation of developing nations. These countries lack the financial resources and the technology to innovate around the causes of global air pollution. Yet their cooperation is critical, given the expected rate of industrial and economic growth in these nations and the associated ramifications for the environment. This is why their participation in the Kyoto Protocol was viewed as particularly significant by the United States and other countries.

Despite the difficulties, there has been progress in recognizing the relevant issues, acknowledging the unknowns, and investigating alternative solutions. Furthermore, scientific research is ongoing in the hope of reaching a general consensus about the implications of global air pollution. That policy development has been tentative, particularly in responding to the threat of global warming, is recognition of how important a balancing of benefits and costs is to that process.

Summary

- Ozone depletion refers to damage to the earth's stratospheric ozone layer caused by certain pollutants.
- Scientists agree that the presence of chlorofluorocarbons (CFCs) in the atmosphere is the most likely explanation for ozone depletion. CFCs are a family of chemicals commonly used in refrigeration, air conditioning, packaging, and insulation and as aerosol propellants.
- The Montreal Protocol of 1987, subsequently revised through a series of amendments, is an international agreement aimed specifically at the problem of ozone depletion. The agreement calls for a full phase-out plan for CFCs, halons, and other ozone-depleting substances, most by 1996.
- Title VI of the 1990 Clean Air Act Amendments is dedicated to protecting the earth's ozone layer. Included are two legislated instruments that explicitly use market incentives to eliminate ozone-depleting substances: an excise tax and a marketable allowance system.
- The excise tax on the production of ozone-depleting substances was designed to be escalating over time. The tax rate per pound was defined as a base dollar amount multiplied by the chemical's ozone depletion potential (ODP), where the base amount was higher for each successive year in the phase-out schedule. The tax is still applicable to imported recycled CFCs.
- Under the allowance system, firms could produce or import ozone depleters only if they held allowances. The number of available allowances was gradually reduced and eventually brought to zero. Allowance exchanges were permitted under strict guidelines. A proposal is currently under review to use an allowance system in the United States to facilitate the phaseout of hydrochlorofluorocarbons.
- According to an EPA-commissioned study conducted by the Rand Corporation, a tradeable permit system achieved the desired ozone-depletion reductions in the most cost-effective manner compared to a technology-based command-and-control approach or a fixed emissions charge.
- Prices of ozone-depleting chemicals signaled the effect of the phaseout and the underlying market adjustments. The excise tax helped to counter the accumulation of excess profits to firms holding the limited number of allowances.

- Global warming is caused by sunlight hitting the earth's surface and radiating back into the atmosphere, where it is absorbed by greenhouse gases (GHGs) such as carbon dioxide (CO_2). This process heats the atmosphere and warms the earth's surface. Any significant disruption to the natural levels of GHGs would have climatological impacts.
- Accumulating CO_2 arises from fossil fuel combustion and widespread deforestation. The resulting global warming is predicted to affect forest ranges, biodiversity, sea levels, agricultural productivity, water and air quality, and human health. The timing and magnitude of these effects are uncertain.
- One agreement generated at the Rio Summit was the U.N. Framework Convention on Climate Change, which became legally binding in March 1994.
- In December 1997, a Conference of the Parties (COP) was held in Kyoto, Japan, to discuss GHG emissions. A key outcome of the meeting was the establishment of emissions targets for developed nations. Discussions continued at a series of COPs through the next several years.
- In March 2001, President Bush announced his opposition to the Kyoto Protocol and took the United States out of the accord. Later in 2001, 178 nations reached an agreement without the United States. The accord calls for 38 industrialized countries to cut their GHG emissions to 5.2 percent below 1990 levels. Credits for carbon-absorbing forestry practices will be allowed, and emissions trading will be permitted on a global scale.
- The U.S. response spanned two presidential administrations. In preparation for the Kyoto meeting in December 1997, President Clinton announced his climate change proposal, which included binding emissions targets, a $5 billion program of tax cuts, research and development for new technologies, and the use of domestic and international emissions trading systems.
- President Bush formulated his climate change plan to act as the U.S. alternative to the Kyoto Protocol. Released in February 2002, the plan's overall goal for the nation is to reduce GHG intensity by 18 percent over the next decade. The plan includes energy tax credits, promotion of new technologies, emission reduction credits, and joint research projects with other nations.
- Market-based instruments have been proposed as strategic options to help mitigate the effects of global warming. Among these are a pollution charge and a tradeable permit system.
- Benefit assessment is critical to developing policy to control global warming. Economists promote market-based policies designed to consider both the benefits and the costs of government controls.
- The environmental problem of global warming can be illustrated by modeling accumulating GHG emissions as a negative externality. To correct the market failure, the external cost must be brought into the market transaction.
- A common type of pollution charge is the gasoline tax. Its applicability is limited because it targets relatively minor emitters of CO_2.
- The carbon tax targets only fossil fuels and more directly taxes the cause of the pollution.
- The Btu tax is a charge based on the energy content of a fuel. Because it is broader based, it shifts the tax burden more equally across the economy.
- An alternative market-based proposal to combat global warming is the use of an international market for permits to emit GHGs, an approach adopted under the Kyoto Protocol.

Key Concepts

ozone layer
ozone depletion
chlorofluorocarbons (CFCs)
ozone depletion potential (ODP)
excise tax on ozone depleters
allowance market for ozone-depleting chemicals
greenhouse gases (GHGs)
global warming
global warming potential (GWP)
carbon sinks
National Action Plan (NAP)
greenhouse gas (GHG) intensity
pollution charge
gasoline tax
Btu tax
carbon tax
corrective tax
tradeable permit system for GHG emissions

Use the Key Concepts listed above to begin your search for additional articles and information using the InfoTrac College Edition database.

Review Questions

1. By restricting CFC production prior to the phase-out deadline, the Montreal Protocol and U.S. regulations increased the price of CFCs. Assume that your employer had been a major CFC producer. Present an economic argument either for or against the restrictions during the phase-out period.
2. Other than financial assistance, how might industrialized countries help developing countries to control ozone depletion?
3. Consider the distributional effects of agricultural productivity due to global warming. Discuss some of the ramifications of this outcome with regard to regional economies, national economies, and world trade.
4. a. Why is it that a carbon tax is preferred to either a Btu tax or a gasoline tax when the objective is to reduce carbon dioxide (CO_2) emissions?
   b. Instead of enacting a carbon tax, assume that Congress decides to provide tax incentives to non-carbon-based energy sources, such as solar and wind power. Would this instrument be cost-effective in reducing CO_2 emissions?
   c. Now suppose that the government chooses to initiate tax incentives (e.g., a tax credit) for these energy alternatives along with the carbon tax. Would this be a more socially optimal solution? Explain briefly.
5. During the 1970s, domestic oil prices rose sharply because of supply restrictions initiated by the OPEC nations. As prices rose, Congress instituted tax incentives for homeowners to substitute away from oil as a heating fuel and move toward alternative sources, such as solar energy. During the 1980s, these incentives were eliminated. What was the underlying motivation of implementing the tax incentives in the first place? Why did Congress remove them?

Additional Readings

Barthold, Thomas A. "Issues in the Design of Environmental Excise Taxes." *Journal of Economic Perspectives* 8 (winter 1994), pp. 133–51.

Benedick, Richard Elliot. *Ozone Diplomacy: New Directions in Safeguarding the Planet.* Cambridge, MA: Harvard University Press, 1991.

Carey, John, and Catherine Arnst. "Greenhouse Gases: The Cost of Cutting Back." *Business Week,* December 8, 1997, pp. 64–66.

Choucri, Nazli, ed. *Global Accord: Environmental Challenges and International Responses.* Cambridge, MA: MIT Press, 1993.

Garbaccio, Richard, with Mun S. Ho and Dale Jorgenson. "Controlling Carbon Emissions in China." *Environment and Development* 2, pt. 4 (October 1999) pp. 493–518.

Goldstein, Andrew, and Matthew Cooper. "How Green Is the White House?" *Time,* April 29, 2002, pp. 30–33.

McKibbin, Warwick J., and Peter J. Wilcoxen. "The Role of Economics in Climate Change Policy." *Journal of Economic Perspectives* 16 (spring 2002), pp. 107–29.

Miller, Clark, and Paul N. Edwards (eds.). *Changing the Atmosphere.* Cambridge, MA: MIT Press, 2001.

Morgenstern, Richard D. "Environmental Taxes: Is There a Double Dividend?" *Environment* 38 (April 1996), pp. 16–20, 32–34.

Muller, Frank. "Mitigating Climate Change: The Case for Energy Taxes." *Environment* 38 (March 1996), pp. 13–20, 36–43.

Nordhaus, William D. "Global Warming Economics." *Science,* November 9, 2001, pp. 1283–84.

Roberts, Paul Craig. "Warning: The Greens May Be Hazardous to Our Economy." *Business Week,* September 29, 1997, p. 22.

Schelling, Thomas C. "The Cost of Combating Global Warming: Facing the Tradeoffs." *Foreign Affairs* 76 (November/December 1997), pp. 8–14.

Schmalensee, Richard, Thomas M. Stoker, and Ruth A. Judson. "World Carbon Dioxide Emissions: 1950–2050." *Review of Economics and Statistics* (February 1998), pp. 15–27.

Shapiro, Michael, and Ellen Warhit. "Marketable Permits: The Case of Chlorofluorocarbons." *Natural Resources Journal* 23 (July 1983), pp. 577–91.

Shogren, Jason F., and Michael A. Toman. "Climate Change Policy." In Paul R. Portney and Robert N. Stavins, eds., *Public Policies for Environmental Protection.* Washington, DC: Resources for the Future, 2000.

Stipp, David. "Science Says the Heat Is On." *Fortune,* December 8, 1997, pp. 126–29.

"Symposia on Global Climate Change." *Journal of Economic Perspectives* 7 (fall 1993), pp. 3–86.

Toman, Michael A. "A Framework for Climate Change Policy." *Resources,* no. 127 (spring 1997).

Related Web Sites

Accelerated Phaseout of Class I Ozone-Depleting Substances in the United States	**http://www.epa.gov/ozone/title6/phaseout/accfact.html**
Benefits of the CFC phaseout	**http://www.epa.gov/ozone/geninfo/benefits.html**
Climate Change Impacts on the United States, National Assessment Synthesis Team (2001)	**http://www.usgcrp.gov/usgcrp/Library/nationalassessment/foundation.htm**
Climate Change Review–Initial Report, U.S. Government, White House (June 2001)	**http://www.whitehouse.gov/news/releases/2001/06/climatechange.pdf**
Energy Star Program	**http://www.energystar.gov/default.shtml**
"Global Climate Change: Market-Based Strategies to Reduce Greenhouse Gases," Parker (March 12, 2002)	**http://www.cnie.org/nle/crsreports/climate/clim-5.pdf**
"Global Climate Change Treaty: The Kyoto Protocol," Fletcher (March 6, 2000)	**http://www.cnie.org/nle/crsreports/climate/clim-3.cfm**
Global warming effects	**http://www.epa.gov/globalwarming/impacts/index.html**

Global warming potentials (GWPs)	**http://www.epa.gov/globalwarming/emissions/national/gwp.html**
Global warming science	**http://www.epa.gov/globalwarming/climate/index.html**
Kyoto Protocol ratification status	**http://unfccc.int/resource/kpstats.pdf**
List of substitutes for ozone-depleting substances based on the U.S. Significant New Alternatives Policy (SNAP) program	**http://www.epa.gov/ozone/snap/lists/index.html**
Montreal Protocol ratification status	**http://www.unep.ch/ozone/ratif.shtml**
Multilateral Fund	**http://www.unmfs.org/general.htm**
Ozone depletion health effects	**http://www.epa.gov/sunwise/uvandhealth.html**
Ozone depletion potentials (ODPs)	**http://www.epa.gov/ozone/ods.html**
Ozone depletion science	**http://www.epa.gov/ozone/science/science.html**
"Ozone Science: The Facts Behind the Phaseout," U.S. EPA, Office of Air and Radiation (February 2002)	**http://www.epa.gov/ozone/science/sc_fact.html**
Ozone treaties, United Nations, Ozone Secretariat	**http://www.unep.ch/ozone/treaties.shtml**
The Plain English Guide to the Clean Air Act, "Repairing the Ozone Layer," U.S. EPA, Office of Air Quality Planning and Standards	**http://www.epa.gov/oar/oaqps/peg_caa/pegcaa06.html#topic6**
President Clinton's Proposal on Global Climate Change, U.S. Government, White House (October 22, 1997)	**http://www.epa.gov/globalwarming/publications/actions/clinton/index.html**
Report on the Supply and Demand of CFC-12 in the United States 1999, U.S. EPA, Stratospheric Protection Division, Office of Air and Radiation (June 1999)	**http://www.epa.gov/ozone/geninfo/sdreport99.html.**
Summary Report for Policymakers, Intergovernmental Panel on Climate Change (IPCC) (2001)	**http://www.ipcc.ch/pub/spm22-01.pdf**
Title VI of the Clean Air Act	**http://www.epa.gov/ozone/title6/index.htm/#titlevc**
United Nations Framework Convention on Climate Change	**http://unfccc.int/index.html**
U.S. Climate Action Report: 2002, U.S. Department of State (May 2002)	**http://www.epa.gov/globalwarming/publications/car/**
U.S. EPA, global warming site	**http://www.epa.gov/globalwarming/**
U.S. EPA, ozone depletion site	**http://www.epa.gov/ozone/**
U.S. position papers on climate change	**http://www.epa.gov/globalwarming/publications/actions/us_position/index.html**

A Reference to Acronyms and Terms in Global Air Quality Control Policy

Environmental Economics Acronyms

MPC_F	Marginal private cost of fossil fuel users
MSC_F	Marginal social cost of fossil fuel users
MEC_F	Marginal external cost of fossil fuel users
MSC_A	Marginal social cost of alternative fuel users
MPC_A	Marginal private cost of alternative fuel users
S_P	Private market supply
S_S	Social market supply

Environmental Science Terms

Btu	British thermal unit
CFCs	Chlorofluorocarbons
CFC-11	Chlorofluorocarbon-11
CFC-12	Chlorofluorocarbon-12
CO_2	Carbon dioxide
GHG	Greenhouse gas
GWP	Global warming potential
HCFCs	Hydrochlorofluorocarbons
CH_4	Methane
N_2O	Nitrous oxide
NO_X	Nitrogen oxides
O_3	Ozone
ODP	Ozone depletion potential
ppbv	Parts per billion by volume
ppmv	Parts per million by volume
SO_2	Sulfur dioxide
VOC	Volatile organic compound

Environmental Policy Acronyms

COP	Conference of the Parties
IPCC	Intergovernmental Panel on Climate Change
NAP	National Action Plan
RIA	Regulatory Impact Analysis
SNAP	Significant New Alternatives Policy
UNFCCC	U.N. Framework Convention on Climate Change

PART 5

The Case of Water

Water is so much a part of what we are and where we live that it is generally taken for granted. Satellite photographs of earth are convincing evidence of the predominance of water on our planet, covering over 70 percent of its surface. Over half of the 3-mile-deep outer layer of the earth is water. All biological organisms depend on water for life: to transport nutrients, to regulate temperature, and to support virtually all biochemical systems that sustain life.[1] Ironically, the abundance of water on earth disguises the fact that *usable* water is scarce, and, as a result, efforts to conserve and protect this natural resource have been less than adequate.

Although most of the earth is covered by water, much of it is seawater and hence unusable for drinking and crop irrigation. According to a review of accumulated scientific research, the estimate of fresh water on the planet is less than one-thirtieth of the earth's entire supply, and very little of this amount is contained in lakes and streams. Most of our freshwater supply—some 77 percent, according to estimates—is trapped in ice and snow. Another 22 percent lies beneath the earth's surface, and most of this (about two-thirds) is not accessible, absent prohibitive costs.[2] Finally, of what fresh water is available and accessible, much of it has been damaged by pollution.

Population growth and industrial development have placed competing demands on water resources. Fortunately, nature provides a powerful mechanism that regularly replenishes water supplies. Yet despite this restorative process, our water supply is not unlimited. Some parts of the world face severe water shortages, and virtually everywhere on earth, industrial activity, improper waste disposal, and human carelessness have damaged lakes and streams and contaminated accessible groundwater supplies.

The dependence of all forms of life on this scarce resource demands that society understand the risks of water pollution and take appropriate action to minimize those risks. However, because water pollution is an externality, corrective action must come about through government intervention. In the United States, water quality control policy has taken decades to develop and, by most accounts, still has much to accomplish. Most argue that national policy is in need of significant reform, although there is disagreement about what revisions should be made and how to implement them. What *is* clear is that water quality laws have not been consistently effective. Although there have been successes, there are still water bodies in decline and there are continuing threats to some drinking water supplies. Furthermore, it appears that national policy objectives are not efficient and that many control instruments in use are not cost-effective.

In this module, we undertake a collection of important tasks: to understand how water resources are threatened by contamination, to examine policy initiatives aimed at the problem, and to analyze the outcome using economic criteria. We begin our study in Chapter 14, which provides a broad overview of the Clean Water Act. Chapter 15 focuses on the primary control instruments used to achieve clean water goals. In Chapter 16, we shift our attention to the Safe Drinking Water Act and the standards used to protect human health.

[1] Lyklema and van Hylckama (1988).

[2] United Nations (1978), as cited by White (1988).

Chapter 14

Defining Water Quality: The U.S. Clean Water Act

Water is a classic example of a natural resource characterized by the absence of property rights. Unless government intervenes, water supplies likely will be overused and contaminated. In the United States, controls on water quality have a long history, dating back to the late nineteenth century, and, remarkably, some of these early laws are still on the books. Of course, much has happened in water quality legislation over the last 100 years or so, and the evolution continues.

Effective water policy depends on a careful appraisal of existing water quality conditions, the setting of appropriate objectives, and the design of effective instruments to bring the two together. As we have learned in previous chapters, these tasks depend on good risk assessment techniques and sound risk management practices. For water policy formulation, however, many problems have impeded risk analysis. Assessing the extent of water pollution has been hampered by inadequate monitoring systems. National objectives have not always been properly motivated by benefit and cost considerations, and, even when they have, reliable estimates have been hard to come by. Furthermore, some policy instruments have been guided more by public and political pressures than by sound environmental management practices. As a consequence, resources have been misallocated, and society has had unrealistic expectations about what federal policy can accomplish.

Beyond these procedural difficulties are the natural complexities of water resources, which present a further challenge to policymakers. Water supplies are heterogeneous, with different chemical, biological, and ecological attributes. They also serve different functions, an important factor in policy decisions. Consider how quality controls on water used for drinking have to be much more stringent than those applied to waters used only for navigation. Geographic location, including proximity to urban or industrial centers, affects the usage of water resources and their vulnerability to contamination. These differences suggest that policy objectives and control instruments must be sensitive to the variations caused by nature and by human activity.

Adding to the challenge is a complex, natural cycle that links water resources, land, and the atmosphere together. This interdependence suggests the need for an integrated approach that acknowledges the transmedia implications. Taken together, these issues add up to an ambitious policy agenda, which we begin to investigate in this chapter.

As a preface to our analysis of water quality policy, we present a general overview of water resources and the sources of contamination that threaten them. This discussion helps to motivate our subsequent investigation of how U.S. policy to protect these resources has evolved to the present day. We then focus on national policy goals under the Clean Water Act and how standards are used to define water quality for the nation. A two-part policy analysis follows. The first is an evaluation of the standard-setting process. The second is a benefit-cost analysis of U.S. water quality initiatives that uses the economic analytical tools we have developed previously. At the end of the chapter is a list of acronyms and terms commonly used in discussing the Clean Water Act and related issues.

Understanding Water Resources for Policy Development

Identifying Water Resources and Their Interdependence

Most of us are cognizant of the fact that water is a significant component of the surface of the earth and its underlying geological layers. In fact, policy specifically addresses two major categories of water resources: **surface water** and **groundwater.**

- **Surface water** refers to all bodies of water that are open to the earth's atmosphere, such as rivers, lakes, oceans, and streams, and also springs, wells, or other collectors that are directly influenced by surface water.
- **Groundwater** refers to the fresh water located beneath the earth's surface, generally in what are called aquifers, which are underground geological formations that supply wells and springs.

surface water Bodies of water open to the earth's atmosphere as well as springs, wells, or other collectors directly influenced by surface water.

groundwater Fresh water beneath the earth's surface, generally in aquifers.

Setting policy to protect and maintain these resources is a major undertaking because these water supplies are so vast and because they are remarkably heterogeneous.

Surface waters represent a highly diverse group of water bodies that support distinct ecological systems, serve different uses, and often face dissimilar sources of pollution. Because of these differences, policymakers in the United States categorize surface waters into the following groups: **open ocean waters, ocean coastal waters, estuaries, inland waters,** and **wetlands,** each of which is described in Table 14.1. Although national policy is aimed at protecting all these waters, specific policy instruments and programs often are aimed at one of these categories.

Groundwater is an enormous resource—by volume over 50 times the annual flow of the earth's surface waters. It is of critical importance, since it is available at virtually every point

Table 14.1 *Surface Waters of the Earth*

Surface Water Type	Description
Open ocean waters	The deep waters beyond the continental shelf and the waters over the continental shelf that are not measurably affected by the input of fresh water from rivers.
Ocean coastal waters	The saltwaters along ocean shorelines, which are less enclosed and more influenced by oceanic processes than estuaries and which generally lie over the continental shelf within the territorial sea.
Estuaries	Regions of interaction between rivers and nearshore ocean waters, where tidal action and river flow mix fresh water and saltwater together. Estuaries include bays, mouths of rivers, salt marshes, and lagoons.
Inland waters	Bodies of fresh water, including rivers, streams, lakes, and reservoirs.
Wetlands	Areas saturated by surface water or groundwater with vegetation adapted for life under those soil conditions. These regions are transitional between aquatic and terrestrial ecosystems and include swamps, bogs, fens, marshes, and estuaries.

Sources: U.S. Congress, Office of Technology Assessment (1987), p. 4, box A; U.S. EPA Office of Communications, Education, and Public Affairs (September 1992).

Figure 14.1 *Hydrologic Cycle*

Groundwater and surface water are linked together by the **hydrologic cycle.** Water is transported from the atmosphere to the surface of the earth through **precipitation.** Some precipitation never reaches the earth's surface but instead is **evaporated** into the atmosphere as water vapor or clouds. Some of it settles on trees and plants where it is absorbed and later returned to the atmosphere through the process of **transpiration.** Water landing on buildings and other man-made structures is also evaporated.

The precipitation that *does* reach the earth's surface can take a number of routes from the point called the **watershed** or the **drainage basin.** It can collect in pools, where it is later evaporated. It also might flow over the surface, where it is collected in lakes and streams, a process known as **runoff.** From there, it may evaporate, soak into the ground through the process of **infiltration,** or flow into the oceans, where some portion of it is evaporated. Finally, precipitation may infiltrate the ground directly and flow into soil and rock formations into the water table, where it is collected within an aquifer, a process called **percolation.**

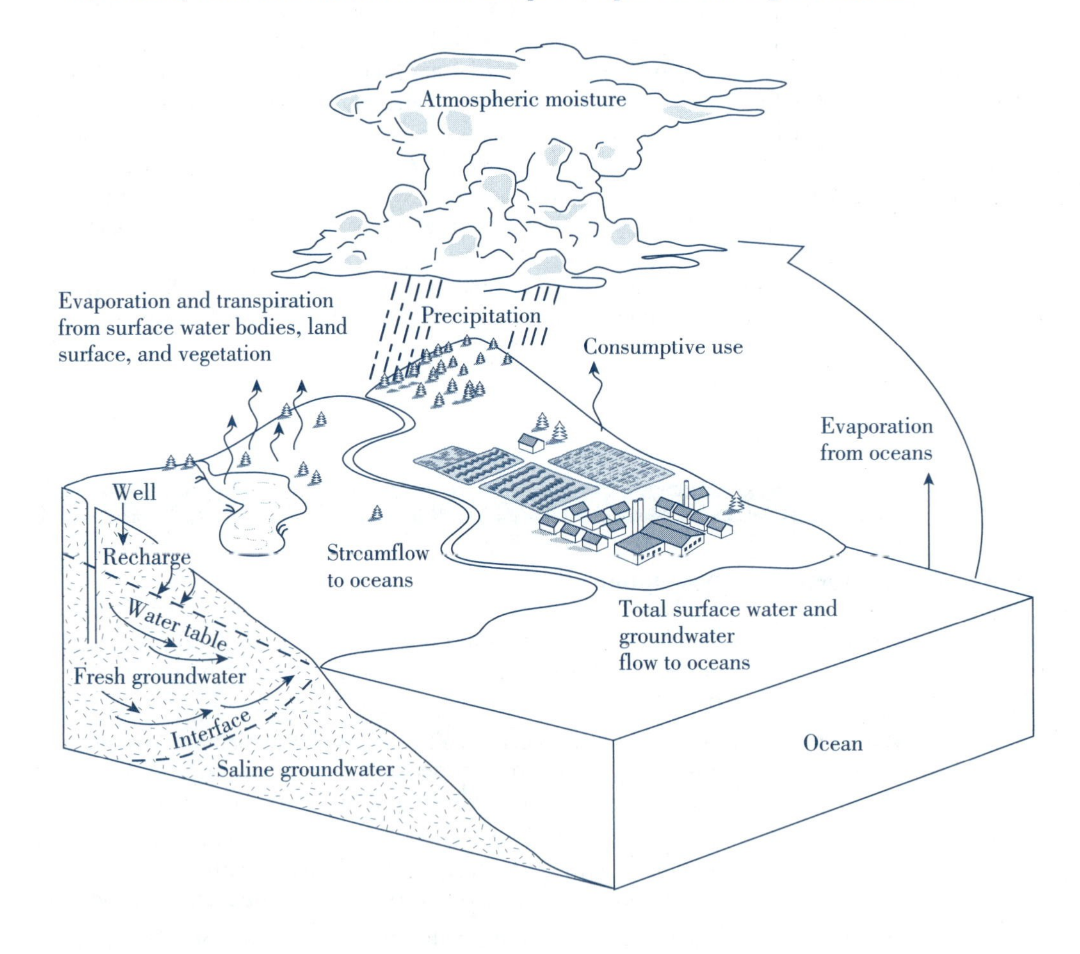

Source: Council on Environmental Quality (1989), p. 21.

on earth and because it is the *only* reliable source for many arid and semiarid regions of the world.[1]

Adding to the complexity of water supplies is the interdependence of all water resources, a phenomenon explained by the **hydrologic cycle.** This cycle models the natural movement of water from the earth's atmosphere to the surface and beneath the ground, and back into the atmosphere through a series of natural processes.[2] This model is depicted in Figure 14.1, which shows how the earth's water supply is continuously in motion to replenish itself.

hydrologic cycle
The natural movement of water from the atmosphere to the surface, beneath the ground, and back into the atmosphere.

Notice that the hydrologic cycle illustrates the natural process that joins together the waters on the earth's surface, those below the ground, and the water vapor in the atmosphere. It also communicates an important message about the ramifications of water contamination. The earth's atmosphere and its water supplies are remarkably interdependent. Pollutants that damage a river or stream can easily be carried to nearby groundwater supplies, and contaminants in the atmosphere can damage surface waters through precipitation.[3] Hence, although surface water and groundwater are in a practical sense distinct water supplies, potential threats to one can have implications for the other because of the hydrologic cycle that ties them together.

Targeting Water Quality Policy

Both surface water and groundwater resources are vulnerable to contamination from a wide variety of sources, some more obvious than others. In fact, most people tend to think only of the discharges from industrial facilities and power plants as the primary sources of water pollution. However, water resources are contaminated by many other sources that are more obscure but nonetheless serious. Conventionally, the source categories of water pollution are characterized as **point sources** and **nonpoint sources.**

- A **point source** is any single identifiable source from which pollutants are released, such as a factory smokestack, a pipe, or a ship.
- A **nonpoint source** is one that cannot be identified accurately and degrades the environment in a diffuse, indirect way over a relatively broad area.

point source
Any single identifiable source of pollution from which pollutants are released, such as a factory smokestack, a pipe, or a ship.

nonpoint source
A source of pollution that cannot be identified accurately and degrades the environment in a diffuse, indirect way over a relatively broad area.

As a group, point sources cover a lot of ground, so policy instruments often distinguish among the following categories:

- **Publicly owned treatment works (POTWs),** which treat wastewaters flowing through sewer systems
- **Industrial facilities,** such as private factories, mills, or other physical plants
- **Combined sewer systems,** which carry both sewage and storm water runoff to waste treatment plants

Nonpoint sources by definition are difficult to identify and hence difficult to control. Examples are land runoff (such as from farms, urban areas, construction sites, and mines), septic systems, landfills, spills, and atmospheric deposition. Only recently have government authorities begun to address these less obvious polluting sources. Hence, control efforts are still in the early stages of development. As an example, Application 14.1 discusses how the Chesapeake Bay has been polluted by both nonpoint and point sources and what is being done to save this important natural resource.

---

[1] For more detail, see Heath (1988).
[2] For a complete discussion of this important water cycle, see U.S. Congress, Office of Technology Assessment (1988).
[3] Recall from Chapter 12 that sulfur dioxide (SO_2) and nitrogen oxide (NO_X) emissions contribute to acid rain, which in turn damages lakes and other surface waters.

Application

14.1 *Deterioration and Restoration of the Chesapeake Bay*

For more than 20 years, the Chesapeake Bay was deteriorating. This popular U.S. estuary, the largest in North America, suffered serious stress from a barrage of polluting sources. Among them were sewage treatment plants, atmospheric deposition, urban and rural runoff, and industrial dischargers. The vulnerability of this majestic water body is explained partly by the large number of major rivers that flow into the bay from Maryland, Virginia, West Virginia, Pennsylvania, and Delaware. Furthermore, its geographic location within the boundaries of Maryland and Virginia is such that it does not benefit from the cleansing forces of the open ocean.

The bay receives an enormous amount of nutrients and suspended solids. Most originate from municipal treatment plants and agricultural runoff, entering from upstream sources. The major effects are declines in submerged vegetation and severe eutrophication, that is, the gradual aging process during which a water body evolves into a marsh and eventually disappears. Because the eutrophication is extreme, hypoxia (a serious depletion of dissolved oxygen) is common in the bay. This condition has been blamed for fish kills that are both ecologically and commercially devastating. Other contaminants, such as heavy metals and organic substances, are linked to industrial and municipal pipelines as well as to urban and rural runoff.

One of the more dramatic consequences of the Chesapeake's contamination is the decline in its productivity. Commercial fishing from this large estuary has always been prolific even when measured on a national scale. In 1985, for example, most of the 815 million pounds of fish caught in Virginia and Maryland came from the bay's waters. For that year, this catch accounted for 13 percent of the national total and was valued at almost $124 million. National attention was drawn immediately to observations that the bay's bountiful marine life was being threatened by pollution. In the last two decades, oyster production has fallen by 99 percent. Severe reductions also have been observed in such species as striped bass, white perch, and the blueback herring.

Official response to the bay's decline began in the 1970s. A seven-year study completed in 1983 reported on the major causes and sources of the deterioration. Later that year, the Chesapeake Bay Agreement was signed by the EPA, Maryland, Virginia, Pennsylvania, the District of Columbia, and the Chesapeake Bay Commission. This historic document authorized the Restoration and Protection Plan, the oldest estuarine program in the United States. The plan is a cooperative agreement among the signatories to clean up the bay and restore its productivity. In 1987, the program was made a part of the provisions of the Water Quality Act, and a new, more specific agreement was drafted and signed. In 1992, the agreement was amended through another in the series of Chesapeake Bay documents, this one aimed at accelerating the restoration efforts to ensure progress toward meeting the program's objectives.

In June 2000, a new bay document, the Chesapeake 2000 Agreement, was signed. With its enactment, officials have formally extended the Chesapeake Bay restoration plan through 2010. Among the 93 commitments comprising the new agreement are to permanently preserve 20 percent of the land in the watershed, to increase public access by 30 percent, and to increase the oyster population tenfold.

These efforts have not gone unrewarded. Since 1985, annual phosphorus loads and nitrogen loads delivered to the Chesapeake Bay from all its tributaries have been reduced by 8.1 million and 52 million pounds, respectively. Furthermore, submerged vegetation has begun to return to the shorelines, oyster harvest declines are starting to reverse, and the population of striped bass is showing signs of increase. For more information, visit **http://www.chesapeakebay.net/info/c2k.cfm.**

http://

Sources: U.S. EPA, Office of Water (December 1995), pp. 330–31; U.S. EPA, Office of Communications, Education, and Public Affairs (April 1992), pp. 32–33; U.S. Congress, Office of Technology Assessment (1987), pp. 21–22, 100–108, 157–64; U.S. EPA (August 1988), p. 66; Chesapeake Bay Program (June 28, 2000); Chesapeake Bay Program (August 22, 2001).

Setting the Policy Agenda

Recognizing the complexity of water resources and their vulnerability to pollution, policymakers have a comprehensive agenda of what must be accomplished through specific initiatives. In the United States, part of the National Water Program includes a three-part goal for water resources as follows:

- All Americans will have drinking water that is clean and safe to drink.
- Effective protection of America's rivers, lakes, wetlands, aquifers, and coastal and ocean waters will sustain fish, plants, and wildlife, as well as recreational, subsistence, and economic activities.
- Watersheds and their aquatic ecosystems will be restored and protected to improve human health, enhance water quality, reduce flooding, and provide habitat for wildlife.[4]

These themes communicate that water quality is important not only to protect human health but also to protect the ecology and to ensure that all uses of water are maintained. Therefore, we should expect water quality to be defined broadly in the law and pollution controls to be instituted from a wide range of perspectives. Indeed, many different laws and special programs are aimed at water pollution, although the most comprehensive is the Clean Water Act. To get a sense of how the federal government became involved in clean water legislation and how today's policy position developed, consider the following evolution of U.S. water quality policy.

Water Quality Legislation in the United States: Overview

Similar to the development of U.S. air policy, government controls on water quality date back to the late 1800s. In the early days, these rulings were aimed solely at protecting human health and were unsophisticated by today's standards. Over time, the legislation evolved into a fairly complex set of laws to protect not only human health but also the natural condition of the earth's water resources and the aquatic life they support. A summary of the evolution is given in Table 14.2, which serves as an outline for the following discussion.[5]

Early U.S. Water Quality Laws

The evolution of U.S. water quality laws began with the Rivers and Harbor Act of 1899. Technically still in force, this legislation prohibits the discharge of refuse into all U.S. navigable waters. Federal legislation on water pollution per se originated nearly half a century later with the Water Pollution Control Act of 1948. Up until this time, water pollution control laws had been enacted by state and local governments. Much like the first national law on air quality control, the role of the federal government in the 1948 act was limited, apparently for the same reason. There was a reluctance to bring federal intervention into what had been perceived as the responsibility of the states.

The legislation was revised in 1956 and again in 1965, but little progress was achieved. Part of the problem was that state governments were being held responsible for such tasks as setting standards, issuing permits, and establishing monitoring and enforcement programs, but they were ill prepared from an implementation standpoint to accomplish what needed to be done. They also lacked an incentive to enforce these regulations. Why? Because state

[4] U.S. EPA, Office of the Chief Financial Officer (September 2000), p. 10, available at **http://www.epa.gov/ocfo/plan/2000strategicplan.pdf**

[5] Much of the following discussion is drawn from Adler, Landman, and Cameron (1993), pp. 5–10; Freeman (1990), pp. 97–108; and Freeman (1978), pp. 45–53.

Table 14.2 ***Evolution of U.S. Water Quality Laws***

Legislative Act	Major Provisions
Rivers and Harbor Act of 1899	Prohibits the discharge of refuse into all U.S. navigable waters without a permit.
Water Pollution Control Act of 1948	Called for the federal government to conduct research, to conduct surveys to study contamination, and to provide loans to municipalities for construction of publicly owned treatment works (POTWs).
Water Pollution Control Act Amendments of 1956	Authorized state governments to set criteria to define water quality; strengthened federal government's role; established a grant program to subsidize construction of POTWs.
Water Quality Act of 1965	Required states to establish ambient water quality standards, devise implementation plans, issue permits to satisfy standards, and set up a monitoring and enforcement program.
Marine Protection, Research, and Sanctuaries Act of 1972	Regulates ocean dumping of materials that may harm health, the marine ecology, or the economic potential of the oceans; also acts as the domestic instrument that implements the London Dumping Convention.
Federal Water Pollution Control Act of 1972	Shifted the primary responsibility for water quality controls from states to the federal government; established the first set of national goals for water quality; authorized the EPA to establish technology-based effluent limitations.
Clean Water Act of 1977	Postponed compliance deadlines for meeting the effluent limitations; established new treatment standards for wastes sent to POTWs; strengthened controls on toxic pollutants.
Water Quality Act of 1987	Established federal subsidies to states for loans to finance POTW construction; required states to set up programs for nonpoint polluting sources.

Sources: Adler, Landman, and Cameron (1993), pp. 5–10; Freeman (1990), pp. 97–108; Freeman (1978), pp. 45–53.

officials recognized that stringent discharge limits would discourage industry from locating in their jurisdictions.[6] In sum, there was little question that further reform was needed.

Evolving Toward Today's Policy Position

Responding to the continuing decline in national water quality and recognizing the failure of delegating virtually all control to state authorities, Congress passed two major pieces of water control legislation in 1972. One was the Marine Protection, Research, and Sanctuaries Act (MPRSA), the primary objective of which was to regulate ocean dumping. Application 14.2 presents a discussion of this and other legislation protecting ocean waters. The other major law enacted in 1972 was the Federal Water Pollution Control Act (FWPCA), which is considered a milestone in the evolution of U.S. water quality legislation. In fact, much of what guides U.S. policy today originated with this act.[7]

It was through the FWPCA that primary responsibility for water quality shifted from states to the federal government. National goals were defined for the first time, including one calling for the elimination of all polluting discharges by 1985. New **technology-based effluent**

[6] Notice how this incentive problem arises — just as it does in states' development of implementation plans to control air pollution.
[7] A brief historical overview of the 1972 law is presented at **http://www.epa.gov/history/topics/fwpca/05.htm.**

Application

14.2 *U.S. Policy to Protect Ocean Waters*

Deliberate ocean dumping and accidental oil spills are two problems that require special mandates. Therefore, Congress passed legislation specifically to protect oceans and coastal waters. Two such laws are the Marine Protection, Research, and Sanctuaries Act (MPRSA) and the Oil Pollution Act (OPA) of 1990.

Marine Protection, Research, and Sanctuaries Act

In 1972, Congress passed the MPRSA, also known as the Ocean Dumping Act. Its major purpose is to regulate ocean dumping of materials that may harm human health, the marine ecology, or the economic potential of the oceans. This act is also the domestic instrument that implements the rulings of the London Dumping Convention, an international agreement on ocean dumping to which the United States is a party.

Under the MPRSA, dumping of certain materials, such as radiological, chemical, and biological warfare agents, high-level radioactive wastes, and medical waste, is explicitly prohibited. The dumping of other wastes, such as dredged materials, is controlled through permitting. The MPRSA also authorizes the Secretary of Commerce to designate certain areas of the marine environment as National Marine Sanctuaries. To learn about these areas, visit **http://www.sanctuaries.nos.noaa.gov/oms/oms.html.**

http://

In 1988, the Ocean Dumping Ban Act was enacted by Congress to amend the MPRSA. Among its mandates was a ruling to end the dumping of sewage sludge and industrial waste by the close of 1991. It also regulates garbage barges and makes it unlawful for medical wastes to be disposed of in any coastal or navigable U.S. waters. More information is available at **http://www.epa.gov/owow/ocpd/marine.html.**

Oil Pollution Act of 1990 (OPA)

Oil spills from tanker accidents or offshore oil drilling platforms have made headlines worldwide. Perhaps the best-known spill in U.S. waters is that of the *Exxon Valdez* on March 24, 1989. The environmental and economic damages of this accident were substantial, estimated in the several billions of dollars. In the aftermath, Congress recognized the need to pass additional legislation to control and respond to oil spills. In 1990, it enacted the OPA, which amends provisions in the Clean Water Act to institute tougher controls on oil spills. One section of this act is dedicated to the Prince William Sound region, where the *Valdez* incident occurred. The rest is aimed at controlling and preventing the threat of oil pollution to all marine environments.

Under the OPA, a National Contingency Plan is established to provide for effective action to remove oil and hazardous substance discharges. To execute this plan, the act instituted a National Planning and Response System, activated in part by response teams formed by the U.S. Coast Guard. Other provisions in the OPA identify those parties responsible for removal costs, damages to natural resources and property, associated loss of profits or earning capacity, and costs of providing public services. Congress also recognized the need to develop preventive initiatives. Among these are provisions to strengthen the standards for obtaining licenses, certificates of registry, and merchant mariners' documents and to clarify the basis under which these various documents may be suspended or revoked. There are also rulings aimed at improving the construction of oceangoing vessels.

Sources: Marine Protection, Research, and Sanctuaries Act; Oil Pollution Act of 1990; U.S. EPA, Office of Water (February 1989); Council on Environmental Quality (January 1993), pp. 34–35.

limitations were to be set within one year by the EPA as the primary instruments of U.S. water quality control.[8] The federal cost share of constructing publicly owned treatment works (POTWs) was increased from 55 to 75 percent, and a National Commission on Water Quality was formed to monitor policy effectiveness and to study any economic and technological im-

[8]In the next chapter, we will investigate how these effluent limits are defined and conduct an economic analysis of their effectiveness.

plications. It soon became apparent, however, that the new law was overly ambitious, falling far short of its objectives.

Making much needed midcourse corrections, the Clean Water Act (CWA) of 1977 extended the compliance deadlines for meeting the effluent limitations. It also strengthened the law on toxic water pollutants. Still more extensions and revisions followed when the act later came up for reauthorization—these in the form of the Water Quality Act of 1987.[9] The 1987 act authorized more federal monies to support POTW construction. It also required states to establish programs aimed at nonpoint polluting sources and authorized federal funding of $400 million to support that effort.

The most recent reauthorization of the Clean Water Act began under the 102d Congress with many hearings held by the Senate and the House in 1991. The review process ignited intense discussion among Senate members, the EPA, and administration officials. In the interim, no analogous bill was introduced in the House of Representatives. An accumulation of disagreements so delayed the procedures that it soon became clear that the 102d Congress would not complete the process. Similar debate continued through the next three sessions of Congress, leaving the reauthorization unresolved to date.[10] In the interim, the EPA continues to work toward achieving the nation's three-part goal to achieve clean and safe water, as outlined earlier.

Policy Objectives Under the Clean Water Act (CWA)

In the United States, as in many industrialized nations, many laws are aimed at protecting water quality, some directed solely at ocean waters, some at drinking water, and some at specific water bodies.[11] Of these, none is more comprehensive than the Clean Water Act.[12] Although its provisions are not consistently stringent for all water resources and for all polluting sources, they do cover both groundwater and surface water, and they address both point and nonpoint sources of contamination.

To understand the far-reaching intent of this major act, look at Table 14.3, which lists the national goals for water quality enacted by Congress. Of these, the three considered most important are the first three, which are referred to as the **zero discharge goal,** the **fishable-swimmable goal,** and the **no toxics in toxic amounts goal,** respectively. These are implemented primarily through technology-based effluent limits as well as the three approaches outlined in the fourth, fifth, and sixth goals stated in the act. These approaches address funding for POTW construction, area-wide waste treatment management, and research and development, respectively. The seventh goal, which calls for the development of nonpoint source pollution programs, was added with the Water Quality Act of 1987.

zero discharge goal
A U.S. objective calling for the elimination of all polluting effluents into navigable waters.

fishable-swimmable goal
An interim U.S. objective requiring that surface waters be capable of supporting recreational activities and the propagation of fish and wildlife.

Zero Discharge Goal

The **zero discharge goal** called for the elimination of all polluting effluents into navigable waters by 1985. The objective was ambitious. It was motivated by the ineffectiveness of states' control efforts prior to 1972 and by evidence that had been accumulating about the deterioration of U.S. water resources. For example, in 1969 the buildup of oil and industrial wastes in Ohio's Cuyahoga River caused it to literally catch fire. Two years later, a report of Ralph Nader's investigation of U.S. water quality, titled *Water Wasteland,* cited incident after incident

[9] Although this act officially changed the title of U.S. legislation, we will follow convention and refer to this legislation as the Clean Water Act.
[10] For more on the reauthorization process, see Knopman and Smith (January/February 1993) and Adler (November 1993).
[11] In Chapter 16, we will analyze U.S. policy to protect drinking water with a particular emphasis on the Safe Drinking Water Act.
[12] The complete text of the Clean Water Act can be accessed online at **http://www.epa.gov/region5/water/cwa.htm.**

Table 14.3 ***Goals and Policy of the U.S. Clean Water Act***

The objective of this Act is to restore and maintain the chemical, physical, and biological integrity of the Nation's waters. In order to achieve this objective it is hereby declared that, consistent with the provisions of this Act—

(1) it is the national goal that the discharge of pollutants into the navigable waters be eliminated by 1985;

(2) it is the national goal that wherever attainable, an interim goal of water quality which provides for the protection and propagation of fish, shellfish, and wildlife and provides for recreation in and on the water be achieved by July 1, 1983;

(3) it is the national policy that the discharge of toxic pollutants in toxic amounts be prohibited;

(4) it is the national policy that Federal financial assistance be provided to construct publicly owned waste treatment works;

(5) it is the national policy that areawide waste treatment management planning processes be developed and implemented to assure adequate control of sources of pollutants in each state;

(6) it is the national policy that a major research and demonstration effort be made to develop technology necessary to eliminate the discharge of pollutants into the navigable waters, waters of the contiguous zone, and the oceans; and

(7) it is the national policy that programs for the control of nonpoint sources of pollution be developed and implemented in an expeditious manner so as to enable the goals of this Act to be met through the control of both point and nonpoint sources of pollution.

Source: Federal Water Pollution Control Act, as amended by the Clean Water Act of 1977, sec. 101(a).

Note: To access the full text of the Clean Water Act online, visit **http://www.epa.gov/region5/water/cwa.htm.**

of damage to U.S. waters caused by pollution.[13] Believing that there was a need for a strong government response with a fixed deadline, Congress wrote the zero discharge goal into law.

The Fishable-Swimmable Goal

The **fishable-swimmable goal** was written as an interim objective to be met until the zero discharge goal could be achieved. According to this objective, by 1983 surface waters were to be capable of supporting recreational activities and the propagation of fish and wildlife. In so doing, this goal established a baseline level of water quality across all states and set a fixed deadline of July 1, 1983, by which the baseline would be achieved.

No Toxics in Toxic Amounts

The **no toxics in toxic amounts goal** is a national policy prohibiting the release of toxic substances in toxic amounts into all water resources. Singling out toxic pollutants responded to the observation of an increasing number of water bodies contaminated by dangerous chemicals. Throughout the 1960s and 1970s, chemical use had been on the rise. The discovery of new synthetic chemicals during that period added uncertainty to the potential threat.

no toxics in toxic amounts goal
A U.S. goal prohibiting the release of toxic substances in toxic amounts into all water resources.

These three goals were intended to be the guiding principles for achieving and maintaining water quality throughout the United States. In truth, none of them were met by the stated deadlines, nor have they since been achieved, although they are still the ultimate targets of policy initiatives.

[13] Zwick and Benstock (1971), as cited in Adler, Landman, and Cameron (1993), pp. 5–6.

Application

14.3 *Ecological and Economic Impacts of Oil Tanker Spills*

All too often, the media report on another oil tanker that has run aground, spilling its hazardous cargo into surrounding ocean waters. As recently as January 1996, a barge off the coast of Rhode Island made national headlines—this one releasing over 700,000 gallons of heating oil into the ocean and damaging some of the region's renowned beaches. The largest spill in history involved 97 million gallons when two tankers collided in 1979 off the coast of Trinidad and Tobago. In the United States, the most damaging spill occurred in March 1989 when the *Exxon Valdez* ran aground in Alaskan waters, and 11 million gallons of oil poured into Prince William Sound.

In January 1993, the second of two major oil spills within five weeks of one another made world headlines. Both occurred off the coast of Europe. The first involved a Greek oil tanker, the *Aegean Sea,* that ran aground in stormy seas off the coast of Spain in December 1992. The tanker split in two and dumped 23 million gallons of oil into the northwestern port of La Coruna. Early damage reports placed a value of $50 million on the destruction. The second incident involved the Liberian-registered tanker *Braer,* carrying 26 million gallons of light crude oil. The 18-year-old vessel lost power just south of the Shetland Islands in the North Sea and ran aground along the rocky coastline. Preliminary damage reports estimated the cost in the hundreds of millions of dollars.

In the aftermath of these accidents, society has had to contend with the environmental damage and the economic losses. Birds, fish, sea otters, and other marine life are killed or found coated with oil, struggling to survive. Coastlines are contaminated as the crude oil makes its way to shore. Some of the losses have market implications, such as regional economic losses in tourism and the destruction of commercially valued fish and shellfish. Even more difficult to assess are damages based on the existence value of the ecology—the ocean itself, the marine life it supports, and natural coastlines.

No one can accurately determine the numbers of birds, fish, and sea mammals destroyed as a result of these accidents. Even if this were possible, there is the daunting problem of assigning a dollar value to ecological quality and biological life. Nonetheless, analysts attempt to assess and monetize these damages along with other market effects. In the *Exxon Valdez* case, between 3,500 and 5,500 sea otters were destroyed. An estimated 30,000 sea birds also were killed, among them the nation's prized bald eagle. Including these losses, dollar damages attributable to the *Valdez* incident have been estimated in the billions of dollars.

Sources: Bai and Allen (January 21, 1996); Miller (January 6, 1993); "Updating Previous Big Spills," (January 6, 1993); Schmidt (January 6, 1993); Raven, Berg, and Johnson (1993), pp. 15, 197.

Identifying Pollutants Under the Clean Water Act

Water can be polluted at virtually any point in the hydrologic cycle and by many different contaminants. The contamination can arise either deliberately through illegal waste disposal or accidentally, such as when oil tankers spill their cargo into the oceans, a problem discussed in Application 14.3. Many pollutants are responsible for degrading the earth's water supplies, ranging from excess plant nutrients to synthetic toxics. A list of the major categories of water pollutants is given in Exhibit A.7 in the Appendix.

Three categories of pollutants were introduced in the Clean Water Act of 1977 and are relevant to national law. These are **toxic pollutants, conventional pollutants,** and **nonconventional pollutants.**

toxic pollutant
A contaminant which, upon exposure, will cause death, disease, abnormalities, or physiological malfunctions.

- A **toxic pollutant** is a contaminant that, upon exposure, will cause death, disease, behavioral abnormalities, genetic mutations, or physiological malfunctions in biological organisms or their offspring.

- A **conventional pollutant** is an identified pollutant that is well understood by scientists and may be in the form of organic waste, sediment, acid, bacteria, viruses, nutrients, oil and grease, or heat. Conventional pollutants include suspended solids, biological oxygen-demanding (BOD) substances,[14] fecal coliform, and pH.
- A **nonconventional pollutant** is the default category for those pollutants not identified as toxic or conventional.

conventional pollutant
An identified pollutant that is well understood by scientists.

nonconventional pollutant
A default category for pollutants not identified as toxic or conventional.

These categories play a role in how the technology-based effluent limits are implemented, an issue discussed in the next chapter. As to their motivation, these classifications came about to allow for tougher controls on toxic contaminants. Classifying a pollutant as toxic is based on its degradability (i.e., its ability to break down into a less complex form), its persistence (i.e., how long it remains in the environment), the presence of affected organisms and their importance, and the nature and degree of effect of the pollutant on such organisms. Among the 65 toxic compounds and families of compounds listed in the Clean Water Act are such substances as benzene, chloroform, lead, mercury, and arsenic. This list translates into 126 individual toxic substances, referred to as **priority pollutants.**[15]

Defining Water Quality: Standard-Setting Under the Clean Water Act

As originally required under the Water Quality Act of 1965, surface water quality is defined by **receiving water quality standards.** These state-established standards have two distinct components:

- **Use designation** for the water body
- **Water quality criteria** to sustain the designated uses

receiving water quality standards
State-established standards defined by use designation and water quality criteria.

The **use designation** identifies the intended purposes of a water body, such as for irrigation or for shellfishing. The **water quality criteria** give the biological and chemical water attributes necessary to sustain or achieve these designated uses, including the **maximum concentration of pollutants** allowed.

use designation
A component of receiving water quality standards that identifies the intended purposes of a water body.

Use Designation[16]

States are authorized to decide the designated beneficial uses for all intrastate water bodies, subject to EPA approval. Among the uses they should consider are public water supplies, propagation of fish and wildlife, recreational activities, and agricultural purposes. More detail is available in Exhibit A.8 in the Appendix. At minimum, designated uses must be sufficient to support swimming and some fishing in order to be consistent with the national **fishable-swimmable goal.** Furthermore, adoption of waste transport or waste assimilation as a designated use is specifically prohibited. This latter ruling refers directly to the assertion that pollution dilution is not a viable substitute for waste treatment.[17]

water quality criteria
A component of receiving water quality standards that gives the biological and chemical attributes necessary to sustain or achieve designated uses.

Use-Support Status

Periodically, state authorities must determine the **use-support status** of a water body by assessing its present condition and comparing it with what is needed to maintain its designated uses. One of five classifications is assigned to characterize use-support status of surface

use-support status
A classification of a water body based on a state's assessment of its present condition relative to what is needed to maintain its designated uses.

[14]This refers to the amount of oxygen needed by microorganisms to decompose organic compounds.
[15]See CWA, sec. 307(a)(1) on toxic pollutants and sec. 304(a)(4) on conventional pollutants; and U.S. EPA, Office of Water Regulations and Standards (September 1988), p. 9.
[16]The following is drawn from 40 CFR 131.10, *Designation of Uses,* and U.S. EPA, Office of Water (March 1994).
[17]U.S. Congress, Congressional Research Service (1972), as cited in Adler (November 1993), p. 159; and Van Putten and Jackson (summer 1986).

Table 14.4 ***Classifications of Use Support for U.S. Surface Water Quality***

Use-Support Level	Water Quality Condition	Definition
Fully supporting	Good	Meets applicable standards, both criteria and designated use.
Threatened	Good	Currently meeting standards, but there are concerns about degradation in the near future.
Partially supporting	Fair	Meeting standards most of the time, but exhibits occasional exceedances.
Not supporting	Poor	Not meeting standards; does not support a healthy aquatic environment and/or prevents some human activities.
Not attainable		Support of one or more designated uses is not attainable because of specific biological, chemical, physical, or economic/social conditions.

Source: U.S. EPA, Office of Water (June 2000), pp. 52–53, 82–83.

waters. As described in Table 14.4, these classifications are **fully supporting, threatened, partially supporting, not supporting,** and **not attainable.** States' findings on use-support status are submitted to the EPA, analyzed, and reported to Congress as part of a biennial National Water Quality Inventory required by the Clean Water Act.[18]

Water Quality Criteria[19]

To ensure that designated uses are achieved and maintained, states must either establish **water quality criteria** subject to EPA approval or adopt criteria set by the EPA itself. The EPA is responsible for developing and publishing criteria that reflect the most current scientific knowledge on all identifiable effects of pollution on health, aquatic life, and welfare. These criteria are pollutant specific and may be expressed as concentrations of pollutants allowed in water (known as **numeric criteria**), in qualitative statements (called **narrative criteria**), or as comments about the overall condition of an aquatic system (called **biocriteria**). All criteria must have a sound scientific basis, and if there are multiple uses designated for a water body, the criteria must be set to support the most sensitive use.

Analysis of Receiving Water Quality Standards

The **receiving water quality standards** are a critical element of U.S. water quality policy because they are linked directly to the nation's objectives, particularly the fishable-swimmable goal. However, the process of establishing these standards has been problematic, as has their reliance on the effluent limitations for implementation.

[18] See, for example, U.S. EPA, Office of Water (June 2000), or access the 1998 National Water Quality Inventory online at **http://www.epa.gov/305b/98report/index.html.**

[19] The following discussion is drawn from 40 CFR 131.11, *Criteria;* CWA, sec. 303; and U.S. EPA, Office of Water Regulations and Standards (September 1988), pp. 8–11.

Absence of Benefit-Cost Analysis in Setting the Standards

When states were called upon to establish receiving water quality standards, the law allowed them to specify different standards for each interstate water body within their jurisdictions. It was implied that they could use an evaluation of benefits and costs in setting these standards, but they were not required by law to do so.[20] Of course, the use designation had to be consistent with national goals, indicating at minimum a water quality level to support fishing and swimming, a goal that was solely benefit based with no consideration for economic costs.

Even in states' reports to the EPA on use-support status, benefit-cost assessment is not being done, even though the law calls for such an analysis. According to the Clean Water Act, states are called on to submit a biennial water quality report in which they must identify waters that have met the fishable-swimmable use designation. This report is supposed to include estimates of the associated costs and benefits of that achievement. The law stipulates that:

> Each State shall prepare and submit . . . a report which shall include . . . an estimate of (i) the environmental impact, (ii) the economic and social costs necessary to achieve the objective of this Act in such state, (iii) the economic and social benefits of such achievement, and (iv) an estimate of the date of such achievement.

Most information on the benefits and costs of water quality improvements is limited to the results of research studies focusing on a single location or a specific water body. A case in point is an academic study of the benefits of water quality improvements to marine sportfishing along the East Coast. For a closer look at this study, which is being supported by the EPA, see Application 14.4.

Efforts such as these may advance researchers' understanding about how best to measure water quality benefits on a comprehensive scale. In the interim, state officials are a long way from assessing the benefit categories recommended by the EPA, such as aesthetic benefits, commercial benefits, and community benefits. For details, refer to Exhibit A.9 in the Appendix. Even if data collection and assessment methods improve, there is currently no explicit mandate to ensure that the benefit-cost decision rule is used to define water quality standards. Hence, there is no assurance that water pollution abatement will be set at an efficient level.

Lack of Consistency with the Technology-Based Effluent Limitations

Another problem with the standard-setting procedure relates to the use of federally mandated effluent limitations to achieve the states' receiving water quality standards. The source of contention is that the link between the water quality standards and the effluent limitations is blurred at best. Why? Because each is motivated differently. Specifically,

- the standards are motivated by water *usage*
- the effluent limits are motivated by *technology*

Because the effluent limits are technology based, they are motivated by what is practical or feasible rather than by benefit-cost analysis or environmental criteria. The intent was to avoid the problem that states had encountered in the past of trying to estimate the relationship between a predetermined water quality level and the pollution reduction needed to achieve it. However, it is precisely this relationship that is needed to map the effluent limits to the water quality standards.

Because the technological limits are applied uniformly within defined groups of point sources, they do not account for varying conditions across water bodies or for the different uses of water bodies designated by states. So, even if all dischargers met the effluent limits,

[20] Freeman (1990), p. 102.

Application

14.4 *How Much Is Cleaner Water Worth to Marine Sportfishing?*

Most water quality research focuses on a particular water body or region as the context of its analysis. Such is the case for a University of Maryland study supported by the EPA that is analyzing the coastal region that extends from New York to south Florida (excluding the Florida Keys). The selection is motivated by the economic activity being studied: marine sportfishing. Eighty percent of all East Coast marine sportfishing takes place in the area targeted by the study. Furthermore, the region is one where active pollution control initiatives and management plans for recreational fisheries are in place.

The objective of this university research is to develop a database and a procedure that can be used to estimate the economic value of two related factors: access to marine sportfishing and changes in the catch rate of various species, where the catch rate is the average number of fish caught per fishing trip at a given site. The link between these two factors and economic benefits is a logical one. Water quality policy reform can improve the catch rate, which in turn will affect fishermen's decisions about where to fish, what species they fish for, whether they fish from the shore or from a boat (called the fishing mode), and even how often they go fishing. By measuring these changes in fishermen's behavior, researchers can make the link to a monetized benefit measure of improved fishing conditions that can be achieved through tougher pollution controls.

To determine catch rates, the analysis uses survey data collected by the National Marine Fisheries Service. Three broad categories of catch rates are defined by type of fish: big game fish (e.g., marlin and tuna), small game fish (e.g., bluefish and mackerel), and bottom fish (e.g., snapper and grouper). Thus far, three different benefit estimates have been calculated:

- A 20 percent increase in the catch rate of small game fish for both fishing modes at all sites would increase the average benefit of each fishing trip by $0.33.
- A 20 percent increase in the catch rate of bottom fish by boat would increase the average benefit per trip by $1.27.
- A 20 percent increase in the catch rate of large game fish would yield an increase in benefits of $1.56 per trip.

By publishing a detailed report of their findings, the university hopes their work will lay the groundwork for other regional analyses. In addition, the model developed by the research will be capable of assessing interregional impacts, such as how changes in fishing conditions in Florida can affect fishing activity in the Chesapeake Bay region.

Source: U.S. EPA, Office of Water (April 1992b), p. 164.

there is no guarantee that water quality as the states have defined it would be achieved. In fact, it is exactly because of this possibility that the law requires states to identify those waters for which the effluent limitations are insufficient. Officially, these waters are to be labeled "water quality limited" and placed in a priority ranking. More stringent controls must then be established.[21] These additional rulings are necessary because the policy instrument (the effluent limitation) is not properly linked to the objective (the water quality standard).

Benefit-Cost Analysis of U.S. Water Quality Control Policy

Water quality is a goal about which there is little debate, at least from a qualitative perspective. The issue is not a zero-one option of whether or not this objective is worth pursuing but rather a determination of the extent to which water contamination should be controlled. As is

[21] See CWA, sec. 303(d)(1)(A), and U.S. GAO (January 1989), pp. 2–3.

always the case in environmental policy, the question is, How clean is clean? Once this fundamental question has been addressed, a secondary issue is to evaluate how this goal, however defined, is to be achieved. Both issues are complex, but they can be addressed with considerable objectivity using benefit-cost analysis and the efficiency criterion. Relying on the careful work of well-respected environmental economists, we can examine the evidence and draw some qualitative, albeit guarded, conclusions.[22]

Benefit-Cost Analysis of the FWPCA of 1972

The enactment of the Federal Water Pollution Control Act (FWPCA) of 1972 marked a major shift in U.S. policy toward stronger controls administered by the federal government. Because of its importance in the development of U.S. policy, the FWPCA has been a common subject of analysis, not always with favorable results.

Part of the criticism is that overall water quality did not improve significantly as a result of the 1972 reforms. However, a review of water quality trend data is insufficient evidence to evaluate the effectiveness of federal policy. First, it is not valid to assume that an observed favorable trend is entirely the result of regulatory controls. Water quality is influenced by other factors, such as use intensity, population growth, and industrial development. Second, even if the trend data *were* indicative of improved water quality and could be attributed to policy, these improvements or benefits must be considered along with the costs society incurred to achieve them.

A number of studies have estimated and monetized the benefits of improved water quality, but these analyses typically focus on a specific water body or a particular region. Furthermore, some examine only a single aspect of water quality, such as recreational use or the value to the commercial fishing industry.[23] Only one analysis has been done that attempts to synthesize the findings of existing research to arrive at a comprehensive benefit measure of improved water quality attributable to the FWPCA of 1972. This landmark study was conducted by A. Myrick Freeman III in 1979 for the Council on Environmental Quality and later revised in 1982.[24] An overview of Freeman's findings follows, with all values expressed in 1984 dollars.

Estimating the Benefits of the FWPCA of 1972

To arrive at a dollar value of the total social benefits (*TSB*) associated with the FWPCA of 1972, Freeman surveys the findings of about 20 existing empirical studies. As shown in Table 14.5, he classifies the benefit measures into four categories: recreational use, nonuser benefits, commercial uses, and diversionary uses. Adjusting for differences in price levels, Freeman arrives at an estimate of annual *TSB* for 1985 of between $5.7 billion and $27.7 billion, with a point estimate of $14.0 billion. This dollar value can be interpreted as the annual benefit of achieving the effluent limitations established by the 1972 legislation or, put another way, of achieving the level of water quality to be attained by 1985.[25]

Estimating the Costs of the FWPCA of 1972

To estimate the comparable costs associated with the FWPCA, Freeman examines two sets of estimates: one provided by the Council on Environmental Quality (CEQ) and one by the EPA.[26] The CEQ data use EPA estimates of the engineering costs of complying with the 1972 act, absent the additional spending associated with the 1977 Amendments. According to these data, the average annual costs over the 1979–1988 period are about $23.2 billion. The EPA data are based upon legislative requirements as of December 1982 and therefore include

[22] To review an EPA benefit-cost analysis of water policy, see U.S. EPA, Office of Water (1994).
[23] A survey of some of these studies is found in Cropper and Oates (June 1992), particularly sec. IV.
[24] Freeman (December 1979) and (1982). Freeman (2000) briefly revisits these findings in a recent discussion of water policy.
[25] The 1985 date is linked to the goals of the FWPCA, specifically, that all polluting discharges were to be eliminated by that year.
[26] Council on Environmental Quality (1980); Freeman (1982); and U.S. EPA (1984), pp. 15–16, table 3, as cited in Freeman (1990), pp. 125–26.

Table 14.5 ***Monetized Annual Benefits of the FWPCA as of 1985 ($1984 billions)***

Category	Range	Point Estimate
Recreation		
Freshwater fishing	$0.7–2.1	$ 1.5
Marine sportfishing	0.1–4.5	1.5
Boating	1.5–3.0	2.2
Swimming	0.3–3.0	1.5
Waterfowl hunting	0.0–0.5	0.2
Subtotal	$2.6–13.1	$ 6.9
Nonuser benefits		
Aesthetics, ecology, property value	$0.7–5.9	$ 1.8
Commercial fisheries	$0.6–1.8	$ 1.2
Diversionary Uses		
Drinking water/health	$0.0–3.0	$ 1.5
Municipal treatment costs	0.9–1.8	1.3
Households	0.2–0.7	0.4
Industrial supplies	0.7–1.4	0.9
Subtotal	$1.8–6.9	$ 4.1
Total	$5.7–27.7	$14.0

Source: Freeman (1982), p. 161, table 8-3, and p. 170, table 9-1.

Note: Estimates are originally presented in Freeman (1982) in 1978 dollars. They have been converted to 1984 dollars using the implicit price deflator.

some of the costs incurred by the 1977 revisions. These estimates show that the average annual costs are $30.8 billion. Freeman drew from both sources and arrived at an estimate of total social costs (*TSC*) in 1985 of between $25 billion and $30 billion, with a midpoint of $27.5 billion.

Benefit-Cost Comparison

Now consider a comparison of Freeman's estimates, assuming they reflect the true benefits and costs to society. The estimated range of values for *TSB* is between $5.7 billion and $27.7 billion, and the comparable *TSC* estimate is between $25.0 and $30 billion. Notice that the two sets of values overlap by only a small amount.

- Estimated range of *TSB:* **$5.7 billion to $27.7 billion ($1984)**
- Estimated range of *TSC:* **$25.0 billion to $30.0 billion ($1984)**

We also can examine the relationship between Freeman's point estimate of *TSB* ($14.0 billion) and the midpoint of his *TSC* estimate range ($27.5 billion).

- Point estimate of *TSB:* **$14.0 billion ($1984)**
- Point estimate of *TSC:* **$27.5 billion ($1984)**

In this case, the *TSC* of water quality control as of 1985 exceeds the associated *TSB*, a finding that indicates an unfavorable outcome. This result is illustrated in Figure 14.2a. Notice that at the abatement level associated with U.S. regulations as of 1985 (A_{1985}), *TSC* is higher

Figure 14.2 *Benefit-Cost Analysis of the U.S. FWPCA of 1972, as of 1985*

(a) This model is based on Freeman's estimates of the total social costs (*TSC*) and total social benefits (*TSB*) as of 1985, based on regulations given in the FWPCA of 1972. As the diagram shows, $TSC > TSB$ in 1985, and abatement level A_{1985} is higher than the efficient level (A_E), where net benefits are maximized.

Total Social Benefits and Total Social Costs

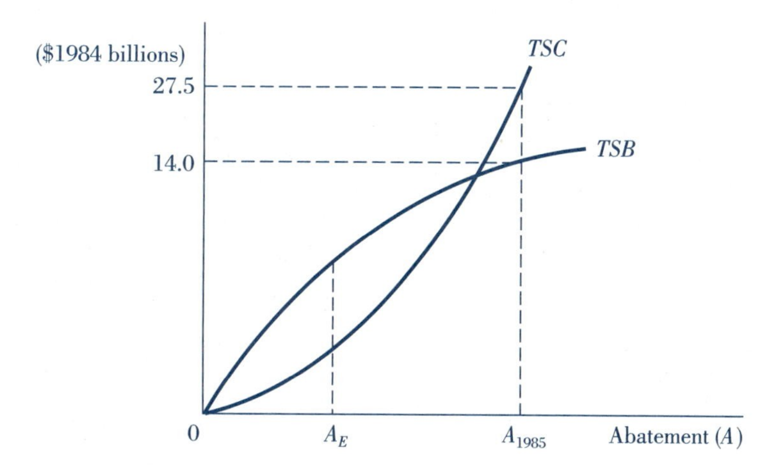

(b) This model shows the comparable relationships to Figure 14.2a, using marginal analysis. Notice that A_E in this depiction corresponds to the point where the marginal social cost (*MSC*) curve intersects the marginal social benefit (*MSB*) curve. In 1985 at abatement level A_{1985}, $MSC > MSB$. Notice that A_{1985} is to the *right* of A_E, suggesting that U.S. policy in effect at that time overregulated water quality.

Marginal Social Benefit and Marginal Social Costs

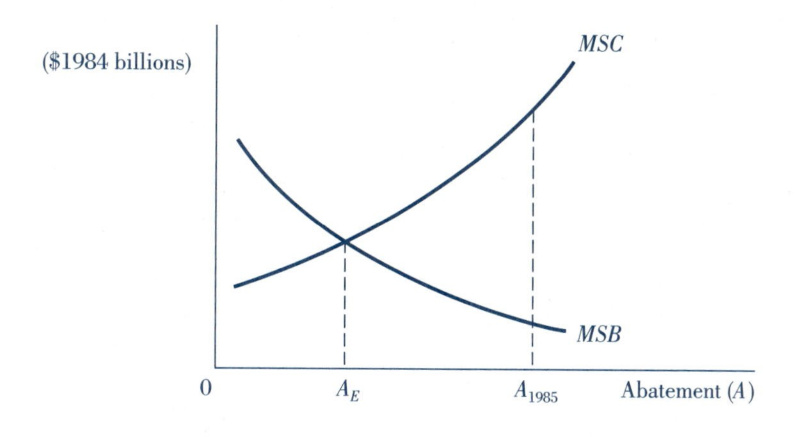

Source: Values depicted based on Freeman (1982), p. 161, table 8-3, and p. 170, table 9-1.

than *TSB*. Also note that A_{1985} is higher than the efficient level (A_E), where net benefits are maximized.

The comparable marginal analysis is shown in Figure 14.2b, where A_E corresponds to the point where the marginal social cost (*MSC*) curve intersects the marginal social benefit (*MSB*) curve. Notice that at A_{1985}, *MSC* exceeds *MSB*, and that this abatement level is to the *right* of A_E.

If Freeman's estimates are reasonable, the findings suggest that as of 1985, the United States was overregulating water quality. The marginal benefit to society was not justified by the marginal cost incurred, and too many resources were allocated to water quality controls. This result is symptomatic of any number of problems. It could be that some of the water quality standards were too stringent, forcing an abatement level that was too expensive for the benefits received. Alternatively, the standards could have been set appropriately but the instruments used to implement them, such as the permitting system or the effluent limits, were not cost-effective.[27]

As in any empirical study, there are always caveats attached to the conclusions. In a subsequent analysis, Freeman (1990) points out several reservations about the values used to support his conclusions. First, the EPA's cost estimates include expenditures associated with the added 1977 controls on toxics, although the associated gains are not captured in the benefit estimate. Second, he argues that the cost estimates suffer from the usual biases associated with the **engineering approach.**[28] Freeman suggests that the cost values are likely inflated, because they do not allow for the potential of more cost-effective decisions at the firm level or cost savings due to technological advances. Furthermore, in a recent publication, Freeman (2000) states that in retrospect he thinks the economic benefits he assigned to sportfishing and boating are too high and that the aesthetic benefits are too low.

An independent commentary on Freeman's work argues further that the benefit estimate would likely be higher if existence and option values were included.[29] Although Freeman does attempt to include some valuation of nonuser benefits, the estimate for this category is highly uncertain because of the methods used in the underlying studies on which it is based.[30] Recognizing the potential inaccuracies while defending his overall conclusion, Freeman argues that, even with a 20 percent *upward* adjustment to the benefit estimate and a 20 percent *downward* adjustment in the cost values, the data still suggest an inefficient outcome.[31]

Freeman's work, though not without limitations, is a valuable analysis in that it draws attention to the potential inefficiencies of the 1972 FWPCA. Taken in the context of the subsequent revisions to this legislation, we need to consider whether the post-1972 reforms accomplished much in terms of improving the efficiency of water quality policy. To examine this question, we next consider the benefits and costs of U.S. water pollution policy in a more current period.

Advances in Benefit-Cost Analysis of U.S. Water Quality Policy

Updated Benefit Estimate of U.S. Water Quality Controls

Assessing benefits continues to be the most difficult task in benefit-cost analysis of environmental policy. Nonetheless, there have been advances in the methods used to estimate environmental benefits. One in particular is the use of the **contingent valuation method**

---

[27] If this were the case, then the cost estimates would lie *above* the true *MSC*.

[28] Recall from Chapter 8 some of the reservations about using this approach, including uncertainty about price movements, availability of raw materials, and energy costs in future time periods.

[29] Howe (September 1991), p. 15.

[30] Freeman (December 1979), pp. 161–62. Look back at Table 14.5. Notice that although the estimated range for this category is within a range of $0.7 billion and $5.9 billion, Freeman uses a conservative point estimate of $1.8 billion.

[31] Freeman also discusses the difficulty of attempting to assess the overall benefit of water quality control given the heterogeneity of surface waters and the sensitivity of water quality indicators to sampling locations and time. He further points out that the benefit estimates used in his analysis measure only the gains from controls on point sources, even though many bodies of water are seriously impaired by nonpoint sources (Freeman, December 1979, pp. ix–x).

Table 14.6 ***Benefit Estimates of Water Quality Improvements: Contingent Valuation Study ($1990)***

Survey Results of Contingent Valuation Study

Water Quality	Estimated Willingness to Pay (average per person per year)
Boatable	$106
Fishable	80
Swimmable	89
Total	$275

Preliminary Estimate of Annual Economic Benefits Based on Survey Results

Economic Benefit to Achieve Swimmable Water Quality from Baseline of Nonboatable Quality	
Range	$24–45 billion
Point Estimate	$29.2 billion

Adjusted Estimate of Annual Economic Benefits of Clean Water

Adjusted to add benefits of commercial usage and marine recreational activities from Freeman (1982)	$39.1 billion
Scaling for increases in real income and changes in attitude about water pollution from 1983 to 1990	$46.7 billion

Source: Drawn from Carson and Mitchell (July 1993).

(CVM), touted for its ability to capture existence value as well as user value of environmental resources. (To review the CVM, refer to the discussion in Chapter 7.)

A study conducted by Carson and Mitchell (1993), of Resources for the Future in Washington, DC, uses the CVM approach to estimate the value of water quality improvements associated with U.S. regulations. The researchers surveyed individuals across the United States in 1983, asking them to assign a dollar value to the minimum levels of water quality specified in the Clean Water Act, that is, for boatable, fishable, and swimmable surface waters.[32]

As indicated in Table 14.6, usable responses helped to form a preliminary nationwide estimate of the total social benefits (*TSB*) of water quality at $29.2 billion ($1990) per year. (All values are expressed in 1990 dollars.) Carson and Mitchell then combine this estimate with elements of Freeman's earlier findings that were not captured by their survey and arrive at an adjusted value of $39.1 billion. Notice that this estimate is considerably higher than what Freeman found, which is $20.1 billion in 1990 dollars when adjusted for inflation and number of households.[33] A final modification is made to adjust for changes in the number of house-

[32] A later study, conducted by Bingham et al. (1998), uses Carson and Mitchell's results to estimate selected benefits for those rivers predicted to actually achieve the swimmable water quality level.

[33] Carson and Mitchell (July 1993) explain that the difference is due to the broader measure of water quality improvement considered in their analysis. Recall that Freeman measures the benefit of achieving the effluent limitations, whereas Carson and Mitchell measure the benefit of achieving the swimmable goal. Also, as Freeman (2000) points out, the baseline in his estimates was 1972 water quality, whereas Carson and Mitchell's was a no-control baseline. Another possibility is that Carson and Mitchell's estimate is higher because the CVM captures the existence value of clean water.

Table 14.7 ***EPA Cost Estimates of Water Quality and Drinking Water Control (1990 billions)***

	Year				
Program	**1980**	**1987**	**1990**	**1995**	**2000**
Water quality[a]	\$27.2	\$41.0	\$46.3	\$57.5	\$68.7
Drinking water	2.4	3.7	4.3	6.4	7.9
Total[b]	\$29.5	\$44.7	\$50.6	\$63.8	\$76.4

Source: U.S. EPA, Office of Policy, Planning, and Evaluation (December 1990), pp. 2-2–2-3, table 2-1.

Note: Estimates are originally given by the EPA in 1986 dollars. They have been adjusted to 1990 dollars using the CPI. Data for 1990 and beyond are projections.

[a]Water quality costs are those pursuant to the Clean Water Act as amended in 1987 and the Marine Protection, Sanctuaries, and Research Act of 1972.

[b]Some totals do not agree with components shown due to rounding.

holds, the general price level, real income, and attitudes about water pollution up to 1990. Carson and Mitchell's final estimate of *TSB* is \$46.7 billion per year as of 1990. This magnitude represents the value of improving water quality from a baseline of nonboatable to swimmable water quality.

Comparable Costs of U.S. Water Quality Controls

For comparison to Carson and Mitchell's (1993) benefit assessment, we can consider cost estimates from a number of government sources. One comprehensive source is the EPA, whose estimates of annualized costs of water quality control are shown in Table 14.7. Notice that the EPA data show a projected annual cost in 1990 of \$50.6 billion. Carson and Mitchell cite the most recently available cost data from the Department of Commerce, which indicate that 1988 annual costs of U.S. water quality control were approximately \$37.3 billion expressed in 1990 dollars.[34] To be conservative, we use both these sources and assume that annualized total social costs (*TSC*) for 1990 are somewhere in the range \$37.3–\$50.6 billion, or about \$44.0 billion as a midpoint estimate.[35]

Updated Benefit-Cost Comparison for U.S. Water Control Policies

We next compare the relative values of the *TSB* and *TSC* estimates (using 1990 dollars). In so doing, we observe that the *TSB* of \$46.7 billion is slightly higher than the comparable *TSC* estimate of \$44.0 billion.

- Point estimate of *TSB*: **\$46.7 billion (\$1990)**
- Point estimate of *TSC*: **\$44.0 billion (\$1990)**

However, Carson and Mitchell (1993) suggest that costs are likely to rise above benefits in the future because of the spending needed to bring all surface waters up to the swimmable level. In any case, the comparison of total magnitudes as of 1990 does not indicate whether the associated abatement level is efficient.

As Figure 14.3 illustrates, the efficient abatement level occurs where net benefits are maximized at point A_E, the point where *TSB* exceeds *TSC* by the greatest distance. Given the available estimates for 1990, we can conclude only that the associated abatement level lies

[34]Bratton and Rutledge (1990), pp. 32–38, as cited in Carson and Mitchell (July 1993).

[35]Although these cost estimates are comprehensive and include costs to all economic sectors, they do not capture the true social costs of water quality control because implicit costs are not included.

Figure 14.3 ***Benefit-Cost Analysis of U.S. Post-1972 Water Quality Policy: Total Social Benefits and Total Social Costs***

The efficient abatement level occurs where net benefits are maximized at point A_E, the point where *TSB* exceeds *TSC* by the greatest vertical distance. Based on available estimates for 1990, the abatement level for post-1972 U.S. regulations lies somewhere between 0 and A_1, where *TSB* is everywhere above *TSC*. Note, however, that there is no reason to assume that it coincides with A_E.

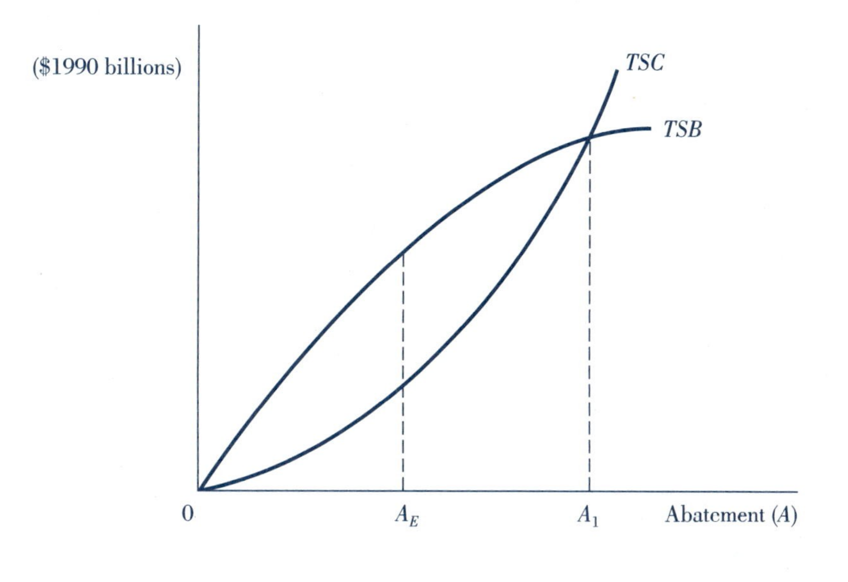

somewhere between 0 and A_1, where *TSB* is above *TSC*. However, there is no reason to assume that this level coincides with A_E.

To determine whether or not this is the case, we need to compare marginal social cost (*MSC*) and marginal social benefit (*MSB*) as of 1990. Although these magnitudes are not directly available, a rough approximation is possible by comparing *incremental* costs and *incremental* benefits over a relevant time period as long as the magnitudes are measured in constant dollars. In this context, the appropriate comparison is the incremental costs and benefits between 1985 and 1990, using Freeman's estimates for 1985 and Carson and Mitchell's estimates for 1990. After converting all estimates to 1990 dollars, these incremental values are as follows:

Incremental control costs for 1985–1990:

1990 control costs (EPA and Dept. of Commerce):	$44.0 billion ($1990)
− 1985 control costs (Freeman):	34.6 billion ($1990)
Incremental costs	$ 9.4 billion ($1990)

Incremental benefits for 1985–1990:

1990 benefits (Carson and Mitchell):	$46.7 billion ($1990)
− 1985 benefits (Freeman):	20.1 billion ($1990)[36]
Incremental benefits	$26.6 billion ($1990)

[36] As noted, this magnitude is the value of Freeman's estimate after being adjusted by Carson and Mitchell for number of households and inflation.

Figure 14.4 **Benefit-Cost Analysis of U.S. Water Control Policy for 1985–1990: Marginal Social Cost and Marginal Social Benefit**

In this model, incremental costs for 1985–1990 are shown using the conventional marginal social cost (*MSC*) curve with an analogous interpretation on the benefit side. Notice that the 1990 abatement level (A_{1990}) occurs at a point *lower* than the efficient level (A_E), where the *MSC* curve intersects the MSB curve.

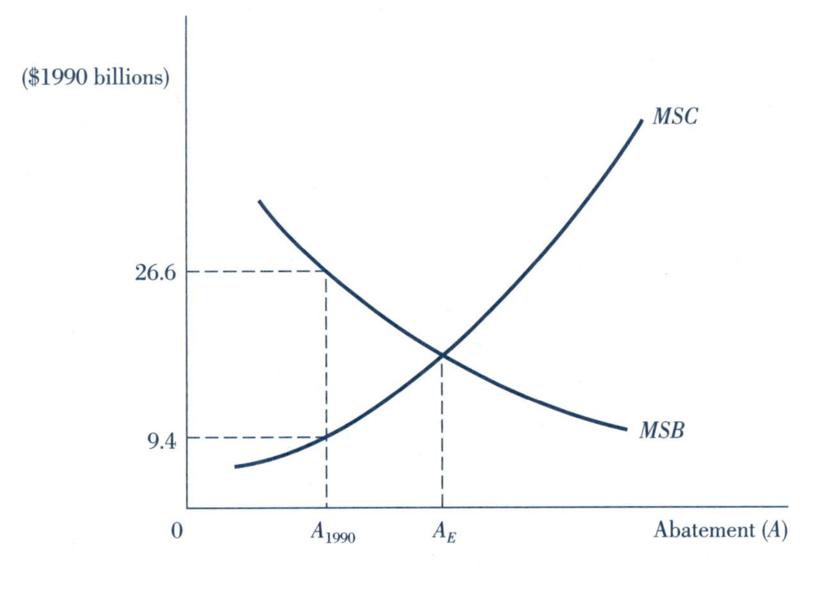

These calculations communicate an interesting result. Because the incremental benefits are greater than the incremental costs, there is reason to believe that the post-1972 revisions to U.S. water legislation have not achieved an efficient abatement level. This result is illustrated in Figure 14.4, where incremental costs are shown using the conventional marginal social cost (*MSC*) curve with an analogous interpretation on the benefit side. Notice that the 1990 abatement level (A_{1990}) is *lower* than the efficient level (A_E), where the *MSC* curve intersects the *MSB* curve. This suggests that more stringent controls on polluting sources may be justified.

This assessment is interesting, but it is important to point out that it is based on crude calculations. For one thing, as Carson and Mitchell assert, most U.S. waters are at a level above the nonboatable quality baseline assumed in the benefit estimate. Hence, costs to achieve swimmable quality are lower than they would be if all water bodies were actually at the baseline level.

Furthermore, the calculations implicitly assume that the values between 1985 and 1990 are directly comparable, which is not the case on the benefit side. Because the 1990 benefit assessment is determined by the CVM, it is a more comprehensive estimate than Freeman's because it includes some measure of existence value. Consequently, Freeman's estimate of $20.1 billion likely undervalues benefits for the 1985 period, which in turn means that incremental benefits likely are biased upward. It can be argued, however, that existence values in 1985 would not be large enough to change the *qualitative* result. Also, Freeman (2000) points out that the two benefit estimates use different baselines of water quality, making the findings not directly comparable. In any case, until more accurate data are available, the conclusion that current U.S. regulations underallocate resources to water quality control is a tentative one at best.

Conclusions

The water on earth has some natural capacity to replenish itself through the hydrologic cycle. Likewise, water bodies have some ability to assimilate certain of the pollutants they receive. However, there is a limit to nature's restorative powers. Furthermore, the hydrologic cycle that contributes to the restoration process is also the mechanism that can spread contamination. Water pollution is pervasive, and virtually every form of human activity and many natural processes contribute to the problem.

Although water pollution is widespread, society has come to recognize the problem and has begun to take action. In the United States, water quality legislation has been expanded and strengthened over the last several decades. This evolution coupled with what are unquestionably ambitious national objectives would seem to suggest that policy officials have made a concerted effort to respond to the country's water pollution problems. Despite these assertions, most argue that whatever legislative progress has been realized, it has happened at too slow a pace. There have been missed deadlines both in establishing effluent limitations and in setting the receiving water quality standards.

Even without the delays that admittedly have slowed the process, economic analyses suggest that the result may not be an efficient solution. If this is the case, valuable resources are not being employed in their best possible use, and society is paying for the misallocation. Accepting that we cannot change the past, the real question is whether the nation can learn from its mistakes. There is still much to be done in revising existing policy and establishing new initiatives, a realization that is an integral part of Congress's recent deliberations to reauthorize the Clean Water Act.

An important motivation in moving national policy forward will be for Congress to consider how much progress has been made toward reaching national water quality objectives and at what measure of costs. Once done, Congress must determine how much of any realized improvement can realistically be attributed to existing policy. Finally, constructive revisions must be proposed, and feasible options formulated as needed. In the next chapter, we will examine the control instruments currently used to achieve national objectives—in a very real sense a textbook exposition of what U.S. policymakers are now undertaking as they continue to develop water quality policy for the nation.

Summary

- The earth's water supply comes from two major sources: surface water and groundwater.
- Surface water and groundwater are linked by the hydrologic cycle, a model that explains the movement of water from the earth's atmosphere to the surface and back into the atmosphere.
- The two primary source categories of water pollution are point sources and nonpoint sources.
- Federal legislation on water pollution began with the Water Pollution Control Act of 1948. The legislation was revised in 1956 and again in 1965, but little progress was achieved.
- In 1972, Congress passed the Marine Protection, Research, and Sanctuaries Act (MPRSA) to regulate ocean dumping. It also enacted the Federal Water Pollution Control Act (FWPCA), which shifted primary responsibility from states to the federal government. Further amendments were accomplished through the Clean Water Act (CWA) of 1977 and the Water Quality Act of 1987.

- The three most important objectives of the Clean Water Act are the zero discharge goal, the fishable-swimmable goal, and the no toxics in toxic amounts goal.
- U.S. law distinguishes among three categories of pollutants: toxic pollutants, conventional pollutants, and nonconventional pollutants.
- Surface water quality is defined in the Clean Water Act by receiving water quality standards. These standards assign a use designation for each water body and identify water quality criteria necessary to sustain designated uses.
- State authorities determine the use-support status of a water body by assessing its present condition and comparing it with what is needed to maintain its designated uses.
- One problem with the receiving water quality standards is that states are not required by law to use a benefit-cost evaluation to establish these standards. Also, the link between receiving water quality standards and effluent limits is unclear because each is motivated differently—one by usage and the other by technology.
- According to Freeman's benefit-cost analysis of the FWPCA, the total social costs of regulatory control as of 1985 exceed the associated total social benefits. This suggests that too many resources were being allocated to water quality controls at that time.
- A more recent analysis by Carson and Mitchell, which uses the contingent valuation method to estimate benefits, found that as of 1990, the total social benefits of U.S. controls exceeded the total social costs.
- Over the 1985–1990 period, crude estimates indicate that the incremental benefits exceeded incremental costs by $17.2 billion, suggesting that current legislation may underregulate polluting sources.

Key Concepts

surface water
groundwater
hydrologic cycle
point source
nonpoint source
zero discharge goal
fishable-swimmable goal
no toxics in toxic amounts goal
toxic pollutant
conventional pollutant
nonconventional pollutant
receiving water quality standards
use designation
water quality criteria
use-support status

Use the Key Concepts listed above to begin your search for additional articles and information using the InfoTrac College Edition database.

Review Questions

1. Use the concept of the hydrologic cycle to explain how contamination of surface waters can also cause degradation of groundwater.
2. Refer to Application 14.1 on the Chesapeake Bay. Discuss both the natural and man-made conditions that are most responsible for the degradation of this water body.
3. Nutrients and pesticides are prevalent causes of water pollution. From an economic perspective, analyze why more has not been done to control agricultural runoff, the major source of these contaminants.
4. Identify two significant trends that characterize the evolution of U.S. water quality policy up through the 1987 Water Quality Act.
5. Using economic analysis, comment on the following statement: The Clean Water Act's zero discharge goal was doomed from the beginning.

6. a. Discuss the major difficulties associated with estimating the benefits of water pollution abatement.
   b. Propose a policy approach that would promote states' use of benefit-cost analysis in setting and evaluating standards.
7. Reconsider the benefit-cost analyses of U.S. clean water policy offered by Freeman (1979) and by Carson and Mitchell (1993). Are their different conclusions justified from a policy perspective? Why or why not? Are the respective estimation methods used by these researchers a factor in your assessment? Explain.
8. Consider the following benefit and cost relationships for mercury abatement (A):

$$MSB = 30 - 0.3A, \qquad MSC = 16 + 0.2A,$$
$$TSB = 30A - 0.15A^2, \qquad TSC = 16A + 0.1A^2,$$

   where A is the percentage of mercury abatement and dollar values are in millions.
   a. Suppose the mercury abatement level was set at 20 percent for 2003. Are net benefits positive or negative?
   b. On the basis of the efficiency criterion, should controls on mercury be tightened or relaxed? Support your response with specific calculations.
9. Visit the Web site at **http://www.epa.gov/305b/98report/statefct.html,** and select a state in the nation. Write a one-page summary of the surface water and groundwater quality in that state. Provide some policy suggestions to improve water quality levels. Be specific.

Additional Readings

Cooper, Christopher. "Fears Linger on 10th Anniversary of Exxon Valdez Spill." *Wall Street Journal,* March 23, 1999, p. B4.

Dahl, T. E. *Wetlands Losses in the United States: 1780's to 1980's.* Washington, DC: U.S. Department of the Interior and U.S. Fish and Wildlife Service, 1990.

Jacobs, J. W., and J. L. Wescoat. "Managing River Resources: Lessons from Glen Canyon Dam." *Environment* 44, no. 2 (2002), pp. 8–19.

Knopman, Debra S., and Richard A. Smith. "20 Years of the Clean Water Act: Has U.S. Water Quality Improved?" *Environment* 35 (January/February 1993), pp. 17–20, 34–41.

Morgan, Cynthia, and Nicole Owens. "Benefits of Water Quality Policies: The Chesapeake Bay." *Ecological Economics* 39 (November 2001), pp. 271–84.

Newman, Alan. "A Blueprint for Water Quality." *Environmental Science and Technology* 27, no. 2 (1993), pp. 223–25.

Renzetti, Steven (ed.). *The Economics of Industrial Water Use.* Northampton, MA: Elgar, 2002.

Rogers, Peter. *America's Waters.* Cambridge, MA: MIT Press, 1993.

Satchell, Michael. "Rape of the Oceans." *U.S. News and World Report,* June 22, 1992, pp. 64–68, 70–71, 75.

U.S. Environmental Protection Agency, Office of Water. *The Quality of Our Nation's Water: 1990.* Washington, DC, June 1992.

U.S. General Accounting Office. *Water Pollution: Improved Coordination Needed to Clean Up the Great Lakes.* Washington, DC, September 1990.

World Health Organization. "Health Hazards of Water Pollution." In David H. Speidel, Lon C. Ruedisili, and Allen F. Agnew, eds., *Perspectives on Water: Uses and Abuses.* New York: Oxford University Press, 1988.

http:// Related Web Sites

Chesapeake 2000 Agreement	**http://www.chesapeakebay.net/info/c2k.cfm**
Clean Water Act	**http://www.epa.gov/region5/water/cwa.htm**
Historical overview of the Federal Water Pollution Control Act of 1972	**http://www.epa.gov/owow/cwa/historytopics/fwpca/05.htm**
Individual fact sheets for states, territories, tribes, and interstate commissions on water quality	**http://www.epa.gov/305b/98report/statefct.html**
	http://www.epa.gov/305b/98report/ch13.pdf
	http://www.epa.gov/305b/98report/ch14.pdf
National Marine Sanctuaries	**http://sanctuaries.nos.noaa.gov/oms/oms.html**
National Water Program: Goals and Objectives	**http://www.epa.gov/water/programs/goals.html**
National Water Quality Inventory: 1998 Report to Congress, U.S. EPA, Office of Water (June 2000)	**http://www.epa.gov/305b/98report/index.html**
Ocean Dumping Ban Act summary	**http://www.epa.gov/owow/ocpd/marine.html**
Strategic Plan, U.S. EPA, Office of the Chief Financial Officer (September 2000)	**http://www.epa.gov/ocfo/plan/plan.htm**
U.S. EPA, Office of Water	**http://www.epa.gov/ow/**
U.S. EPA, Office of Wetlands, Oceans, and Watersheds	**http://www.epa.gov/owow/**

A Reference to Acronyms and Terms in Water Quality Control Policy

Environmental Economics Acronyms

CVM	Contingent valuation method
TSB	Total social benefits
TSC	Total social costs
MSB	Marginal social benefit
MSC	Marginal social cost

Environmental Science Terms

BOD	Biological oxygen demand
NO_X	Nitrogen oxide
SO_2	Sulfur dioxide

Environmental Policy Acronyms

CWA	Clean Water Act
CEQ	Council on Environmental Quality
FWPCA	Federal Water Pollution Control Act
MPRSA	Marine Protection, Research, and Sanctuaries Act
OPA	Oil Pollution Act of 1990
POTWs	Publicly owned treatment works

Chapter 15

Improving Water Quality: Controlling Point and Nonpoint Sources

As we discussed in Chapter 14, U.S. water policy has a fairly long history and one that has been characterized by a series of significant revisions. Yet few would debate the need for further reform. Although U.S. water quality has measurably improved, some water bodies continue to deteriorate and most are still threatened by contamination—particularly the pollution from nonpoint sources. Furthermore, there is concern about certain of the policy instruments in use. A common criticism is an overreliance on uniform technology-based limits and a lack of economic incentives to get the job done. An investigation of these and other issues form the agenda for this chapter.

Our analysis will cover the following aspects of U.S. water quality policy:

- Effluent limitations and permits to control point sources,
- Funding programs for publicly owned treatment works (POTWs), and
- Nonpoint source control policy.

In each case, we present an overview of the control approach followed by an analytical evaluation. Where appropriate, the evaluation will be based on the economic criteria of efficiency and cost-effectiveness. We conclude the chapter with a discussion of several market-based instruments to achieve water quality, some of which are already in use on a limited scale. A list of acronyms and terms is provided at the end of the chapter for reference.

Controlling Point Sources: Effluent Limitations

As evidence began to mount in the 1960s and early 1970s that U.S. waters were deteriorating, Congress realized that federal legislation needed a major overhaul. The result was a command-and-control policy dominated by uniform pollution standards, a response not unlike the initial reaction to America's air quality problems.

Overview of the Effluent Limits and National Permits

Effluent Limitations

Instituted through the Federal Water Pollution Control Act (FWPCA) of 1972, **technology-based effluent limitations** are the primary control instruments through which U.S. water quality objectives are to be achieved. Following a command-and-control approach, these standards limit the amount of contaminants that may be released into surface waters by point sources. As technology advances, these EPA-established regulations are to be revised to identify available control measures and practices. Over time, the intent is for effluents to be completely eliminated to meet the zero discharge goal.

technology-based effluent limitations Standards to control discharges from point sources based primarily on technological capability.

All point sources are subject to these end-of-pipe effluent limits, which vary by the type of polluting source, the age of the facility, and sometimes by the contaminant released. For example, **publicly owned treatment works (POTWs)** must meet a special set of what are known as secondary treatment standards.[1] For **indirect industrial dischargers** that release

[1] Secondary treatment refers to the second step in the waste treatment process in which bacteria consume the organic elements of waste.

their effluents to these facilities, standards are established for the amount of pretreatment that must be done. Still other types of technology-based limits are defined for **direct industrial dischargers** that release their wastes into surface waters without any intermediary. A summary of these different technology-based effluent limits is given in Exhibit A.10 in the Appendix, although the distinctions are sometimes more apparent than real.[2] Also, keep in mind that although there are different standards for different types of polluting sources, they are applicable *uniformly* to all individual polluters within each designated category.

Permit System

Once determined, pollution limits for direct industrial dischargers and for POTWs are communicated through a permitting system, called the **National Pollutant Discharge Elimination System (NPDES),** administered by the EPA. In simplest terms, this system prohibits any direct discharges into navigable waters without an NPDES permit.[3] These permits state precisely what the effluent limitations are as well as the requirements for monitoring and reporting. Though the EPA bears the overall responsibility of the program, states may administer NPDES permits as long as certain federal requirements are satisfied. Currently, 44 states and one territory have assumed this responsibility.[4]

National Pollutant Discharge Elimination System (NPDES)
A federally mandated permit system used to control effluent releases from direct industrial dischargers and POTWs.

Defining effluent limits for direct industrial dischargers is a monumental task, because the limits must reflect technological differences across industry groups, such as steel mills, pesticide manufacturers, and fertilizer manufacturers. In fact, problems associated with these standards caused serious delays that slowed the nation's progress toward achieving its water quality objectives. For these and other reasons, these standards have been the subject of the most criticism by policy analysts, and hence we examine them more closely.

Technology-Based Effluent Limitations for Direct Industrial Dischargers

Many factors are considered in setting the effluent limits for commercial facilities, but the most important is the technological capability of each industry, meaning what can be achieved by using certain pollution control equipment. Because of this, these standards are called **technology-based effluent limitations.** However, polluting sources are allowed to choose the means by which the limit is achieved. This means that these effluent limitations are more accurately termed **performance-based standards.**

These industry-specific standards are further delineated by the age of the polluting source, much as is done for air quality standards, with differences made between new and existing sources. According to the Clean Water Act, a **new source** is one whose construction was begun after proposed regulations have been announced. For new industrial point sources, the effluent limitations are more stringent than those for existing sources, based on what is called the best available demonstrated control technology. In fact, wherever practical, these limits are to explicitly prohibit effluent discharges. For **existing sources,** two types of limits are applicable, based on the type of pollutant released: one for controlling nonconventional and toxic pollutants (called best available technology economically achievable) and one applicable to conventional pollutants (called best conventional control technology). These limits are considered minimum or baseline controls, allowing for more stringent standards to be used if needed to achieve the desired water quality for a particular water body. Refer to Exhibit A.10 in the Appendix for detail.

In setting the standards for nonconventional and toxic pollutants, the technology level for each industry is determined by a number of factors, including processes used, costs, energy requirements, and any impact on the environment beyond water quality. The determinants for the standards on conventional pollutants add a consideration for the reasonableness of the relationship between the associated benefits and costs.

---

[2] More information about these limitations is available online at **http://cfpub.epa.gov/npdes/techbasedpermitting/effguide.cfm?program_id=15.**

[3] Permits are not required for indirect dischargers, whose effluents are released to permitted POTWs. More information on the NPDES program in available online at **http://cfpub.epa.gov/npdes/.**

[4] U.S. EPA, Office of Water (June 2001), p. 2.

Analysis of Effluent Limitations on Point Sources

The command-and-control approach to U.S. water quality policy has been the subject of some debate. Much of the criticism has been lodged against the technology-based effluent limits, which have been blamed for the lack of progress achieved by U.S. policy. Among the recognized problems are

- Delays in establishing the standards,
- Vague statutory definitions,
- Meeting the zero discharge goal, and
- Lack of economic decision rules.

These problems contribute to an inefficient solution to the U.S. water pollution problem. Even accepting the inefficiency, the evidence strongly suggests that the effluent limitations are not cost-effective.

Administrative Delays

Under the 1972 law, industrial point sources were to have achieved an initial phase of compliance with standards called best practicable technology by 1977. This was to be followed by a second phase that called for attainment with different standards (best available technology economically achievable) by 1983. As it turned out, neither of these deadlines was met, with much of the blame pinned on the EPA. Although the agency was to have determined all the best practicable technology standards within one year of the act's enactment, it failed by a wide margin. The one-year time frame may have been overly ambitious, but it is harder to defend why the EPA had not defined these standards by 1977, the date by which polluters were to have achieved compliance with them. In any case, the first phase deadline had to be postponed to 1983. The second phase was not only delayed but also redefined through a new set of standards for conventional pollutants, with the best available technology economically achievable standards retained specifically for toxics and nonconventional pollutants.

Over the past 25 years, the EPA's track record has been slow in developing and revising effluent guidelines, which are the basis for the federal limitations.[5] Such delays are particularly worrisome for toxic effluents, which pose relatively high environmental risks, some of which are inequitably distributed, as discussed in Application 15.1. What caused the delays? Part of the answer lies in how the standards were originally defined in the law.

Imprecise and Inconsistent Definitions

A fundamental problem with the effluent limitations is that they are not aligned with the nation's objectives. The basis of the standards is what is technologically feasible for each group of polluting sources, as opposed to what is necessary to achieve water quality. In fact, the potential inadequacy of these standards is implied by the statutes themselves. Modifications to the effluent limits are allowed in the form of **water quality-related limitations.** These are to be met by every polluting source if the desired level of water quality is not being achieved, even if a source is already satisfying the technology-based limits.

water quality-related limitations
Modified effluent limits to be met if the desired water quality level is not being achieved, even if polluters are satisfying the technology-based limits.

Although this may seem reasonable, it nonetheless speaks to the deficiency of the effluent limits. The fact that modifications might be needed calls into question the characterization of these technology-based limitations as "best." The law acknowledges that these so-called best standards might not be sufficient to achieve water quality and further requires that some polluting sources abate beyond what has been determined as the best available or best practicable technology.

[5] In January 1992, the Natural Resources Defense Council filed litigation against the EPA, which led to the EPA entering a Consent Decree affecting the Effluent Guidelines Program. The decree requires the EPA to adhere to a schedule for devising new regulations and to form an advisory task force (U.S. EPA, Office of Water, May 3, 2002).

15.1 *Toxic Fish Consumption: Are the Risks Equitable?*

Application

Two distinct motivations affect an individual's decision to fish and consume the catch. One is purely recreational—fishing for sport. The other is subsistence—a means to sustain a regular food supply. Individuals engaging in either or both types of fishing are at greater risk of exposure to water pollution than the general population. It is true that those who fish purely for pleasure can minimize their risk by traveling to less polluted waters or by electing not to consume their catch. However, those who rely on fish and shellfish as part of their regular diet have limited options and therefore bear the highest relative risk of exposure to contaminated waters.

Researchers have tried to identify which population groups are most vulnerable to toxic contamination through fish consumption. Some studies suggest a relationship between race or ethnic origin and per capita fish consumption. For example, survey data indicate that people of color consume more fish than the rest of the population. Other studies find that Asians and African Americans eat more seafood than whites. Also, some American Indians comprise one of the largest subsistence fishing communities in the United States. Unfortunately, research on this group of Americans has been frustrated by a lack of data. The problem is that most studies use a sample of only licensed fishermen. Such data do not include reservation-based American Indians because according to treaty rights, these individuals do not have to obtain a state fishing license.

To illustrate the risk implications of such consumption differences, we consider the findings from a recent survey study of Detroit residents. According to this investigation, the average daily fish consumption across the entire sample is 18.3 grams per person per day. This statistic by itself is disconcerting, because the standard used in the state of Michigan to regulate point source toxic releases is based on a much lower consumption rate of 6.5 grams per person per day. However, the equity implications are even more troubling. The highest consumption rate across all defined population groups is for off-reservation American Indians, estimated at 24.3 grams per person per day. The comparable statistics for African Americans, other minorities, and whites, are 20.3 grams, 19.8 grams, and 17.9 grams, respectively. These data suggest that the risk of exposure to contaminated waters is not shared equally across population groups. Even worse, the extent of the inequality may be far greater than the estimates imply, because reservation-based American Indians are unavoidably excluded from the analysis.

If researchers could determine the cause of the relatively high consumption rate for American Indians, it might be possible to make inferences about the on-reservation population. It is true that some tribes have a cultural tradition of fishing-based economies, but it is also believed that American Indians living on reservations depend on subsistence fishing for economic reasons. If this hypothesis is correct, then both cultural and economic factors play a role in fish consumption rates. Thus, it may be that reservation-based American Indians consume fish at a higher rate than the available data suggest and therefore face a high health risk from toxic exposure.

Sources: West (1992); West (March/April 1992); U.S. EPA, Office of Communications, Education, and Public Affairs (February 22, 1994a); "EPA: Fish Aren't Safe to Eat from 46 Waterways" (November 20, 1992).

Still more problems arise in attempting to establish and implement these limitations for designated categories of point sources. Although the use of technology-based limits appears to be an objective and scientific approach to standard-setting, the law does not identify clearly what level of technology is appropriate. Phrases such as "best practicable," "best available," and "best conventional" are not exact terms. Without more careful guidelines and objective decision rules, officials are left with the difficult task of trying to infer what these terms mean and how to implement them in practice.[6]

[6] Clean Water Act, sec. 302; Freeman (1990), pp. 106–107.

Meeting the Zero Discharge Goal

In retrospect, the zero discharge goal established in 1972 was overly ambitious. Furthermore, its call for the complete elimination of water-polluting effluents is likely an inefficient objective. Nonetheless, zero discharge was the target set by Congress under federal law.[7] The Clean Water Act also requires the EPA to review existing standards and successively advance them toward a zero limit as new technology becomes available. Yet, the EPA's effort in this regard has been something less than satisfactory. In the last 20 years, in only a few instances (among them onshore oil and gas wells) has this limit been imposed.

Although there is no comprehensive way to determine which of the effluent guidelines should have been moved to a zero discharge limit, there are indications that this may have been the case in certain instances. One example is the set of effluent standards for the organic chemicals, plastics, and synthetic fibers industry. Even though the EPA knew that certain industrial plants had the capacity to achieve the zero discharge goal, it did not impose a zero effluent limit. The Natural Resources Defense Council challenged the EPA's position. Yet even after a federal appeals court required the agency to reassess its decision, the EPA did not change its ruling. Instead, it defended its position on the basis of differences across plants and its lack of resources to investigate further.[8]

Absence of Economic Decision Rules

Economists and other policy analysts are quick to point out the absence of economic decision rules in guiding the specification of the effluent standards. To understand the implications of this deficiency, consider the following observations:

- The law does not mandate that the standards be set to maximize net benefits, which prevents the attainment of an **efficient** level of abatement.
- The standards are applied *uniformly* across dischargers within identified industrial groups, which impedes a **cost-effective** outcome.

Lack of an Efficiency Criterion

The Clean Water Act allows for many factors to be considered in setting the effluent standards beyond technological feasibility, including economic consequences. However, the provisions do little more than list the relevant factors, offering no guidance as to how they are to be used in decision making. This is problematic because many determinants are itemized in the law and they cross over from engineering and scientific criteria to economic considerations.

In the definition of the standards for conventional pollutants, there is a troubling lack of precision in the reference to benefits and costs. According to law, officials are to consider the "reasonableness of the relationship between the costs of attaining a reduction in effluents and the effluent reduction benefits derived" (CWA, sec. 304b–4-B). This is a far cry from setting abatement levels at the point where marginal benefits and marginal costs are equal.

Absent from the standards on nonconventional and toxic pollutants is any reference at all to economic benefits. For these limits, only the cost of achieving the mandated effluent reduction is among the list of determinants.

Without congressional guidelines for how all determinants, including economic ones, are to be measured and weighted relative to one another, the standards are left to discretion and subjective judgment. The result? Many disagreements and legal battles have ensued among industrial sources, government officials, and environmental groups, delaying the process on which much of the force of the law depends.[9] Beyond these general problems is the absence of any opportunity for the standards on nonconventional and toxic pollutants to achieve an efficient level of abatement and only a remote one for the limits on conventional pollutants.

[7] For a more detailed discussion about the zero discharge goal, see Van Putten and Jackson (summer 1986).
[8] Adler, Landman, and Cameron (1993), pp. 143–44.
[9] Freeman (1978), p. 55.

Table 15.1 ***Cost-Effectiveness of Market-Based Controls: Some Quantitative Studies***

Investigators and Year	Command-and-Control Approach	Geographic Area	Ratio of Command-and-Control Cost to Least-Cost
Johnson (1967)	Equal proportional treatment	Delaware Estuary	3.13 at 2 mg/l 1.62 at 3 mg/l 1.43 at 4 mg/l
O'Neil (1980)	Equal proportional treatment	Lower Fox River, Wisconsin	2.29 at 2 mg/l 1.71 at 4 mg/l 1.45 at 6.2 mg/l
Eheart, Brill, and Lyon (1983)	Equal proportional treatment	Willamette River, Oregon	1.12 at 4.8 mg/l 1.19 at 7.5 mg/l
Eheart, Brill, and Lyon (1983)	Equal proportional treatment	Delaware Estuary in Pennsylvania, Delaware, and New Jersey	3.00 at 3 mg/l 2.92 at 3.6 mg/l
Eheart, Brill, and Lyon (1983)	Equal proportional treatment	Upper Hudson River, New York	1.54 at 5.1 mg/l 1.62 at 5.9 mg/l
Eheart, Brill, and Lyon (1983)	Equal proportional treatment	Mohawk River, New York	1.22 at 6.8 mg/l
Faeth (2000)	Equal treatment	Minnesota River Valley	2.7 at 1 ppm/l
Faeth (2000)	Equal treatment	Rock River, Wisconsin	1.74 at 1 mg/l
Faeth (2000)	Equal treatment	Saginaw Bay, Michigan	5.9 at 1 mg/l

Source: U.S. EPA, Office of Policy, Economics, and Innovation, Office of the Administrator (January 2001), p. 26, table 3-3, citing Tietenberg (1985) and the original sources given in the table.

Notes:

mg/l = milligrams per liter

ppm/l = parts per million per liter

Cost-Ineffective Decision Making

As we have discussed in previous chapters, in instances where the law prevents the use of the efficiency criterion, a "second-best" economic solution is to select cost-effective policy instruments to achieve an objective. **Cost-effectiveness** requires that abatement levels be set to achieve equal marginal abatement cost (*MAC*) levels across all polluters. However, under the Clean Water Act, the *uniformity* of the effluent limits likely prevents such an outcome. Although the law allows the limitations to be industry specific, the associated effluent reductions must be achieved by all polluting sources within each industry group, regardless of firm-level differences in resource availability or technological expertise. If all dischargers must achieve the same standard, the only way the *MAC*s would be equivalent across polluters is if these polluting sources were identical. Because this clearly is not the case, the standards impose higher costs to society than is necessary.

Just how significant is the problem? Unfortunately, there is no research that provides a comprehensive answer to this question. However, some economic studies have attempted to quantify the cost implications of command-and-control instruments for specific water bodies. Table 15.1 summarizes the findings of some of these. Each analysis determines a ratio of the cost of implementing a command-and-control approach to that of the least-cost market-based method. In each case, the ratio is significantly greater than 1, meaning that the command-and-control approach is relatively more costly than the use of economic incentives.

Underlying the problem of uniform standards is the lack of a reward system for efficient abaters to reduce effluents beyond legal limits. In fact, it has been argued that the structure of the effluent limits acts as a market disincentive for technological innovation.[10] Based on the law, if a discharger were to develop a new technology to reduce effluents more efficiently, the limits would be tightened based on the innovative discovery. This response would impose higher abatement costs on all effluent dischargers, including the innovating entity. Furthermore, because the limits are used to establish performance-based rather than technology-based standards, the law would not require polluters to use the new technology. Thus, there is no market opportunity and hence no monetary reward for new innovation. The result is that most dischargers tend to employ the same technology used to set the limits in the first place, even though they are not required to do so. The rationale is to avoid the potential of an official compliance inquiry that might arise if an alternative technology were used.

Waste Treatment Management and the POTW Program

An important and sometimes controversial aspect of the Clean Water Act is the federal funding authorized by Congress to support the construction of publicly owned treatment works (POTWs). POTWs are potentially significant sources of water contamination. Among the pollutants released by these facilities are pesticides, heavy metals, viruses, and bacteria. Unless municipal wastewater is properly treated, the associated pollution threatens groundwater and surface water drinking supplies, aquatic life, recreational opportunities, and the overall health and stability of ecosystems.

Responding to the potential risks, the Clean Water Act requires POTWs to satisfy technology-based secondary treatment standards and calls for federal appropriations to support this mandate. In fact, the fourth policy goal of the act specifically states:

> It is the national policy that Federal financial assistance be provided to construct publicly owned waste treatment works. (CWA, sec. 101a–4)

The Pre-1987 Federal Grant Program

federal grant program
Provided major funding from the federal government for a share of the construction costs of POTWs.

Prior to 1987, one titled section of the Clean Water Act was devoted to waste treatment management. The voluminous and highly detailed set of provisions outlined the **federal grant program** for the construction of POTWs throughout the nation. Administered by the EPA, the program authorized billions of dollars of federal monies to municipalities each year, starting in 1973 with $5 billion and ending in 1990 with $1.2 billion. Before its demise, the grant program allocated over $60 billion for POTW construction.[11] The federal share of construction costs was set at a maximum of 75 percent until 1984, when it was reduced to 55 percent.

Shift to the State Revolving Fund (SRF) Program in 1987

State Revolving Fund (SRF) program
Establishes state lending programs to support POTW construction and other projects.

As part of the 1987 Clean Water Act reauthorization, capitalization grants to each state are authorized to establish a revolving fund for water pollution projects. States are to provide 20 percent in matching funds to support the effort. This **State Revolving Fund (SRF) program,** which replaced the federal grants, is targeted to provide loans for POTW construction as well as for other environmental projects. All 50 states and Puerto Rico have established SRF programs, and all have been operating them for more than a decade. As of 2001, more than $30 billion in cumulative assistance has been made available to fund over 9,500 projects.[12]

[10] Freeman (1978), p. 57.
[11] U.S. EPA, Office of Water (June 28, 2002).
[12] U.S. EPA, Office of Water (May 2001), pp. 3, 6. For more detail on the SRF program, visit **http://www.epa.gov/owm/cwfinance/cwsrf.**

Analysis of the POTW Funding Program

Most understand the motivation behind the federal subsidy of POTW construction, given the potential health and ecological threat of inadequate waste treatment. Yet the issue is not whether the intent of the program is well founded but rather whether federal subsidies are an effective instrument to improve national water quality. To organize a study of this important component of U.S. water quality regulation, we consider three primary questions:

1. What is the relationship between the allocation of federal monies and any observed improvement in waste treatment across the United States?
2. To what extent is **efficiency** served in the federal aid plan, and what incentives or disincentives are present?
3. What are the **equity** implications in terms of who actually pays for this funding?

Assessing the Accomplishments Attributable to Federal Subsidies

To argue that federal subsidies have been effective because municipal waste treatment has progressed measurably is falsely motivated. The problem is that such an argument implicitly assumes that the federal program is fully responsible for any observed improvement. In fact, the evidence suggests that this is not true at all.

It *is* true that an increasing proportion of the population is served by facilities using at least secondary treatment. In 1960, of all people served by a wastewater facility this proportion was only 4 percent. By 1980 the proportion rose to 73 percent. As of 1988, 138 million Americans were being served by at least secondary treatment, an 85 percent increase over 1977. More recent data show that as of 1996, over 150 million Americans were being served by modern waste treatment facilities, as indicated in Figure 15.1.[13] Yet, research shows that most of the federal grant monies only *displaced* local funds that would have been allocated to POTW construction had the national program not been in place.

[13] Council on Environmental Quality (1982), p. 295, table A-61, as cited by Freeman (1990), p. 136; U.S. EPA (December 1990), p. 2; U.S. EPA, Office of Water (January 1995), as cited by Anderson, Lohof, and Carlin (August 1997), p. 7–36, fig. 7-3.

Figure 15.1 ***U.S Population Served by Modern Sewage Facilities***

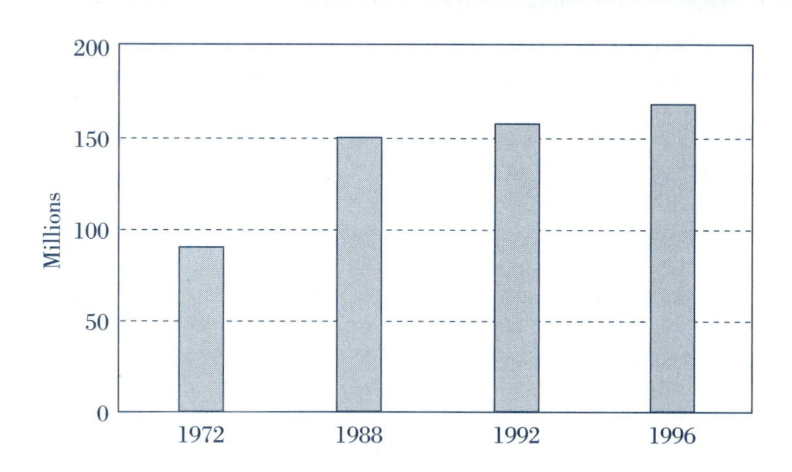

Source: U.S. EPA, Office of Water (January 1995), pp. 6–14.

A 1984 empirical study found the rate of fund displacement to be 67 percent, meaning that every additional dollar of federal money *permanently* displaced 67 cents of municipal spending on POTW construction.[14] This means that roughly two-thirds of every dollar in federal grant money acted only as a *substitute* for local funding, with only a third going to incremental investment in POTW construction.

The significance of this result is that the observed improvement in wastewater treatment cannot be attributed fully to the grant program. Because only a portion of the federal monies was incremental to what would have been spent at the local level, only a fraction of the improved water quality can be linked to the federal aid program. See Application 15.2 for more on this important observation and the 1984 study.

Inefficiencies in the Federal Grant Program

That the Federal Grant Program achieved something less than a dollar-for-dollar improvement is an important realization, but it is only part of the story. It turns out that the well-intended plan was plagued by a number of inefficiencies. One was that the original program, as established in 1972, did not ensure that federal funds were allocated where incremental benefits were expected to be highest, such as in areas suffering from the worst pollution problems. This misallocation was not addressed until Congress passed the 1977 amendments, which required states to establish priority lists of facilities most in need of financial assistance.

Another problem is that the federal aid package was directed toward proper *construction* of new POTWs yet lacked any provision to encourage *maintenance* and *operation* of these facilities once they had been built. In the 1970s, the EPA's annual inspections of POTWs revealed that only 30–50 percent were operating satisfactorily. Similar findings were reported by the U.S. General Accounting Office (GAO).[15]

Lack of Incentives

Most, if not all, of the grant program's inefficiencies were due to one important void: a lack of incentives. Fundamental to the program's design was the share of costs to be absorbed by the federal government. Recall that this share was set at a 75 percent maximum until 1984. Such a large proportion shifted most of the spending away from local governments, leaving them little incentive to minimize costs in the building of treatment facilities. In fact, with the addition of special aid from federal and state governments, some municipalities had to raise as little as 5 percent of the overall construction costs. Consequently, some POTWs are larger and more elaborate than necessary, motivated by a desire to attract industry and promote growth. Such excess capacity is a waste of economic resources and has been blamed for the observed poor operating performance of some facilities.[16]

Policy Response

Responding to such inefficiencies, the U.S. government made several changes to the program. Grant awards were cut to $2.4 billion per year through 1985, a significant reduction from the range in prior years of $4–$6 billion. Starting in 1985, the federal cost share was reduced to 55 percent. Tighter restrictions on grant authorizations were instituted, such as limiting funds to servicing the needs of existing rather than projected population levels. This latter revision was an attempt to discourage the deliberate construction of excess capacity as a way to promote local economic growth.[17] Finally, the Water Quality Act of 1987 altered the basic premise of the program from outright construction aid to funding for state-administered revolving loans.

Notice that these changes shift more of the cost burden to local governments. In so doing, municipalities must become more self-sufficient in developing revenue sources to fund wastewater treatment and therefore should make more cost-effective decisions. Furthermore, con-

[14]Jondrow and Levy (1984), p. 176.

[15]See Council on Environmental Quality (1976), p. 18, and U.S. EPA (November 1974), p. 16, reported in U.S. EPA (1975), p. IV-37.

[16]U.S. Congress, Congressional Budget Office (CBO) (1985a), pp. ix, 12.

[17]See U.S. Congress, CBO (1985a), p. 3. To learn more about the inherent incentive toward excess capacity, see Freeman (1978), pp. 63–64.

15.2 *The POTW Grant Program: Displacement of Local Funding*

Application

In a 1984 empirical study, James Jondrow and Robert A. Levy reported some interesting findings about the U.S. government's POTW grant program. Their investigation shows that every additional dollar of federal grants awarded *permanently* displaced 67 cents of municipal spending. The analysis further estimates that 28 cents of every dollar of unspent federal budget authority *temporarily* displaced (i.e., postponed) local expenditures on POTW construction. This latter form of displacement is associated mainly with the early phase of the federal program. At that point, local authorities had an incentive to delay expenditures while the federal program was getting under way or while awaiting federal approval of a proposed project.

According to Jondrow and Levy, these estimates translate to the following dollar values for 1973 at the start of the POTW grant program:

	\$1,050 million in federal grants
–	\$ 714 million of municipal expenditures *permanently* displaced
	\$ 336 million
–	\$1,400 million of municipal expenditure *temporarily* displaced
–	\$1,064 million net addition to POTW construction

Based on these calculations, the total displacement of municipal spending on sewer treatment in 1973 was approximately 200 percent. Thus, the federal program actually caused a *reduction* in overall funding for POTW construction of over \$1 billion.

According to the researchers, the high proportion of displacement in 1973 is explained mainly by the large amount of temporary displacement that is more likely to occur in the initial stages of the program. Hence, this displacement would be expected to decline as the program became more established. In fact, conducting an analogous exercise for 1980, Jondrow and Levy find that temporary displacement was much lower for the period. Nonetheless, total displacement was still a significant amount, estimated at \$2.5 billion, or 64 percent of the total grant expenditures for that year.

Related statistics bear out the findings of Jondrow and Levy. According to a report conducted by the Congressional Budget Office (CBO), the \$18 billion federal grant program authorized in 1972 marked a major shift of the financial burden for POTW construction from local governments to the federal level. From 1970 to 1977, the annual federal outlay under the grant program rose from \$0.5 billion to \$6.0 billion, with average local spending falling from \$4.0 billion to \$1.5 billion.

Collectively, these data suggest that the POTW grant program did not contribute dollar for dollar to improvements in U.S. water quality. Given the enormous financial investment in the program, this conclusion cannot be taken lightly. In simplest terms, this outcome is a classic case of misallocated resources. One can only hypothesize about the gain in environmental quality that could have been achieved had these billions of dollars been spent more wisely.

Sources: Jondrow and Levy (1984); U.S. Congress, CBO (1985a), p. 4.

cern for higher sewer fees prompts local residents to take a more active role in ensuring that local officials carefully consider costs. According to a statistical analysis done by the U.S. Congressional Budget Office (CBO), increases in the lifetime local cost share of up to about 50 – 60 percent do lead to efficiency improvements measured in terms of declines in lifetime unit costs.[18] Finally, by placing responsibility closer to the funded project, decisions can better reflect state and local needs.

[18] For more detail on this study, consult U.S. Congress, CBO (1985a), chap. 2.

Equity Implications

Beyond the efficiency issues, equity considerations are also associated with the POTW program. Some municipalities had not yet been funded when the grant program was eliminated, placing a relatively higher cost burden on those communities. The inequity is greater for smaller, rural communities that are unable to take advantage of scale economies. The shift to the State Revolving Fund (SRF) program may have provided some measure of offset, however, because state-managed loans can be tailored through interest rates or grace periods to accommodate lower income or wealth levels of certain local communities.[19]

On a broader scale, the grant program apparently did provide for a more equitable tax burden across the national population. A 1980 study showed that federal cost sharing reduces the burden imposed on the lowest income families from 0.6 percent to 0.4 percent of family income and increases the comparable burden of highest income families from 0.3 percent to 0.8 percent.[20] It is not yet apparent as to how the allocation of the tax burden has been affected by the shift to the SRF program.

Controlling Nonpoint Sources[21]

In 1987, the federal government added a seventh policy goal to the Clean Water Act, calling for the development of programs to control nonpoint polluting sources:

> It is the national policy that programs for the control of nonpoint sources of pollution be developed and implemented in an expeditious manner so as to enable the goals of this Act to be met through the control of both point and nonpoint sources of pollution. (CWA, sec. 101a–7)

This addition to the national list of objectives was in response to the growing awareness that these diffuse sources are major contributors to surface water and groundwater pollution. In fact, the EPA asserts that nonpoint source pollution is the greatest source of water quality problems in the United States.

Nonpoint Source Management Program

Nonpoint Source Management Program
A three-stage, state-implemented plan aimed at nonpoint source pollution.

Enacted officially as Section 319 of the Water Quality Act of 1987, the **Nonpoint Source Management Program** was launched as a three-stage plan to be implemented by states with federal approval and financial assistance.[22]

- The first step of the program calls for states to prepare assessment reports in which they identify waters that cannot achieve or maintain water quality standards without some action taken toward nonpoint sources. The states must also identify categories of nonpoint sources or specific point sources responsible for the problem.
- In the second stage, states must develop management programs in which they designate **best management practices (BMP)** to reduce pollution from every identified category, subcategory, or individual nonpoint source, taking into account the impact on groundwater resources. As of 2001, all states and territories and over 70 American Indian tribes have full EPA approval of their management programs.
- The final step calls for implementation of these programs over a multiyear time period.

best management practices (BMP)
Strategies other than effluent limitations to reduce pollution from nonpoint sources.

[19] U.S. Congress, CBO (1985a), pp. 54, 59–60.
[20] Gianessi and Peskin (February 1980).
[21] The following discussion is drawn from the Clean Water Act, sec. 319, Nonpoint Source Management Programs, and U.S. EPA, Office of Water (March 1994), pp. 247–48. For online information about the EPA's Nonpoint Source Pollution Control Program, visit **http://www.epa.gov/owow/nps/**.
[22] See **http://www.epa.gov/owow/nps/cwact.html** for more detail on sec. 319, including access to the full text.

To support states' efforts, federal grants are available for up to 60 percent of the total costs incurred. In 1991, the EPA issued final guidance on how these funds are to be awarded and managed. This guidance encourages states to focus on high-priority activities, such as dealing with high-risk nonpoint problems, promoting comprehensive watershed management, and protecting sensitive and ecologically significant waters, such as wetlands, estuaries, and scenic rivers. From fiscal year 1990 through 2001, federal appropriations totaled over $1.3 billion.[23]

Devising and Updating a Framework

In 1989, the EPA developed its *Nonpoint Source Agenda for the Future* to help define national goals for nonpoint source pollution and to find appropriate mechanisms to achieve them. Over time, state, tribal, and local programs were developed and implemented. In 1995, representatives from the EPA and state officials began a series of discussions to devise a new framework for strengthening existing nonpoint source programs. The collaborative effort, sponsored by the Association of State and Interstate Water Pollution Control Administrators (ASIWPCA), resulted in a new Section 319 program and grant guidance signed by both the EPA and the ASIWPCA. Application 15.3 has more on this national effort.

Watershed Approach

As is evident in Application 15.3, watershed management has become an important element of nonpoint source control policy. Refer back to Figure 14.1 to see that a **watershed** refers to the land areas that drain into a particular water body. By attending to the watershed, as opposed to just a specific lake or stream, public officials gain a better sense of the overall environmental conditions in an area and can better identify factors that negatively affect those conditions. In January 2002, President Bush announced that over $20 million in his 2003 budget would be allocated to a new initiative aimed at watershed protection. According to the EPA administrator, the monies are to be used to fund efforts in up to 20 of the nation's most highly valued watersheds.[24]

Analysis of Controls on Nonpoint Sources

There is no debate that the emphasis of water pollution control in the United States has been on point sources. Through 1987, policy officials focused their energies on developing standards, issuing permits, and monitoring these more visible and obvious sources of contamination. Over time, it became increasingly apparent that these efforts were insufficient. Nonpoint source pollution from land runoff and atmospheric deposition was continuing to threaten surface water and groundwater supplies, and existing policy was simply not designed to deal with it.

In 1994, then EPA administrator Carol Browner characterized the extent of the problem as follows:

> The single greatest remaining threat to America's rivers, lakes, and estuaries is polluted runoff, sometimes called nonpoint source pollution. Silt, pesticides, fertilizer, and other pollutants are carried off farms, suburban lawns, industrial plants, and city streets into water bodies whenever it rains.[25]

Since the 1987 provisions delegate much of the implementation to state governments, we need to consider the reasonableness of this decision. What are the pros and cons? We also

[23] U.S. EPA, Office of Water (February 2002).
[24] U.S. EPA (January 25, 2002).
[25] Browner (summer 1994), p. 6.

Application

15.3 *The Nonpoint Source Program: A Collaborative Effort*

To help define objectives for nonpoint source programs and ways to implement them, the EPA drafted its *Nonpoint Source Agenda for the Future* in 1989. This agenda outlined a comprehensive approach to controlling nonpoint sources and achieving the policy goals of the Clean Water Act. Over the next several years, state and local officials used this national guidance to develop and implement nonpoint source programs. In 1995, the EPA decided it was time to open a series of discussions with state officials to find ways to strengthen existing programs. The discussions continued for over a year and resulted in a new Section 319 program and grant guidance, which was signed by both the EPA and the Association of State and Interstate Water Pollution Control Administrators (ASIWPCA).

The new approach exemplifies the common commitment by the EPA and ASIWPCA to upgrade states' nonpoint source management programs by integrating nine program elements aimed at achieving and maintaining beneficial water uses. Specifically, these nine elements call for states to:

- Include in their programs explicit short- and long-term goals, objectives, and strategies to protect surface and ground waters;
- Strengthen partnerships with appropriate state, interstate, tribal, regional, and local entities, private sector groups, citizen groups, and federal agencies;
- Use a balanced approach emphasizing statewide nonpoint source programs and on-the-ground management of watersheds where waters are impaired or threatened;
- Abate known water quality impairments from nonpoint source pollution and prevent significant threats to water quality from nonpoint source activities;
- Identify watersheds impaired by nonpoint source pollution and important unimpaired waters that are at risk, and develop and implement watershed implementation plans;
- Review, upgrade, and implement all program elements required by the Clean Water Act and establish flexible approaches to achieve and maintain beneficial water uses;
- Identify federal lands and activities not managed consistently with state nonpoint source objectives and seek EPA assistance as needed;
- Manage and implement their nonpoint source programs efficiently and effectively; and
- Periodically review and evaluate their nonpoint source management programs and revise them at least every 5 years.

At a national nonpoint source meeting sponsored by the EPA and the ASIWPCA in April 2000, states and the EPA launched a new State-EPA Nonpoint Source Partnership. The partnership provides a framework for a cooperative effort aimed at identifying, prioritizing, and solving nonpoint source pollution problems. Eight work groups have been formed, each focusing on a particular need: watershed planning and implementation, rural nonpoint sources, urban nonpoint sources, nonpoint source grants management, nonpoint source capacity building and funding, information transfer and outreach, nonpoint source results, and nonpoint source monitoring. The expectation is that the information yielded by these work groups will assist states in implementing their nonpoint source programs more effectively.

Sources: U.S. EPA, Office of Water (June 2000), pp. 247–249; U.S. EPA, Office of Water (July 24, 2002).

must evaluate what the federal government is doing vis-à-vis what is needed to support states' efforts.[26]

Delegating Control to the States: The Pros

Variability of Nonpoint Source Pollution

One factor in support of delegating nonpoint source control to state governments is the variability of nonpoint source pollution. Not only are these sources difficult to identify and isolate, but also the extent of the associated damage is unpredictable. A major source of the problem is land runoff from farms, city streets, mines, construction sites, etc. Because runoff is influenced by precipitation, the resulting contamination is affected by many exogenous factors, such as weather, geological patterns, and soil conditions. Not only are these factors uncontrollable, but they also vary considerably from location to location. Logically, the use of broad-based, uniform controls is likely to be ineffective and difficult to implement.

Land Use Practices

Another factor influencing the coordination of effort between state and federal authorities is the relationship between nonpoint source pollution and land use practices. Land use practices include such activities as agriculture, mining, forestry, and urban development, all of which historically have been controlled by local governments. It therefore becomes a politically sensitive issue when the federal government sets policy that dictates how a state or local community is to use its own land resources.

A good example is environmental policy affecting forestry, an industry that can harm the ecology but also often sustains local or regional economies. In some sense, the 1987 nonpoint source program attempts to address the interests of both sides by giving states the major responsibility of policy design and implementation, while assigning oversight to the EPA through its review of states' assessments and programs.

Delegating Control to the States: The Cons

Information Deficiencies

Although there is some logic to controlling nonpoint pollution close to the source, state governments often lack information to carry out their responsibilities. Data on the extent of water contamination are inadequate, in large part because of ineffective monitoring systems. Although the Clean Water Act requires states to assess surface waters and report the findings every two years, the law does not specify what proportion of these waters is to be included in the biennial review.[27]

Poor Monitoring Systems

Adding to the dilemma is the fact that most state assessment data are collected from monitoring systems designed to detect point source pollution. Therefore, they do not give an accurate picture of the damage from nonpoint sources. Insufficient monitoring data not only hinders environmental assessment but also impedes the evaluation of states' management programs and their designation of best management practices (BMP). Without good monitoring systems, changes in contaminant levels cannot be properly measured. Hence, incentives to propose new plans or revisions to existing ones are limited at best.

Inconsistent Controls

Finally, there is the potential problem of inconsistent pollution controls when state governments are in charge of policy implementation. Such inconsistencies are problematic because water contamination in one state can flow downstream into another state's jurisdiction. Anticipating this possibility, Congress provided for intervention by the EPA to arrange for

[26] Much of the following discussion is drawn from U.S. GAO (October 1990), chap. 2. Another comprehensive examination of policy on nonpoint source pollution is U.S. EPA, Office of Policy, Planning, and Evaluation (June 29, 1992). A review of the EPA's 12 findings in this source is given by Adler, Landman, and Cameron (1993), pp. 188–89.
[27] U.S. GAO (July 1991), pp. 21–22.

interstate management conferences. If a state's water body is not meeting standards because of nonpoint sources in another jurisdiction, that state can petition the EPA to arrange a conference to solve the problem.

Analyzing the Federal Role in Nonpoint Source Controls

Although states have the primary responsibility for implementing nonpoint source controls, they require support from the federal government on a number of fronts. The problem is that policy resources dedicated to water quality have long been focused on point sources, and consequently some adjustments must be made at the federal level. To illustrate, we consider two areas in need of attention:

- Allocation of resources between point and nonpoint source pollution policy
- Coordination of nonpoint source programs with other federal programs.

Resource Allocation

Because of the historical emphasis on point source water pollution, federal funds are needed to fill information voids about nonpoint source pollution. For example, states need water quality criteria to develop standards suited to these diffuse sources of impairment. Unfortunately, the EPA's experience with establishing criteria for point source pollutants is not directly applicable, so the task is expected to be time intensive. Similarly, efforts must be made to improve the nation's monitoring technologies, or protocols as they are called. Here again, the work to date has focused on measuring contaminants released by point sources. It now must be extended to establishing better protocols aimed at the more complex problem of nonpoint source pollution. Attending to these needs will require substantial resources.

Table 15.2 shows the relative spending on point versus nonpoint source pollution control for selected years starting in 1972 and projected through 2000. The data clearly indicate the national emphasis on point source control. Notice also that the proportion of total spending dedicated to nonpoint source pollution shows a decline from 6.2 percent in 1972 to 1.7 percent in 2000. Part of the reason is that the EPA is charged with much greater responsibility for controlling point sources. However, two other observations are more difficult to explain.

First, the EPA has not been aggressive in using the funds that *have* been authorized by Congress for nonpoint sources. Through 1991, the EPA had requested only $22 million of the $400 million approved for Section 319 programs, and through 1993 the accumulated requests totaled $200 million. Second, the proportion of funding across the two types of sources is not supported by relative risk analysis. One study showed that, although health risks from point and nonpoint source pollution are comparable, nonpoint sources pose a much greater risk to ecosystems.[28] As discussed in previous chapters, policy controls do not always follow scientific findings on actual risk but instead are based on public perceptions of risk. Hence, there is a need to improve public awareness about nonpoint source pollution.

Coordination with Other National Programs

Another issue to be resolved at the federal level is the conflict between water quality objectives and other regulations. This becomes an issue for government programs that support such industries as agriculture and forestry, which are major contributors to nonpoint source pollution.

Consider, for example, the agricultural commodity programs operated by the U.S. Department of Agriculture (USDA). These programs are designed to stabilize and support crop prices, which in turn protects the incomes of the nation's farmers. Over two-thirds of U.S. croplands are enrolled in these programs. To operationalize the price protection, an acreage base is established for a given program crop along with a crop yield based on that acreage,

[28] The issue of funding priorities between point and nonpoint sources is discussed in U.S. GAO (October 1990), pp. 49–53, and by Adler, Landman, and Cameron (1993), p. 256. Findings on relative risks between the two sources are given in U.S. EPA (August 1989).

Table 15.2 ***Annualized Expenditures on Point Versus Nonpoint Source Pollution Control ($1986 millions)***

	Year				
Program	**1972**	**1980**	**1987**	**1995**	**2000**
Point source	$8,543 (93.8%)	$22,116 (97.2%)	$33,642 (97.7%)	$47,300 (98.1%)	$56,604 (98.3%)
Nonpoint source	567 (6.2%)	647 (2.8%)	779 (2.3%)	893 (1.9%)	959 (1.7%)
Water quality total[a]	$9,110 (100%)	$22,763 (100%)	$34,421 (100%)	$48,194 (100%)	$57,563 (100%)

Source: U.S. EPA, Office of Policy, Planning, and Evaluation (December 1990), p. 3-3, table 3-3.

Notes: Relative proportions given in parentheses.

[a]Total excludes expenditures on drinking water controls.

both of which are determined from historical values. Farmers participating in this program cannot plant crops other than the so-called program crop, nor can they plant more than their base acreage in that crop. Such supply restrictions help to maintain crop prices, but they also promote practices that can lead to water pollution.

A case in point is the incentive for farmers to specialize in certain program crops, because benefits are based on historical production levels. The problem is that repeated plantings of the same crop year after year deplete the soil and make plantings more vulnerable to pests. Consequently, farmers tend to rely more on agrichemicals, such as fertilizers and pesticides, both of which contribute to agricultural runoff—the leading source of pollution in U.S. rivers, streams, and lakes.[29]

Proposals for Reform: Using the Market

On balance, federal officials have been slow to react to the inefficiency and cost-ineffectiveness of U.S. water quality policy. There are, however, indications that this trend might be changing. Following the lead of several European countries, the United States has begun to consider the potential cost savings of using market-based approaches to water quality policy. In fact, some local governments are currently experimenting with these types of incentive-based policy tools.

To analyze these economic instruments, we consider two market approaches to point source pollution control: effluent fees and tradeable effluent permits. We then extend the analysis to the more complex case of nonpoint source control by studying product charges and effluent reduction trading within a bubble.

Market Approaches to Point Source Pollution

Effluent Fees

As discussed in Chapter 5, an **effluent fee** is a charge based on the discharge of pollution. For an effluent charge to provide an incentive for pollution reduction, it must be based on either the volume or the type of effluent released. For example, a **volume-based effluent fee**

volume-based effluent fee
A fee based on the quantity of pollution discharged.

[29]Specifically, agricultural runoff contributes 59 percent of the water quality problems in impaired rivers and streams and 31 percent in impaired lakes (U.S. EPA, Office of Water, June 2000, pp. 62, 88).

Application

15.4 *Germany's Effluent Charge System*

Like most advanced countries, Germany experienced substantial economic growth and industrialization during the 1960s. By the early 1970s, this surge of development had taken its toll on Germany's natural environment, particularly its water resources. In some regions, the contamination was so severe that it was impeding customary uses of some water bodies.

Germany's first line of defense was to create the Cabinet Committee for Environmental Protection, which was established in 1971. By design, this body was responsible for coordinating all environmental activities at the federal level of government. In one of its first official documents, the committee advocated the use of market-based instruments. Among its recommendations was the implementation of an effluent charge to help restore Germany's water quality. As proposed, this charge was to be levied on every discharger in an amount equal to the incremental damages caused by its effluents—precisely the kind of solution economists endorse.

Response to the committee's recommendation for an effluent charge was mixed. Support came from some international organizations, such as the Organisation for Economic Co-operation and Development (OECD) and the 1972 World Environmental Conference sponsored by the United Nations. However, the Länder (i.e., all the states of the Federal Republic of Germany) opposed the shift to an economic approach. Instead, they supported a more moderate transition with an integration of market-based instruments into the country's existing command-and-control structure. There was also strong opposition from most of Germany's industrial sector. Ultimately, a revised initiative was imposed, which integrated an effluent charge system into Germany's existing command-and-control regulatory framework.

In September 1976, the German government passed the Effluent Charge Law, which combined a discharge fee with a permit structure similar to the U.S. system. Implemented in 1981, the new law requires each German state to levy an effluent charge on all direct dischargers of such pollutants as settleable solids, mercury, and cadmium. One of the unique elements of this landmark legislation is a market-based incentive that reduces a polluter's charge liability if it complies with federally mandated minimum standards. Studies indicate that this incentive has had its intended qualitative effect, at least in some German cities and towns. Some municipalities and industrial dischargers claim that this aspect of the new law was the primary motivation for their increased investment in waste treatment.

From a financial perspective, the effluent charge adds to government revenues. There is also anecdotal evidence that this instrument can provide significant cost savings. Germany's Council of Experts on Environmental Questions claims that an effluent charge can achieve a given level of water quality for about 33 percent less money than if uniform standards were used. Finally, there are indications that the effluent charge may stimulate innovation in pollution abatement. According to the OECD, Germany's clean water technology market has grown considerably and is now the largest segment of its environmental protection market nationwide.

Sources: Brown and Johnson (October 1984); OECD (1989).

is imposed on a per unit basis so that polluters pay higher amounts for larger quantities of discharges. Similarly, to discourage the release of more damaging effluents, such as those containing toxics, the government can use a **pollutant-based effluent fee,** meaning that the fee is higher for discharges containing more harmful substances. Countries such as Australia, Belgium, France, and Germany have instituted effluent charges to control water contamination.[30] Application 15.4 discusses Germany's initial experience with this market-based instrument. Worldwide use of effluent fees is still somewhat limited. There are complications

pollutant-based effluent fee
A fee based on the degree of harm associated with the contaminant being released.

[30] OECD (1999a), pp. 22–23.

in setting the level of the fee, and because of the complexity, administrative costs can be prohibitive.

To illustrate these issues, suppose the government sets a per unit effluent fee for releases of some toxic water pollutant. The outcome would be that each polluting source would abate up to the point where its marginal abatement cost (*MAC*) equaled the fee. Because all *MAC*s would be equal, the effluent charge would yield a cost-effective solution. This is shown in Figure 15.2 for two hypothetical firms, each facing different *MAC* curves but the same marginal effluent fee (*MEF*). Firm 1 faces MAC_1 and abates A_1 units of effluent, and Firm 2 faces MAC_2 and abates A_2 units. Although the abatement amounts are different, the associated level of *MAC* for each firm is the same.

Recognize, however, that the combined abatement level achieved by all firms would not be efficient unless the marginal social benefit (*MSB*) of abatement were equal to the associated marginal social cost (*MSC*) of abatement. Recall that the *MSC* is the horizontal sum of all the individual *MAC*s (MAC_{mkt}) plus the government's marginal cost of monitoring and enforcement (*MCE*). Identifying the *MSB* and the *MSC* in practice is difficult, and the costs of collecting the necessary data would likely be prohibitive.

Even if this task were somehow accomplished, the solution would be efficient only in the aggregate and not for the specific water bodies affected. Unless the marginal benefits and costs at each pollution site were identical, the fee would not yield a water body–specific efficient outcome. For example, the fee would effectively overregulate polluters in regions where the associated marginal benefits of abatement were relatively low, such as in sparsely populated locations, and would underregulate polluters where the conditions were reversed, holding all else constant.

Such an outcome is shown in Figure 15.3, where it is assumed for simplicity that the *MSC* is the same for two regions but that there are different *MSB* curves: MSB_{low} in a low population

Figure 15.2 *Cost-Effectiveness of a Per Unit Effluent Fee*

Suppose that the government sets a per unit effluent fee for releases of some toxic water pollutant. The decisions of two hypothetical firms are shown. Each firm faces a different *MAC* curve but the same marginal effluent fee (*MEF*). Firm 1 faces MAC_1 and abates A_1 units of effluent, and Firm 2 faces MAC_2 and abates A_2 units. Although the individual abatement levels are different, the associated *MAC* levels incurred by each firm is the same.

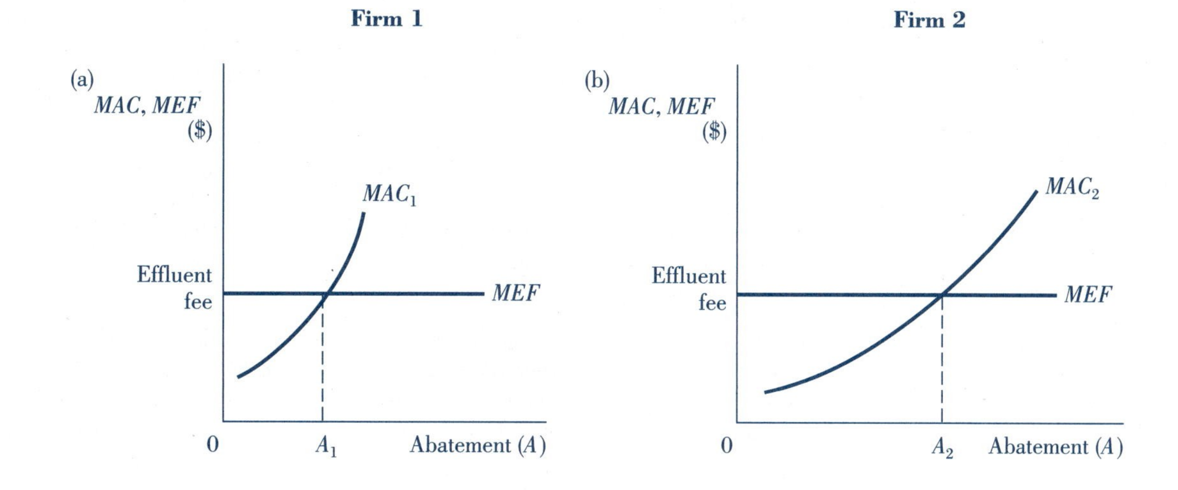

Figure 15.3 ***Inefficiency of a National Per Unit Effluent Fee***

In the model shown, the *MSC* curve is the same for two regions, but there are different *MSB* curves: MSB_{low} in a low population area and MSB_{high} in a high or densely populated area. If the per unit effluent fee is set at *MEF*, each region will abate A_0 units. Notice that A_0 is above the efficient level for the sparsely populated region (A_{low}) and below the efficient level for the densely populated region (A_{high}). If the objective were to achieve efficiency at the regional level, regulatory officials would have to establish a unique effluent fee for every location corresponding to the abatement level where the respective *MSB* and *MSC* curves were equal.

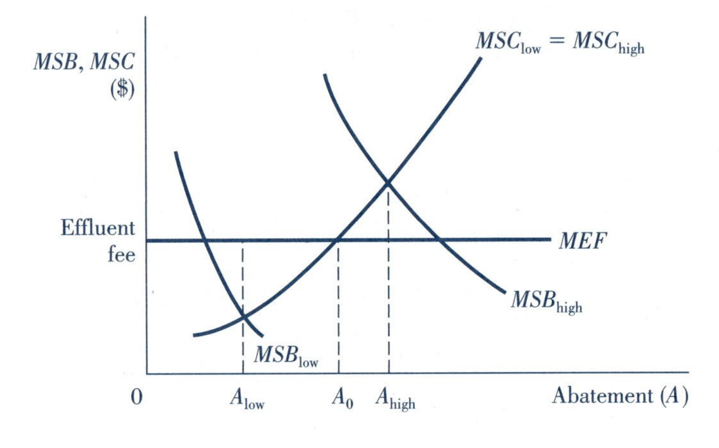

area and MSB_{high} in a high population area. If the per unit effluent fee is set at *MEF*, then each region will abate A_0 units. Notice, however, that A_0 is *above* the efficient level for the sparsely populated region (A_{low}) and *below* the efficient level for the densely populated region (A_{high}). Hence, if the objective were to achieve efficiency at the regional level, regulatory officials would have to impose a unique effluent fee for every location corresponding to the abatement level where the respective *MSB* and *MSC* curves are equal.

The data required to implement such a complex plan coupled with the sophisticated monitoring needed would make this approach highly impractical, at least on a national scale. However, there are opportunities for states to institute such programs, and in fact some have begun to do so. Officials in a number of states, such as Florida, Connecticut, and New York, have instituted volume-based fees, whereas California, Indiana, Louisiana, and others have set fees that vary with volume and toxicity.[31] A similar approach is being used by some POTWs. These initiatives are extensions of the permit fees implemented through the National Pollutant Discharge Elimination System (NPDES). Normally, these charges are independent of the type or amount of effluent being discharged. Thus, the associated *MEF* is zero, offering no incentive for polluting sources to abate.

Tradeable Effluent Permits

tradeable effluent permit market
The exchange of rights to pollute among water-polluting sources.

An alternative economic instrument is the establishment of a **tradeable effluent permit market.** This market could be implemented using the existing NPDES, by issuing each permit as a set of tradeable, per unit rights to pollute. As discussed in Chapter 5, there are potential gains from trading as long as *MACs* differ across polluting sources. If a discharger does

[31] Duhl (December 1993), p. 10, as cited in U.S. EPA, Office of Policy, Economics, and Innovation (January 2001), p. 36, table 4-2.

Table 15.3 ***Examples of Point-to-Point Tradeable Effluent Permit Markets***

State	Water Body	Pollutant
Wisconsin	Fox River	Biological oxygen demand (BOD), nutrients
California	South San Francisco Bay	Copper
Colorado	Cherry Creek	Phosphorus
Florida	Tampa Bay	Nitrogen, total suspended solids

Source: U.S. EPA, Office of Water (June 15, 2002).

Notes: BOD, or biological oxygen demand, measures the amount of oxygen needed by microorganisms to decompose organic compounds. Therefore, BOD rises as the waste level in a water body rises. For more information on effluent trading in watersheds, visit **http://www.epa.gov/owow/watershed/tradelinks.html,** which lists external links to Web sites for specific trading programs.

not need to release all the effluents allowed, it can hold the excess rights for future use or sell them to other dischargers. If it needs to release more than allowed by law, it must either abate more or buy the rights it needs from polluters holding an excess. Low-cost abaters will have excess rights, which they would be willing to sell at any price greater than or equal to their *MAC.* High-cost abaters would be willing to buy these rights as long as the selling price is less than or equal to their own *MAC.* Market forces will establish a price for permits such that all firms abate to the point where the associated level of *MAC* is equal across firms—a cost-effective solution.

Table 15.3 lists some of the applications of point-to-point tradeable effluent permit markets, including a frequently cited program designed to clean up the Fox River. In 1981, Wisconsin officials instituted this plan for one type of pollutant, biological oxygen demand (BOD) based effluents, being released to the Fox River. The initial allocation of permitted discharges was based on historical data and was valid for a five-year period. Officials expected that unequal *MACs* among polluting sources would provide the impetus for active trading of effluent reduction credits. One researcher estimated that this market potentially could yield a cost savings of about $7 million. However, the savings have not materialized, because only one trade has taken place.[32] Why? Several conditions may justify the absence of trading in this case.

For one thing, participants in the program still had to meet the federally established technology-based effluent limits. Because such standards constrain independent decision making, they likely limit the incentive mechanism of the trading program. Furthermore, two of the major classes of dischargers involved (namely, pulp and paper mills and municipal waste treatment plants) do not operate in competitive markets, which also dampens market forces. Finally, the key participants in the program did not fully support the plan, further limiting its potential for success.[33] These observations suggest that market conditions and existing regulatory constraints must be considered in the design of a successful trading program.

Market Approaches to Nonpoint Source Pollution

Using a market approach to reduce nonpoint source pollution is more complex because the problem itself is more difficult to identify. The dilemma becomes apparent if we reconsider such a pricing instrument as the effluent fee in this context. Implementation is problematic, because land runoff and atmospheric deposition arise from a number of sources. Since the contribution of each polluting source is indeterminate, an effluent fee cannot be implemented

[32] U.S. EPA, Office of Policy, Planning, and Evaluation (July 1992), p. 5-15, citing O'Neil et al. (December 1983) on the cost-savings estimation.
[33] Hahn (spring 1989).

fairly, at least not in the usual sense. If policymakers were to impose a uniform fee on all known contributors to a runoff problem, the result would be inequitable. The fees would be passed on to consumers of the goods produced at these sites through higher prices, and those individuals would bear most of the costs. However, the benefits of the reduced runoff would accrue primarily to individuals located near the waterway.[34] Because of these difficulties, indirect approaches are a viable alternative. Two examples are a **product charge** levied on the commodity whose usage adds to a known runoff problem and **effluent reduction trading** within a bubble.

Product Charge

product charge
A fee added to the price of a pollution-generating product based on its quantity or some attribute responsible for pollution.

A commonly cited example of using a **product charge** to control runoff is the use of a **fertilizer tax.** Because agricultural runoff is a major contributor of nonpoint source pollution, such a market instrument appears to have merit. Imposing a tax on fertilizer causes its effective price to rise, which in turn should reduce the quantity demanded. In theory, the optimal tax is one that covers the marginal external cost (*MEC*) associated with fertilizer use in the production of agricultural products. In practice, however, the operative issue is whether the demand response to the elevated effective price is sufficient to measurably reduce the runoff problem.

According to a 2001 EPA report, at least 46 states impose such a tax, but the rates apparently are too low to have much of an effect on fertilizer consumption. Tax rates range from less than $1.00 to $4.00 per ton—a very small proportion of the product's price, which is anywhere from $150 to $200 per ton. With taxes levied at no higher than a 2.5 percent rate (and usually much lower than that), the decline in consumption has been negligible.[35]

Even if fertilizer tax rates were elevated, the effect on quantity demanded would likely not be sufficient to significantly mitigate the associated pollution problem. Because fertilizer is an important input in crop production, its demand is likely to be price inelastic. In fact, studies of the use of such fees in Europe show that even with a tax rate of 50 percent, fertilizer usage is not significantly reduced.[36] Ultimately, the incremental benefit of these taxes tends to be not in the decline in fertilizer use but in the associated revenues that can be used to enhance environmental protection and fund research. About $14 million are raised each year from these taxes in the United States.[37]

Effluent Reduction Trading Within a Bubble

effluent reduction trading policy
Establishes an abatement objective for a watershed and allows sources to negotiate trades for rights to pollute.

effluent reduction credits
Tradeable permits issued if a polluter discharges a lower level of effluents than what is allowed by law.

effluent allowances
Tradeable permits issued up-front that give a polluter the right to release effluents in the future.

An alternative market-based approach to nonpoint source pollution is the use of an **effluent reduction trading policy** within a bubble (sometimes called a bowl, to signify its use for a watershed). The premise is exactly the same as what underlies a bubble to control air pollution. In simple terms, an overall abatement objective is established for a watershed, and then each polluting source is assigned an effluent limit. Once done, these sources are allowed to negotiate trades among themselves for rights to pollute, issued as **effluent reduction credits** or **effluent allowances. Effluent reduction credits** are issued if a polluter discharges *less* than what is permitted by law. **Effluent allowances** are issued up-front and give the polluter the right to release pollution in the future. (To review the distinction between credits and allowances, see Chapter 5.)

Polluting sources have an incentive to trade as long as their *MAC*s are not equal at current abatement levels. Ultimately, the market will establish a price for each allowance or credit such that all polluters abate to the point where their respective *MAC* levels are equal. This approach has been used to allow trading between point and nonpoint source polluters and between nonpoint sources, as shown in Table 15.4.[38] In theory, such trades should bring about

[34] Harrington, Krupnick, and Peskin (January–February 1985).
[35] U.S. EPA, Office of Policy, Economics, and Innovation (January 2001), p. 51.
[36] U.S. EPA, Office of Water (January 1992), p. 196.
[37] U.S. EPA, Office of Policy, Planning, and Evaluation (July 1992), p. 3-5. For a discussion of specific usage of such product charges, see U.S. EPA, Office of Water (January 1992), pp. 196–97.
[38] Case studies on some of these projects can be viewed online at **http://www.epa.gov/owow/watershed/hotlink.htm.**

Table 15.4 ***Examples of Point-to-Nonpoint and Nonpoint-to-Nonpoint Tradeable Effluent Permit Markets***

State	Water Body	Pollutant	Type of Trading
Colorado	Dillon Reservoir	Phosphorus	Point to nonpoint, nonpoint to nonpoint
Colorado	Boulder Creek	Ammonia, nutrients	Point to nonpoint
Colorado	Chatfield Basin	Phosphorus	Point to nonpoint
Maryland	Wicomico River	Phosphorus	Point to nonpoint
New York	Long Island Sound	Dissolved oxygen	Point to nonpoint
North Carolina	Tar-Pamlico Basin	Nitrogen, phosphorus	Point to nonpoint
Ohio	Honey Creek Watershed	Phosphorus	Point to nonpoint
Tennessee	Boone Reservoir	Nutrients	Point to nonpoint
Washington	Chehalis River Basin	Biological oxygen demand (BOD)	Point to nonpoint

Source: U.S. EPA, Office of Water (June 15, 2002).

efficiency gains, since abatement costs are typically much higher for point sources than they are for nonpoint sources. Perhaps the most frequently cited example of a point-nonpoint trading program is one designed to clean up the Dillon Reservoir.[39]

The Dillon Reservoir is an important resource to Colorado, because it supplies Denver residents with more than half their water supply. Over time, residential and industrial development in the area caused a serious pollution problem. The reservoir was receiving significant amounts of phosphorus, which was contributing to its decline. Officials tracked the damaging discharges and determined that over half was coming from nonpoint sources. The remainder was coming from 4 municipal treatment plants, 16 small treatment plants, and an industrial facility. Although these point sources were being controlled by discharge limits, they faced substantial marginal costs trying to achieve them. More important, the large municipal treatment facilities faced much higher marginal abatement costs than the nonpoint sources in the watershed.

The Dillon Reservoir problem was a textbook case of unequal marginal abatement costs (*MACs*) across polluting sources where cost savings could be realized through a trading scheme.[40] With the EPA's support, officials in the area instituted a point-nonpoint source trading program. Initial allocations of phosphorus discharge limits were set, and trades were allowed between point and nonpoint sources at a 2 to 1 ratio.[41]

Despite the classic conditions at the Dillon Reservoir, active trading has been minimal. There are, however, some logical explanations for this outcome. For one thing, point sources

[39] See U.S. EPA, Office of Policy, Planning, and Evaluation (July 1992), pp. 5-15, 5-16, and U.S. EPA, Office of Policy, Economics, and Innovation (January 2001), p. 102, for more information on the Dillon Reservoir project, from which much of the following discussion is drawn.

[40] An EPA study suggests that there was a potential for over $1 million in cost savings. See Elmore et al. (1984), cited by Harrington, Krupnick, Peskin (January–February 1985). For a discussion of why these savings might not be realized, see Hahn (spring 1989).

[41] A trading ratio of 2 means that point sources are allowed 1 unit of effluent for every 2 they purchase from nonpoint sources. A ratio greater than 1 favors point source reductions and discourages trades. Conversely, a ratio less than 1 favors reductions by nonpoint sources and encourages trades. For an analysis of these ratios and an overview of point-nonpoint source trading, see Letson (spring 1992). A more rigorous study is given by Malik, Letson, and Crutchfield (November 1993).

in the Dillon area have become more efficient abaters over time. As they have, their abatement costs have declined, which in turn lessens their incentive to trade for pollution rights. Also, the region's economic development has begun to slow down, which necessarily lessens the amount of discharges and hence the need for market trades. Finally, the 2 to 1 trading ratio, although well intentioned, discourages trading, because point sources are allowed only 1 unit of effluent for every 2 credits they buy. Interestingly, a few trades have been proposed between nonpoint sources at the Dillon Reservoir region, an outcome that was not anticipated by the program planners.[42]

It is still too soon to predict the long-term success of such trading programs, because they are still relatively new. Nonetheless, the EPA estimates that potential compliance cost savings from effluent trading markets could be significant, within a range of $658 million to $7.5 billion. Of this amount, trading among point sources is expected to save $8.4 million to $1.9 billion, with point-nonpoint source trades estimated to save $611 million to $5.6 billion.[43] Hence, the experience at the Dillon Reservoir may be valuable to other officials planning to establish similar markets and achieve some measure of cost savings in their own communities.

Conclusions

An ongoing source of debate is the nation's dependence on a command-and-control approach to achieving water quality and the methods it uses to guide critical decisions. Since 1972, U.S. water control policy has been rooted in the use of technology-based effluent limitations. There have been documented delays in setting these limits, which in turn have slowed states' efforts in achieving receiving water quality standards. Beyond the time problem, there is a lack of guidance in the law as to how the technological and economic determinants of these limits are to be considered relative to one another. An efficient abatement level is not likely, given the absence of any mandate to use benefit-cost analysis in defining these standards. Furthermore, because the effluent standards are applied uniformly within major groups of polluters, a cost-effective solution is also unlikely.

Another costly policy decision has been the federal funding of POTW construction. These funds have served primarily to displace rather than supplement local spending, suggesting that observed improvements in waste treatment are not directly attributable to the billions of dollars spent at the federal level for this program. In addition, by removing much of the cost burden from municipalities, local officials had little incentive to make cost-conscious decisions in building waste treatment facilities. Policymakers have recognized the problem and have revised the law in an attempt to restore the incentive mechanisms removed by the long-term grant program.

Beyond this dependence on command-and-control instruments, there are gaps in the overall policy approach. Nonpoint polluting sources, such as agricultural and urban runoff, are major contributors to water contamination, yet the United States has only begun to address this highly complex issue. The 1987 amendments initiated the Nonpoint Source Management Program, which recently has been updated through a collaborative effort involving state and federal entities. At issue is whether or not there are sufficient resources allocated to improving monitoring technologies and to advancing the knowledge base in nonpoint source pollution.

In the past, the federal government virtually ignored market incentives in implementing water quality regulations. Economic instruments such as effluent fees and permit trading programs may be viable alternatives to the nation's current standards-based approach. Perhaps the use of economic incentives will bring about significant costs savings and cleaner water for the nation. Clearly, U.S. water quality policy development has been and continues to be an

[42] U.S. EPA, Office of Policy, Planning, and Evaluation (July 1992), pp. 5-15, 5-16. The reader should consult Harrington, Krupnick, and Peskin (January–February 1985), particularly pp. 29–30, for a clear explanation of why trading might be more likely among nonpoint sources and other determinants of trading activity.
[43] U.S. EPA (n.d.), as cited in U.S. EPA, Office of Policy, Economics, and Innovation (January 2001).

evolutionary process. Environmental controls must be continuously evaluated and revised to reflect new technologies, changes in natural conditions, and innovative proposals for better solutions.

Summary

- Point sources are subject to technology-based effluent limits that vary by the type of polluting source, the age of the facility, and sometimes the type of contaminant released.
- Effluent limits for direct industrial dischargers and for publicly owned treatment works (POTWs) are communicated through the National Pollutant Discharge Elimination System (NPDES).
- The effluent limitations for new industrial point sources are more stringent than those for existing sources.
- For existing industrial point sources, two types of limits are applicable, based on the type of pollutant released: limits for controlling nonconventional and toxic pollutants and limits for controlling conventional pollutants.
- There have been delays in developing and revising the guidelines that serve as the basis for the effluent limitations. These limitations are not aligned with U.S. objectives, because they are based on what is technologically feasible rather than on what is needed to achieve water quality.
- The EPA has been criticized for not imposing zero effluent limits where achievable, in accordance with the nation's zero discharge goal.
- The Clean Water Act does not mandate that the standards be set to maximize net benefits, which prevents an efficient solution. Also, they are applied uniformly across dischargers within identified groups, which disallows a cost-effective outcome.
- Before the 1987 revisions, the Clean Water Act outlined a federal grant program for the construction of POTWs. This was replaced in 1987 with the State Revolving Fund (SRF) program, which authorized capitalization grants to states for setting up a revolving fund for POTW projects.
- A criticism of the POTW grant program is that most of the federal monies only displaced local funding for new construction. Also, federal aid was aimed only at construction of POTWs, with no provisions to ensure proper operation and maintenance. An economic issue is that the federal cost share shifted most of the expenditures away from local governments, leaving them with little incentive to minimize costs.
- Under the Nonpoint Source Management Program initiated in 1987, states must prepare assessment reports, develop management programs designating best management practices (BMP), and implement the programs.
- In 1995, EPA and state officials collaborated to strengthen existing nonpoint source efforts and devised a new Section 319 program and grant guidance.
- Factors supporting states' responsibility for nonpoint sources are the location-specific nature of nonpoint source pollution and the political sensitivity of federal controls on land use. On the opposite side are information deficiencies that hinder states' ability to carry out their responsibilities and the potential problem of inconsistent control efforts.
- At the federal level, far less funding has been allocated to nonpoint source pollution than to point source contamination. There also have been conflicts between water quality objectives and the goals of other regulations and programs, such as those directed by the USDA.
- The United States has begun to consider the potential cost savings of using market-based approaches to water quality policy. Some local governments are experimenting with these types of incentive-based policy tools.

- An effluent charge can be based on either the volume or the type of effluent released. Australia, Belgium, France, and Germany have instituted effluent charges to control water contamination.
- Tradeable effluent permit markets could be implemented using the existing NPDES by issuing the permit as a set of tradeable, per unit rights to pollute. Potential gains from trading exist as long as marginal abatement costs differ across polluting sources.
- Product charges can be levied on a commodity whose usage adds to a known runoff problem, such as fertilizers that contribute to agricultural runoff. An alternative instrument is an effluent reduction trading policy within a bubble.

Key Concepts

technology-based effluent limitations
National Pollutant Discharge Elimination System (NPDES)
water quality–related limitations
federal grant program
State Revolving Fund (SRF) program
Nonpoint Source Management Program
best management practices (BMP)
volume-based effluent fee
pollutant-based effluent fee
tradeable effluent permit market
product charge
effluent reduction trading policy
effluent reduction credits
effluent allowances

Use the Key Concepts listed above to begin your search for additional articles and information using the InfoTrac College Edition database.

Review Questions

1. a. Evaluate the use of technological attainability as the primary determinant of the effluent limitations.
   b. Economically analyze the use of *uniform* technology-based effluent limitations.
2. Discuss any incentive and disincentive implications of the federal assistance programs for POTW construction.
3. As an alternative to standards, one policy proposal is the use of permit trading among point sources of water pollution. Give the major reason why this is advantageous (a) from an economic perspective and (b) from an environmental perspective.
4. According to a former EPA administrator,

   > We've done the easy part of controlling pollution at the end of the pipeline. For the first time ever, we are tackling the hard part—the control of polluted runoff, which is the biggest remaining barrier we face in keeping the nation's waters clean.[44]

   Working as part of the EPA's team, your assignment is to give a clear, objective presentation of a market-based approach to reduce nonpoint source pollution. Include in your discussion the theoretical issues and any practical concerns that must be addressed before implementing your proposal.
5. a. To help fight the problem of nonpoint source pollution associated with agricultural runoff, your state is contemplating charging an annual fee of $500 to every seller of pes-

[44]U.S. EPA, Office of Communications, Education, and Public Affairs (February 22, 1994b).

ticides. If this fee is to achieve an efficient solution, state specifically according to externality theory (a) what the $500 fee must represent and (b) in which market.

b. Illustrate graphically, labeling where and how the fee is imposed.

6. Assume for simplicity that there are two identified point sources discharging chemical wastes into a local water body. Currently, each source releases 30 units of effluent, for a total of 60 units. To improve water quality, suppose that the government sets an aggregate abatement standard of 30 units. The two polluters' abatement cost functions are

$$\text{Point source 1:} \quad TAC_1 = 10 + A_1^2, \quad MAC_1 = 2A_1.$$

$$\text{Point source 2:} \quad TAC_2 = 20 + 2A_2^2, \quad MAC_2 = 4A_2.$$

a. Suppose the government allocates the abatement responsibility equally across the two point sources so that each source must abate 15 units of effluent. Graphically illustrate this policy, and explain why this abatement allocation does not yield a cost-effective solution. Support your answer numerically.

b. What cost condition is required for the government's abatement allocation to be cost-effective?

c. Suppose that, instead of using an abatement standard, the government institutes an effluent fee of $40 per unit of pollution. How many units of pollution would each point source abate? Is the $40 fee a cost-effective strategy for meeting the 30-unit abatement standard? Explain.

Additional Readings

Downing, Donna, and Stuart Sessions. "Innovative Water Quality-based Permitting: A Policy Perspective." *Journal of Water Pollution Control Federation* 57 (May 1985), pp. 358–65.

Joeres, Erhard F., and Martin H. David, eds. *Buying a Better Environment: Cost-Effective Regulation Through Permit Trading.* Madison, WI: University of Wisconsin Press, 1983.

Knopman, Debra S., and Richard A. Smith. "20 Years of the Clean Water Act: Has U.S. Water Quality Improved?" *Environment* 35 (January/February 1993), pp. 17–20, 34–41.

Krupnick, Alan J. "Reducing Bay Nutrients: An Economic Perspective." *Maryland Law Review* 47 (1988), pp. 452–80.

Magat, Wesley A., and W. Kip Viscusi. "Effectiveness of the EPA's Regulatory Enforcement: The Case of Industrial Effluent Standards." *Journal of Law and Economics* 33 (October 1990), pp. 331–60.

Oates, Wallace E., and Dianna L. Strassmann. "Effluent Fees and Market Structure." *Journal of Public Economics* 24 (1984), pp. 29–46.

O'Neil, William B. "The Regulation of Water Pollution Permit Trading Under Conditions of Varying Streamflow and Temperature." In E. Joeres and M. David, eds., *Buying a Better Environment: Cost-Effective Regulation Through Permit Trading,* pp. 219–31. Madison, WI: University of Wisconsin Press, 1983.

Smith, Stephen. *Green Taxes and Charges: Policy Practices in Britain and Germany.* London, UK: Institute for Fiscal Studies, November 1995.

U.S. Environmental Protection Agency, Office of Water. *The Clean Water State Revolving Fund and the Clean Water Action Plan.* Washington, DC, March 1998.

U.S. General Accounting Office. *Water Pollution: Improved Monitoring and Enforcement Needed for Toxic Pollutants Entering Sewers.* Washington, DC, April 1989.

http:// Related Web Sites

Clean Water Act, Section 319	**http://www.epa.gov/owow/nps/cwact.html**
Draft Framework for Watershed-Based Trading, U.S. EPA, Office of Water (May 1996)	**http://www.epa.gov/owow/watershed/framwork.html**
Effluent Limitations Guidelines and Standards	**cfpub.epa.gov/npdes/techbasedpermitting/effguide.cfm?program_id=15**
Effluent Trading Case Studies	**http://www.epa.gov/owow/watershed/hotlink.htm**
Effluent Trading in Watersheds Policy Statement, U.S. EPA, Office of Water (May 14, 2001)	**http://www.epa.gov/owow/watershed/tradetbl.html**
Effluent trading links	**http://www.epa.gov/owow/watershed/tradelinks.html**
NPDES permit program	**http://cfpub.epa.gov/npdes/**
Protecting and Restoring America's Watersheds, U.S. EPA, Office of Water (June 2001)	**http://www.epa.gov/owow/protecting/**
SRF Program and Construction Grants Program	**http://www.epa.gov/owm/cwfinance/cwsnf**
State-EPA Nonpoint Source Partnership	**http://www.epa.gov/owow/nps/partnership.html**
U.S. EPA, Nonpoint Source Pollution	**http://www.epa.gov/owow/nps/**
U.S. EPA, Office of Wastewater Management	**http://www.epa.gov/owm**
U.S. EPA, Office of Water	**http://www.epa.gov/ow/**
U.S. EPA, Office of Wetlands, Oceans, and Watersheds	**http://www.epa.gov/owow/**
U.S. EPA, Watershed Protection	**http://www.epa.gov/owow/watershed**
Watershed Approach Framework	**http://www.epa.gov/owow/watershed/framwork.html**
Watershed Information Network	**http://www.epa.gov/win/**

A Reference to Acronyms and Terms in Water Quality Control Policy

Environmental Economics Acronyms

MAC	Marginal abatement cost of a single polluter
MAC_{mkt}	Marginal abatement cost of all polluters
MCE	Marginal cost of enforcement
MEC	Marginal external cost
MEF	Marginal effluent fee
MSB	Marginal social benefit of abatement
MSB_{low}	Marginal social benefit of abatement in an area with a low population
MSB_{high}	Marginal social benefit of abatement in an area with a high population
MSC	Marginal social cost of abatement

Environmental Science Terms

BOD	Biological oxygen demand
mg/l	Milligrams per liter
ppm/l	Parts per million per liter

Environmental Policy Acronyms

ASIWPCA	Association of State and Interstate Water Pollution Control Administrators
BMP	Best management practices
CWA	Clean Water Act
FWPCA	Federal Water Pollution Control Act
NPDES	National Pollutant Discharge Elimination System
POTWs	Publicly owned treatment works
SRF	State Revolving Fund

Chapter 16

Protecting Drinking Water: The U.S. Safe Drinking Water Act

In 1993, 400,000 residents of Milwaukee, Wisconsin, became ill from a waterborne disease that was transmitted through the city's drinking water. Ultimately, more than 40 people lost their lives. Beyond the human suffering, an estimated $37 million in wages and productivity was lost. How could a city's water supplies become so polluted? What went wrong? There are several theories—human error, aging facilities, illegal discharges, and monitoring failures among them. Another is that the disease outbreak was a result of poorly defined federal controls. The claim is that regulations divert attention and resources to lower priority problems, which in this case caused officials to miss the warning signs of the contamination.[1]

Whatever the reason, the Milwaukee incident sparked national concern about the potential for similar problems across America. It also triggered skepticism and criticism of U.S. drinking water policy. What regulations are in place to protect drinking water supplies? How effective is this part of U.S. water quality control policy?

Drinking water supplies depend on ground- and surface water resources, both of which are protected by the Clean Water Act. However, this act requires a level of water quality to support aquatic life and recreational uses—a level that calls for far less stringent standards than those necessary to support safe drinking. Furthermore, although the Clean Water Act is comprehensive, its protection of groundwater resources is limited. Groundwater is the source of drinking water for nearly half of the U.S. population and for 99 percent of rural Americans.[2] Hence, regulations beyond those in the Clean Water Act are necessary to minimize the risks of contaminated drinking water.

In this chapter, we focus on the nation's drinking water policy, which is governed by the Safe Drinking Water Act. Like the Clean Water Act, this legislation is based primarily on a standards-based, command-and-control policy approach—a characteristic we evaluate from an economic perspective. We begin with a brief review of the origin and evolution of drinking water standards and laws. From this point, we discuss the objectives of the Safe Drinking Water Act and the standards used to define drinking water quality for the nation. Once done, we analyze the federal standard-setting process, the performance of state and local governments in meeting these standards, and the cost implications for society. We then consider the economics of pricing water supplies. Once again, a list of acronyms and terms is provided at the end of the chapter.

The Evolution of U.S. Safe Drinking Water Legislation

Throughout the early history of the United States, deaths caused by diseases such as cholera and typhoid fever were not uncommon. For many years, no one realized that these diseases were caused by bacteria living in water supplies used for drinking. It was not until 1854 that an epidemiologist named John Snow linked cholera to contaminated water. Another 30 years passed before specific bacteria were identified as the cause of waterborne disease. Armed with this evidence, the federal government enacted the Interstate Quarantine Act of 1893. As

[1] Velma Smith (summer 1994).

[2] Solley, Pierce, and Perlman (1988), as cited in U.S. EPA, Office of Water (June 2000), p. 158.

Table 16.1 *Evolution of U.S. Drinking Water Legislation*

Legislation	Major Provisions
Interstate Quarantine Act of 1893	Authorized the surgeon general to issue regulations to prevent the spread of disease; led to the first U.S. water regulation passed in 1912 prohibiting the use of a common drinking cup on interstate carriers.
Public Health Service Act of 1912	Provided for the U.S. Public Health Service to set bacteriological standards for drinking water.
Revisions to Public Health Standards in 1925, 1942, 1946, 1962	Strengthened existing standards; required more stringent testing procedures; established maximum allowable concentrations for certain substances.
Safe Drinking Water Act of 1974	Authorized the EPA to establish drinking water standards; controlled underground injection activities and protected sole-source aquifers.
Safe Drinking Water Amendments of 1986	Accelerated procedures for the standard-setting process; provided greater protection of groundwater sources of drinking water; banned the use of lead in public drinking water systems.
Safe Drinking Water Amendments of 1996	Integrates risk assessment and benefit-cost analysis into standard-setting procedures; authorizes a $1 billion per year Drinking Water State Revolving Fund; promotes prevention through source water protection and better management.

Sources: Larson (1989); McDermott (1973); Dzurik (1990); U.S. Congress, OTA (October 1984), chap. 3, pp. 64–75; Wolf (1988), chap. 4, pp. 133–37; U.S. Congress (August 6, 1996).

Table 16.1 shows, this was the first federal law in the evolution of U.S. drinking water controls. This act authorized the surgeon general to issue regulations to prevent the spread of disease from any foreign country to the United States or from one state to another. In 1912 the first water-related regulation was passed—a ruling that outlawed the use of a common drinking cup on interstate carriers.[3]

Setting Standards to Protect Drinking Water

Realizing that banning common drinking cups was of little value if the water itself was contaminated, federal authorities passed the Public Health Service Act of 1912—the first U.S. law to specifically call for drinking water health standards. As authorized, these standards were limited in scope—targeting only contaminants capable of spreading communicable waterborne diseases. Hence, in 1914 the Public Health Service set bacteriological standards for drinking water. Recognizing the potential health risks of other types of substances, the standards were expanded in 1925 to consider chemical characteristics like lead and copper. Over the next several decades, more revisions followed, each time adding more structure to the law and strengthening existing standards and controls.[4]

In the 1960s, there was a turn of events that was to change the direction of U.S. drinking water policy. In 1963, the Public Health Service formed an advisory committee to evaluate the existing drinking water standards and make appropriate recommendations. In 1967, the committee recommended that toxic contaminants be addressed—an advisory that would

[3] McDermott (1973), as cited in Larson (1989).
[4] McDermott (1973), as cited in Larson (1989); Dzurik (1990), p. 61.

have led to limits on such chemicals as chlordane and DDT. However, it was argued that setting standards on toxics went beyond the jurisdiction of the Public Health Service, whose authority was limited to controlling communicable diseases.

Although the responsibility may have been misplaced, the concern was appropriate. Chemical usage was on the rise, pesticides were being detected in groundwater supplies, and the general public was becoming alarmed. A federal committee was formed in 1969 to evaluate the 1962 standards and make recommendations. It completed its work in 1971—one year after the EPA had been established. The newly established agency reviewed the committee's report, which ultimately became the cornerstone of new regulations. These were issued under the Safe Drinking Water Act (SDWA) of 1974, at which point the responsibility of protecting drinking water shifted from the Public Health Service to the EPA.[5]

The Safe Drinking Water Act (SDWA) of 1974[6]

Capitalizing on the broader jurisdiction of the EPA, the Safe Drinking Water Act (SDWA) was aimed at protecting drinking water from *any* contaminant that could threaten human health or welfare—not just bacteria responsible for communicable disease. Primary standards were to be established for organic and inorganic chemicals, radionuclides, and microorganisms identified as having adverse effects on public health. Secondary standards, which were actually just guidelines, were to be defined to protect welfare by controlling such characteristics as taste and odor.

Not unlike other early environmental laws, the SDWA of 1974 was not without problems that ultimately caused major delays. It originally called for interim health standards to be set within 180 days. These were to be revised through a coordinated effort between the EPA and the National Academy of Sciences (NAS), at which time they would become final regulations. These regulations were to establish pollution limits, called maximum contaminant levels (MCLs), that the NAS believed would protect human health. The complex procedures coupled with the EPA's alleged lack of initiative delayed the process considerably. In fact, by the mid-1980s, relatively few contaminants had been regulated by the EPA, and most interim standards still had not been revised. These procedural problems and uneven compliance by public water utilities led to the 1986 Amendments.

The Safe Drinking Water Act Amendments of 1986

The 1986 Amendments expanded federal controls on drinking water and corrected some of the failings of the original SDWA. To accelerate procedures for the standard-setting process, maximum contaminant levels (MCLs) were required for a list of 83 contaminants according to a strict timetable over the 1987 to 1989 period. Responding to growing concerns about lead contamination, the amendments prohibited all future use of lead pipe and solder in public drinking water systems, now known as the "lead ban." Other revisions attempted to provide better protection of underground drinking water supplies.

Although the SDWA of 1986 addressed some important concerns, most agreed that further changes were needed. Numerous proposals for reform were discussed as part of the most recent reauthorization of this act. Ultimately, Congress passed the SDWA Amendments of 1996—the law currently in force.

Drinking Water State Revolving Fund (DWSRF) Authorizes $1 billion per year from 1994 to 2003 to finance infrastructure improvements.

The Safe Drinking Water Act Amendments of 1996[7]

In an attempt to correct existing problems with the nation's safe drinking water legislation, President Clinton signed into law the 1996 SDWA Amendments in August 1996. An important element of this legislation is the establishment of a **Drinking Water State Revolving Fund (DWSRF).** Recognizing the success of the POTW State Revolving Fund program for

[5] Larson (1989).

[6] This discussion is drawn from U.S. Congress, Office of Technology Assessment (OTA) (October 1984), vol. 1, chap. 1, pp. 8–9, and chap. 3, pp. 64–75; Wolf (1988), chap. 4, pp. 133–37.

[7] For online information on the implementation of the SDWA Amendments of 1996, visit **http://www.epa.gov/ogwdw/sdwa/sdwa.html.**

publicly owned treatment works (POTWs), Congress initiated a similar plan for drinking water as part of these new amendments. This fund authorizes $1 billion per year for the 1994 to 2003 period to finance infrastructure improvements.[8]

Another important change addresses standard-setting procedures. The new amendments repeal the provisions that required the EPA to design and implement standards for 25 additional contaminants every 3 years. Replacing these rulings is a new requirement that drinking water standards be based on sound risk assessment and benefit-cost analysis. In fact, the law specifically states [emphasis added]:

> in considering the appropriate level of regulation for contaminants in drinking water, *risk assessment,* based on sound and objective science, and *benefit-cost analysis* are important analytical tools for improving the efficiency and effectiveness of drinking water regulations to protect human health.[9]

This represents a significant revision, integrating an economic decision rule into the standard-setting process. There are also new provisions promoting prevention through better management and source water protection rather than relying solely on remediation.

Directives of the Safe Drinking Water Act

Interestingly, the SDWA has no statutory objectives per se. However, according to the House of Representatives, the primary purpose of the act is:

> to assure that water supply systems serving the public meet minimum national standards for protection of public health.[10]

To accomplish this goal, the SDWA of 1974 authorized three key **directives:**

- To establish drinking water health standards applicable to public water systems
- To control underground injection activities that may threaten drinking water[11]
- To designate **sole-source aquifers** (i.e., those that are the only source of drinking water for a given area) to protect aquifer recharge areas[12]

sole-source aquifers Underground geological formations containing groundwater that are the only supply of drinking water for a given area.

The 1986 Amendments added another directive:

- To protect groundwater supplies used for drinking

Relating the SDWA to Other Laws

It is important to have some perspective about how these directives relate to those of other U.S. water quality laws. First, compared to the Clean Water Act, which has a comprehensive span of control, notice that the Safe Drinking Water Act (SDWA) is more focused. Its aim is to define, monitor, and enforce whatever standards are needed to assure that water drawn from the tap is safe for human consumption.[13] Also, note the phrase "drawn from the tap," suggesting that bottled water is not regulated through the SDWA. Assuring the potability of this

[8] To learn more about the DWSRF, visit **http://www.epa.gov/ogwdw/dwsrf.html.**
[9] Pub. L. 104-182, § 3.7, 110 Stat. 1613 (codified at 42 U.S.C. § 300f).
[10] U.S. Congress, OTA (October 1984), p. 74, citing U.S. House of Representatives (1974).
[11] This directive controls various underground wells, such as those used to inject wastes or fluids from oil production into the ground, to assure that these materials do not contaminate underground drinking water supplies.
[12] U.S. Congress, OTA (October 1984), p. 74.
[13] For more information about water drawn from the tap, consult U.S. EPA, Office of Water (July 1997), which is also available on the Internet at **http://www.epa.gov/ogwdw/wot/wot.html.**

Application

16.1 *Who Regulates the Quality of Bottled Water?*

Sales of bottled drinking water have grown into a thriving $4–$5 billion market. To a large extent, these impressive revenues reflect consumers' uncertainty about the quality of public drinking water. Some are concerned about aesthetic issues, and others about more serious threats like lead contamination. Whatever the cause, the response of many consumers is to substitute bottled water for tap water.

Implicit in households' decisions to buy bottled water is the assumption that this substitute commodity is of higher quality than tap water. For this assumption to be valid, it must be the case that bottled water is subject to more stringent standards than is public drinking water or that suppliers of bottled water are closer to compliance with some set of universal standards than their public counterparts. Is either of these scenarios correct? To answer this question, we need to compare the regulations affecting private and public drinking water supplies.

To begin, the two types of water supplies are controlled by different entities. Public drinking water supplies are controlled by the Safe Drinking Water Act through standards administered by the EPA and state authorities. Bottled water quality falls within the jurisdiction of the Food and Drug Administration (FDA). The reason is that bottled water is considered a "food" and, as such, is part of the FDA's charge.

To assure consumer safety, the FDA requires that bottled water products be produced in compliance with FDA Good Manufacturing Practices. All products must be clean and safe for human consumption and must be processed and distributed under sanitary conditions. Furthermore, according to a 1978 agreement, the FDA must adopt the EPA's public drinking water standards for bottled water. The FDA also has its own standards, dealing with aesthetics and health concerns. As a further precaution, domestic bottled water producers who distribute their products interstate are subject to periodic, unannounced site inspections by the FDA and by state health officials. There is also some measure of self-regulation within the industry.

What does all this mean to the consumer? It is true that there are federal and state regulations in place to protect the quality of bottled water, and indeed there is consensus that bottled water is safe for human consumption. However, there is no reason to assume that it is safer to drink than ordinary tap water, since the same health standards are applicable to both.

Beyond this issue, there is also debate about whether bottled water is being unfairly or deceptively marketed. For example, it has been alleged that some bottled water is nothing more than ordinary tap water. Yet, its price can be as much as 1,000 times higher than the price of public water supplies. Further allegations have been made about misleading information about the characteristics or sources of bottled water products. To address these problems, the FDA established new labeling requirements for bottled water that will provide consistency to such designations as "mineral," "distilled," or "sterile" and will require the identification of bottled water drawn from municipal supplies. Truth in labeling is not a new concept, but its application to the increasingly popular bottled water industry may be long overdue.

Sources: U.S. EPA, Office of Drinking Water (n.d.); Kirchhoff (January 1, 1993); U.S. EPA (March 1991); Sullivan (August 14, 1995).

water falls within the jurisdiction of the Food and Drug Administration—a matter discussed in Application 16.1.

Second, it is important to realize that the goals of the SDWA are not independent of the Clean Water Act, which controls *all* water resources, including those used for drinking. Since the rulings of the Clean Water Act are not stringent enough to assure the safety of drinking water, the SDWA imposes the tougher standards needed to assure potability. However, it does rely implicitly on the Clean Water Act to achieve a baseline level of water quality and to control the effluents of polluting sources.

Figure 16.1 Groundwater Contaminants Prioritized by States

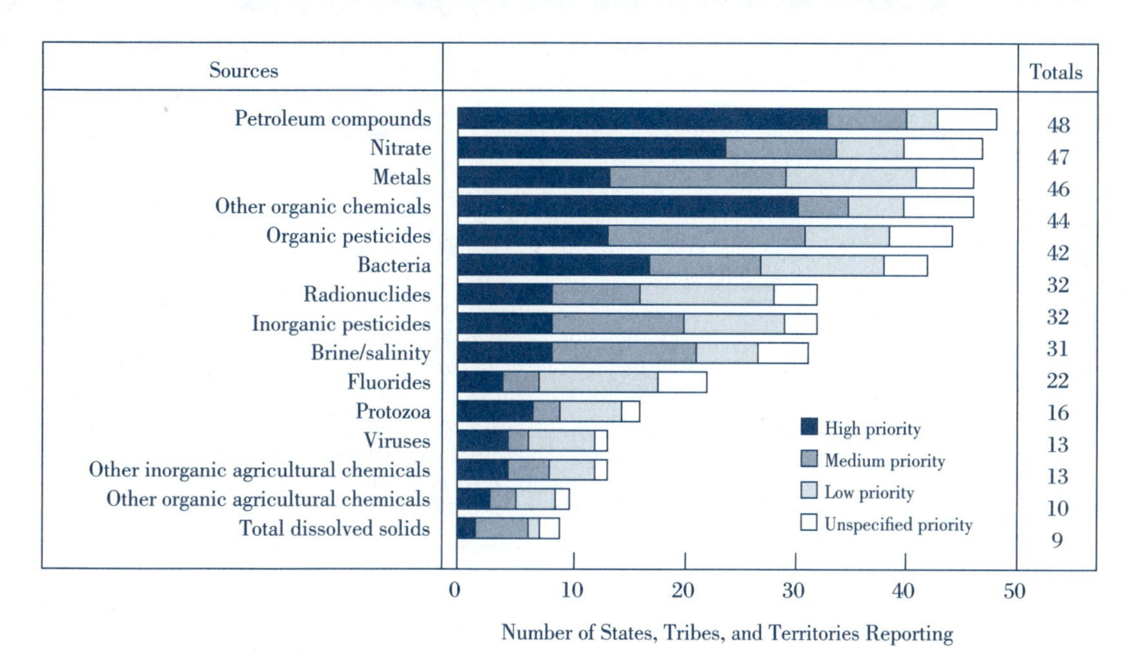

Source: 1994 Section 305(b) reports submitted by states, tribes, and territories, as presented in U.S. EPA, Office of Water (December 1995), p. 108, fig. 6.6.

Third, the groundwater protection added in 1986 is indicative of the fragmentary U.S. approach to controlling groundwater contamination. There are many statutes implemented by different government agencies aimed at this important natural resource—arguably an ineffective approach with no unified oversight.[14] This outcome evolved from the false belief that groundwater was naturally protected from contamination because of its location within layers of soil and rock. In the 1970s, this assumption was called into question when pesticides were discovered in U.S. groundwater supplies. Since then, every state in the nation has discovered some amount of pollutants in its groundwater.[15]

The most commonly observed groundwater contaminants are shown in Figure 16.1. Public concern prompted a call for more aggressive legislative controls, and existing laws were amended in an attempt to address the problem. However, no single legislative act has ever been passed whose main objective is to protect groundwater resources.[16]

Pollutants Controlled Under the Safe Drinking Water Act

The Safe Drinking Water Act controls all types of contaminants that may threaten human health and not just those linked to communicable disease. Its rulings refer to "contaminants" as any physical, biological, or radiological substances in water. The 1986 Amendments were much more specific, listing 83 contaminants for which drinking water standards are to be set,

[14] Among the laws with some provisions to protect groundwater are the Comprehensive Environmental Response, Compensation, and Liability Act; the Federal Insecticide, Fungicide, and Rodenticide Act; the Resource Conservation and Recovery Act; and the Toxic Substances Control Act. For more detail on this important issue, the reader should consult U.S. Congress, OTA (October 1984), vol. 1, chap. 3.

[15] U.S. EPA, Office of Pesticides and Toxic Substances (October 1991), pp. 1–2; U.S. Congress, OTA (October 1984), vol. 1, p. 3.

[16] U.S. Congress, OTA (October 1984), vol. 1, pp. 63, 73–75.

Application

16.2 *Pesticides and Other Agrichemicals in U.S. Groundwater*

Surveys conducted by federal, state, and local governments in the late 1970s revealed some troubling statistics about the pollution of U.S. groundwater. As more contamination was discovered, monitoring efforts intensified, as did research endeavors to determine the magnitude of the problem. Recognizing a need to assimilate the accumulating data, the EPA established the Pesticides in Ground-Water Data Base, using information gathered from some 150 monitoring studies. In 1988, the agency published the results in the *Interim Ground-Water Data Base Report.* One of its findings was that 46 different pesticides had been found in the groundwater supplies of 26 states—all of which were believed attributable to conventional agricultural activities. Although the database was valuable in generally assessing the problem, it nonetheless was a collection of information gleaned from diverse sources. Hence, the EPA saw the need to establish a nationally sponsored baseline of groundwater contamination data from which to monitor progress over time. It was this recognition that motivated the National Pesticide Survey of Drinking Water Wells conducted between 1988 and 1990.

The national survey analyzed samples from 1,349 drinking water wells across all 50 states. In November 1990, the agency released the survey's Phase I report, which asserted that approximately 10.4 percent of the community wells and 4.2 percent of the private wells in the sample had a detectable amount of at least one pesticide. Based on these sample data, the EPA inferred that nearly 10,000 community drinking wells and about 446,000 private domestic wells in the United States were contaminated by at least one pesticide. The survey report went on to indicate, however, that less than 1 percent of private wells and no community drinking water wells had contamination levels high enough to threaten human health.

Phase II of the EPA's survey report was completed in 1992 and is a statistical analysis of the findings in Phase I. A number of statistical associations were identified, including a link between pesticide contamination and agricultural activity. The analysis also found that the transport of chemicals to well water is affected by numerous factors, including precipitation and the proximity of surface waters to drinking water wells.

Overall, the EPA contends that while pesticide contamination of groundwater exists, most of the nation's supplies are relatively safe, at least in the short term. In those local areas where this is not the case, the EPA asserts that every effort must be made to address the problem and the associated health risks. The agency does acknowledge, however, that detectable levels of pesticides in well water on a widespread scale may suggest a threat to groundwater quality in the long run. In this regard, the survey data establish the baseline needed to help monitor changes in contaminant levels over time. Such trend data should be useful in formulating policy initiatives and designing national strategies to protect U.S. groundwater supplies.

Sources: U.S. EPA, Office of Pesticides and Toxic Substances (September 1990); U.S. EPA, Office of Pesticides and Toxic Substances (October 1991); U.S. EPA, Office of Water, Office of Pesticides and Toxic Substances (winter 1992).

priority contaminants
Pollutants for which drinking water standards are to be established based on specific criteria.

with new contaminants to be drawn from a list of **priority contaminants** devised by an advisory group, published, and updated regularly.

The following are among the criteria for selecting **priority contaminants:**

- The contaminant must be known or expected to occur in a public water system.
- The contaminant may have an expected adverse effect on human health.

Consideration of which substances are identified on the priority list shall include but not be limited to contaminants identified in the Comprehensive Environmental Response, Compensation, and Liability Act (also known as Superfund), and pesticides registered under the Federal Insecticide, Fungicide, and Rodenticide Act. More on the problem of pesticides and other agricultural chemicals in drinking water is given in Application 16.2.

Still more changes came with the 1996 Amendments, calling for **risk assessment** and **benefit-cost analysis** to govern which contaminants are to be regulated and the standard-setting process itself. In addition to developing new rules and guidance for priority contaminants, a new infrastructure has been established for future decisions, called the **National Contaminant Occurrence Database (NCOD).** This is a collection of data on both regulated and unregulated contaminants that may occur in U.S. public water systems. The purpose of the NCOD is to facilitate identification and selection of contaminants to be controlled in the future. For more information, visit **http://www.epa.gov/ncod/** on the Internet.

Setting Standards to Define Safe Drinking Water

The most important directive of the SDWA is to set standards that define drinking water quality for the nation. Under the law, there are two types of standards—**primary standards** to protect human health and **secondary standards** to protect public welfare.[17] The distinction is an important one, both in how each is defined and in how each is implemented.

Establishing National Primary Drinking Water Regulations (NPDWRs)

The emphasis of the SDWA is clearly on the primary or health standards for drinking water. These are applicable only to public water systems and are to be implemented *uniformly* throughout the country. More formally, these standards are called **National Primary Drinking Water Regulations (NPDWRs).** Each regulation consists of three parts:

- **Maximum contaminant level goal (MCLG)**
- **Maximum contaminant level (MCL)**
- **Best available technology (BAT)** for public water supply treatment[18]

National Primary Drinking Water Regulations (NPDWRs)
Health standards for public drinking water supplies that are implemented uniformly.

Setting the Goal: The MCLG

One of the key elements of a primary drinking water regulation is the **maximum contaminant level goal (MCLG).** This defines the level of a pollutant at which no known or expected adverse health effects occur, allowing for an adequate margin of safety. MCLGs are based on data obtained in the risk assessment process discussed in Chapter 6. Of particular relevance is the evidence on carcinogenicity of water contaminants. The MCLG for known or probable carcinogens is zero, meaning that no amount of such a contaminant is allowed in public drinking water. For any substance that is not a carcinogen, the MCLG is set according to the established reference dose (RfD) for that contaminant. An RfD is an estimate of the amount of a pollutant to which humans can be exposed over a lifetime without harm.[19]

A key point to remember is that the MCLG is *not* an enforceable standard. Instead, it serves as a target or objective toward which the primary standard is to be aimed.

maximum contaminant level goal (MCLG)
A component of an NPDWR that defines the level of a pollutant at which no known or expected adverse health effects occur, allowing for a margin of safety.

Setting the Standard: The MCL

Once the target or MCLG is established, the primary standard is set. The primary standard gives the **maximum contaminant level (MCL)** allowed in drinking water. It is to be set as close to the MCLG as is feasible, where feasibility is defined through the BAT. More formally, an MCL is the highest permissible level of a contaminant in water delivered to any user of a public system. It is expressed as an **action level** measured in milligrams per liter (mg/l).

Unlike the MCLGs, the MCLs are federally enforceable. Determination of the MCLs was changed by the 1996 Amendments. Specifically, the new law requires a published

maximum contaminant level (MCL)
A component of an NPDWR that states the highest permissible level of a contaminant in water delivered to any user of a public system.

action level
The manner in which MCLs are expressed, generally in milligrams per liter.

[17] Note the consistency between the meaning of the primary and secondary drinking water standards and that of the primary and secondary ambient air quality standards in the United States.
[18] Each regulation must also outline requirements for monitoring, reporting, and public notification (U.S. EPA, Office of Water, May 1992).
[19] As discussed in Chapter 6, an RfD is expressed in milligrams of a pollutant per body weight per day.

Table 16.2 *National Primary Drinking Water Standards for Selected Contaminants*

Contaminant	MCL[a]	Health Effects
Organic chemicals		
Atrazine	0.003	Reproductive and cardiovascular effects
Benzene	0.005	Anemia; increased cancer risk
Carbon tetrachloride	0.005	Liver problems; increased cancer risk
Chlordane	0.002	Liver or nervous system problems; increased cancer risk
Heptachlor	0.0004	Liver damage; increased cancer risk
Styrene	0.1	Liver, kidney, and circulatory problems
Vinyl chloride	0.002	Increased cancer risk
Inorganic chemicals		
Arsenic	0.01	Skin damage; circulatory system problems; increased cancer risk
Asbestos	7 MFL[b]	Increased risk of developing benign intestinal polyps
Cadmium	0.005	Kidney damage
Fluoride	4.0	Bone disease
Lead	TT[c] Action level = 0.015	Delays in physical development in infants and children; kidney problems and high blood pressure in adults
Mercury	0.002	Kidney damage
Radionuclides		
Beta particle and photon emitter	4 mrem/yr[d]	Increased cancer risk
Radium 226/228	5 pCi/l[e]	Increased cancer risk
Microbiological		
Turbidity	TT[c]	Higher levels of disease-causing microorganisms
Legionella	TT[c]	Legionnaire's disease (pneumonia)
Viruses	TT[c]	Gastrointestinal illness

Source: U.S. EPA, Office of Water, Office of Ground Water and Drinking Water (April 17, 2002).

[a]MCLs are measured in milligrams per liter (mg/l) unless otherwise noted.

[b]MFL = million fibers per liter, with fiber length > 10 microns.

[c]TT means treatment technique requirement in effect.

[d]mrem/yr = millirems or 1/1000 rem per year.

[e]pCi/l = picocuries per liter.

determination as to whether the benefits of the MCL are justified by the cost. Table 16.2 lists the MCLs for selected contaminants and their associated health effects.[20]

best available technology (BAT) A treatment technology that makes attainment of the MCL feasible, taking cost considerations into account.

Defining Treatment Technologies

The law requires that each national primary drinking water regulation identify the treatment technology that makes attainment of the MCL feasible. This is characterized as the **best available technology (BAT)** observed under field conditions, taking account of cost considerations. Since this technology is not a requirement per se, the primary water quality regulations are more accurately characterized as **performance-based standards.**

[20]For a complete listing of contaminants and their standards, visit **http://www.epa.gov/safewater/mcl.html.**

Current Status of the National Primary Drinking Water Regulations (NPDWRs)

As of 2002, NPDWRs have been announced for 69 organic and inorganic chemicals, 7 microorganisms, 7 disinfectants, and 4 radionuclides.[21] All standards are to be reviewed at least once every five years and must be amended whenever enhanced health protection is possible through changes in technology, treatment, or other means. For example, more stringent primary standards for lead in drinking water were announced by the EPA in 1991. As a consequence, public water supplies in many U.S. communities were found to be exceeding the lead action level. For more on the implications of this outcome, see Application 16.3.

Establishing National Secondary Drinking Water Regulations (NSDWRs)

Protection of public welfare is the statutory objective of secondary drinking water standards. More to the point, these standards deal with contaminants that so impair aesthetics and other non-health-threatening characteristics like odor and taste that a substantial number of individuals may be forced to discontinue use of the public water system. Referred to as **secondary maximum contaminant levels (SMCLs),** these standards serve as guidelines to protect public welfare and are *not* enforceable by the federal government. Furthermore, unlike the primary standards, the secondary standards are not uniform, since they may vary with geographic or other conditions. Exhibit A.11 in the Appendix gives the U.S. SMCLs as of 2002 and the associated effects of each listed contaminant. The current list of SMCLs is also available online at **http://www.epa.gov/safewater/mcl.html.**

secondary maximum contaminant levels (SMCLs)
National standards for drinking water that serve as guidelines to protect public welfare.

Analysis of U.S. Safe Drinking Water Policy

In analyzing safe drinking water policy in the United States, it is important to understand not only the current legislation enacted through the 1996 Safe Drinking Water Act Amendments, but also the pre-1996 regulations. Doing so helps to explain the motivation behind some of the recent revisions in the law.

Prior to the 1996 Amendments, several aspects of the Safe Drinking Water Act were criticized by environmental economists and policy analysts, particularly in light of reported incidents of contaminated drinking water supplies. One important concern centered on the standard-setting process at the federal level and the potentially inefficient outcome it generates. Were contaminant levels being properly determined? If not, what were the ramifications? Another key issue was the subpar performance at the state and local levels in meeting the standards and maintaining adequate treatment facilities. Was the less than satisfactory track record the result of ill-defined standards, poor enforcement procedures, inadequate resources, or some combination of factors? To begin, let's consider the standard-setting process that forms the fundamental basis of the Safe Drinking Water Act.

The Federal Role: Setting the Standards

Efficiency Implications

The statutory goals of the Safe Drinking Water Act (SDWA), that is, the maximum contaminant level goals (MCLGs), are defined as the level of a pollutant at which no adverse human health effects occur with a "margin of safety." Allowing for an adequate margin of safety seems to suggest a contaminant level where no individual would be harmed. Given the variability of human sensitivities, this seems to imply that at least some MCLGs must be set at zero. Notice that there is no mention of feasibility or cost considerations in setting these goals. They are purely **benefit based.**

Since the maximum contaminant levels (MCLs) are aimed at these goals, they too were benefit based prior to the 1996 Amendments. Although the MCLs are to consider "feasibil-

[21] U.S. EPA, Office of Water, Office of Ground Water and Drinking Water (April 17, 2002).

Application

16.3 *Strengthening Controls on Lead Contamination of Water Supplies*

In 1991, the EPA tightened the lead standard for drinking water to a maximum contaminant level (MCL) of 0.015 mg/l. Announcing the new standard, then EPA administrator William K. Reilly stated:

> Today's action will reduce lead exposure for approximately 130 million people. . . . We estimate approximately 600,000 children will have their blood lead content brought below our level of concern because of these standards. (U.S. EPA, Office of Communications and Public Affairs, May 7, 1991)

According to the EPA's initial 1992 survey based on test data from 6,400 large water systems, approximately 13 percent did not meet the new lead standard. Of those cities with populations greater than 50,000, the following were found to have the highest lead content in their public drinking water supplies:

City	Lead Levels in 1992 [parts per billion (ppb)]
Charleston, South Carolina	165
Utica, New York	160
Newton, Massachusetts	123
Columbia, South Carolina	114
Medford, Massachusetts	113
Chicopee, Massachusetts	110
Yonkers, New York	110
Waltham, Massachusetts	76
Brookline, Massachusetts	72
Taylor, Michigan	69

The EPA estimates that only about 1 percent of water systems with elevated lead levels will have to treat source waters. The reason is that most of the lead content in drinking water comes from the public water delivery system and not from the water supply itself. The contamination arises from corrosion of lead pipe service lines and plumbing solder. The "lead ban," which was legislated as part of the Safe Drinking Water Amendments of 1986, prohibits the future use of lead pipes, solder, or flux. Nonetheless, some older homes built before 1930 have lead water pipes, and some newer homes built before the lead ban was enacted have copper pipes joined with lead-based solder. Consequently, in areas with a highly corrosive water supply, lead can leach out of these pipes or solder joints, contaminating the water that reaches residents.

To combat this problem, treatment technologies have been developed to reduce water corrosivity. Congress also added more controls by passing the Lead Contamination Control Act of 1988, which calls for the repair or removal of water coolers that are not lead free and bans the future sale and manufacture of these potential sources of contamination.

Beyond these comprehensive measures, the EPA is also using a public education program to communicate simple procedures households can follow to reduce their risk of lead exposure. The EPA advisory recommends that all tap water be allowed to run for about a minute before using it for cooking or drinking as a means to flush out any lead sediment from pipes. For homes in areas where the new action level is exceeded, further precautions are recommended, such as running tap water for several minutes before using and avoiding the use of hot tap water for drinking or cooking. These precautionary measures coupled with the strengthened federal regulations should lessen considerably the risk of lead exposure that threatens human health.

Sources: U.S. EPA, Office of Communications and Public Affairs (May 7, 1991); Noah (May 12, 1993); U.S. EPA, Office of Water (July 1988).

Figure 16.2 ***Inefficiency of Benefit-Based Primary Drinking Water Standards***

If the maximum contaminant level (MCL) is set to maximize social health benefits, the resulting abatement level is A_1, corresponding to the point where the *TSB* curve reaches its highest level. In this case, A_1 is *above* the efficient level of abatement (A_E). Hence, the benefit-based standard would overregulate this contaminant, imposing unnecessarily high costs on society.

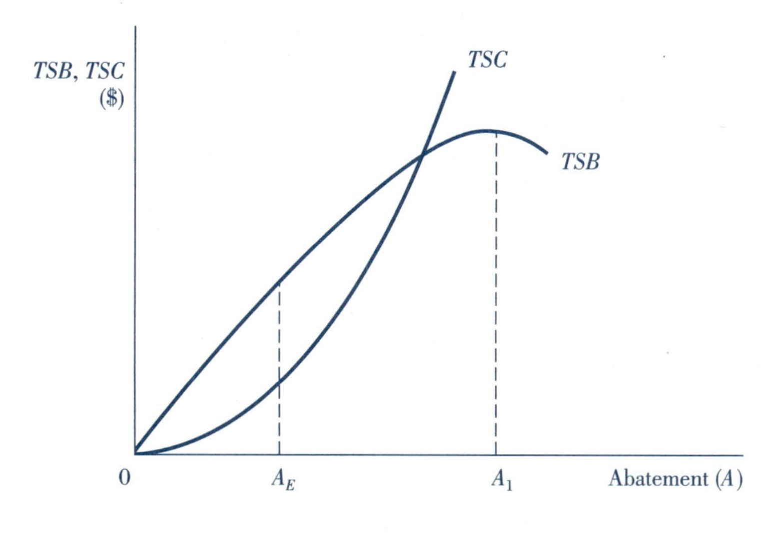

ity" defined through technological availability, there was no legal requirement that the associated marginal costs of achieving the standard be balanced with the expected marginal gains. As is the case with any such regulation, the absence of an explicit balancing of costs and benefits at the margin means that efficiency is not being used to regulate drinking water. And given the language of the MCLG, there was potential for overregulation of some contaminants and inflated costs before the 1996 Amendments.

Consider Figure 16.2, which models the total social health benefits (*TSB*) and total social costs (*TSC*) of abating some hypothetical drinking water contaminant beyond the efficient level. If the MCL is set to maximize social health benefits, the resulting abatement level is A_1, corresponding to the point where the *TSB* curve reaches its highest level. In this case, A_1 is *above* the efficient level of abatement (A_E). Hence, the benefit-based standard would overregulate this contaminant, imposing unnecessarily high costs on society.

All of this explains the importance of a paragraph in the 1996 Amendments specifying precisely how the MCLs are to be determined. According to this paragraph, at the time a new NPDWR is proposed, the EPA must publish a determination identifying whether or not the benefits of the MCL justify the costs. Effectively, this means that the EPA must conduct an **Economic Analysis (EA)** for any proposed NPDWR. The benefit-cost analysis is to be based on specific health risk reduction and cost analysis used in risk assessment for any proposed MCL.[22]

Economic Analysis (EA)
A requirement under Executive Order 12866 that calls for information on the benefits and costs of a "significant regulatory action."

[22] To learn more about the EPA's use of economic considerations in setting drinking water standards, visit **http://www.epa.gov/ogwdw/ria/riadoc.html.**

Table 16.3 ***Trend Data and Projections on Control Costs for Safe Drinking Water***

Year	Annualized Control Costs ($1986 millions)
1986	$2,979
1987	3,111
1988	3,250
1989	3,415
1990	3,587
1991	3,926
1992	4,319
1993	4,586
1994	4,917
1995	5,350
1996	5,684
1997	5,949
1998	6,264
1999	6,491
2000	6,571

Source: Drawn from U.S. EPA, Office of Policy, Planning, and Evaluation (December 1990), pp. 2-2–2-3, table 2-1.

Note: These cost data assume full implementation at a 7 percent discount rate.

Cost Implications

A common criticism of U.S. water quality policy is the rise in treatment costs that have come about since the SDWA was enacted. Much of the increase in regulatory costs is due to the more stringent requirements that were called for in the 1986 Amendments. According to estimates based on analyses done in late 1986 and early 1987, the annual costs of implementing these amendments were about $2.5 billion.[23]

A more recent report projected annualized control costs through the year 2000. These estimates are shown in Table 16.3 and assume full implementation of the SDWA prior to the 1996 Amendments. Notice how real costs are expected to rise over time—an outcome that may impede the EPA's ability to implement the law. The financial burden that these regulations place on state and local authorities is an important concern, particularly since many public water systems are already unable to comply with the law due to resource limitations.

What these findings imply is that both social costs and benefits are important considerations in the regulatory process. The premise is rooted in economic theory and recommended in President Clinton's Executive Order 12866 and previously in Executive Order 12291 issued by President Reagan. Under both presidential directives, the benefits and costs associated with a major regulatory action must be assessed and made public.

Regulatory Impact Analysis (RIA)
A requirement under Executive Order 12291 that called for information about the potential benefits and costs of a major federal regulation.

Establishing or revising primary drinking water standards is an example of a government action subject to this requirement. In 1991 when the lead standard was revised, Reagan's Executive Order was in force, so the estimated benefits and costs were presented within a **Regulatory Impact Analysis (RIA),** which is summarized in Application 16.4. The 2001 revision to the arsenic standard was initially proposed under the Clinton administration and was later accepted by the Bush administration. The key results of this recent **Economic Analysis (EA)** are considered next.

[23] U.S. GAO (June 1990), p. 53, table 4.1.

16.4 Regulatory Impact Analysis (RIA) for the New Lead Standard in Drinking Water

Application

In June 1991, the EPA announced a maximum contaminant level goal (MCLG) of zero for lead and a more stringent maximum contaminant level (MCL) of 0.015 mg/l. This new primary standard lowered the allowable lead level in drinking water from its former limit of 50 parts per billion (ppb) to 15 ppb. Regulations for water treatment also were instituted.

Taken together, these regulations were expected to have a substantial financial impact on the regulated community—in excess of $100 million per year. Hence, by law the new rulings were subject to Executive Order 12291 and had to be accompanied by a **Regulatory Impact Analysis (RIA)** with a full description of the associated benefits and costs. A summary of the estimates, stated as annualized values, is given below.

According to the RIA, the expected incremental benefits of the tougher lead standard include reductions in damages to health and materials. For health benefits, the qualitative assessment was described as a reduction in the exposure of 130 million people to lead in drinking water and a reduction in the blood lead level of an additional 570,000 children to below 10 micrograms per deciliter.

To monetize these benefits, they were translated into avoided medical costs and estimated as within a range of $2.87 billion and $4.54 billion annually. Incremental material benefits include such gains as longer pipe life, reduced leakages, extended life of water-using appliances, and fewer repairs. The estimated value of these benefits was stated as $500 million per year. Combining these values, the aggregate incremental benefits associated with tightening the lead standard in drinking water were estimated at between $3.4 billion and $5.0 billion per year.

On the cost side, the RIA itemized annual expenditures for treatment, monitoring, education, and implementation. The highest costs were assessed for treatment procedures at between $390 million and $680 million per year. In the aggregate, the incremental costs of implementing the new lead action level were estimated at between $500 million and $790 million per year.

Based on the benefit and cost estimates, the EPA determined that the annualized net benefits of tightening the standard for lead in drinking water were between $2.9 billion and $4.2 billion. Because the new standard was proposed under the pre-1996 law, the RIA results did not dictate the new action level. However, they did communicate in economic terms the net gain to society associated with the regulatory decision.

Benefits	
Health (based on avoided medical costs)	
From corrosion control and source water treatment:	$2.8–$4.3 billion per year
From replacement of lead service lines:	$70–$240 million per year
Material	
Accruing to households and water systems:	$500 million per year
Incremental benefits	$3.4–$5.0 billion per year
Costs	
Treatment, implementation, education costs	
Treatment costs:	$390–$680 million
Monitoring costs:	$40 million
Education costs:	$30 million
State implementation costs:	$40 million
Incremental costs	$500–$790 million per year
Net Benefits	
Net benefits	$2.9–$4.2 billion per year

Sources: U.S. EPA, Office of Water, Office of Ground Water and Drinking Water (May 1991); *U.S. Federal Register* 56, no. 110 (June 7, 1991).

Economic Analysis (EA) for the New Arsenic Standard in Drinking Water[24]

For some 60 years, the MCL for arsenic had been set at 50 parts per billion (ppb). Arsenic has long been associated with various health effects, but more recent scientific research suggests that the risks may be greater than originally believed. Specifically, some studies indicate that high doses of arsenic in drinking water may lead to several types of fatal cancers. Citing such health concerns, in 1996 Congress directed the EPA to review the arsenic standard and issue a revision by January 2001. In May 2000, the EPA proposed a change in the MCL from 50 ppb to 5 ppb. Objections from industry and municipal water authorities prompted the EPA to change the proposed standard to 10 ppb, a decision made in January 2001 by the Clinton administration, just before President Bush took office.

Concerned that the decision to set the arsenic standard at 10 ppb may have been rushed, the Bush administration initially rescinded the Clinton standard. The newly appointed EPA administrator, Christine Todd Whitman, then called for a comprehensive review of the scientific and economic evidence used by the Clinton administration in its decision. Among the participants were the National Academy of Sciences (NAS), the National Drinking Water Advisory Council, and the EPA's own Science Advisory Board (SAB). Although there were concerns about the underlying methodology used and the resulting benefit and cost estimates that had been cited in the requisite **Economic Analysis (EA),** Whitman ultimately decided there was sufficient scientific and economic evidence to warrant the new arsenic standard. Hence, after much debate and political scrutiny, the MCL of 10 ppb originally proposed under President Clinton ultimately was accepted by the Bush administration.

Beyond the political and scientific complexities involved in this decision, the EA for the new standard proved to be particularly significant. This analysis was necessary because the change in the standard was expected to have a substantial financial impact on the regulated community—in excess of $100 million per year. A brief overview of the benefit and cost estimates cited in the EA follows, with a summary of the findings given in Table 16.4.

Incremental Benefits of the New Arsenic Standard

Citing numerous epidemiological studies, the EA for the new arsenic standard indicates that its major incremental health benefit is the number of lung and bladder cancer cases avoided. For the reduction in fatal cases, the value of a statistical life of $6.1 million ($1999) is used to monetize these health benefits, based on current EPA guidance. Avoided nonfatal cancer cases are monetized using a willingness-to-pay value of $607,162. Other health gains, such as reductions in skin, kidney, and liver cancers, are addressed in the analysis but not monetized. Collectively, the estimated health benefits for the 10-ppb standard for arsenic are reported to be between $139.6 million and $197.7 million ($1999) per year.

Incremental Costs of the New Arsenic Standard

In assessing incremental costs of the new standard, the EA identifies annual expenditures associated with compliance, monitoring, implementation, and enforcement, as shown in Table 16.4. Treatment procedures account for the highest proportion of estimated costs at $200.6 million per year. In sum, the estimated incremental costs of complying with and enforcing the new arsenic standard are reported to be $205.6 million ($1999) per year.

Net Benefits of the New Arsenic Standard

According to the overall findings of the EA, the annualized net benefits associated with the more stringent arsenic standard in drinking water are between −$66 million and −$7.9 million ($1999) per year. Based on these estimates, the Bush administration's concern that the cost of complying with the 10-ppb standard might not have been justified by the benefits may have been warranted. However, since a substantial number of potential benefits were not

[24]This discussion is drawn from Seelye (November 1, 2001) and U.S. EPA, Office of Water, Office of Ground Water and Drinking Water (December 2000). This report is also available online at **http://www.epa.gov/safewater/ars/econ_analysis.pdf.**

Table 16.4 ***Estimated Benefits and Costs of the New Arsenic Standard ($1999): Results from the Economic Analysis (EA)***

	Incremental Benefits
Health benefits	
Descriptive assessment: Avoided cases of bladder and lung cancer. Avoided premature deaths were valued at $6.1 million per statistical life. Avoided nonfatal cancer cases were monetized using a willingness-to-pay value of $607,162.	
Monetized annual value:	$139.6–$197.7 million per year
Additional benefits (not monetized): Avoided cases of skin, liver, kidney, and prostate cancer; avoided cardiovascular and pulmonary effects and psychological effects of knowing that the drinking water is safer to consume.	
ANNUALIZED BENEFITS	$139.6–$197.7 million per year
	Incremental Costs
Systems costs (assuming a 7% discount rate)	
Descriptive assessment: Treatment costs; monitoring and administrative expenses.	
Monetized annual value:	
Treatment costs:	$200.6 million per year
Monitoring/administrative costs:	$ 3.8 million per year
State costs	
Monetized Annual Value	$ 1.2 million per year
ANNUALIZED COSTS	$205.6 million per year
	Net Benefits
ANNUALIZED NET BENEFITS	−$66.0 million to −$7.9 million per year

Source: U.S. EPA, Office of Water, Office of Ground Water and Drinking Water (December 2000).

monetized in the EA, the EPA decided that the actual benefits would likely be higher than the estimates. Hence, the agency argued that the true incremental benefits would justify the incremental costs and announced the new standard in October 2001.

The State and Local Role: Compliance and Enforcement

Since the Safe Drinking Water Act relies heavily on a command-and-control approach, its success depends critically on sound monitoring and enforcement procedures that assure compliance with the primary standards. Public water systems are required to periodically sample water supplies, test them in an approved laboratory, and report the data to state authorities. In turn, these data are to be analyzed to determine if the system is meeting its monitoring responsibilities and if it is complying with the national standards. States are responsible for enforcement if any violations are noted, giving priority to those systems designated as significant noncompliers (SNCs).[25] The EPA must step in if the state does not properly enforce the law.

[25] The classification as an SNC is based on either the frequency of violations or the severity of the violations, such as the extent to which a system exceeds the MCL. See U.S. GAO (June 1990), p. 14.

Compliance Problems

Although these procedures may seem to be in order, an investigation conducted by the U.S. General Accounting Office (GAO) found that the extent of compliance by public water systems had been overstated by the EPA.[26] Part of the problem lies in the EPA's classification of SNCs versus other noncompliers. The classification system was put into place to focus scarce resources on the worst offenders. However, according to the GAO, the SNC criteria are such that many serious violators are not being confronted. Deficiencies were identified at every level of government, specifically the local water system, the state, and the EPA. Problems were discovered in the sampling and testing procedures conducted by community water systems. The GAO also found that states were not consistently reporting violations to the EPA. Since the EPA's data management system relies on state tracking systems for compliance information, its reporting of these data is sometimes inaccurate and overstated.

Enforcement Failures

Another failing in the process is the inadequate enforcement action taken by both state authorities and the EPA, despite the added stringency provided by the 1986 Amendments. In its review of enforcement procedures in six states, the GAO found that states took timely and appropriate action in only 24 of the 95 SNC violations committed by 75 public water systems. Most notable was the observation that many SNCs remained in noncompliance for several years. In fact, 46 of the SNC cases reviewed, or nearly half, had met the SNC criteria for more than 4 years.

Analyzing the Issues

Why is the compliance record so poor? Some of the reasons are procedural, having to do with how states and the EPA respond to violations of federal law. However, even if these administrative issues were corrected, there would still be unresolved compliance issues due to both technological and economic considerations. A recent analysis done by the Natural Resources Defense Council reports that severely outmoded treatment facilities and failing water distribution systems characterize the majority of public water systems across the nation.[27] These technological problems are exacerbated by the absence of watershed protection approaches, which forces an overreliance on treatment techniques.

With regard to economic considerations, the GAO argues that the SDWA imposes what it calls "staggering costs" upon public water systems faced with contamination problems. This is particularly true for small systems that are unable to finance the costs of corrective action and to compete for external funding. Of the 75 systems examined by the GAO, two-thirds served 500 or fewer people, and 87 percent served 3,300 or less.[28] This economic dilemma is worse in those cases where water prices are set below cost.

The funding problem helps to explain the motivation for the State Revolving Fund (SRF) authorized by the 1996 Amendments. Among the fund's objectives are to finance infrastructure improvements, to assist small and disadvantaged communities, and to encourage pollution prevention as a means to ensure safe drinking water.[29] In retrospect, it seems that a balancing of benefits and costs may have avoided some of the costing difficulties public officials face. Another likely explanation is the lack of economically sound pricing practices for drinking water supplies—an issue we examine more closely.

Economic Principles in Pricing Water Supplies

At all levels of government, the costs of implementing the SDWA present a challenge. Not only are these costs inflated because of inefficient decision making, but in many communities they are not properly reflected in the pricing of water supplies. Basic economic theory illus-

[26] U.S. GAO (June 1990).
[27] Cohen and Olson (March 1994).
[28] U.S. GAO (June 1990), p. 46.
[29] U.S. EPA, Office of Water, Office of Ground Water and Drinking Water (May 8, 2002).

Figure 16.3 *International Water Prices in 1996 (U.S. dollars per cubic meter)*

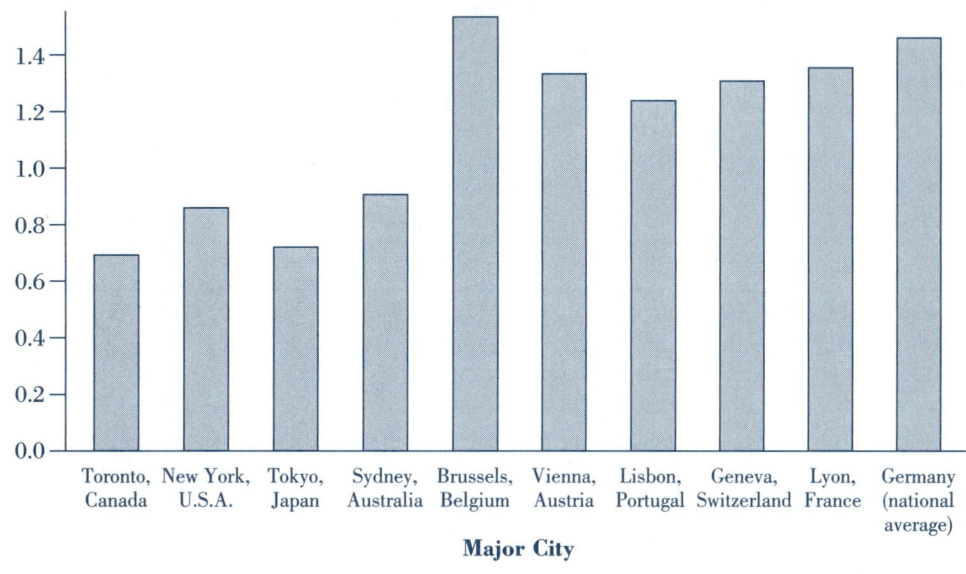

Source: OECD (1999b), p. 77.

trates that resources are misallocated if the price of a good or service is not equal to the associated marginal social cost (*MSC*) of production. In this context, society's marginal cost of *current* water usage includes the explicit costs incurred by water supply facilities plus the opportunity costs of forgone *future* consumption. All too often, water is priced at a fixed rate that is independent of use and therefore independent of rising marginal costs. The result is overconsumption. If present consumption is higher than its allocatively efficient level, future supplies will be adversely affected. The 1996 Amendments recognize the importance of efficient water use and explicitly require the EPA to establish guidelines to encourage water conservation.[30]

An International Comparison

An international comparison of water consumption patterns reveals two very interesting findings. First, U.S. per capita consumption is higher than it is in most nations around the world. For example, in 1990 the average American consumed approximately 2,161 cubic meters of water. Contrast this with the same statistic for Canada, which is reported at about 1,500 cubic meters, or for all of Europe, estimated at 750 cubic meters.[31]

A second important observation is that water prices vary considerably across countries, as shown in Figure 16.3. Notice how cities in European countries face much higher prices for water than urban centers in the United States and Canada. And European nations have lower consumption rates than their North American counterparts. Of course, this outcome is precisely what the Law of Demand predicts—higher prices are associated with lower quantities

[30] To learn more about the nation's water efficiency program and guidelines for water conservation plans, visit **http://www.epa.gov/owm/water-efficiency/**.
[31] World Resources Institute (1992a), p. 102.

Table 16.5 Percentage Use of Various Pricing Structures in 1995

Pricing	Percentage of Community Water Systems
Uniform rate	49.0
Declining block	16.0
Increasing block	11.0
Peak period	0.9
Separate flat fee	15.3
Combined flat fee	10.0
Other	8.2

Source: U.S. EPA, Office of Water, Office of Ground Water and Drinking Water (January 1997), p. 15.

Notes: Percentages do not sum to 100% because some systems use more than one rate structure.

Definitions are as follows:

Uniform rate: A price that is a fixed amount per unit of water used.

Declining block: A pricing scheme designed such that the per unit price declines as water use increases. Units are typically referred to as blocks, where each block represents a specific quantity of water.

Increasing block: A pricing scheme designed such that the per unit price increases as water use increases.

Flat fee: A fixed fee paid monthly or annually that is independent of actual water use.

consumed. What these data convey is that water use *is* price sensitive—an important observation given that some locations in the United States employ pricing policies that actually encourage inefficient water use. By examining these pricing practices in more detail, we can identify the inherent weaknesses and consider some economically sound alternatives.

Pricing Practices of U.S. Water Utilities

flat fee pricing scheme
Pricing water supplies such that the fee is independent of water use.

uniform rate (or flat rate) pricing structure
Pricing water supplies to charge more for higher water usage at a constant rate.

Table 16.5 provides data on the actual pricing practices of water utilities based on a 1995 survey of community water systems.[32] Notice that a considerable percentage uses a **flat fee pricing scheme,** which refers to pricing that is independent of water use. Such an approach would be efficient only if the associated marginal costs were zero. Nearly half employ a **uniform rate (or flat rate) pricing structure,** which does charge more for higher usage but at a constant rate. This structure would be efficient only if marginal costs were constant and equal to the uniform rate being charged. Neither of these pricing structures reflects the rising marginal social costs (*MSC*) of water provision, and thus they act as a disincentive for consumers to economize on water usage.

The other major pricing methods represented in the table are the **declining block** and **increasing block pricing structures.** Each of these allows for changes in per unit prices for different blocks of consumption levels, but obviously each works in the opposite direction of the other. Both pricing structures are illustrated graphically in Figure 16.4. Since they are motivated differently, we need to examine them more closely and consider how each affects water usage.

declining block pricing structure
A pricing structure in which the per unit price of different blocks or quantities of water declines as usage increases.

Declining Block Pricing Structure

Referring back to the data in Table 16.5, the **declining block pricing structure** is used by 16 percent of the surveyed water systems. Its usage appears to have its basis in the typical utility's cost structure. Throughout the United States, most water utilities tend to have high fixed costs and relatively low variable costs.[33] In an attempt to recover these high fixed costs, some utilities encourage higher consumption levels with declining block pricing, as shown in

[32] U.S. EPA, Office of Water, Office of Ground Water and Drinking Water (January 1997), p. 15.
[33] U.S. EPA, Office of Policy, Planning, and Evaluation (March 1991), pp. 4–7.

Figure 16.4 ***Alternative Pricing Structures of Water***

(a) In the United States, most water utilities incur relatively low variable costs but high fixed costs. In an attempt to recover these fixed costs, utilities use declining block pricing to encourage higher consumption. In so doing, they are able to exploit available scale economies and incur lower average costs.

(b) An increasing block pricing structure provides an economic incentive for more conservative water usage. Benefit-cost analysis is implicitly reflected in this approach. As each additional block of water is used, higher marginal costs are considered along with the marginal benefits of consumption.

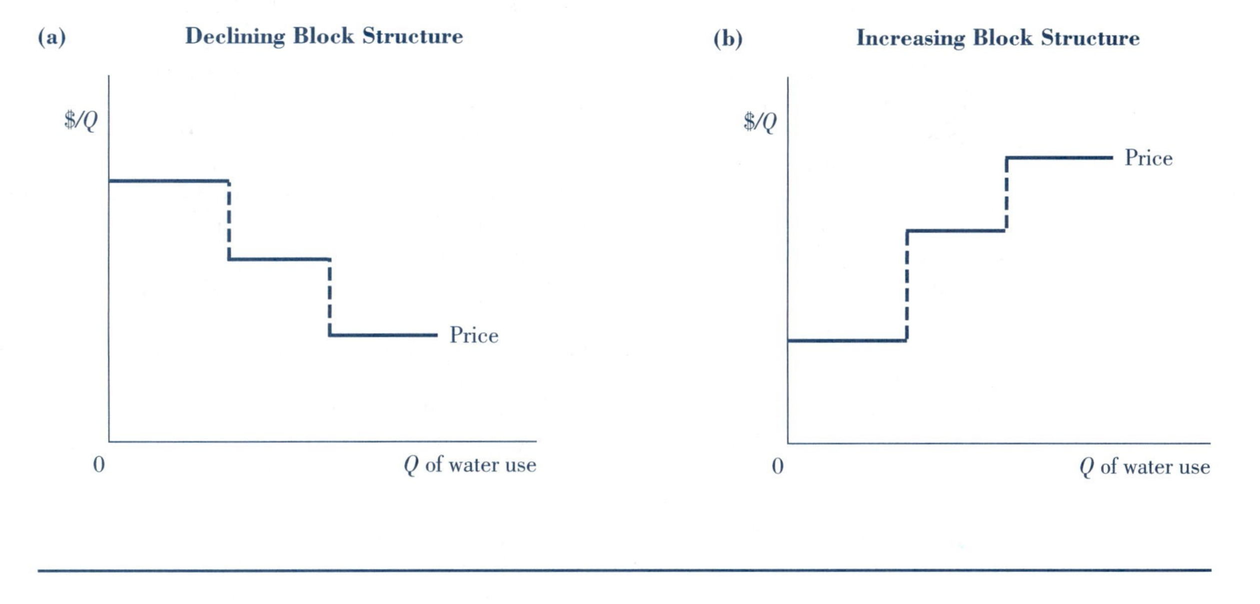

Figure 16.4a. In so doing, they can exploit available scale economies and incur lower average costs.

From the utility's perspective, the declining block price structure has merit. But from society's point of view, the pricing scheme is inefficient. Although this inefficiency arises for several reasons, the major culprit is that suppliers are incorrectly using average cost pricing as opposed to a marginal cost pricing framework.

Increasing Block Pricing Structure

Looking again at Table 16.5, notice that one of the least common pricing practices is **increasing block pricing.** As illustrated in Figure 16.4b, the customer pays a higher per unit price as water consumption increases—a practice that provides an economic incentive for more conservative water usage. Notice how benefit-cost analysis is implicitly reflected in this approach. As each additional block of water is used, higher marginal costs are considered along with the marginal benefits of consumption. Of course, unless the increasing price blocks accurately reflect society's marginal cost, the result will not be efficient. Also, there are implementation costs to be considered in instituting this pricing structure, since metering is necessary to track usage. This factor explains why some smaller utilities favor a uniform rate system despite its inherent inefficiencies.

increasing block pricing structure
A pricing structure in which the per unit price of different blocks of water increases as water use increases.

Based on this fundamental analysis, it is argued that the EPA and other government authorities should attempt to structure water pricing to reflect the associated marginal cost to society. A further consideration is to find ways to make such a pricing scheme financially feasible for all types of water utilities, including smaller facilities. Although this would be a challenging undertaking, it could make a measurable difference in both current and future water quality.

Conclusions

For over 25 years, the Safe Drinking Water Act has been the legislation responsible for protecting human health from contaminated tap water. Like most environmental laws, this one has been the subject of debate and criticism. Confidence in national drinking water policy has weakened in recent years. Right or wrong, most Americans had been fairly complacent about the quality of tap water. However, reports such as the EPA's 1992 survey on lead levels and the 1993 Milwaukee incident have made both private citizens and public officials question the effectiveness of U.S. drinking water policy.

Prior to the 1996 Amendments, the structure of the SDWA was markedly simple—a direct application of the command-and-control approach. It established contaminant limits designed to assure that public water supplies were safe for human consumption and relied on state and local authorities to make certain that these standards were met. From an economic perspective, such an approach is flawed. The standards were benefit based, with no requirement for a balancing of marginal benefits and costs. Furthermore, there were no incentives to encourage compliance or to stimulate technological advance to meet the standards in a cost-effective manner. Instead, achieving safe drinking water relied on constant monitoring and enforcement—procedures that are difficult and costly to implement.

The recently enacted SDWA Amendments of 1996 address some of these shortcomings. A major set of revisions deals with integrating risk assessment and benefit-cost analysis into the setting of drinking water standards. This is a major shift away from the strict command-and-control approach that characterized this legislation until now. There is also a new Drinking Water State Revolving Fund (DWSRF), which is aimed at helping states and local communities improve the infrastructure needed to ensure clean drinking water supplies. Another important reform is the integration of preventive programs to protect and manage sources of drinking water supplies.

Because water resources are so diverse and at the same time interconnected, policy formulation is not straightforward, nor is the revision process. Both the Clean Water Act and the Safe Drinking Water Act continue to be rooted strongly in the command-and-control approach with standards providing the legislative muscle. But there has been some integration of market-based incentives into this policy structure and gradual changes to launch initiatives to *prevent* water contamination and lessen the dependence on end-of-pipe treatment. These are nontrivial revisions, as are the problems confronting the nation at this phase of its water quality policy development.

Summary

- The Interstate Quarantine Act of 1893 was the first federal law in the evolution of U.S. drinking water standards. The Public Health Service Act of 1912 called for standards to prevent the spread of communicable waterborne diseases. These were set in 1914 and revised over the next several decades.
- The Safe Drinking Water Act (SDWA) of 1974 was aimed at protecting drinking water from any contaminant that could threaten human health or welfare. It established primary standards to protect human health and secondary standards to protect welfare.
- The 1986 Amendments expanded federal controls on drinking water and corrected some of the failings of the original SDWA.
- The 1996 Amendments were signed into law in August 1996. Among the chief revisions are the integration of risk assessment and benefit-cost analysis into standard setting, a new Drinking Water State Revolving Fund (DWSRF) to improve infrastructure, and efforts to encourage prevention through better management and source water protection.
- The primary purpose of the Safe Drinking Water Act is to assure that public water systems meet minimum standards to protect human health.

- The 1986 Amendments identified 83 contaminants for which drinking water standards were to be set.
- Under U.S. law, there are National Primary Drinking Water Regulations (NPDWRs) aimed at protecting human health. Each regulation consists of a maximum contaminant level goal (MCLG), a maximum contaminant level (MCL), and a specification of the best available technology (BAT) for public water supply treatment.
- The MCLG defines the level of a pollutant at which no known or expected adverse health effects occur, allowing for an adequate margin of safety.
- The MCL is the highest permissible level of a contaminant in water delivered to any user of a public system. Determination of the MCLs was changed by the 1996 Amendments such that a published determination must be made as to whether the benefits of the MCL are justified by the costs.
- The BAT represents the feasible treatment technology capable of meeting the standard, taking account of cost considerations.
- National Secondary Drinking Water Regulations (NSDWRs) establish secondary maximum contaminant levels (SMCLs) for pollutants that impair aesthetics and other characteristics like odor and taste. These are nonenforceable federal guidelines aimed at protecting public welfare.
- Enforcement of the primary drinking water standards is shared between the EPA and state governments. The major responsibility is assigned to state governments, provided that certain criteria are met.
- Prior to the 1996 Amendments, the Safe Drinking Water Act (SDWA) did not call for a balancing of benefits and costs in setting the maximum contaminant levels (MCLs). Revisions in the 1996 Amendments specify that when a new NPDWR is proposed, the EPA must publish a determination identifying whether or not the benefits of the MCL justify the costs.
- According to a government study, the extent of compliance by public water systems has been overstated by the EPA, and enforcement action has been inadequate.
- Many U.S. communities use improper pricing practices of water supplies, such as a flat fee, uniform rate, or declining block structure, none of which account for rising marginal costs. The result is overconsumption of water resources.

Key Concepts

Drinking Water State Revolving Fund (DWSRF)
sole-source aquifers
priority contaminants
National Primary Drinking Water Regulations (NPDWRs)
maximum contaminant level goal (MCLG)
maximum contaminant level (MCL)
action level
best available technology (BAT)
secondary maximum contaminant levels (SMCLs)
Economic Analysis (EA)
Regulatory Impact Analysis (RIA)
flat fee pricing scheme
uniform rate (or flat rate) pricing structure
declining block pricing structure
increasing block pricing structure

Use the Key Concepts listed above to begin your search for additional articles and information using the InfoTrac College Edition database.

Review Questions

1. In what ways did the Safe Drinking Water Act Amendments of 1986 best address the shortcomings of the law? In what ways did they fail?
2. a. Explain the practical difference between the MCLG and the MCL.
   b. Draw the relationship of marginal social benefit and marginal social cost for the MCLG for lead, assuming it is set at an efficient level. Intuitively explain your model.
3. Support or refute Congress's decision to make the secondary drinking water standards *not* enforceable by the federal government.
4. According to the EPA, uncontrolled disposal of pesticide residues and containers contributes significantly to groundwater contamination and hence threatens drinking water supplies. Design and evaluate a deposit/refund system for pesticide containers.
5. Determine the water pricing structure that is used in your hometown. Using the criterion of allocative efficiency, defend the current practice or propose an alternative pricing structure.

Additional Readings

Bergstron, John C., Kevin J. Boyle, and Gregory L. Poe, eds. *The Economic Value of Water Quality.* Northampton, MA: Elgar, 2001.

Dziegielewski, Benedykt, and Duane D. Baumann. "Tapping Alternatives: The Benefits of Managing Urban Water Demands." *Environment* 34 (November 1992), pp. 6–11, 35–41.

Ferguson, Tim W. "Socialized Water." *Forbes,* March 11, 1996, pp. 68–69.

Gordon, Wendy. "Federal Protection of Ground Water." In David H. Speidel, Lon C. Ruedisili, and Allen F. Agnew, eds., *Perspectives on Water: Uses and Abuses,* pp. 326–29. New York: Oxford University Press, 1988.

Howitt, Richard E., Dean E. Mann, and H. J. Vaux Jr. "The Economics of Water Allocation." In Ernest A. Engelbert with Ann Foley Scheuring, eds., *Competition for California Water,* pp. 136–62. Berkeley, CA: University of California Press, 1982.

Innes, Robert, and Dennis Dory. "The Economics of Safe Drinking Water." *Land Economics* 77 (February 2001), pp. 94–117.

Levin, R. "Lead in Drinking Water." In R. D. Morgenstern, ed., *Economic Analyses at EPA: Assessing Regulatory Impact,* pp. 205–32. Washington, DC: Resources for the Future, 1997.

MacLeish, William H. "Water, Water, Everywhere, How Many Drops to Drink?" *World Monitor* 13 (December 1990), pp. 54–58.

Mazari, Marisa. "Potential for Groundwater Contamination in Mexico City." *Environmental Science and Technology* 27, no. 5 (1993), pp. 794–802.

Mohl, Bruce. "Testing the Water." *Boston Globe,* February 26, 1996, pp. 25–26.

Newman, Alan. "A Blueprint for Water Quality." *Environmental Science and Technology* 27, no. 2 (1993), pp. 223–25.

Tomsho, Robert. "Cities Reclaim Waste Water for Drinking." *Wall Street Journal,* August 8, 1994, pp. B1, B6.

Um, M. J., S. J. Kwak, and T. Y. Kim. "Estimating Willingness to Pay for Improved Drinking Water Quality Using Averting Behavior Method with Perception Measure." *Environmental and Resource Economics* 21, no. 3 (2002), pp. 285–300.

Waxman, Henry. "Amending the Safe Drinking Water Act: View from Congress." *EPA Journal* 20 (summer 1994), pp. 32–33.

http:// Related Web Sites

Arsenic in Drinking Water Rule: Economic Analysis, U.S. EPA, Office of Ground Water and Drinking Water	**http://www.epa.gov/safewater/ars/econ_analysis.pdf**
Drinking Water and Health Fact Sheets	**http://www.epa.gov/ogwdw/hfacts.html**
Drinking Water Regulations and Guidance	**http://www.epa.gov/ogwdw/regs.html**
Drinking Water State Revolving Fund (DWSRF)	**http://www.epa.gov/ogwdw/dwsrf.html**
Economic Considerations in Drinking Water Standard Setting	**http://www.epa.gov/ogwdw/ria/riadoc.html**
Community Water Systems Survey, vol. 1, Overview, U.S. EPA, Office of Water, Office of Ground Water and Drinking Water (January 1997)	**http://www.epa.gov/ogwdw000/cwssvr.html**
Water on Tap: A Consumer's Guide to the Nation's Drinking Water, U.S. EPA, Office of Water (July 1997)	**http://www.epa.gov/ogwdw/wot/wot.html**
National Drinking Water Contaminant Occurrence Database	**http://www.epa.gov/ncod/**
National Primary Drinking Water Regulations (NPDWRs)	**http://www.epa.gov/safewater/mcl.html**
Safe Drinking Water Act	**http://www.epa.gov/ogwdw/sdwa/sdwa.html**
U.S. EPA, Office of Ground Water and Drinking Water	**http://www.epa.gov/safewater/**
National Secondary Drinking Water Regulations (NSDWRs)	**http://www.epa.gov/safewater/mcl.html**
U.S. Geological Survey, Water Use in the United States	**http://water.usgs.gov/watuse/**
Water Efficiency Program	**http://www.epa.gov/own/water-efficiency/**

A Reference to Acronyms and Terms in Drinking Water Quality Control

Environmental Economics Acronyms

MSC	Marginal social cost
TSB	Total social benefits
TSC	Total social costs

Environmental Science Terms

MFL	Million fibers per liter
mg/l	Milligrams per liter
ppb	Parts per billion
RfD	Reference dose

Environmental Policy Acronyms

BAT	Best available technology
DWSRF	Drinking Water State Revolving Fund
EA	Economic Analysis
FDA	Food and Drug Administration
MCL	Maximum contaminant level
MCLG	Maximum contaminant level goal
NAS	National Academy of Sciences
NCOD	National Contaminant Occurrence Database
NPDWRs	National Primary Drinking Water Regulations
NSDWRs	National Secondary Drinking Water Regulations
RIA	Regulatory Impact Analysis
SAB	Science Advisory Board
SDWA	Safe Drinking Water Act
SMCL	Secondary maximum contaminant level
SNC	Significant noncomplier

PART 6

The Case of Solid Wastes and Toxic Substances

In 1978, the state of New York urged 1,000 families to leave their homes and ordered an emergency evacuation of 240 others—all residents of a Niagara Falls community known as Love Canal. Built on a site that 30 years earlier had been a chemical dumping ground, the ill-fated Love Canal was eventually declared a disaster area by President Carter. In 1982, the residents of Times Beach, Missouri, learned that their groundwater and soil contained dangerously high levels of dioxins—the result of contaminated road oil that had been used in the town 11 years earlier. Times Beach was completely evacuated. These events and countless others awakened society to the potential risks of solid waste pollution and exposure to toxic chemicals.

The extent of these problems is not completely known, and efficient solutions appear to be just as elusive. Not only are current waste generation and disposal practices at issue, but also the damages caused by waste mismanagement of the past. In a broad sense, there are two interrelated issues to be addressed—excess waste generation and the use of toxic substances that eventually become part of the waste stream. Chemicals leaking from buried waste and mismanaged disposal sites have caused damage to human health and the ecology—some of it irreparable. Less severe, but nonetheless cause for concern, is the growth rate of municipal trash generation—millions of tons of bottles, cans, food scraps, and the like. In many communities, residents and public officials are still searching for cost-effective ways to collect and dispose of the accumulating heap. What makes these problems all the more troubling is that the environmental damage cuts across all media—land, water, and air.

Policy development aimed at solid waste pollution got a late start in the United States, and most argue that current legislation is inadequate. In the 1960s and 1970s, U.S. environmental policy was firmly focused on controlling air and water pollution, but little in the way of substantive federal legislation was passed during this period to manage and reduce the solid waste stream. Why the lack of initiative? For one thing, population centers had remained fairly compact for a long time and seemingly were able to manage the wastes being generated. Managing solid wastes was viewed as a local responsibility and not one that required national action. Furthermore, the problems associated with the treatment and disposal of chemical wastes, particularly synthetics, did not accelerate until the 1970s.[1]

Regardless of the logic of these explanations, there is one root cause that underlies all of them. Society failed to recognize the significance of solid waste as an environmental and health risk. The result? Policymakers are having to play catch-up with a problem that, at least until recently, has been advancing in magnitude and severity.

In this module, we assess this environmental and social dilemma and analyze the solutions brought forth by government. In Chapter 17, we focus on hazardous waste pollution—the risks it poses to society and the federal legislation designed to minimize those risks. In Chapter 18 we examine the challenges of managing nonhazardous wastes and the policy efforts of both the federal and lower levels of government. Our focus shifts in Chapter 19, where we analyze federal laws aimed at controlling pesticides and other toxic substances *before* they are introduced into commerce and eventually enter the waste stream. Not unlike our analyses of air and water degradation, the tools of economic analysis are used throughout to assess government's effectiveness in responding to environmental risks.

[1] U.S. EPA, Office of Solid Waste and Emergency Response (October 1985).

Chapter 17

Managing Hazardous Solid Waste and Waste Sites

An official order to evacuate homes and businesses following the discovery of a hazardous waste leak is a chilling reminder of the potential risks of solid waste pollution. All too often, there are accounts of strange odors emanating from basements, tainted water supplies, or health symptoms for which doctors have no diagnosis—reports that are eventually linked to a waste accident or cover-up from years past. Once-thriving communities in the United States like Love Canal, New York, and Times Beach, Missouri, became ghost towns because of the toxic effects of hazardous waste. Why did the problem reach such a level before something was done?

By the time federal policies were formulated in the United States, the nation was already suffering from the ill effects of mismanaged hazardous wastes that had been accumulating for decades. At the same time, the growth rate of the waste stream was rising due to urban development, industrialization, population growth, and increasing chemical usage. There was also a glaring lack of information about the magnitude of the problem. In fact, some argue that we still do not have a complete picture of the extent and severity of solid waste pollution in this country.

What we do know is that the damage extends to all environmental media, which means that the effects are widespread, difficult to control, and costly to society. Accepting this characterization, it should not be surprising to learn that there are several major legislative acts aimed at the dilemma. While some deal with the damage from past mismanagement, others attempt to control the present solid waste stream.

Investigating this multidimensional control approach is the objective of this chapter. Our aim is to explain the policy position of the United States in dealing with hazardous waste. The operative issue is whether or not existing policy is effective and, if not, whether there are viable alternatives that should be implemented. We begin with an overview of the hazardous waste problem and a brief discussion of how the federal response evolved. We then focus on the two major U.S. laws aimed at hazardous waste—the Resource Conservation and Recovery Act (RCRA) and Superfund. Here, the goal is to explain the overall intent and structure of these laws so that they can be analyzed from an economic perspective. A list of acronyms is provided for reference at the end of the chapter.

hazardous solid wastes
Any unwanted materials or refuse capable of posing a substantial threat to health or the ecology.

waste stream
A series of events starting with waste generation and including transportation, storage, treatment, and disposal of solid wastes.

Characterizing the Hazardous Waste Problem[1]

Hazardous solid wastes are any unwanted materials or refuse capable of posing a substantial threat to health or the environment. A diverse range of substances fall into this broad category, so it is useful to consider the hazardous constituents of wastes as belonging to subgroups like pesticides and synthetic chemicals, as listed in Table 17.1. Despite their toxicity, these substances serve important functions in productive activity. Often, the properties that make them useful in production are precisely those that make them dangerous to society once they enter the **waste stream.** The **waste stream** refers to the series of events starting with

[1] Following the literature and government data reporting, we have omitted from this discussion any reference to medical wastes and radioactive wastes.

Table 17.1 *Types of Hazardous Substances*

Hazardous Substance	Description
Acids and bases	Acids have a low pH, between 0 and 6, and bases have a high pH, between 8 and 14. Highly acidic or basic substances are corrosive, which means that discarded acids and bases can corrode the containers in which they are stored and leach into land and nearby waterways.
Heavy metals	Heavy metals are highly toxic and do not readily break down in the human body. Examples include lead, arsenic, cadmium, barium, copper, and mercury.
Reactives	Reactives, which include explosives and flammables, behave violently when combined with air or water. Common examples are petroleum or natural gas by-products.
Synthetic organic chemicals	Synthetic organic chemicals are man-made substances that are hydrocarbon based, some created from relatively new processes like chemical splicing and molecular engineering. Two important subcategories are solvents and pesticides.
Solvents	Solvents are liquids that can dissolve or disperse one or more substances. Examples include ethylene, benzene, pyridine, and acetone.
Pesticides	Pesticides are used to prevent, destroy, or repel any unwanted form of plant or animal life. Some are highly resistant, which allows them to persist in the environment for a long time without the need for frequent reapplication. Some are nearly insoluble in water, making their use as insecticides on crops resistant to rainfall. Many are acutely toxic.

Sources: Drawn from Phifer and McTigue (1988), chap. 3; Epstein, Brown, and Pope (1982), pp. 14–26; U.S. EPA (December 1990), p. 120.

waste **generation** and proceeding though the **transportation, storage, treatment,** and **disposal** of these materials.[2] Environmental and health risks arise both because of excess generation and improper management of these materials once they enter the waste stream.

Magnitude and Source of the Problem

Extent of the Problem

In the United States, hazardous waste generation has been estimated at about 173 million tons per year. This translates to about 0.66 tons per person annually. Nations everywhere must deal with the risks of accumulating hazardous wastes. In Italy, for example, the reported annual level of hazardous wastes is 2.7 million tons, or about 0.04 tons per person, and in Mexico, the level is estimated at 8.0 million tons, or 0.10 tons per person.[3] Third world countries also are confronting the problem as they evolve into more industrialized economies. These developing nations also have to contend with wastes deliberately dumped inside their borders by firms from advanced countries seeking to avoid the costs of and regulatory constraints on disposal.[4] There is no doubt that the problem of hazardous waste is worldwide, and there is no question that there are serious risks in ignoring it.

[2] To read more about the waste stream, visit the EPA's Office of Solid Waste Web site at **http://www.epa.gov/epaoswer/osw/**, and for more detail about hazardous wastes, visit **http://www.epa.gov/osw/hazwaste.htm**.
[3] OECD (1999b), pp. 168–70, 317–18. These data refer to wastes to be controlled according to the Basel Convention on the Control of Transboundary Movements of Hazardous Wastes and Their Disposal. Note that international comparisons cannot be made with precision, given differences in definitions and estimation methods across nations.
[4] Kharbanda and Stallworthy (1990), p. 105.

Application

17.1 *The Hazardous Waste Site Called Love Canal*

In the late 1880s, an entrepreneur named William T. Love began excavating a canal that would connect the upper and lower ends of the Niagara River in upper New York state. Envisioned as the future site of a major industrial complex, the Love Canal project was designed to make use of the enormous supply of hydroelectric power that would be provided naturally by the 280-foot drop between the two ends of the river. Before the plan was finished, alternating current (AC) was developed, and the once futuristic Love Canal project lost its appeal. Love's visionary complex was never finished, but one segment of the partially constructed canal remained. It simply filled with water and for a time was used for swimming.

In the 1940s, Hooker Chemical and Plastics Corporation, through agreement with the canal's owner, began using the site to dump chemical wastes. Ultimately buying the canal to use as a dump site, Hooker discarded over 21,000 tons of chemicals into the Love Canal between 1942 and 1952. Among the toxic wastes were benzene derivatives, dioxins, and trichlorophenol (TCPs) contaminated with a highly potent carcinogenic substance called TCDD. According to one source, TCDD is so lethal that less than three ounces could kill all the residents of New York City (Epstein, Brown, and Pope 1982, p. 93).

Some years later, the postwar construction boom that characterized America in the 1950s prompted the city of Niagara Falls to condemn the properties surrounding the canal to make room for a new school and a residential area. In 1953, Hooker signed the canal over to the city for $1 in exchange for a release from any liability for damage associated with the dump site. Despite the danger, the canal was filled in, and construction proceeded. The school opened in 1955, and hundreds of homes were built.

Buyers of the new homes were unaware of the chemical dump site. Soon after moving in, they began to complain about fumes coming from the former dump site and chemical burns their children got from playing in nearby fields. Lawns and gardens refused to grow, and pools of black liquid began to surface in backyards. In the mid-1970s, the same black sludge started to seep into basements after heavy rains. Finally, in 1977 after numerous complaints, the city called in a consultant. The findings confirmed the residents' worse fears—toxic chemicals had leached from the canal and contaminated the groundwater and surface water.

One resident, Lois Gibbs, took action, suspecting that her son's history of health problems might have been caused by the chemical leaks. What followed was a complex and bitter struggle for Gibbs and the other Love Canal residents. Despite mounting evidence, the state's response to the problem was less than adequate. Gibbs and the other residents formed the Love Canal Homeowners Association Inc. and took matters into their own hands. Every level of government eventually became involved along with public agencies, health officials, researchers, and the media. In 1979, the Department of Justice filed a series of lawsuits against Hooker Chemical, the city of Niagara Falls, the city's board of education, and the Niagara County Health Department. In May 1980, President Carter declared Love Canal a disaster area, and that summer $15 million in grants and loans was offered to New York to purchase new homes for the relocated families.

To learn more about Love Canal, visit **http://ublib.buffalo.edu/libraries/projects/lovecanal** at the State University of New York at Buffalo.

Sources: Epstein, Brown, and Pope (1982), chap. 5; Gibbs (1982); Griffin (January/February 1988).

Table 17.2 ***Hazardous Waste Generation in the United States by Major Industry***

Industry	Percentage of Hazardous Waste Generation
Chemical	79.0
Petroleum refinery	7.0
Transportation equipment	1.0
Fabricated metals	1.0
Primary metals	1.0
Electrical equipment	0.4
National security	0.4
Other	9.0

Source: U.S. EPA (August 1988), p. 80, fig. L-2, from U.S. EPA, Office of Solid Waste (n.d.).

Note: Excludes radioactive wastes.

All environmental media—the atmosphere, groundwater and surface waters, and soil—are vulnerable to hazardous waste contamination. Ocean pollution, soil contamination, disease, fish kills, and livestock loss are among the potential damages. The health and ecological effects can be severe and long term, particularly from exposure to persistent pollutants like PCBs and DDT that bioaccumulate in the environment.

Water and soil contamination are the most prevalent type of damage linked to hazardous waste sites. Contamination of drinking water can occur when rainwater absorbs pollutants as it runs through disposal sites. The resulting leachate can pollute groundwaters and surface waters—damage that is costly and sometimes impossible to rectify. Surrounding soil can be polluted in the same way. Food crops grown in such contaminated soil absorb pollutants that are later ingested by humans and animals.[5] Particularly vulnerable are households living near hazardous waste sites. One of the most often cited cases of an entire community affected by the contamination is Love Canal, New York. A brief account of the story is given in Application 17.1.

Sources of Hazardous Waste

Just who is responsible for generating all these toxic waste materials? It turns out that every sector of the economy—households, industry, government, and institutions—contributes to the hazardous waste stream.[6] While none of these sources should be ignored, there is no question that most hazardous wastes are generated by industry. In the United States, 99 percent of industrial hazardous wastes comes from so-called large-quantity generators, those producing over 2,200 pounds per month. As shown in Table 17.2, the chemical and petroleum industries are responsible for generating most of the hazardous waste in the United States, with 79 percent attributable to the former.[7]

From an economic perspective, any adverse effects associated with the use and disposal of hazardous substances are **negative externalities.** Hence, policy is needed to regulate markets where the problem arises and internalize the external costs. In the United States, the early development of solid waste policy was anything but aggressive. Most argue that

[5]Council on Environmental Quality (March 1992), p. 338, table 90, citing U.S. EPA (1991).
[6]For example, U.S. households dispose of an estimated 385 million gallons of used motor oil each year—literally 35 times more oil than the *Valdez* spilled into Alaskan waters in 1989.
[7]U.S. EPA (August 1988), p. 80.

this segment of environmental legislative history should be characterized as having been reactive—even passive at times. This lack of spirited attention to solid waste problems by federal officials was matched by the same absence in the private sector. The environmentalism of the 1960s was preoccupied with fouled rivers and streams, urban smog, and threatened wildlife, paying little attention to the accumulating trash and the risks of chemical wastes.[8]

Evolution of U.S. Solid Waste Policy

It is often suggested that the U.S. government did virtually nothing in the way of controlling solid wastes until after the Love Canal saga made national news in the late 1970s. This suggestion on face value is erroneous. The primary piece of federal legislation on solid wastes, the Resource Conservation and Recovery Act, had been passed by Congress two years prior to that infamous event. Furthermore, the draft version of the Superfund legislation dealing with abandoned and uncontrolled hazardous waste sites had been completed in October 1978—only two months after the nation first learned about Love Canal and well before the related publicity that continued into 1980.[9]

The point is that the United States did have a national policy agenda on solid waste in the 1970s, but the legislation had passed with little scrutiny and was not fully responsive to the magnitude of the problem. Hence, while Love Canal was not the catalyst for introducing federal solid waste legislation, this event and others like it motivated public officials and private citizens to give the matter more attention. Several federal laws and amendments speak to this evolution, as shown in Table 17.3. An overview of the chronology that ties together these laws and their primary objectives follows.

Federal Recognition of the Solid Waste Problem

Solid Waste Disposal Act (SWDA) of 1965

U.S. policy on waste control originated with the Solid Waste Disposal Act (SWDA), passed in 1965. Its primary objectives were to provide financial assistance to state and local governments for planning waste management programs and to initiate a national research plan aimed at finding better disposal methods.

Resource Recovery Act of 1970

In 1970, the SWDA of 1965 was amended by the Resource Recovery Act. More significant in intent, this act marked a slight shift in emphasis from disposal to conservation by encouraging recycling and technological development to reduce waste generation. However, the emphasis continued to be on nonhazardous waste and the use of land disposal in local communities.[10]

Developing Policy to Control the Risks of Hazardous Wastes

Resource Conservation and Recovery Act (RCRA) of 1976

"cradle-to-grave" management system A command-and-control approach to regulating hazardous solid wastes through every stage of the waste stream.

In 1976, the Resource Conservation and Recovery Act (RCRA) was passed by Congress as amendments to the original Solid Waste Disposal Act. The hallmark of RCRA was a distinct set of regulations within the Act that established a **"cradle-to-grave" management system** for controlling hazardous wastes. The metaphor refers to a policy strategy for controlling, managing, and tracking these wastes throughout every stage of the waste stream, as shown in

[8] One important exception is Rachel Carson's *Silent Spring* (1962), which warned of the risks of pesticide chemicals.
[9] Landy, Roberts, and Thomas (1990), p. 140.
[10] U.S. EPA, Office of Solid Waste and Emergency Response (October 1985).

Table 17.3 ***Overview of U.S. Legislation on Hazardous Solid Wastes***

Legislative Act	Major Provisions
Solid Waste Disposal Act (SWDA) of 1965	Provided financial assistance to state and local governments in planning solid waste disposal programs; initiated a national research plan aimed at finding better waste disposal methods.
Resource Recovery Act of 1970	Encouraged recycling and technological controls to reduce waste at the generation point; continued to promote land disposal of wastes in local communities and to emphasize nonhazardous solid waste.
Resource Conservation and Recovery Act (RCRA) of 1976	Represented the first U.S. official position on hazardous waste control; established a cradle-to-grave management system for hazardous waste; delegated the administration of nonhazardous waste mainly to state governments.
Hazardous and Solid Waste Amendments of 1984	Reauthorized RCRA and broadened federal control; shifted emphasis from land disposal to waste reduction and toward improved treatment technologies for hazardous wastes; elevated standards for hazardous waste facilities.
Comprehensive Environmental Response, Compensation, and Liability Act (CERCLA) of 1980 (known as Superfund)	Called for the preparation of a national inventory of hazardous waste site information from which to identify sites posing the greatest threat to health and the ecology; placed these sites on the National Priorities List (NPL); designated a fund of $1.6 billion to clean up NPL sites and pay for associated damages.
Superfund Amendments and Reauthorization Act (SARA) of 1986	Reauthorized CERCLA; increased the fund to $8.5 billion to be financed primarily from feedstock taxes; called for federal action on 375 sites within a five-year period; promoted permanent cleanup technologies.

Sources: Wolf (1988), chaps. 6 and 7; U.S. EPA, Office of Solid Waste and Emergency Response (October 1985); Resource Conservation and Recovery Act of 1976; Hazardous and Solid Waste Amendments of 1984; Comprehensive Environmental Response, Compensation, and Liability Act of 1980; Superfund Amendments and Reauthorization Act of 1986.

Figure 17.1. A separate set of provisions delegated the administration of nonhazardous waste primarily to state governments.[11]

Hazardous and Solid Waste Amendments of 1984

Congress reauthorized and strengthened RCRA through the Hazardous and Solid Waste Amendments of 1984.[12] The new provisions reinforced the evolving policy shift away from land disposal and toward waste reduction efforts and better treatment technologies. Standards for hazardous waste facilities were strengthened with new safety requirements such as groundwater monitoring equipment and leachate collection systems. Control was broadened by adding two groups of facilities to those regulated under RCRA—owners and operators of certain underground storage tanks and all small-quantity waste generators. Stringent rulings

[11] The provisions applicable to hazardous waste are given in Subtitle C of RCRA. In the next chapter, we will investigate and analyze the rulings in Subtitle D of RCRA aimed at nonhazardous municipal solid wastes.
[12] Despite the change in title, conventional practice is to use the acronym RCRA to refer to this legislation.

Figure 17.1 ***The "Cradle-to-Grave" Management System Under RCRA: Controlling, Managing, and Tracking the Hazardous Waste Stream***

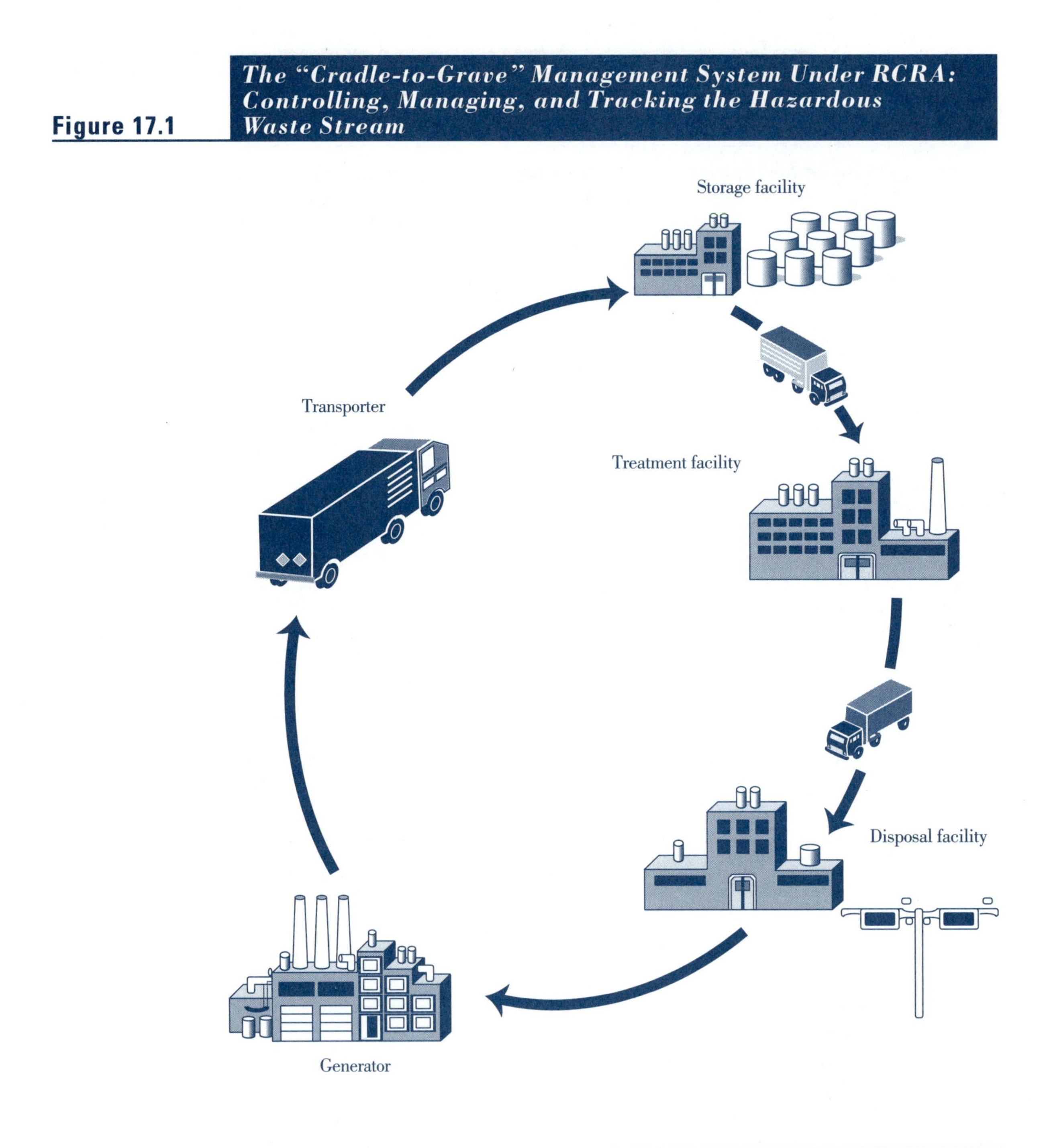

Source: U.S. EPA, Office of Solid Waste (November 1986), p. 11.

were imposed on underground storage tanks that house potentially dangerous substances. Application 17.2 has more on these rulings and the potential risks of underground storage tanks.

Beyond RCRA's controls on *current* solid waste management, guidelines and financial support were needed to address the damages caused by *past* contamination and to clean up abandoned and uncontrolled hazardous waste sites. On the heels of the Love Canal dilemma, the

Application

17.2 *Hazardous Wastes from Leaking Underground Storage Tanks*

Current estimates indicate that there are over 5 million underground storage tanks containing petroleum products or other hazardous substances in the United States. About 2 million of these are regulated under RCRA. The practice of burying storage tanks was done as a safety measure to prevent fires. Ironically, these tanks now have become the source of a serious environmental hazard. Some 400,000 of the underground storage tanks in the United States are believed to be leaking—a number expected to rise in the future. These tanks are so large, even a pinhole can lead to a serious waste problem. About 49 percent of the regulated storage tanks in the United States are filled with petroleum and owned by gasoline stations, with another 47 percent owned by industries that store petroleum for their own use, such as airports, farms, golf courses, and firms with large trucking fleets. The remaining 4 percent are filled with other chemicals stored for industrial use.

A major source of concern is that the inventory of buried tanks is aging, making leakages more likely and increasingly so as time goes on. It turns out that many of these tanks were placed in the ground in the 1950s and 1960s during the oil boom of that era. Studies have indicated that about one-third of fuel storage tanks in the United States are over 20 years old or of unknown age. Unfortunately, many of these older vessels were constructed of bare steel, making them vulnerable to corrosion. Adding to the dilemma are all the abandoned storage tanks left by defunct gas stations that failed during the OPEC crisis of the 1970s. In these cases, it is feared that the tanks may not have been closed properly. Since so much time has passed, determining ownership and responsibility for any discovered leaks is difficult at best.

According to the EPA, the primary cause of leaks is corrosion of steel tanks. Other important causes are spills and overfills. Spills occur when the hose that delivers the contents is pulled from the entry pipe before it has completely drained, and overfills result from trying to add more product than the tank can hold. Piping failures and loose fittings for attachments such as pumps and vent lines can also be the cause of leakages.

Leaks from underground storage tanks or their piping can contaminate groundwater and soil, poison crops, cause fires or explosions, and harm human health. Human exposure can occur through direct contact with contaminated soil or water or via inhalation of toxic emissions. Aside from breathing dangerous vapors released into a nearby building, inhalation also can occur during showering with contaminated water when volatile components are released.

Recognizing the environmental hazard, Congress has responded on a number of fronts. Perhaps the most important move was adding Subtitle I to RCRA as part of the 1984 Amendments. This section of the law required the EPA to develop and implement a comprehensive plan to address underground storage of petroleum and other hazardous substances. Simultaneously, Congress banned the use of steel tanks as underground storage devices starting in 1985. A year later, Congress established the Leaking Underground Storage Tank (LUST) Trust Fund, which would be financed by a 0.1 cent tax on each gallon of motor fuel. Since its inception, $2.6 billion has been added to the LUST Trust Fund, which is appropriated by Congress for the EPA. Acting through its regional offices, the EPA disperses about 85 percent of these monies to states to fund their assistance with the administration, oversight, and enforcement of each cleanup action.

Sources: U.S. EPA (August 1988), pp. 102–105; U.S. EPA (December 1990), p. 16; U.S. EPA, Office of Solid Waste and Emergency Response (January 1992); U.S. EPA, Office of Solid Waste and Emergency Response, Office of Underground Storage Tanks (March 25, 2002); U.S. EPA, Office of Solid Waste and Emergency Response, Office of Underground Storage Tanks (February 14, 2002).

Comprehensive Environmental Response, Compensation, and Liability Act (CERCLA), or Superfund as it is more commonly known, was passed in 1980.

Comprehensive Environmental Response, Compensation, and Liability Act (CERCLA) of 1980

Under CERCLA, the federal government was authorized to clean up contaminated sites and to recover damages from parties identified as responsible. Past and present owners of hazardous waste site facilities were required to notify the EPA of conditions at these sites by 1981. This baseline information along with data collected on an ongoing basis would comprise a national inventory of hazardous waste sites called the **Comprehensive Environmental Response, Compensation, and Liability Information System (CERCLIS).** This system is used to identify those sites posing the greatest threat to human health and the ecology—a classification referred to as the **National Priorities List (NPL).**

Comprehensive Environmental Response, Compensation, and Liability Information System (CERCLIS)
A national inventory of hazardous waste site data.

The first "Superfund" of $1.6 billion was established to finance the cleanup of priority sites and to pay for damages. As it turned out, CERCLA was a national failure on two counts:

1. The $1.6 billion Superfund was used to remediate only eight of the thousands of contaminated sites across the nation.[13]
2. The CERCLIS inventory did not assess the full magnitude of the problem, as was its intended purpose.[14]

Superfund Amendments and Reauthorization Act (SARA) of 1986

These failures were the primary motivation for the revision and reauthorization of CERCLA through the Superfund Amendments and Reauthorization Act (SARA), passed in 1986. SARA raised the Superfund to $8.5 billion. This was to be financed primarily from **feedstock taxes,** including $2.75 billion from a petroleum tax and $1.4 billion from a tax on raw chemicals. Prior to their expiration at the end of 1995, these taxes raised some $1.5 billion per year.[15] Although there have been attempts to reauthorize the tax, none have yet materialized.

feedstock taxes
Taxes levied on raw materials used as productive inputs.

An important reform accomplished through SARA was the promotion of *permanent* cleanup technologies to replace short-term solutions of burying toxic wastes or transferring them from one site to another. Another was a requirement to initiate federal action on 375 sites within a five-year period. Despite these changes, Superfund continues to be the source of political debate and controversy.

Recent Policy and Brownfields[16]

In an effort to squelch some of the controversy, President Bush signed a bill in January 2002 dealing with certain Superfund-related concerns. The bill, officially called the Small Business Liability Relief and Brownfields Revitalization Act, combines two issues put forth by the Senate and the House of the 107th Congress. One of these addresses the extensive liability associated with Superfund (to be discussed later in the chapter). Specifically, the new legislation protects from liability households and small businesses that dispose only municipal solid waste at Superfund sites.

brownfields
Abandoned or underutilized industrial sites where redevelopment is discouraged by actual or perceived contamination.

The second element of the bill provides for a brownfields abatement program with a budget allocation averaging $250 million per year. **Brownfields** are abandoned or underutilized industrial sites that are less contaminated than Superfund sites, but their actual or perceived contamination deters development. (Visit **http://www.epa.gov/epahome/hi-brownfields.htm** for more information.) Attention to brownfields is said to be coincident with the view that it might be time to bring the Superfund program to a close, assuming it has dealt with the most

[13] Wolf (1988), p. 227.
[14] U.S. GAO (December 1987), pp. 2–3.
[15] Reisch (November 23, 1998).
[16] Information on the new legislation discussed here is drawn from Reisch (March 21, 2002).

seriously contaminated sites. This type of sentiment along with concerns about the program's costs and associated litigation is partly responsible for congressional failures to reauthorize this legislative act.[17]

Controlling Hazardous Wastes: RCRA[18]

To minimize the risks of hazardous waste pollution, efforts must be made to reduce the quantity and toxicity of the waste stream. Most agree that two general approaches are needed to achieve these broad-based objectives:

- **Source reduction,** which is a preventive strategy aimed at the generation stage
- **Waste management,** which refers to strategies to control those wastes that cannot be eliminated

source reduction
Preventive strategies to reduce the quantity of any hazardous substance, pollutant, or contaminant released to the environment at the point of generation.

waste management
Control strategies to reduce the quantity and toxicity of hazardous wastes at every stage of the waste stream.

While RCRA's objectives make a general reference to preventive strategies, there is little question that management of the waste stream is the primary emphasis. A command-and-control framework characterizes the implementation plan, just as is true for clean air and water initiatives.

Primary responsibility for controlling hazardous wastes is assigned to the federal government—mainly through the EPA. However, Congress provided little guidance as to *how* the agency should design and implement its newly drafted policy position effected by the 1976 RCRA. This absence of legislative muscle is often blamed for what became long delays in launching the program and in deflecting opposition from those most affected by the new regulations. Just as critical to the policy challenge was the uncertainty about the extent of hazardous waste pollution. Absent the obvious sources that had become national news, no one—not even major federal agencies—had a complete picture of the problem. The EPA's response was to devise a command-and-control approach intended to give structure to what must have seemed little more than ordered chaos.

The "Cradle-to-Grave" Management Approach[19]

The RCRA program emerged as the "**cradle-to-grave" management system** —a multipronged command-and-control approach to regulating hazardous waste. Its four major components are:

- **Identification** of hazardous wastes
- A **national manifest system** for tracking and monitoring the movement of wastes
- A **permit system** for **treatment, storage, and disposal facilities (TSDFs)**
- Development of **standards** for TSDFs[20]

Identification of Hazardous Wastes[21]

A fundamental element of the management system is a procedure for identifying wastes considered to be hazardous and therefore subject to federal regulations. The rules are highly detailed, since an error of omission would allow a dangerous material to escape federal control.

---

[17]To read more about this reauthorization process during the 105th and 106th Congresses, see Reisch (November 23, 1998) and Reisch (October 30, 2000), respectively, each of which can be accessed online at **http://cnie.org/nle/crsreports/waste/waste-17.cfm** and at **http://cnie.org/nle/crsreports/waste/waste-28.cfm**.
[18]For more information on RCRA, visit **http://www.epa.gov/rcraonline/**.
[19]Much of the following discussion is drawn from U.S. EPA, Office of Solid Waste (November 1986).
[20]U.S. Congress, OTA (1983), chap. 7.
[21]See 40 CFR 261, Identification and Listing of Hazardous Waste.

Figure 17.2 *U.S. Characteristic and Listed Wastes in 1999*

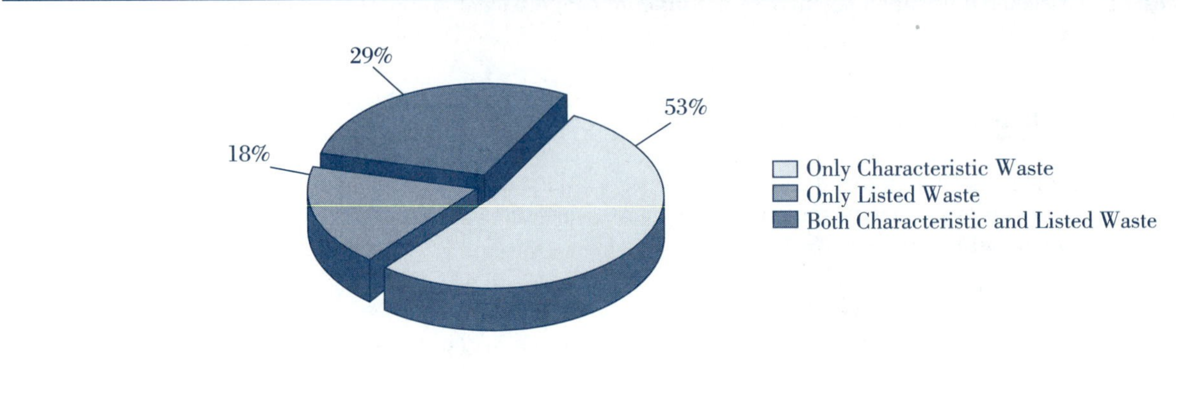

Source: U.S. EPA, Office of Solid Waste and Emergency Response (June 2001), p. 1-13, exhibit 1.9.

Essentially, a waste is considered hazardous under the law if it falls into one of two defined categories:

characteristic wastes Hazardous wastes identified as those exhibiting certain characteristics that imply a substantial risk.

listed wastes Hazardous wastes that have been preidentified by government as having met specific criteria.

- **Characteristic wastes,** which are those with characteristics or attributes that imply a substantial risk. In the United States, these attributes are ignitability, corrosivity, reactivity, and toxicity.
- **Listed wastes,** which are those preidentified by the EPA as having met certain criteria such as the presence of toxic or carcinogenic constituents.

The proportions of hazardous waste in each category based on 1999 data are shown in Figure 17.2. Using these categories, waste generators are responsible for determining which of their wastes are hazardous and assuring that these are managed in compliance with federal law.

The National Manifest System for Tracking Wastes

manifest A document used to identify hazardous waste materials and all parties responsible for its movement from generation to disposal.

If a generator chooses to transfer any hazardous wastes offsite for treatment, storage, or disposal, that movement is tracked by the **national manifest system.** Once the wastes are ready for transport, the generator must prepare a document called a **manifest** that identifies the hazardous material and all the parties responsible for its movement. The document remains with the shipment from the point of generation through to its final disposal. The objective is to assure that dangerous waste materials are accounted for through every phase of the waste cycle, reducing the opportunity for illegal dumping.

The Permit System

permitting system A control approach that authorizes the activities of TSDFs according to predefined standards.

While the tracking system monitors the location of hazardous waste, a **permitting system** controls its management at the **treatment, storage, and disposal facilities (TSDFs).**[22] Every TSDF must obtain a permit to operate. The permitting process is stringently controlled, since its purpose is to ensure that TSDFs meet federal standards, many of which are aimed at protecting groundwater from waste contamination.

[22] For basic information on the activities of TSDFs, visit **http://www.epa.gov/epaoswer/osw/tsd.htm.** To learn more about the permitting of hazardous wastes, see **http://www.epa.gov/epaoswer/hazwaste/permit/prmtguid.htm.**

Standards

Two sets of standards control the waste management practices of TSDFs. **General regulatory standards** apply to *all* types of TSDFs and control generic functions like inspections, emergency plans, and participation in the manifest program. There are also **technical regulatory standards** that outline procedures and equipment requirements for specific types of waste facilities. For example, storage facilities must follow careful procedures to avoid leakages into the surrounding environment. Disposal sites must use technologies such as cover systems, liners, and leakage-detection systems.

Although these four components of the "cradle-to-grave" system provide controls over the hazardous waste stream, the program is inadequate in one important respect—it does not indicate nor even suggest a hierarchy of preferred methods for managing these wastes. Prior to subsequent revision, this omission meant that virtually nothing was being done to slow the overuse of land at the disposal stage.[23] From a broader perspective, it also meant that national policy was not supporting preventive strategies to reduce the size of the hazardous waste stream. Responding to this inadequacy, Congress passed the Hazardous and Solid Waste Amendments of 1984. Unlike their predecessor, these amendments outline in detail how hazardous wastes are to be controlled and explicitly set forth restrictions on land disposal. In so doing, the law prioritizes hazardous waste treatment technologies, which is intended to provide an incentive for source reduction and hence prevention of hazardous waste pollution.

Moving Toward Pollution Prevention

Several references in the 1984 Amendments speak to the policy shift away from land disposal and toward preventive solutions. The law's general provisions were revised to include the following mandate:

> reliance on land disposal should be minimized or eliminated, and land disposal, particularly landfill and surface impoundment, should be the least favored method for managing hazardous wastes.[24]

Other rulings specifically prohibit land disposal of untreated hazardous wastes except under certain conditions. These controls are intended to encourage facilities to find alternatives that prevent pollution. To reinforce the emphasis on prevention, the 1984 law added the following to RCRA's statement of objectives:

> The Congress hereby declares it to be the national policy of the United States that, wherever feasible, the generation of hazardous waste is to be reduced or eliminated as expeditiously as possible.[25]

More about pollution prevention will be discussed in subsequent chapters. Information on this issue is also available on the Internet at **http://www.epa.gov/p2/**.

Analysis of U.S. Hazardous Waste Policy

The nation's "cradle-to-grave" management system is a classic example of a command-and-control policy approach. As such, it lacks incentives, which decreases the likelihood of an efficient or cost-effective outcome. To investigate this hypothesis, we consider the following aspects of the RCRA program:

- Identification of hazardous waste
- Standard-setting process

[23] For more detail, see Fortuna and Lennett (1987), chap. 9.
[24] 42 U.S.C. § 6901(b)(7).
[25] 42 U.S.C. § 6902(b).

- Manifest system
- 1984 land restrictions

Risk-Based Uniform Rules of Identification

Central to RCRA's control approach is the identification of hazardous wastes. Two elements of this process merit consideration. One is the absence of **risk-benefit analysis** in defining the identifying categories, and the other is the **uniformity** in how these categories are implemented.

Risk-Based Criteria

To begin, the law requires that hazardous waste be identified according to characteristics and criteria that are **risk based.** Although the rulings attempt to accurately identify the risks of exposure to hazardous waste, they include no provision for balancing these risks with the benefits provided by these materials before they enter the waste stream. That being the case, all wastes satisfying these criteria are regulated with the same stringency regardless of how they originated. Consequently, a waste material associated with a product that is vital in use is treated under the law in exactly the same way as one generated from an output of relatively low value to society.[26]

Applying Criteria Uniformly

Another problem is that the identifying criteria are applied **uniformly.** There are no qualifiers to allow for differences in the degree of toxicity across various waste constituents. Effectively, the procedure sets up a zero-one decision-making process that translates to a set of uniform standards. If a waste falls into one of the predefined categories, it is hazardous; if not, it is nonhazardous by default. Also, the law does not allow for any flexibility in determining how much of a given substance poses a hazard to society. Yet there can be wide variation in the risk of exposure to the same quantity of different hazardous substances.[27] From an economic perspective, ignoring this variation is **allocatively inefficient.** Since there are known differences in risk across hazardous waste constituents, it is also true that the marginal benefits of abating these constituents are different. If marginal abatement costs are the same, society's welfare could be improved by allocating more resources to abating relatively high-risk substances and fewer to those posing a lower risk. Yet there is no provision in the law for such flexibility.

Benefit-Based Uniform Standards

Benefit Based

RCRA controls hazardous waste management primarily through standards imposed on the generators, transporters, and treatment, storage, and disposal facilities (TSDFs). According to the law, these standards are to be established "as may be necessary to protect human health and the environment."[28] Such a mandate clearly defines the standard-setting process as **benefit based** with no consideration for economic costs.

The lack of cost considerations is particularly problematic for standards with long-run implications. Examples include the technical standards for TSDFs relating to groundwater monitoring, closure, and postclosure procedures. Standards on postclosure procedures guide a facility's activities for 30 years after a facility is closed. The purpose is to prevent active hazardous waste facilities from becoming problem sites after they cease operation—undoubtedly a response to the nation's concerns with abandoned and uncontrolled sites.[29] Al-

[26] It is interesting to note that certain wastes are exempt from this identification process, such as those associated with the production of crude oil and natural gas. The importance of these products suggests that a risk-benefit assessment may have been used to justify these exemptions. It is also true that the initial RCRA was passed during a period of great uncertainty about U.S. oil supplies.
[27] For more detail, see U.S. Congress, OTA (1983), chap. 6.
[28] RCRA, sec. 3004(a).
[29] U.S. Congress, OTA (1983), pp. 283–85.

though the intent seems to be well motivated, the long-term nature of these standards assures that compliance costs will be significant. Yet costs are not allowed to influence the standard-setting process. As long as this is the case, the standards so fundamental to federal policy likely will not be set to achieve **allocative efficiency.**

Uniform Standards

Another criticism of the standards-based approach is that with few exceptions, it offers states no flexibility in how they administer RCRA's hazardous waste program. While state officials can submit their own plans, these are approved only if they are equivalent to the national program or consistent with those applicable in other states.[30] There is virtually no provision for site-specific differences—either in the facilities themselves or in the types of wastes they handle. The lack of consideration for such differences means that the criterion of **cost-effectiveness** likely will not be met.

Failures of the Manifest System

Benefit Based

One of the more prominent elements of RCRA's "cradle-to-grave" approach is the manifest system. Congress's intent in establishing this elaborate tracking procedure was to deter illegal disposal and reduce the adverse effects associated with such practices.[31] The system is a definitive command-and-control instrument. There are no incentives, no pricing mechanisms, and no options. It was a no-nonsense response to a tough problem, and in that sense the approach was understandable, albeit inefficient. Why? Because its objective is solely **benefit based**—to halt the damage caused by mismanaged hazardous wastes. Characteristic of the RCRA program, economic considerations play no role in its design or implementation.

Limited in Scope

Beyond the efficiency implications, a more fundamental issue is the effectiveness of the manifest plan. It turns out that despite its elaborate design, the tracking system has not been a major factor in controlling hazardous wastes because of its limited scope. For all its complexity, the tracking system applies only to wastes that are transported off-site by the generator. According to estimates, these represent only 4–5 percent of the nation's hazardous waste stream. Most large firms manage wastes on their own premises. It is smaller businesses, particularly those in crowded urban areas, that rely on other facilities to treat and dispose of their wastes.[32] It is true that this proportion should increase over time due to the 1984 revision to make small-quantity generators subject to RCRA. Nonetheless, it is expected to remain a small fraction of the total hazardous waste stream.

High Compliance Costs

Finally, the high compliance costs associated with the tracking program set up the potential for a perverse outcome. To understand this claim, consider the following. Hazardous waste generators that typically rely on off-site facilities have three options available to them:

- Legally dispose of wastes by complying with the manifest system
- Evade the law and illegally discard wastes
- Reduce waste generation

Assuming that generators wish to maximize profits, they must consider the relative costs of these options in their decision making.

When the United States instituted the manifest system, it elevated the costs of legal disposal, making both source reduction and illegal disposal relatively cheaper and hence more

---

[30] RCRA, sec. 3006(b).
[31] U.S. GAO (February 22, 1985), p. 3.
[32] Wirth and Heinz (May 1991), p. 47; U.S. EPA, Office of Solid Waste (November 1986), p. 9.

attractive options from a profitability perspective. The key question is, Which of these two options will firms pursue? While there is some anecdotal evidence that "midnight dumping" has diminished since RCRA was passed, there is no definitive answer to this critical question.[33] Until such evidence is available, there is no reason to dismiss the possibility that this well-intended program might promote the very practice it sought to deter—illegal hazardous waste disposal.

From nearly every perspective, RCRA's track record during its early years was less than impressive. There were major delays in implementation, more than a few policy reversals, several potential loopholes, and many questions about its effectiveness. These observations were largely responsible for Congress's enactment of the hard-hitting 1984 Amendments. Attempting to solve a litany of problems, these Amendments continue to rely upon command-and-control policy instruments. There is still a glaring absence of economic incentives in the law, though the land restrictions may have been an exception.

Market Implications of the 1984 Land Restrictions

As the nation's hazardous waste program was evolving in the late 1970s, it became apparent that existing policy was doing nothing to halt the overuse of land disposal. One reason why landfilling had become so prevalent was that it was cheaper than other disposal options. This differential is due in part to more significant scale economies for landfilling relative to alternative practices, particularly waste incineration. For example, landfills face unit costs of $70 to $260 per metric ton, while incineration facilities incur much higher unit costs of $1,169 to $2,104 per ton of solid materials.[34]

The problem is that these values are private costs and do not account for the **external costs** of environmental damages. If these latter costs were internalized, prices would be higher and any misallocation corrected. An appropriate economic solution would be to use a pricing instrument that captures the social costs of all hazardous waste services. In 1984, policymakers chose instead to use a command-and-control approach—a restriction on landfilling. This policy decision does have economic consequences, though the net effect is not clear-cut.

Restricting the land disposal of hazardous waste elevates the costs of providing these services, which should be reflected in higher prices to waste generators. Just as in our analysis of the manifest system, we need to ask how generators are likely to respond. As profit maximizers, they would consider the relative costs of all available options. Again, they could reduce the amount of waste they generate or they could maintain their waste level and seek an alternative waste management practice. The difference here is that alternative waste management does not necessarily mean illegal disposal. Waste generators could use less landfilling by employing a treatment method like incineration to reduce the volume of wastes to be disposed of. Although the options are complex, we can gain some insight by modeling the two *legal* options facing the waste generator:

- Substituting treatment for disposal
- Using preventive strategies to achieve source reduction

By way of illustration, we use incineration as the representative treatment method.

Pre-1984 Market Equilibrium

The two relevant markets are shown in Figure 17.3—land disposal in panel (a) and incineration in panel (b). Marginal private benefit (*MPB*) curves represent the decisions of waste generators, which are identical to marginal social benefit (*MSB*) functions, assuming no consumption externalities. On the supply side, marginal private cost (*MPC*) curves denote the

[33] Wirth and Heinz (May 1991), p. 47.

[34] Peretz and Solomon (April 1995) and a survey by Environmental Information Ltd. in Krukowski (1995), as cited in Sigman (2000), p. 225, table 7-3.

Figure 17.3 ***Impact of Land Restrictions Under RCRA's 1984 Amendments***

The higher costs of the 1984 land restrictions shift up the *MPC* and *MSC* curves for land disposal, decreasing the private market solution from L_0 to L_1. Because production is lower, external costs are smaller, declining from area *abc* to area *def.* If the waste generator uses less landfilling because it uses source reduction, the net result of the 1984 rulings is an improvement for society measured by this external cost reduction. But if the generator does not reduce its waste level, then the decrease in land disposal must have been achieved by using some alternative practice, such as incineration. Increased use of incineration shifts rightward the *MSB* curve in that market, which increases output from I_0 to I_1. As a result, external costs increase from area *ghi* to area *gjk.* Unless the decline in external costs in the landfilling market outweighs the increase in the incineration market, the 1984 land restrictions would achieve no net reduction in the external costs borne by society.

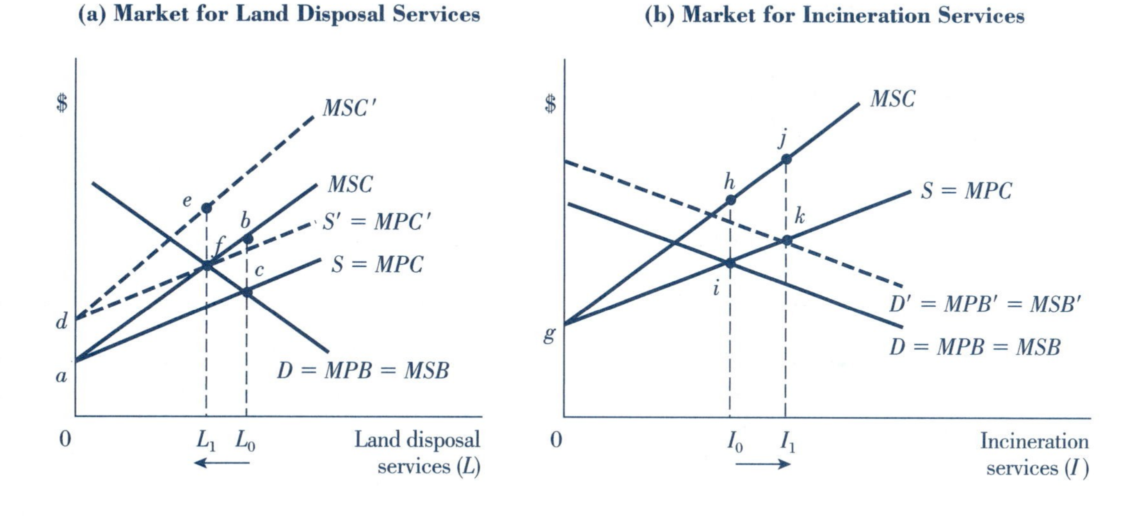

decision making of the respective waste service facilities. The *MPC* curves are distinct from the marginal social cost (*MSC*) curves to indicate that **negative externalities** are associated with production in each market, with each marginal external cost (*MEC*) implied by the vertical distance between each pair of *MPC* and *MSC* curves.

Prior to the 1984 land restrictions, private equilibrium in each market is determined by the intersection of the *MPC* and *MPB* curves. In the landfill market, the pre-1984 equilibrium output level is L_0. The total measure of external costs associated with that output level is the area between the *MPC* and *MSC* curves up to that point, or triangular area *abc.* In the incinerator market, the initial equilibrium output level is I_0 and area *ghi* represents the analogous measure of external costs in that market.

Post-1984 Market Equilibrium

Now consider the effect of the 1984 land restrictions. These rulings lead to higher operating costs in the land disposal market. In Figure 17.3a, the *MPC* curve shifts up to *MPC'*, and the *MSC* curve rises by the same amount to *MSC'*. The result is a decline in landfill use from L_0 to L_1. Because production is lower, the associated external costs are smaller as well,

Table 17.4 ***Changes in Hazardous Waste Generation and Waste Management Costs Associated with the 1984 RCRA Amendments***

	1983[a]	1990 with No Waste Reduction[b]	Percent Change from 1983	1990 with Waste Reduction[c]	Percent Change from 1983
Waste generation (thousands of metric tons)	265,595	280,364	+ 5.6	229,141	−13.7
Annual costs for waste management ($1983 millions)	$5,779	$11,201	+93.8	$8,429	+45.9

Source: U.S. Congress, CBO (1985b), p. 48, table 13, and p. 51, table 15.

[a]Assumes full compliance with RCRA before the 1984 Amendments.

[b]Assumes no waste reduction efforts by industry in response to the 1984 RCRA Amendments. Projection of waste levels in 1990 based on growth in industrial output levels forecasted by the CBO generation model. Decreases in waste quantities therefore result from declining levels of industrial and waste-producing activities.

[c]Assumes waste reduction efforts by industry.

declining to area *def.* Notice that the land restrictions do not eliminate the market inefficiency, since there is nothing in these rulings that causes external costs to be internalized by the market participants. It is the case, however, that the total external costs associated with land disposal have declined. Hence, if waste generators use less landfilling because they pursue source reduction, the 1984 rulings should reduce the health and ecological damages caused by hazardous wastes, holding all else constant. However, if generators do *not* reduce their level of wastes, then the decrease in land disposal must have been achieved by using some alternative waste management practice. Assuming the alternative is the use of incineration, we need to consider how the effects in that market affect the net result.

Increased use of incineration is shown in Figure 17.3b as a shift rightward of the *MSB* curve to *MSB'*, which increases private equilibrium output from I_0 to I_1. This outcome was predicted by the U.S. Congressional Budget Office (CBO), which forecasted that the 1984 restrictions would increase the quantity of waste incinerated from 2.7 million metric tons (MMT) in 1983 to more than 11 MMT by 1990.[35] Notice however that at the higher equilibrium level, external costs have increased to area *gjk*. What can we conclude from this observation? Unless the decline in external costs in the landfilling market outweighs the increase in the incineration market, the 1984 land restrictions would achieve no net reduction in the external costs borne by society.

Obviously, a key issue is to determine the extent to which source reduction is undertaken in response to the 1984 RCRA land rulings. The outcome is difficult to predict. In part, the problem is that there are many factors that influence a firm's decision to reduce waste—an issue discussed in Application 17.3. Despite the inherent difficulties, the CBO conducted an analysis of how the 1984 RCRA Amendments might be expected to affect waste generation levels. Table 17.4 presents the CBO's aggregate estimates of pre- and post-1984 waste generation both with and without source reduction. If there were no waste reduction undertaken, aggregate hazardous waste generation is estimated to increase from 265,595 thousands of metric tons in 1983 to as much as 280,364 in 1990—a 5.6 percent increase. At the other extreme is the best possible outcome, which assumes sufficient incentives to encourage waste

[35]U.S. Congress, CBO (1985b), p. 49, table 14.

17.3 *INFORM's Study of Industrial Source Reduction Activities*

Application

What motivates a firm to reduce its waste generation? Once the decision is made, how do firms determine the best course of action? Answers to questions like these are usually known only to industry insiders and corporate decision makers. Fortunately, independent research on these issues recently has been conducted by INFORM, a nonprofit organization located in New York. The results of this 10-year effort provide new evidence on what factors determine industry's efforts to engage in **source reduction** activities.

To better understand the motives of source reduction and the eventual outcome, INFORM used a case study approach to examine pollution prevention strategies and the results achieved by a sample of 29 highly diverse organic chemical plants. An initial survey was conducted during 1985. A follow-up study was later undertaken to document changes instituted between 1985 and 1990 by the 27 plants that had remained in operation. To assure a representative depiction of the industry, the sample of plants was drawn from three states: California, New Jersey, and Ohio. Similarly, INFORM made certain to include different types of facilities distinguished by such characteristics as size, degree of centralization, type of production process, and output produced.

INFORM's analysis revealed important information about what motivates source reduction activities. The most commonly reported reason for initiating a source reduction action was given as the costs and overall problems of waste disposal. The increasing burden of regulation was the second most common response and appears to be an increasingly important factor in these decisions based on the five-year trend. Other determinants cited by the survey participants were liability issues, process costs, worker safety, and community relations.

INFORM's comprehensive case study approach also reported on the eventual outcome of the plants' source reduction actions. The 1985 survey found that all activities yielded a waste reduction of approximately 7 million pounds per year. A dramatic improvement was documented for 1990—a combined reduction in wastes for all activities of 180 million pounds per year. More insight can be gained by examining other data from INFORM's 1990 report that quantify the effectiveness of source reduction, as shown in the accompanying table.

To access fact sheets, research, and other information from INFORM Inc., visit their Web site at **http://www.informinc.org/.**

http://

Source: U.S. EPA, Office of Pollution Prevention (October 1991), pp. 63–66.

Factor	Amount
Average reduction	68 percent
Average increase in product yields	6.8 percent
Average quantity reduced/activity/year	2,405,621 lbs
Total quantity reduced during the study period	180,421,596 lbs
Average dollars saved/activity/year	$484,616
Total dollars saved during the study period	$28,592,367
Average dollars spent/activity	$946,612
Total dollars spent during the study period	$48,277,197
Average payback period	12.2 months[a]
Average length of time to implement	8.8 months

[a]This average payback period is based on only those activities where both dollars spent and dollars saved were reported. Also, the three highest figures for the payback period (i.e., 889, 70, and 48 years) were excluded because they were much higher than other values reported.

Application

17.4 *Using the Market to Control Hazardous Wastes*

Although the feedstock tax imposed under CERCLA may appear to have been motivated by the incentives of a product charge, its true intent was to provide a reliable source of funds for the nation's remediation program. Beyond this revenue-raising initiative, the use of a tax to reduce hazardous waste generation remains exclusive to state governments. While some states had introduced waste levies before the Superfund tax was effected, most imposed their own version of a pollution charge after the feedstock tax became law. Typically, this instrument was implemented as a **waste-end charge.**

According to a 1998 survey, 30 states were identified as actively taxing hazardous wastes at or beyond the point of generation. Technically, such a waste-end charge is imposed to encourage waste reduction. However, this motivation is not universally apparent. In fact, revenue raising appears to be the overriding influence in some instances. Many states use the revenues to finance monitoring costs or cleanup expenditures at sites where damages have occurred. Some add the waste tax revenues to their general funds, forcing environmental needs to compete for these funds along with all other state programs.

The characteristics of these market instruments vary considerably from state to state. Some states impose the tax directly on the generator at the first stage of the waste stream. Others levy the charge at the post-generation phase by taxing the facility operator, such as a storage or treatment facility. Revenue motivations aside, taxing the generator offers a stronger incentive for waste reduction than does imposing the charge on a facility engaged in treatment, storage, or disposal. Some levy higher taxes on landfilling than on incineration, which should discourage land disposal.

The rate of the tax also differs widely. For example, the rate of tax imposed by Vermont and California is more than $100 per ton for land disposal. Six other states levy a tax that is greater than $50 per ton. Across all states, including those using no tax, the average rate is $21 per ton. More refined data would show even more state-specific differences, such as varying rate schedules based on waste characteristics like the type of waste and/or whether the waste was generated outside state boundaries. And some states allow an exemption for on-site disposal. If these differences properly reflect local economic and environmental conditions, the use of a hazardous waste tax at the state level may be far superior to any uniform charge imposed at the national level.

Sources: U.S. Congress, CBO (1985b), p. 82; U.S. Congress, CBO, Office of Technology Assessment (September 10, 1984); Reese (1985); Hoerner (1998); U.S. EPA, Office of Policy, Economics, and Innovation (January 2001).

reduction. In this case, aggregate waste generation is estimated to decline by about 14 percent between 1983 and 1990, even allowing for waste growth associated with higher production levels.[36]

Table 17.4 also presents the estimated costs associated with these waste generation magnitudes. With no waste reduction activity, aggregate costs rise from $5,779 million in 1983 to $11,201 million in 1990, measured in 1983 dollars—nearly a doubling of real expenditures to comply with the new regulations. If these estimates are reasonable, it is likely that firms will initiate waste reduction activities. If so, under the most favorable assumptions, the CBO estimates that the increase in 1990 compliance costs will be limited to about $2.7 billion.[37]

These findings suggest that at least one aspect of U.S. hazardous waste policy—the land restrictions—may provide an economic incentive to help achieve the national goal of reduc-

[36] U.S. Congress, CBO (1985b), p. 48, table 13.

[37] The dollar expenditures would increase even with the decline in generation because of more stringent requirements for treatment and disposal methods in the 1984 Amendments. U.S. Congress, CBO (1985b), pp. 50–51, table 15.

ing hazardous waste generation. Absent this single element of the law, U.S. policy relies on a command-and-control approach that fails to achieve either efficiency or cost-effectiveness. This is a disconcerting observation, particularly since other nations and even some state governments have used market instruments successfully to help control hazardous waste pollution.

Market Instruments in Hazardous Waste Control Policy

Market instruments attempt to use the price mechanism to make polluters confront the full costs of their actions and respond accordingly. Since there are significant risks associated with hazardous waste pollution, the external costs can be enormous. Hence, if price were somehow made to reflect these costs, polluters—or in this context hazardous waste generators—would have a powerful incentive to engage in source reduction activities.

A conventional market approach to waste control policy is the use of a **pollution charge,** which can be implemented as a **waste-end charge.** A waste-end charge is so named because it is effected at the time of disposal based on the quantity of waste generated. This type of economic instrument is used internationally. For example, the Czech Republic, Denmark, Finland, and Hungary are among the nations that charge a fee on hazardous wastes.[38] In each case, the waste-end charge elevates the effective price of waste management services. If the charge is high enough, it should *discourage* waste generation and *encourage* source reduction.

pollution charge
A fee that varies with the amount of pollutants released.

waste-end charge
A fee implemented at the time of disposal based on the quantity of waste generated.

In the United States, similar economic approaches have been employed by state governments in designing and implementing their hazardous waste programs—an issue discussed in Application 17.4. Despite the international evidence and the experience of state governments, federal hazardous waste policy has employed no analogous market instruments to date.

Managing Uncontrolled Hazardous Waste Sites: CERCLA[39]

By most accounts, the Comprehensive Environmental Response, Compensation, and Liability Act of 1980 (CERCLA), more commonly known as Superfund, is considered a landmark legislative act because its intent is solely remedial rather than preventive. Despite its precedent-setting intent and a $1.6 billion fund, the 1980 law failed to execute on its commitments. The 1986 Amendments attempted to address these failures, but many problems still remain. To understand why, we need to investigate the basic elements of this national policy.

Response and Cleanup

Superfund gives the federal government broad authority to respond whenever there is a release or the threat of a release of a hazardous substance, which it identifies as:[40]

- One that has been so designated under the Water Pollution Control Act, the Solid Waste Disposal Act, the Federal Water Pollution Control Act, the Clean Air Act, or the Toxic Substances Control Act
- One that may present substantial danger to health, welfare, or the environment when released.[41]

Notification of any release exceeding amounts specified in the law (1 pound unless stated otherwise) is to be made to the National Response Center (NRC). Failure to do so is punishable by a fine or imprisonment.

[38] OECD (1999a), pp. 20–21, table 3.3.
[39] For more information on CERCLA, visit **http://www.epa.gov/superfund/action/law/cercla.htm.**
[40] CERCLA, sec. 101(14) and sec. 102(a).
[41] Since these descriptions are purposefully comprehensive, CERCLA itemizes certain exclusions, among them petroleum, natural gas, natural gas liquids, liquified natural gas, or synthetic gas usable for fuel.

Figure 17.4 *Superfund Cleanup Process*

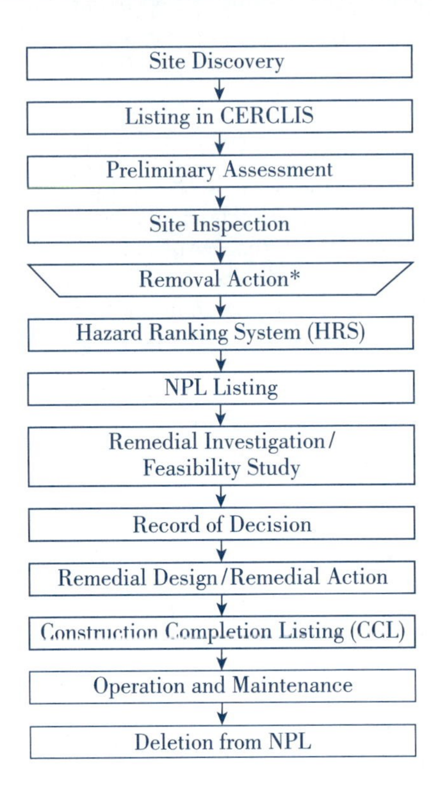

Source: U.S. EPA, Office of Solid Waste and Emergency Response (December 11, 2000).

*Removal Action is in a different shape to signify that it can occur whenever deemed necessary.

Superfund cleanup process
A series of steps used to determine and implement the appropriate response to threats posed by the release of a hazardous substance.

removal actions
Official responses to a hazardous substance release aimed at restoring immediate control.

remedial actions
Official responses to a hazardous substance release aimed at achieving a more permanent solution.

To implement these response actions, a comprehensive set of regulations was established, called the National Contingency Plan (NCP). The NCP outlines the procedures to put Superfund into action. The resulting steps in the **Superfund cleanup process** are summarized in Figure 17.4, with details provided in Exhibit A.12 in the Appendix.[42] By executive order, response activities are delegated to the EPA and are to be carried out through removal actions or remedial actions.

Removal Actions

Removal actions are responses aimed at restoring immediate control to a release site. An example is an emergency cleanup of a chemical spill at a waste site or at the scene of an accident involving a transporter of hazardous chemicals. These activities are undertaken only as short-term measures.[43]

Remedial Actions

Remedial actions are long-term responses aimed at finding a more permanent solution to a release site. Several steps are involved, making the process very time intensive. Following the official identification of a problem, called site discovery, a complete assessment is made of all available information, and an inspection is ordered to sample soil, groundwater, and surface water and to document the site layout. The findings ultimately are evaluated using a risk rank-

[42] More information about this process is also available online at **http://www.epa.gov/superfund/action/process/sfproces.htm.**
[43] U.S. EPA (August 1987); U.S. EPA, Office of Solid Waste and Emergency Response (May 1992).

ing system called the **Hazard Ranking System (HRS),** which assigns a numerical score to a site based on its inherent risks. If the site receives a 28.50 score or higher out of a 100-point maximum, it is placed on the **National Priorities List (NPL).**[44] As of 2002, 1,479 sites have been placed on the NPL. Only if a site is listed on the NPL can remedial actions be undertaken by the federal government. The rest are relegated to state and local officials for an appropriate response.

National Priorities List (NPL)
A classification of hazardous waste sites posing the greatest threat to health and the ecology.

Once a site is placed on the NPL, a feasibility study is done, which involves a complete examination of the site and an appraisal of available cleanup options. This can take from 18 to 30 months and can cost up to $1 million. The action plan selected and support for the decision are documented and used in efforts to recover funding from parties responsible for the release. Design work follows to adapt the action plan to the specific attributes of the site. This stage takes another 12 to 18 months and adds another $1 million on average to the tally. Finally, the actual site cleanup or remedial action begins. On average, the total cost is $25 million per site and can take years to accomplish.[45]

Compensation, Liability, and Enforcement

Superfund gives the EPA authority to force those parties responsible for a hazardous substance release to correct the problem and pay for the damage. This is accomplished either through an administrative order or by bringing a civil action against the responsible party. According to current law, so-called **potentially responsible parties (PRPs)** include any existing or former owner or operator of a hazardous waste facility as well as those involved, however remotely, in the disposal, treatment, or transport of hazardous substances to a contaminated site. The extent of the financial accountability is similarly extensive—to include all costs of removal or remedial action and any damages to natural resources or human health.[46] From an economic perspective, the intent is to force guilty parties to **internalize the externality** of whatever market transaction contributed to the release.

potentially responsible parties (PRPs)
Any current or former owner or operator of a hazardous waste facility and all those involved in the disposal, treatment, or transport of hazardous substances to a contaminated site.

Emergency Planning

Title III of the Superfund Amendments and Reauthorization Act of 1986 is a free-standing piece of legislation known as the Emergency Planning and Community Right-to-Know Act. The purpose of this act is to inform citizens about the existence and potential release of hazardous substances and to provide a planning system for emergencies. To develop the planning system, each state government must devise and implement a comprehensive plan to deal with a hazardous substance release.

A related objective of Title III is to keep the public informed about the production of hazardous substances and their discharge into the environment. To accomplish this goal, Title III has a number of reporting requirements. One of these calls for annual toxic chemical release reports from facilities that manufacture, process, or use specific chemicals in an amount that exceeds a mandated threshold. Data from these reports are used to compile the **Toxics Release Inventory (TRI),** a national database about releases of hazardous substances.[47] All data are published annually by the EPA and are available to the general public. Currently, information is collected on more than 600 chemicals and chemical categories.

Toxics Release Inventory (TRI)
A national database that gives information about hazardous substances released into the environment.

An Analysis of Superfund

In concept, the Comprehensive Environmental Response, Compensation, and Liability Act (CERCLA) is the most unusual of all U.S. environmental laws. Unlike RCRA and other environmental legislation like the Clean Air Act, Superfund is not a regulatory program. Rather, its aim is to clean up the damage at uncontrolled or abandoned hazardous waste sites, using

[44] U.S. EPA, Office of Emergency and Remedial Response (June 1992), p. 9. To view an online listing by state of NPL sites, visit **http://www.epa.gov/superfund/sites/npl/npl.htm.**
[45] U.S. EPA, Office of Solid Waste and Emergency Response (May 1992), p. 6; U.S. EPA, Office of Emergency and Remedial Response (June 1992), pp. 10–12.
[46] CERCLA, sec. 107(a).
[47] More information about the TRI is available on the Internet at **http://www.epa.gov/tri/.**

Table 17.5 *Inventory of CERCLIS and NPL Sites: 1980–2000*

Year	Cerclis (cumulative number)	NPL[a] (cumulative number)
1980–1990	33,371	1,236
1991	35,108	1,245
1992	36,869	1,275
1993	38,169	1,321
1994	39,099	1,360
1995[b]	15,622	1,374
1996	12,781	1,210
1997	9,245	1,194
1998	9,404	1,192
1999	9,237	1,217
2000	9,297	1,226

Source: U.S. EPA, Office of Emergency and Remedial Response (April 2000), as cited by Council on Environmental Quality (1998), p. 312, table 8.9 and updated online; U.S. EPA, Office of Solid Waste and Emergency Response (April 1997).

[a]The number of NPL sites reflects the cumulative total of proposed, final, and deleted NPL sites as of the end of each fiscal year.

[b]The CERCLIS data in 1995 reflect the removal of more than 24,000 sites from the Superfund inventory, which was done as part of the EPA's Brownfields Economic Redevelopment Initiative to help promote redevelopment of these properties. These sites are placed in an archived database of CERCLIS sites, a process that is ongoing and expected to involve about 1,000 sites per year. Over 30,000 sites have been archived as of 2002.

liability standards to identify the so-called potentially responsible parties and make them pay for their actions.[48] While the overall intent is beyond reproach, the implementation has produced mixed results.

Assessing Superfund's Performance

Table 17.5 gives inventory data on identified CERCLIS and NPL sites for the 20-year period between 1980 and 2000. Year-to-year CERCLIS assessments are difficult because over 24,000 sites were removed from the CERCLIS inventory in 1995 and archived. This adjustment was part of the EPA's Brownfields Economic Redevelopment Initiative to help promote redevelopment of these properties. These sites have been placed in an archived database of CERCLIS sites, a process that is ongoing and expected to involve about 1,000 sites per year.[49] Hence, the NPL inventory is simpler to assess over time.

Based on the NPL data, Superfund's Remedial Program has moved at a snail's pace by most accounts. As of 2002, a total of 1,479 sites had been placed on the NPL. Of this total, 810, or 55 percent, have been placed on the **Construction Completion List (CCL),** which means that any physical construction work is finished and the site qualifies for deletion.[50] To date, only 258 of the 1,479 sites, or 17 percent, have been officially deleted. Further, it is argued that even the small number of removed sites is an inflated measure of Superfund's progress, since the EPA has not in all cases properly addressed the potential of recurring problems.[51] This argument relates to the selection of an appropriate risk management strategy for dealing with these sites.

If **benefit-cost analysis** guided the decision making, the high costs of cleanup would be used to determine how far remedial actions should go in cleaning up a site. Opponents of such

[48]For an interesting discussion of how CERCLA is part of an important trend in U.S. environmental policy, see Humphrey and Paddock (1990).

[49]U.S. EPA, Office of Solid Waste and Emergency Response (April 1997).

[50]The CCL was established by the EPA in 1993 to help simplify site categorization, but it has no legal significance. See **http://www.epa.gov/superfund/action/process/ccl.htm** for further information and to view the current CCL by state.

[51]See U.S. Congress, OTA (1985).

an approach argue that the associated risks of a hazardous release should be reduced to zero or close to it, implying that a pure risk standard ought to be employed. Both arguments are firmly entrenched in the issue of how clean is clean. Until this is resolved, significant amounts of resources may be wasted. This very issue explains some of the controversy surrounding the EPA's ordering of General Electric Company to dredge PCBs from the Hudson River, a matter discussed in Application 17.5.

What's Wrong with Superfund?

Superfund's lack of success cannot be blamed on any single factor. There are many reasons that explain its generally disappointing track record, all of which tend to fall into two major categories. The first is the lack of information about the extent of the problem, both at the outset of the program and throughout its implementation. The second source of failure is linked to the structure of the program and the absence of incentives to advance the remediation process.

Information Problems[52]

A major obstacle in addressing the nation's hazardous waste site problem was that the EPA had to start at ground zero, with very little data on the extent of the problem. As information first began to accumulate, the agency learned that the problem was far worse than it had anticipated. There also has been a contention that all the significant problems still have not been identified. At issue is to determine how much of this information gap is within the EPA's control.

A major criticism of the Superfund program is that there has been insufficient federal control, direction, and financial support for state programs aimed at identifying potential Superfund sites. While all states are required under RCRA to submit an inventory of any sites ever used for hazardous waste disposal, the methods used to comply with this requirement are not consistent. Furthermore, there is a lack of initiative in how states identify sites in need of remedial action, relying mainly on private citizens' complaints. The EPA has been criticized for not providing sufficient guidance or funding to rectify this problem, choosing to allocate its budget solely to cleanup activities.

Beyond these data problems, there is also a lack of knowledge about the most effective technologies to accomplish the complex cleanup task. The many phases of remedial action are time intensive and costly. Presumably, experience will help to shorten the time horizon. In the interim, the EPA is continuing its cleanup efforts and at the same time pursuing cost recovery settlements from responsible parties to help defray Superfund's enormous expense.

Lack of Incentives

Like the provisions of RCRA, the Superfund legislation is distinguished by the conspicuous absence of market mechanisms. Even the feedstock taxes used to fund the $8.5 billion Superfund were aimed mainly at raising revenues rather than encouraging decisions to improve the environment. Since these taxes were levied on oil and certain raw chemicals, it may seem as if they are product charges aimed at deterring the use of these hazardous substances. However, Congress's decision to target these materials was motivated by the need for a substantial and fairly reliable tax base from which to collect much-needed revenues.[53] Despite the statutory label of "environmental taxes," these levies offered little incentive to diminish hazardous waste pollution.

Other incentive issues arise in how liability is established under Superfund. In the law itself as well as in how the courts have inferred Congress's intent, the determination of who is responsible for damages and to what extent is all-encompassing and unyielding. Motivated by the knowledge that the $8.5 billion fund would not be nearly enough to pay for the massive

---

[52] Much of the following discussion is drawn from U.S. GAO (December 1987), chap. 3, p. 14, table 2.1; and U.S. EPA (March 1987), cited on p. 22 of this GAO report.

[53] SARA added another revenue source in 1986, calling for a new tax based on a mandated percentage of corporate income. U.S. EPA, Office of Policy, Planning, and Evaluation (July 1992), p. 3-5; CERCLA, Title II; and SARA, pt. 7, sec. 59A.

Application

17.5 *G.E. Ordered to Dredge the Hudson River: Do the Benefits Justify the Costs?*

In one of the more controversial decisions of its history, the EPA in 2001 ordered General Electric Company (GE) to dredge polychlorinated biphenyls (PCBs) from a 38-mile length of New York's Hudson River. GE, one of the largest and most well respected companies in the world, did indeed release 1.1 million pounds of PCBs into the river. The huge conglomerate did so legally starting in 1940 up until PCB use was banned in 1976. Like other manufacturers at the time, GE used the chemicals because of their superlative insulating properties. Nonetheless, the company caused the environmental damage and now is being ordered to clean up the mess. So why the controversy?

From a benefit-cost perspective, the five-year dredging project makes sense only if the marginal costs of abatement are balanced by the associated marginal benefits. The explicit costs alone are estimated to be $460 million, which must be paid by GE under the Superfund law. The expected benefits are the reduced health and ecological damages from PCB exposure. The problem is that there are a number of issues surrounding the dredging order that are in dispute, and any one of these could easily upset the benefit-cost balance needed to achieve efficiency.

A major issue is the possibility that dredging up the PCBs and "resuspending" them will cause further environmental harm to the Hudson River and its ecosystems. Instead, goes the argument, the river should be left alone, since any risk of exposure diminishes as layers of silt naturally form on top of the chemicals. There is some evidence that this natural cleansing may be occurring, since measured PCB levels in fish drawn from the Hudson have been declining, though unevenly. In response to the potential problem, EPA administrator Christine Todd Whitman is establishing performance standards by which to evaluate various stages of the project, so that it can be halted if necessary. However, if the dredging does worsen environmental conditions, the EPA order could actually yield negative benefits to society.

A second source of controversy is that the dredging order is intended to reduce the carcinogenic risk of PCB exposure—an expected benefit of abatement. However, a recent study casts doubt on this expectation. The original decision by Congress to ban PCBs followed the results of a study conducted by Renate Kimbrough, who was then with the Centers for Disease Control and Prevention. Her findings indicated that rats that ingested large amounts of PCBs developed liver cancer. However, more than 20 years later, in a study funded in part by GE, this same scientist along with two colleagues, Martha Doemland and Maurice LeVois, report that no association could be found between exposure to PCBs and death by cancer or any other disease. The two scientists studied 7,035 people over an average of 31 years who worked in the very two GE plants that released the PCBs into the Hudson River. Their findings were published in the *Journal of Occupational and Environmental Medicine*, a peer-reviewed publication, in March 1999. If these new findings are accurate, the benefits of abating PCBs in the Hudson would be greatly diminished.

Lastly, and quite perversely, the segment of society that would be expected to gain the most from pollution abatement—the nearby residents—has been the most vocal in *opposing* the idea. Why? Because the massive dredging effort with its unsightly equipment and truckloads of oily sludge will wreak havoc on property values, quality of life, and local tax revenues. These are added costs of the project that are difficult to quantify but would likely be nontrivial. In a proper benefit-cost analysis, these should be estimated and added to the direct abatement costs of nearly a half-billion dollars.

Taken together, these concerns suggest that even the explicit costs of the Hudson dredging, let alone the implicit costs, may not be justified by the expected benefits. If so, allocative efficiency will not be achieved, suggesting that resources used for this massive project would be better utilized in some alternative effort.

Sources: Cohen (December 12, 2001); Cohen (January 7, 2002); Murray (December 5, 2001); U.S. EPA, Region 2 (December 6, 2000).

cleanup, Congress and the courts use whatever legislative and judicial muscle is necessary to impose the "polluter-pays principle." While the intent was to achieve swift action against the guilty parties and advance the remediation process, the plan backfires because it sets up an unexpected and perverse incentive.

The problem is rooted in how the law identifies a potentially responsible party (PRP) and in how the liability is defined. In establishing legal responsibility, the courts use the concepts of strict liability and joint and several liability.[54] **Strict liability** means that individuals can be held liable even if negligence is not proven. **Joint and several liability** means that a single party found liable for damages can be held responsible for *all* associated costs even if that party's actual contribution to the damages is minimal.[55] The use of such tough legal standards was intended to save the government the complex and time-intensive process of proving negligence and then determining what proportion of costs each PRP must pay.[56] There was also the expectation that the use of joint and several liability would encourage an identified PRP to name all other responsible parties at a given site.

strict liability
The legal standard that identifies individuals as responsible for damages even if negligence is not proven.

joint and several liability
The legal standard that identifies a single party as responsible for all damages even if that party's contribution to the damages is minimal.

Unfortunately, these expectations proved to be off the mark. Recognizing the potential to be held accountable for *all* costs at a named waste site, few PRPs were willing to come forward with information. Furthermore, because joint and several liability assigns the entire costs of cleanup to a single party, there was a strong financial incentive for a liable party to use delay tactics or to incur the relatively lower costs of litigation to fight the charges.[57] As a consequence, the well-intended legal standards have led to long delays and a diversion of resources from cleanup to litigation proceedings. In fact, a 1985 study puts the estimate of legal expenses at about 55 percent of the total dollars expended on remediation.[58]

Conclusions

Most would agree that U.S. hazardous waste policy has broadened and strengthened considerably over the past two decades. As the evolution continues, we observe public officials beginning to integrate preventive initiatives into the overall policy solution. By addressing the generation of hazardous waste *before* it becomes a problem, the nation can avoid some of the associated health and ecological risks, not to mention the high costs of correcting the degradation. Yet in spite of these favorable assessments and the acknowledgment that some progress has been made, there are still questions about the efficiency and cost-effectiveness of current U.S. policy.

With 20/20 hindsight, it is apparent that the federal government should have acted sooner to respond to the risks of hazardous waste pollution. But such an assessment assumes full information—a luxury the United States clearly did not have. Confronted with tough problems, a marked lack of data, and a very angry constituency, Congress sought policy instruments that they thought would be direct, uncompromising, and capable of achieving a swift resolution. Given this motivation, it is not surprising that the major laws on hazardous solid waste are couched within a command-and-control framework.

Understanding this motivation, however, does not change the fact that such a policy approach came at some measure of sacrifice—a loss of efficiency and cost-effectiveness. Exacerbating the problem, many of Congress's intentions were not carried out as planned. Not only

[54] Interestingly, these legal concepts are not written into Superfund's provisions. They were initially in the bill submitted before Congress but were removed as part of an agreement between the House of Representatives and the Senate before signing it into law. However, the courts took the initiative and reinstated these notions based on what they perceived to be Congress's intent (Mazmanian and Morell, 1992, p. 36).

[55] It is also true that the use of these liability standards under Superfund extends well beyond what is imposed under traditional common law. For example, causality does not have to be shown to establish liability under Superfund, as is the case under common law. Hence, there is no need to show proof that a defendant's wastes are those that caused the damage. All that is necessary is evidence that the defendant is a PRP. See Anderson (1989), pp. 427–32, for more detail.

[56] For an interesting article that discusses the issue of cost allocation in detail, see Butler et al. (March 1993).

[57] Mazmanian and Morell (1992), pp. 36–37.

[58] Light (July 1985), p. 10205.

did most of the initiatives lack incentive mechanisms, but many were set in motion without a full understanding of the extent of the problem. As such, they were destined for failure. Such was the fate of the Superfund legislation, a unique body of law that continues to be a source of contention among government officials, industry, and private citizens.

Midcourse corrections might save the nation considerable expense in executing its policy missions and could restore the level of environmental risk to an acceptable level more quickly. The strategy has to be directed both at repairing the damage from the past and seeking ways to reduce and more efficiently manage the present waste stream. Congress must decide a new course of action to reauthorize the Superfund law. If nothing else, policymakers seem to have recognized the need for reform and the importance of using risk management strategies in all environmental initiatives—even in the absence of extensive information. Such an awareness may be significant to what appears to be a critical phase in the evolution of U.S. hazardous waste policy.

Summary

- Hazardous solid wastes are any unwanted materials or refuse capable of posing a substantial threat to health or the environment.
- In the United States, hazardous waste generation has been estimated at about 173 million tons per year. This translates to about 0.66 tons per person annually. Industry is responsible for the largest proportion. Chemical producers generate about 79 percent of all the hazardous wastes in the United States.
- U.S. policy on solid wastes originated with the Solid Waste Disposal Act (SWDA), enacted in 1965. This act was amended in 1970 by the Resource Recovery Act and later by the Resource Conservation and Recovery Act (RCRA) in 1976. RCRA represented the nation's first official policy position on hazardous waste control.
- Congress reauthorized and strengthened RCRA through the Hazardous and Solid Waste Amendments of 1984.
- The Comprehensive Environmental Response, Compensation, and Liability Act (CERCLA), or Superfund, was passed in 1980. This act was aimed at identifying and cleaning up the nation's worst inactive hazardous waste sites and at recovering damages from responsible parties.
- CERCLA was revised in 1986 through the Superfund Amendments and Reauthorization Act (SARA), which brought the Superfund budget up to $8.5 billion.
- A recent bill signed by President Bush provides some liability relief under Superfund to households and small businesses. It also funds a new program aimed at abating brownfields.
- Under RCRA, a command-and-control hazardous waste program was devised, commonly referred to as the "cradle-to-grave" management approach. Its four components are the identification and listing of hazardous waste; a national manifest system; a permit system for treatment, storage, and disposal facilities (TSDFs); and the setting of standards for these facilities.
- Several references in the 1984 Amendments suggest a policy shift away from land disposal and toward more preventive solutions.
- One problem with RCRA is that the characteristics and criteria used to identify hazardous waste are risk based with no provision to consider the benefits of these materials before they enter the waste stream. Another problem is that the identifying categories are applied uniformly with no qualifiers to allow for differences in the degree of toxicity across various waste constituents or differences in quantity.
- The standards applicable to TSDFs are benefit based with no consideration for economic costs. Hence, it is not likely that allocative efficiency will be achieved. Furthermore, with

few exceptions, the standards are applied uniformly. There is virtually no provision for site-specific differences, likely disallowing a cost-effective solution.

- The manifest system is a command-and-control instrument that is benefit based in motivation. Economic considerations played no role in its design or implementation. It has not been a major factor in reducing risk because of its limited scope.
- The land restrictions imposed with the 1984 Amendments caused land disposal facilities to face higher unit costs and elevate their prices. In response, generators could either initiate waste reduction or alter their waste management practices. Although these rulings may reduce the external costs borne by society, inefficiency is not eliminated.
- One market-based approach to hazardous waste management is a waste-end charge levied at the time of disposal. Despite evidence that such an approach can be effective, federal hazardous waste policy has employed no analogous market instruments to date.
- Under CERCLA, the federal government can undertake a response action whenever there is an actual or potential release of a hazardous substance that may present imminent danger to public health or welfare. The National Contingency Plan (NCP) outlines procedures to implement these activities as removal actions or remedial actions carried out through the Superfund cleanup process. By executive order, response activities are delegated to the EPA and are carried out through removal actions or remedial actions.
- As part of a remedial action, a site is evaluated using a risk-ranking system that assigns a numerical score to the site based on its inherent risks. If the site receives a 28.50 score or higher out of a 100-point maximum, it is placed on the National Priorities List (NPL).
- The EPA has the authority to force those parties responsible for a hazardous substance release to correct the problem and pay for the damage. This is accomplished either through an administrative order or by bringing a civil action against the so-called potentially responsible parties (PRPs).
- Title III of SARA is a free-standing piece of legislation known as the Emergency Planning and Community Right-to-Know Act. Its purpose is to inform citizens about potential releases of hazardous substances and to provide a planning system for emergencies. One of its reporting requirements calls for data that are used to form the Toxics Release Inventory.
- Superfund's provisions have produced mixed results. One problem is the lack of information about the extent of the problem. The second source of failure is the lack of incentives to enhance the remediation process. The use of strict liability and joint and several liability has led to delays and the diversion of resources from cleanup to litigation proceedings.

Key Concepts

hazardous solid wastes
waste stream
"cradle-to-grave" management system
Comprehensive Environmental Response, Compensation, and Liability Information System (CERCLIS)
feedstock taxes
brownfields
source reduction
waste management
characteristic wastes
listed wastes
manifest
permitting system
pollution charge
waste-end charge
Superfund cleanup process
removal actions
remedial actions
National Priorities List (NPL)
potentially responsible parties (PRPs)
Toxics Release Inventory (TRI)
strict liability
joint and several liability

Use the Key Concepts listed above to begin your search for additional articles and information using the InfoTrac College Edition database.

Review Questions

1. Consider the various provisions in RCRA that discourage land-based waste disposal. Identify and explain one aspect of the law that is command-and-control in approach and one that is incentive based.
2. Propose an alternative method of identifying hazardous waste that is more efficient than the one mandated under RCRA without compromising the objective of risk reduction. Support your proposal with a well-defined risk management strategy.
3. Use benefit-cost analysis to qualitatively evaluate the Superfund remedial action program.
4. Recommend an incentive-based reform that would improve states' identification of hazardous waste sites.
5. a. An ongoing debate about the Superfund program is the determination of the optimal abatement level, characterized as the "how clean is clean" problem. Choose one of the risk management strategies, and propose how it might be used to resolve this issue.
   b. In your view, why has there been no movement to initiate such a proposal?

Additional Readings

Been, Vicki. "Unpopular Neighbors: Are Dumps and Landfills Sited Equitably?" *Resources* (spring 1994), pp. 16–19.

Fullerton, Don, and Seng-Su Tsang. "Should Environmental Costs Be Paid by the Polluter or Beneficiary? The Case of CERCLA and Superfund." *Public Economics Review* (June 1996), pp. 85–117.

Gerrard, Michael B. *Whose Backyard, Whose Risk.* Cambridge, MA: MIT Press, 1994.

Hamilton, James, and W. Kip Viscusi. "How Costly Is 'Clean'? An Analysis of the Benefits and Costs of Superfund Site Remediations." *Journal of Policy Analysis and Management* 18 (winter 1999), pp. 2–27.

Harper, Richard K., and Stephen C. Adams. "CERCLA and Deep Pockets: Market Responses to the Superfund Program." *Contemporary Economic Issues* 14 (January 1996), pp. 107–15.

Hoffman, Andrew J. "An Uneasy Rebirth at Love Canal." *Environment* 37 (March 1995), pp. 4–9, 25–31.

Landy, Marc K., Marc J. Roberts, and Stephen R. Thomas. *The Environmental Protection Agency: Asking the Wrong Questions.* New York: Oxford University Press, 1990.

Levenson, Howard. "Wasting Away: Policies to Reduce Trash Toxicity and Quantity." *Environment* 32 (March 1990), pp. 10–15, 31–36.

Levinson, Arik. "State Taxes and Interstate Hazardous Waste Shipments." *American Economic Review* 89 (June 1999), pp. 666–77.

National Academy of Sciences. *Reducing Hazardous Waste Generation: An Evaluation and a Call for Action.* Washington, DC: National Academy Press, 1985.

Probst, Katherine N., David M. Konisky, Robert Hersh, Michael B. Batz, and Katherine D. Walker. *Superfund's Future: What Will It Cost?* Washington, DC: Resources for the Future, July 2001.

Ritter, Don. "Challenging Current Environmental Standards." *Environment* 37 (March 1995), pp. 11–12.

Schoenbaum, M. "Environmental Contamination, Brownfields Policy, and Economic Redevelopment in an Industrial Area of Baltimore, Maryland." *Land Economics* 78, no. 1 (2002), pp. 72–87.

Sigman, Hilary. "The Effects of Hazardous Waste Taxes on Waste Generation and Disposal." *Journal of Environmental Economics and Management* 30 (March 1996), pp. 199–217.

U.S. Congress, Office of Technology Assessment (OTA). *Serious Reduction of Hazardous Waste.* Washington, DC: U.S. Government Printing Office, September 1986.

Viscusi, W. Kip, and James T. Hamilton. "Are Risk Regulators Rational? Evidence from Hazardous Waste Decisions." *American Economic Review* 89 (September 1999), pp. 1010–27.

http:// Related Web Sites

Brownfield site information	http://www.epa.gov/epahome/hi-brownfields.htm
CERCLA information	http://www.epa.gov/superfund/action/law/cercla.htm
Construction Completions List	http://www.epa.gov/superfund/action/process/ccl.htm
Hazardous waste data	http://www.epa.gov/epaoswer/hazwaste/data
Hazardous waste information	http://www.epa.gov/osw/hazwaste.htm
The Hazardous Waste Permitting Process: A Citizen's Guide, U.S. EPA, Office of Solid Waste and Emergency Response (June 1996)	http://www.epa.gov/epaoswer/hazwaste/permit/prmtguid.htm
INFORM Inc.	http://www.informinc.org/
Leaking Underground Storage Tank (LUST) Trust Fund	http://www.epa.gov/swerust1/ltffacts.htm
Love Canal	http://ublib.buffalo.edu/libraries/projects/lovecanal
NPL site listing by state	http://www.epa.gov/superfund/sites/npl/npl.htm
Pollution prevention	http://www.epa.gov/p2
RCRA Online	http://www.epa.gov/rcraonline
Superfund Program	http://www.epa.gov/superfund/index.htm
Superfund Reauthorization Issues in the 105th Congress, Reisch (November 23, 1998)	http://cnie.org/nle/crsreports/waste/waste-17.cfm
Superfund Reauthorization Issues in the 106th Congress, Reisch (October 30, 2000)	http://cnie.org/nle/crsreports/waste/waste-28.cfm
Superfund site cleanup process	http://www.epa.gov/superfund/action/process/sfproces.htm
Toxics Release Inventory (TRI)	http://www.epa.gov/tri/
TSDF activities	http://www.epa.gov/epaoswer/osw/tsd.htm
Underground Storage Tank Program	http://www.epa.gov/swerust1/overview.htm
U.S. EPA, Office of Solid Waste	http://www.epa.gov/epaoswer/osw/

A Reference to Acronyms and Terms in Hazardous Waste Control Policy

Environmental Economics Acronyms

MPB	Marginal private benefit
MSB	Marginal social benefit
MEC	Marginal external cost
MPC	Marginal private cost
MSC	Marginal social cost

Environmental Science Acronyms

MMT	Millions of metric tons
PCBs	Polychlorinated biphenyls

Environmental Policy Acronyms

CERCLA	Comprehensive Environmental Response, Compensation, and Liability Act
CERCLIS	Comprehensive Environmental Response, Compensation, and Liability Information System
CCL	Construction Completions List
HRS	Hazard Ranking System
LUST	Leaking underground storage tank
NCP	National Contingency Plan
NPL	National Priorities List
NRC	National Response Center
PRP	Potentially responsible party
RCRA	Resource Conservation and Recovery Act
SARA	Superfund Amendments and Reauthorization Act
SWDA	Solid Waste Disposal Act
TRI	Toxics Release Inventory
TSDFs	Treatment, storage, and disposal facilities

Chapter 18

Managing Municipal Solid Waste

municipal solid waste (MSW) Nonhazardous wastes disposed of by local communities.

Every local community has to deal with collecting and disposing of what most of us offhandedly call trash. **Municipal solid waste (MSW),** as it is more formally termed, is the collection of cans, bottles, food scraps, newspapers, lawn clippings, and old furniture that characterizes everyday living. How could such a mundane matter as everyday trash become an issue? Mainly because society has seen it as exactly that—a routine part of living that merits no particular attention. As long as the unsightly pile of old newspapers and garbage leaves the street corner, most citizens give the matter little thought—that is, until some public official proposes a new landfill site on the next block. The typical reaction is a negative one—dubbed the "not in my backyard," or "NIMBY," syndrome. This response coupled with the growth trend in municipal waste generation and an aging disposal system has left many communities with a difficult problem.

As part of the nonhazardous waste stream, municipal solid wastes pose no *direct* threat to human, animal, or plant life. Nonetheless, there are risks to society and the ecology if too much is generated or if it is improperly managed. Like many other countries around the world, the United States has depended on landfills to dispose of its MSW. Unsanitary conditions at these sites can contaminate water and soil with disease-spreading bacteria. A more serious risk is the release of toxic substances into the environment. Toxic contamination can arise from natural decomposition processes or the presence of household or industrial hazardous substances mixed in with municipal refuse. There is also the risk of atmospheric pollution caused by gases released from waste decomposition or from the incomplete combustion of incinerated wastes.

Given the potential damages, public policies are needed to control waste management practices, to find ways to recover and reuse waste materials, and to develop new technologies. In the United States, much of the responsibility for these policies is delegated to state and local governments with some measure of federal oversight. Such an approach allows for more flexibility in devising waste management programs than is the case for hazardous waste controls. In some communities, innovative policies have been developed to deal with MSW, including the use of market instruments to achieve environmental goals.

In this chapter, we consider the risks of MSW pollution and analyze the policy response to these risks. To understand the motivation of government programs and regulations, we begin by characterizing the generation and composition of this waste stream. Here, we use international comparisons to provide context to our discussion of U.S. data. Once done, we outline the delegation of responsibilities between federal and state governments given by the Resource Conservation and Recovery Act (RCRA). We then present a market model of MSW management services and explore potential sources of inefficiency. This discussion prepares the way for an economic analysis of various market instruments used by state and local governments to manage the MSW stream. At the end of the chapter is a list of commonly used acronyms and terms.

Table 18.1 ***Trend Data on U.S. Annual MSW Generation from 1960 to 1999 by Weight***

	1960	1970	1980	1990	1999
Total MSW generated by weight (millions of tons)	88.1	121.1	151.6	205.2	229.9
Total population (thousands)	179,979	203,984	227,255	249,907	272,691
Per capita MSW (pounds per day)	2.68	3.25	3.66	4.50	4.62
Gross domestic product (GDP) per capita (chained $1996)	13,148	17,446	21,521	26,834	32,512

Source: U.S. EPA, Office of Solid Waste and Emergency Response (July 2001), p. 2, tables ES-1 and ES-2; Council of Economic Advisers, (January 2001), table B-31 (U.S. Department of Commerce).

Characterizing Municipal Solid Waste[1]

Observing a Trend

Many U.S. communities have been observing an increase in the size of the MSW stream, a phenomenon that is only partly linked to population growth. As shown in Table 18.1, per capita generation rates are rising, though they reflect, at least in part, rising economic growth. Nonetheless, most argue that the characterization of Americans as a "disposable society" may be well placed. In-built product obsolescence, persuasive advertising that encourages excessive consumption, and reliance on prepared and therefore heavily packaged food products are all contributing factors. Despite the apparent growth trend in MSW generation, the development of better waste management practices has lagged behind.

For years, the United States and most European countries have depended primarily on landfills to dispose of MSW, a practice that at least for a time seemed a reasonable solution. In the 1980s, reports of a "capacity crisis" warned that many municipal landfills were running out of space and others were being closed for improper waste management practices. More recently, counterarguments have emerged, stating that the so-called capacity crisis in the United States was exaggerated, occurring in only certain locations. These reports point out that the rash of dump closings has involved mainly small sites and that increased recycling and a greater use of waste incineration have contributed to what is now a surplus of landfill space in some areas. Indeed, some communities have been able to negotiate lower contract prices with trash management firms, suggesting that the reported glut in disposal capacity may be real—at least in some regions. Yet some major waste companies continue to argue that landfill space is declining.

A look at recent data helps to explain the disparate views. Recent EPA data do show an ongoing dependence on landfilling in the United States, with more than 57 percent of MSW landfilled in 1999. It is also the case that the number of landfills in the nation has been declining. For example, in 1989 there were 7,379 landfills in the United States, and in 1999 the number fell sharply to 2,216. However, since the size of the average landfill has been

[1] To access the most recent characterization of municipal solid waste in the United States and various data, visit **http://www.epa.gov/epaoswer/non-hw/muncpl/.**

Figure 18.1 ***Proportion by Weight of Products and Materials Generated in U.S. Municipal Solid Waste in 1999***

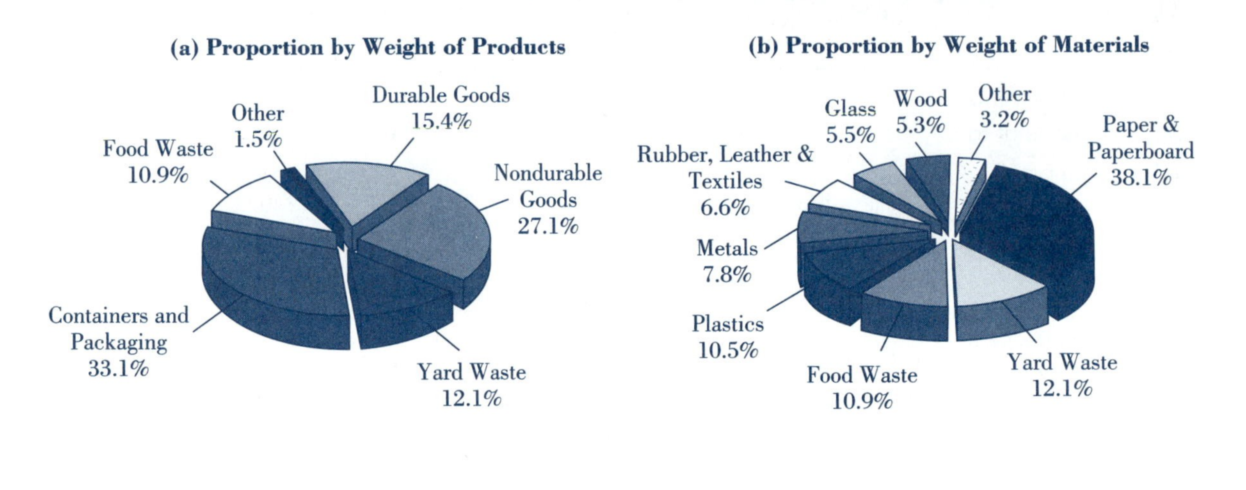

Source: U.S. EPA, Office of Solid Waste and Emergency Response (July 2001), p. 8, fig. ES-4, and p. 7, fig. ES-3.

increasing, overall capacity has remained fairly level. The EPA asserts that capacity is not a problem in the United States but that regional dislocations do occur on occasion. For example, two states currently report having less than 5 years of landfill capacity left.[2]

In any case, there is virtually no debate that the MSW disposal system in the United States is aging. According to a report from the U.S. Congress, Office of Technology Assessment, 70 percent of MSW landfills in the United States began operation before 1980. Many lack proper controls such as waterproof cover systems, suitable liners to prevent leaching, or leachate collection systems. The absence of well-engineered controls creates a potential for serious environmental problems, such as methane gas explosions, the accumulation of bacteria, and contamination of groundwater and surface waters. These concerns are exacerbated when even small proportions of hazardous waste are mixed in with nonhazardous materials.[3]

Composition of MSW in the United States[4]

To get a sense of what comprises MSW in the United States, we can examine its composition both by the types of products being discarded and the kinds of materials entering the waste stream.

Product Groups

product groups
Categories in the MSW stream identified as durable goods, nondurable goods, containers and packaging, and other wastes.

The designated **product groups** are durable goods (e.g., appliances, furniture, and tires), nondurable goods (e.g., magazines, clothing, and household cleansers), containers and packaging, and other wastes, such as yard trimmings and food waste. Figure 18.1a gives the relative proportions by weight of these product groups based on the 229.9 million tons of MSW generated in 1999. Notice that packaging and containers account for a third of this total, or 76 million tons.

Within the durable goods product group, one area of interest is the disposal of consumer electronics, which includes video, audio, and Information Age equipment such as televisions,

[2] Data from *BioCycle* (selected issues, 1989–2000) and Goldstein (April 2000), as cited in U.S. EPA, Office of Solid Waste and Emergency Response (July 2001), p. 106, table 28.
[3] U.S. Congress, OTA (1989), p. 284.
[4] Much of the following discussion is drawn from U.S. EPA, Office of Solid Waste and Emergency Response (July 2001).

Table 18.2 ***Trend Data on MSW Generated in the United States: 1960–1999 by Materials***

	Millions of Tons[a]					Percent of Total[a]				
	1960	**1970**	**1980**	**1990**	**1999**	**1960**	**1970**	**1980**	**1990**	**1999**
Materials in products										
Paper and paperboard	30.0	44.3	55.2	72.7	87.5	34.0	36.6	36.4	35.4	38.1
Glass	6.7	12.7	15.1	13.1	12.6	7.6	10.5	10.0	6.4	5.5
Metals	10.8	13.8	15.5	16.6	17.8	12.3	11.4	10.2	8.1	7.8
Plastics	0.4	2.9	6.8	17.1	24.2	0.4	2.4	4.5	8.3	10.5
Rubber and leather	1.8	3.0	4.2	5.8	6.2	2.1	2.5	2.8	2.8	2.7
Textiles	1.8	2.0	2.5	5.8	9.1	2.0	1.7	1.7	2.8	3.9
Wood	3.0	3.7	7.0	12.2	12.3	3.4	3.1	4.6	6.0	5.3
Other[b]	0.1	0.8	2.5	3.2	4.0	0.1	0.6	1.7	1.6	1.7
Subtotal	54.6	83.3	108.9	146.5	173.6	62.0	68.8	71.8	71.4	75.5
Other wastes										
Food wastes	12.2	12.8	13.0	20.8	25.2	13.8	10.6	8.6	10.1	10.9
Yard trimmings	20.0	23.2	27.5	35.0	27.7	22.7	19.2	18.1	17.1	12.1
Miscellaneous inorganic wastes	1.3	1.8	2.3	2.9	3.4	1.5	1.5	1.5	1.4	1.5
Subtotal	33.5	37.8	42.8	58.7	56.3	38.0	31.2	28.2	28.6	24.5
Total MSW generated by weight	88.1	121.1	151.6	205.2	229.9					

Source: U.S. EPA, Office of Solid Waste and Emergency Response (July 2001), p. 29, table 1.

[a]The sum of any given column may not add to the total due to rounding.

[b]The category "Other" in products is primarily associated with disposable diapers. Also included are electrolytes and other materials in batteries not classified as plastics or metals.

camcorders, CD players, personal computers, computer monitors, and fax machines. Rising sales of these products along with their relatively short economic lives has contributed to their growth in the waste stream. The EPA began to estimate the quantity of consumer electronics in MSW starting in 1999, reporting 1.8 million tons for this new subcategory. Tracking this segment of MSW is relevant not only because of the rising volume but also because much of this equipment includes toxic materials like lead. Hence, efforts to recover, reuse, and remanufacture these products are becoming increasingly important. In 1999, only 9 percent of the 1.8 million tons generated were recovered for recycling.[5]

Materials Groups

The **materials groups** identified in MSW are paper and paperboard, yard trimmings, glass, metals, plastics, textiles, rubber and leather, wood, and other miscellaneous wastes. Proportions by weight for these categories are shown in Figure 18.1b. According to these data, paper and paperboard represent the largest proportion by weight of discarded materials in 1999, accounting for 38.1 percent of the total.

materials groups Categories in the MSW stream identified as paper and paperboard, yard trimmings, glass, metals, plastics, textiles, rubber and leather, wood, and other miscellaneous wastes.

Composition data by materials convey information about how manufacturing decisions ultimately affect waste accumulation. Table 18.2 gives trend data on the materials content of the nation's MSW for the 1960–1999 period. Notice that the generation of paper and paperboard

[5]For more information, see appendix C of U.S. EPA, Office of Solid Waste and Emergency Response (July 2001) or the National Recycling Coalition's "Electronics Recycling Initiative" available at **http://www.nrc-recycle.org/resources/electronics/index.htm.**

Application

18.1 *The Facts on Recycling Plastics*

Over the last several decades, plastic use in the United States has risen by over 10 percent per year, which translates to a substantial growth rate of plastic wastes. In 1960, plastic wastes totaled 390,000 tons, which grew to 24.2 million tons in 1999. Relative to the entire MSW stream in 1999, plastic wastes represent 10.5 percent of total generation by weight, but by volume, they account for a much higher proportion. Containers and packaging are by far the largest proportion of plastic wastes, representing 11.2 million tons in the United States in 1999, or 46 percent of the plastics waste stream.

Beyond the sheer quantity of plastic wastes, these materials also can endanger the environment. According to the EPA, most of the wastes collected during harbor surveys and beach cleanups are plastics. In addition to the aesthetic degradation of such littering, disposal of plastics in surface waters threatens virtually all forms of marine life. Plastics also contain such additives as colorants, stabilizers, and plasticizers, some of which include toxics like cadmium and lead. Reportedly, 28 percent of the cadmium and 2 percent of the lead found in MSW arise from plastics.

Exacerbating the problem, the recovery rate for plastic wastes in the United States has been poor. In 1985, only 0.1 million tons, or less than 1 percent, of all plastic wastes generated was recovered. In 1999, the figure rose to 1.4 million tons, or roughly 5.6 percent of the total. Why is the overall recovery rate for plastics so poor? The bottom line is that recycled plastics have to compete with virgin materials. And for any recycled product to be competitive, all three steps in the recycling process—collection, separation of materials, and manufacture of new products—must be executed efficiently.

One of the key issues in producing recycled plastic is the resin content. Plastics are made from a variety of different types of resins. Currently, there appear to be sizable and lucrative markets for products made from single resins but not for commodities produced from mixed plastics, which command a lower market value. The problem is, even though many products are made of only a single resin, all of these end up together in the waste stream. To achieve a homogeneous collection of a particular resin, different kinds of plastics have to be identified and separated after collection—a costly step in the recycling process. An important problem is that plastic wastes often are not easily identified, even by experts.

Hence, in order for plastic waste recovery rates to improve, cost-effective methods must be developed to more easily identify different types of plastics and to more readily separate them into batches of single resins. Until these technologies are developed, recycling will continue to focus on easily recognizable plastic wastes that accumulate in large amounts. This explains why soft drink bottles and milk containers account for the greatest majority of recycled plastic. It also explains why many packaging manufacturers label their containers with a recycling symbol and a code number that identifies the resins that were used.

http://

For current information on plastics and plastics recycling, visit **http://www.plasticsresource.com/recycling/index.html,** prepared by the American Plastics Council, Inc.

Sources: Resource Integration Systems, Ltd., and Waste Matters Consulting (Portland, OR) (December 1990), pp. 7–9; U.S. EPA, Office of Solid Waste and Emergency Response (February 1990); U.S. EPA, Office of Solid Waste and Emergency Response (July 2001), pp. 41–43, 81.

has grown fairly steadily over time, rising from 30 million tons, or 34 percent of the total in 1960, to 87.5 million tons, or 38.1 percent in 1999. Paper wastes are procyclical, meaning that the amount varies directly with the economic business cycle.

The fastest growing segment of MSW in the United States is plastics, which has risen steadily from 0.4 million tons in 1960 to 24.2 million tons in 1999. This growth rate is just as impressive when viewed on a proportionate basis, at less than 1 percent of the 1960 total MSW generated by weight to 10.5 percent of the total in 1999. These data explain why there has been a surge of interest in recycling plastics, an issue discussed in Application 18.1. Recovery rates for plastics and the other components of MSW as of 1999 are shown in Figure 18.2.

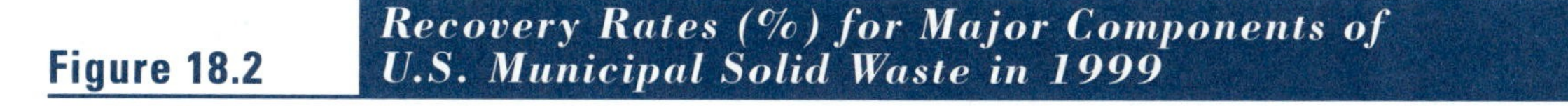
Figure 18.2 *Recovery Rates (%) for Major Components of U.S. Municipal Solid Waste in 1999*

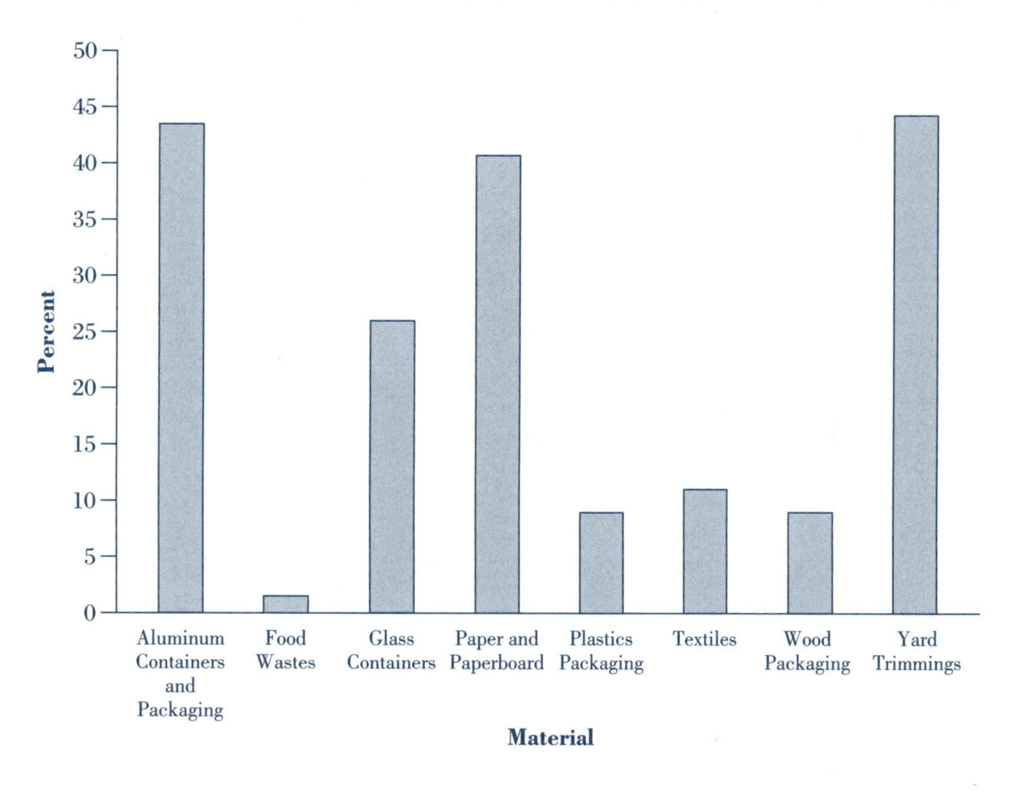

Source: U.S. EPA, Office of Solid Waste and Emergency Response (July 2001), p. 9, table ES-5.

International Comparisons

Generation

Proper management of the MSW stream is a universal objective. Table 18.3 shows per capita MSW generation (measured in kilograms per year) and gross national product (GNP) for various nations around the world. Notice that the major industrialized nations are at the upper end of MSW generation rates. What is also apparent from the data is that Americans are among the highest MSW generators in the world. Notice how the U.S. data compare with other industrialized nations, such as France and Japan. For example, Japan's per capita rate of 400 kilograms pales in comparison to the U.S. statistic.

Some of the international differences are attributable to the amount of packaging used by producers. For example, in 1988, it was estimated that the average American generated 463 pounds of packaging waste. Citizens of both the European Community and Japan are said to have produced at least 25 percent less than that amount.[6] These variations also may reflect cultural preferences, environmental awareness, economic conditions, or government regulations. Of course, statistical comparisons across nations are not totally reliable. Estimation methods and even the definitions used for MSW are often different. Nonetheless, even

[6]McCarthy (winter/spring 1993).

Table 18.3 ***Per Capita MSW Generation and Gross National Product (GNP) in Selected Countries***

Country	Waste (kilograms per capita)	1998 GNP ($ per capita)
United States	720	29,240
Australia	690	20,640
Iceland	650	16,180
Norway	600	34,310
France	590	24,210
Denmark	560	33,040
Ireland	560	18,710
Canada	500	19,170
Italy	460	20,090
Japan	400	32,350
Greece	370	11,740
Poland	320	3,910
Czechoslovakia	310	5,150
Mexico	310	3,840

Source: OECD (1999b), as cited in U.S. Census Bureau (2001); The World Bank, as cited in U.S. Census Bureau (2000).

Note: Waste figures are for 1998 or latest available year.

assuming some measure of inaccuracy, the United States is clearly at the high end of worldwide MSW generation rates.

Recycling Rates

There is also considerable variability in recycling rates across nations. In the United States, the national MSW recycling rate for 1999 was 27.8 percent.[7] Japan and many European countries have practiced recycling for much longer than the United States, and their national averages reflect their experience. Japan in particular is touted as a nation with one of the highest national recycling rates for municipal wastes. Recent data show that Japan recycles over 50 percent of its paper and glass waste materials. Other nations with land constraints similar to Japan's such as the Netherlands exhibit relatively high recycling rates as well.[8] However, direct international data comparisons must be adjusted for political, cultural, economic, and demographic distinctions that affect the recyclability of manufactured products and the effectiveness of recycling programs.

The Policy Response: An Overview

As mentioned in the previous chapter, there is a set of provisions within the Resource Conservation and Recovery Act (RCRA) that applies to nonhazardous waste, including MSW.[9] For the most part, the responsibility of nonhazardous waste management is assigned to states with supervision and support provided by the federal government. This delegation of authority implies that policymakers view MSW controls as a responsibility best undertaken close to the source. Not every community faces the same problems, and even among those that do, the degree of environmental risk can vary considerably across localities. Factors such as town pop-

[7]To learn more about recycling in the United States, visit **http://www.epa.gov/epaoswer/non-hw/muncpl/recycle.htm.**

[8]OECD (1999b), pp. 173–74.

[9]These provisions are given in subtitle D of RCRA. For online information on this section of the law, visit **http://www.epa.gov/epaoswer/osw/non-hw.htm.**

18.2 State Recycling Goals: New Jersey's Solid Waste Initiative

Application

Following the lead of states like Oregon and Rhode Island, New Jersey enacted a mandatory recycling law in 1987. The legislation calls for a statewide recycling rate of 15 percent for the first year and a 25 percent rate for each year thereafter, with an ultimate objective of 65 percent by 2000. To implement the law, all of New Jersey's 21 counties had to develop and submit a recycling plan as part of their solid waste plans. The law specifies that at least three materials must be recycled. While it does not mandate what these three must be, it is generally assumed that they will be newspapers, aluminum cans, and glass containers.

The statewide recycling efforts are to be financed by a $1.50 per ton facilities surcharge, which officials estimate will generate approximately $12 million in revenues each year. These funds will be allocated to the program participants. While this revenue sharing provides some incentive for counties and municipalities to participate in the program, New Jersey authorities took further steps to assure the program's success. For example, several incentive mechanisms were established, including a 50 percent tax credit for industries purchasing new recycling equipment. There is also a requirement that 45 percent of the state's paper purchases be spent on recycled paper to promote markets for recyclables. In addition, authorities have established an infrastructure throughout the state with recycling coordinators positioned within each county and municipality to develop and implement local recycling plans and to communicate important information to program participants.

Despite the complexities of orchestrating a recycling program in a state with an 8.4 million population and 567 municipalities, New Jersey has enjoyed remarkable success thus far. During 1989, just two years after the recycling law went into effect, about 20 percent of its residential and commercial solid waste and almost 39 percent of its entire solid waste stream were being recycled—wastes that would have found their way to New Jersey's limited landfills were it not for the 1987 law. By 1996, the state's recycling rate rose to 45 percent. Although New Jersey did not meet its target of a 65 percent recycling rate in 2000, it did achieve a statewide rate of 53 percent that year.

Sources: U.S. EPA, Office of Solid Waste (January 1989), pp. 20–22; National Solid Wastes Management Association (1990); Heumann (August 1997); State of New Jersey, Department of Environmental Protection (March 13, 2002).

ulation, age of a landfill, and the proximity of groundwater and surface water sources to landfills are among the conditions affecting the outcome.

States' Responsibilities

According to RCRA, states are to develop their own waste management plans, but these must meet certain federal requirements. For example, state plans must include provisions to close or upgrade existing open dumps and to prohibit the establishment of new ones. They also must require that nonhazardous wastes either be used for resource recovery or be properly disposed of, such as in a sanitary landfill. Some flexibility is allowed to account for state- or region-specific factors that may affect the MSW stream.

RCRA also requires states to establish whatever regulatory powers they need to comply with the law. Typically, states pass their own legislation to meet federal requirements and to fulfill their responsibilities as outlined in their state plans. Some have used incentive-based approaches such as providing grant monies for their cities and towns to set up waste management programs. Others have passed laws calling for local governments to set up recycling plans. One example is New Jersey's recycling law, discussed in Application 18.2. Some have passed more stringent laws mandating that local governments and disposal facilities recycle—among them, Connecticut, Oregon, Rhode Island, Washington, and Wisconsin.[10]

[10] U.S. EPA, Office of Solid Waste and Emergency Response (November 1989), pp. 16–17.

In support of such efforts, in October 1991 President George H. W. Bush signed Executive Order 12780, titled *The Federal Recycling and Procurement Policy.* This directive required all federal agencies to step up their recycling efforts and to foster the development of markets for recycled products.[11] With a similar intent, President Clinton issued Executive Order 12873 in October 1993, titled *On Federal Acquisition, Recycling, and Waste Reduction,* which called for all printing and writing paper to contain at least 20 percent recovered paper. This executive order was subsequently replaced with Executive Order 13101, which raised the target to 30 percent. While these responses show initiative, they fall short of what some other nations are doing. For example, in the same year that President Bush signed his executive order, Germany passed a new ordinance that is far more aggressive in its approach toward recycling, as Application 18.3 explains.

Federal Responsibilities

Based on RCRA's provisions for nonhazardous wastes, the federal government must provide financial and technical assistance to states in designing and implementing their waste management plans. It also must encourage states to conserve resources and assist them in finding ways to maximize the use of recoverable resources. The EPA is responsible for establishing minimum criteria for sanitary landfills and other land disposal sites.[12] Facilities not meeting the criteria are considered open dumps and must be closed or upgraded in accordance with specific rules.

In September 1991, new regulations were issued to establish tougher standards for land disposal sites starting in 1993. These new rules, accessible at **http://www.epa.gov/epaoswer/non-hw/muncpl/safedis.htm,** are effective throughout the useful life of a facility plus 30 years after its closure. In addition to the risk reduction provided by these tougher rules, the revisions are expected to encourage source reduction and recycling on a national scale.[13] New regulations also were issued for waste-to-energy incineration facilities. These establish controls on various air emissions and assure proper combustion conditions. These rulings are of special importance, since there are about 100 municipal waste-to-energy facilities currently in operation in the United States.[14] Prior to this time, the EPA had established guidelines, but no regulatory controls per se, for MSW incineration facilities.

The Current Policy Direction

Because states are responsible for their own nonhazardous waste plans, they have the flexibility to develop cost-effective programs. As of 2002, many states have developed bona fide plans, following the EPA's **integrated waste management system.** This system promotes using a combination of techniques and programs aimed at **source reduction, recycling, combustion,** and **land disposal**—in that order. To understand how states are responding to this initiative, we need to develop a general market model of MSW services. This provides the analytical tool with which to assess the cost-effectiveness and efficiency of various state programs.

integrated waste management system
An EPA initiative to guide state MSW plans that promotes using source reduction, recycling, combustion, and land disposal.

Modeling the Market for MSW Management Services

In the market for MSW services, the relevant commodity is actually a combination of several distinct activities—the collection, transportation, and disposal of municipal solid waste. Based on this output definition, we model the market for MSW services in Figure 18.3, using

[11] Council on Environmental Quality (March 1992), pp. 112–14.
[12] U.S. EPA, Office of Solid Waste (November 1986).
[13] U.S. EPA, Office of Solid Waste and Emergency Response (November 1989), pp. 18–19, 108–109; Council on Environmental Quality (March 1992), pp. 112–13.
[14] U.S. EPA, Office of Solid Waste and Emergency Response (July 2001), p. 103.

18.3 *Germany's Green Dot Program*

Application

A commonly cited example of a national recycling plan is Germany's packaging ordinance. According to one press report, this 1991 law, the German Packaging Ordinance of 1991, and the Amendment of 1998, is the most ambitious recycling program in the world. The progressive mandate set recycling targets of 72 percent of glass, tinplate, and aluminum and 64 percent of cardboard, paper, plastic, and composites to be met by July 1, 1995. To achieve these ambitious objectives, Germany's ordinance stipulates that industry is to be responsible for the collection and recycling of all its packaging. If businesses' efforts fail to meet the statutory targets, the German government will institute costly deposits on essentially all packaging.

Responding to these tough requirements, firms collaborated to form a private, nonprofit company, called Duales System Deutschland AG (DSD), meaning a dual system, to provide collection and recycling services to consumers. In 1997, DSD became a publicly limited company with approximately 600 shareholders. More information on the company is available at their Web site, **http://www.gruener-punkt.de/en/home.php3.** Through its network, DSD assures that used packaging will be collected and sent to a proper recycling facility. To facilitate the collection phase, DSD distributes yellow bins and bags for all packaging waste other than glass containers, which continue to be collected via drop-off plans already in place.

Of course, such an undertaking doesn't come cheap. DSD's start-up costs were estimated to be in the neighborhood of $10 billion, while its operating expenses were expected to be $1 billion per year. To finance these fixed and variable costs, DSD sells to participating companies the right to use a "green dot" on their packages, a symbol that guarantees that the packaging is eligible for the services provided by DSD. The price of a green dot varies with the amount of packaging on each product and its recyclability. At the start of the program, the average price was about one cent per package.

Thus far, reported statistics indicate progress has been made. For example, by April 1992, both domestic- and foreign-based firms had purchased 5,000 licenses for DSD's green dots to assure the use of its services for some 40 billion packaging units. By 2000, the company sold more than 19,000 licenses. Consumers have responded just as vigorously—even more so than was forecasted by analysts. Original predictions were that the Green Dot Program would result in the collection of about 100,000 tons of plastics per year, but these estimates were far off the mark. In 2000, German households accumulated nearly 5.7 million tons of packaging waste bearing the green dot.

While response to Germany's new ordinance has been favorable, the Green Dot Program that supports it is not without problems. DSD lost more than $300 million during its first full year of operation. A further problem involves the potential for fraud in the program. Apparently, some companies are putting the green dot symbol on their products without paying for it. In fact, one DSD spokesperson claims that approximately one-third of the green dots that appear on products are fraudulent. Finally, there is concern by the European Union Commission that certain conditions associated with the contractual obligations of using the green dot are not compatible with a unified marketplace, and hence changes in the licensing arrangements may need to be made.

Sources: Shea (July/August 1992); Cairncross (March–April 1992); "German Recycling Is Too Successful" (July 27, 1993); "German Packaging Regulations" (May 23, 2002); Duales System Deutschland AG (May 23, 2002).

http://

Figure 18.3 ***Market for Municipal Solid Waste Services***

This model illustrates the market for MSW management services. The hypothetical demand (or *MPB* curve) and the supply (or *MPC* curve) determine the competitive equilibrium price of MSW services (P_C) and the equilibrium quantity (Q_C). The impact of Congress's decision to tighten federal controls on landfill disposal causes the *MPC* curve to shift upward to *MPC'*, elevating equilibrium price to P_C' and causing a decline in quantity to Q_C'. The decrease in quantity of MSW services may mean that generators are using a source reduction strategy. It might also be the case that the generation rate is the same but generators are recycling more of their wastes. Finally, they might be maintaining the same generation rate *and* the same recycling rate but be engaging in illegal disposal to get rid of some of their wastes.

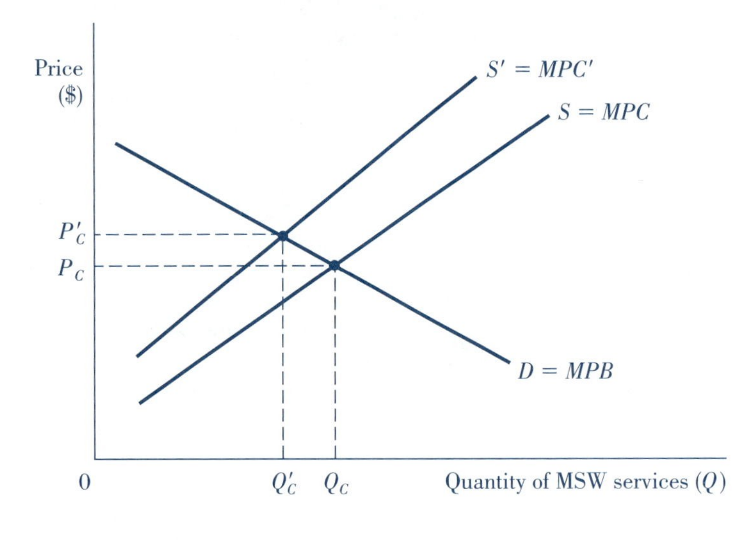

some hypothetical demand (*D*) or marginal private benefit (*MPB*) curve and some supply (*S*) or marginal private cost (*MPC*) curve. Together, these curves determine the competitive equilibrium price or fee for MSW services (P_C) and the equilibrium quantity (Q_C).

Supply of MSW Services

The supply side of the MSW services market represents the production decisions of private firms under contract with cities and towns or those of municipalities who provide these services directly to the community. In this market, the costs of production reflect the expenses of operating a fleet of disposal trucks, managing an approved landfill or incinerator, and labor. Under the usual assumptions about production and diminishing returns, the *MPC* curve exhibits a positive slope. Among the nonprice factors affecting supply in this market are land availability and government regulation.

Consider, for example, the impact of Congress's decision to tighten federal controls on landfill disposal. These tougher standards led to the closing of several hundred landfills that were unable to meet the law's new requirements. This outcome along with the NIMBY syndrome limited the availability of landfill space in some locations. As a result, landfill prices, or **tipping fees** as they are called, have risen. As Figure 18.3 shows, this outcome is modeled as an upward shift of the *MPC* curve, which causes an increase in price to P_C' and a decline in equilibrium quantity to Q_C'.

tipping fees
Prices charged for disposing of wastes in a facility such as a landfill.

Demand for MSW Services

The demand side of this market represents the purchasing decisions of MSW generators. In this context, the quantity response to changes in price takes on an important meaning about how wastes are managed. To understand this, reconsider the decline in quantity from Q_C to Q_C' due to the regulation-induced price increase described in Figure 18.3. How might generators change their behavior to achieve this reduction in quantity? One possibility is that they use a source reduction strategy and produce less trash. Another is that they generate the same amount of trash but demand fewer services because they recycle or reuse. Finally, they might maintain the same generation rate *and* the same recycling rate but engage in illegal disposal to avoid the higher price of MSW services.[15]

Which of these options is chosen depends on their availability to the generator and the prices of these options relative to the price of MSW services. Recognizing the natural market response of demanders to higher priced MSW services, a local community can encourage recycling by offering a cost-effective program to its residents. An example of such of program is the provision of curbside instead of drop-off recycling services. In 1998, there were over 9,300 active curbside recycling programs in the United States, serving about 52 percent of the population.[16] In such cases, the opportunity cost of recycling to the waste generator is reduced.[17] In the absence of such a program, some generators may be motivated to illegally dispose of their wastes.

The demand, or *MPB*, for municipal waste services also responds to certain nonprice changes. For example, more affluent individuals tend to generate larger amounts of trash, since they purchase more products and replace them more frequently. Thus, demand for MSW services likely would shift to the right as the income of a community rises, holding all else constant. Another nonprice determinant of demand is tastes and preferences. As generators become more environmentally responsible, we would expect their demand for these services to decline as they adjust their purchases toward products with less packaging. In sum, waste generators in each community likely face a uniquely shaped demand curve that responds somewhat predictably to both price and nonprice changes.

If actual MSW markets behaved in accordance with this model *and* if there were no externalities, we could conclude that MSW markets achieve an efficient solution where *MPC* = *MPB*. However, it turns out that these conditions typically are violated in actual MSW markets. The resulting resource misallocation is an important issue that merits further investigation.

Resource Misallocation in the Market for MSW Services

There are two distinct problems that arise in most MSW services markets, and both are associated with the supply side of the market:

- Pricing of MSW services does not properly reflect the rising *MPC* associated with increases in production levels.
- Production of MSW services gives rise to negative externalities.

Flat Fee Pricing of MSW Services

In most communities, suppliers charge a **fixed fee** per household or commercial establishment for MSW services. Since the fee is the same regardless of the quantity of waste generated at each location, this type of pricing scheme is known as a **flat fee pricing system.** Notice that the waste generator is charged nothing for any additional containers of trash beyond the first one, which means that the price does not reflect rising *MPC*. Demanders effectively pay a marginal price of zero and hence have no incentive to reduce wastes. More formally, the market price under this scenario is being determined *as if* the *MPC* is zero.

fixed fee or flat fee pricing system
Pricing MSW services independent of the quantity of waste generated.

[15] Jenkins (1993), pp. 4–6.
[16] U.S. Census Bureau (1999), and *Biocycle* (various issues, 1999) (1998 data), as cited in U.S. EPA, Office of Solid Waste and Emergency Response (July 2001), p. 99, table 25.
[17] See, for example, Duggal, Saltzman, and Williams (summer 1991) and Callan and Thomas (fall 1997).

Figure 18.4 Modeling a Flat Fee Pricing System for MSW Services

In a **flat fee pricing system,** the waste generator is charged nothing for any additional containers of trash beyond the first one, which means that the price does not reflect rising *MPC*. Instead, demanders are effectively paying a marginal price of zero and hence have no incentive to reduce wastes. More formally, the market price is being determined *as if* the *MPC* is zero or coincident with the horizontal axis. Under such a scenario the market equilibrates where $MPC = MPB = 0$, at Q_0. Compared to the competitive equilibrium, too many resources are being allocated to MSW services.

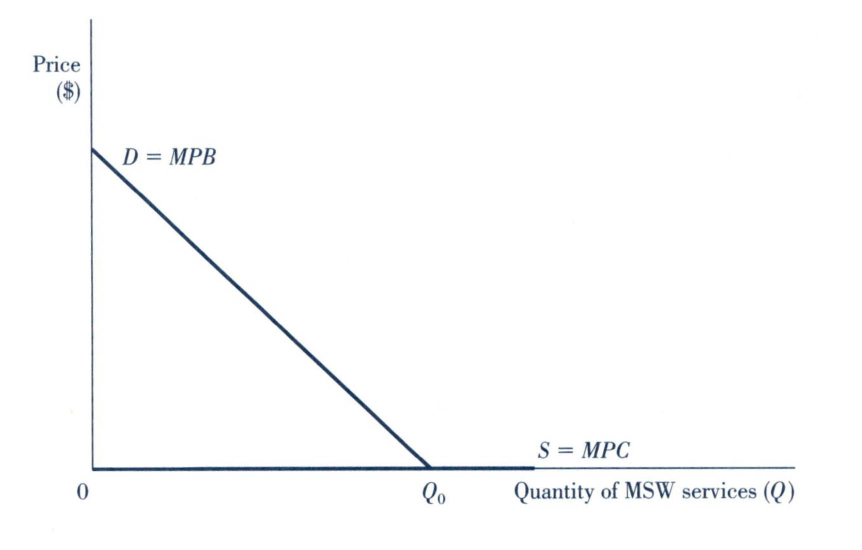

This phenomenon is illustrated in Figure 18.4, where the *MPC* curve is shown coincident with the horizontal axis. Notice that under such a scenario, the market equilibrates where $MPC = MPB = 0$ at Q_0. Compared to the competitive equilibrium, too many resources are being allocated to MSW services. In communities where MSW services are publicly provided, the price mechanism is further dampened because the fee is not explicitly communicated but rather collected through property taxes. But whether these services are publicly or privately provided, the use of a fixed fee has serious efficiency implications.[18]

Negative Externalities

Even if waste generators are charged a fee equal to the *MPC*, there still may be a resource misallocation if production of MSW services gives rise to negative externalities. Such an outcome is not uncommon in waste management markets. For example, Application 18.4 reports on the negative externalities associated with tire disposal and a new technology that may help reduce these effects. In the MSW services market, external costs may be due to groundwater contamination, air pollution from incineration, or impairment of aesthetics.

As we have illustrated in other contexts, such a production externality is captured by a marginal external cost (*MEC*) curve, which must be added to the *MPC* to identify the marginal social cost (*MSC*) of producing the good. Assuming no externalities on the demand side, the marginal social benefit (*MSB*) of consumption is the same as the *MPB* function. The efficient

[18] Wirth and Heinz (May 1991), pp. 48–49.

18.4 *Garbage or Resource? The Case of Scrap Tires*

Application

One person's trash is another person's treasure—an appropriate commentary on the market for recyclables. Some solid waste can be returned to productive use, provided there is an active market for the recycled material. Encouraging waste generators to recycle is important, but it is only part of the story. Recycling sets up a reliable supply of inputs that presumably can be used in the production of some new commodity. However, in order for this market to be viable, there must be a healthy demand side of buyers willing and able to purchase these recycled materials at the going market price. If not, society can expect a trade-off of one problem for another—a waste heap for a glut of unwanted recycled materials.

An interesting case is the market for scrap tires. Current estimates place the total number of discarded tires at approximately 3 billion—a nontrivial disposal problem that is increasing at the rate of 273 million per year in the United States alone. Recycling of scrap tires has grown significantly over the years. For example, in 1990, 83 percent of discarded tires found their way to either a landfill or a tire stockpile. Contrast this with the fact that by 2001, only 24 percent of all scrapped tires were disposed in this manner.

What contributed to the dramatic change in the way scrap tires are treated? The answer is based on economic principles. Predictably, the enormous supply of used tires provided a powerful incentive to find some productive use for this burdensome contribution to the solid waste stream. Most experts agree that the use of scrapped tires as an energy source makes good sense from an energy efficiency standpoint. Their heat value exceeds that of coal, at about 12,000 to 16,000 Btu per pound. However, environmental and economic factors have been major stumbling blocks to expanding the use of scrap tires for energy recovery. Because of both air quality and aesthetic impairment, there is resistance to siting an energy facility that is dependent on scrap tires for fuel. Until recently, most tire-to-energy technologies have been relatively dirty, resulting in atmospheric emissions that are unacceptable. But new technologies have lead the way in making the burning of scrap tires for energy recovery both environmentally and economically feasible. One such technology patented by a New Mexico firm, Titan Technologies, appears to hold promise.

Titan's innovative process bakes 6-inch squares of tire chips instead of burning whole tires or tire shreds. According to company officials, the process uses a closed system of catalytic drums and requires relatively low temperatures—about 450° Fahrenheit. The result is a conversion process that yields by-products much less damaging to the environment. Furthermore, Titan operates at a large scale—much larger than conventional waste-to-energy facilities. Exploiting scale economies allows Titan to achieve lower unit costs and to market its energy at a competitive price. This "buy-back rate" is critical to the viability of this particular waste-to-energy market. However, to be a major player in the market, Titan must continue to find ways to lower unit costs. If this economic feasibility is established, Titan's technology may help further reduce the number of scrap tires that are stockpiled, landfilled, or illegally dumped.

Sources: Metz (December 28, 1993); U.S. EPA, Office of Solid Waste and Emergency Response (October 1991); U.S. EPA, Office of Policy, Planning, and Evaluation (March 1991), pp. 2-13–2-15; U.S. EPA, Office of Solid Waste and Emergency Response (May 9, 2002).

Figure 18.5 ***Allocative Inefficiency in Private Markets for MSW Services: Presence of a Negative Externality***

A resource misallocation arises if production of MSW services gives rise to a negative externality. The production externality is captured by a marginal external cost (*MEC*) curve, which is added to the *MPC* to identify the marginal social cost (*MSC*) of producing the good. Assuming no externalities on the demand side, the marginal social benefit (*MSB*) function is the same as the *MPB*. The efficient solution is determined where *MSC* equals *MSB*. Notice that the efficient output level occurs at Q_E, which is *lower* than the private market outcome, Q_C.

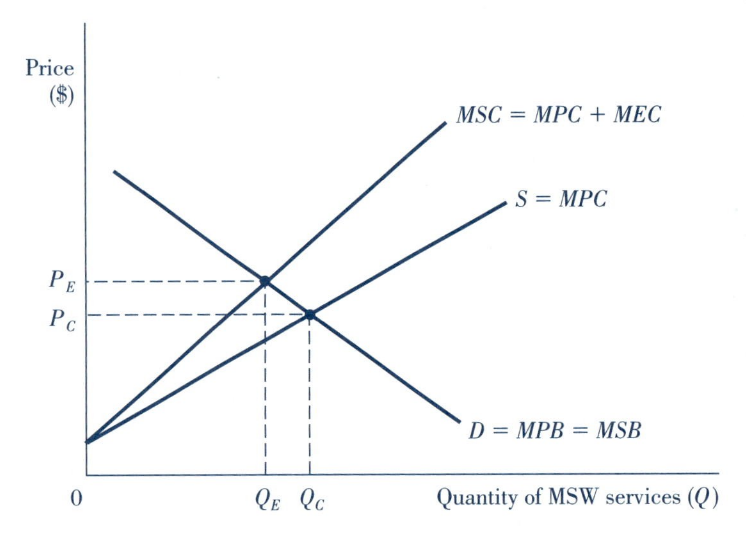

solution is determined where the *MSC* equals the *MSB*, as illustrated in Figure 18.5. (To avoid confounding the effect of the externality with that of charging a fixed fee, we show the *MPC* as a positively sloped curve.) Notice that the efficient output level is Q_E, which is *lower* than the private market outcome, Q_C.

To summarize, we observe that private MSW markets typically do not yield an efficient outcome. The use of fixed fees distorts the signaling mechanism of price, effectively removing the incentive to economize on trash generation. The private market's disregard for external costs leads to a further distortion. Singularly or in combination, these factors contribute to an overallocation of resources to waste services and an overproduction of municipal trash. Is there any solution to these problems? Actually, a number of pricing schemes can be used to correct the inefficiency.

Market Approaches to MSW Policy[19]

In recent years, some communities have begun to institute market-based policies aimed at reducing the problems associated with MSW. Of interest are three approaches that directly exploit the signaling mechanism of price. These are:

[19]Much of the following discussion is drawn from Wirth and Heinz (May 1991), pp. 48–65, and U.S. EPA, Office of Policy, Economics, and Innovation (January 2001), chaps. 4 and 5.

Figure 18.6 **Modeling a Waste-End Charge to Restore Efficiency in the Market for MSW Services**

This model illustrates how a **waste-end charge** can be used to achieve efficiency in the market for MSW services. Because such a charge varies with the quantity of waste, it avoids the market distortion caused by a flat fee pricing scheme. To achieve efficiency, the waste-end charge must be set to cover the *MSC* at the efficient equilibrium, where *MSB* equals *MSC*. In the model shown, the appropriate per unit fee would be set equal to P_E.

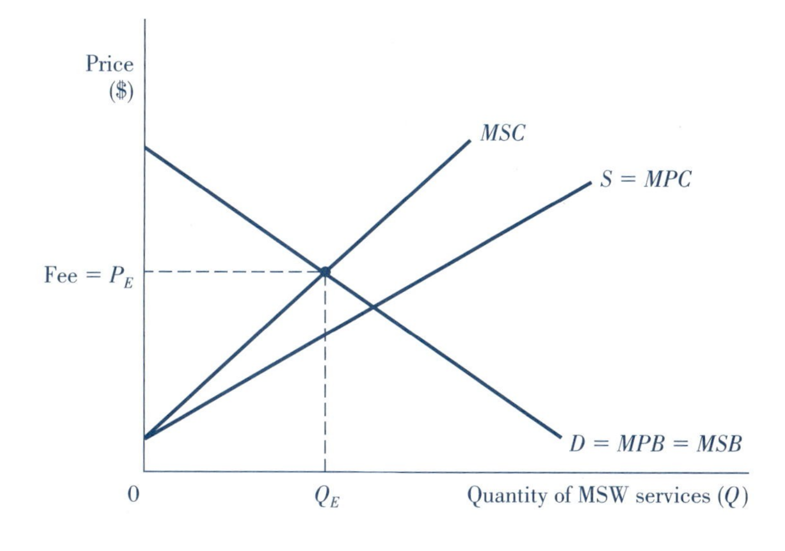

- **Back-end or waste-end charges**
- **Front-end or retail disposal charges**
- **Deposit/refund systems**

Back-End or Waste-End Charges

As discussed in the previous chapter, a **waste-end charge** is so named because it is implemented at the time of disposal based on the quantity of waste generated. Because such a charge varies with the quantity of waste, it avoids the market distortion caused by a flat fee pricing scheme. To achieve efficiency, the waste-end charge must be set to cover the *MSC* at the efficient equilibrium. This solution is illustrated in Figure 18.6, with the appropriate per unit fee shown as P_E. Notice that this fee is a *price per unit of waste* to be paid by all generators, which is very different from charging a *price per household*, as is done under a flat fee system.

In practice, programs that use waste-end charges are called **unit pricing schemes,** also known as **pay-as-you-throw (PAYT) programs,** to indicate that prices for MSW services are charged on a per-unit-of-waste basis. The units may be measured either by weight or by volume generated, though the latter is more commonly used at the present time. The same price can be charged for each unit, called **flat rate pricing,** or the price can vary, called **variable rate pricing.**[20] Perhaps the best-known example of a community having

back-end or waste-end charge
A fee implemented at the time of disposal based on the quantity of waste generated.

unit pricing scheme
A common designation for the use of a waste-end charge.

flat rate pricing
A unit pricing scheme that charges the same price for each additional unit of waste.

variable rate pricing
A unit pricing scheme that charges a different price for each additional unit of waste.

[20] To learn more about unit pricing programs in specific communities, visit **http://www.epa.gov/payt/comm.htm.**

implemented unit pricing is the city of Seattle. In fact, Seattle's program has come to be the prototype upon which other communities have modeled their own MSW unit pricing plans. The program is implemented by having each MSW generator register with the city for the number and volume of trash containers it expects to use in a week. Based on the registration, a fee is calculated using a schedule of volume-based prices.[21]

By moving from a fixed fee schedule to a unit pricing program, two important market incentives are restored. First, unit pricing communicates to generators that increased waste services are associated with a nonzero marginal cost of production, ideally the marginal social cost. Effectively, this provides an incentive to economize on waste generation by using a source reduction strategy. Second, by elevating the price of collection and disposal, the *relative* price of alternative approaches declines. Given that recycling and disposal are substitute activities, the change in relative prices may encourage more recycling, which means less disposal. There are also favorable equity implications, since unit pricing does not force small MSW generators to subsidize larger ones, as is the case with a flat fee system. Finally, while there is some indication that unit pricing can encourage illegal disposal, the evidence does not indicate that this activity is increasing dramatically where such programs are in use.[22]

Beyond theoretical arguments, there is a fair amount of anecdotal and empirical evidence that unit pricing is a viable policy instrument. For example, it is commonly noted that Seattle officials observed a decline in landfill disposal since unit pricing was introduced. Also, surveys conducted in New York and New Hampshire suggest that unit pricing programs motivate consumers to identify ways to reduce waste generation, including changing their purchasing decisions.[23]

More formal evidence is presented by Robin Jenkins (1993), who conducted a rather extensive statistical study on unit pricing. The results show that the effect of switching from a fixed fee to a volume-based fee of 50 cents per container is a decrease in waste per person per day of 0.2 pounds. This translates to an estimated decline of 3,650 tons per year for a community of 100,000 people or 18,250 tons annually for a community of 500,000 people. Another study of communities in California and the Midwest found that disposal decreased by 6–50 percent following the launch of a variable rate pricing scheme, where the reductions were positively related to the price level charged (Miranda and Aldy, 1996).

As the evidence accumulates, more and more communities have followed Seattle's lead and have moved to unit pricing programs. Current estimates indicate that over 4,100 communities in 42 states are using this form of MSW pricing.[24] In fact, some local authorities have enhanced the Seattle prototype to improve its efficiency. For example, one of the problems with Seattle's "per can" pricing plan is that the preregistered number of receptacles on which the fee is based is not always an accurate forecast of actual trash generation. Consequently, generators can be charged a full rate even if some receptacles are half-full or not used at all. Recognizing this potential inefficiency, some communities have instituted more flexible programs whereby generators purchase stickers for receptacles of various sizes. Then, they choose how many receptacles they need each period and simply apply the prepaid stickers to the receptacles used.[25] Application 18.5 discusses this so-called **bag-and-tag approach,** which has met with some success.

bag-and-tag approach
A unit pricing scheme implemented by selling tags to be applied to waste receptacles of various sizes.

Front-End or Retail Disposal Charges

front-end or retail disposal charge
A fee levied on a product at the point of sale designed to encourage source reduction.

An alternative pricing scheme aimed at the MSW problem is the use of a **front-end or retail disposal charge.** In contrast to a waste-end charge imposed on *wastes* at the point of *disposal,* a retail disposal charge is levied on *products* at the point of *sale.* Since such a charge is instituted at the pregeneration stage, its objective is to encourage pollution prevention

[21] Seattle Solid Waste Utility (1988), as cited in Wirth and Heinz (May 1991), pp. 50–51.
[22] See, for example, Fullerton and Kinnaman (September 1996) and Skumatz and Zach (1993).
[23] U.S. EPA, Office of Policy, Economics, and Innovation (January 2001), p. 44.
[24] Skumatz (March 21, 1996), as cited in U.S. EPA, Office of Policy, Economics, and Innovation (January 2001), p. 41.
[25] Wirth and Heinz (May 1991), p. 51.

18.5 *Bag-and-Tag Systems: Another PAYT Approach*

Application

Although the city of Seattle has received national acclaim for its innovative unit pricing waste program, the market-based plan has not been without problems. One in particular is that per can pricing schemes are based on residents' forecasts of their waste generation rate, which may or may not be accurate. As a result, some generators end up paying for disposal services on a full can of trash, even if it is half empty or not used at all. This potential inefficiency has prompted some communities to design modified versions of the Seattle prototype, called **bag-and-tag systems.**

Under these bag-and-tag plans, the advantages of unit pricing are maintained, but residents have the flexibility to assure that they pay only for the wastes they generate. One way these systems are implemented is by providing collection and disposal services only to trash placed in specially designated containers sold by the municipality. Program participants purchase in advance a supply of these trash bags, typically available in various sizes and priced accordingly. In each collection period, residents use only as many of the prepaid trash bags as they need. A variation of the system operates from the same premise except that stickers or tags are sold by the municipality, and these are then affixed to residents' own trash containers. In either case, the inefficiency of the per can pricing plan is resolved while maintaining the advantages of unit pricing. A few examples will illustrate how this approach works in practice.

In the late 1980s, High Bridge, New Jersey, instituted a tag pricing system, charging residents an annual fee for 52 stickers. An annual charge per household was set, which covered the program's administrative costs plus the disposal cost of one container per week for each household. Restrictions on maximum weight and volume are imposed on a per sticker basis. To provide greater flexibility to residents, additional stickers can be purchased from the town or traded between low-use and high-use residents. In its first 10 months of operation, High Bridge officials reported a 24 percent decrease in waste tonnage.

At about the same time as High Bridge got its plan under way, the town of Perkasie, Pennsylvania, began selling town-issued trash bags to implement its unit pricing program. To distinguish between low and high waste generators, Perkasie uses a differential charge based on bag size. A 20-pound trash bag commands a price of $0.80, while the larger 40-pound container is sold for $1.50. All of Perkasie's trash bags have a tree logo to distinguish them from receptacles used by nonparticipants. Like the High Bridge plan, residents pay for services only on the quantity of waste generated in a given period and not on a predetermined and likely inaccurate forecast. A 40 percent decline in waste tonnage was observed after Perkasie's first full year of operation.

In 1993, officials in San Jose, California, launched a hybrid variable rate program called a cart/sticker system. Local residents subscribe to designated cart sizes, as in the Seattle plan, and pay specified fees for the carts they select. However, if more disposal services are needed, households can use 32-gallon bags, as long as each bag bears a designated sticker that can be purchased for $3.50 each. This combination method adds needed flexibility to the Seattle subscription method, yet still adheres to the pay-as-you-throw (PAYT) principle. An added incentive for residents to economize on disposal is provided by San Jose's curbside recycling services offered at no additional charge to residents.

For information, data, and case studies on these and other PAYT programs, visit **http://www.epa.gov/payt/index.htm.**

http://

Sources: Wirth and Heinz (May 1991), pp. 49–53; U.S. EPA, Office of Policy, Planning, and Evaluation (March 1991), pp. 2.7–2.12; Tregarthen (September/October 1989), p. 17; U.S. EPA, Office of Policy, Economics, and Innovation (January 2001), pp. 41–46.

through source reduction.[26] This includes motivating manufacturers to seek out product designs and packaging that are more environmentally responsible.

From a practical perspective, local conditions tend to dictate when a retail disposal charge might be more appropriate than a waste-end charge. For example, a waste-end pricing scheme should be coordinated with a recycling program to deter illegal disposal. Hence, if a community has not instituted such a plan, a better option might be to use a front-end charge. A similar argument applies to communities where there is a predominance of multiunit residences, making waste-end fees difficult to implement. Finally, a front-end charge can be used to complement a waste-end charge system to discourage the use of products that yield large amounts of waste or wastes that pose a particular threat to health or the environment. Examples include products like motor oil, batteries, and automobile tires.

To better understand the environmental and economic consequences of retail disposal charges, consider the market for household batteries. Batteries are made with heavy metals like mercury or cadmium, so their disposal poses an environmental risk to society. Battery consumers do not consider the external effects, so the *MPB* effectively overstates the true marginal benefits of consumption. This means that the *MSB* curve is actually *below* the *MPB* curve, as shown in Figure 18.7.[27] In the absence of government intervention, equilibrium output is determined where *MPB* equals *MPC*, or Q_C. However, the efficient output level is lower, determined by the intersection of the *MSC* and *MSB* curves, or Q_E. (For simplicity, we assume that no production externalities exist, so *MSC* = *MPC*.)

To correct the inefficiency, government can impose a retail disposal charge on batteries at the point of sale equal to the difference between the *MSB* and the *MPB* at Q_E. In Figure 18.7, we model the imposition of the charge as a shift up of the *MSC* by distance *xy*, resulting in an effective price to the consumer of P_R.[28]

Retail disposal charges are being used both internationally and domestically. For example, Austria imposes these charges on refrigerators and air conditioners, and Finland uses them for crude oil and tires. While part of the objective is to diminish the damage caused by disposal of these products, these charges are also used to generate revenue, which may in turn help to fund cleanup efforts.[29] Domestically, retail disposal charges are imposed on a multitude of products—many by state governments. Table 18.4 lists some common applications. Most of the target products and materials are those expected to generate high external costs. For example, Texas imposes a $3.50 charge on truck tires, and Rhode Island charges $0.10 per gallon of antifreeze.

Deposit/Refund Systems

deposit/refund system
A market instrument that imposes an up-front charge to pay for potential damages and refunds it for returning a product for proper disposal or recycling.

An alternative market instrument that can mitigate MSW pollution is a **deposit/refund system,** which is actually a two-part pricing scheme. It imposes an up-front charge for potential damages caused by improper disposal, and it allows for a refund of that charge at the end of the product cycle if the consumer takes proper action to avoid those damages. Typically, the consumer must return the product or its container for recycling or safe disposal. The formal

[26] While source reduction is also possible from using unit pricing, it is not assured. All that can be said with certainty is that unit pricing schemes encourage a decline of waste *disposal.* But this may be achieved through an increase in recycling with no change in the waste generation rate. Conversely, front-end or retail disposal charges directly encourage a decline in waste *generation* by reducing the quantity demanded of a product *before* it enters the waste stream.

[27] The externality is not modeled as an *MEC* added to the *MPC,* because in this case it is not associated with the production side of the market. Rather, the externality arises from consumption or the demand side. Hence, the externality must be *deducted* from the *MPB* curve to show that the true benefits to society, the *MSB,* are reduced by the amount of the environmental damage. For a more rigorous and detailed analysis of this type of externality, see Tresch (1981).

[28] It matters not whether the retail disposal charge is levied on the seller or the buyer of the product, since economic incidence (i.e., the party who ultimately bears the burden of the fee) is not determined by statutory incidence (i.e., the party who is initially charged the fee). For more detail, the reader can consult any good principles of microeconomics text.

[29] OECD (1999a), pp. 34–35, table 3.15.

Figure 18.7 ***Modeling a Retail Disposal Charge to Correct a Consumption Externality: Market for Household Batteries***

The disposal of batteries poses an environmental risk to society because they are made with heavy metals. Battery consumers do not consider the external effects, so the *MPB* overstates the true marginal benefits of consumption. Hence, the *MSB* curve is actually *below* the *MPB* curve. In the absence of government intervention, equilibrium output is Q_C, where *MPB* intersects *MPC*. The efficient output level is lower, at the intersection of *MSC* and *MSB*, or Q_E. (We assume no production externalities, so *MSC* = *MPC*.) To correct the inefficiency, government can impose a retail disposal charge equal to the difference between *MSB* and *MPB* at Q_E. We can model the charge as a shift up of the *MSC* by distance *xy*, resulting in an effective price of P_R.

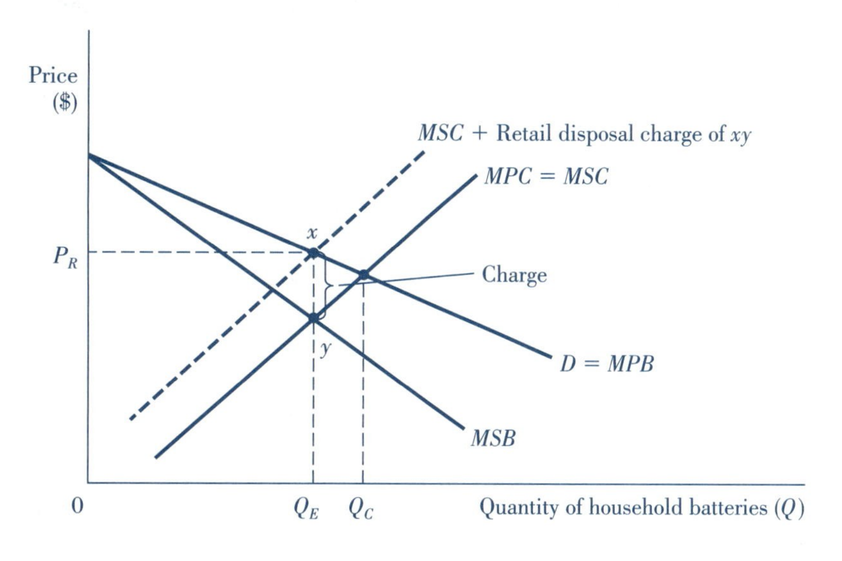

Table 18.4 ***Examples of U.S. Retail Disposal Taxes***

Governing Entity	Waste	Amount of Tax
34 states	Passenger car tires	\$0.25 to \$2.00 each
46 states	Fertilizers	Generally less than \$1.00 per ton
Federal government	Tires	\$0.15 to \$0.50 per pound[a]
Nebraska	Fertilizers	\$4.00 per ton
North Carolina	White goods	\$3.00 per item
Rhode Island	Lubricating oil	\$0.05 per quart
	Antifreeze	\$0.10 per gallon
	Organic solvents	\$0.25 per gallon
	Tires	\$0.50 per tire
Texas	Truck tires	\$3.50 per tire

Source: U.S. EPA, Office of Policy, Economics, and Innovation (January 2001), pp. 50–52.

[a]An additional fixed fee is associated with higher weight tires.

Table 18.5 *Examples of State Deposit/Refund Programs*

State	Product	Amount of Deposit
Arizona	Batteries	$5.00
Arkansas	Batteries	$10.00
California	Beverage	$0.025 for < 24 oz. $0.05 for > 24 oz.
Connecticut	Batteries	$5.00
	Beverage	$0.05 minimum
Delaware	Beverage	$0.05
Iowa	Beverage	$0.05
Maine	Batteries	$10.00
	Beverage	$0.05–$0.15
Massachusetts	Beverage	$0.05
Michigan	Beverage	$0.05–$0.10
New York	Beverage	$0.05
Oregon	Beverage	$0.03–$0.05
Vermont	Beverage	$0.05–$0.15
Washington	Batteries	$5.00 minimum

Sources: U.S. EPA, Office of Policy, Economics, and Innovation (January 2001), pp. 57–66; Battery Council International (May 23, 2002).

model of this system was presented in detail in Chapter 5, so it is not repeated here. More relevant to this discussion is the conceptual relationship between this instrument and the other two pricing schemes. Notice that the deposit is imposed at the time of purchase exactly like a retail disposal charge. The refund attempts to reduce disposal and encourage recycling much like a waste-end charge.[30]

This innovative economic instrument has gained acceptance both domestically and internationally. In the United States, deposit/refund systems are aimed primarily at beverage containers and lead-acid batteries, as Table 18.5 indicates. Similar applications are quite common in other countries as well. Among the nations that use deposit/refund systems for beverage containers are Australia, Canada, Denmark, Mexico, and Sweden. Other countries, including Denmark, Korea, and Mexico, use a deposit/refund system for batteries. Norway and Sweden use this instrument on motor vehicles. In these countries, new car buyers pay a deposit at the time of purchase and receive a larger refund if the automobile is returned to a designated recovery site when it is no longer wanted.[31]

Conclusions

Municipal solid wastes—the simple by-products of production and consumption—have become a local public issue in communities everywhere. Today, most agree that former claims of a landfill crisis were exaggerated. Nonetheless, the world was listening and, in the absence of 20/20 hindsight, responded to what it heard. Private citizens began to change their behavior—avoiding products with excess packaging and participating in local recycling programs, and they turned to government for broader solutions.

[30] See Fullerton and Kinnaman (July 1995) relating to this issue.
[31] OECD (1999a), pp. 39–43.

In retrospect, many argue that people responded to false signals—a crisis that did not exist and unfounded predictions of doom and gloom. This assertion is probably true, at least to a point. However, it is wrong to conclude that there were no solid waste issues to resolve and no environmental risks to manage. Open dumps had become a health hazard, and most of society had been ignorant of the potential problems of excess waste generation and poor waste management practices. Having said this, a more relevant question is whether the response itself—both by private citizens and by government—however motivated, was and is appropriate.

In the United States, Congress passed laws to restrict the use of land for waste disposal and to replace the nation's open dumps with modern sanitary landfills. It then delegated the responsibility for devising local waste management programs to the states, encouraging plans that promoted waste reduction and recycling. The land restrictions and the landfill rulings added costs to local waste management, and some states responded with market instruments that reflected these costs.

What was the result of these policy moves? At this stage, the most accurate assessment is that these efforts have yielded mixed results. Some waste generators are pursuing source reduction activities, while others are illegally disposing of their wastes. Some states and local governments have developed successful, cost-effective recycling centers, while others are struggling with expensive curbside programs that have placed a heavy tax burden on their residents. Add to this the series of debates about the viability of markets for recycled materials. Netting it out, these observations imply that municipal waste policies are still evolving, and public officials at all levels of government need to consider some needed reforms. Much can be learned from observing market responses to regulatory decisions and revisions. Rising prices for waste disposal and gluts of recycled materials are important signals to policymakers about how and where to make strategic changes. If these signals are not ignored, future policy should approach more efficient and cost-effective solutions to the problem of municipal solid waste.

Summary

- The composition of municipal solid waste (MSW) can be characterized both by the types of products being discarded and the kinds of materials entering the waste stream. In 1999, containers and packaging accounted for the highest proportion by weight of all discarded products in the U.S. municipal solid waste stream. The fastest growing segment of MSW in the United States is plastics.
- Internationally, the major industrialized nations are among the highest MSW generators. Some of the international differences are attributable to the amount of packaging used by producers, cultural preferences, environmental awareness, economic conditions, or government regulations.
- One set of provisions in the Resource Conservation and Recovery Act (RCRA) is concerned with the management of nonhazardous wastes, including the MSW stream. For the most part, this section of the law assigns the responsibility of nonhazardous waste management to states with supervision and support provided by the federal government.
- Under RCRA, states are required to develop their own waste management plans, but these must meet certain requirements to receive federal approval. They also must establish whatever regulatory powers they need to comply with RCRA, such as enacting legislation to meet federal requirements.
- The federal government must provide financial and technical assistance to states in designing and implementing their waste management plans. The EPA must establish minimum criteria for sanitary landfills and other land disposal sites.
- The EPA actively encourages state authorities to use an integrated waste management system, which promotes source reduction, recycling, combustion, and land disposal, in that order.

- Resources are misallocated in private MSW services markets that use a flat fee pricing system, which charges a fixed fee per household. The inefficiency occurs because the constant price does not properly reflect rising *MPC* and because production of MSW services gives rise to a negative externality.
- A waste-end charge is implemented at the time of disposal based on the quantity of waste generated. In practice, programs using waste-end charges are referred to as unit pricing schemes to indicate that prices for MSW services are charged on a per-unit-of-waste basis.
- Retail disposal charges are levied on the product at the point of sale. Since this charge is imposed at the pregeneration stage, its objective is to encourage pollution prevention through source reduction.
- A deposit/refund system imposes an up-front charge for potential damages caused by improper disposal and allows for a refund of that charge at the end of the product cycle if the consumer takes proper action to avoid those damages.

Key Concepts

municipal solid waste (MSW)
product groups
materials groups
integrated waste management system
tipping fees
fixed fee or flat fee pricing system
back-end or waste-end charges
unit pricing scheme
flat rate pricing
variable rate pricing
bag-and-tag approach
front-end or retail disposal charge
deposit/refund system

Use the Key Concepts listed above to begin your search for additional articles and information using the InfoTrac College Edition database.

Review Questions

1. a. Summarize a specific municipal solid waste problem in your hometown or one that recently has been reported in the media. Use the specific stages of the waste stream in your discussion.
   b. What policies have been proposed or implemented to address the problem you describe? Analyze these policies from an economic perspective.
2. Using what you have learned about market-based incentives, suggest a policy initiative that would discourage land disposal in the United States.
3. Refer to Application 18.2 on New Jersey's recycling program. Identify which elements of the program are command and control in approach and which are market based.
4. Using the information in Application 18.3, compare Germany's Green Dot Program to U.S. national initiatives on recycling. In your view, which is likely to be more effective in the long run? Explain.
5. Diagram a general model of MSW management services. Show the simultaneous effect of the federal restrictions on landfills *and* rising consumer awareness of the benefits of "green packaging." Assuming a private competitive market, predict the effect on the price and quantity of MSW services.
6. Consider the following model of a hypothetical market for MSW management services:

$$MPB = 25 - 2Q,$$
$$MPC = 4 + Q,$$
$$MEC = 0.5Q,$$

where Q is the number of trash containers per household per month.

a. Quantitatively determine the effect of the resource misallocation due to (i) the presence of the negative externality, and (ii) the use of a flat fee pricing system in the presence of a negative externality.
b. Support your answer to part (a) with a graphical model.
c. Determine the dollar value of a waste-end charge that would restore efficiency to this market. Explain your answer intuitively.

7. a. Contrast the Seattle per can pricing program with Perkasie's bag-and-tag approach both from an environmental and an economic perspective.
b. Why might weight-based unit pricing for MSW management be more advantageous than volume-based programs?

Additional Readings

Alster, Norm, and William Echikson. "Are Old PCs Poisoning Us?" *Business Week,* June 12, 2000, pp. 78, 80.

Bailey, Jeff. "Curbside Recycling Soothes the Soul, But Cost Is High." *Wall Street Journal,* January 19, 1995, pp. A1, A8.

Boerner, Christopher, and Kenneth Chilton. "False Economy: The Folly of Demand-Side Recycling." *Environment* 36 (January/February 1994), pp. 6–15, 32–33.

Brown, Kelly M., Ronald Cummings, Janusz R. Mrozek, and Peter Terrebonne. "Scrap Tire Disposal: Three Principles for Policy Choice." *Natural Resources Journal* 41, no. 1 (2001), pp. 9–22.

Callan, Scott J., and Janet M. Thomas. "Adopting a Unit Pricing System for Municipal Solid Waste: Policy and Socioeconomic Determinants." *Environmental and Resource Economics* 14 (December 1999), pp. 503–18.

Fullerton, Don, and Thomas C. Kinnaman, eds. *The Economics of Household Garbage and Recycling Behavior.* Northampton, MA: Elgar, 2002.

Fullerton, Don, and Wenbo Wu. "Policies for Green Design." *Journal of Environmental Economics and Management* 36, no. 2 (1998), pp. 131–48.

Hickman, H. Lanier, ed. *Principles of Municipal Solid Waste Management.* Silver Spring, MD: Solid Waste Association of North America, 1996.

Judge, R., and A. Becker. "Motivating Recycling: A Marginal Cost Analysis." *Contemporary Policy Issues* 11 (July 1993), pp. 58–68.

Miranda, Marie Lynn, and Joseph E. Aldy. "Unit Pricing of Residential Municipal Solid Waste: Lessons from Nine Case Study Communities." *Journal of Environmental Management* 52, (January 1998), pp. 79–93.

Miranda, Marie Lynn, Jess W. Everett, Daniel Blume, and Barbeau A. Roy Jr. "Market-Based Incentives and Residential Municipal Solid Waste." *Journal of Policy Analysis and Management* 13 (fall 1994), pp. 681–98.

Palmer, Karen, and Margaret Walls. "Optimal Policies for Solid Waste Disposal and Recycling: Taxes, Subsidies, and Standards." *Journal of Public Economics* 65, no. 2 (1997), pp. 193–205.

Pellow, David Naguib. *Garbage Wars: The Struggle for Environmental Justice in Chicago.* Cambridge, MA: MIT Press, 2002.

Repetto, Robert, Roger C. Dower, Robin Jenkins, and Jacqueline Geoghegan. *Green Fees: How a Tax Shift Can Work for the Environment and the Economy.* Washington, DC: World Resources Institute, 1992.

Ueta, K., and H. Koizumi. "Reducing Household Waste: Japan Learns from Germany." *Environment* 43 (November 2001), pp. 20–32.

Related Web Sites

Duales System Deutschland AG	**http://www.gruener-punkt.de/en/home.php3**
Electronics Recycling Initiative Project	**http://www.nrc-recycle.org/resources/electronics/index.htm**
Landfill regulations	**http://www.epa.gov/epaoswer/non-hw/muncpl/safedis.htm**
Lead-acid battery laws	**http://www.batterycouncil.org/states.html**
MSW in the United States	**http://www.epa.gov/epaoswer/non-hw/muncpl/**
Municipal Solid Waste Fact Book	**http://www.epa.gov/epaoswer/non-hw/muncpl/factbook**
National Recycling Coalition	**http://www.nrc-recycle.org**
National Solid Wastes Management Association	**http://www.nswma.org**
Pay-As-You-Throw home page	**http://www.epa.gov/payt/index.htm**
Pay-As-You-Throw individual programs	**http://www.epa.gov/payt/comm.htm**
Plastics and plastics recycling	**http://www.plasticsresource.com/recycling/index.html**
RCRA Subtitle D	**http://www.epa.gov/epaoswer/osw/non-hw.htm**
Recycling	**http://www.epa.gov/epaoswer/non-hw/muncpl/recycle.htm**
U.S. EPA, Office of Solid Waste	**http://www.epa.gov.epaoswer/osw/**

A Reference to Acronyms and Terms in Municipal Solid Waste Policy

Environmental Economics Acronyms

MPB	Marginal private benefit
MSB	Marginal social benefit
MEC	Marginal external cost
MPC	Marginal private cost
MSC	Marginal social cost

Environmental Science Term

Btu	British thermal unit

Environmental Policy Acronyms

MSW	Municipal solid waste
NIMBY	"Not in my backyard"
PAYT	Pay-as-you-throw
RCRA	Resource Conservation and Recovery Act

Chapter 19

Controlling Toxic Chemicals: Production, Use, and Disposal

Toxic substances are extensively regulated in the United States. Air toxics, for example, are controlled under the Clean Air Act. The discharge of hazardous pollutants into waterways is governed by the Clean Water Act, and the Resource Conservation and Recovery Act (RCRA) regulates the hazardous solid waste stream. Notice how all these federal laws deal with the *release* of toxics into the environment. In this chapter, we study legislation aimed at controlling chemical substances *before* they become residuals and enter the waste stream—when they are formulated and produced.

Approximately 5 percent of the U.S. economy is engaged in the production of chemicals and allied products. About 75,000 chemical substances are produced in this country, with new ones introduced every year.[1] While this increasing reliance on chemicals has added to the variety, durability, and usefulness of products available to society, there also has been a cost. Certain of these chemicals, such as polychlorinated biphenyls (PCBs) and the pesticide dichloro-diphenyl-trichloroethane (DDT), pose a threat to human health and ecosystems. And in some cases the risks are not fully understood until long after the damage has been done.

Recognizing the potential risks of chemical usage and the impact of toxic residuals in the waste stream, Congress passed the Federal Insecticide, Fungicide, and Rodenticide Act (FIFRA) and the Toxic Substances Control Act (TSCA). These two laws, each from a different vantage point, control the production, distribution, and consumption of pesticides and other chemicals. By instituting controls before these substances are introduced into commerce, those that pose an unacceptable risk can be restricted in use or even banned. In so doing, the toxicity of the hazardous waste stream should be abated and the associated environmental risks reduced.

Notice that this approach is preemptive because it eliminates the need to treat and dispose of some amount of hazardous residuals. The underlying premise is that the waste stream is part of a larger cycle that originates with the design and development of products. Hence, regulating hazardous substances at the pregeneration phase of the waste stream is an effective way to prevent pollution.

In this chapter, our objectives are to understand the intent of FIFRA and TSCA, to analyze their implementation, and to evaluate their effectiveness using economic criteria and decision rules. To organize our investigation, we begin with an overview of FIFRA—its evolution and its current regulatory structure. Following this presentation, we conduct an analysis of this act and its overall control approach. An analogous treatment is used to study TSCA. At the conclusion of these two parallel discussions, we conduct a comprehensive analysis of U.S. policies on solid waste and toxic substances that have been the focus of this module. Risk-benefit analysis is the primary approach used in this summary evaluation. A list of acronyms and terms is provided at the end of the chapter.

[1] U.S. EPA (August 1988); U.S. EPA, Office of Pollution Prevention and Toxics (May 9, 2002).

Overview of U.S. Pesticide Controls: FIFRA

Brief Retrospective[2]

Federal Insecticide, Fungicide, and Rodenticide Act (FIFRA) of 1947

Congress enacted the Federal Insecticide, Fungicide, and Rodenticide Act (FIFRA) in 1947 and charged the Department of Agriculture with its administration. Its primary aim was to regulate pesticide labeling to protect against fraudulent products. In 1970, the newly established EPA assumed responsibility for FIFRA, and two years later Congress passed amendments to the act. These amendments changed virtually every aspect of the original law, shifting its focus to health and environmental protection. A key revision was the requirement for **registration** of new pesticides and the reevaluation and **reregistration** of those already on the market. Pesticide registration continues to be the chief regulatory instrument of federal controls on pesticides.

Over time, further revisions were made. Part of the reform has dealt with requiring more sophisticated test data from registrants to determine if better safeguards are needed in labeling, packaging, or formulations. Not unlike other major policies, U.S. pesticide initiatives have sometimes proven to be overly ambitious. In fact, of the approximately 600 pesticides needing reregistration under FIFRA, the EPA had issued **registration standards** for only about 185 of them by the late 1980s. These standards are important, since each includes a review of available data, a listing of information needed for reregistration, and a statement of the current regulatory position on the pesticide.

FIFRA Amendments of 1988

In 1988, Congress passed the FIFRA Amendments. These revisions improved registration and reregistration procedures and authorized a fee system to help pay for the reregistration process. The amendments also strengthened the EPA's control over the storage, transportation, and disposal of pesticides, including labeling of both the pesticide and its container. Added to the EPA's responsibilities was to regulate the use, disposal, refill, and reuse of pesticide containers. Congress also rescinded one of the agency's responsibilities—that of accepting stocks of suspended and canceled pesticides and disposing of them at the government's expense. Instead, the EPA can require registrants and distributors to handle a recall. Application 19.1 reports on the disposal dilemma that arose following the U.S. cancellation of ethylene dibromide (EDB).

Food Quality Protection Act (FQPA) of 1996

In August 1996, President Clinton signed into law the Food Quality Protection Act (FQPA) of 1996, which amended FIFRA as well as the Federal Food, Drug, and Cosmetic Act (FFDCA), another piece of legislation that regulates pesticides—this one by establishing tolerances (i.e., maximum levels allowed) for pesticide residues in food. The major thrust of the Food Quality Protection Act is to provide a more consistent set of controls by establishing a single health-based standard for all pesticides in all foods. It also gives special attention to pesticide risks faced by infants and children. The new standard requires that all tolerances be "safe." Under the law, safety refers to "a reasonable certainty that no harm will result from aggregate exposure."[3] By imposing one standard, the FQPA eliminates an inconsistency problem posed by one of the provisions of the FFDCA, known as the Delaney Clause, which is discussed in Application 19.2.[4]

---

[2] The following section is drawn from U.S. EPA, Office of Pesticide Programs (December 13, 2000); U.S. EPA, Office of Pesticides and Toxic Substances (December 1988); U.S. EPA (August 1988), pp. 113–19; and Wolf (1988), p. 154. For an online review of FIFRA, visit **http://www.epa.gov/pesticides/fifra.htm.**

[3] Pub. L. 104-170, § 405(b)(2)(A)(ii), 110 Stat. 1489 (1996).

[4] Council on Environmental Quality (1997), pp. 94–96, Kimm (January–March 1993). To read more about the FQPA of 1996, visit **http://www.epa.gov/oppfead1/fqpa/.**

Application

19.1 *Pesticide Cancellation: The Case of Ethylene Dibromide*

In 1948, a new pesticide was introduced to commercial markets—ethylene dibromide (EDB). Its effectiveness was borne out by its extensive use on fruits, vegetables, and stored grain. Over 35 years later, it was learned that EDB caused cancer in laboratory animals and was found to be linked to soil and groundwater contamination. In 1984, the EPA halted most uses of the product. All supplies of the toxic EDB immediately became unusable and unwanted material—effectively becoming part of the hazardous waste stream. Since the EDB cancellation predated the 1988 FIFRA Amendments, the EPA was bound by law to dispose of any of the unused pesticide if requested to do so. It also had to compensate users and manufacturers for any costs incurred from the emergency action.

As a consequence of the cancellation, the EPA was left with a huge stockpile of EDB wastes and no known method of safe disposal. Pending a decision as to how the disposal would be accomplished, the pesticide was held in storage. In 1986, an ominous discovery was made. Drums filled with EDB in a Missouri warehouse were found to be leaking. News of the incident alerted the general public to the serious problems of pesticide waste disposal.

The EPA had planned to neutralize the EDB stock through a chemical process that had been tested only in laboratories, but the plan had been delayed by mechanical problems. When the agency attempted to remove the EDB from the leaking drums, which by that time had begun to corrode, toxic vapor emissions were discharged into the atmosphere. The EPA spent $1.5 million before it determined that the chemical treatment it had been considering for EDB was not viable for all formulations. It then looked into incineration that could somehow accommodate the corrosivity of the substance. That option was expected to cost between $6 million and $8 million. By 1990, all stocks of EDB had finally been disposed of.

In terms of understanding the risks of managing pesticide wastes, there are likely no clearer examples than those that arise when pesticides are canceled on a mass scale. In addition to the EDB suspension, the United States has had to deal with other emergency cancellations, including one for 2,4,5-T/silvex and another for dinoseb. Dinoseb, an herbicide and crop desiccant, was banned in 1986. Expected disposal costs for this pesticide were estimated at over $100 million.

Beyond these emergency actions, the United States also has canceled the registrations of a number of pesticides and halted the use of certain toxic, inert ingredients in pesticide formulations. Although the EPA no longer must accept supplies of suspended and canceled pesticides, registrants and distributors have to be prepared to handle a recall. The potential difficulties of finding safe disposal methods for canceled pesticides and the risks of interim storage underscore the importance of risk assessment within the registration process and the need to develop safer, less toxic substitutes.

Sources: U.S. EPA (August 1988), pp. 118, 128–33; U.S. EPA (December 1990), pp. 18–20.

19.2 *Solving the Delaney Clause Dilemma*

Application

The Delaney Clause was perhaps the most controversial piece of environmental legislation ever passed in the United States. Named after its sponsor, former U.S. Representative James Joseph Delaney of New York, the 1958 law effectively banned the use of any food additive found to cause cancer—no matter how negligible the degree of risk was. Under Section 409 of FFDCA dealing with processed foods, pesticide residues are considered food additives. Hence, the Delaney Clause directly affected the use of pesticides based on the residue that remained after processing. As it turned out, the provision had even more far-reaching implications because the EPA elected to coordinate its decisions on pesticide residues in raw food with those applicable to processed foods. In fact, in a federal appeals court decision in July 1992, it was ruled that the EPA could not allow any pesticide residues in processed food that exceeded what was allowed on raw food if the pesticide posed *any* measure of carcinogenic risk—no matter how small.

Some of the controversy associated with the Delaney Clause arose because of a statute in Section 409 called the flow-through provision. In essence, this provision said that processed foods would be considered safe if the residues remaining were not in excess of what had been established for raw foods under Section 408. Because of the relationship between these two sections in the FFDCA, the EPA coordinated its policies on setting tolerances for residues on raw food with those applicable to processed foods. Through this "coordination policy," the EPA would not approve the use of a pesticide for raw foods under Section 408, if this use could have resulted in a violation under Section 409 for processed foods.

A problem arose because some pesticides concentrate during processing, and the resulting concentration could have caused the remaining residue to exceed the approved tolerance level for raw foods. If so, the pesticide then became subject to separate approval under Section 409 and the Delaney Clause. Thus, if the pesticide was associated with even the smallest measure of carcinogenic risk, its use was banned. It was prohibited from use not only for processed food but, because of EPA's coordination policy, for raw food as well.

This coordination policy came under fire because of the inconsistencies it caused. Tolerance levels under Section 408 were to be set using risk-benefit analysis, but those under Section 409, because of the Delaney Clause, were risk based. Furthermore, pesticides with negligible carcinogenic risks would be banned from use if they concentrated during processing, yet pesticides with higher carcinogenic risks could have been used on raw foods because, if they did not concentrate during processing, they did not require a Section 409 ruling.

The Food Quality Protection Act (FQPA) of 1996 solved the paradox by amending both FIFRA and FFDCA. First, it establishes a single standard applicable to both raw and processed food, avoiding the use of different decision rules under two laws. Second, it defines safety to include all risks—not just carcinogenic risks. After decades of debate and controversy, the Delaney dilemma seems to have been resolved.

Sources: Kimm (January–March 1993); Council on Environmental Quality (January 1993), pp. 87 and 424, table 100; Council on Environmental Quality (1997), pp. 94–96.

Controlling New Pesticides Through Registration

pesticide registration A formal listing of a pesticide with the EPA that must be approved, based on a risk-benefit analysis, before it can be sold or distributed.

risk-benefit analysis An assessment of risks of a hazard along with the benefits to society of not regulating that hazard.

A major objective of FIFRA and its amendments is to accomplish proper **registration** of any pesticide before it can be distributed or sold. Registration is subject to EPA approval based upon a **risk-benefit analysis.** The EPA must assure that the benefits of using the substance outweigh the associated health and ecological risks. Hence, the law implicitly identifies those pesticides associated with unreasonable adverse effects by denying their registration. Currently, there are about 20,000 pesticide products registered or licensed for use in the United States.[5]

Based on the law, a pesticide may not be registered if any of the following hold:

- Its composition does not warrant its proposed claims.
- Its labeling does not comply with the law.
- It will not perform its intended function without causing unreasonable adverse effects on the environment.
- It cannot be used in widespread practice without generating unreasonable adverse environmental effects.

Advances in risk assessment are helping officials reduce the probability of making uninformed registration decisions and prevent the use of any pesticide posing an unreasonable risk to society.

pesticide tolerances Legal limits on the amount of pesticide remaining as a residue on raw agricultural products or in processed foods.

At the time of registration, the EPA sets legal limits, or **tolerances,** on the amount of a pesticide that may remain as a residue on raw or processed food without causing an unacceptable health risk. In effect, the registration acts as a license for use, and the tolerances specify the conditions under which that use is approved. Tolerances for most foods are enforced by the Food and Drug Administration (FDA) according to the provisions of the Federal Food, Drug, and Cosmetic Act (FFDCA).

Controlling Existing Pesticides Through Reregistration[6]

pesticide reregistration A formal reevaluation of a previously licensed pesticide already on the market.

Among the more important aspects of the 1988 FIFRA Amendments and the FQPA of 1996 are the rulings dealing with **reregistration** of previously licensed pesticides. As a result of this reevaluation process, many unsafe pesticides have been restricted in use or banned. Examples include agricultural uses of chlordane and virtually all uses of DDT and ethylene dibromide (EDB). As harmful chemicals like these are restricted or eliminated, industries and consumers must find alternative, and presumably less toxic, pesticides. Consequently, some amount of hazardous pollution should be prevented instead of having to be remediated after the fact.

At the time of the 1988 amendments, approximately 600 groups of pesticide active ingredients were targeted for reregistration. This represents a major undertaking, and the 1988 amendments were intended to accelerate what had been an inordinately slow procedure, setting deadlines for the various phases in the process. The steps in the five-phase reregistration process are shown in Table 19.1. In the final phase, the EPA issues a **Reregistration Eligibility Decision (RED)** document for each pesticide, which gives the full results of the agency's review. Over 200 REDs have been completed thus far.

Further changes to the reregistration program came with the enactment of the FQPA of 1996. Under this law, the EPA is required to use reregistration as a means to reassess all established tolerances to make certain that the new safety standard established by the

[5] U.S. EPA, Office of Prevention, Pesticides, and Toxic Substances (December 1999), p. 11. For online information on pesticide registration, visit **http://www.epa.gov/pesticides/chemreg.htm.**

[6] Much of the following discussion is drawn from U.S. EPA, Office of Pesticides and Toxic Substances (June 1991), and U.S. EPA, Office of Pesticide Programs (December 13, 2000).

Table 19.1 ***The Five-Phase Pesticide Reregistration Process***

Phase	Timetable	Description
Phase I: List active ingredients	Completed by October 1989	The EPA published lists of pesticide active ingredients subject to reregistration and inquired whether registrants intended to seek registration.
Phase II: Declare intent and identify studies	Completed in 1990	Registrants were to notify the EPA of intent to reregister, commit to providing necessary new studies, and pay the first installment of the reregistration fee.
Phase III: Summarize studies	Completed in October 1990	Registrants were required to summarize existing studies, flag those indicating adverse effects, recommit to satisfying all data requirements, and pay the final installment of the reregistration fee.
Phase IV: EPA review and data call-in	Completed in 1993	The EPA reviewed registrants' submissions and required them to satisfy any unfulfilled data requirements within 4 years.
Phase V: Reregistration decisions	Expected to be completed in 2002	The EPA conducts a comprehensive investigation of all data submitted and decides whether each pesticide is eligible for reregistration. The results are presented in a document called a Reregistration Eligibility Decision (RED).

Source: U.S. EPA, Office of Pesticide Programs (December 13, 2000).

Note: The status of the reregistration of any pesticide is accessible on the Internet at **http://www.epa.gov/pesticides/reregistration/status.htm.**

FQPA is met. Thus far, this reassessment has been completed for 435 groups, or over 70 percent of pesticide active ingredients. As the work continues, the status of any individual pesticide review can be accessed on the Internet at **http://www.epa.gov/pesticides/reregistration/status.htm.** The entire reassessment process is expected to be completed in 2002. By this time, a new reregistration review program, also put in place by the FQPA, will have started. This review program calls for the EPA to reassess every pesticide on a 15-year cycle.

Based on original estimates, the reregistration process was expected to cost about $250 million. At issue is whether these costs are economically justified. The fact is, controlling pesticides poses a tough problem for policymakers and for society as a whole. As a group of substances, pesticides can contribute positively to agriculture and other industries. Yet exposure to these substances can harm the ecology and cause serious health problems like cancer, birth defects, and neurological impairments. Both sets of factors need to be considered in a risk-benefit analysis. Hence, the $250 million expense to administer the reregistration process may be warranted if it facilitates government's ability to conduct such an analysis and properly manage whatever risks are justified by the associated social benefits.

Analysis of FIFRA

Risk-Benefit Analysis Under FIFRA

No pesticide may be distributed or sold unless it has been registered with the EPA, and registration is denied for any substance associated with "unreasonable adverse effects on the environment." Thus, the statutory definition of this phrase implicitly identifies the standard used to evaluate pesticides:

> The term "unreasonable adverse effects on the environment" means any unreasonable risk to man or the environment, taking into account the economic, social, and environmental costs and benefits of the use of any pesticide.[7]

Notice that, unlike most U.S. environmental laws, FIFRA explicitly calls for a consideration of the costs and benefits of pesticide use in determining unreasonable risk. This clearly suggests the need for **risk-benefit analysis.** That is, the risks of using a pesticide measured in terms of health and ecological effects should be weighed against the associated benefits, such as increased crop yields. It turns out, however, that risks are the dominant factor in the EPA's registration decisions.

New Pesticides

For **new pesticides,** risks are evaluated from data on health and environmental effects submitted by manufacturers as part of the registration application. Hence, the onus of proving that risks are not "unreasonable" lies with the producer. If risks are found to be negligible, the EPA generally assumes that benefits exist based on the manufacturer's willingness to absorb the high cost of registration. If risks are found to be greater than negligible, the manufacturer typically has to formulate a risk-reduction strategy or show that the benefits exceed the risks.[8]

Existing Pesticides

When the EPA conducts a special review of an **existing pesticide,** a formal benefit analysis is done. In these instances, benefits are measured by determining biological gains, such as changes in agricultural yields, and then these effects are monetized.

Problems in Risk Assessment

At least in principle, the nation's approach to pesticide control appears to have merit. So why is U.S. pesticide policy the subject of intense debate? Probably the best answer is that there is much uncertainty about the associated risks, despite the widespread exposure. Many scientists question the EPA's methods of assessing pesticide risks. Some argue that the risks are underestimated, citing the agency's lack of attention to inert ingredients or the cumulative effects from using multiple pesticides in combination. Others assert that the opposite is true, claiming that findings from animal bioassays overstate the potential harm to humans. Beyond these issues, there is also concern about how tougher pesticide controls would affect agricultural productivity. Without viable pesticide substitutes, there could be serious implications for world food supplies.

New Policy Direction

In 1993, a report by the National Academy of Sciences (NAS) triggered what appears to be a major shift in U.S. pesticide policy and perhaps a resolution to some of the issues. What the NAS found was that children are not being sufficiently protected from pesticide risks by U.S.

---

[7] FIFRA, sec. 2(bb).

[8] Generally, manufacturers try to reduce the risks or withdraw the registration application. U.S. GAO (March 1991), pp. 3, 9.

policy. At the core of this assessment is that the smaller size of children and their diets make them more vulnerable to pesticide risks. Almost immediately, the Clinton administration announced its intention to develop a new pesticide plan—a joint effort by the EPA, the FDA, and the Department of Agriculture.

Ultimately launched in 1994 as the **Pesticide Environmental Stewardship Program (PESP),** the plan's major objective is to *reduce* pesticide use—not just regulate it. This voluntary program operates by establishing partnerships with pesticide users and implementing preventive strategies. More information is available online at **http://www.epa.gov/oppbppd1/PESP/.** Among the program's goals is the promotion of **Integrated Pest Management (IPM)**—a collection of methods to foster more selective use of pesticides and greater reliance on natural deterrents. In support of this objective is an incentive-based proposal to encourage chemical manufacturers to develop alternative products.[9]

Integrated Pest Management (IPM)
A combination of control methods aimed at encouraging more selective use of pesticides and greater reliance on natural deterrents.

Overview of U.S. Legislation on Toxic Substances: TSCA

Chemicals are an integral part of modern society and represent a major industry in the U.S. economy. Used properly, most of these substances contribute positively to the quality of life. Yet exposure to chemicals can pose significant risks to society. Unfortunately, some of the health risks of chemical exposure, particularly long-term and permanent effects, were not known until after these substances had been introduced into commerce and had become widely used. An important example is the chemical group polychlorinated biphenyls (PCBs), substances formerly used as insulators in electrical transformers and as lubricants and dye carriers in paints, inks, and dyes. The toxicity of PCBs was not known until after these substances had leaked into the environment and began to accumulate in the tissues of animals and fish, thereby entering the food chain. The discovery led to a 1976 ban on their manufacture, processing, distribution, and use in the United States, except in completely enclosed electrical equipment.[10]

Policy Response to Chemical Risks[11]

Responding to the problem of chemical risks, Congress enacted the Toxic Substances Control Act (TSCA) in 1976. One of the act's primary objectives is to identify and control chemical substances that present a risk to health or the environment *before* they are introduced into commerce. By confronting the environmental and health risks of chemicals at the premanufacture stage, the government has an opportunity to prevent pollution rather than having to correct problems after the fact.

TSCA also provides a way to monitor the risks of chemicals already on the market and to take action as needed. Since so little was known about most of the chemical substances in use, Congress also authorized the compilation of an inventory of all chemicals commercially produced or processed in the United States between 1975 and 1979. Using data collected from manufacturers and importers, the government published the first **TSCA inventory** in 1979, which contained information on more than 62,000 chemicals. Today, there are about 75,000 chemicals in the TSCA inventory, and the EPA estimates that about 15,000 of these are presently in commerce. More information on the inventory, including how to access it, is available at **http://www.epa.gov/opptintr/newchems/invntory.htm.**

TSCA inventory
A database of all chemicals commercially produced or processed in the U.S.

TSCA gives the EPA authority to gather information on chemical risks from producers, to require testing on existing chemicals, and to review most newly introduced chemicals before

[9] U.S. EPA, Office of Pesticide Programs (April 22, 2002); "White House to Seek Reduced Use of Pesticides" (September 21, 1993).
[10] U.S. EPA (August 1988), p. 120.
[11] Much of the following discussion is drawn from U.S. EPA, Office of Toxic Substances (June 1987), pp. 2–8, and U.S. EPA (August 1988), pp. 112–27. For online access to the full text of TSCA, visit **http://www.law.cornell.edu/uscode/15/ch53.html.**

Table 19.2 ***Selected Laws Governing the Control of Substances Exempted Under TSCA***

Legislation	Agency	Substance or Control
Atomic Energy Act	Nuclear Regulatory Commission	Nuclear waste disposal, nuclear energy production
Federal Insecticide, Fungicide, and Rodenticide Act	Environmental Protection Agency	Pesticides
Federal Food, Drug, and Cosmetic Act	Food and Drug Administration	Foods, food additives, drug additives, cosmetics
Safe Drinking Water Act	Environmental Protection Agency	Controls contaminants in drinking water supplies
Occupational Safety and Health Act	Occupational Safety and Health Administration	Controls hazards found in the workplace
Clean Air Act	Environmental Protection Agency	Controls the emissions of hazardous air pollutants
Clean Water Act	Environmental Protection Agency	Controls the discharge of hazardous pollutants into surface waters
Hazardous Materials Transportation Act	Department of Transportation	Regulates the transport of hazardous materials
Marine Protection, Research, and Sanctuaries Act	Environmental Protection Agency	Regulates waste disposal at sea
Resource Conservation and Recovery Act	Environmental Protection Agency	Regulates hazardous waste generation, storage, transport, treatment, and disposal
Comprehensive Environmental Response, Compensation, and Liability Act	Environmental Protection Agency	Provides for cleanup of inactive and abandoned hazardous waste sites
Surface Mining Control and Reclamation Act	Department of the Interior	Regulates environmental aspects of mining and reclamation

Source: U.S. EPA, Office of Solid Waste (November 1986).

they are manufactured and made available for use on a broad scale. Depending on its findings, the EPA takes appropriate action. The regulatory response can range from requiring warnings during production and distribution to banning the chemical's manufacture. Risk controls may be imposed at any stage in the chemical's life cycle, including its manufacture, processing, distribution in commerce, usage, and disposal.

To avoid the inefficiency and potential inconsistency of dual controls, TSCA acknowledges and accounts for the mandates of other laws dealing with toxic substances and the agencies that implement them. Some of these are listed in Table 19.2. Under the law, the EPA is to coordinate its control activities with those of other agencies, such as the Food and Drug Administration (FDA), the U.S. Department of Agriculture (USDA), and the Occupational Safety and Health Administration (OSHA). As a chemical risk is discovered, the EPA must consider whether other agencies have investigated that risk, and if so, it must determine if existing laws adequately address the problem. In fact, the EPA must use other laws it administers to reduce risk, such as the Clean Water Act, the Clean Air Act, and RCRA, *before* implementing TSCA's provisions. Finally, eight product categories are explicitly exempted from TSCA, most of which are regulated under other laws.[12]

Controlling the Introduction of New Chemicals

Unlike the nation's pesticide law, TSCA does not use a registration procedure to control toxic chemical use. Instead, it requires manufacturers to notify the government at least 90 days before they intend to produce or import any **new chemical,** which is one that is not listed in the TSCA inventory. The notification is made to the EPA via a **premanufacture notice (PMN).** The PMN provides information about the chemical's characteristics, the expected exposure to workers, its intended use, and the health and ecological effects. Upon receipt of the PMN, the EPA has a 90-day review period to evaluate the risks and respond. If the risks are too high or, in the absence of full information, are expected to be too high, the EPA may restrict usage of the chemical permanently or, if more data are needed for a complete risk analysis, temporarily. The determination of unreasonable risk is based on test results of exposure effects, such as cancer or birth defects.[13]

new chemical
Any substance not listed in the TSCA inventory of existing chemicals.

premanufacture notice (PMN)
An official notification made to the EPA by a chemical producer about its intent to produce or import a new chemical.

Over 24,500 PMNs were submitted to the EPA between 1979 and 1996. Of this total, the EPA has restricted or prohibited the production or use of more than 725 chemicals pending further information. Over 850 additional chemicals have been controlled through voluntary agreements with producers, and almost 1,000 have been withdrawn or suspended from the review process by manufacturers in response to EPA concerns. In the aggregate, over 10 percent of the chemicals submitted for review since 1979 have been suspended, restricted, or withdrawn from production or use.[14]

Controlling Existing Chemicals in Use

Other provisions of TSCA deal with evaluating the risks of some 75,000 **existing chemicals,** which are those in the TSCA inventory. The law requires manufacturers to notify the EPA if any chemical is found to present a substantial risk to human health or the environment. When this occurs, the EPA's Office of Toxic Substances evaluates the information and takes action, ranging from labeling requirements to outright bans. To get a sense of the activity level, consider that between 1980 and 1991, about 5,000 of these notifications were received by the EPA. In more than 550 instances, producers altered their processes to reduce the release of the substances, and the use of the chemical was temporarily or permanently halted. One of the more significant examples of such an action was the banning of certain asbestos-containing products in the late 1980s.[15]

existing chemical
A substance listed in the TSCA inventory.

Analysis of TSCA

Risk-Benefit Analysis Under TSCA

TSCA's statutory objectives are to obtain data on chemical risks and to regulate chemicals posing an "unreasonable risk" to human health or the environment. Interestingly, "unreasonable risk" is not explicitly defined in the law. However, a House of Representatives report

[12] The eight product groups are tobacco, nuclear material, firearms and ammunition, food, food additives, drugs, cosmetics, and pesticides.
[13] U.S. EPA, Office of Pollution Prevention (October 1991), p. 142; U.S. EPA, Office of Toxic Substances (June 1987), pp. 3–4; TSCA, sec. 4(b)(2)(A).
[14] U.S. EPA, Office of Prevention, Pesticides, and Toxic Substances (May 16, 2001).
[15] U.S. EPA (August 1988), p. 114; *U.S. Federal Register* 54, pp. 29460–513 (July 12, 1989), as cited in U.S. EPA, Office of Pollution Prevention (October 1991), p. 143.

sheds some light on what is implied, stating that the determination of "unreasonable risk" involves the following:

> balancing the probability that harm will occur and the magnitude and severity of that harm against the effect of proposed regulatory action on the availability to society of the benefits of the substance or mixture.[16]

Notice how this determination supports the use of **risk-benefit analysis** in the approval process—just as is the case for pesticides. Like FIFRA, TSCA empowers the EPA to ban or restrict the use of toxic chemicals that do not pass the risk-benefit test. Such actions should effectively encourage the substitution of alternative, less dangerous substances, which in turn lessens the toxicity of the hazardous waste stream.

A major difference between TSCA and FIFRA is the process used for review of new substances. While FIFRA calls for extensive test data as part of a complex registration process, TSCA requires only a 90-day advance notice of intent to produce a new chemical. Testing is done only upon formal request by the EPA. It has been argued that this regulatory difference properly reflects the relative magnitude of risks between toxic substances and pesticides. Pesticides generally pose a greater risk since they are biologically active. Hence, the more stringent controls and higher costs of introducing a new pesticide into commerce may be justified by the inherently greater risk potential.[17]

Bias Against New Chemical Introductions[18]

A common observation about TSCA's effectiveness is that the EPA has been slow to develop data on **existing chemicals** despite the evidence that such information is critical. Ironically, much of the delay seems to be caused by how TSCA is written, imposing command-and-control procedures that are time intensive and possibly needless. For example, the EPA literally must write a new regulation each time it wishes to test a chemical substance. Another contributing factor is the agency's current use of a single chemical review process rather than one aimed at a group of similar substances.

Overall, the analysis of existing chemicals is complex and bogged down with time-intensive procedures—markedly different from the relatively straightforward rules in place for reviewing new substances. The result? Not only do these regulations delay the development of an important database, but they also create a bias against the introduction of **new chemicals.** The government is much more efficient in its testing of new substances than it is for those already on the market. Consequently, chemical producers wishing to avoid such procedures and the associated costs can continue to sell existing and possibly more dangerous substances. Hence, certain of TSCA's rulings appear to generate the perverse outcome of deterring the development of new, safer chemicals.

New Policy Direction[19]

Although TSCA's premanufacture rulings for new chemicals have been effective in encouraging substitution of safer chemicals for toxic ones, there has been less success in fostering the same activity for existing chemicals. Recognizing the weaknesses in certain of TSCA's provisions, the United States is taking steps to develop chemical programs aimed at **pollution prevention** to complement existing policy. One such effort is the **Green Chemistry Program.** This initiative, with domestic and international partners in industry, academia, research centers, trade organizations, national laboratories, and others, promotes the development and application of innovative chemical technologies to achieve pollution prevention by minimizing or even eliminating the production of hazardous substances.

Green Chemistry Program
An initiative that promotes the development and application of innovative chemical technologies to achieve pollution prevention.

[16] Dominguez (1977), p. 5.9.
[17] Shapiro (1990), pp. 213–14.
[18] This discussion is drawn from Shapiro (1990), pp. 223–24, 232–36.
[19] This discussion is drawn mainly from U.S. EPA, Office of Pollution Prevention and Toxics (August 2001), and U.S. EPA, Office of Solid Waste and Emergency Response (December 1998).

Figure 19.1 ***Selected Partners in the Green Chemistry Program***

Category	Partners
Industry	The BF Goodrich Company The Dow Chemical Company Dow Corning Corporation E.I. DuPont de Nemours Eastman Kodak Company
International Organizations	Center for Green Chemistry (Australia) Organisation for Economic Co-operation and Development (OECD) Green & Sustainable Chemistry Network (Japan) Royal Australian Chemistry Institute
Academia	National Autonomous University of Mexico University of Massachusetts, Boston University of South Alabama
Environmental Groups	Environmental Defense
Scientific Organizations	American Chemical Society International Union of Pure and Applied Chemistry National Research Council National Science Foundation
Trade Associations	American Petroleum Institute American Chemistry Council Society of the Plastics Industry
National Laboratories	Los Alamos National Laboratory National Renewable Energy Laboratory
Research Centers	Center for Process Analytical Chemistry at the University of Washington Green Chemistry Institute National Environmental Technology at the University of Massachusetts, Amherst

Source: U.S. EPA, Office of Pollution Prevention and Toxics (October 21, 2001).

To accomplish its objective, the program actively supports research in environmentally benign chemistry along with other activities such as public conferences and meetings. The program includes four major projects, which are Green Chemistry Research, the Presidential Green Chemistry Challenge, Green Chemistry Education, and Scientific Outreach. For more information and updates, visit **http://www.epa.gov/greenchemistry/.** Among the program's voluntary partners are the Dow Chemical Company, the American Petroleum Institute, the University of Massachusetts, Boston, and the National Science Foundation, as shown in Figure 19.1.

Extended Product Responsibility (EPR) A commitment by all participants in the product cycle to reduce any life-cycle environmental impacts of products.

Another preventive approach being used to reduce chemical risk is **Extended Product Responsibility (EPR),** also known as **product stewardship.** This refers to committed responsibility by all participants in the product cycle—manufacturers, retailers, consumers, and disposers—to find ways to reduce a product's impact on the environment. Though its objectives are comprehensive, an important element of this strategic framework is to reduce the toxicity of inputs used in production. An example of how EPR works is the team approach adopted by Nortel Networks Corporation with its primary chemical supplier. To prevent pollution, the two firms drafted a supply contract that includes incentives for both companies to reduce chemical use. More examples and further information are available online at **http://www.epa.gov/epr.**

Economic Analysis of U.S. Solid Waste and Toxics Policy

National policy on toxic chemicals is closely linked to solid waste controls. As we have observed, FIFRA and TSCA attempt to control the production and use of potentially harmful substances *before* they are released into the environment. If such initiatives are successful, there is less risk when these materials enter the waste stream. Given the connection, it is appropriate to evaluate the allocation of resources to both approaches and determine if the allocation makes sense from a risk management perspective. While there are many factors to consider, the major issues can be addressed by examining the evidence on the costs and benefits of these policy efforts.

Cost Analysis of U.S. Solid Waste and Toxics Policy[20]

On the cost side of the issue, much attention has been drawn to the enormous expenditures associated with RCRA and Superfund. Given the complex problems addressed by these laws, high costs are not surprising. The issue is whether or not the costs are justified by the benefits. As we suggested in Chapter 17, the benefit-based approach of both of these laws coupled with the lack of market incentives suggests that the high costs of implementing these laws may not be defensible on economic grounds. To accommodate a broader objective, we now use a different perspective and evaluate these costs relative to those associated with TSCA and FIFRA. Ultimately, we want to determine if the comparison makes sense in terms of the social benefits each set of laws is expected to achieve.

Direct costs associated with U.S. solid waste and toxic control policy for selected years are presented in Table 19.3.[21] The dollar values represent costs incurred by all economic sectors—the private sector, the EPA, other federal agencies, and state and local governments. All values are expressed in millions of 1986 dollars.[22] Notice there are two major cost categories for solid wastes—those associated with RCRA and those dealing with CERCLA or Superfund. The data for RCRA are further delineated by specific provisions, that is, those for nonhazardous solid waste, those for hazardous waste, and the rulings for underground storage tanks. Cost data on chemical control policy are separated into those associated with toxic substances under TSCA and those for pesticides governed under FIFRA.

Solid Waste Control Costs

Over the period shown in Table 19.3, the aggregate cost data on solid waste controls exhibit a strong positive trend, rising from approximately $8.4 billion in 1972 to $19.1 billion in 1987—an increase of 126 percent for the 15-year period. Based on projections, this trend was expected to continue at least until 2000.

[20]The following discussion is drawn mainly from U.S. EPA, Office of Policy, Planning, and Evaluation (December 1990).

[21]Since implicit costs are not available, we use the second-best approach of examining explicit or direct costs, recognizing that the magnitude likely undervalues the true social costs.

[22]These pollution control costs are estimates based on full implementation of federal regulations, using survey data on historical expenditures for years up to 1987 and *ex ante* estimates of forthcoming regulations not in place as of 1987.

Table 19.3 ***Pollution Control Costs for U.S. Solid Waste and Chemicals Programs ($1986 millions)***

Program	1972	1980	1987	1995	2000
RCRA total	$8,436	$13,612	$18,409	$32,468	$38,055
Nonhazardous solid waste	8,436	13,612	16,683	20,338	22,302
Hazardous solid waste	*	*	1,725	9,210	12,062
Underground storage tanks	*	*	1	2,920	3,691
CERCLA (Superfund)	*	*	683	4,690	8,093
Solid waste total	**$8,436**	**$13,612**	**$19,092**	**$37,158**	**$46,148**
TSCA	*	429	365	1,119	1,234
FIFRA	92	461	453	1,353	1,658
Chemicals total	**$92**	**$889**	**$818**	**$2,472**	**$2,892**

Source: U.S. EPA, Office of Policy, Planning, and Evaluation (December 1990), p. 3-4, table 3-4, and p. 3-5, table 3-5.

Note: Data for 1995 and 2000 are projections. An asterisk means the program was not in existence.

For the most part, this general rise in costs reflects the strengthening of the U.S. policy position since the mid-1970s, particularly for hazardous wastes. Consider, for example, the high costs of implementing the "cradle-to-grave" approach mandated by RCRA—a financial commitment that was increased by the 1984 Amendments. By 1987, national costs of controlling hazardous wastes had grown to $1.7 billion, and the projections indicate continued growth. Based on forecasts, these costs would rise to $12.1 billion by 2000. A similar trend was expected for the costs of regulating underground storage tanks. Notice that expenditures were $1 million in 1987 but projected to reach nearly $3.7 billion by 2000.

Control costs also have been on the rise for nonhazardous waste, though the trend has been much more gradual. Federal laws on nonhazardous wastes have been in effect for a longer period of time and have undergone fewer revisions than those governing hazardous wastes. Furthermore, although these costs consistently represent the largest allocation of funding for any solid waste program, the great majority of them are incurred by local governments and private entities with only a small proportion directly linked to federal laws.

Superfund costs pale in comparison to those associated with RCRA, but the growth rate is higher than for any other solid waste program. Looking again at Table 19.3, notice that costs to fully implement the Superfund program in 2000 were projected at $8 billion—nearly 12 times the expenditure in 1987. In relative terms, the proportion of total solid waste control costs allocated to Superfund has risen steadily from 3.6 percent in 1987 to a projected 17.5 percent in 2000. There is, however, considerable debate about the accuracy of Superfund cost projections.

Although the EPA's cost estimates for the Superfund program are based on many factors, none of these is certain. One of the key elements affecting the projections is the expected number of National Priority List (NPL) sites to be remediated. Another uncertain data point is the expected cost of cleaning up each of these sites. Such per unit estimates are averages that do not capture the variability across thousands of sites throughout the nation. If the EPA's estimate of $25 million per remedial action is accurate, aggregate costs could be over $30 billion. Finally, there is a common assertion that most Superfund cost projections are undervalued because they typically do not include litigation costs, which are expected to be significant.

Chemical Control Costs

Referring again to Table 19.3, notice that aggregate control costs for both FIFRA and TSCA are much smaller than those incurred for solid wastes. However, they do exhibit a strong growth pattern. Total expenditures show a dramatic increase from $92 million in 1972 to nearly nine times that level, or $818 million, in 1987. And cost projections suggest a considerable growth rate up to 2000.

On a less aggregated level, the growth rate of costs is fairly consistent for the two national programs under TSCA and FIFRA. Control expenditures under TSCA are directly related to the increasing number of premanufacture notices (PMNs) received each year. Based on current data, the EPA reviews between 2,000 and 2,500 of these each year—much higher than the number submitted in the 1970s and early 1980s.[23] Projections indicate that this increase in new chemical introductions will continue.

National controls on pesticides have a much longer history, dating back to 1947. Rising costs to implement FIFRA reflect increasingly tougher laws on registration, reregistration, and safety requirements mandated through recent revisions. The EPA anticipates that private costs will continue to rise as monies are allocated to research and development, cancellations and suspensions of pesticides, and compliance with more stringent farmworker safety and pesticide applicator requirements.

In sum, the costs of U.S. policy on solid waste and toxic substances are significant and rising over time. It is also true that the rate of increase is higher than for any other major environmental program. Based on EPA estimates, total expenditures for solid waste grew by nearly 67 percent from 1985 to 1990 and for chemicals by about 104 percent in the same period. These are much higher than the comparable growth rates for air and water control costs at 19 percent and 28 percent, respectively. These observations may mean that standard setting and the selection of policy instruments to execute solid waste and toxics control have missed the mark—at least from an efficiency perspective. Indeed this assessment has been suggested by some. However, such an assertion is unwarranted without considering whether these relatively high costs are justified by comparable levels of social benefits.

Benefit Analysis of U.S. Solid Waste and Toxics Policy

As the government launches initiatives to control solid wastes and toxic substances, society's risk of exposure to these hazards should be reduced, and the accompanying adverse effects should decline. In theory, the associated benefits can be assessed by measuring and monetizing the reduction in damages to human health and the ecology. Unfortunately, however, no comprehensive benefit analysis has been undertaken for the major U.S. programs on solid wastes and toxics. Even specific benefit measures are lacking. According to one article, of the tens of thousands of chemicals registered with the EPA, health studies have been conducted on only about 10,000, and of these, only about 1,000 have been studied for acute effects.[24] There is an even greater void in formal assessments of ecological benefits. These have particular relevance to Superfund, given its emphasis on compensation for damages to natural resources.[25] Why has so significant an issue escaped the rigors of formal analysis?

Part of the explanation is that national policy on solid waste and toxics is not as well established as air and water quality controls. The United States got a late start in initiating these policies, and there were long delays in implementation once the rulings were in place.[26] Beyond the timing issues, it is also true that certain of the expected benefits of these initiatives are difficult if not impossible to identify. Consider, for example, the manifest program under RCRA. An important expected benefit is the reduction in damages due to a lower incidence of illegal hazardous waste disposal. But it is not possible to measure a decrease in the number of illegal disposals when such activity is not observable.

Even those data points that *are* observable are difficult to measure given the sheer magnitude of the relevant variables. There are over 39,000 identified CERCLIS sites (including those that have been archived), over 70,000 registered chemicals, over 25,000 pesticide products in use, over 5 million underground storage tanks, hundreds of hazardous waste land disposal facili-

[23] U.S. EPA, Office of Pollution Prevention and Toxics (March 1997), as cited in Chemical Industry Archives (November 1, 2001).

[24] Stranahan (February/March 1990).

[25] For more specific information on the nature of ecological risks, see U.S. EPA, Region 5 (May 1991).

[26] Recall that although FIFRA was enacted in 1947, it did not address environmental issues until its revision in 1972.

Table 19.4 ***Risk Rankings for Selected Environmental Problems***

Problem	Expert Risk Ranking				Public Risk Ranking, Overall
	Cancer Risk	Noncancer Health Risk	Ecological Effects	Welfare Effects	
Active hazardous waste site	2	Low	Low	Medium	High
Inactive hazardous waste site	2	Low	Medium	Medium	High
Nonhazardous waste site: municipal	3	Medium	Medium	Medium	Not ranked
Nonhazardous waste site: industrial	2	Medium	Medium	Low	Not ranked
Accidental releases of toxics	4	High	Medium	Low	High
Releases from storage tanks	3	Low	Low	Low	Not ranked
Pesticide residues	1	High	High	Minor	Moderate
Other pesticide risks	2	Medium	High	Medium	Moderate
New toxic chemicals	2	Not ranked	Not ranked	Minor	Not ranked
Consumer product exposure	1	High	Not ranked	Minor	Low
Worker chemical exposure	1	High	Not ranked	Minor	Moderate

Source: U.S. EPA, Office of Policy Analysis, Office of Policy, Planning, and Evaluation (February 1987).

Note: Cancer rankings are on a scale of 1 through 5, with 1 being the highest relative risk and 5 being a "not assessed" or "no risk" rank.

ties, and thousands of municipal landfills. These statistics help to explain the uncertainty in even defining the extent of the problem, much less assessing the progress achieved by policy.

Using Comparative Risk Analysis to Assess Benefits

In the absence of quantitative benefit measures, **comparative risk analysis** can be used to gain some insight about expected social benefits. By examining how the risks of exposure to various environmental hazards compare with one another, we can make inferences about the relative benefits that accrue to society from regulating these hazards and reducing the associated risk. These qualitative inferences can shed some light on whether or not the associated control costs are justified, assuming that risk-based priorities guide the nation's policy agenda.

comparative risk analysis
An evaluation of relative risk.

In this context, the objective is to compare the benefits of solid waste policies and those of chemical control policies to alternative mandates by analyzing the risks these regulations attempt to lessen. Two landmark studies of risk ranking have been conducted in recent years. The findings of the first investigation, titled *Unfinished Business: A Comparative Assessment of Environmental Problems,* are used to support the following analysis.[27]

The Risk-Ranking Study

To conduct the comparative risk study, a task force designated by the EPA defined the universe of environmental issues as a set of 31 problems. For each of these, the team of experts assessed four types of risks: cancer risks, noncancer risks, ecological effects, and welfare effects. The objective was to determine the **relative risk** of each problem within each category. The risk-ranking process was based on available quantitative risk data, subsequent interpretation, and expert judgment. Ultimately, the task force established a priority ranking of environmental problems that could be used to guide national policy. Findings also were compared to an ordinal ranking of the public's perception of environmental risks based on a survey conducted by the Roper Organization. Both the expert ranking and the public's ranking for selected environmental problems are presented in Table 19.4. The overall observations and recommendations of the analysis are given in Table 19.5.

[27] U.S. EPA, Office of Policy Analysis, Office of Policy, Planning, and Evaluation (February 1987). The conclusions of a follow-up study are reported in U.S. EPA, Science Advisory Board (September 1990).

Table 19.5 ***General Observations and Recommendations by the EPA Task Force on Relative Environmental Risks***

1. No environmental problems were ranked consistently "high" or "low" across all four risk types.
2. The project has developed a useful tool to help set priorities.
3. Risks and EPA's current program priorities do not always match. In part, these differences seem to be explainable by public opinion on the seriousness of different environmental problems.
   - Areas of high risk/low EPA effort
     Radon, indoor air pollution, stratospheric ozone depletion, global warming, accidental releases of toxics, consumer and worker exposure to chemicals, nonpoint sources of water pollution, and "other" pesticide risks.
   - Areas of medium or low risk/high EPA effort
     Active (RCRA) and inactive (Superfund) hazardous waste sites, releases from storage tanks, and municipal nonhazardous waste.
4. Statutory authorities do not match neatly with risks.
5. National rankings do not necessarily reflect local situations—local analyses are needed.
6. Some chemicals show up as major concerns in multiple problem areas, notably lead, chromium, formaldehyde, solvents, and some pesticides.
7. More research is needed in several areas. The general weakness of exposure data is a special problem because exposure is such an important determinant of risk. In addition, specific data on the different types of risks and environmental problems are often lacking.
8. EPA should now study other areas important to setting priorities.

Source: U.S. EPA, Office of Policy Analysis, Office of Policy, Planning, and Evaluation (February 1987), pp. 94–100.

As the risk rankings in Table 19.4 indicate, it appears that the federal government's environmental priorities do not align consistently with the expert risk ranking—an observation that seems to be explained in part by how the general public perceives environmental risk. This observation is described more fully in the list of recommendations given in Table 19.5, specifically item 3. Notice how the task force identifies environmental hazards of high risk where EPA's control efforts are low and hazards of low or medium risk where the comparable effort is high. Every hazard in this latter category is a solid waste issue—active (RCRA) and inactive (Superfund) hazardous waste sites, releases from storage tanks, and municipal nonhazardous wastes. According to this observation, the large amount of resources allocated to solid waste control does not appear to be justified by relative risk analysis.

Now, examine the list of environmental hazards in the "high risk/low EPA effort category." Notice that there are several chemical exposure hazards identified in this category—consumer and worker exposure to chemicals, accidental releases of toxics, and "other" pesticide risks. (This last category includes runoff, leaching, and air deposition associated with the use of pesticides and agricultural chemicals.) The same logic applies to this observation except in the opposite direction. Based on relative risk analysis, it appears that too few resources are being used to control pesticides and other chemicals.

Evaluating the Evidence

If we consider the results of the environmental risk analysis along with the national cost data discussed earlier, we can make three important observations about U.S. policy on solid waste and toxic chemicals:

- There is reason to believe that economic resources are being misallocated across environmental risks.

- National policy appears to be motivated at least in part by public perception of risk.
- Inadequate data and uncertainty limit the policy analysis.

Each of these observations is discussed briefly below.

Misallocation of Resources

Given an overall policy objective to reduce environmental risk, the benefits of regulation are appropriately measured by the risk reduction achieved. Based on **benefit-cost analysis,** resources should be allocated to balance the incremental benefits, or the reduction in risk, with the incremental costs of achieving them. Yet the results of the risk-ranking study suggest that the high costs of implementing RCRA and Superfund are not justified by the expected risk reduction, which is low relative to other environmental hazards.

Put another way, the comparative risk analysis suggests that resources may be overallocated to solid waste pollution control.[28] Of course, if too many resources are being allocated to solid waste problems, then it must be the case that too few are being used to satisfy other needs. It may be that the misallocation is between government and the private sector, meaning that public policies are overfunded. Indeed, this is a common criticism. However, such a broad assertion is difficult to verify at best. What we can consider is how the allocation of resources *among* environmental public policies might be improved, based on the existing distribution between private and public uses.

Reexamine the list of "high risk/low EPA effort" problems identified by the task force in Table 19.5. Of particular relevance to the present investigation are the chemical-related hazards included in this list. The implication is that government should intensify its efforts under TSCA and FIFRA where there is greater opportunity for risk reduction. In fact, of the chemical-related problems in the "high risk/low EPA effort" category, worker exposure to chemicals is tied for first in the cancer risk ranking and ranked as "high" in the noncancer health risk category. Assuming the risk rankings are accurate, society would gain by transferring some resources away from those hazards where the risks are relatively low and reallocating them toward problems where the risks are higher. In theory, this transfer process should continue until the marginal benefit of reducing risk is the same across all environmental hazards in accordance with the **equimarginal principle of optimality.**[29]

Although the evidence is far from complete, it seems reasonable to argue that the current standards specified by RCRA and Superfund may have erred on the side of being too stringent and that the controls under TSCA and FIFRA may be too lax. Hence, the costs of achieving these statutory standards likely cannot be justified on economic grounds.

Public Perception and Policy Making

Since it appears that allocative efficiency is not governing national priorities on the environment, we ought to consider what might be influencing these decisions. Referring back to Table 19.4, notice that there are major differences between the risk ranking of experts and that of the general public. Furthermore, the intensity of EPA efforts to combat environmental problems is more consistent with public *perception* of risk than it is with *actual* environmental risk. In fact, this assertion was one of the major conclusions of the EPA report, as Table 19.5 indicates.

---

[28] There is an important caveat here. As the task force indicates in their final observations, the "low risk/high EPA effort" areas may be indicating that the risks are low *because* of the intensified national effort. If so, some continuance of control efforts may be needed to maintain reduced risk. The issue hinges on whether the marginal benefit of further controls is justified by the marginal costs. In any case, the findings call for further investigation of how national resources are being allocated across competing needs.

[29] Notice that such a reallocation is based on relative rankings of just regulated environmental problems. Hence, the result would optimize the environmental risk reduction possible from government efforts based on a given distribution of resources. A general solution could be determined only from a benefit-cost analysis across all competing uses and across all economic sectors. Were this empirically possible, one might determine that government uses too many resources for all of its programs, as is commonly suggested.

Application

19.3 *Hazardous Waste Sites: Are the Risks Disproportionate?*

Unlike the effects of acid rain that can be felt by people living hundreds of miles away from the polluting source, the risks of exposure to hazardous waste tend to be localized. For example, the health effects of groundwater contamination caused by a mismanaged waste site are confined to nearby residents. On that basis alone, we know that these environmental risks are not evenly distributed across the nation. But we need to look beyond this simple observation. The operative issue is to determine if these higher risk communities are dominated by any racial, ethnic, or income groups. If so, policy officials have to consider how to address the environmental inequity.

http://

In 1990, the EPA formed the Environmental Equity Workgroup to study the evidence on whether certain socioeconomic groups face a disproportionate degree of environmental risk. In 1994, the agency formed the Office of Environmental Justice, which has a home page on the Internet at **http://www.epa.gov/compliance/environmentaljustice.** Motivating these efforts is the notion that risk-based priorities should guide the direction of U.S. environmental policy. Simply put, society can gain more from environmental policy initiatives if resources are allocated to high-risk regions of the country and high-risk population groups.

As part of its two-year mission, the EPA's workgroup investigated available information on the characteristics of populations living near existing and abandoned hazardous waste facilities. Among other findings, the EPA workgroup discovered that there is a lack of good data on environmental health effects sorted on the basis of socioeconomic factors like race and income. Nonetheless, it did find that minorities and low-income groups appear to have a greater exposure to certain environmental threats, one of which is hazardous waste sites. Why are these groups more strongly represented in areas where these potentially dangerous sites are located? The reasons are complex, but among the determinants are land use decisions, historical residence patterns, and politics.

In its 1992 report, the EPA workgroup referenced a 1983 study conducted by the U.S. General Accounting Office (GAO), which focused on the socioeconomic characteristics of communities near four hazardous waste landfills in eight southeastern states. According to the GAO's findings, three out of four of the landfills cited are located in communities where more than 50 percent of the population is black. At one of these, blacks represent 90 percent of the local population. The income data are also important, indicating a fairly large proportion of the local population living below the poverty level—most of whom are black. In fact, for two of the sites, 100 percent of the families living in poverty are black.

Another study was conducted in 1987 by the United Church of Christ to investigate these same issues on a national scale. The results showed that minorities are more likely to live in communities where hazardous waste facilities are located. Specifically, minorities account for over 37 percent of the population in communities where one of the five largest hazardous waste landfills is located or where there is more than one treatment, storage, or disposal facility. This research did not find as strong an association for socioeconomic status as the GAO analysis discovered.

In a recent discussion of this issue, Sigman (2000) examined several articles that study the location of hazardous waste facilities and various minority groups. The overall conclusion in this commentary is that the results are mixed. Some studies find that these facilities are disproportionately located in minority communities, while others determine that no such relationship exists once income or other characteristics are accounted for.

Sources: U.S. EPA, Office of Policy, Planning, and Evaluation (June 1992); U.S. GAO (1983); United Church of Christ, Commission for Racial Justice (1987); Sigman (2000).

Consider, for example, the experts' risk ranking of active hazardous waste sites. The experts classify the associated cancer risk in category 2 of 5, adding that fewer than 100 cases of cancer per year should result nationwide.[30] They also give the problem a low ranking for noncancer risks and for ecological effects. Yet national efforts to control these sites through RCRA is high, and so too is the public perception of the associated risks. A similar argument applies to the mismatch of expert risk ranking for inactive hazardous waste sites relative to the intensity of national effort and the associated costs to control them. Yet again, the public perceives the associated risks as high.

Limitations of the Risk-Ranking Study

There are important and in some cases disturbing findings associated with the EPA's landmark risk analysis. However, it is critical to understand not only what the study accomplishes but also what its limitations are. As the task force points out in its report, risk assessment data are not complete, and methods for assessing noncancer and ecological risks are not well defined. There is a need for further research to advance scientific knowledge about environmental hazards and the implications of short- and long-term exposure. In the interim, given the limited data and the scientific uncertainty, evaluations about U.S. policy on solid waste and toxics cannot be made without reservation.

Another qualification of the risk-ranking analysis is that some intangible aspects of risk were not considered. For example, the study did not account for the extent to which certain risks are voluntary, nor did it consider any risks to the existence value of natural resources. Furthermore, the task force acknowledges that its findings were not adjusted for **environmental equity.** In fact, one of the concluding observations is that national risk rankings are not always indicative of local conditions and that these must be considered in setting environmental priorities at nonfederal levels of government.[31] Some environmental problems affect certain geographic regions or specific segments of the population more than others. Hence, analyses based on overall or national risks tend to obscure the true implications of these hazards. A case in point is the risk of exposure to hazardous waste facilities—an instance of environmental inequity discussed in Application 19.3.

Even accepting the study's limitations, the risk-ranking analysis strongly suggests the need for further investigation of how the federal government sets priorities for environmental policy initiatives. In the context of solid waste and toxics, it appears that the costs incurred by society to develop and implement comprehensive legislation like RCRA and Superfund relative to those associated with FIFRA and TSCA are not justified by comparative risk analysis. Further research is needed, particularly in benefit assessment, to support and expand upon this hypothesis.

Conclusions

Recognizing the potential risks of chemical exposure and the enormous costs of remediation efforts, the government has developed legislation aimed at controlling hazardous substances *before* they become a part of the waste stream. Although many legislative acts contribute to this effort, two of the more significant are the Federal Insecticide, Fungicide, and Rodenticide Act (FIFRA) and the Toxic Substances Control Act (TSCA). Both of these are designed to prevent the production and use of substances posing an unacceptable risk to society and to monitor the risks of those that have been introduced to the marketplace.

Although many pesticides and toxic substances are associated with known health and ecological effects, both can, and do, contribute to society's well-being. Hence, it is appropriate that decision making under FIFRA and TSCA be guided by risk-benefit analysis. There are,

[30] U.S. EPA, Office of Policy Analysis, Office of Policy, Planning, and Evaluation (February 1987), p. 73.
[31] U.S. EPA, Office of Policy Analysis, Office of Policy, Planning, and Evaluation (February 1987), pp. 7, 97.

however, practical problems that have hindered the realization of what this strategy can achieve. For example, registration decisions for new pesticides are based on risk data submitted by manufacturers, but benefits are assumed to justify those risks based on the registrant's willingness to pay the registration fee. There are similar flaws in the implementation of TSCA, particularly those provisions applicable to chemicals already on the market.

New programs have been developed to complement these laws and to integrate more incentives to encourage the production and use of safer substitutes. The direction is one that fosters pollution prevention to displace some of the reliance on treatment and cleanup. Such a shift in emphasis is supported by comparative risk analysis, particularly in light of the costs to support RCRA and Superfund relative to FIFRA and TSCA. According to this preventive approach, society has to be made to rethink how market decisions ultimately affect the size and toxicity of the waste stream. Firms must adjust how their products are designed, manufactured, and packaged, and households need to modify consumption and consider the external costs of their market decisions. Pollution prevention is an important theme in the ongoing development of environmental policy—one of several issues discussed in the concluding module of the text.

Summary

- Congress enacted the Federal Insecticide, Fungicide, and Rodenticide Act (FIFRA) in 1947. In 1972, Congress passed amendments that shifted the focus of the law to health and environmental protection. Additional amendments in 1988 were aimed at improving registration and reregistration procedures.
- In 1996, the Food Quality Protection Act (FQPA) was enacted, which amended FIFRA, as well as the Federal Food, Drug, and Cosmetic Act (FFDCA), another piece of legislation that regulates pesticides.
- A major objective of FIFRA is to register all pesticides before they are distributed or sold. Registration is subject to EPA approval based on a risk-benefit analysis. At the time of registration, the EPA must set tolerances on the amount of a pesticide that may remain as a residue on food without causing an unacceptable health risk.
- The dominant factor in pesticide registration decisions is a risk evaluation, based on data submitted by manufacturers. If risks are found to be negligible, the EPA assumes that benefits exist based on the manufacturer's willingness to absorb the registration costs.
- In 1994, the United States launched the Pesticide Environmental Stewardship Program (PESP), a voluntary program that establishes partnerships with pesticide users and implements preventive strategies. Among its goals is the promotion of Integrated Pest Management (IPM), which fosters more selective use of pesticides and greater reliance on natural deterrents.
- In 1976, Congress enacted the Toxic Substances Control Act (TSCA). One of the act's primary objectives is to control chemicals that present a risk to health or the environment before they are introduced into commerce. TSCA also provides for the monitoring and regulation of chemicals already on the market.
- TSCA requires chemical producers to notify the government before they intend to produce or import any new chemical, using a premanufacture notice (PMN). The EPA reviews each PMN to evaluate the risks and respond. If the risks are too high, the EPA may restrict usage of the chemical.
- For existing chemicals, TSCA requires manufacturers to notify the EPA if any chemical is found to present a substantial risk to human health or the environment.
- The United States is taking steps to develop chemical programs aimed at pollution prevention that will complement existing policy. One such effort is the Green Chemistry Program, which promotes the design of chemical products and processes that minimize or

eliminate the production of hazardous substances. Another is Extended Product Responsibility (EPR), which refers to a commitment by all participants in the product cycle to find ways to reduce a product's impact on the environment.

- Aggregate cost data on solid waste controls exhibit a strong positive trend between 1972 and 1987, which reflects the strengthening of U.S. policy since the mid-1970s. Aggregate control costs for FIFRA and TSCA are much smaller but have increased dramatically over the same period.
- According to an EPA risk-ranking study, federal resources might be overallocated to solid waste control and underallocated to the regulation of toxic substances and pesticide use. These findings imply that the social costs of RCRA and Superfund relative to those of FIFRA and TSCA may not be justified by comparative risk analysis.

Key Concepts

pesticide registration
risk-benefit analysis
pesticide tolerances
pesticide reregistration
Integrated Pest Management (IPM)
TSCA inventory
new chemical
premanufacture notice (PMN)
existing chemical
Green Chemistry Program
Extended Product Responsibility (EPR)
comparative risk analysis

Use the Key Concepts listed above to begin your search for additional articles and information using the InfoTrac College Edition database.

Review Questions

1. a. Critically compare and contrast the policy approaches of TSCA and FIFRA.
   b. In your view, which of these is more effective in preventing pollution? Explain.
2. Suggest two market-based instruments that would support the Pesticide Environmental Stewardship Program. Explain.
3. In 1976, the United States banned the manufacture of PCBs. Propose an alternative policy that would effectively reduce society's exposure to these cancer-causing substances. Support your proposal by discussing two of its strong points. Realistically, why might opponents argue against your idea?
4. a. If you were to plan an empirical study of the benefits of RCRA and Superfund, how would you categorize the primary social benefits?
   b. How, if at all, would your answer differ if you were to do the same type of study for FIFRA and TSCA? Explain.
5. Critically evaluate the EPA's risk-ranking study. Cite one major accomplishment of this analysis and one major failure. Support your answer from a policy perspective.

Additional Readings

Cropper, Maureen L., William N. Evans, Stephen J. Berard, Maria M. Ducla-Soares, and Paul R. Portney. "The Determinants of Pesticide Regulation: A Statistical Analysis of EPA Decision Making." *Journal of Political Economy* 100 (February 1992), pp. 175–97.

Geiser, Kenneth. *Materials Matter: Toward a Sustainable Materials Policy.* Cambridge, MA: MIT Press, 2001.

Geyelin, Milo. "Pollution Suits Raise Charges of Racism." *Wall Street Journal,* October 29, 1997, p. B3.

Hamilton, James T. "Testing for Environmental Racism: Prejudice, Profits, Political Power?" *Journal of Policy Analysis and Management* 14, no. 1 (1995), pp. 107–32.

Macauley, Molly K., Michael D. Bowes, and Karen L. Palmer. *Using Economic Incentives to Regulate Toxic Substances.* Washington, DC: Resources for the Future, 1992.

Mendeloff, John M. *The Dilemma of Toxic Substance Regulation.* Cambridge, MA: MIT Press, 1988.

Schierow, Linda-Jo. *Risk Analysis and Cost-Benefit Analysis of Environmental Regulations.* CRS Report 94-961 ENR. Washington, DC: Congressional Research Service, Library of Congress, December 2, 1994.

Thorton, Joe. *Pandora's Poison: Chlorine, Health, and a New Environmental Strategy.* Cambridge, MA: MIT Press, 2000.

U.S. Congress, OTA. *Green Products by Design: Choices for a Cleaner Environment.* Washington, DC: U.S. Government Printing Office, October 1992.

Weber, Peter. "A Place for Pesticides?" *World Watch* (May/June 1992), pp. 18–25.

Wilson, James D. "Resolving the 'Delaney Paradox.'" *Resources,* no. 123 (fall 1996), pp. 14–17.

http:// Related Web Sites

Extended Product Responsibility (EPR)	**http://www.epa.gov/epr**
Federal Insecticide, Fungicide, and Rodenticide Act (FIFRA)	**http://www.epa.gov/pesticides/fifra.htm**
Food Quality Protection Act (FQPA) of 1996	**http://www.epa.gov/oppfead1/fqpa/**
National Pesticide Information Center	**http://npic.orst.edu/**
Pesticide Environmental Stewardship Program (PESP)	**http://www.epa.gov/oppbppd1/PESP/**
Pesticide Registration	**http://www.epa.gov/pesticides/chemreg.htm**
Pesticide Reregistration Fact sheet	**http://www.epa.gov/oppfead1/trac/factshee.htm**
Pesticide Reregistration Status	**http://www.epa.gov/pesticides/reregistration/status.htm**
Toxic Substances Control Act (TSCA)	**http://www.law.cornell.edu/uscode/15/ch53.html**
TSCA Inventory	**http://www.epa.gov/opptintr/newchems/invntory.htm**
U.S. EPA Chemical Registry System	**http://www.epa.gov/crs**
U.S. EPA Chemical Testing and Information home page	**http://www.epa.gov/opptintr/chemtest/**
U.S. EPA Green Chemistry Program	**http://www.epa.gov/greenchemistry/**
U.S. EPA, Office of Environmental Justice	**http://www.epa.gov/compliance/environmentaljustice**
U.S. EPA, Office of Pesticide Programs	**http://www.epa.gov/pesticides/**
U.S. EPA, Office of Pollution Prevention and Toxics	**http://www.epa.gov/opptintr/index.html.**

A Reference to Acronyms and Terms in Toxic Substances Policy Control

Environmental Science Acronyms

DDT	Dichloro-diphenyl-trichloroethane
EDB	Ethylene dibromide
PCBs	Polychlorinated biphenyls

Environmental Policy Acronyms

CERCLA	Comprehensive Environmental Response, Compensation, and Liability Act
CERCLIS	Comprehensive Environmental Response, Compensation, and Liability Information System
EPR	Extended Product Responsibility
FDA	Food and Drug Administration
FFDCA	Federal Food, Drug, and Cosmetic Act
FIFRA	Federal Insecticide, Fungicide, and Rodenticide Act
FQPA	Food Quality Protection Act
IPM	Integrated Pest Management
NAS	National Academy of Sciences
NPL	National Priorities List
OSHA	Occupational Safety and Health Administration
PESP	Pesticide Environmental Stewardship Program
PMN	Premanufacture notice
RCRA	Resource Conservation and Recovery Act
RED	Reregistration Eligibility Decision
TSCA	Toxic Substances Control Act
USDA	U.S. Department of Agriculture

PART 7

Global Environmental Management

As society became more aware of environmental damage and the associated risks, it recognized that economic advance and industrialization were largely responsible for the pollution problems it confronted. Such a realization seemed to suggest an unacceptable trade-off between competing social objectives—economic prosperity and environmental quality. Economic growth is not a goal that many consider to be optional, nor are advances such as air travel, electricity, computer technology, and telecommunications. Nonetheless, most saw the need for policy initiatives that would improve environmental quality and reduce the risk of exposure to hazards like urban smog, polluted waterways, and abandoned hazardous waste sites.

The thrust of the initial policy response was to use end-of-pipe controls on the pollutants released into the environment. Although this approach has achieved some success, public officials have been concerned about the high costs, the unattained policy goals, and the conflict between economic growth and environmental protection. In response, policy revisions are being drafted and new programs launched in nations around the world. Incentives are displacing some command-and-control instruments, integrated programs are replacing some pollutant- and media-specific initiatives, and preventive efforts are superseding treatment as a waste management option. Taken together, these trends speak to a global transition in policy development toward broader, more long-term solutions to environmental damage.

In retrospect, it is easy to be critical of what has at times been a painstakingly slow policy development process. But, in truth, some of the accomplishments in environmental management have been a direct result of learning—sometimes the hard way—from mistakes and failures. Believing that experience is perhaps the best teacher, we have reason to be optimistic about this new direction of environmental policy both nationally and internationally. The wisdom of that experience coupled with scientific advances and technological innovation can bring about significant change. Perhaps more important, environmental objectives have broadened to consider the future along with the present and to accommodate global interests along with national and local needs. Such is the fundamental premise of sustainable development—a goal that integrates economic prosperity with environmental preservation, as a legacy to future generations.

In this final module, we examine environmental management from a global perspective, with an eye toward the objective of sustainable development and selected strategies being used to achieve that goal. Although there is much that has yet to be defined, several themes are beginning to emerge. Among these are efforts to advance environmental literacy, to redefine policy toward prevention, to reconcile international trade objectives with environmental goals, and to develop domestic and international agreements that facilitate common objectives.

To organize our discussion, in Chapter 20 we present an overview of sustainable development and the implications of population growth on the environment. We also examine the framework established by worldwide conferences, the intent of key international agreements to control transboundary pollution, and the interaction of international trade and environmental issues. In Chapter 21 we deal with implementation of these efforts, discussing evolving themes that are becoming long-run environmental approaches for many nations—industrial ecology and pollution prevention. The materials balance model is reintroduced to underscore the importance of these approaches, and strategic initiatives to operationalize them are discussed as well. Mindful of what has been accomplished and what remains to be done, we explore these themes that characterize what many view as an important transition in global environmental management.

Chapter 20

Sustainable Development: International Trade and International Agreements

Direct regulation of economic activity has characterized much of the evolution of environmental policy in most nations around the world. While this command-and-control approach has contained the problem, most recognize that reliance on so-called end-of-pipe controls does not adequately address the long-term implications of environmental damage. Furthermore, this approach does not accommodate the broader objective of **sustainable development**—achieving environmental quality *and* economic prosperity, which is particularly difficult for third world or developing nations.

The comprehensive goal of sustainable development calls for fundamental changes in how society makes market decisions—both in production and consumption. The challenge is to achieve economic prosperity but alter market activity so that natural resources and the environment are protected. Effecting changes of this magnitude calls for a different policy approach than one that relies on rules and limits, control instruments that often run counter to the polluter's market incentive. If society is to sustain a long-term commitment to preserving the earth, there has to be a motivation to do so beyond the avoidance of penalties for regulatory noncompliance.

Logically, the motivation should be consistent with economic incentives. The premise is that economic growth and environmental quality can be reinforcing rather than competing objectives. Perceptions must be changed to recognize that resource conservation and pollution abatement can enhance private as well as social interests. Communication must be improved—both within and among nations—to share information about technologies, inputs, and processes that can protect the environment without diminishing profitability. If successful, cooperation should displace what is sometimes an adversarial position between the public and private sectors of society, and there should be less reliance on costly monitoring and enforcement procedures.

In this chapter, we explore the fundamentals of this transitional phase in environmental policy, focusing on objectives and international agreements. We begin by discussing the concept of sustainable development and explore its motivation by studying the implications of economic growth on environmental quality. We then expand our analysis to examine selected international conferences and agreements to determine how these arrangements facilitate achieving the dual objectives of economic growth and the realization of environmental quality. Finally, we assess international trade issues and examine how trade objectives sometimes conflict with environmental goals.

Sustainable Development as a Global Objective

Redefining Environmental Objectives

Environmental Quality Revisited

Over the last several decades of environmental policy development, the clear focus has been to achieve **environmental quality** by reducing anthropogenic pollution to a level that is "acceptable" to society. Policymakers have struggled with the tough issue of *how clean is clean* and the challenge of devising cost-effective instruments to achieve whatever quality level has been set. Through revisions, political debate, and a growing social consciousness, there has

environmental quality
A reduction in anthropogenic contamination to a level that is "acceptable" to society.

been progress and some needed adjustments in policy development. Yet there also has been an increasing awareness of the need for change that is more fundamental than periodic revisions of legislative provisions—a redefinition of environmental policy objectives.

Moving Toward Sustainable Development

A consensus is forming that both public and private decision making should be driven by a broader goal that is global in scope and dynamic in perspective. One such goal is **sustainable development**—managing the earth's resources such that their long-term quality and abundance is ensured for future generations.[1] This objective, posed as a pressing issue at the 1992 Earth Summit, is also commonly referred to as **intergenerational equity.** In any case, what this goal makes clear is that environmental policy must consider the long run. Just as it is foolhardy to pursue economic growth without regard for the environmental implications over time, so too is it irrational to pursue environmental objectives that ignore the future economic consequences. An example of the potential conflict is discussed in Application 20.1, which focuses on the controversy surrounding government subsidies to support the timber industry.

sustainable development
Management of the earth's resources such that their long-term quality and abundance is ensured for future generations.

Sustainable Development in Practice

To be fair, not all the commentary on sustainable development or sustainable growth is entirely favorable. One common criticism, expressed by Nobel Laureate Robert Solow (1991) and others, is that the concept itself is vague and hence not a good notion to guide policy. Furthermore, and more important, the idea of each generation leaving the earth as they found it for future generations is not feasible and in some instances not even desirable. Consider that such a premise, followed literally, would mean that society should engage in no construction activity and should use no depletable resources, such as copper or oil. Notwithstanding a rational recognition of scarce resources, such inactivity does not make sense.

As an alternative, Solow (1991) argues that sustainability ought to be considered as "an obligation to conduct ourselves so that we leave to the future the option or the capacity to be as well off as we are" (p. 132). Even at that, there is still the difficulty of attempting to forecast both the preferences and the technological capability of future generations—no easy task by any measure. Solow also asserts, though somewhat controversially, that sustainability need not mean preservation of a particular species or tract of land. Rather, substitutability ought to be allowed in applying the notion of intergenerational equity, or distributional equity over time, to policy prescriptions. Lastly, he argues that, if policy is to be guided by this idea of intergenerational equity, to be consistent, it must also consider *intra*generational equity, with clear effort expended toward reducing poverty today.

Note that the practicality of Solow's ideas does not dispel the importance of recognizing the obligation to leave future generations with comparable capabilities to today's society. Moreover, it does not minimize the challenge of doing so in the face of population growth. This reality is, of course, more of an issue for developing nations. So we need to consider the ramifications of growth a bit more carefully.

Understanding the Implications of Economic and Population Growth on the Environment

Sustainable development is based on the premise that economic growth and environmental quality must be reconciled. Statistical estimates on worldwide population and income growth help to explain why this is important.

What the Data Imply

According to one source, per capita income levels have to grow by at least 2 percent per year to reduce world poverty and close the gap between the rich and the poor. This communicates

[1] To learn more about this objective, visit the Web site of the United Nations Division for Sustainable Development at **http://www.un.org/esa/sustdev/.**

Application

20.1 *The Potential Conflict Between Economic Gain and Environmental Quality*

One federal subsidy that has been the subject of public debate and media attention is the below-cost timber sale program. Administered by the U.S. Forest Service, this program grants companies the right to cut timber on federal lands at reduced costs. Environmentalists believe that this federal program has encouraged the practice of clear-cutting, causing severe soil erosion and threatening ecosystems and the already tenuous future of endangered species. Characterized as the controversy over the plight of the northern spotted owl, the much publicized debate actually has much broader concerns. Among these is the allegation that federal subsidies foster the use of virgin resources, effectively discouraging the use of recycled materials.

Beyond the environmental implications, the timber subsidy program also has been questioned on economic grounds. In its sale of public timber, the Forest Service is not obligated to recover the federal government's cost of cultivating and marketing the valuable timber. Consequently, although its enormous assets would give it a top-five position in the *Fortune 500*, its pricing policies would classify it as bankrupt in the private sector. In practice, the sale of publicly owned timber is subsidized according to one of three methods: (1) charging a sale price based on the industry's ability to pay versus the timber's market value, (2) assigning a price based solely on the government's cost, or (3) using public funds to finance access roads for harvesting.

Responding to strong arguments on a number of fronts, the Clinton administration suggested the possibility of reducing or even eliminating the timber subsidy program, but the suggestion never materialized. More recently, proposals have been made in the Senate to cut spending for the program but have been defeated thus far. Nonetheless, opposition to the program is firm, so such legislative proposals will likely be reintroduced. Yet the eventual outcome of limiting the forest subsidy plan is not entirely predictable.

The EPA has identified several possible scenarios, each of which would have very different effects on the economy and on national recycling objectives. First, the timber industry might be able to maintain current production levels by reducing other operating costs to counter the effect of the subsidy reduction. As a result, there would be no effect on alternative markets or on the nation's recycling efforts. Second, it could encourage more efficient use of existing supplies and an overall reduction in lumber consumption. Third, the cost effect could precipitate a market shift away from domestic suppliers toward lower cost foreign producers. Finally, there may be a substitution effect toward other domestic products, including recycled lumber materials.

While no one can forecast the future of the timber subsidy or the related market implications, most observers argue that such programs are in need of reform. Tax preferences such as these, however well intended, create market distortions. In this case, the subsidies embedded in the below-cost timber sales undervalue our nation's forests, setting up an artificial signal to use virgin materials over recycled substitutes. This distortion has important implications. Overharvesting of national forests jeopardizes the long-term supply of timber, threatens the future of biological species, and endangers the overall quality of the environment.

Sources: Wirth and Heinz (May 1991), pp. 77–86; O'Toole (1988); U.S. Congress, OTA (1989), chap. 5; U.S. EPA, Office of Policy, Planning, and Evaluation (March 1991), pp. 5-33–5-39; Kranish (April 3, 1993); Lazaroff (Septem-ber 14, 1999).

the importance of achieving economic growth. Furthermore, world population is growing at about 1.7 percent each year, a rate that is expected to decline only very slowly. Recognizing that economic growth has an effect on the environment, these data imply that the associated environmental impact per unit of income must decline at a rate between 3.5 percent and 4 percent per year to avoid further pollution and natural resource depletion.[2] To generalize

[2] Nitze (April–June 1993).

this assertion, we can characterize the relationship among population growth, income growth, and the environment at a point in time as follows:[3]

$$\text{Environmental Impact} = \text{Income per capita} \times \text{Environmental impact per unit of income} \times \text{Population.}$$

Developing countries face more urgent conditions, struggling to advance economically to accommodate a rapidly rising population and at the same time confronted with the environmental contamination that these pursuits have exacerbated. For example, recent projections estimate an annual population growth through 2050 of 4.7 percent for Ethiopia, 3.4 percent for Nigeria, 2.7 percent for Pakistan, and 1.5 percent for Bangladesh.[4] As population grows, so too does the demand for goods and services. A case in point is the expected increase in commercial energy demand for China and India, which is expected to double by 2015 due to population increases, economic growth, and rising demand for modern products.[5] If an economy's productive capacity cannot accommodate the population growth, shortages arise and resources are misused in an attempt to compensate for the imbalance.[6]

The Environmental Kuznets Curve

The implications of the relationship between growth and environmental quality have motivated numerous studies. Some of these examine the possibility of a technical relationship or pattern between economic development (commonly measured as income per capita) and environmental degradation. Known as the **environmental Kuznets curve,** this pattern has been hypothesized as an inverted U shape, as shown in Figure 20.1.

environmental Kuznets curve Models an inverted U shaped relationship between economic growth and environmental degradation.

Such a model suggests that early stages of industrialization are associated with rising levels of pollution, when growth is a greater priority than natural resource protection and when environmental controls are lenient or virtually nonexistent. It further implies that more advanced development is linked to a shift in focus in the opposite direction with increasing concern for environmental quality and a concomitant strengthening of environmental regulation. The implications of this theory are far-reaching and have motivated debate among empirical researchers, as Application 20.2 explains. That said, a consensus does seem to be forming that the trade-off between economic growth and environmental quality may not be as severe as once feared. This in turn suggests that progress in reconciling these goals may be realized.

Global Framework for Sustainable Development

Sustainable development is intended to be a global objective, the benefits of which should accrue to all segments of society and to all nations. Because of its intent and its pervasive implications, it calls for a collaborative effort from all stakeholders. Fundamental to such an effort is communication on a global scale, which can be facilitated at conferences that set an agenda and establish a framework for achieving growth and improving the environment.

United Nations Conference on Environment and Development (UNCED)[7]

More commonly called the Rio Summit, the U.N. Conference on Environment and Development (UNCED) was a 12-day worldwide forum held in Rio de Janeiro, Brazil, in June 1992.

[3] This relationship is just an identity, which is more apparent when the equation is rewritten using ratios, as follows:

$$\text{Environmental Impact} = \frac{\text{Income}}{\text{Population}} \times \frac{\text{Environmental Impact}}{\text{Income}} \times \text{Population.}$$

[4] United Nations, Population Division (1996), as cited in World Resources Institute (1998), pp. 244–45, table 7.1.
[5] U.S. Department of Energy, Energy Information Administration (April 1997), p. 115, as cited in World Resources Institute (1998), p. 37.
[6] For a classic article illuminating the environmental implications of population growth, see Hardin (1968).
[7] The following discussion is drawn from Reilly (September/October 1992); Sessions (April–June 1993a), p. 12; and Parson, Haas, and Levy (October 1992).

Figure 20.1 ***The Environmental Kuznets Curve***

The environmental Kuznets curve illustrates a theory that there is a technical relationship, or pattern, between economic development and environmental degradation. As shown in the graph, this expected pattern can be depicted as an inverted U shape. Such a model suggests that early stages of industrialization are associated with rising levels of pollution, when growth is a greater priority than natural resource protection and when environmental controls are lenient or virtually nonexistent. It further implies that more advanced development is linked to a shift in focus in the opposite direction with increasing concern for environmental quality and a concomitant strengthening of environmental regulation.

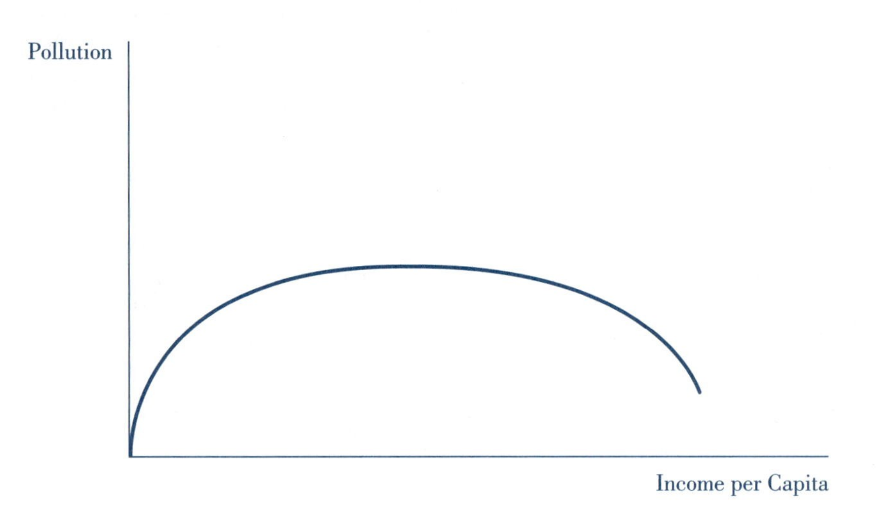

Thousands of delegates from over 170 nations attended the event to discuss issues and concerns dealing with sustainable development. Among the major documents produced from the Rio Summit are *Agenda 21* and the *Rio Declaration.* A brief description of each follows.

Agenda 21

This 40-chapter document is a voluntary action plan, outlining the course for worldwide progress toward sustainable development. Table 20.1 summarizes its four primary sections collectively intended to guide decision making into the 21st century. Online access to the text of this document is available at **http://www.un.org/esa/sustdev/agenda21text.htm.** Major issues covered within the 900 pages of *Agenda 21* include

- Financing for developing countries
- Conservation and sustainable development for forests
- Prevention and minimization of hazardous and solid wastes
- Strategies to address ocean pollution and the protection of marine life
- International cooperation for technological advance
- Risk assessment and management of toxic chemicals

Perhaps the most important achievement of the Rio Summit, *Agenda 21* is the result of 2½ years of negotiations to reach an international consensus on environmental issues.

20.2 *Economic Growth and Environmental Quality: The Environmental Kuznets Curve*

Application

An interesting hypothesis in the environmental economics literature is that the relationship between pollution and economic development might look like an inverted U. Known as the **environmental Kuznets curve (EKC),** this relationship implies that as a society begins to develop, it focuses more on growth than on the environment. Put another way, there is a trade-off between growth and environmental quality in the early stages of a society's development. As a society becomes more advanced, it recognizes the importance of preserving the environment and has the resources to begin implementing environmental policy. Hence, at some point the trade-off ceases to exist. Beyond that point, environmental quality may actually improve with further growth.

Not surprisingly, the EKC has prompted debate among environmental economists. Even among those who accept the basic notion of a trade-off between growth and environmental quality, there is uncertainty about the extent of the trade-off and the point at which the trade-off ends. Simply put, exactly where is the turning point of the inverted U, and what is the shape of the curve up to and beyond that point? Such questions are part of the motivation for empirical research in this area.

As the findings have begun to accumulate, a divergence of opinions has become apparent. A recent article by Dasgupta et al. (2002) presents an overview of the literature, which seems to suggest three different views about the shape and position of the EKC. Each of these is shown in the accompanying figure. For discussion purposes, these are identified as **conventional, pessimistic,** and **optimistic.**

The **conventional** view depicts the standard inverted U shape. In general, empirical results show that the turning point occurs between \$5,000 and \$8,000 income per capita. For example, Grossman and Krueger (1995) estimate this point as falling between \$4,000 and \$6,000 for air pollution measured by sulfur dioxide (SO_2) and particulate matter. Yet, according to Stern (1998), who reviewed the literature, such findings seem to arise only for certain pollutants.

Less prevalent is the **pessimistic** view, which suggests that beyond a critical income level, the best a nation can do is maintain some existing level of pollution. In fact, some findings suggest that the EKC may not ever flatten out but continues to increase with economic growth.

At the other extreme is the **optimistic** view, which implies that the trade-off is not as severe (i.e., the curve is flatter), does not persist (i.e., the turning point occurs at lower levels of income), and the extent of environmental damage is lower at each level of development (i.e., the EKC is everywhere below its conventional counterpart). This suggests that the EKC might be shifting downward. Dasgupta et al. (2002) indicate that this depiction is the most likely. Possible causes include increasing effectiveness of environmental regulation, improvements in abatement technology, and greater public awareness of pollution and its effects. Of course, there are no certainties, but if this perspective continues to be validated empirically, there is reason to be more confident that the trade-off between growth and environmental quality, even in the early phases of economic development, can be measurably reduced.

Sources: Dasgupta et al. (winter 2002); Stern (May 1998); Grossman and Krueger (May 1995).

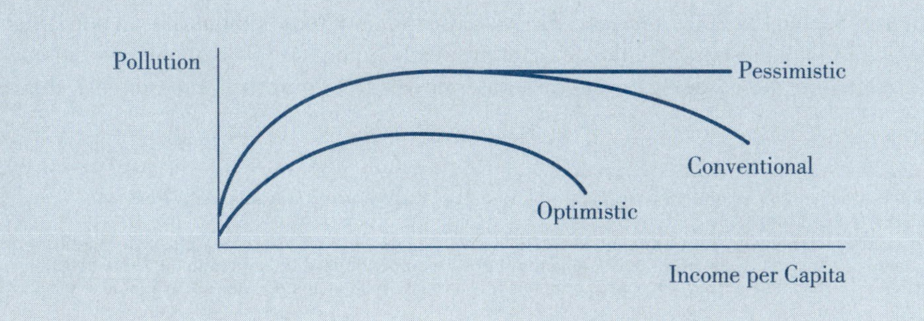

Table 20.1 ***Summary of the Rio Summit's Agenda 21***

Section	Description
Section 1: Social and Economic Dimensions	Includes recommended actions on sustainable development cooperation, poverty, consumption, demographics, health, human settlements, and integration of the environment and development in decision making.
Section 2: Conservation and Management of Resources for Development	Includes chapters on atmospheric protection, land resources, deforestation, desertification and drought, mountains, agriculture, biological diversity, biotechnology, oceans, freshwater resources, toxic chemicals, hazardous wastes, solid wastes, and radioactive wastes.
Section 3: Strengthening the Role of Major Groups	Includes ways to increase the participation in sustainable development efforts of major social groups: women, youth, indigenous peoples, nongovernmental organizations, local authorities, trade unions, business and industry, scientific and technological communities, and farmers.
Section 4: Means of Implementation	Includes chapters on financial resources; technology transfer, cooperation, and capacity building; science; education; public awareness and training; institutional arrangements; legal instruments and mechanisms; and information collection, analysis, and dissemination.

Source: Sessions (April–June 1993b).

Rio Declaration on Environment and Development

The *Rio Declaration* outlines 27 principles to act as guidelines for achieving global environmental quality and economic development.[8] For example, Principle 4 requires that environmental protection be an integral part of development, and Principle 8 calls for a reduction in "unsustainable patterns of production and consumption." Reportedly, the declaration is a compromise from what was originally anticipated. The preparatory meetings were intense, often immersed in seeking a balance among the views of developing and industrialized countries. So fragile was this agreement that it was adopted in Rio without further negotiation for fear that further discussion would jeopardize what was left of the intended Earth Charter.

To monitor progress on agreements made at the world conference, the Commission on Sustainable Development (CSD) was created in December 1992. The mandate, membership, and activities of the intergovernmental CSD are available at **http://www.un.org/esa/sustdev/csd.htm.**

World Summit on Sustainable Development: Johannesburg 2002[9]

In August–September 2002, ten years after the Rio Summit, tens of thousands of participants converged in Johannesburg, South Africa, at the World Summit on Sustainable Development. Heads of state, national delegates, and representatives from industry, trade unions, the sci-

[8] The full text of the *Rio Declaration* is available online at **http://www.un.org/documents/ga/conf151/aconf15126-1annex1.htm.**

[9] Information on the Johannesburg Summit was obtained from the official United Nations Web site for the Johannesburg Summit 2002, available at **http://www.johannesburgsummit.org/index.html** and from the United Nations Environmental Programme at **http://www.unep.org/wssd/.** More information and updates are available at these sites.

entific community, and other so-called Major Groups, attended this major event. The objective was to renew worldwide interest in sustainable development and to assess progress achieved since the 1992 Rio Summit.

Though the original goals of the Rio Summit remain in place, a key focus of the 2002 summit was to improve implementation of *Agenda 21.* Specifically, officials identified five characteristics that the Johannesburg Summit should possess. It should be:

- A summit of implementation: bridging the gaps between commitment and realization
- A summit of partnership: strengthening collaborations between governments, business, nongovernmental organizations, and others
- A summit of responsible prosperity: committing to the elimination of poverty and equitable distribution of globalization gains
- A summit of integration: assimilating environmental, economic, and social development policies
- A summit of concrete action: setting targets with a timetable and measurable benchmarks

Intended to maximize the effectiveness of the world summit, a series of Preparatory Committee Meetings were held in the months prior to the 2002 meeting at the national, regional, and global levels to prepare documents and engage in dialogues with the various stakeholders. Results of these meetings helped to define and motivate the global forum in August–September 2002.

International Agreements to Control Transboundary Pollution

Some pollution problems are transboundary, such as acidic deposition, ozone depletion, global warming, and some surface water pollution. Hence, contamination in one nation can travel beyond its borders—a type of **international externality.** In these cases, formal treaties must be negotiated and agreed to by all affected countries. Within these agreements are often rulings specifically aimed at the distinct problems of developing nations and the need for financial and technical support from their wealthier and more advanced counterparts. Certain of these agreements we have discussed in other contexts, while others are introduced here for the first time.

international externality
A spillover effect associated with production or consumption that extends to a third party outside the market in another nation.

Montreal Protocol and Amendments

Originally signed by 24 countries in September 1987, the Montreal Protocol on Substances that Deplete the Ozone Layer is an important example of international cooperation aimed at environmental protection. Strengthened by the 1990 London Amendments, the 1992 Copenhagen Conference, the Montreal Amendments in 1997, and the Beijing Amendments in 1999, this treaty is aimed at phasing out **chlorofluorocarbons (CFCs)** and other ozone-depleting substances. A summary of the phase-out deadlines, as they evolved over time, is presented in Table 20.2. As of 2002, over 180 nations have ratified the Montreal Protocol.[10]

chlorofluorocarbons (CFCs)
A family of chemicals that scientists believe contributes to ozone depletion.

Recognizing the special problems of developing nations and at the same time the importance of gaining their support, industrialized countries set up a $160 million Interim Multilateral Fund to assist them in transitioning toward CFC substitute technologies. Developing countries also were given a 10-year grace period to meet the targets of the agreement. The

[10] Current ratification status of the Montreal Protocol and subsequent amendments is available at **http://www.unep.ch/ozone/ratif.shtml.**

Table 20.2 Montreal Protocol and Amendments Aimed at Phasing Out Ozone Depleters

	1987 Montreal Protocol	1990 London Amendments	1992–1999 Up Through the Beijing Amendments
CFCs	50% cut by 2000 (5 types covered)	Phase out by 2000 (15 types covered)	Phase out by 1996 (15 types covered)
Carbon tetrachloride[a]	No controls	Phase out by 2000	Phase out by 1996
Halons	Freeze at 1986 levels in 1992	Phase out by 2000	Phase out in 1994
HCFCs[b]	No controls	Phase out by 2040	Freeze in 2004 at 1989 levels; phase out by 2020
Methyl bromide[c]	No controls	No controls	Freeze in 1995 at 1991 levels; phase out by 2005
Methyl chloroform[d]	No controls	Phase out by 2005	Phase out by 1996

Sources: UN Environment Programme cited in Dumanoski (November 26, 1992); Council on Environmental Quality (January 1993), p. 139; Parker (July 12, 2000).

Note: For more detail on the Montreal Protocol or any of the subsequent amendments, visit the Ozone Secretariat at the United Nations Web site at **http://www.unep.ch/ozone/index.shtml.** Deadlines shown are for developed nations; developing nations generally were given grace periods that extended the phase-out deadlines. After the London Amendments, there were the 1992 Copenhagen Conference, the 1997 Montreal Amendments, and the 1999 Beijing Amendments.

[a] Carbon tetrachloride is a solvent used in CFC production.

[b] HCFCs are CFC substitutes.

[c] Methyl bromide is a pesticide.

[d] Methyl chloroform is a solvent.

halons
A major group of ozone depleters with long atmospheric lifetimes.

Multilateral Fund became permanent in 1992.[11] Cooperation from the developing world is critical, since CFC production in these nations rose by nearly 2.5 times between 1986 and 1995, with CFC consumption increasing by 40 percent over that same period.[12] Most of this increase is attributable to a few rapidly growing countries, such as Brazil, India, Mexico, and especially China. A case in point is China's production of **halons,** a major group of ozone depleters, which rose from 4,000 metric tons in 1991 to 2.5 times that amount, or 10,000 metric tons, just 4 years later.[13]

Important accomplishments for the Montreal Protocol were the addition of China to the group of signatories in 1991 and the addition of India in 1992. Both nations have high economic growth rates and very large populations, which means that their commitments to the protocol are critical. Specifically, China's 1998 population is estimated at 1.26 billion and is projected to grow to 1.52 billion by 2050. For India, the 1998 population is estimated at 0.98 billion, with a forecast of 1.53 billion by 2050.[14]

---

[11] As of 2001, 32 industrialized nations had contributed a total of $1.3 billion to the Multilateral Fund. For updates and further information, visit **http://www.unmfs.org/general.htm.**

[12] Oberthür (1997), pp. vi, 35, as cited in World Resources Institute (1998), p. 178.

[13] United Nations Environment Programme (1996), p. 8, table 1-1, as cited in World Resources Institute (1998), p. 178.

[14] United Nations, Population Division (1996), as cited in World Resources Institute (1998), pp. 244–45, table 7.1.

U.N. Framework Convention on Climate Change (UNFCCC)

Adopted by more than 150 nations, including the United States, the U.N. Framework Convention on Climate Change (UNFCCC) established a baseline for a global cooperative response to climate change. In March 1994, the treaty became legally binding, following its ratification by the requisite number of 50 countries in 1993. Among the first 10 nations to ratify the agreement was China. This nation will need strong support from other nations to contain its rising consumption of coal and associated CO_2 emissions, predicted to triple over the next 35 years.[15]

A key provision of the UNFCCC, and one that prompted much contention, was a commitment by each signatory to implement a national strategy to limit the release of **greenhouse gases (GHGs).** Each strategy was to support the common objective of reducing emissions to 1990 levels by 2000. At a subsequent 1995 meeting in Berlin, some participants continued to support the commitment, while others argued it was inadequate. Further discussion centered on the establishment of specific timetables and emissions targets. Major carbon dioxide (CO_2) generators like Japan and the United States were pressured to make stronger pledges to the climate control treaty, but lobbyists from these countries expressed concerns about what such commitments might mean to international competitiveness and domestic jobs. The conference produced little more than an agreement to conduct further negotiations, which was called the Berlin Mandate.[16]

greenhouse gases (GHGs)
Gases collectively responsible for the absorption process that naturally warms the earth.

Kyoto Protocol

At the third Conference of the Parties (COP) held in Japan in December 1997, the Kyoto Protocol was formulated. A principal result of the protocol was the setting of binding GHG emission targets for developed nations. These emissions limits were to become effective when 55 nations ratified the agreement. To assure that the ratification represented the major GHG emitters, the 55 nations had to include developed nations that were responsible for at least 55 percent of worldwide emissions. At this point, the United States had committed to reduce GHG releases by 7 percent below their 1990 levels by the period 2008–2012, and the European Union had agreed to a larger cut of 8 percent. These limits were to be achieved using emissions trading among nations.

Although the United States signed the protocol in November 1998 during the fourth COP in Buenos Aires, its action was merely symbolic. The president did not submit the protocol to the Senate for ratification because of growing opposition to the treaty on Capitol Hill. A major concern was that the mandated emission limits would significantly hurt American industry and hence the economy. Moreover, the United States was holding firm to its commitment not to ratify the Kyoto Protocol until developing nations adopted binding emissions limits along with their industrialized counterparts.

Debate and discussion continued over the next several years at a series of COPs held around the world. Before the seventh session of the COP convened in Morocco, President Bush took the United States out of the Kyoto treaty in March 2001. This decision was significant for a number of reasons, not the least of which is the fact that the United States is responsible for about one-fourth of worldwide GHG emissions. This makes ratification of the protocol difficult, given the requirement that ratifying nations must include developed countries responsible for at least 55 percent of emissions.

In spite of the U.S. decision, 178 countries reached an international climate change accord in 2001, with details decided at the Morocco meeting in November 2001. The final Kyoto Protocol calls for 38 developing nations to reduce GHG emissions to 5.2 percent below their 1990 levels. No such targets were imposed on developed countries.[17] In mid-2002, the European Union and Japan ratified the treaty, bringing the total number of ratifying nations to 44.

---

[15]Council on Environmental Quality (January 1993), pp. 142–43; World Resources Institute (1994), p. 202; Chandler, Makarov, and Dadi (September 1990), p. 125, as cited in World Resources Institute (1994), p. 78.
[16]"New Emissions Plan Sought" (April 8, 1995).
[17]Revkin (July 24, 2001); Winestock (July 24, 2001); Fialka (November 12, 2001).

To date, about 36 percent of developed nations' aggregate emissions are represented, which means that most remaining industrialized nations must ratify to reach the requisite 55 percent target. Hence, it is of some interest that in June 2002, Australia decided to follow the United States and pulled itself out of the accord.[18] Ratification status of the Kyoto Protocol can be viewed online at **http://unfccc.int/resource/kpstats.pdf.**

London Dumping Convention (LDC)

Ocean dumping of certain wastes, including radioactive wastes, is prohibited by the Convention on the Prevention of Marine Pollution by Dumping of Waste and Other Matter, commonly referred to as the London Dumping Convention (LDC). Originally decreed in November 1972, the LDC now has been ratified by 75 nations, including the United States.[19] In fact, all nations that have recently engaged in ocean dumping of radioactive wastes are contracting parties to the LDC. A discussion of some of the initiatives brought before the convention and the pending 1996 Protocol to the London Convention is the subject of Application 20.3.

U.S.-Canada Air Quality Agreement

acid rain
Arises when sulfuric and nitric acids mix with other airborne particles and fall to the earth as precipitation.

In March 1991, The United States–Canada Air Quality Agreement was finalized to combat the problem of **acid rain** and visibility impairment. The agreement calls for each country to commit to national emissions caps on sulfur dioxide (SO_2) and nitrogen oxides (NO_X). The bilateral provisions also include directives to establish a forum for addressing other transboundary air quality issues. Among these are rulings for each nation to do research to improve understanding of common air pollution problems and stipulations to facilitate a regular information flow about monitoring, effects of air pollutants, and control methodologies.

The U.S. commitment to these concerns is confirmed by the 1990 Clean Air Act Amendments, which refer specifically to Canada's acid rain program. The statutes call for regular reporting to Congress on the SO_2 and NO_X emissions of all Canadian provinces participating in Canada's control program. Access to recent progress reports as well as to the text of the agreement is available online at **http://www.epa.gov/airmarkt/usca/index.html.**[20]

International Trade and Environmental Protection

International trade negotiations have always been the subject of political and economic debate. Although there are known economic advantages to trade, exchange between nations is nonetheless a complex undertaking. Trade negotiations are rarely, if ever, totally distinct from political objectives and national defense issues. The gains from trade can easily become clouded by protectionist attitudes, fueled by differences in product safety regulations, labor laws, and nationalism. In recent years, international trade discussions have triggered controversy of another sort—the potential conflict between the associated gains from trade and a decline in environmental quality.

Overview of the Controversy: Free Trade Versus Protectionism

Free Trade

Proponents of free trade argue that nations should trade with one another because there are tangible gains to be realized. Though not intended to be exhaustive, the following list identifies some of the chief benefits of international trade:

- Greater consumer choice
- Higher worldwide output
- Efficiency gains from specialization
- International political stability from forming trading partnerships
- More competition and lower prices in a global marketplace

[18] "EU Ratifies Kyoto Protocol" (May 31, 2002); Revkin (June 6, 2002); UNFCCC, (June 4, 2002a; June 4, 2002b).
[19] For more detail on the origin and content of the LDC, see National Advisory Committee on Oceans and Atmosphere (April 1984).
[20] For an economic analysis of this agreement, see Menz (fall 1995).

20.3 *The London Dumping Convention (LDC): An International Agreement on Ocean Dumping Rules*

Application

Signed over 30 years ago, the London Dumping Convention (LDC) is an international agreement to control the practice of dumping wastes at sea. It is the only such convention to which the United States is a signatory nation. The LDC has been accepted by 79 nations, among them the major industrialized countries that generate most of the wastes being dumped into the sea. To view a list of the signatories, visit **http://www.londonconvention.org/PartiesToLC.htm.** The LDC controls ocean dumping through various means. It bans the ocean disposal of so-called "black-listed" substances like mercury and cadmium. The disposal of "gray-listed" or less damaging substances is controlled through special permits, and the dumping of any other substances is managed through a general permitting system.

http://

One of the convention's most significant achievements was a ban on the disposal of high-level nuclear waste in ocean waters. Another was a 10-year voluntary moratorium on the disposal of low-level nuclear waste, which was agreed to by the signatories in 1983. At the LDC's 1990 annual meeting, a consensus was reached to phase out ocean dumping of all industrial wastes by January 1, 1996. The phaseout was to be achieved without transferring the damage to some other environmental media. Each member nation was to be responsible for enforcing this resolution and for prosecuting any of its ships found to be in violation. At this same session, the delegates proposed that an international agreement be drafted to control land-based ocean pollution. This action is in direct response to the fact that more than 80 percent of ocean pollution is generated on land.

In a 1992 meeting of 44 of the 75 LDC signatories, the assembly agreed to update the rules of the convention. There were three major issues that were considered for revision: the disposal of low-level radioactive waste, the dumping of industrial waste, and the incineration at sea of industrial waste and sewage sludge. Several countries, namely Denmark, Norway, and Iceland, submitted their collective proposal to ban all of these activities on a permanent basis. In November 1993, 37 of the signatories voted to permanently ban the ocean disposal of low-level radioactive wastes. Although there were no votes against, 5 nations abstained—Belgium and four nuclear powers, China, France, Russia, and the United Kingdom. Of greatest concern was Russia's abstention, since this nation, just one month prior, had dumped over 237,000 gallons of low-level radioactive waste in the Sea of Japan. However, in 1993, following the completion of various technical, scientific, and other studies, all parties agreed to ban the dumping of low-level radioactive wastes, effective February 1994.

This review of convention rules continued and was completed when the parties adopted the 1996 Protocol to the London Convention. This accord, which explicitly acknowledges international agreements executed at the Rio Summit, has not yet been ratified by the requisite number of signatories. If and when ratification is achieved, the 1996 Protocol will replace the London Dumping Convention. In any case, as the international community continues to restrict or eliminate waste disposal at sea, nations must develop alternatives that will allow the advance of industrial activity without simply transferring the environmental damage from the earth's oceans to some place on land.

Sources: U.S. Congress, OTA (1987), pp. 73, 149; "Industrial Nations Ban Dumping Waste at Sea" (November 3, 1990); "44 Nations Agree to Update Rules on Dumping" (November 14, 1992); Lederer (November 13, 1993); "Russia to Continue Dumping Nuclear Waste at Sea" (October 19, 1993); International Maritime Organization, Office for the London Convention 1972 (March 25, 2001).

International trade models have been developed to illustrate these gains and help sort out which nations should export which products, but these are beyond the scope and intent of this discussion. What *is* relevant is to ask why there are strong opponents to trade, given that these gains from trade exist.

The Protectionist View

protectionism
Fostering trade barriers, such as tariffs or quotas, to protect a domestic economy from foreign competition.

On the opposite side of the coin are those who support **protectionism,** arguing that nations are better off without engaging in international trade. Several arguments are commonly offered to support this position, such as concerns that trade can threaten national security, create unfair competition, lose jobs to nations with cheap labor, and limit the growth of infant industries. Though seemingly plausible, there are strong counterarguments to each, which are adopted by most economists. Nonetheless, protectionist arguments are used to justify trade barriers like quotas and tariffs.

Though the protectionist view is not new, concerns about environmental quality now have been added to its arsenal of arguments. In part, these concerns arise from the disparity of environmental standards among trading nations—a disparity that is particularly striking between developing nations and their more advanced trading partners. To better understand this issue, we need to explore exactly how environmental policies can tangibly influence trading arrangements among nations.

International Trade and Environmental Objectives

A common concern about international trade is that lenient labor laws and relatively low wage rates in less developed nations can adversely affect employment in more advanced countries. Similar apprehensions have arisen about differences in environmental regulations between trading partners. The relevant issue is that production costs are lower in countries that have more lenient environmental standards. This cost differential gives producers in these nations a competitive advantage over those in countries with more stringent controls. Predictably, such disparate regulations are more commonly found between developing and advanced trading partners.

Another source of controversy is the quality and desirability of imports produced in nations with lax regulations on such issues as toxic chemical use, fuel efficiency, and coal consumption. Beyond any cost considerations, these imports would be associated with environmental externalities that can extend beyond national borders. For example, importing goods produced with high-sulfur coal has negative implications for global air quality, which identifies the market failure as an **international externality.** At issue is whether demand for such products should be implicitly augmented as a result of an international agreement.

As a counter to this concern, trade advocates argue that the economic gains from trade will help poorer nations afford the costly cleanup of what is in many cases severe environmental pollution. In the case of transboundary pollution, the benefits from this response may accrue to bordering countries or those located downstream along a common river. An improved economy in all likelihood will also provide the financial support to design and implement more comprehensive environmental policy, as suggested by the **environmental Kuznets curve.** These arguments are consistent with the premise of sustainable development—that economic prosperity, in this context enhanced through trade, can bring about improvements in environmental quality.

These and other issues were part of the lengthy and often contentious negotiations associated with the North American Free Trade Agreement (NAFTA), the latest round of the General Agreement on Tariffs and Trade (GATT), and the World Trade Organization (WTO) agreements.

International Trade Agreements and the Environment

North American Free Trade Agreement (NAFTA).[21]

Following a difficult series of negotiations, many of which centered on environmental issues, NAFTA finally was reached by the United States, Mexico, and Canada in 1992 and approved by Congress in 1993. In fact, because of the anticipated environmental impact of this treaty, in particular the adverse effects of Mexico's lax environmental regulations, a federal court case was filed that could have put an end to this major trade accord. An overview of the issues and the court's decision is discussed in Application 20.4.

Among NAFTA's provisions are the following, which are aimed at economic advance and environmental quality:

- Explicit language asserting the signatories' commitment to sustainable development
- Agreement among the three nations to implement NAFTA in accordance with the aim of environmental protection and not to lower health, safety, or environmental standards to attract investment
- Consensus to aim for congruence of each country's respective environmental regulations while preserving each nation's right to select a level of environmental quality that it deems appropriate
- Agreement that NAFTA dispute settlement panels will solicit environmental experts for advice on factual issues as needed

The full text of NAFTA is accessible on the Internet at **http://www.nafta-sec-alena.org/english/nafta.index.htm.**

http://

Consistent with NAFTA, the United States and Mexico also implemented a 1992–1994 Integrated Environmental Plan for the border region. The 2,000-mile border has been a source of concern because of the concentration of the now infamous *maquiladora* factories positioned there for access to American markets. Effluent releases into the Rio Grande River and air emissions far exceed U.S. standards. The plan, supported by $208 million from Congress and $160 million from the Mexican government, was aimed at improving water quality, monitoring air pollution, tracking hazardous waste, developing enforcement cooperation, and promoting pollution prevention.

Today, there is an ongoing binational effort aimed at some of these same border issues. Launched in 1996 as a five-year plan, the U.S.-Mexico Border XXI Program was designed to deal specifically with the most challenging air and water quality problems in border regions. The 10 U.S.-Mexico border states worked on plans to address environmental pollution, and the EPA dedicated $22 million to border tribes to help improve water and wastewater systems. The program was also responsible for the collection of continuous air monitoring data in three high-priority areas. In the last year of Border XXI, officials began planning a successor program based on input from a series of roundtable discussions with various constituencies. The new program should be finalized near the end of 2002.

As to how NAFTA actually affected the flow of trade, consider the following summary statistics. In NAFTA's first six years, U.S. exports to Mexico and Canada rose 78 percent to $253 billion. These exports supported an estimated 2.7 million American jobs in 1999, which represents a 34 percent increase (translating to 685,000 new jobs) over 1993, the year before NAFTA took effect. Two-thirds of U.S. exports to Mexico and nearly all of those to Canada now enter duty-free, and the average tariff imposed by Mexico on U.S. exports has dropped to below 2 percent.

[21] This discussion is drawn from Council on Environmental Quality (January 1993), pp. 52–53; U.S. Trade Representative, World Regions (June 6, 2002); U.S. Trade Representative (June 10, 2002); North American Commission for Environmental Cooperation (June 6, 2002); and U.S. EPA (May 10, 2002).

Application

20.4 *How NAFTA Was Affected by the National Environmental Policy Act*

The National Environmental Policy Act (NEPA) is a broad-based set of rulings that outlines the U.S. overall position on environmental issues and requires that the environmental impact of any public policy decision be formally considered. To implement this latter requirement, NEPA calls for an Environmental Impact Statement (EIS) to accompany proposals for legislation or major federal actions. According to NEPA sec. 102(1)(c), an EIS must specify the following:

(i) The environmental impact of the proposed action,
(ii) Any adverse environmental effects which cannot be avoided should the proposal be implemented,
(iii) Alternatives to the proposed action,
(iv) The relationship between local short-term uses of man's environment and the maintenance and enhancement of long-term productivity, and
(v) Any irreversible and irretrievable commitments of resources which would be involved in the proposed action should it be implemented.

Despite the specific rulings in NEPA, not all federally proposed actions have complied with the law, and in some cases lawsuits have been filed. For example in 1997, 102 cases were filed against government agencies or departments for failure to file an EIS or for incomplete documents. But of the hundreds of cases filed to date, none captured greater world attention than one involving the North American Free Trade Agreement (NAFTA).

In a stunning federal court decision handed down in 1993, a judge ruled that the filing of an EIS was required for negotiations of NAFTA between the United States, Mexico, and Canada. The case originally had been brought before the District Court of Washington, DC, in August 1992 as *Public Citizens v. Office of the U.S. Trade Representative* for failure to file an EIS for NAFTA and for the Uruguay Round of negotiations under the General Agreement on Tariffs and Trade (GATT). Among the plaintiffs in the case was the Sierra Club, a major lobbying environmental group. The district court dismissed the case, and a subsequent court of appeals ruling stated that the plaintiffs had failed to identify a final agency action. Responding to President Bush's announcement of an agreement on NAFTA on September 12, 1992, a new court action was filed—this time arguing the need for an EIS only for NAFTA.

The 1993 decision by Judge Charles B. Richey stirred up another round of debate centered on the economic and environmental implications of open trade among the three nations. Judge Richey was assertive about his position, writing in his decision:

> An impact statement is essential for providing Congress and the public with the information needed to assess the present and future environmental consequences of . . . the NAFTA. (Gosselin, p. 1)

Critics complained that filing an EIS could take years—a delay they believed could mean the end of NAFTA. Responding to Judge Richey's ruling, then U.S. trade representative Mickey Kantor stated that the government would appeal the decision. Interestingly, in September 1993, Judge Richey's ruling was reversed—not because of the anticipated delays but because NAFTA is a presidential action not covered by NEPA. And, in December 1993, President Clinton signed NAFTA into law.

Sources: Council on Environmental Quality (1998), p. 355; Dumanoski (July 1, 1993); Gosselin (July 1, 1993); York (September 25, 1993).

To address the environmental impact of the increased trade, the North American Commission for Environmental Cooperation (CEC) was formed among the three trading nations in 1993. This international organization complements NAFTA and is aimed at protecting and improving the natural environment. Since 1993, the CEC has engaged in a number of environmental efforts aimed at such issues as biodiversity, pollutant reporting, chemical management, and trade-environmental issues. One of the organization's current projects is aimed at increasing the market share of green goods and services in North America and finding ways to use more market-based instruments in these markets.

General Agreement on Tariffs and Trade (GATT)[22]

Originally executed in 1947, GATT was a major international treaty on foreign trade. Its avowed purpose was to reduce tariffs and other trade barriers. The 107 signatories, mostly developing countries, met periodically for negotiations called rounds, the last being the Uruguay Round. This round began in 1986 and was signed in December 1993. The associated negotiations, much like those preceding the signing of NAFTA, considered environmental protection issues. However, most environmentalists say that the long-awaited agreement did not respond adequately to their concerns.

Environmentalists' opposition to GATT was stronger than it was to NAFTA. Reportedly, the intense debate was fueled by a 1991 GATT panel ruling that the United States could not block imports of tuna from Mexico harvested under conditions yielding an unacceptable incidental kill rate of dolphins.[23] However, the arguments that ultimately pitted open traders against environmentalists during the GATT round run broader and deeper than this single incident.

Fundamentally, environmentalists are wary of any agreement that promotes economic growth and hence increases the potential for ecological damage. Furthermore, they are concerned about how GATT rulings that deter trade restrictions run counter to environmental aims. A few examples will illustrate. A nation cannot use countervailing duties on imports from a country whose environmental regulations are lower, and therefore less costly to producers, than its own. An import cannot be restricted based solely on the exporter's use of a pollution-generating input or production method. Relatively high environmental standards in one country can be viewed as restrictions on free trade.

Based on the results of the Uruguay Round, countries are required to use the least trade restrictive measures to achieve environmental goals, and in most cases international standards should be employed rather than national ones. Economically, the concern is that such a universal set of guidelines may become a ceiling instead of a floor. Nations might recognize the incentive to keep standards and enforcement relatively low as a way to keep compliance costs down and attract foreign investment. In any case, once this pattern emerges, other countries may follow suit to maintain export competitiveness.

Another important outcome of the Uruguay Round was the formation of the World Trade Organization (WTO), which is the successor to GATT. However, the multilateral trading system that was developed and established under GATT continues under the WTO.

World Trade Organization (WTO)[24]

Formed in 1995 as the successor to GATT, the World Trade Organization (WTO) is an international organization aimed at facilitating trade between nations. Consisting of 144 member nations and customs territories as of 2002, the organization performs many functions, such as:

- Administering trade agreements
- Facilitating trade negotiations and resolving trade disputes
- Overseeing national trade policies
- Helping developing nations in trade policy matters

Aggregate data suggest that reduced trade barriers due in part to the WTO and its predecessor have markedly improved trade flows. For example, over the 1994–1998 period, U.S. exports increased by 36 percent with more than 1.4 million new jobs created as a consequence. More information is available at the WTO's Web site, **http://www.wto.org.**

http://

[22] This discussion is drawn from Council on Environmental Quality (January 1993), pp. 52–53; Davis (January 10, 1994); Noah (December 16, 1993); Cough (April–June 1993); and "The Greening of Protectionism" (February 27, 1993).

[23] According to GATT provisions, import restrictions may be allowed to protect biological life only in the importing country, not the exporting nation.

[24] This discussion is drawn from World Trade Organization (June 4, 2002, June 6, 2002); U.S. Trade Representative, WTO and Multilateral Affairs (March 2, 2000; April 12, 2000).

The WTO generally makes decision by consensus among the entire membership, over 57 percent of which are developing or least developed nations. Its high-level decision-making body is the Ministerial Conference, which meets at least once every two years. The multilateral trading system established under GATT is referred to as the set of WTO agreements (composed of some 30,000 pages), which are essentially rules of trade established through negotiation among its members. Some 90 percent of all world trade is accounted for by the WTO. Its annual budget is about $85 million.

That the WTO has a role to play in environmental matters was initially established at the Rio Summit. The *Rio Declaration* specifically asserts that a multilateral trading system can make an important contribution to protect the environment and achieve sustainable development. Indeed, called for by the 1994 Ministerial Decision on Trade and Environment at the end of the Uruguay Round, the WTO established a Committee on Trade and Environment (CTE). The committee's directive is:

- To identify the relationship between trade measures and environmental measures to foster sustainable development
- To recommend any necessary changes to the multilateral trading system

The WTO is careful to assert that its purpose is not to act as an environmental protection agency. Rather its competency regarding the environment is limited to trade matters and to the effect of environmental policy on trade flows. To accomplish its goals, the organization fosters coordination and cooperation among nations to address environmental issues and supports the identification of market access opportunities that will aid developing countries in achieving sustainable development.

Conclusions

While some argue that the realization was long overdue, society has begun to recognize the importance of achieving a balance between economic growth and the preservation of natural resources. Although population growth in advanced countries is stable or even declining, the same cannot be said of developing nations. Based on current trends, world population is estimated to rise to 9.4 billion by 2050, an increase of about 3.4 billion.[25] Attempts to increase production to provide for this growth will place inordinate stress on the ecology and the earth's stock of resources.

Acknowledging and understanding the relevant issues is an important first step, but the real challenge is in establishing appropriate goals and in taking responsible action to achieve them. Critical to meeting this challenge is a true understanding of the interdependence between economic activity and nature. Understanding this connection is a precondition for effective policy development and informed decision making, both of which are essential elements of global environmental management. If sustainable development is to be achieved, there must be a cooperative and educated effort from industry, private citizens, and public officials at all levels of government and around the world. As this process unfolds, certain strategies are evolving that may be effective in realizing long-term environmental objectives. In the next chapter, we examine these strategies to learn how the United States and other countries are attempting to implement the common goal of sustainable growth and development.

[25] United Nations, Population Division (1996), as cited in World Resources Institute (1998), p. 141.

Summary

- A consensus is forming that public and private decision making should be driven by a goal that is global in scope and dynamic in perspective. One such goal is sustainable development—managing the earth's resources such that their long-term quality and abundance is assured for future generations.
- Sustainable development is based on the premise that economic growth and environmental quality must be reconciled. Estimates on worldwide population and income growth indicate that environmental impact per unit of income must decline to offset the effect of anticipated economic and population growth.
- The environmental Kuznets curve suggests there is a technical relationship or pattern between economic development and environmental degradation that graphs as an inverted U shape.
- The Rio Summit in 1992 dealt with the objective of sustainable development. To monitor progress of the world conference, the Commission on Sustainable Development (CSD) was created in December 1992. Among the major documents produced from the Rio Summit were *Agenda 21* and the *Rio Declaration.*
- In August–September 2002, tens of thousands of participants converged in Johannesburg, South Africa, at the World Summit on Sustainable Development.
- When pollution problems are transboundary, the contamination can generate an international externality. In such cases, formal treaties must be negotiated among all affected countries.
- Strengthened by the 1990 London Amendments, the 1992 Copenhagen Conference, the 1997 Montreal Amendments, and the 1999 Beijing Amendments, the Montreal Protocol is aimed at phasing out chlorofluorocarbons (CFCs) and other ozone-depleting substances.
- The U.N. Framework Convention on Climate Change (UNFCCC), which became legally binding in 1994, established a baseline for a global cooperative response to climate change.
- At the third Conference of the Parties (COP) held in Japan in December 1997, the Kyoto Protocol was formulated. A principal result was the setting of binding greenhouse gas emission targets for developed nations. Before the seventh session of the COP convened in Morocco, President Bush took the United States out of the Kyoto treaty.
- Ocean dumping of certain wastes is prohibited by the London Dumping Convention (LDC) of 1972. The LDC has been ratified by 75 nations, including the United States.
- The United States–Canada Air Quality Agreement was finalized in March 1991 to combat the problem of acid rain and visibility impairment. Under this accord, each country must commit to national emissions caps on sulfur dioxide (SO_2) and nitrogen oxides (NO_X).
- A common concern about international trade is that lenient labor laws, relatively low wage rates, and lenient environmental standards can give international competitors an unfair advantage. Another source of controversy is the quality and desirability of imports produced in nations with lax regulations on toxic chemical use and fuel efficiency. Such issues have affected trade negotiations on a global scale.
- Following difficult negotiations, many of which centered on environmental issues, NAFTA was reached by the United States, Mexico, and Canada in 1992. To address the environmental impact of the increased trade, the Commission for Environmental Cooperation (CEC) was formed in 1993.
- Originally executed in 1947, the General Agreement on Tariffs and Trade (GATT) was a major international treaty on foreign trade, whose purpose was to reduce tariffs and other trade barriers. Environmentalists' opposition to GATT was stronger than it was to NAFTA.

- Formed in 1995 as the successor to GATT, the World Trade Organization (WTO) is an international organization aimed at facilitating trade. Called for by the 1994 Ministerial Decision on Trade and Environment, the WTO established a Committee on Trade and Environment (CTE).

Key Concepts

environmental quality
sustainable development
environmental Kuznets curve
international externality
chlorofluorocarbons (CFCs)
halons
greenhouse gases (GHGs)
acid rain
protectionism

Use the Key Concepts listed above to begin your search for additional articles and information using the InfoTrac College Edition database.

Review Questions

1. Critically discuss the following statement: "Without a well-enforced command-and-control regulatory structure, society will not take the necessary steps toward a sustainable future."

2. Summarize in your own words Robert Solow's view of using substitutability in applying intergenerational equity to environmental policy. Do you agree with this view? Why or why not?

3. a. Visit the UNFCCC's Web site at **http://unfccc.int/resource/kpstats.pdf,** and determine the current ratification status of the Kyoto Protocol.

b. Explain the dilemma faced by advanced nations deciding about ratification, given that the United States is no longer a party to this agreement.

4. In your view, does NAFTA advance or hinder the achievement of sustainable development? Explain.

5. a. Critically evaluate the provision resulting from the Uruguay Round that requires that the least trade restrictive measures be used to achieve environmental goals. Include in your answer both the environmental and economic implications.

b. Support or refute the use of international environmental standards among trading partners.

Additional Readings

Abaza, Hussein, and Andrea Baranzini, eds. *Implementing Sustainable Development.* Northampton, MA: Elgar, 2002.

Bohringer, Christoph. "Climate Politics from Kyoto to Bonn: From Little to Nothing?" *Energy Journal* 23, no. 2 (2002), pp. 51–72.

Chambers, P. E., and R. A. Jensen. "Transboundary Air Pollution, Environmental Aid, and Political Uncertainty." *Journal of Environmental Economics and Management* 43 (January 2002), pp. 93–112.

Chichilnisky, Graciela. "What Is Sustainable Development?" *Land Economics* 73 (November 1997), pp. 467–91.

Commoner, Barry. "Economic Growth and Environmental Quality: How to Have Both." *Social Policy* 15 (summer 1985), pp. 18–26.

Dean, Judith, ed. *International Trade and the Environment.* Burlington, VT: Ashgate, 2001.

Deere, Carolyn L., and Daniel C. Esty, eds. *Greening the Americas: NAFTA's Lessons for Hemispheric Trade.* Cambridge, MA: MIT Press, 2002.

Lesser, Jonathan A., and Richard O. Zerbe Jr. "What Can Economic Analysis Contribute to the Sustainability Debate?" *Contemporary Economic Policy* 13 (July 1995), pp. 88–100.

Lofdahl, Corey L. *Environmental Impacts of Globalization and Trade: A Systems Study.* Cambridge, MA: MIT Press, 2002.

Lopez, Ramon. "Corruption, Pollution, and the Kuznets Environmental Curve." *Journal of Environmental Economics and Management* 40 (September 2000), pp. 137–50.

Obasi, Godwin O. P. "Embracing Sustainability Science: The Challenges for Africa." *Environment* 44 (May 2002), pp. 8–19.

O'Riordan, Timothy, William C. Clark, Robert W. Kates, and Alan McGowan. "The Legacy of Earth Day: Reflections at a Turning Point." *Environment* 57 (April 1995), pp. 6–15, 37–42.

Pezzey, John C. V., and Michael A. Toman, eds. *The Economics of Sustainability.* Burlington, VT: Ashgate, 2002.

Sampson, Gary P. "The Environmentalist Paradox: The World Trade Organization Challenges." *Harvard International Review* 23 (winter 2002), pp. 56–62.

Swanson, Timothy M., and Sam Johnston. *Global Environmental Problems and International Environmental Agreements.* Northampton, MA: Elgar, 1999.

Taylor, Jerry. *Policy Analysis: NAFTA's Green Accords—Sound and Fury Signifying Little.* Washington, DC: Cato Institute, November 17, 1993.

Wettestad, Jorgen. "Clearing the Air: Europe Tackles Transboundary Pollution." *Environment* 44 (March 2002), pp. 32–40.

Related Web Sites

http://

Agenda 21 (full text)	**http://www.un.org/esa/sustdev/agenda21text.htm**
Kyoto Protocol ratification status	**http://unfccc.int/resource/kpstats.pdf**
London Dumping Convention (LDC) signatories	**http://www.londonconvention.org/PartiesToLC.htm**
Montreal Protocol ratification status	**http://www.unep.ch/ozone/ratif.shtml**
Multilateral Fund	**http://www.unmfs.org/general.htm**
North American Commission for Environmental Cooperation	**http://www.cec.org**
North American Free Trade Agreement (NAFTA) (full text)	**http://www.nafta-sec-alena.org/english/nafta.index.htm**
North American Free Trade Agreement (NAFTA) (comprehensive information)	**http://www.ustr.gov/regions/whemisphere/nafta.shtml**
Rio Declaration (full text)	**http://www.un.org/documents/ga/conf151/aconf15126-1annex1.htm**
United Nations Commission on Sustainable Development (CSD)	**http://www.un.org/esa/sustdev/csd.htm**
United Nations Framework Convention on Climate Change (UNFCCC)	**http://unfccc.int/**
United Nations Ozone Secretariat	**http://www.unep.ch/ozone/index.shtml**
United Nations Sustainable Development Web site	**http://www.un.org/esa/sustdev/**
U.S.–Canada Air Quality Agreement (full text and progress reports)	**http://www.epa.gov/airmarkt/usca/index.html**
U.S. EPA, Smart Growth Web site	**http://www.epa.gov/livability/**
U.S.–Mexico Border Program	**http://www.epa.gov/usmexicoborder/**
World Summit on Sustainable Development: Johannesburg 2002	**http://www.johannesburgsummit.org/index.html** **http://www.unep.org/wssd/**
World Trade Organization (WTO)	**http://www.wto.org**
WTO Committee on Trade and Environment	**http://www.wto.org/english/tratop_e/envir_e/issu1_e.htm**

Chapter 21

Sustainable Approaches: Industrial Ecology and Pollution Prevention

As illustrated in the previous chapter, **sustainable development** is an ambitious pursuit. Fundamental to its achievement are substantive changes in how market activity is undertaken within the circular flow and in how environmental policy is drafted. The key is to modify behavior so that economic growth and environmental protection can become reinforcing rather than competing goals. Although the needed changes will take time to orchestrate, several initiatives being developed in nations around the world are moving public and private efforts in this direction.

Acting as a multidisciplinary framework for certain of these initiatives is the concept of **industrial ecology.** This framework fosters a systems approach to environmental management and recognizes the interaction of ecological systems with industrial systems. The goal is to minimize the impact of economic activity on the environment. In so doing, industrial ecology integrates economic objectives with environmental goals—precisely the motivation of sustainable development. Consequently, it advocates environmental strategies that are different from the more traditional command-and-control approach. In particular, industrial ecology promotes the development of industrial ecosystems in which residuals or wastes from one manufacturing process are used as inputs in the production of another good.

A related and more familiar approach to achieving sustainable development is **pollution prevention.** As the name implies, pollution prevention refers to initiatives that reduce or eliminate wastes rather than deal with them at the end of a product cycle. These types of initiatives avoid unnecessary and costly abatement and are considered necessary to achieving sustainable growth and development. Though their underlying aims are different, pollution prevention and industrial ecology share a common view that end-of-pipe policy controls are not sufficient for achieving long-term goals.

These approaches and associated programs that make them operable are the focus of this chapter. We begin with an overview of industrial ecology. To motivate the discussion and to provide a relevant context, we reintroduce the materials balance model presented originally in Chapter 1. This model clearly explains the relationship between industrial activity and nature, which is at the core of industrial ecology. We then move on to study pollution prevention—what it means in practice and how it is being supported internationally. Lastly, we explore a number of voluntary programs currently used in the United States and in other nations to operationalize these approaches and move toward sustainable development.

Industrial Ecology: A Systems Approach to Sustainable Development

As a society, we are beginning to understand that end-of-pipe pollution controls are inadequate for achieving long-run environmental objectives. It is simply not enough to clean up the effects of pollution after the damage has been done. What this means is that we have to use a broader perspective when assessing the environmental effects of consumption and production and that firms must adopt sustainable business practices.

What Is Industrial Ecology?[1]

An increasingly pervasive notion in environmental policy development is that the entire life cycle of a product, including all the materials and energy flows, ought to be considered in efforts to improve the environment. This conviction underlies an emerging discipline known as **industrial ecology.** Though there is no standardized definition, industrial ecology can be thought of as a multidisciplinary systems approach to the flow of materials and energy between industrial processes and the environment. It is considered a systems approach because it supports the integration of ecological systems and industrial systems. Moreover, it promotes conceptualizing the production process itself as an industrial ecosystem, which is something we will discuss later in the chapter.

industrial ecology
A multidisciplinary systems approach to the flow of materials and energy between industrial processes and the environment.

In any case, the primary purpose of industrial ecology is to promote the use of recycled wastes from one industrial process as inputs in another. It also supports optimal materials flows, or dematerialization of output, which means efficient use of materials and energy in production. Notice that this approach to environmental protection is centered on industry and is implemented over an interdependent group of firms.

Based on its conceptual arguments, industrial ecology is intimately linked with sustainable development. In fact, some argue that industrial ecology is the means by which a society promotes sustainable development or, at the very least, that the achievement of sustainable development is its ultimate purpose. It is also considered a management tool, since it directly addresses industrial production. In any case, since this field deals directly with the flow of materials in an economic system, it is appropriate to examine the materials balance model in this context.

Revisiting the Materials Balance Model

The dynamic relationship between economic activity and nature can be better understood by positioning the circular flow model within a larger framework to generate the **materials balance model.** Originally introduced in Chapter 1, this model illustrates the link between economic activity and the natural environment—a relationship that is critical to achieving sustainable development. As shown in Figure 21.1, this relationship is defined by the flow of resources from nature to an economic system and the return flow of **residuals** from economic activity back to the environment.

materials balance model
Positions the circular flow within a larger schematic to show the connections between economic decision making and the natural environment.

residual
The amount of a pollutant remaining in the environment after a natural or technological process has occurred.

Throughout our study of environmental economics, the primary focus has been on the flow of residuals and the potential damages associated with their release to nature. Indeed, the conventional policy approach uses end-of-pipe command-and-control instruments aimed at limiting the amount and toxicity of residuals released to the environment *after* they are generated. Implicit in this approach is a perception that the flow of materials through the economy is linear and open.

Linear or Open Materials Flow: Cradle to Grave

As the graphic in Figure 21.2 illustrates, a **linear flow of materials** assumes that materials run in one direction, entering an economic system as inputs and leaving as wastes or residuals. This "cradle-to-grave" flow emphasizes use, waste generation, and disposal, which coincides with policies aimed at abating contaminating residuals only at the end of the flow. For example, conventional clean air policy efforts try to control harmful residuals released into the atmosphere, such as sulfur dioxide and particulate matter. Although such policy efforts can and do improve environmental quality in the short run, they do not adequately address the long-run consequences. To understand why, think about the dynamics of the materials balance model in the context of policy initiatives.

linear flow of materials
Assumes that materials run in one direction, entering an economic system as inputs and leaving as wastes or residuals.

First, nature's capacity to convert matter and energy is limited. Although residuals can be converted to other forms that can flow back into productive use, this process is not without bound. Hence, policy decisions affecting resource use and environmental damage have

[1] This discussion is drawn from Garner and Keoleian (November 1995) and Erkman (1997).

Figure 21.1 *Revisiting the Materials Balance Model: Implications for Sustainable Development*

The materials balance model shows the dynamic relationship between economic activity and nature. Resources flow from nature to an economic system, and residuals from economic activity represent the return flow back to nature. Conventional policy uses end-of-pipe instruments aimed at controlling the residual flow. These policies do not adequately address the long-run consequences of intertemporal trade-offs between one generation and the next. Society must find ways to reconcile economic growth and environmental quality, which is the underlying premise of sustainable development.

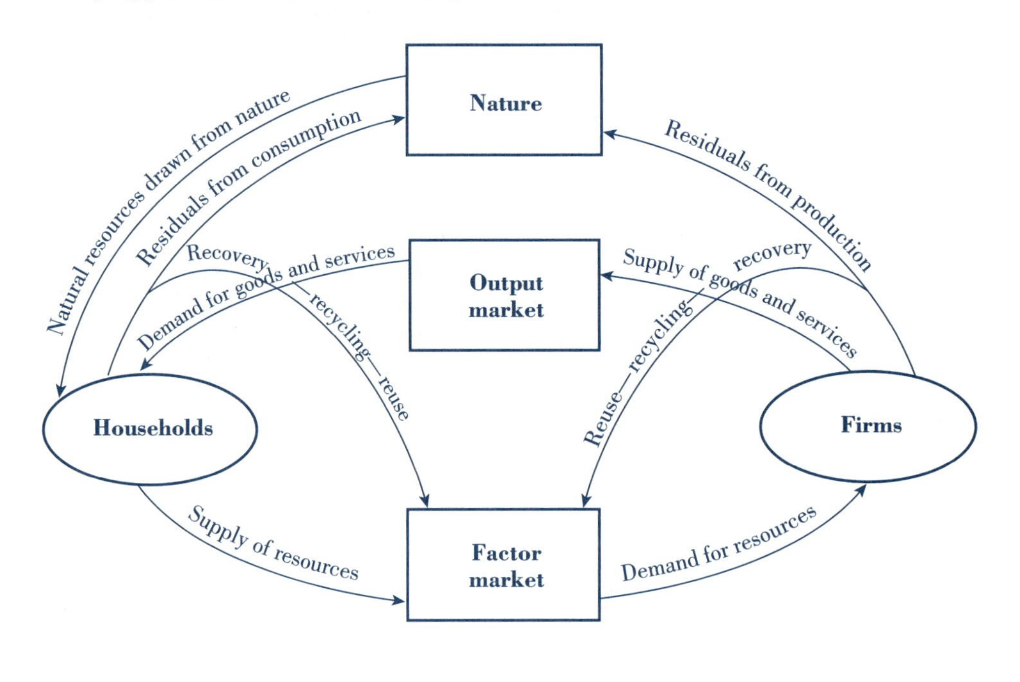

Source: Adapted from Kneese, Ayres, and D'Arge (1970).

implications for future generations. Look again at Figure 21.1. Government programs to promote recovery and reuse of residuals shown by the inner flows in the figure have helped delay their release but not eliminate them.

Second, end-of-pipe policy controls take the form of abatement or remediation after the damage has been done. Such methods draw resources away from other productive activities—resources that ultimately are added to the residual flow. Think about the use of raw materials and energy needed to clean up a hazardous waste site or to dredge a polluted waterway. These resources, once expended, go back into the residual flow.

Third, command-and-control initiatives generally operate at cross purposes with private market incentives and the broader social goal of economic growth. This observation points to the potential for intertemporal trade-offs between one generation and the next. The pursuit of economic development today can so harm the natural environment as to leave future generations unable to continue that progress. In the context of the materials balance model, we know that population growth and economic development increase the flow of resources into market activity and the return flow of residuals back to nature. These changes can adversely affect the well-being of future generations, if present decision making is not mindful of the long run. On the other hand, restricting the pursuit of economic gains to protect the environment deprives today's society as well as future generations of a higher standard of living.

Figure 21.2 ***Conventional Linear Perspective of Materials Flow***

The aim of performance- or technology-based standards is to control the amount of polluting residuals released to the environment. Implicit in this approach is a perception that the flow of materials through the economy is linear and open. As the graphic shows, this view assumes that materials run in one direction, entering an economic system as inputs and leaving as wastes or residuals. Hence, the policy focus is on abating contaminating residuals at the end of the flow.

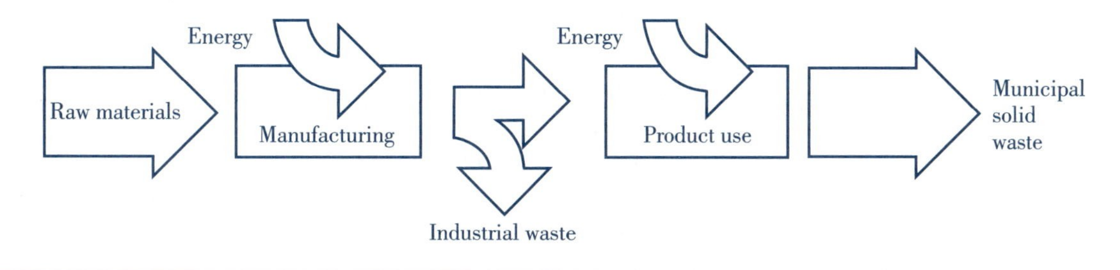

Source: U.S. Congress, OTA (October 1992), as cited in Gibbons (September/October 1992), p. 31.

Figure 21.3 ***Moving Toward a Closed System of Materials Flow***

This model of the materials flow suggests a more comprehensive approach to environmental protection than controlling residuals *after* they have been generated. Instead, this perspective illustrates how economic activity can be altered *throughout* the cycle of production and consumption to reduce the associated environmental impact. Product design, manufacturing processes, and energy use can be modified to achieve a cyclical or closed flow of materials.

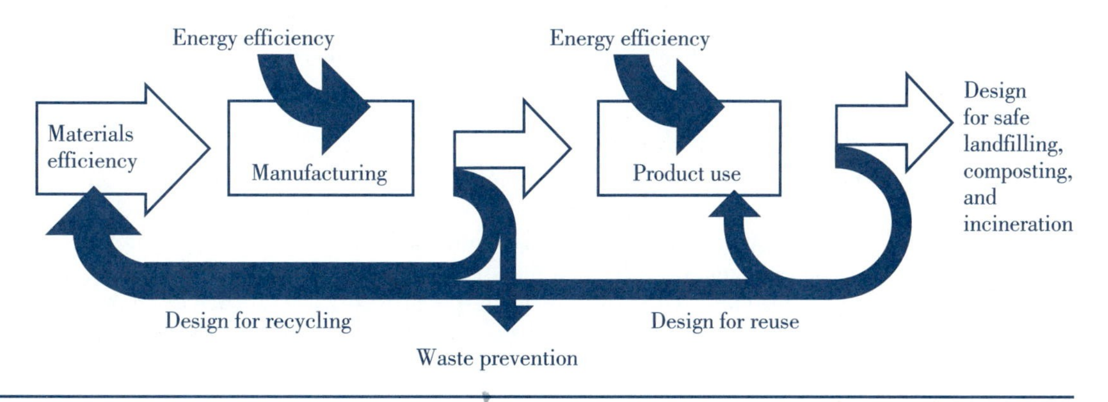

Source: U.S. Congress, OTA (October 1992), as cited in Gibbons (September/October 1992), p. 31.

Cyclical or Closed Materials Flow: Cradle to Cradle

Part of the solution is to recognize that the flow of materials through an economy need not be linear. Instead, product design, manufacturing processes, and energy use can be modified to achieve a **cyclical, or closed, flow of materials,** as shown in Figure 21.3. Notice that this model, considered a "cradle-to-cradle" flow, suggests a much broader approach to environmental protection than controlling residuals *after* they have been generated. Waste prevention, energy and resource efficiency, and design for reuse or recycling are among the options suggested by this paradigm. In sum, the key implication of a cyclical system is that economic

cyclical flow of materials
Assumes that materials run in a circular pattern in a closed system, allowing residuals to be returned to the production process.

activity can be altered *throughout* the cycle of production and consumption to achieve a reduction in the associated environmental impacts.

Life Cycle Assessment (LCA)

life cycle assessment (LCA) Examines the environmental impact of a product or process by evaluating all its stages from raw materials extraction to disposal.

To put the notion of a cyclical materials flow into practice, a **life cycle assessment (LCA)** should be undertaken. This is an analytical tool that examines the environmental impact of a product by evaluating all the stages of a product or process, from raw materials extraction to final disposal. Once done, steps can be taken to make improvements in production to lessen environmental risk.

Though still evolving, there is some consensus that a life cycle assessment has four key components:[2]

- **Goal Definition and Scoping:** Describe the product or process to be assessed, the context for the assessment, and the environmental effects to be reviewed
- **Inventory Analysis:** Identify and quantify resource usage and environmental releases
- **Impact Analysis:** Assess the human and ecological effects based on the inventory analysis
- **Interpretation:** Evaluate the results of the inventory and impact analyses, and select the preferred product or process

More information is available at the National Risk Management Research Laboratory's Life Cycle Assessment (LCA) Web site at **http://www.epa.gov/ORD/NRMRL/lcaccess .index.htm.**

That life cycle assessments are becoming increasingly important is evidenced by the development of databases and software to support the effort on a wide scale. In addition, the International Organization for Standardization (ISO) in Geneva, Switzerland, is working to formalize the methods used in this process for nations around the world.[3] In fact, a major component of the new **ISO 14000 standards** for environmental management addresses life cycle assessment. These are voluntary international standards for environmental management aimed at providing countries with a common approach to environmental issues. Read more about the ISO 14000 standards in Application 21.1.

ISO 14000 standards Voluntary international standards for environmental management.

Industrial Ecosystems

industrial ecosystem A closed system of manufacturing whereby the wastes of one process are reused as inputs in another.

The real-world outcome of implementing a closed system is the formation of an **industrial ecosystem** in which residuals from one or more manufacturing processes are reused as inputs for others.[4] In its most fundamental form, wastes from one process, such as emissions or wastewater, are collected and then redirected to an entirely different process to produce another good or service. The transfers generally take place through a collaborative effort among different companies. Though seemingly esoteric, such eco-industrial parks, as they are also known, are becoming more commonplace and are operating or being developed in Asia, Europe, and the United States. Perhaps the most well known example is the Kalundborg, Denmark, ecosystem, which is discussed in Application 21.2.

Following the Kalundborg example, a number of similar industrial ecosystems are operating or are under development in the United States. In Choctaw County, Mississippi, the notion of creating an industrial ecosystem became a reality during the construction of an industrial park. The idea emerged from one element of the project—a power plant that uses lignite (a low grade of coal) for fuel. Developers began to consider the environmental effects of the plant on the surrounding community. As they did, they began to identify manufacturing pro-

[2] U.S. EPA and Science Applications International Corporation (2001).
[3] U.S. EPA, Science Advisory Board (April 4, 2002).
[4] See Frosch (December 1995) and Frosch and Gallopoulos (September 1989) for more detail.

21.1 *ISO 14000 International Standards on Environmental Management*

Application

With increasing numbers of companies competing and operating in global markets, there is a growing need to develop cooperative international partnerships as well as formal agreements on trade, economic, and political issues. Disparate regulations and practices across nations can pose barriers to economic activity if not properly addressed. A case in point is the variability in environmental standards across countries, which has often been a source of debate during trade negotiations. In response to this issue, the International Organization for Standardization (ISO), a private international body founded in 1947 in Geneva, Switzerland, developed voluntary standards for environmental management. These standards are commonly known as the **ISO 14000** series. (Note that ISO is not an acronym for the International Organization for Standardization. Rather, ISO is from the Greek *isos,* meaning equal.)

ISO has developed over 13,000 standards for many technical specifications in engineering and business, including information technology, aircraft, and quality management. Each series of standards is aimed at providing consistency and homogeneity of products and technology, based on international consensus. The expected outcome of utilizing these voluntary standards is greater clarity in global market activity, which in turn should facilitate international business and trade flows.

Development of the ISO 14000 series in environmental management emerged from a consultation process begun in 1991, involving 20 nations, 11 international organizations, and over 100 experts in environmental issues. This effort combined with ISO's commitment to sustainable development led to the formation of a technical committee in environmental management standards in 1993. The work yielded the new ISO 14000 series, elements of which were published between 1996 and 1998.

Key areas addressed by the ISO 14000 series are

- **Environmental Management Systems (EMS):** Specifies requirements and guidance for establishing an EMS aimed at producing a high-quality product with minimal environmental impact.
- **Environmental Performance Evaluations:** Gives guidance on establishing a baseline and monitoring performance.
- **Environmental Auditing:** Sets procedures and criteria for an EMS audit.
- **Life Cycle Assessment (LCA):** Provides methodology, principles, and a framework for conducting an LCA study.
- **Environmental Labeling:** Offers guidance on terminology and procedures for declaring the environmental aspects of a product.

Achieving certification to the ISO 14000 standards is intended to benefit the individual firm as well as the international business community at large. According to the International Organization for Standardization, firms complying with ISO 14000 can expect to realize lower waste management costs, energy savings, an improved public image, and a model to help them monitor their progress in environmental performance. For more on the ISO 14000 series, visit the International Organization for Standardization at **http://www.iso.org.**

http://

Sources: International Organization for Standardization (1998); International Organization for Standardization (June 4, 2002); EnviroSense (December 20, 1995).

cesses that could use some of the waste products created by the power facility. The result is what officials there are calling an ecoplex. Clay from the lignite mining process to fuel the power plant will be used by a brick manufacturer. The fly ash from the combustion of lignite will be used to make wallboard or manufacturing cement. And any residual heat will be used by a greenhouse nursery, which in turn will exchange its wastes with a local fish farm.[5]

[5] U.S. EPA, Office of Solid Waste and Emergency Response (March 2000), p. 11. To read about other U.S. industrial ecosystems, visit **http://www.smartgrowth.org/library/eco_ind_case_intro.html.**

Application

21.2 *Industrial Symbiosis in Kalundborg, Denmark: When a Bad Becomes a Good*

The generation and disposal of production wastes has become a national concern and, as a consequence, the motivation for a concept called the **industrial ecosystem.** This strategic approach to industrial manufacturing is based on the premise that the conventional input-output process can be perceived as a closed system, such that the residuals generated from one production process are used as raw inputs for another. The link between processes generally occurs among several different entities. Obviously, this shift toward "closing the loop" in manufacturing calls for close coordination among the industrial participants.

The concept of an industrial ecosystem is in keeping with the broader concept of sustainable development and reconciling the achievement of economic growth and environmental preservation. As such, it is directed toward two major objectives. The first is to employ energy and raw materials optimally to assure that each productive input achieves maximum efficiency. The second is to facilitate waste reduction by minimizing the generation of nonreusable wastes that occur as part of the manufacturing function.

Consider the experience of a group of five entities and a municipality cooperating within a closed manufacturing system in Denmark. The participants are the municipality of Kalundborg, an oil refinery, a coal-fired power plant, a biotechnology and pharmaceutical plant, a soil remediation firm, and a wallboard manufacturer. Just as the concept of an industrial ecosystem suggests, these participants are linking their manufacturing processes together to save resources, energy, and the environment. Just how does this cooperative relationship work?

Three waste by-products are produced by the coal-fired electrical power plant: surplus steam, fly ash, and surplus heat. In a conventional manufacturing environment, these wastes would simply be disposed of. However, as a result of establishing a closed-loop manufacturing environment, the surplus steam is sold to the local biotechnology plant and to the oil refinery for use in their respective manufacturing processes. This arrangement helps to conserve water supplies, which must be pumped from a lake seven miles away. Similarly, the surplus heat is sold by the utility to the city of Kalundborg for its heating needs. The fly ash associated with the utility's pollution control equipment is actually a composition of limestone. This is used by the building industry. Gypsum generated from the same process is sold to a wallboard manufacturer, which uses it in place of natural gypsum. Sludge that is generated as a residual from the local water treatment plant is used as a nutrient by the soil remediation company.

As a consequence of this industrial ecosystem, some amount of industrial wastes is eliminated, and more efficient resource use is achieved. The Kalundborg case clearly illustrates how industrial ecology can be applied in a real-world context. This innovative and successful system has drawn worldwide attention, and now other industrial ecosystems are operating or are under development. Arguably, this is a mutually beneficial relationship for all participants that can help an economy move toward sustainable development.

http://

For more detail about the Kalundborg system, visit the Web site of the Symbiosis Institute in Denmark at **http://www.symbiosis.dk/tsi_uk.htm.** To learn about some of the industrial ecosystems operating in the United States, go to **http://www.smartgrowth.org/library/eco_ind_case_intro.html.**

Sources: Frosch and Gallopoulos (September 1989); Smith and Woodruff (May 11, 1992); Schmidheiny (1992), pp. 104–106; Symbiosis Institute (June 6, 2002).

Research in this arena should facilitate more projects along this line. Scientists from the Massachusetts Institute of Technology and Eneco Inc., a Salt Lake City firm, have developed a new technology that can transform heat pollution into electricity in a way that is cost-effective and efficient. The main contribution of their work is the improvement in the conversion rate to 17 percent from the typical 10 percent in such thermoelectric devices, with expectations of achieving an even higher rate. Scientists point out that the theoretical maximum is 50 percent. Hence, the new technology should make the process efficient enough to be adopted for widespread use. Production of a marketable prototype device is anticipated within a year.[6]

Pollution Prevention

There is an old adage that says, "An ounce of prevention is worth a pound of cure." This time-honored advice is now being used to help achieve sustainable development. Recent policy initiatives in nations around the world have begun to emphasize preventive measures that eliminate the release of contaminating residuals and hence avoid having to deal with the associated adverse effects after the fact.

What Is Pollution Prevention?

Pollution prevention, commonly referred to as **P2,** is a long-term approach aimed at reducing the amount or the toxicity of residuals released to the environment. Operationally, it refers to practices that reduce or eliminate wastes or residuals at their source. Based on this definition, pollution prevention promotes a shift from end-of-pipe controls to front-end reduction strategies. Effectively, this implies that residual generation is a strategic variable and not a given to be dealt with after the fact. In the context of the materials balance model, preventive strategies change how economic activity is undertaken so that less contaminating residuals are released to nature.

pollution prevention (P2)
A long-term strategy aimed at reducing the amount or toxicity of residuals released to nature.

Strong support for pollution prevention is offered by Barry Commoner, a scientist and director of the Center for Biology of Natural Systems at Queens College. He writes:

> The impact of a pollutant on the environment can be remedied in two general ways: either the activity that generates the pollutant is changed to eliminate it; or, without altering the activity, a control device is added that traps or destroys the pollutant before it enters the environment . . . . The few real improvements have been achieved not by adding controls or concealing pollutants but by simply eliminating them. . . . This suggests an addition to the informal environmental laws: If you don't put something into the environment, it isn't there.[7]

Commoner supports his argument with references to such preventive policy initiatives as the elimination of leaded gasoline in the United States and the ban on DDT effected in 1972.

Comparing Pollution Prevention to Industrial Ecology

What pollution prevention (P2) and industrial ecology have in common is their support for a sustainable solution, using something other than end-of-pipe controls. In that sense, both views adopt the **cyclical flow of materials,** and both employ **life cycle assessment (LCA)** as an important tool to guide solutions. For these reasons, there is a real temptation to position P2 under the broad umbrella of industrial ecology. In fact, some research papers do exactly this or even use the two concepts interchangeably. However, although comparisons are

[6] Chang (November 27, 2001).
[7] Commoner (1992), pp. 42–43.

made difficult by the lack of standardized definitions for each, there are some clear differences between the two concepts. It is well worth the effort to consider some of these differences to help clarify the perspective of these two strategies.

In a very illuminating and readable paper on the subject, Oldenburg and Geiser (1997) offer the following observations. First and probably foremost, P2 promotes risk reduction through minimizing or eliminating wastes, while industrial ecology argues in favor of using wastes as inputs in other production processes. In this same regard, P2 solutions are aimed mainly at the single firm (though other sectors also participate), while industrial ecology is aimed at a network of businesses. This also means, from a benefit-cost perspective, that the potential gains of industrial ecology have to be shared. Second, and not entirely unrelated to the first, recycling is not viewed as a preventive solution, while it essentially *is* the solution in industrial ecology. Third, P2 proposals generally assume some direction or oversight by government, while proponents of industrial ecosystems tend not to mention government intervention. And fourth, P2 considers efficiency as a potential tool to achieve its objectives, while industrial ecology tends to view efficiency of resource use and materials flows as an end in itself.

source reduction
Preventive strategies to reduce the quantity of any hazardous substance, pollutant, or contaminant released to the environment at the point of generation.

toxic chemical use substitution
The use of less harmful chemicals in place of more hazardous substances.

source segregation
A procedure that keeps hazardous waste from coming in contact with nonhazardous waste.

raw materials substitution
The use of productive inputs that generate little or no hazardous waste.

changes in manufacturing processes
The use of alternative production methods to generate less hazardous by-products.

product substitution
The selection of environmentally safe commodities in place of potentially polluting products.

Objectives and Techniques in Pollution Prevention

While pollution prevention promotes the reduction of residuals or wastes generated by all segments of society, its emphasis has been on the industrial sector—the major source of hazardous pollution. In that regard, there are two major preventive objectives for industry. One of these is **source reduction,** which refers to preventive strategies aimed at reducing the quantity of any hazardous substance, pollutant, or contaminant released to the environment at the point of generation. The other objective is **toxic chemical use substitution,** which is the practice of using less harmful chemicals in place of more hazardous substances.[8] A number of techniques have been proposed to help industry achieve these objectives.[9] Among these are

- Source segregation
- Raw materials substitution
- Changes in manufacturing processes
- Product substitution

The first of these, **source segregation,** refers to any process that keeps hazardous waste from coming in contact with nonhazardous waste. This relatively simple and inexpensive method is widely used by many industries. By keeping the two types of waste materials separate, the accumulation of hazardous waste is not needlessly increased by contaminated nonhazardous waste. These efforts help generators lower their costs of managing wastes and enhance social benefits by reducing the risk of exposure.

Two other methods, **raw materials substitution** and **changes in manufacturing processes,** also hold promise for helping to prevent pollution. Both can be used by industry during the production phase of its operations. By using inputs that generate little or no hazardous waste, the size of the residual flow can be dramatically reduced. Likewise, alternative manufacturing processes can be sought, or existing ones altered, to generate less hazardous byproducts. Finally, **product substitution,** an adjustment made *after* the production phase, refers to the selection of environmentally safe commodities over their potentially polluting substitutes.

[8]U.S. EPA, Office of Pollution Prevention (October 1991), p. 7.
[9]This discussion is drawn from U.S. EPA, Office of Solid Waste (November 1986), p. 19.

Figure 21.4 ***Pollution Prevention Hierarchy Under the Pollution Prevention Act of 1990***

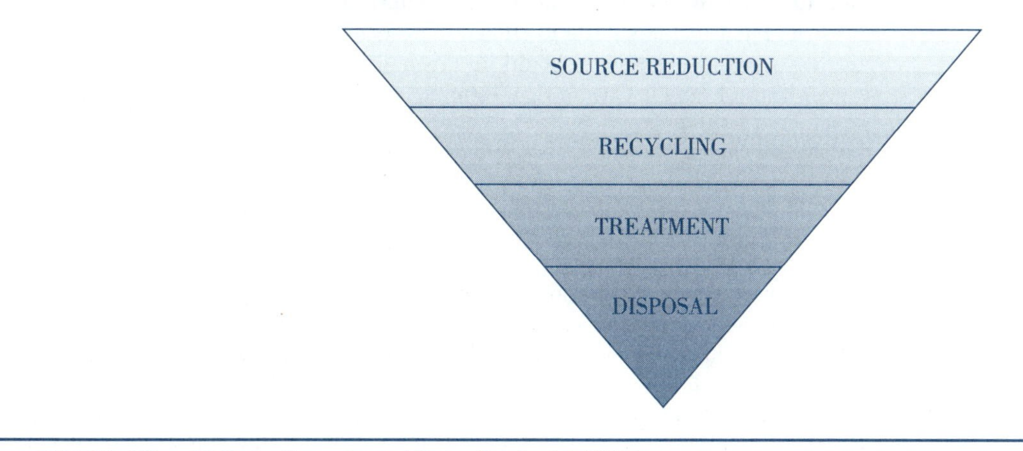

Source: U.S. EPA, Office of Pollution Prevention and Toxics (October 10, 2001b).

National Legislation Promoting Pollution Prevention

United States P2 Legislation

Beginning with the National Environmental Policy Act (NEPA) of 1969, the federal government in the United States has integrated pollution prevention into U.S. environmental legislation. One of the purposes of NEPA is "to promote efforts which will prevent or eliminate damage to the environment and the biosphere."[10]

To formalize pollution prevention as an integral part of U.S. policy, Congress passed the Pollution Prevention Act of 1990. Its provisions outline the priorities for national environmental policy as follows:

> The Congress hereby declares it to be the national policy of the United States that
>
> - pollution should be prevented or reduced at the source whenever feasible;
> - pollution that cannot be prevented should be recycled in an environmentally safe manner, whenever feasible;
> - pollution that cannot be prevented or recycled should be treated in an environmentally safe manner whenever feasible; and
> - disposal or other release into the environment should be employed only as a last resort and should be conducted in an environmentally safe manner.[11]

Notice that this legislation explicitly ranks pollution prevention as the primary objective, with management of residuals as a secondary objective to be achieved through recycling, treatment, and disposal—in that order. This hierarchy is depicted in Figure 21.4. The reference in the law to recycling as a secondary objective has been the source of some controversy, since some have inferred from this that recycling is a solution supported by pollution prevention. However, since recycling only postpones waste disposal, it is not a true preventive solution.[12]

[10]Pub. L. 91-190, sec. 2 (1970), 83 Stat. 852 (codified as 42 U.S.C. § 4321).
[11]Pollution Prevention Act of 1990, 42 U.S.C. § 13101(b). To access the complete text of this legislation online, visit **http://www.epa.gov/opptintr/p2home/p2policy/act1990.htm.**
[12]The exception is in-process recycling, which allows materials to be reused in the same process.

P2 objectives are also supported by other U.S. policies. For example, source reduction is explicitly listed as the first priority in the EPA's integrated waste management system that guides states' solid waste programs. Also, the Federal Insecticide, Fungicide, and Rodenticide Act (FIFRA) promotes toxic chemical use substitution by empowering the EPA to cancel or restrict the use of any pesticide found to pose an unreasonable risk. In so doing, markets for less harmful pesticides should be created. Similar rulings are given in the Toxic Substances Control Act (TSCA) for nonpesticide chemicals.

P2 Legislation in Other Nations

In addition to the United States, national laws promoting pollution prevention are prevalent in other countries as well, particularly in industrialized nations. A few examples will illustrate.

The European Union (EU) has developed a set of rules on permitting for all member nations, called the Integrated Pollution Prevention and Control (IPPC) Directive of 1996. This directive refers specifically to minimizing pollution in the EU and called for each of the 15 members to adjust their national legislation in accordance with the directive by October 1999.[13]

Environment Canada says that the cornerstone of the Canadian Environmental Protection Act (CEPA) of 1999 is pollution prevention. Similar to the United States, CEPA defines pollution prevention as:

> The use of processes, practices, materials, products, substances or energy that avoid or minimize the creation of pollutants and waste and reduce the overall risk to the environment or human health. (c. 33)

CEPA's preamble specifically states that the Canadian government is dedicated to making pollution prevention a national objective and implementing it as the priority approach to protecting the environment. The law also gives the Minister of the Environment additional authority to move toward sustainable development and to require pollution prevention plans from certain facilities. Also authorized in the law is the establishment of a national pollution prevention clearinghouse.[14]

Australia developed a National Strategy for Ecologically Sustainable Development (ESD), endorsed in 1992 by Heads of Government, which outlines a framework for the promotion of ESD throughout the country. Included is a discussion of waste minimization and management in which the stated challenge specifically refers to avoiding hazardous waste generation.[15] Australia also has adopted the preventive strategy of **cleaner production** as defined by the United Nations Environment Programme (UNEP). Included in Australia's rationale for adopting cleaner production is what it terms a preventive approach, in which it argues:

cleaner production
A preventive environmental strategy applied to products and processes to improve efficiency and reduce risk.

> It is cheaper and more effective to prevent environmental damage than attempt to manage or fix it. Prevention requires going upstream in the production process to prevent the source of the problem instead of attempting damage control downstream. Pollution prevention replaces pollution control.[16]

It is also the case that, while some nations have not yet established specific P2 laws, they have begun to move toward preventive objectives. For example, environmental laws enacted in Bahrain and Jordan encourage firms to adopt P2 procedures and decrease the generation of

---

[13] European Commission (April 26, 2002).
[14] CEPA 1999; Environment Canada (March 2000). For updates and further information, visit the CEPA Environmental Registry at **http://www.ec.gc.ca/ceparegistry.** The Web site for the Canadian Pollution Prevention Information Clearinghouse (CPPIC) is **http://www3.ec.gc.ca/cppic/en/index.cfm.**
[15] Australia, Ecologically Sustainable Development Steering Committee (December 1992).
[16] Australia and New Zealand Environment and Conservation Council (March 1999), p. 15. For further information, visit Environment Australia online at **http://www.erin.gov.au/index.html.**

waste. Similarly, China makes specific reference to cleaner production and pollution prevention in several of its environmental laws, including those dealing with air and water pollution.[17]

Corporate Experience with Pollution Prevention[18]

Recognizing the cost implications of environmental laws and observing the trend toward pollution prevention, some forward-thinking firms decided to respond in a progressive manner rather than simply complying with new mandates. One of the first U.S. companies to adopt this approach was 3M Corporation, which launched its Pollution Prevention Pays Program in 1975. Through this "3P Program," as it is known, the corporate giant designed an employee-based plan aimed at finding ways to reduce the company's pollution while maintaining or improving financial performance. Most of 3M's prevention strategies are based on technological improvements, such as product reformulation, process modification, equipment redesign, and waste recovery for resale.

Over the course of the company's 3P Program, over 4,700 separate projects have been launched by employees around the world. According to corporate officials, the payoffs from these efforts have been substantial. Since the inception of the 3P Program, 807,000 tons of pollution have been prevented through 1999. Furthermore, these preventive initiatives have yielded cost savings to the company of over $827 million.[19]

As the advantages of pollution prevention have become known, some corporations are forming cooperative arrangements within the private business sector. Other collaborative efforts are the result of corporations joining forces with environmental groups. Perhaps the best-known example is the joint effort between McDonald's and an environmental group, Environmental Defense—a coalition that began in 1990 with the formation of the Waste Reduction Task Force. The task force developed a Waste Reduction Action Plan (WRAP), completed in 1991, that eventually grew into over 100 separate projects over the subsequent 8-year period.

Ultimately, the task force devised a 42-step action plan, formulated a corporate environmental policy, and found ways to integrate waste reduction efforts throughout McDonald's organization. While most strategies were aimed at recycling through the McRecycle USA initiative, the task force also devised pollution prevention plans. Most of these have been aimed at reducing McDonald's product packaging, accomplished in large part by reducing the thickness of cups, napkins, trashcan liners, and carry-out bags. By 1998, paper use by the fast-food giant was reduced by 5.2 million pounds per year. Over the 1990 to 1999 period, as a result of reducing product packaging, McDonald's decreased its packaging wastes by 150,000 tons.[20]

Following the example of McDonald's, 3M, and other corporate leaders in environmentalism, many other companies have implemented their own pollution prevention programs with favorable results. By way of example, consider the selected results from corporate P2 efforts given in Table 21.1.

At a more aggregate level, a number of trade associations also have developed pollution prevention programs for their members. The American Petroleum Institute, for example, devised a set of 11 environmental principles for petroleum facilities, one of which specifically sets a goal for each member to reduce emissions and waste generation. The institute also publishes reports to communicate trends in waste generation and management throughout the industry and to update members on waste minimization practices.[21] A similar plan has been devised for the chemical industry, a major contributor to environmental pollution. This plan, called the

---

[17] United Nations, Environment Programme (1999); State Economic and Trade Commission (June 14, 2002).

[18] General information for the following discussion is drawn from U.S. EPA, Office of Pollution Prevention (October 1991), chap. 3.

[19] More on 3M's corporate environmental programs is available at the company's environmental Web page at **http://www.3m.com/about3m/environment/index.jhtml.**

[20] U.S. EPA, Office of Solid Waste and Emergency Response (spring 1992), p. 1, 3; U.S. EPA, Office of Solid Waste and Emergency Response (October 1993), p. 13; McDonald's Corporation (May 1999); McDonald's Corporation (June 18, 2002).

[21] To learn more, visit the API's home page at **http://api-ec.api.org** and click on Environmental Commitment.

Table 21.1 ***Results of Selected Corporate Pollution Prevention (P2) Programs***

Company	Major Results
Eastman Kodak Company	Compared to 1997 levels, achieved a 42 percent emissions reduction of 30 priority chemicals and a 12 percent reduction in carbon dioxide emissions by 2001; reduced annual worldwide manufacturing energy use by 18 percent in 2001 and annual water usage at manufacturing plants by 23 percent; relative to a 1999 baseline, achieved a 58 percent average reduction of heavy metals (including lead and mercury) in Kodak products.
E. I. Du Pont de Nemours and Company	In 2001, reduced worldwide greenhouse gas emissions by 63 percent relative to 1990; global emissions of air toxics and carcinogens in 2001 declined by 76 percent and 87 percent, respectively, relative to 1987 levels; hazardous wastes in 2001 were 37 percent below 1990 levels.
Herman Miller Inc.	Adopted a packaging reduction program and eliminated over 370 tons of wood pallet waste, nearly 270 tons of corrugated cartons, and nearly 8 tons of polystyrene packaging filler in 1999; reduced landfill use by 65 percent relative to 1994.
McDonald's Corporation	Over the 1990–1999 period, decreased packaging wastes by 150,000 tons. By 1998, reduced paper use by 5.2 million pounds per year.
3M Corporation	Between 1990 and 2000, reduced volatile organic compound (VOC) emissions by 88 percent, water pollutants by 82 percent, and solid wastes by 24 percent. Of these reductions, pollution prevention activities accounted for 55 percent of the VOC reduction, 43 percent of the effluent reduction, and 76 percent of the solid waste reduction.

Sources: 3M Corporation (June 18, 2002); E. I. Du Pont de Nemours and Company (2001); Eastman Kodak Company (2001); McDonald's Corp. (May 1999); McDonald's Corp. (June 18, 2002); U.S. EPA, Office of Solid Waste and Emergency Response (March 2000), pp. 9, 12.

Responsible Care Program, was developed by the American Chemistry Council (ACC) (formerly the Chemical Manufacturers Association) and is the subject of Application 21.3.

Using Economic Analysis to Implement Pollution Prevention

Firms must determine when preventive approaches should be employed, to what extent, and through which strategy. Economic criteria and their associated decision rules are just as applicable to guiding the use of preventive strategies as they are to any other environmental policy instrument.

Cost-Effectiveness Criterion

Whether or not a firm adopts pollution prevention as part of its business strategy will depend to a large extent on the associated cost relative to other options, such as the use of treatment or abatement technologies. The goal is to select the least-cost option to achieve **cost-effectiveness** and enhance profit. Similarly, if the firm determines that pollution prevention *is* a cost-effective option, it then must select among the available strategies to implement it using the cost-effectiveness criterion as a guide. If, for example, the firm can access nontoxic raw materials more cheaply than converting its production processes, it will pursue that less costly strategy.

Given the importance of relative costs in this decision-making process, government can help promote pollution prevention through various means. For example, it could encourage the development of inexpensive nontoxic chemical substitutes and cost-effective production changes through grants or subsidies. It might also develop ways to enhance the communication of effective pollution prevention strategies to industry. Another approach is for government to

21.3 *The Chemical Industry's Response to Pollution Prevention: The Responsible Care® Program*

Application

As a major contributor to the nation's hazardous waste stream, the chemical industry developed a pollution prevention program that signals the industry's commitment to a cleaner environment. In 1988, the U.S. industry's trade alliance, the American Chemistry Council (ACC) (then called the Chemical Manufacturers Association), launched its Responsible Care® Program as a comprehensive chemical management plan to be adopted by its membership. Great Britain's counterpart to the ACC, the Chemical Industries Association, also joined the ranks of trade associations working to prevent pollution. Their Responsible Care Program came on line in March 1989—the first such plan in Europe. For each nation, the objectives of their Responsible Care Programs are simple but powerful—to protect the health and safety of the workforce and to protect the environment from contamination.

Following the lead of the Canadian Chemical Producers Association, which had established its plan in 1985, the U.S. program ushered in a series of principles to guide its members in reducing both waste generation and dangerous releases into the environment. Currently, the American Chemistry Council (2002) adheres to the following 10 guiding principles:

- To seek and incorporate public input regarding our products and operations.
- To provide chemicals that can be manufactured, transported, used and disposed of safely.
- To make health, safety, the environment and resource conservation critical considerations for all new and existing products and processes.
- To provide information on health or environmental risks and pursue protective measures for employees, the public and other key stakeholders.
- To work with customers, carriers, suppliers, distributors, and contractors to foster the safe use, transport and disposal of chemicals.
- To operate our facilities in a manner that protects the environment and the health and safety of our employees and the public.
- To support education and research on the health, safety and environmental effects of our products and processes.
- To work with others to resolve problems associated with past handling and disposal practices.
- To lead in the development of responsible laws, regulations, and standards that safeguard the community, workplace and environment.
- To practice Responsible Care® by encouraging and assisting others to adhere to these principles and practices.

Since the program's inception, improvements in the industry's environmental, health, and safety performance have been achieved. Furthermore, the program has helped the less than favorable public image of the industry. However, according to ACC president Fred Webber, more needs to be done to "put a more positive face on the chemical industry." Using the Responsible Care Program as a foundation, the ACC's current strategy is to communicate to the public the benefits of the industry without being defensive. This is being accomplished by more aggressive public relations activities, such as sending the message about "better living through chemistry." However, to gain public approval over the long run, the industry must remain committed to the program's guiding principles.

Sources: American Chemistry Council (June 18, 2002); U.S. EPA, Office of Pollution Prevention (October 1991), pp. 40–41; Chemical Manufacturers Association (September 1990), pp. 9–15; Simmons and Wynne (1993); Schmitt (July 5–July 12, 2000).

assist firms in making accurate cost assessments about pollution prevention strategies. This is precisely the motivation of **environmental accounting.** In the United States, the Environmental Accounting Project was launched in 1992. This project is a cooperative effort among government, business, and other stakeholders aimed at improving managerial accounting and capital budgeting practices to more fully incorporate environmental considerations. In so doing, the project expects that more businesses will recognize the advantages of using preventive technologies and processes.[22]

Efficiency Criterion

It is also the case that **benefit-cost analysis** can help determine the extent to which a given preventive strategy should be employed. Qualitatively, we can identify the associated benefits to the firm as abatement cost savings, the avoidance of regulatory penalties for noncompliance, and the gain in revenues associated with presenting an environmentally responsible image to consumers. On the cost side, the firm has to assess the search costs of identifying available options, the administrative costs in analyzing these options, engineering design expenses, and the expenditures to implement the strategy, such as retooling or capital investment. To achieve the **allocatively efficient** level at which a pollutant is eliminated, the decision rule is the same as in other contexts. The firm should find the point where the marginal cost of doing so is exactly offset by the marginal benefit.[23]

Strategic Initiatives and Programs

Sustainable development and approaches like pollution prevention and industrial ecology call for significant change in how consumers and firms make decisions and in particular how firms design, produce, and market their products. If businesses perceive these changes as counter to their self-interest, they will be met with resistance at best. To foster a positive evolution in thinking and business planning, governments are working with industry and environmental groups to share information, to foster technological development, and to find solutions that satisfy both environmental and business objectives.

Although the perspective of sustainable development is global, initiatives promoting sustainability have to begin within national borders. Just as is the case for international agreements, these domestic initiatives require cooperation, but of a different sort. Often, the collaborative effort is launched by government but includes stakeholders from other economic sectors, and typically such efforts are based on voluntary partnerships. In the United States, such arrangements are generally launched by the EPA. An overview of some of these programs is given in Table 21.2.[24] Of course, such programs are not exclusive to the United States. Three important initiatives that are emerging in many nations are **Extended Product Responsibility (EPR), Design for the Environment (DfE),** and the **Green Chemistry Program.** A brief discussion of these efforts, which are interrelated, follows.

Extended Product Responsibility (EPR)
A commitment by all participants in the product cycle to reduce any life-cycle environmental impacts of products.

Extended Product Responsibility (EPR)[25]

Extended Product Responsibility (EPR) refers to efforts aimed at identifying and reducing any life-cycle environmental impacts of products. Sometimes known as product stewardship, EPR is closely linked to sustainable development and is preventive in orientation. The underlying premise of EPR is that *all* participants in the product chain—designers, manufacturers, distributors, consumers, recyclers, remanufacturers, and disposers—are re-

---

[22] U.S. EPA, Office of Pollution Prevention and Toxics (October 1993). Visit **http://www.epa.gov/opptintr/acctg/** for more information on the EPA's Environmental Accounting Project.

[23] For an economic analysis of pollution prevention and the hierarchy given in the Pollution Prevention Act of 1990, see Helfand (October 1994).

[24] Visit **http://www.epa.gov/partners/** to learn more about the many partnership programs sponsored by the EPA, collectively called Partners for the Environment, or go to the Web site on Industry Partnerships at **http://www.epa.gov/epahome/industry.htm.**

[25] This discussion is drawn from U.S. EPA, Office of Solid Waste and Emergency Response (December 1998); and from U.S. EPA, Office of Solid Waste (August 29, 2001).

Table 21.2 ***Selected Voluntary Programs for Pollution Prevention in the United States***

Program	Year Launched	Objective
33/50 Program	1991	To reduce releases and transfers of 17 priority chemicals by 33 percent by 1992 and by 50 percent by 1995 compared to a 1988 baseline. Completed in 1995.
Environmental Accounting Project (EAP) **(http://www.epa.gov/opptintr/acctg)**	1992	To improve corporate understanding of environmental costs and how to integrate them into business operations.
Design for the Environment (DfE) **(http://www.epa.gov/dfe)**	1992	To help companies integrate environmental considerations into product and process design.
Green Chemistry Program **(http://www.epa.gov/greenchemistry)**	1992	To support chemical technologies that reduce or eliminate generation of hazardous substances.
Pesticide Environmental Stewardship Program (PESP) **(http://www.epa.gov/oppbppd1/PESP/)**	1994	To promote pesticide risk reduction through improved pesticide stewardship.
Indoor Air Quality (IAQ) **(http://www.epa.gov/iaq)**	1995	To promote low-cost, easy approaches to reducing risks associated with indoor air quality.
Adopt Your Watershed **(http://www.epa.gov/adopt)**	1994	To facilitate citizen involvement in protecting local watersheds.
Energy Star **(http://www.energystar.gov/default.shtml)**	1994	To promote energy efficiency through new building and product design and practices.
WasteWise **(http://www.epa.gov/wastewise)**	1994	To support organizations in finding ways to eliminate costly municipal solid waste.
Extended Product Responsibility (EPR) **(http://www.epa.gov/epr)**	Not available	To promote identifying and reducing any life-cycle environmental impacts of products.

Sources: U.S. EPA, Office of Policy, Economics, and Innovation (January 2001), table 10.3; information from the Web pages listed in table.

http://

sponsible for a product's effect on the environment. This is a more extensive approach than **extended *producer* responsibility** adopted by some Asian countries, Europe, and certain of the Canadian provinces, which places essentially all the onus of product responsibility on the firm.

Under extended product responsibility (EPR), all players in a product cycle are expected to participate. Consumers, for example, should make responsible market decisions when choosing a product. They can vote with their dollars for companies that use less product packaging or less energy. Of course, manufacturers typically have the largest role to play, and under EPR guidelines, they are expected to devise ways to make production decisions with the environment in mind by addressing the following:

- Raw materials selection: reducing the amount and toxicity of raw materials used
- Production impacts: seeking opportunities to reduce waste and energy consumption
- Product use: designing products to use fewer resources and less waste
- Products at end of life: offering take-back programs whereby used products are collected for recycling or **remanufacturing**

Some countries have federal laws to require certain of these EPR approaches, such as the German Packaging Ordinance of 1991 (discussed in Chapter 18), which requires businesses to collect and recycle their own packaging. In the United States, EPR is essentially a voluntary program, though some state programs mandate EPR-like efforts, such as requiring manufacturers to take back certain types of used batteries. But even in the absence of federal laws, some firms are participating in EPR because it is in their best interest to do so. For example, firms engage in **remanufacturing** because it is lucrative as well as being good for the environment. See Application 21.4 for more information.

remanufacturing Collection, disassembly, reconditioning, and reselling of the same product.

Design for the Environment (DfE)[26]

Design for the Environment (DfE) An initiative that promotes the use of environmental considerations along with cost and performance in product design and development.

Based on the engineering notion of "design for X," where X might be any desirable product attribute such as durability, **Design for the Environment (DfE)** is an initiative that promotes the use of environmental considerations along with cost and performance in product design and development. This effort directly supports a **cyclical flow of materials,** inherently recognizing that all stages of the product life cycle are important to environmental protection. Look back at Figure 21.3 and note the flows labeled "Design for recycling" and "Design for reuse." These flows are important elements of a DfE program and have been adopted by a number of major corporations like 3M, BMW, Dell Computer, and Hewlett-Packard.

Hewlett-Packard launched its DfE program in 1992. Among its goals are to design products and packaging that use fewer resources and less energy, to develop products that facilitate disassembly for reuse or recycling, and to lower emissions and other wastes from its manufacturing processes. Part of BMW's DfE program includes efforts to develop technologies for reducing vehicle weight, making cars more fuel efficient. One such technology is called carbon fiber reinforced polymers (CFRP), which according to BMW, shows the greatest potential for weight-saving of automobile bodies. The German automaker claims that CFRP, which is also corrosion resistant and strong, is 50 percent lighter than steel and 30 percent lighter than aluminum.[27]

The DfE program in the United States is voluntary and hence relies on partnerships formed between the EPA and industry. Among the partnered industries involved in this effort to date are printing, computer display screens, dry cleaning, and automotive refinishing. The idea is to provide decision makers with the information needed to make knowledgeable decisions that integrate environmental considerations with business decisions. See Table 21.3 for an overview on what is involved in the DfE process. Among the expected benefits of using DfE are lower health and ecological risks, a reduced regulatory burden, improved communication and collaboration among stakeholder organizations, and improved market opportunities.

Green Chemistry Program[28]

Green Chemistry Program An initiative that promotes the development and application of innovative chemical technologies to achieve pollution prevention.

Another emerging trend in environmental protection is the development of **Green Chemistry Programs,** also known as benign chemistry or sustainable chemistry programs. A **Green Chemistry Program** promotes the research, development, and application of innovative chemical technologies to achieve pollution prevention in ways that are both scientifically grounded and cost-effective. Like DfE initiatives, green chemistry recognizes the importance of a product's life cycle in pollution prevention. By finding alternatives to hazardous chemicals that are safer or less toxic, health and ecological risks are reduced, production processes are safer, and final products pose less of a threat at the end of their economic lives. To realize their objectives, green chemistry programs support research, education, conferences, and international activities in environmentally benign chemistry.

[26] This discussion is drawn from U.S. EPA, Office of Pollution Prevention and Toxics (October 10, 2001a); and U.S. EPA, Office of Prevention, Pesticides, and Toxic Substances (March 2002).
[27] U.S. EPA, Office of Solid Waste and Emergency Response (October 1998), p. 5; BMW Group (2001), p. 43.
[28] This discussion is drawn from U.S. EPA, Office of Pollution Prevention and Toxics (December 2001); U.S. EPA, Office of Pollution Prevention and Toxics (August 2001).

21.4 Remanufacturing: A Lucrative Approach to Pollution Prevention

Application

In an effort to be environmentally responsible, improve corporate image, and save costs, some firms are participating in a practice known as **remanufacturing.** Combined with a take-back program, remanufacturing refers to a process whereby a used product is collected, disassembled (sometimes called demanufacturing), reconditioned to "as good as new" working condition, reassembled, and resold.

How is this different from recycling? According to remanufacturing industry expert Robert T. Lund, of Boston University, remanufacturing attempts to capture the value-added of energy, labor, and manufacturing when the good was originally produced. Recycling, on the other hand, captures only the recoverable raw materials in the product, which represent a much smaller proportion of the good's value. Energy savings are also substantial. Lund estimates that remanufacturing saves 120 trillion Btu of energy each year on a worldwide scale, which translates to about $500 million in energy savings.

Though not all products are good candidates for this process, there are many remanufactured goods on the market today, including automotive parts, toner cartridges, photocopiers, personal computers, tires, telephones, and medical equipment. In the United States, the largest single remanufacturer is the U.S. Department of Defense, which refurbishes weaponry and military equipment on a regular basis to cut costs. Both small and large companies also participate in this process.

One example is Xerox Corporation, which began its remanufacturing program by encouraging its customers to return used copy machines. Once the products are in house, Xerox employees begin the disassembly process, sorting parts good enough to meet remanufacturing criteria. These are then used to build new products, using the same assembly lines set up for new goods. Interestingly, the firm's leasing program facilitated the effort, since it called for products to be returned at the end of the lease period. Instead of sitting idly in warehouses, the used goods became obvious candidates for remanufacture. The next advance occurred when Xerox used life cycle assessment and adapted its design program to facilitate remanufacturing.

Today, the Connecticut-based firm takes back printers, toner bottles, and other products as well. These efforts reduce waste generation, preventing the pollution normally associated with disposal or incineration. Because of its remanufacturing program, Xerox saved over 1.1 million pounds of plastic and 88 million pounds of metal in 1997 alone.

Another interesting application of remanufacturing is being undertaken by Miller SQA, a subsidiary of the well-known, environmentally conscious Herman Miller Inc., one of the largest office furniture makers in the United States. Miller SQA (which stands for "simple, quick, and affordable") was launched from the parent company's buy-back program. Buy backs were encouraged by offering discounts on new purchases to customers who returned their old furniture. Today, 50–75 percent of the used furniture is remanufactured and sold by Miller SQA in its *As New* product line. These products must meet the same quality standards as new Miller furniture.

Remanufacturing has grown into a substantial market, estimated to consist of over 70,000 firms with annual aggregate sales of about $53 million. These data are not surprising given the estimated cost savings associated with this process. A case in point is Ford Motor Company, which estimates that it has saved $1.2 million since 1991 by participating in a return program with a toner cartridge remanufacturer. On a more aggregate level, estimates suggest that remanufacturing is 40–65 percent less costly than building a new product. All of this underscores how both economic gains and environmental benefits are powerful motivations for continued growth in remanufacturing.

Sources: U.S. EPA, Office of Solid Waste and Emergency Response (October 1998); U.S. EPA, Office of Solid Waste and Emergency Response (May 1997b); Ginsburg (April 16, 2001); The Remanufacturing Institute (June 19, 2002), citing Lund in Klein and Miller (1993).

Table 21.3 ***The Design for the Environment (DfE) Process***

Steps in the DfE Process	Description
Identify	Identifying available technologies, products, and processes that can perform a particular function as well as any pollution prevention opportunities.
Assess	Assessing and evaluating the risk, performance, and cost tradeoffs of alternative processes and emerging technologies.
Disseminate	Disseminating this information to the entire industry.
Promote	Promoting the use of cleaner technologies, safer chemicals, and other findings through incentives and other mechanisms.

Source: U.S. EPA, Office of Pollution Prevention and Toxics (October 10, 2001a).

In 1998, the Advisory Group of Risk Management for the Organisation for Economic Cooperation and Development (OECD) gave its support for a new sustainable chemistry project for OECD member countries. Among the nations electing to coordinate the effort are Germany, Italy, Japan, Mexico, and the United States. Ultimately, a steering group for the OECD Sustainable Chemistry Initiative was formed, consisting of 40 representatives from 10 nations. The group developed a detailed work plan to implement what it believes are the highest priority activities. These include supporting the research and development of innovative sustainable chemistry technologies, disseminating information related to sustainable chemistry, and integrating principles of sustainable chemistry into chemistry education.

Just as DfE and EPR programs rely on partnering with various economic sectors, so too do green chemistry initiatives rely on collaboration with business, academia, government agencies, research centers, laboratories, and other relevant organizations. In the United States, the EPA administers the nation's green chemistry program under its DfE program, since the two initiatives are complementary. It has established partnerships with companies like Eastman Kodak and the Dow Chemical Company, with academic institutions including the University of Massachusetts, Boston, and with trade associations like the American Petroleum Institute and the American Chemistry Council. Today, the U.S. Green Chemistry Program consists of four major projects: Green Chemistry Research, the Presidential Green Chemistry Challenge, Green Chemistry Education, and Scientific Outreach, which are briefly described in Table 21.4.

Disseminating Information and Technology on a Global Scale

Strategic advances and related environmental technologies are important to achieving economic growth while preserving and protecting the natural environment. However, they are generally developed and implemented within industrialized nations and hence need to be communicated globally, particularly to third world countries. In order for this to occur, there must be an infrastructure for sharing knowledge, and there must be an awareness of environmental issues.

Technology Transfer

technology transfer The advancement and application of technologies and strategies on a global scale.

It is often argued that **technology transfer** is critical to consistent progress toward sustainable development. This concept refers to the advancement and application of technologies and management strategies throughout the world. Environmental technologies cover a broad range of products and services. Included are the so-called dark green technologies that deal with the control, abatement, monitoring, and remediation of pollution. There are also more in-

Table 21.4 *Major Projects of the U.S. Green Chemistry Program*

Project	Description
Green Chemistry Research	Provides grant funding to support fundamental research in green chemistry; in 1992 and 1994 EPA signed a Memorandum of Understanding with the National Science Foundation (NSF) to jointly fund these efforts.
Presidential Green Chemistry Challenge	Recognizes exceptional accomplishments in benign chemistry through an annual awards program.
Green Chemistry Education	Supports educational endeavors, such as course development for training professional chemists in industry through a collaborative formed with the National Pollution Prevention Center (NPPC), the Partnership for Environmental Technology Education, and the American Chemistry Society.
Scientific Outreach	Supports various outreach projects, including publishing in scientific journals, distributing computational databases, and organizing important workshops or scientific meetings.

Source: U.S. EPA, Office of Pollution Prevention and Toxics (August 2001).

direct but nonetheless important advances, dubbed light green technologies. These refer to strategies or production changes that benefit the environment, even though that is not their primary intent. An example is the use of electronic mail in place of hard-copy memos, a change that reduces waste generation.[29] Effecting technology transfer relies on a number of interdependent factors—among them, research, physical capital investment, communication, financial resources, and perhaps most important, education.

Environmental Literacy

In order for knowledge and technology to be effectively disseminated, people everywhere need to be made aware of environmental risks and the importance of responding to those risks in a responsible way. The promotion of environmental education across the globe has grown over time and was an important theme at the 1992 Earth Summit in Rio. *Agenda 21* specifically references the importance of education, public awareness, and training in helping to implement the global agenda.[30] The aim is to advance **environmental literacy** across all regions of the world. This can be achieved through communication and education within and among nations. Without this awareness, society will not understand the need for change, will tend not to support it, and may be unwilling to participate in the process.

environmental literacy
Awareness of the risks of pollution and natural resource depletion.

Conclusions

Over the past two decades, many nations have made measurable progress in moving closer to what have become universal environmental objectives, specifically, environmental quality, biodiversity, and sustainable development. With greater emphasis on pollution prevention, product life cycle, and environmental literacy, many believe the progress can continue over the long run. That said, there is still much more to be done. Aggregate statistics disguise those regions where environmental pollution is particularly severe. There are entire nations in Asia, Africa, South America, and Central America that face extremely poor air quality, contaminated drinking water, and serious degradation of their land resources.

[29] "What Is Environmental Technology" (fall 1994).
[30] Parson, Haas, and Levy (October 1992).

The need for setting a global environmental agenda and implementing it is more than apparent. Identifying an appropriate course of action is not easy, and it takes time. This collective effort calls for cooperation from every market sector and from every nation, a process that is now under way. As this important transition evolves, society must move forward to restore what can be repaired, to launch initiatives to prevent further degradation, and to educate people everywhere about the importance and fragility of the natural environment. This commitment was advanced at the 1992 Earth Summit in Rio and is articulated in the preamble of *Agenda 21:*

> Humanity stands at a defining moment in history. We are confronted with a perpetuation of disparities between and within nations, a worsening of poverty, hunger, ill health, and illiteracy, and the continuing deterioration of the ecosystems on which we depend for our well-being. However, integration of environment and development concerns and greater attention to them will lead to the fulfilment of basic needs, improved living standards for all, better protected and managed ecosystems and a safer, more prosperous future.[31]

Summary

- Industrial ecology is a multidisciplinary systems approach to the flow of materials and energy between industrial processes and the environment. Its main purpose is to promote the use of recycled wastes from one industrial process as inputs in another.
- A linear flow of materials assumes that materials run in one direction, entering an economic system as inputs and leaving as residuals. This "cradle-to-grave" flow emphasizes use, waste generation, and disposal. Most national policy focuses on abating residuals at the end of the flow, not adequately addressing long-run consequences.
- Product design, manufacturing processes, and energy use can be modified to achieve a cyclical flow of materials, or a "cradle-to-cradle" approach. A life cycle assessment (LCA) can be used to examine the environmental impacts of a good at all product stages.
- The real-world outcome of implementing a closed system is the formation of an industrial ecosystem, whereby residuals from one or more manufacturing processes are reused as inputs for others. The most well-known example is the Kalundborg, Denmark, ecosystem.
- Pollution prevention (P2) is a long-term approach aimed at reducing the amount or toxicity of residuals released to the environment.
- Two major preventive objectives are source reduction and toxic chemical use substitution. Among the techniques that can help achieve these objectives are source segregation, raw materials substitution, changes in manufacturing processes, and product substitution.
- National laws promoting pollution prevention are prevalent, particularly in industrialized nations. To formalize pollution prevention as an integral part of U.S. policy, Congress passed the Pollution Prevention Act of 1990. The European Union (EU) has developed a set of rules for all member nations called the Integrated Pollution Prevention and Control (IPPC) Directive of 1996.
- The economic criteria of cost-effectiveness and allocative efficiency, along with their associated decision rules, can be used to guide a firm's use of preventive strategies.
- To support sustainable development and the use of preventive approaches, various initiatives are emerging in many countries, including Extended Product Responsibility (EPR), Design for the Environment (DfE), and Green Chemistry.

---

[31] United Nations, Division for Sustainable Development, ch. 1.1 (June 29, 2000).

- Design for the Environment (DfE) promotes the use of environmental considerations along with cost and performance in product design and development.
- A Green Chemistry Program promotes the research, development, and application of innovative chemical technologies to achieve pollution prevention in ways that are scientifically grounded and cost-effective.
- Critical to consistent progress toward sustainable development is technology transfer, which refers to the advancement and application of technologies and management strategies across economic sectors throughout the world.
- Environmental literacy, achieved through communication and education, is part of an effective strategy to preserve and protect the earth's resources.

Key Concepts

industrial ecology
materials balance model
residual
linear flow of materials
cyclical flow of materials
life cycle assessment (LCA)
ISO 14000 standards
industrial ecosystem
pollution prevention (P2)
source reduction
toxic chemical use substitution
source segregation
raw materials substitution
changes in manufacturing processes
product substitution
cleaner production
Extended Product Responsibility (EPR)
remanufacturing
Design for the Environment (DfE)
Green Chemistry Program
technology transfer
environmental literacy

Use the Key Concepts listed above to begin your search for additional articles and information using the InfoTrac College Edition database.

Review Questions

1. Assume you are a city planner working on a new industrial park and contemplating the use of an industrial ecosystem. Discuss the major advantages and disadvantages of an industrial ecosystem that you would consider in making your decision.
2. Choose a product that negatively affects the environment, assuming a linear flow of materials. Then, use a cyclical materials flow approach, and conduct a hypothetical life cycle assessment (LCA), pointing out at least two preventive initiatives that would reduce environmental risk.
3. a. Identify the economic incentives that motivate private firms to engage in pollution prevention activities.
   b. How might the government devise policy initiatives to exploit these natural incentives?
4. Visit the Web site of Environmental Defense, and review the list of their recent alliance partnerships at **http://www.environmentaldefense.org/alliance/partners index.html.** Select one, and summarize the cooperative efforts between Environmental Defense and a private firm. Identify some of the potential environmental and economic benefits associated with that partnership.
5. Extended Product Responsibility (EPR) assumes that all participants in the product cycle play a role in finding ways to reduce environmental risk. Identify the specific role played by the average consumer in this effort.
6. Environmental technology is argued to be an important element in society's effort to achieve sustainable development.

http://

a. Choose a specific market-based instrument that likely would encourage the advance of dark green technologies *and* improve U.S. exports of these goods and services. Explain using economic analytical tools.
b. Now propose a different market-based instrument that would foster the use of more light green technologies domestically and internationally.

Additional Readings

Alberini, A., and K. Segerson. "Assessing Voluntary Programs to Improve Environmental Quality." *Environmental and Resource Economics* 22 (2002), pp. 157–84.

Allen, Scott. "The Greening of McDonald's." *Boston Globe,* January 24, 2000, pp. C1, C3.

Anastas, Paul T., and Joseph J. Breen. "Design for the Environment and Green Chemistry: The Heart and Soul of Industrial Ecology." *Journal of Cleaner Production* 5 (1997), pp. 97–102.

Ayres, Robert U., and Leslie W. Ayres, eds. *A Handbook of Industrial Ecology.* Northampton, MA: Elgar, 2002.

Beardsley, Dan, Terry Davies, and Robert Hersh. "Improving Environmental Management." *Environment* 39 (September 1997).

Bylinsky, Gene. "Manufacturing for Reuse." *Fortune,* February 6, 1995, pp. 102–12.

DeSimone, Livio D., and Frank Popoff. *Eco-Efficiency: The Business Link to Sustainable Development.* Cambridge, MA: MIT Press, 1997.

Flynn, Julia, Zachary Schiller, John Carey, and Ruth Coxeter. "Novo Nordisk's Mean Green Machine." *Business Week,* November 14, 1994, pp. 72–75.

Geiser, Kenneth. *Materials Matter: Toward a Sustainable Materials Policy.* Cambridge, MA: MIT Press, 2001.

Kneese, Allen V. "Industrial Ecology and 'Getting the Prices Right.'" *Resources* 130 (winter 1998).

Köhn, Jörg, John Gowdy, and Jan van der Straaten, eds. *Sustainability in Action.* Northampton, MA: Elgar, 2001.

Lesourd, Jean-Baptiste, and Steven G. M. Schilizzi. *The Environment in Corporate Management: New Directions and Economic Insights.* Northampton, MA: Elgar, 2002.

Martin, Shelia A., Robert A. Cushman, Keith A. Weitz, Aarti Sharma, and Richard C. Lindrooth. "Applying Industrial Ecology to Industrial Parks: An Economic and Environmental Analysis." *Economic Development Quarterly* 12 (August 1998), pp. 218–237.

O'Brien, Mary. *Making Better Environmental Decisions: An Alternative to Risk Assessment.* Cambridge, MA: MIT Press, 2000.

Oldenburg, Kirsten U., and Joel S. Hirschhorn. "Waste Reduction: A New Strategy to Avoid Pollution." *Environment* 29 (March 1987), pp. 16–20, 39–45.

Preston, Lynelle. "Sustainability at Hewlett-Packard: From Theory to Practice." *California Management Review* 43 (spring 2001), pp. 26–37.

Tibbs, Hardin. *Industrial Ecology: An Environmental Agenda for Industry.* Emeryville, CA: Global Business Network, 1993.

Related Web Sites

Canada's National Office of Pollution Prevention	**http://www.ec.gc.ca/nopp/english/index.cfm**
Canadian Environmental Protection Act (CEPA) Environmental Registry	**http://www.ec.gc.ca/ceparegistry**
Design for the Environment (DfE)	**http://www.epa.gov/dfe/**
Environment Australia	**http://www.erin.gov.au/index.html**
EnviroSense: Partners for the Environment	**http://es.epa.gov/partners**
Extended Product Responsibility	**http://www.epa.gov/epr/**
Green Chemistry Program	**http://www.epa.gov/opptintr/greenchemistry**
Industrial ecology, Center of Excellence for Sustainable Development	**http://www.sustainable.doe.gov/business/indeco.shtml**
Industrial Ecosystem Case Studies	**http://www.smartgrowth.org/library/eco_ind_case_intro.html**
International Organization for Standardization (ISO)	**http://www.iso.org/iso/en/ISOOnline.frontpage**
Life-Cycle Assessment (LCA), National Risk Management Research Laboratory	**http://www.epa.gov/ORD/NRMRL/lcaccess.index.htm**
McDonald's Corporation environmental Web page	**http://www.mcdonalds.com/countries/usa/community/environ/index.html**
Pollution Prevention Act of 1990	**http://www.epa.gov/opptintr/p2home/p2policy/act1990.htm**
Pollution Prevention Information Clearinghouse (CPPIC), Canada	**http://www3.ec.gc.ca/cppic/en/index.cfm**
Pollution Prevention Information Clearinghouse (PPIC), United States	**http://www.epa.gov/opptintr/library/ppicindex.htm**
Symbiosis Institute, Kalundborg, Denmark	**http://www.symbiosis.dk/tsi_uk.htm**
3M Corporation's environmental Web page	**http://www.3m.com/about3m/environment/index.jhtml**
U.S. EPA, Environmental Accounting Project (EAP)	**http://www.epa.gov/opptintr/acctg/**
U.S. EPA, Office of Pollution Prevention and Toxics	**http://www.epa.gov/opptintr/p2home/**
U.S. EPA, Partners for the Environment	**http://www.epa.gov/partners/**
U.S. Office of Industrial Technologies	**http://www.oit.doe.gov**
World Business Council for Sustainable Development	**http://www.wbcsd.org**

Appendix

Exhibits for Chapters 7, 8, 10, and 12–17

Exhibit A.1 *Classification of Biases Encountered in Contingent Valuation Studies (from Chapter 7)*

General Biases	
Strategic:	An individual may have an incentive *not* to reveal his or her true preferences about an environmental good when responding to questions about willingness to pay (WTP). This bias may arise from the free-ridership problem typically associated with public goods.
Information:	If there is insufficient information about the commodity being valued, the individual's WTP response may not be equivalent to their actual WTP.
Hypothetical:	Because the market is hypothetical, the respondent may view the questions as unrealistic and respond with an equally unrealistic estimate of WTP.
Survey Instrument–Related Biases	
Starting point:	Some survey instruments use predefined ranges of values to guide responses. The starting points of these ranges can influence the respondent's answers about WTP.
Payment vehicle:	To make the responses more factual, the survey questions about value are often tied to a specific payment vehicle, such as an increase in taxes or an adjustment on a utility bill. The selection of the payment vehicle used in the survey may influence how an individual responds to questions about WTP.
Procedural Biases	
Sampling:	Problems may arise due to the specific sampling procedure used by the researcher.
Interviewer:	The respondent's answers may be influenced by the individual asking the questions.

Source: Adapted from Smith and Desvousges (1986), p. 73, fig. 4–1.

Exhibit A.2 *Pollution Abatement Expenditures for Selected Industries (from Chapter 8)*

As regulatory controls have been put in place, industry has had to absorb the costs of abatement technology. The accompanying figures present abatement cost data for a selection of U.S. industries over the 1975–1994 period. The magnitude of these costs confirms that private industry has a vested interest in how environmental policy is formulated. Some industries are more directly affected than others. Look at the expenditures for the chemical and allied products group. These account for the largest industrial share of abatement expenditures—a direct reflection of the amount and toxicity of pollution released by this industry to all environmental media.

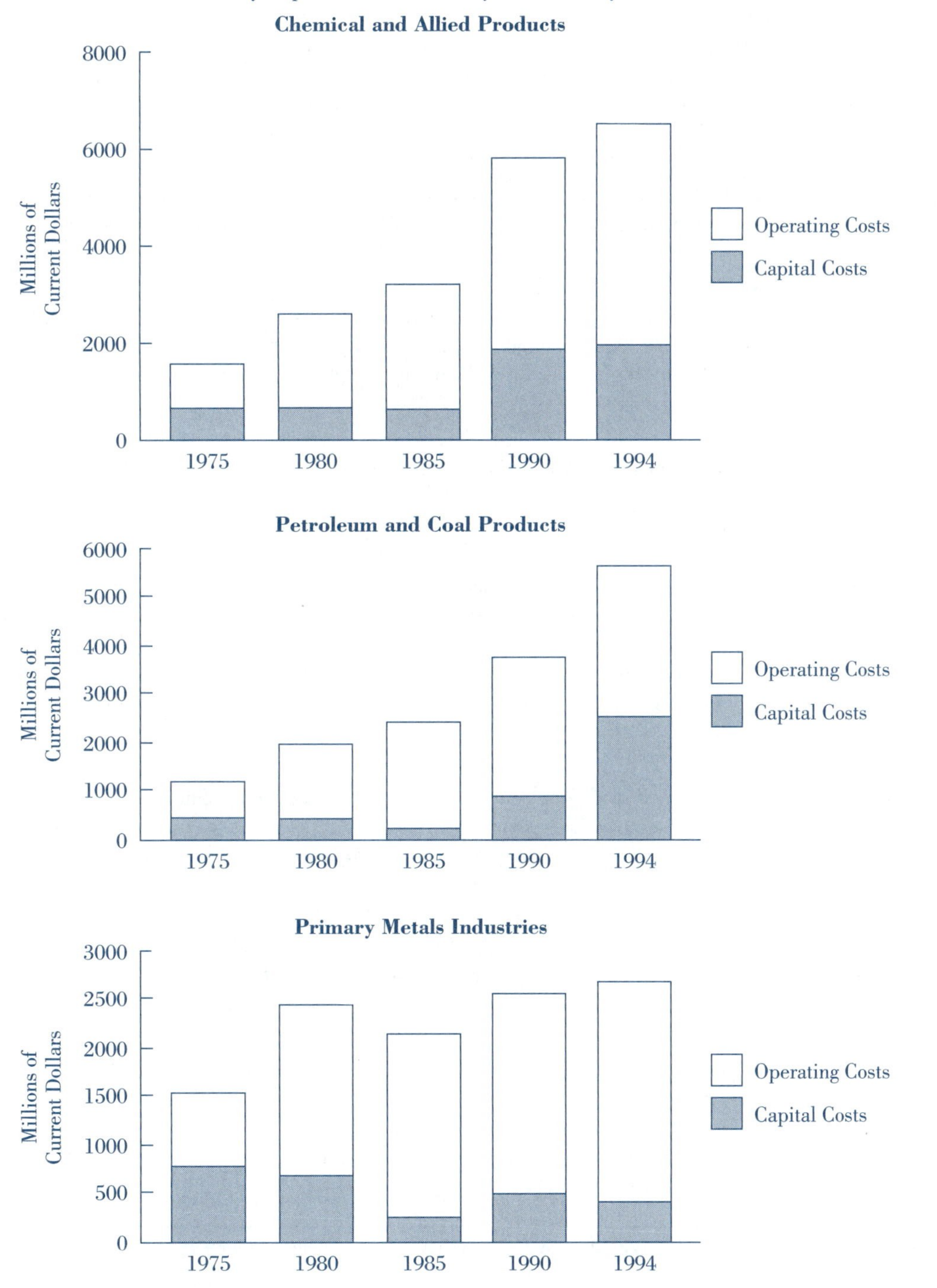

Source: U.S. Department of Commerce, Bureau of the Census, various years, as cited in Council on Environmental Quality (1997), table 13.

Exhibit A.3 *Overview of the Clean Air Act Amendments of 1990 (from Chapter 10)*

Title I. Provisions for Attainment and Maintenance of National Ambient Air Quality Standards

Amends and extends provisions for achieving the National Ambient Air Quality Standards (NAAQS), and regulates changes in State Implementation Plans (SIPs) for nonattainment areas.

Title II. Provisions Relating to Mobile Sources

Establishes more stringent emissions controls and extends the vehicle useful life for most emissions standards to 10 years or 100,000 miles. New provisions address the use of reformulated gasoline, oxygenated fuels, and alternative fuels in certain nonattainment areas. Prohibits leaded gasoline after December 31, 1995.

Title III. Hazardous Air Pollutants

Lists 189 toxic pollutants as the basis for setting emissions controls. Maximum Achievable Control Technology (MACT) standards are to be established for all listed source categories within a 10-year period.

Title IV. Acid Deposition Control

Establishes a market-based allowance program to achieve a permanent 10-million-ton reduction in sulfur dioxide emissions by January 1, 2000. A national cap on sulfur dioxide emissions of 8.9 million tons from utilities is established beginning January 1, 2000. Mandates a 2-million-ton reduction in nitrogen oxide emissions.

Title V. Permits

Requires states to establish permit programs to facilitate compliance with the new amendments. Requires all major sources of air pollution to have federal permits to operate.

Title VI. Stratospheric Ozone Protection

Requires that all Class I and II substances contributing to ozone depletion be identified and listed by the EPA. Outlines a phase-out program for both classifications. Allowances are to be issued by the EPA and a procedure established to facilitate trading that will result in a net reduction in production and consumption of ozone-depleting substances.

Title VII. Provisions Relating to Enforcement

Strengthens enforcement by an increased range of civil and criminal penalties for violations and by the establishment of new authorities.

Title VIII. Miscellaneous Provisions

Establishes programs to control air pollution from sources on the Outer Continental Shelf. Authorizes negotiations with Mexico to improve air quality in border regions. Allocates funds to support studies aimed at evaluating sources of visibility impairment and to assess progress in visibility in Class I areas.

Title IX. Clean Air Research

Requires the EPA to establish research programs associated with air pollution. Reauthorizes the National Acid Precipitation Assessment Program (NAPAP) and modifies and expands its responsibilities. Specifically mandates cost-benefit analysis as the decision rule to be used by the NAPAP in implementing the new amendments, the first such directive in any major environmental legislation.

Title X. Disadvantaged Business Concerns

Requires the EPA to assure that not less than 10 percent of funding for research relating to the 1990 Amendments be made available to disadvantaged firms.

Title XI. Clean Air Employment Transition Assistance

Establishes a training and employment services program for eligible workers who have been laid off or terminated due to compliance with the 1990 Amendments.

Sources: Council on Environmental Quality (March 1992), pp. 12–17; U.S. EPA, Office of Air and Radiation (November 15, 1990); U.S. EPA, Office of Air and Radiation (November 30, 1990).

Note: For an online overview of the 1990 Clean Air Act Amendments, visit **http://www.epa.gov/oar/caa/overview.txt.**

Exhibit A.4 EPA Benefit-Cost Analysis of the Clean Air Act for 1990–2010 (from Chapter 10)

Comparison of Benefits and Costs ($1990 millions)

	Titles I Through V				
	Annual Estimates		Present Value Estimate,	Title VI, Present Value Estimate,	All Titles, Total
Cost or Benefit	**2000**	**2010**	**1990–2010**	**1990–2165**	**Present Value**
Monetized direct costs	$19,000	$27,000	$180,000	$27,000	$210,000
Monetized direct benefits	71,000	110,000	690,000	530,000	1,200,000
Net benefits	$52,000	$83,000	$510,000	$500,000	$1,000,000
Benefit/cost ratio	4:1	4:1	4:1	20:1	6:1

Source: U.S. EPA, Office of Air and Radiation (November 1999), p. 105, table 8-4.

Note: Dollar estimates shown are central estimates, or mean values. A discount rate of 5 percent was used in the calculations.

Exhibit A.5 Technology-Based Standards and Market-Based Programs for PSD and Nonattainment Areas (from Chapter 12)

Area	New or Modified Sources	Existing Sources
PSD[a] Areas Standards	***Best Available Control Technology (BACT):*** BACT refers to an emission limitation based on the maximum degree of pollution reduction that the permitting authority deems achievable on a case-by-case basis, accounting for energy, environmental, and other economic impacts and costs.	***Best Available Retrofit Technology (BART):*** BART is applicable only to emissions that might negatively affect visibility in a PSD area and must consider factors such as compliance costs, existing control technology, and the energy and environmental impacts.
Market-Based Program	***Netting:*** Netting refers to the use of emissions trading among points within a given source for the same type of pollutant, such that any increase in emissions from a modified source is exactly matched by a reduction in emissions from another point within that same source.	***Bubble policy:*** The bubble policy allows a plant or plant complex to measure its emissions of a single pollutant as an average of all emission points emanating from that plant or complex, as if captured within an imaginary bubble.
Nonattainment Areas Standards	***Lowest Achievable Emission Rate (LAER):*** LAER refers to an emissions rate that reflects the most stringent emission limitation outlined in a given SIP[b] for the same type of source, or the most stringent limitation achieved in practice for such a source category, whichever is more stringent.	***Reasonably Available Control Technology (RACT):*** RACT refers to an emissions standard based on the use of technology deemed to be practically available, accounting for technological and economic feasibility, and determined on a case-by-case basis.
Market-Based Program	***Offset plan:*** An offset plan provides for any emissions associated with a new or modified major stationary source to be more than countered by the overall reductions achieved by existing sources.	***Bubble policy:*** The bubble policy allows a plant or plant complex to measure its emissions of a single pollutant as an average of all emission points emanating from that plant or complex, as if captured within an imaginary bubble.

[a]PSD means prevention of significant deterioration.

[b]SIP stands for State Implementation Plan.

Exhibit A.6 *Predictions of Climate Response According to an Expert Panel (from Chapter 13)*

The following table gives a summary of the conclusions of an expert panel convened by the National Academy of Sciences on the possible climate responses to increased levels of greenhouse gases. For details, the full report should be consulted.

Climate Response	Definition	Prediction
Large stratospheric cooling	Increased levels of CO_2 and other trace gases combined with reduced heating from diminished ozone will cause a significant temperature reduction in the upper stratosphere.	Virtually certain
Global mean surface warming	If the level of CO_2 doubles, global mean surface warming over the long term is predicted to be between 1.5° and 4.5° C.	Very probable
Global mean precipitation increase	The heating of the earth's surface will cause elevated evaporation and therefore greater global mean precipitation, although some regions may actually have diminished rainfall.	Very probable
Reduction of sea ice	As the earth's temperature increases, the warmer temperature will cause melting of sea ice.	Very probable
Polar winter surface warming	The air at the polar surface may warm as much as 3 times the global average due to the reduction in sea ice.	Very probable
Summer continental dryness/warming	Several, but not all, studies show that summer continental dryness will result primarily from earlier termination of winter storms.	Likely in the long term
Rise in global mean sea level	An elevated mean sea level is considered probable due to the thermal expansion of the seas and the melting of ice.	Probable

Sources: National Research Council (1987) as cited in U.S. EPA, Office of Policy, Planning, and Evaluation, Office of Research and Development (December 1989), p. xxvi.

Exhibit A.7 *Major Categories of Water Pollutants (from Chapter 14)*

Category	Description	Examples
Plant nutrients	Substances that promote the growth of aquatic plant life; high levels of nutrients in a water body indicate enrichment or eutrophication, which is the fertilization of a water body.	Nitrate and phosphate compounds
Sediment and other suspended solids	Small particles of solid pollutants suspended in water bodies that resist removal by conventional means.	Organic and inorganic particles, including soil and silt
Pathogens	Microorganisms that can cause disease in humans, animals, and plants.	Bacteria, viruses, parasites
Inorganic chemicals	Chemical substances of mineral origin that do not contain carbon.	Acids, road salt
Heavy metals	Metallic elements with high atomic weights that tend to accumulate in the food chain.	Mercury, chromium, cadmium, arsenic, lead
Organic compounds	Synthetics and animal- or plant-produced substances containing primarily carbon and hydrogen, nitrogen, or oxygen.	Pesticides, plastics, oil, gasoline, detergents
Thermal modification	Elevated temperature of water bodies caused by the discharge of heated water from industrial processes and electricity generation.	Heated water
Radioactive substances	Substances that emit ionizing radiation during decay.	Radio-isotopes

Sources: U.S. EPA, Office of Communications, Education, and Public Affairs (September 1992); Raven, Berg, and Johnson (1993).

Exhibit A.8 *Individual Beneficial Uses for Water Bodies as Recommended by the EPA (from Chapter 14)*

Use	Description
Aquatic life support	The water body provides suitable habitat for protection and propagation of desirable fish, shellfish, and other aquatic organisms.
Drinking water supply	The water body can supply safe drinking water with conventional treatment.
Fish consumption	The water body supports a population of fish free from contamination that could pose a human health risk to consumers.
Shellfish harvesting	The water body supports a population of shellfish free from toxicants and pathogens that could pose a human health risk to consumers.
Primary contact recreation, swimming	People can swim in the water body without risk of adverse human health effects (such as catching waterborne diseases from raw sewage contamination).
Secondary contact recreation	People can perform activities on the water (such as canoeing) without risk of adverse human health effects from ingestion or contact with the water.
Agriculture	The water quality is suitable for irrigating fields or watering livestock.
Groundwater recharge	The surface water body plays a significant role in replenishing groundwater, and surface water supply and quality are adequate to protect existing or potential uses of groundwater.
Wildlife habitat	Water quality supports the water body's role in providing habitat and resources for land-based wildlife as well as aquatic life.
Culture[a]	Water quality supports the water body's role in tribal culture and preserves the waterbody's religious, ceremonial, or subsistence significance.

Source: U.S. EPA, Office of Water (June 2000), pp. 12–13.

[a]Tribes may designate their waters for special cultural and ceremonial uses.

Exhibit A.9 *Classification of Economic Benefits of Water Quality Recommended by the EPA (from Chapter 14)*

Benefit Type	Description
INTRINSIC BENEFITS	
No current use by the individual	
Community benefits	Biocentric satisfaction of knowing that an ecological community is sustained for its own sake.
Existence benefits	Vicarious enjoyment from the knowledge that others are using the resource. Stewardship interest in providing an opportunity for others to use the resource in the future.
Potential future use by the individual	
Option benefits	Interest in securing an option to participate in an activity or use the resource at some point in the future.
CURRENT BENEFITS	
Indirect use by the individual	
Aesthetic benefits	Conditions enhance the characteristics of current adjoining fixed amenities, such as lakeside property.
Recreational benefits	Conditions enhance the characteristics of current adjoining transitory activities, such as hiking and photography.
Structural ecosystem benefits	Conditions maintain the functional ecosystem processes, such as stable climate and purification of land, air, and water.
Direct use by the individual	
Recreational benefits	Conditions enhance the characteristics of current water-contact activities, such as boating, swimming, and fishing.
Commercial benefits	Conditions enhance the characteristics of current production processes and activities.
Extractive commercial uses	Production processes where water is a medium for other goods (e.g., commercial fishing, medical industries).
Commercial navigation	Examples include dams, canals, and ports.
Agricultural irrigation	Water used as an input to production of crops.
Industrial processes	Water itself is used as an input to production (e.g., processing, cooling, and steam generation).
Municipal water	Water used for drinking, washing, and fire protection.

Source: Adapted from Desvousges and Smith (April 1993), p. 3-2, fig. 3-1, as cited in U.S. EPA, Office of Water (April 1992b), p. 167, fig. 12-1.

Exhibit A.10 ***Summary of Effluent Limits Specified in the Clean Water Act (from Chapter 15)***

Standard	Polluting Source	Pollutant
Best practicable technology currently available (BPT) (phased out; replaced by BAT and BCT)	Direct industrial dischargers	EPA-specified water pollutants
Best available technology economically achievable (BAT)	Direct industrial dischargers (existing sources)	Toxic and nonconventional pollutants
Best conventional control technology (BCT)	Direct industrial dischargers (existing sources)	Conventional pollutants
Best available demonstrated control technology (BADCT)	Direct industrial dischargers (new sources)	Toxic, conventional, and nonconventional pollutants
Pretreatment standards	Indirect industrial dischargers (existing and new sources)	Pollutants not susceptible to POTW treatment or potentially damaging to POTW operations[a]
Secondary treatment standards	POTWs[a]	Oxygen-demanding substances and suspended solids

Note: More information about these effluent limitations is available online at **http://cfpub.epa.gov/npdes/techbasedpermitting/effguide.cfm?program_id515.**

http://

[a]POTWs stand for publicly owned treatment works.

Exhibit A.11 *National Secondary Drinking Water Regulations (NSDWRs): U.S. Secondary Maximum Contaminant Levels (SMCLs) (from Chapter 16)*

Substance	SMCL[a]	Contaminant Effects
Aluminum	0.05–0.2	discoloration
Chloride	250	taste; pipe corrosion
Color	15 color units	aesthetic
Copper	1.0	taste; staining of porcelain
Corrosivity	Noncorrosive	aesthetic and health related (corrosive water can leach pipe materials, like lead, into water)
Fluoride	2.0	dental fluorosis (brownish discoloration of the teeth)
Foaming agents	0.5	aesthetic
Iron	0.3	taste; staining of laundry
Manganese	0.05	taste; staining of laundry
Odor	3 threshold odor numbers	aesthetic
Hydrogen and hydroxyl ions	pH 6.5–8.5	water is too corrosive
Silver	0.10	argyria (discoloration of the skin)
Sulfate	250	taste; laxative effects
Total dissolved solids	500	taste; possible relationship between low hardness and cardiovascular disease; indicator of corrosivity (related to lead levels in water); can damage plumbing and limit effectiveness of soaps and detergents
Zinc	5	taste

Sources: U.S. EPA, Office of Water (April 1992a); U.S. EPA, Office of Water (December 1991); U.S. EPA, Office of Water, Office of Ground Water and Drinking Water (April 17, 2002).

[a] SMCLs are expressed in mg/L unless otherwise noted.

Exhibit A.12 *The Superfund Cleanup Process (from Chapter 17)*

Step	General Description
Site discovery	Hazardous substance release is identified and reported to the EPA.
Listing in CERCLIS	Site is listed in CERCLIS, which inventories and tracks releases to provide information to all response agencies.
Preliminary assessment	Preliminary assessment is done to determine if an emergency removal action is necessary and to establish site inspection priorities.
Site inspection	On-site investigations are conducted to determine the extent of a release or potential for release. This typically involves sample collection and may include the installation of groundwater monitoring wells.
Removal action[a]	A short-term federal response is executed to prevent or mitigate damage from a hazardous release or threat of release.
Hazard Ranking System (HRS)	Information from the site assessment is used in the HRS to evaluate the environmental hazards at the site and assign a numerical score based on the site's inherent risks.
NPL Listing[b]	Site is added to the NPL as a national priority for long-term cleanup and therefore becomes eligible for federal cleanup funding.
Remedial investigation / feasibility study	In the remedial investigation, data are collected to assess risk and support the selection of response options. The feasibility study involves the development, evaluation, and selection of a remedial action.
Record of decision	An outline of cleanup actions planned for the site is given in the record of decision.
Remedial design	The set of technical plans and specifications for the cleanup actions are presented based on the record of decision.
Remedial action	Construction and other tasks at the site are executed to implement the selected solution.
Construction Completion Listing (CCL)	Physical construction of all cleanup remedies has been finished, all immediate threats have been addressed, and all long-term threats are under control.
Operation and maintenance	Activities at the site are undertaken to ensure that cleanup methods are working properly.
Deletion from NPL	The EPA and the state jointly determine that all appropriate response actions have been implemented and that no additional remedial actions are needed. A Notice of Final Action to Delete is published in the *Federal Register.* If no significant adverse comments are made within 30 days, the site is deleted from the NPL.

Source: U.S. EPA, Office of Solid Waste and Emergency Response (December 11, 2000).

Note: Visit **http://www.epa.gov/superfund/action/process/sfproces.htm** for more details on this process.

[a]A removal action may occur at any step in the process.

[b]NPL stands for National Priorities List.

References

Abdalla, Charles. 1990. "Measuring Economic Losses from Ground Water Contamination: An Investigation of Household Avoidance Cost." *Water Resources Bulletin* 26 (June), pp. 451–63.

Ackerman, B., and W. Hassler. 1981. *Clean Coal/Dirty Air.* New Haven, CT: Yale University Press.

Adler, Jerry, and Mary Hager. 1992. "How Much Is a Species Worth?" *National Wildlife* (April/May), pp. 4–14.

Adler, Robert W. 1993. "Revitalizing the Clean Water Act." *Environment* 35 (November), pp. 4–5, 40.

Adler, Robert W., Jessica C. Landman, and Diane M. Cameron. 1993. *The Clean Water Act: 20 Years Later.* Washington, DC: Island Press.

Alexander, Michael. 1992. "The Challenge of Markets: The Supply of Recyclables Is Larger than the Demand." *EPA Journal* 18 (July/August), pp. 29–33.

Allen, Scott. 1992. "Boston Harbor's Waters Have Started to Heal: Cleanup Helping to Shed 'Dirtiest' Label." *Boston Globe*, September 6, p. 1.

———. 1993. (January). "State Touts Give-and-Take on Clean Air." *Boston Globe*, January 8, p. 1.

———. 1993. (March). "Rights to Pollute Given Up." *Boston Globe*, March 20, p. 35.

———. 1993. (September). "Mass. Firms May Now Trade Clean Air Credits." *Boston Globe*, September 29, p. 26.

Alliance for Responsible CFC Policy. 1986. *Montreal Protocol: A Briefing Book.* Rosslyn, VA: Alliance for Responsible CFC Policy, December.

Alternative Fluorocarbons Environmental Acceptability Study. 2000. *Production, Sales, and Atmospheric Release of Fluorocarbons Through 1996* (an Internet-accessible dataset). Available at **http://ceq.eh.doe.gov/nepa/reports/statistics/tab11x3.html** (last updated May 15, 2000).

Alvarez, Lizette, and David Barboza. 2001. "Support Grows for Corn-Based Fuel Despite Critics." *New York Times*, July 23, p. A1.

American Chemistry Council. 2000. "Responsible Care Practitioners Site: Guiding Principles." Available at **http://www.americanchemistry.com** (accessed June 18, 2002).

Anderson, Frederick R. 1989. "Natural Resource Damages, Superfund, and the Courts." *Environmental Affairs* 16, pp. 405–57.

Anderson, Robert C., Andrew Q. Lohof, and Alan Carlin. 1997. *The United States Experience with Economic Incentives in Environmental Pollution Control Policy.* Washington, DC, August.

Andrews, Richard N. L. 1984. "Economics and Environmental Decisions, Past and Present." In V. Kerry Smith, ed., *Environmental Policy Under Reagan's Executive Order: The Role of Benefit-Cost Analysis*, pp. 43–85. Chapel Hill, NC: University of North Carolina Press.

Armstrong, Larry. 2001. "Hybrids Are Headed for Main Street." *Business Week*, July 9, p. 34.

Arndt, Michael. 2001. "Maybe It's Not So Hard Being Eco-Friendly." *Business Week*, April 16, p. 128F.

Arnst, Catherine. 1997. "Loophole in the Ozone Pact." *Business Week*, September 29, pp. 41–42.

Arnst, Catherine, and Gary McWilliams. 1997. "The Black Market vs. the Ozone." *Business Week*, July 7, p. 128.

Atkinson, Scott E., and Donald H. Lewis. 1974. "A Cost-Effectiveness Analysis of Alternative Air Quality Control Strategies." *Journal of Environmental Economics and Management* 1, pp. 237–50.

Australia, Ecologically Sustainable Development Steering Committee. 1992. *National Strategy for Ecologically Sustainable Development (NSESD).* Australia: December.

Australia and New Zealand Environment and Conservation Council. 1999. *Towards Sustainability: Achieving Cleaner Production in Australia.* Australia: Environment Australia, March.

Bai, Matt, and Scott Allen. 1996. "Oil Spill Spreads off Rhode Island Coast." *Boston Globe*, January 21, pp. 1, 24.

Baker, Stephen. 1990. "Mexico's Motorists Meet the Smog Patrol." *Business Week*, June 25, p. 100.

Ball, Jeffrey. 2000. "U.S. Auto Makers to Rev Up Output of 'Hybrid' Vehicles." *Wall Street Journal*, October 24, p. B4.

Bartik, Timothy. 1988. "Evaluating the Benefits of Nonmarginal Reductions in Pollution Using Information on Defensive Expenditures." *Journal of Environmental Economics* 15, pp. 111–27.

Battery Council International. 2002. "Summary of U.S. State Lead-Acid Battery Laws: October 2000." Available at **http://www.batterycouncil.org/states.html** (accessed May 23, 2002).

Baumol, W. J., and W. E. Oates. 1975. *The Theory of Environmental Policy.* Englewood Cliffs, NJ: Prentice-Hall.

Beckerman, Wilfred. 1990. "Global Warming: A Sceptical Economic Assessment." In Dieter Helm, ed., *Economic Policy Towards the Environment*, pp. 52–85. Cambridge, MA: Blackwell Publishers.

Biers, Dan. 1994. "China Creates Environmental Nightmare in Industrial Rush." *Brockton Enterprise*, June 16, p. 11.

Bingham, Tayler H., Timothy R. Bondelid, Brooks M. Depro, Ruth C. Figueroa, A. Brett Hauber, Suzanne J. Unger, and George L. Van Houtven. 1998. *A Benefits*

Assessment of Water Pollution Control Programs Since 1972. Revised draft report to the U.S. Environmental Protection Agency. Research Triangle Park, NC: Research Triangle Institute.

Block, Debbie Galante. 1993. "CD Jewel Box Only, or Alternatives Too?" *Tape-Disc Business* (June), p. 12.

Blomquist, Glenn C. 1979. "Value of Life Saving: Implications of Consumption Activity." *Journal of Political Economy* 87 (June), pp. 540–58.

BMW Group. 2001. *Sustainable Value Report 2001/2002.* Munich, Germany: BMW Group.

Bowker, James, and John R. Stoll. 1988. "Use of Dichotomous Choice Nonmarket Methods to Value the Whooping Crane Resource." *American Journal of Agricultural Economics* 70 (May), pp. 372–81.

Boyle, Kevin J., and Richard C. Bishop. 1987. "Valuing Wildlife in Benefit-Cost Analyses: A Case Study Involving Endangered Species." *Water Resources Research* 23, pp. 615–16.

Bradsher, Keith. 2001. "Detroit's Answer to 'Hybrid' Cars." *New York Times,* February 20, p. C1.

Brajer, Victor, and Jane V. Hall. 1992. "Recent Evidence on the Distribution of Air Pollution Effects." *Contemporary Policy Issue* 10 (April), pp. 63–71.

Bratton, D., and G. L. Rutledge. 1990. "Pollution Abatement and Control Expenditures, 1985–1988." *Survey of Current Business* 70, pp. 32–38.

Brock, James W. 2001. "The Automobile Industry." In Walter Adams and James Brock, eds., *The Structure of American Industry,* pp. 114–36. Upper Saddle River, NJ: Prentice-Hall.

Brookshire, D. S., and T. D. Crocker. 1981. "The Advantages of Contingent Valuation Methods for Benefit-Cost Analysis." *Public Choice* 36, pp. 235–52.

Brown, Gardner M., Jr., and Ralph W. Johnson. 1984. "Pollution Control by Effluent Charges: It Works in the Federal Republic of Germany, Why Not in the U.S." *Natural Resources Journal* 24 (October), pp. 929–66.

Browner, Carol M. 1994. "The Administration's Proposals." *EPA Journal* 20 (summer), pp. 6–9.

Burtraw, Dallas, and Paul Portney. 1991. "Environmental Policy in the U.S." In Dieter Helm, ed., *Economic Policy Towards the Environment,* pp. 289–320. Cambridge, MA: Blackwell Publishers.

"Bush Won't Drop Utility Pollution Lawsuits." 2002. *New York Times,* January 15.

Butler, John C., III., Mark W. Schneider, George R. Hall, and Michael E. Burton. 1993. "Allocating Superfund Costs: Cleaning Up the Controversy." *Environmental Law Reporter* 23 (March), pp. 10,133–144.

Cahan, Vicky. 1989. "Waste Not, Want Not? Not Necessarily." *Business Week,* July 17, pp. 116–17.

Cairncross, Frances. 1992. "How Europe's Companies Reposition to Recycle." *Harvard Business Review* (March–April), pp. 34–45.

"California's Smog Market: Right to Pollute." 1993. *The Economist* (October 30), 1993, p. 77.

Callan, Scott J., and Janet M. Thomas. 1997. "The Impact of State and Local Policies on the Recycling Effort." *Eastern Economic Journal* 23 (fall), pp. 411–23.

Canadian Institute of Chartered Accountants. 1992. *Environmental Auditing and the Role of the Accounting Profession.* Toronto Canadian Institute of Chartered Accountants.

Carson, Rachel. 1962. *Silent Spring.* Boston, MA: Houghton Mifflin.

Carson, Richard T., and Robert Cameron Mitchell. 1988. *The Value of Clean Water: The Public's Willingness to Pay for Boatable, Fishable, and Swimmable Quality Water.* Discussion Paper 88–13. La Jolla, CA: University of California at San Diego.

———. 1993. "The Value of Clean Water: The Public's Willingness to Pay for Boatable, Fishable, and Swimmable Water." *Water Resources Research* 29 (July), pp. 2245–54.

"'Cash for Clunkers' to Cut Pollution." 1992. *EPA Journal* 181 (May/June), p. 2.

"CFC-Free A/C." 1993. *AAA World* (March/April), p. 23.

Chandler, William, Alexei Makarov, and Zhou Dadi. 1990. "Energy for the Soviet Union, Eastern Europe, and China." *Scientific American* 263 (September), p. 120.

Chang, Kenneth. 2001. "A Practical Way to Make Power from Wasted Heat." *New York Times,* November 27, p. C4.

Chemical Industry Archives. 2001. "The Most Poorly Tested Chemicals in the World." Available at **http://www.chemicalindustryarchives.org/factfiction/testing.asp** (last updated November 1, 2001).

Chemical Manufacturers Association. 1990. *Improving Performance in the Chemical Industry.* Washington DC: CMA. September.

Chesapeake Bay Program. 2000. "New Chesapeake Bay Agreement Signed." Press release. Annapolis, MD: Chesapeake Bay Program, June 28.

———. 2001. "Chesapeake 2000 and the Bay: Where Are We and Where Are We Going?" Fact Sheet. Annapolis, MD: Chesapeake Bay Program, August 22.

"China's Environmental Outlook Bleakens." 2002. Available at **http://www.cnn.com/2002/WORLD/asiapcf/east/01/10/china.enviroment/index.html** (accessed January 10, 2002).

Clawson, Marion, and Jack L. Knetch. 1966. *Economics of Outdoor Recreation.* Washington, DC: Resources for the Future.

Cline, William R. 1992. *Global Warming: The Benefits of Emission Abatement.* Paris: OECD.

"Coal Dependence Has China Looking for Pollution Solution." *Brockton Enterprise,* November 3, 1996, p. 6.

Coalition for Environmentally Responsible Economies. 1989. "The *Valdez* Principles." Boston, MA: CERES.

Coase, Ronald H. 1960. "The Problem of Social Cost." *Journal of Law and Economics* 3 (October), pp. 1–44.

Cogan, Douglas G. 1988. *Stones in a Glass House: CFCs and Ozone Depletion.* Washington, DC: Investor Responsibility Research Center.

Cohen, Bonner R. 2001. "EPA Will Destroy Hudson River to Save It." *Wall Street Journal,* December 12, p. A18.

———. 2002. "EPA's Dredging Scheme Will Wreak Havoc on the Hudson River." *Human Events* 58 (January 7), p. 10.

Cohen, Brian A., and Erik D. Olson. 1994. *Victorian Water Treatment Enters the 21st Century.* Washington, DC: NRDC, March.

Cohn, Laura, John Carey, and Ann Therese Palmer. 2001. "Is the EPA Sandbagging Business? *Business Week,* June 11.

Committee on the Science of Climate Change, and the National Research Council. Division on Earth and Life Sciences, 2001. *Climate Change Science: An Analysis of Some Key Questions.* Washington, DC: National Academy Press.

Commoner, Barry. 1992. *Making Peace with the Planet.* New York: The New Press.

Cook, Elizabeth. 1996. *Making a Milestone in Ozone Protection.* Washington, DC: World Resources Institute.

Cormier, William. 1992. "Smog Eats Mexico City's Monuments." *Brockton Enterprise,* November 23, p. 9.

Cough, Paul. 1993. "Trade-Environment Tensions: Options Exist for Reconciling Trade and Environment." *EPA Journal* 19 (April–June), pp. 28–30.

Council of Economic Advisers. 1982. *The Annual Report of the Council of Economic Advisers: Economic Report of the President.* Transmitted to the U.S. Congress in February 1982. Washington, DC: U.S. Government Printing Office.

———. 2001. *Economic Report of the President.* Washington, DC: U.S. Government Printing Office, January, Table B-31. Available at **http://w3.access.gpo.gov/usbudget/fy2002/sheets/b31.xls.**

Council on Environmental Quality. 1976. *Environmental Quality—1976.* Washington, DC: U.S. Government Printing Office.

———. 1980. *Environmental Quality—1980.* Washington, DC: U.S. Government Printing Office.

———. 1982. *Environmental Quality—1982.* Washington, DC: U.S. Government Printing Office.

———. 1989. *Environmental Trends.* Washington, DC: U.S. Government Printing Office.

———. 1992. *Environmental Quality, 22nd Annual Report.* Washington, DC: U.S. Government Printing Office, March.

———. 1993. *Environmental Quality, 23rd Annual Report.* Washington, DC: U.S. Government Printing Office, January.

———. 1997. *Environmental Quality, 25th Anniversary Report.* Washington, DC: U.S. Government Printing Office.

———. 1998. *Environmental Quality, 1997 Report: The World Wide Web.* Washington, DC: U.S. Government Printing Office.

Cox, Meg. 1991. "Music Firms Try Out 'Green' CD Boxes." *Wall Street Journal,* July 25, p. B1.

Crandall, Robert W. 1984. "The Political Economy of Clean Air: Practical Constraints on White House Review." In V. Kerry Smith, ed., *Environmental Policy Under Reagan's Executive Order: The Role of Benefit-Cost Analysis.* Chapel Hill, NC: University of North Carolina Press.

Crookshank, Steven. 1994. *Air Emissions Banking and Trading: Analysis and Implications for Wetland Mitigation Banking.* Discussion Paper 074. Washington, DC: American Petroleum Institute.

Cropper, Maureen L., and Wallace E. Oates. 1992. "Environmental Economics: A Survey." *Journal of Economic Literature* 30 (June), pp. 675–740.

Cummings, Ronald G., David S. Brookshire, and William D. Schulze. 1986. *Valuing Environmental Goods: An Assessment of the Contingent Valuation Method.* Totowa, NJ: Rowman and Allanheld.

Daley, Beth. 2001. "Harbor Cleanup Falls Short at Beaches: Sewage Discharge Fouling Boston's Cleanup Success." *Boston Globe,* September 10, p. A1.

Daley, Beth, and Ellen Barry. 2001. "Five Indicted in Coolant Smuggling Scheme." *Boston Globe,* August 1, p. B1.

Dardis, Rachel. 1980. "The Value of a Life: New Evidence from the Marketplace." *American Economic Review* 70 (December), pp. 1,077–82.

Dasgupta, Susmita, Benoit Laplante, Hua Wang, and David Wheeler. 2002. "Confronting the Environmental Kuznets Curve." *Journal of Economic Perspectives* 16 (winter), pp. 147–68.

Davies, Terry. 1995. "Congress Discovers Risk Analysis." *Resources,* no. 118 (winter), pp. 5–8.

Davis, Bob. 1994. "U.S. Is Hoping to Blend Environmental World Trade Issues at Morocco Meeting." *Wall Street Journal,* January 10, p. A9.

Desvousges, William H., and V. Kerry Smith. 1993. *Benefit-Cost Assessment Handbook for Water Programs, Volume 1.* Prepared for the Environmental Protection Agency, Economic Analysis Division, by Research Triangle Institute, Office of Policy, Planning and Evaluation, Research Triangle Park, NC, April.

Deutsch, Claudia H. 2001. "Together at Last: Cutting Pollution and Making Money." *New York Times*, September 9, p. 3.1.

Dickerson, Kenneth R. 1991. "At the Gasoline Pump." *EPA Journal* 17 (January/February), pp. 48–50.

Dolin, Eric Jay. 1992. "Boston Harbor's Murky Political Waters." *Environment* 34 (July/August), pp. 7–11, 26–33.

Dominguez, George, ed. 1977. *Guidebook: Toxic Substances Control Act.* Cleveland: CRC Press.

Doneski, D. 1985. "Cleaning up Boston Harbor: Fact or Fiction?" *Boston College Environmental Affairs Law Review* 12 (spring), p. 567.

Duales System Deutschland AG. 2002. "Dual System." Available at **http://www.gruener-punkt.de/en/frames.php3?choice=ds** (accessed May 23, 2002).

Duggal, V. G., C. Saltzman, and M. L. Williams. 1991. "Recycling: An Economic Analysis." *Eastern Economic Journal* 17 (summer), pp. 351–58.

Duhl, Joshua. 1993. *Effluent Fees: Present Practice and Future Potential.* Discussion Paper 075. Washington, DC: American Petroleum Institute, December.

Dumanoski, Dianne. 1992. "Nations Act to Speed Phaseout of Ozone-Depleting Chemicals." *Boston Globe*, November 26, p. 26–27.

———. 1993. "Activists Cite Trade Rules as a Force Shaping Nature." *Boston Globe*, July 1, p. 13.

Dunlap, Riley E., and Lydia Saad. 2001. "Only One in Four Americans Are Anxious About the Environment: Most Favor Moderate Approach to Environmental Protection." Gallup News Service. Available at **http://www.gallup.com/poll/Releases/Pr010416.asp** (April 16).

Dzurik, Andrew A. 1990. *Water Resources Planning.* Savage, MD: Rowman & Littlefield.

Eastman Kodak Company. 2001. *2001 Health, Safety, and Environment Annual Report.* Rochester, NY: Eastman Kodak Company.

Eheart, Wayland, E. Downey Brill Jr., and Randolph M. Lyon. 1983. "Transferable Discharge Permits for Control of BOD: An Overview." In Erhard F. Joeres and Martin H. David, eds., *Buying a Better Environment: Cost-Effective Regulation Through Permit Trading.* Madison, WI: University of Wisconsin Press.

E. I. Du Pont de Nemours and Company. 2001. *Sustainable Growth: 2001 Progress Report.* Wilmington, DE: E. I. Du Pont de Nemours and Company.

Ellerman, A. Denny, Paul L. Joskow, Richard Schmalensee, Juan Pablo Montero, and Elizabeth Bailey. 2000. *Markets for Clean Air: The U.S. Acid Rain Program.* Cambridge: Cambridge University Press.

Elmore, T., J. Jaksch, D. Downing, M. Podar, B. Morrison, B. Zander, and S. Sessions. 1984. *Trading Between Point and Nonpoint Sources: A Cost-Effective Method for Improving Greater Quality—The Case of Dillon Reservoir.* Washington, DC: U.S. Environmental Protection Agency.

Environment Canada. 2000. *A Guide to the Canadian Environmental Protection Act, 1999.* Ottawa, Canada: Minister of Public Works and Government Services, March.

Environmental Information Ltd. 1990. *Industrial and Hazardous Waste Management Firms.* Minneapolis: Environmental Information Ltd.

"Environmentalist Predicts Marked Change in U.S. Policy." 1992. *Boston Globe*, December 7.

EnviroSense. 1995. "Partners for the Environment: EPA Standards Network." Available at **http://es.epa.gov/partners/iso/iso.html** (last updated December 20, 1995).

"EPA: Fish Aren't Safe to Eat from 46 Waterways." *Brockton Enterprise*, November 20, 1992.

"EPA's Whitman Sees Overhaul of Pollution-Control Program." *Wall Street Journal Online*, March 20, 2002.

"EPA Warns of Exposure to Smoke." *Boston Globe*, July 23, 1993, p. 9.

Epstein, Richard A. 2001. "Through the Smog: What the Court Actually Ruled." *Wall Street Journal* (interactive edition), March 1.

Epstein, Samuel S., Lester O. Brown, and Carl Pope. 1982. *Hazardous Waste in America.* San Francisco: Sierra Club Books.

Erkman, S. 1997. "Industrial Ecology: An Historical View." *Journal of Cleaner Production* 5, pp. 1–10.

"EU Ratifies Kyoto Protocol; All Eyes Now on Russia." 2002. *Wall Street Journal Online*, May 31.

European Commission. 2002. "The IPPC Directive." Available at **http://europa.eu.int/comm/environment/ippc/** (last updated April 26, 2002).

Faeth, Paul. 2000. *Fertile Ground: Nutrient Trading's Potential to Cost-Effectively Improve Water Quality.* Washington, DC: World Resources Institute.

Farman, J. C., B. G. Gardiner, and J. D. Shanklin. 1985. "Large Losses of Total Ozone in Antarctica Reveal Seasonal ClO_x/NO_x Interaction." *Nature* 315, pp. 207–10.

Fialka, John J. 1997. "Gore Faces Cool Response to Issue of Global Warming." *Wall Street Journal*, August 27, p. A18.

———. 1997. "Global-Warming Treaty Is Approved." *Wall Street Journal*, December 11, p. A2.

———. 2001. "Kyoto Treaty Moves Ahead Without U.S." *Wall Street Journal*, November 12, p. A4.

Fialka, John J., and Jackie Calmes. 1997. "Clinton Proposes Global-Warming Plan." *Wall Street Journal*, October 23, pp. A2, A6.

Fialka, John J., and Geoff Winestock. 2001. "Future of Kyoto Protocol Minus U.S. Is Uncertain." *Wall Street Journal*, July 16, pp. A2, A8.

Fisher, Ann, Dan Violette, and Lauraine Chestnut. 1989. "The Value of Reducing Risks of Death: A Note on New Evidence." *Journal of Policy Analysis and Management* 8, no. 1, pp. 88–100.

Fletcher, Susan R. 2000. "Global Climate Change Treaty: Summary of the Kyoto Protocol." *CRS Report for Congress.* Washington, DC: Committee for the National Institute for the Environment, March 6.

Fortuna, Richard C., and David J. Lennett. 1987. *Hazardous Waste Regulation: The New Era.* New York: McGraw-Hill.

"44 Nations Agree to Update Rules on Ocean Dumping." 1992. *Boston Globe,* November 14, p. 61.

Freeman, A. Myrick, III. 1978. "Air and Water Pollution Policy." In Paul R. Portney, ed., *Current Issues in U.S. Environmental Policy,* pp. 12–67. Baltimore: Johns Hopkins University Press for Resources for the Future.

———. 1979. *The Benefits of Air and Water Pollution Control: A Review and Synthesis of Recent Estimates.* Report prepared for the Council on Environmental Quality, December.

———. 1979. "Hedonic Prices, Property Values, and Measuring Environmental Benefits: A Survey of the Issues." *Scandinavian Journal of Economics* 81, pp. 154–73.

———. 1982. *Air and Water Pollution Control: A Benefit-Cost Assessment.* New York: Wiley.

———. 1990. "Water Pollution Policy." In Paul R. Portney, ed., *Public Policies for Environmental Protection.* Washington, DC: Resources for the Future, pp. 97–149.

———. 2000. "Water Pollution Policy." In Paul R. Portney and Robert N. Stavins, eds., *Public Policies for Environmental Protection,* pp. 169–213. 2d Edition. Washington, DC: Resources for the Future.

———. 2002. "Environmental Policy Since Earth Day I: What Have We Gained?" *Journal of Economic Perspectives* 16 (winter), pp. 125–46.

Frosch, Robert A. 1995. "Industrial Ecology: Adapting Technology for a Sustainable World." *Environment* 37 (December), pp. 16–24, 34–37.

Frosch, Robert A., and Nicholas E. Gallopoulos. 1989. "Strategies for Manufacturing." *Scientific American* 261 (September), pp. 144–52.

Fullerton, Don, and Thomas C. Kinnaman. 1995. "Garbage, Recycling, and Illicit Burning or Dumping." *Journal of Environmental Economics and Management* 29 (July), pp. 78–91.

———. 1996. "Household Responses to Pricing Garbage by the Bag." *American Economic Review* 86 (September), pp. 971–84.

Garner, Andy, and Gregory A. Keoleian. 1995. *Industrial Ecology: An Introduction.* Ann Arbor, MI: National Pollution Prevention Center for Higher Education, November.

"German Packaging Regulations." 2001. *Commercial Angles Newsletter,* September. Available at **http://www.commercialangles.com/articles/green_dot.htm** (accessed May 23, 2002).

"German Recycling Is Too Successful." 1993. *Brockton Enterprise,* July 27.

Gianessi, Leonard P., and Henry M. Peskin. 1980. "The Distribution of the Cost of Federal Water Pollution Control Policy." *Land Economics* 56 (February), pp. 85–102.

Gibbons, John. 1992. "Moving Beyond the 'Tech Fix.'" *EPA Journal* 18 (September/October), pp. 29–31.

Gibbs, Lois Marie. 1982. *The Love Canal: My Story.* Albany: State University of New York Press.

Ginsburg, Janet. 2001. "Once Is Not Enough." *Business Week,* April 16, pp. 128B–128D.

Goddard, Walter E. 1986. *Just-in-Time: Surviving by Breaking Tradition.* Essex Junction, VT: Oliver Wight.

Goldstein, Nora. 2000. "The State of Garbage in America, Part I." *BioCycle* 41, (April), pp. 32–39.

Gosselin, Peter G. 1993. "Environmental Ruling Blocks Free Trade Pact." *Boston Globe.* July 1, p. 1.

"'Greener' Cooling." 1993. *Consumer Reports* (March), p. 133.

"The Greening of Protectionism." 1993. *The Economists,* February 27, pp. 25–28.

Griffin, Melanie L. 1988. "The Legacy of Love Canal." *Sierra* (January/February), pp. 26–28.

Grossman, Gene M., and Alan B. Krueger. 1995. "Economic Growth and the Environment." *Quarterly Journal of Economics* 110 (May), pp. 353–78.

Gruenspecht, Howard, and Robert Stavins. 2002. "A Level Field on Pollution at Power Plants." *Boston Globe,* January 26, p. A15.

Haas, Peter M., Marc A. Levy, and Edward A. Parson. 1992. "Appraising the Earth Summit: How Should We Judge UNCED's Success?" *Environment* 34 (October), pp. 6–11, 26–33.

Hahn, Robert W. 1989. "Economic Prescriptions for Environmental Problems: How the Patient Followed the Doctor's Orders." *Journal of Economic Perspectives* 3 (spring), pp. 95–114.

Hahn, Robert W., and Gordon L. Hester. 1989. "Marketable Permits: Lessons for Theory and Practice." *Ecology Law Quarterly* 16, pp. 361–406.

Hahn, Robert W., and Roger Noll. 1982. "Designing a Market for Tradeable Emissions Permits." In Wesley Magat, ed., *Reform of Environmental Regulation,* pp. 119–46. Cambridge, MA: Ballinger.

Hall, Alan. 2001. "Corn for Fuel: Not Such a Hot Idea?" *Business Week,* September 25.

Halvorsen, Robert, and Michael G. Ruby. 1981. *Benefit-Cost Analysis of Air-Pollution Control.* Lexington, MA: D. C. Heath.

Hardin, Garrett. 1968. "The Tragedy of the Commons." *Science* 162, pp. 1,243–48.

Harrington, Winston, Alan J. Krupnick, and Henry M. Peskin. 1985. "Policies for Nonpoint-Source Water Pollution Control." *Journal of Soil and Water Conservation* (January–February), pp. 27–32.

Harrison, D., Jr., and D. Rubinfeld. 1978. "Hedonic Housing Prices and the Demand for Clean Air." *Journal of Environmental Economics and Management* 5 (March), pp. 81–102.

Haveman, Robert H., and Burton A. Weisbrod. 1975. "The Concepts of Benefits in Cost-Benefit Analysis: With Emphasis on Water Pollution Control Activities." In Henry M. Peskin and Eugene P. Seskin, eds., *Cost Benefit Analysis and Water Pollution Policy.* Washington, DC: The Urban Institute.

Hazilla, Michael, and Raymond J. Kopp. 1990. "Social Cost of Environmental Quality Regulations: A General Equilibrium Analysis." *Journal of Political Economy* 98, no. 4, pp. 853–73.

Heath, Jenifer. 1993. "Moynihan Bill Sets Goals." *Water Environment and Technology* (July), pp. 28–30.

Heath, Ralph C. 1988. "Ground Water." In David H. Speidel, Lon C. Ruedisili, and Allen F. Agnew, eds., *Perspectives on Water: Uses and Abuses.* New York: Oxford University Press.

Helfand, Gloria E. 1994. "Pollution Prevention as Public Policy: An Assessment." *Contemporary Economic Policy* 12, October, pp. 104–13.

Heritage, John, ed. 1992. "Environmental Protection—Has It Been Fair?" *EPA Journal* 18 (March/April).

Heumann, Jenny M. 1997. "State Recycling Programs: A Waste Reduction Emphasis." *Waste Age* (August).

Hilsenrath, Jon E. 2001. "Environmental Economists Debate Merit of U.S.'s Kyoto Withdrawal." *Wall Street Journal* (interactive edition), August 7.

Hoerner, Andrew J. 1998. *Harnessing the Tax Code for Environmental Protection: A Survey of State Initiatives.* Washington, DC: World Resources Institute.

Hof, Robert D. 1990. "The Tiniest Toxic Avengers." *Business Week,* June 4, pp. 96, 98.

Holtermann, S. E. 1972. "Externalities and Public Goods." *Economica* 39, (February).

Hong, Peter. 1992. "John Henry: A Power Broker for Cleaner Power." *Business Week,* July 20.

Hong, Peter, and Michele Galen. 1992. "The Toxic Mess Called Superfund." *Business Week,* May 11, pp. 32–34.

Houghton, J. T., L. G. Meira Filho, B. A. Callander, N. Harris, A. Kattenberg, and K. Maskell, eds. 1996. *Climate Change 1995: The Science of Climate Change.* Contribution of Working Group I to the Second Assessment of the Intergovernmental Panel on Climate Change. Cambridge: Cambridge University Press.

Howe, Charles W. 1991. "An Evaluation of U.S. Air and Water Policies." *Environment* 33 (September), pp. 10–15, 34–36.

Humphrey, Hubert H., III, and LeRoy C. Paddock. 1990. "The Federal and State Roles in Environmental Enforcement: A Proposal for a More Effective and More Efficient Relationship." *Harvard Environmental Law Review* 14, pp. 7–44.

ICF Resources International. 1989. *Economic, Environmental, and Coal Market Impacts of SO_2 Emissions Trading Under Alternative Acid Rain Control Proposals.* Report prepared for the Regulatory Innovations Staff, Office of Policy, Planning and Evaluation, U.S. Environmental Protection Agency.

"Industrial Nations Ban Dumping Waste at Sea." 1990. *Boston Globe,* November 3, p. 7.

International Energy Agency. "Selected Statistics." Available at **http://www.iea.org/statist/index.htm.**

International Maritime Organization, Office for the London Convention 1972. 2001. "A Brief Description of the London Convention 1972 and the 1996 Protocol." Available at **http://www.londonconvention.org/London_Convention.htm,** last revised March 25, 2001.

International Organization for Standardization. 1998. *ISO 14000: Meet the Whole Family.* Geneva, Switzerland: ISO.

———. 2002. "ISO Online." Available at **http://www.iso.org** (accessed June 4, 2002).

IPCC (Intergovernmental Panel on Climate Change). 2001. *Climate Change 2001: The Scientific Basis.* Contribution of Working Group I to the Third Assessment Report of the Intergovernmental Panel on Climate Change (IPCC). Cambridge: Cambridge University Press.

"Japanese Environmentalist Says Gas Too Cheap in U.S." 1992. *Brockton Enterprise,* February 2.

Jenkins, Robin. 1993. *The Economics of Solid Waste Reduction: The Impact of User Fees.* Brookfield, VT: Edward Elgar Publishing.

Johansson, Per-Olov. 1991. "Valuing Environmental Damage." In Dieter Helm, ed., *Economic Policy Towards the Environment,* pp. 111–36. Cambridge, MA: Blackwell Publishers.

Johnson, Edwin L. 1967. "A Study in the Economics of Water Quality Management." *Water Resources Research* 3, pp. 291–305.

Jondrow, James, and Robert A. Levy. 1984. "The Displacement of Local Spending for Pollution Control by Federal Construction Grants." *American Economic Review* 74, no. 2, pp. 174–78.

Kemp, David D. 1990. *Global Environmental Issues: A Climatological Approach.* New York: Routledge.

Kerr, Richard A. 1997. "Greenhouse Forecasting Still Cloudy." *Science* 276 (May 16), pp. 1,040–42.

Kharbanda, O. P., and E. A. Stallworthy. 1990. *Waste*

Management: Towards a Sustainable Society. New York: Auburn House.

Kiel, K. A. 1995. "Measuring the Impact of the Discovery and Cleaning of Identified Hazardous Waste Sites on House Values." *Land Economics* 71, no. 4, pp. 428–35.

Kimm, Victor J. 1993. "The Delaney Clause Dilemma." *EPA Journal* 19 (January–March), pp. 39–41.

Kinsman, Art. 1992. "Fueling the Future." *AAA World* (April), pp. 2F–2H.

Kirchhoff, Sue. 1993. "FDA Proposes Bottled Water Measure. Up to Tap Standards." *Boston Globe,* January 1, p. 71.

Kirkpatrick, David. 1990. "Environmentalism: The New Crusade." *Fortune,* February 12, pp. 44–52.

Klein, Janice A., and Jeffrey G. Miller, eds. 1993. *The American Edge: Leveraging Manufacturing's Hidden Assets.* New York: McGraw-Hill.

Kneese, Allen V., Robert U. Ayres, and Ralph C. D'Arge. 1970. *Economics and the Environment: A Materials Balance Approach.* Washington, DC: Resources for the Future.

Knepper, Mike. 1993. "Recycling: Investment in the Future." *BMW Magazine* (January), pp. 66–69.

Knopman, Debra S., and Richard A. Smith. 1993. "Legislative Status of the U.S. Clean Water Act," *Environment* 35 (January/February), p. 18.

Kohlhase, Janet E. 1991. "The Impact of Toxic Waste Sites on Housing Values." *Journal of Urban Economics* 30, pp. 1–26.

Kopp, Raymond J., and Alan J. Krupnick. 1987. "Agricultural Policy and the Benefits of Ozone Control." *American Journal of Agricultural Economics* 69 (December), pp. 956–62.

Kranish, Michael. 1993. "Energy Tax Plan Changes Sought." *Boston Globe,* February 19, p. 57–58.

———. 1993. "Clinton Treads into Forests Dispute." *Boston Globe,* April 3, p. 3.

Krimsky, Sheldon, and Dominic Golding. 1991. "Factoring Risk into Environmental Decision Making." In Richard A. Chechile and Susan Carlisle, eds., *Environmental Decision Making: A Multidisciplinary Perspective.* New York: Van Nostrand Reinhold.

Krukowski, John. 1995. "Survey: Thermal Prices End Free-Fall?" *Pollution Engineering* 7, no. 2.

Krupnick, Alan. 1986. "Costs of Alternative Policies for the Control of Nitrogen Dioxide in Baltimore." *Journal of Environmental Economics and Management* 13, pp. 189–97.

Krupnick, Alan, and Virginia McConnell, with Matt Cannon, Terrell Stoessell, and Michael Batz. 2000. *Cost-Effective NO_X Control in the Eastern United States.* Discussion Paper 00-18. Washington, DC: Resources for the Future.

Krutilla, John, V. 1967. "Conservation Reconsidered." *American Economic Review* 57, pp. 777–86.

Kuhn, Thomas S. 1970. *Structure of Scientific Revolutions.* Chicago: University of Chicago Press.

Lancaster, Kelvin J. 1966. "A New Approach to Consumer Theory." *Journal of Political Economy* 78, pp. 311–29.

Landy, Marc K., Marc J. Roberts, and Stephen R. Thomas. 1990. *The Environmental Protection Agency: Asking the Wrong Questions.* New York: Oxford University Press.

Larson, Charles D. 1989. "Historical Development of the National Primary Drinking Water Regulations." In Edward J. Calabrese, Charles E. Gilbert, and Harris Pastides, eds., *Safe Drinking Water Act: Amendments, Regulations, and Standards,* pp. 3–15. Chelsea, MI: Lewis Publishers.

Lashof, D., and D. Tirpak, eds. 1989. *Policy Options for Stabilizing Global Climate: Draft Report to Congress.* Washington, DC: U.S. Environmental Protection Agency, Office of Policy, Planning, and Evaluation, February.

Lave, Lester B. 1982a. "Methods of Risk Assessment." In Lester B. Lave, ed., *Quantitative Risk Assessment in Regulation,* pp. 23–54. Washington, DC: Brookings Institution.

———, ed. 1982b. *Quantitative Risk Assessment in Regulation.* Washington, DC: Brookings Institution.

Lawrence, Jennifer. 1991. "Mobil." *Advertising Age,* January 29, p. 12.

Lazaroff, Catherine. 1999. "Senate Votes for Subsidized National Forest Timber Sales." *Environmental News Service,* September 14.

Lederer, Edith M. 1993. "Ban on Atomic Dumping at Sea Voted; 4 Nuclear Powers Abstain." *Boston Globe,* November 13, p. 7.

Lee, David. 1992. "Ozone Loss: Modern Tools for a Modern Problem." *EPA Journal* 18 (May/June), pp. 16–18.

Lee, Martin R. 1995. *Environmental Protection: From the 103d to the 104th Congress.* CRS Issue Brief for Congress. Washington, DC: Committee for the National Institute for the Environment, January 3.

Letson, David. 1992. "Point/Nonpoint Source Pollution Reduction Trading: An Interpretive Survey." *Natural Resources Journal* 32 (spring), pp. 219–32.

Light, Alfred R. 1985. "A Defense Counsel's Perspective on Superfund." *Environmental Law Reporter* 15 (July), pp. 10,203–207.

Liroff, Richard A. 1986. *Reforming Air Pollution Regulation: The Toil and Trouble of EPA's Bubble.* Washington, DC: Conservation Foundation.

Lyklema, J., and T. E. A. van Hylckama. 1988. "Water Something Peculiar." In David H. Speidel, Lon C. Ruedisili, and Allen F. Agnew, eds., *Perspectives on Water: Uses and Abuses.* New York: Oxford University Press.

Main, Jeremy. 1991. "The Big Cleanup Gets It Wrong." *Fortune,* May 20, pp. 95–96, 100–101.

Malik, Arun S., David Letson, and Stephen R. Crutchfield. 1993. "Point/Nonpoint Source Trading of Pollution Abatement: Choosing the Right Trading Ratio." *American Journal of Agricultural Economics* 75 (November), pp. 959–67.

Maloney, Michael T., and Bruce Yandle. 1984. "Estimation of the Cost of Air Pollution Control Regulation." *Journal of Environmental Economics and Management* 11, pp. 244–63.

Mandel, Michael J. 1989. "The Right to Pollute Shouldn't Be for Sale." *Business Week,* May 22, p. 158G.

Manly, Lorne. 1992. "It Doesn't Pay to Go Green When Consumers Are Seeing Red." *Adweek,* March 23, pp. 32–33.

Marchant, Gary E., and Dawn P. Danzeisen. 1989. "'Acceptable' Risk for Hazardous Air Pollutants." *Harvard Environmental Law Review* 13, no. 2, pp. 535–58.

Marshall, Alfred. 1930 [1890]. *Principles of Economics,* 8th ed. London: Macmillan.

Marshall, Jonathan. 1992. "How Ecology Is Tied to Mexico Trade Pact." *San Francisco Chronicle,* February 25, p. A8.

Mathtech Inc. 1983. *Benefit and Net Benefit Analysis of Alternative National Ambient Air Quality Standards for Particulate Matter,* vol. 1. Prepared for U.S. Environmental Protection Agency, Economic Analysis Branch, Office of Air Quality Planning and Standards. Research Triangle Park, NC: U.S. Environmental Protection Agency, March.

Mazmanian, Daniel, and David Morell. 1992. *Beyond Superfailure: America's Toxics Policy for the 1990s.* Boulder, CO: Westview Press.

McCarthy, James E. 1993. "Packaging Waste: How the United States Compares to Other Countries." *Reusable News* (winter/spring), p. 3.

McDermott, H. 1973. "Federal Drinking Water Standards—Past, Present, and Future." *Water Well Journal* 27, no. 12, pp. 29–35.

McDonald's Corp. 1999. *McDonald's Waste Reduction Action Plan (WRAP): Status Report.* Oakbrook, IL: McDonald's Corp., May.

———. 2002. "Tenth Anniversary of EDF/McDonald's Alliance: Fact Sheet 1990–1999." Available at **http://www.mcdonalds.com/countries/usa/community/environ/info/decade/decade.html** (accessed June 18, 2002).

McGartland, Albert M. 1984. "Marketable Permit Systems for Air Pollution Control: An Empirical Study." Ph.D. dissertation, University of Maryland.

Menz, Fredric C. 1995. "Transborder Emissions Trading Between Canada and the United States." *Natural Resources Journal* 35 (fall), pp. 803–19.

Metz, Robert. 1993. "N. M. Firm Is Ready to Roll into the Tire Recycling Industry." *Boston Globe,* December 28, p. 34.

"Mexico's Bad Air Exacts Financial Toll, Official Says." 1990. *Boston Globe,* May 30, p. A5.

Miller, Krystal. 1992. "Big 3 to Cooperate on Efforts to Meet Clean-Air Rules." *Wall Street Journal,* June 9, p. B9.

Miller, William. 1993. "Oil Spill Threatens North Sea Wildlife." *Boston Globe,* January 6.

Mills, Edwin S. 1978. *The Economics of Environmental Quality.* New York: Norton.

Miranda, Marie Lynn, and Joseph E. Aldy. 1996. *Unit Pricing of Residential Municipal Solid Waste: Lessons from Nine Case Study Communities.* Report EE-0306, prepared for the Office of Policy, Planning, and Evaluation, U.S. Environmental Protection Agency, March.

Mitchell, Robert Cameron, and Richard T. Carson. 1989. *Using Surveys to Value Public Goods: The Contingent Valuation Method.* Washington, DC: Resources for the Future.

Moore, Michael J., and W. Kip Viscusi. 1988. "Doubling the Estimated Value of Life: Results Using New Occupational Fatality Data." *Journal of Policy Analysis and Management* 7 (spring), pp. 476–90.

Moynihan, Daniel Patrick. 1993. "A Legislative Proposal: Why Not Enact a Law That Would Help Us Set Sensible Priorities?" *EPA Journal* 19 (January–March), pp. 46–47.

Mullen, John K., and Fredric C. Menz. 1985. "The Effect of Acidification Damages on the Economic Value of the Adirondack Fishery to New York Anglers." *American Journal of Agricultural Economics* 67 (February), pp. 112–19.

Murray, Matt. 2001. "EPA Orders Dredging of PCBs from the Upper Hudson River." *Wall Street Journal* (interactive edition), December 5, 2001.

National Academy of Sciences. 1983. *Risk Assessment in the Federal Government: Managing the Process.* Washington, DC: U.S. Government Printing Office.

National Academy of Sciences and National Academy of Engineering. 1974. *Air Quality and Automobile Emission Control,* vol. 4, *The Costs and Benefits of Automobile Emission Control.* Washington, DC: U.S. Government Printing Office.

National Acid Precipitation Assessment Program (NAPAP). 1991. *1990 Integrated Assessment Report.* Washington, DC: NAPAP Office of the Director, November.

National Advisory Committee on Oceans and Atmosphere. 1984. *Nuclear Waste Management and the Use of the Sea.* Washington, DC: U.S. Government Printing Office, April.

National Assessment Synthesis Team. 2001. *Climate Change Impacts on the United States: The Potential Consequences of Climate Variability and Change.* Report for the U.S. Global Change Research Program. Cambridge: Cambridge University Press.

National Commission on Air Quality. 1981. *To Breathe Clean Air.* Washington, DC: U.S. Government Printing Office, March.

National Recycling Coalition. 2002. "Electronics Recycling Initiative." Available at **http://www.nrc-recycle.org/resources/electronics/index.htm** (accessed May 23, 2002).

National Research Council. 1979. *Carbon Dioxide and Climate: A Scientific Assessment.* Washington, DC: National Academy Press.

———. 1987. *Current Issues in Atmospheric Change.* Washington, DC: National Academy Press.

National Research Council, Committee on Passive Smoking, Board on Environmental Studies and Toxicology. 1986. *Environmental Tobacco Smoke: Measuring Exposures and Assessing Health Effects.* Washington, DC: National Academy Press.

National Solid Waste Management Association. 1990. *Recycling in the States: Mid-Year Update 1990.* Washington, DC: NSWMA, October.

Newcomb, Peter. 1991. "'Ban the Box.'" *Forbes,* May 13, p. 70.

"New Emissions Plan Sought." 1995. *Boston Globe,* April 8, p. 2.

"Newspaper Recycling Booming." 1995. *Brockton Enterprise,* July 11, p. 18.

Nitze, William A. 1993. "Stopping the Waste: Technology Itself Is Not the Problem." *EPA Journal* 19 (April–June), pp. 31–33.

Noah, Timothy. 1993. "Environmental Groups Say Deal Poses Threats." *Wall Street Journal,* December 16, p. A10.

———. 1993. "EPA Declares 'Passive' Smoke a Human Carcinogen." *Wall Street Journal,* January 6, p. B1.

———. 1993. "EPA Finds Unsafe Lead Levels in Water." *Wall Street Journal,* May 12, p. B6.

Nordhaus, William D. 1999. "The Future of Environmental and Augmented National Accounts: An Overview." *Survey of Current Business* (November), pp. 45–49.

North American Commission for Environmental Cooperation. 2002. "Supporting Environmental Protection and Conservation Through Green Goods and Services." Available at **http://www.cec.org/programs_projects/trade_environ_econ/** (accessed June 6, 2002).

Nussbaum, Bruce, and John Templeman. 1990. "Built to Last—Until It's Time to Take It Apart." *Business Week,* September 17, p. 102.

Oates, Wallace E., Paul R. Portney, and Albert M. McGartland. 1989. "The Net Benefits of Incentive-Based Regulation: A Case Study of Environmental Standard-Setting." *American Economic Review* 75, pp. 1,223–42.

Oberthür, Sebastian. 1997. *Production and Consumption of Ozone-Depleting Substances, 1986–1995.* Bonn, Germany: Deutsche Gesellschaft für Technische Zusammenarbeit.

OECD (Organisation for Economic Co-operation and Development). 1989. *Economic Instruments for Environmental Protection.* Paris: OECD.

———. 1992. *Global Warming: The Benefits of Emission Abatement.* Paris: OECD.

———. 1999a. *Economic Instruments for Pollution Control and Natural Resources Management in OECD Countries: A Survey.* Paris: OECD.

———. 1999b. *OECD Environmental Data: Compendium 1999.* Paris: OECD.

———. 2001. *OECD Environmental Outlook.* Paris: OECD.

Ohnuma, Keiko. 1990. "Missed Manners." *Sierra* (March/April), pp. 24–26.

Oldenburg, Kirsten U., and Kenneth Geiser. 1997. "Pollution Prevention and/or Industrial Ecology?" *Journal of Cleaner Production* 5, nos. 1–2, pp. 103–108.

O'Neil, William B. 1980. "Pollution Permits and Markets for Water Quality." Ph.D. dissertation, University of Wisconsin, Madison.

O'Neil, William B., Martin H. David, Christina Moore, and Erhard F. Joeres. 1983. "Transferable Discharge Permits and Economic Efficiency: The Fox River." *Journal of Environmental Economics and Management* 10, (December), pp. 346–55.

O'Toole, Randal. 1988. *Reforming the Forest Service.* Washington, DC: Island Press.

Ozone Transport Commission. 2002. *Joint OTC-EPA Report Demonstrates Cost-Effective Air Pollution Reductions.* Washington, DC, April 4.

Palmer, A. R., W. E. Mooz, T. H. Quinn, and K. A. Wolf. 1980. *Economic Implications of Regulating Chlorofluorocarbon Emissions from Nonaerosol Applications.* Report R-2524-EPA, prepared by the Rand Corporation for the U.S. Environmental Protection Agency, June.

Parker, Larry. 2000. *Stratospheric Ozone Depletion: Implementation Issues.* CRS Issue Brief for Congress. Washington, DC: Committee for the National Institute for the Environment, July 12.

———. 2002. *Global Climate Change: Market-Based Strategies to Reduce Greenhouse Gases.* CRS Issue Brief for Congress. Washington, DC: Committee for the National Institute for the Environment, March 12.

Parrish, Michael. 1994. "GM Signs On to Environmental Code of Conduct." *Los Angeles Times,* February 4, pp. D1, D4.

Parson, Edward A., Peter M. Haas, and Marc A. Levy. 1992. "A Summary of the Major Documents Signed at the Earth Summit and the Global Forum." *Environment* 34 (October), pp. 12–15, 34–36.

Patton, Dorothy E. 1993. "The ABCs of Risk Assessment: Some Basic Principles Can Help People Understand Why Controversies Occur." *EPA Journal* 19 (January–March), pp. 10–15.

Peretz, Jean H., and Jeffrey Solomon. 1995. "Hazardous Waste Landfill Costs on Decline, Survey Says." *Environmental Solutions* (April), pp. 21–24.

Perl, Lewis, and Frederick C. Dunbar. 1982. "Cost Effectiveness and Cost-Benefit Analysis of Air Quality Regulations." *American Economic Review* 72 (May), pp. 208–13.

Phifer, Russell W., and William R. McTigue Jr. 1988. *Handbook of Hazardous Waste Management.* Chelsea, MI: Lewis Publishers.

Pianin, Eric, and John Mintz. 2001. "EPA Seeks to Narrow Pollution Initiative." *Washington Post,* August 8, p. A1.

Pigou, A. C. 1952. *The Economics of Welfare,* 4th ed. London: Macmillan.

Pollack, Andrew. 2001. "California Adopts Plan for Electric Cars." *New York Times,* January 27, p. A10.

Porter, Richard C. 1983. "Michigan's Experience with Mandatory Deposits on Beverage Containers." *Land Economics* 59, pp. 177–94.

Portney, Paul R. 1990a. "Air Pollution Policy." In Paul R. Portney, ed., *Public Policies for Environmental Protection,* pp. 27–96. Washington, DC: Resources for the Future.

———. 1990b. "Economics and the Clean Air Act." *Journal of Economic Perspectives* 4 (fall), pp. 173–81.

Preuss, Peter W., and William H. Farland. 1993. "A Flagship Risk Assessment: EPA Reassesses Dioxin in an Open Forum." *EPA Journal* 19 (January–March), pp. 24–26.

Protzman, Ferdinand. 1993. "Germany's Push to Expand the Scope of Recycling." *New York Times,* July 4.

Pryde, Lucy T. 1973. *Environmental Chemistry: An Introduction.* Menlo Park, CA: Cummings.

Putnam, Hayes, and Bartlett Inc. 1987. *Economic Implications of Potential Chlorofluorocarbons Restrictions: Final Report.* December 2.

"Quietly, EPA Drops Some Tobacco Research." 1993. *Boston Globe,* January 7, p. 3.

Quirk, James, and Katsuki Terasawa. 1991. "Choosing a Government Discount Rate: An Alternative Approach." *Journal of Environmental Economics and Management* 20, pp. 16–28.

Radin, Charles A. 1995. "With China's 'Miracle,' Pollution Surges." *Boston Globe,* January 2, pp. 47, 50.

Raeburn, Paul, and Gail DeGeorge. 1997. "You Bet I Mind." *Business Week,* September 15, p. 90.

Ramanathan, V. 1988. "The Greenhouse Theory of Climate Change: A Test by an Inadvertent Global Experiment." *Science* 240, pp. 293–99.

Ramanathan, V., R. D. Cess, E. F. Harrison, P. Minnis, B. R. Barkstrom, E. Ahmad, and D. Hartmann. 1989. "Cloud-Radioactive Forcing and Climate: Results from the Earth Radiation Budget Experiment." *Science* 243, pp. 57–63.

Raul, Alan Charles, and Stephen F. Smith. 1999. "Judicial Oversight Can Restrain Regulators' Use of Junk Science." *Legal Backgrounder,* January 8.

Raven, Peter H., Linda R. Berg, and George B. Johnson. 1993. *Environment.* New York: Saunders College Publishing and Harcourt Brace Jovanovich.

Reese, Craig E. 1985. "State Taxation of Hazardous Materials." *Oil and Gas Tax Quarterly* 33, pp. 502–26.

Reidy, Chris. 1996. "Economics of Recycling Paper Take a Tumble." *Boston Globe,* July 24, pp. A1, A16.

Reifenberg, Anne, and Allanna Sullivan. 1996. "Rising Gasoline Prices: Everyone Else's Fault." *Wall Street Journal,* May 1, pp. B1, B8.

Reilly, William K. 1992. "The Road from Rio." *EPA Journal* 18 (September/October), pp. 11–13.

Reisch, Mark. 1998. *Superfund Reauthorization Issues in the 105th Congress.* CRS Issue Brief for Congress. Washington, DC: Committee for the National Institute for the Environment, November 23.

———. 2000. *Superfund Reauthorization Issues in the 106th Congress.* CRS Issue Brief for Congress. Washington, DC: Committee for the National Institute for the Environment, October 30.

———. 2002. *Superfund and Brownfields in the 107th Congress.* CRS Issue Brief for Congress. Washington, DC: Committee for the National Institute for the Environment, March 21.

Remanufacturing Institute. 2002. "Frequently Asked Questions." Available at **http://www.reman.org/frfaqust.htm** (accessed June 19, 2002).

Repetto, Robert. 1992. "Accounting for Environmental Assets." *Scientific American,* June, pp. 94–100.

———. 1992. "Earth in the Balance Sheet: Incorporating Natural Resources in National Income Accounts." *Environment* 34 (September), pp. 12–20, 43–45.

Resource Integration Systems Ltd. and Waste Matters Consulting (Portland, Oregon). 1990. *Decisionmaker's Guide to Recycling Plastics.* Prepared for the Oregon Department of Environmental Quality, Solid Waste Reduction and Recycling Section, and the U.S. Environmental Protection Agency, Region X, Solid Waste Program, December.

Revkin, Andrew C. 2001. "178 Nations Reach Climate Accord; U.S. Only Looks On." *New York Times,* July 24, p. A1.

———. 2002. "World Briefing: Australia." *New York Times,* June 6, p. A6.

Roach, Fred, Charles Kolstad, Allen V. Kneese, Richard Tovin, and Michael Williams. 1981. "Alternative Air Quality Policy Options in the Four Corners Region." *Southwest Review* 1, pp. 44–45.

Rosen, Harvey S. 2002. *Public Finance.* Boston, MA: Irwin–McGraw-Hill.

Ross, Karl. 1993. "Some Foul Air in Puerto Rico." *Boston Globe,* January 7, p. 12.

Rowen, Henry S., and John P. Weyant. 1997. "The Greenhouse Follies." *Wall Street Journal* (interactive edition), December 2, 1997.

Ruff, Larry. 1970. "The Economic Common Sense of Pollution." *The Public Interest* 19, (spring), pp. 69–85.

"Russia to Continue Dumping Nuclear Waste at Sea." 1993. *Boston Globe,* October 19, p. 2.

Samuelson, Paul A. 1954. "The Pure Theory of Public Expenditure." *Review of Economics and Statistics* 36, pp. 387–89.

———. 1955. "Diagrammatic Exposition of a Theory of Public Expenditure." *Review of Economics and Statistics* 37, pp. 350–56.

———. 1958. "Aspects of Public Expenditure Theory." *Review of Economics and Statistics* 40, pp. 332–38.

Sassone, Peter G., and William A. Schaffer. 1978. *Cost-Benefit Analysis: A Handbook.* New York: Academic Press.

"SCAQMD Moves to Relieve Power Plants from Soaring NO_X Emission Credit Costs." 2001. *McGraw-Hill's Utility Environment Report,* New York: McGraw-Hill, January 26.

Scheuplein, Robert. 1993. "Uncertainty and the 'Flavors' of Risk." *EPA Journal* 19, (January–March), pp. 16–17.

Schmalensee, Richard, Paul L. Joskow, A. Denny Ellerman, Juan Pablo Montero, and Elizabeth M. Bailey. 1998. "An Interim Evaluation of Sulfur Dioxide Emissions Trading." *Journal of Economic Perspectives* 12 (summer), pp. 53–68.

Schmidheiny, Stephan. 1992. *Changing Course: A Global Business Perspective on Development and the Environment.* Cambridge, MA: MIT Press.

Schmidt, William E. 1993. "Aground, Tanker Spills Oil on Shetland." *New York Times,* January 6, p. A3.

Schmitt, Bill. 2000. "Responsible Care: Making New Connections." *Chemical Week* 162 (July 5–12), pp. 38–40.

Schrage, Michael. 1992. "In Tokyo, 'Just-in-Time' Deliveries in Need of Administrative Guidance." *Boston Globe,* March 22.

Schulze, William D., and David S. Brookshire. 1983. "The Economic Benefits of Preserving Visibility in the National Parklands of the Southwest." *Natural Resources Journal* 23 (January), pp. 763–72.

Seattle Solid Waste Utility. 1988. *Seattle Solid Waste Utility Rate Sheet and Customer Reply Card.* Seattle, WA: Seattle Solid Waste Utility.

Seattle Solid Waste Utility, Public Information Department. 1991. *Municipal Solid Waste Management Program Description.* Seattle, WA: Seattle Solid Waste Utility.

Seelye, Katharine Q. 2001. "EPA to Adopt Clinton Arsenic Standard." *New York Times,* November 1, p. A18.

Seitz, Fred, and Christine Plepys. 1995. *Monitoring Air Quality in Health People 2000.* Statistical Notes 9. Hyattsville, MD: National Center for Health Statistics.

Seskin, Eugene P. 1978. "Automobile Air Pollution Policy." In Paul R. Portney, ed., *Current Issues in U.S. Environmental Policy,* pp. 68–104. Baltimore: Johns Hopkins University Press.

Sessions, Kathy. 1993a. "Products of the Earth Summit." *EPA Journal* 19 (April–June), p. 12.

———. 1993b. "What's in Agenda 21?" *EPA Journal* 19 (April–June), p. 14.

Shalal-Esa, Andrea. 1993. "Tobacco Industry Sues EPA over Secondhand Smoke Report." *Boston Globe,* June 23, p. 41.

Shapiro, Michael. 1990. "Toxic Substances Policy." In Paul R. Portney, ed., *Public Policies for Environmental Protection,* pp. 195–241. Washington, DC: Resources for the Future.

Shea, Cynthia Pollock. 1992. "Getting Serious in Germany." *EPA Journal* 18 (July/August), pp. 50–52.

Shirouzu, Norihiko. 2001. "Ford, Toyota Are in Talks to Develop New Electric-Gasoline Hybrid Vehicle." *Wall Street Journal* (interactive edition), November 30, 2001.

Sigman, Hilary. 2000. "Hazardous Waste and Toxic Substance Policies." In Paul R. Portney and Robert N. Stavins, eds., *Public Policies for Environmental Protection,* pp. 215–59. 2d ed. Washington, DC: Resources for the Future.

Simmons, Peter, and Brian Wynne. 1993. "Responsible Care: Trust, Credibility, and Environmental Management." In Kurt Fischer and Johan Schot, eds., *Environmental Strategies for Industry: International Perspectives on Research Needs and Policy Implications,* pp. 207–26. Washington, DC: Island Press.

Skumatz, Lisa. 1996. Economic Research Associates. Unpublished memorandum, to whom it may concern March 21.

Skumatz, Lisa, and Phillip Zach. 1993. "Community-Level Adoption of Variable Rates: An Update." *Resource Recycling.*

Smith, Adam. 1937 [1776]. *An Inquiry into the Nature and Causes of the Wealth of Nations.* New York: Random House.

Smith, Emily T. 1989. "Developments to Watch: Aerosol Cans That Run on Compressed-Air Power." *Business Week,* June 12, p. 63.

Smith, Emily T., Vicki Cahan, Naomi Freundlich, James E. Ellis, and Joseph Weber. 1990. "The Greening of Corporate America: 'Sometimes You Find

That the Public Has Spoken and You Get on with It.'" *Business Week,* April 23, pp. 96–103.

Smith, Emily T., and David Woodruff. 1992. "The Next Trick for Business: Taking a Cue from Nature." *Business Week,* May 11, pp. 74–75.

Smith, V. Kerry, and William H. Desvousges. 1985. "The Generalized Travel Cost Model and Water Quality Benefits: A Reconsideration." *Southern Economic Journal* 52 (October), pp. 371–81.

———. 1986. *Measuring Water Quality Benefits.* Norwell, MA: Kluwer-Nijhoff.

Smith, V. Kerry, William H. Desvousges, and Matthew P. McGivney. 1983. "Estimating Water Quality Benefits: An Econometric Analysis." *Southern Economic Journal* 50 (October), pp. 422–37.

Smith, V. Kerry, and John V. Krutilla, eds. 1982. *Explorations in Natural Resource Economics.* Baltimore: John Hopkins University Press.

Smith, Velma. 1994. "Disaster in Milwaukee." *EPA Journal* 20 (summer), pp. 16–18.

Snyder, Adam. 1992. "The Color of Money." *SuperBrands,* pp. 30–31.

Solley, Wayne B., Robert R. Pierce, and Howard A. Perlman. Estimated Use of Water in the United States in 1995. U.S. Geological Survey Circular 1200. Denver: U.S. Geological Survey, 1998. Available at **http://water.usgs.gov/watuse/pdf1995/html/.**

Solow, Robert M. 1991. "Sustainability: An Economist's Perspective." Paper presented at the Eighteenth J. Seward Johnson Lecture to the Marine Policy Center, Woods Hole Oceanographic Institution, Woods Hole, Massachusetts, June 14, 1991. In Robert N. Stavins, ed. *Economics of the Environment.* NY: Norton, pp. 131–38.

"Some Statutory Mandates on Risk." 1993. *EPA Journal* 19 (January–March), p. 15.

South Coast Air Quality Management District. 1992. *Regional Clean Air Incentives Market.* Diamond Bar, CA: SCAQMD.

Spofford, Walter O., Jr. 1984. *Efficiency Properties of Alternative Source Control Policies for Meeting Ambient Air Quality Standards: An Empirical Application to the Lower Delaware Valley.* Discussion Paper D-118. Washington, DC: Resources for the Future, February.

State Economic and Trade Commission. 2002. "Cleaner Production in China: Environmental Legislation." Available at **http://www.chinacp.com/eng/cnenvleg.html** (accessed June 14, 2002).

State of New Jersey, Department of Environmental Protection. 2002. "2000 Generation, Disposal, and Recycling Rates in New Jersey." Available at **http://www.state.nj.us/dep/dshw/recycle/00munrts.htm** (last updated March 13, 2002).

"State Your Claim." 1992. *EPA Journal* 18 (July/August), p. 10.

Stavins, Robert N. 1998. "What Can We Learn from the Grand Policy Experiment? Lessons from SO_2 Allowance Trading." *Journal of Economic Perspectives* 12 (summer), pp. 68–88.

Stern, Arthur C. 1982. "History of Air Pollution Legislation in the United States." *Journal of the Air Pollution Control Association* 32 (January), pp. 44–61.

Stern, D. I. 1998. "Progress on the Environmental Kuznets Curve?" *Environment and Development Economics* 3 (May), pp. 175–98.

Stiglitz, Joseph E. 1988. *Economics of the Public Sector.* New York: Norton.

Stranahan, Susan Q. 1990. "It's Enough to Make You Sick." *National Wildlife* (February/March), pp. 8–15.

"Study: Dioxin Health Threat Much Worse Than Suspected." *Brockton Enterprise,* September 12, 1994.

Sullivan, R. Lee. 1995. "Snob Water." *Forbes,* August 14, p. 192.

Symbiosis Institute. 2002. "Industrial Symbiosis: Trading Byproducts." Available at **http://www.symbiosis.dk/isik_uk.htm** (accessed June 6, 2002).

Tanner, James. 1992. "Carbon Tax to Limit Use of Fossil Fuels Becomes Embroiled in Global Politics." *Wall Street Journal,* June 9, p. A2.

Taylor, Jeffrey. 1992. "Auction of Rights to Pollute Fetches About $21 Million." *Wall Street Journal,* March 31, p. A6.

———. 1993. "CBOT Plan for Pollution-Rights Market Is Encountering Plenty of Competition." *Wall Street Journal,* August 24, pp. C1, C6.

Taylor, Jeffrey, and Rose Gutfeld. 1992. "CBOT Selected to Run Auction for Polluters." *Wall Street Journal,* September 25, p. C1.

Taylor, Jeffrey, and Dave Kansas. 1992. "Environmentalists Vie for Right to Pollute." *Wall Street Journal,* March 26, p. C1.

Thaler, Richard, and Sherwin Rosen. 1976. "The Value of Life Savings." In Nester Terleckyj, ed., *Household Production and Consumption.* New York: Columbia University Press.

3M Corporation. 2002. "Leading Through Innovation: Our Environment." Available at **http://www.3m.com/about3M/environment/index.jhtml** (accessed June 18, 2002).

Tietenberg, T. H. 1985. *Emissions Trading: An Exercise in Reforming Pollution Policy.* Washington, DC: Resources for the Future.

———. 1996. *Environmental and Natural Resource Economics.* New York: Harper Collins.

Tregarthen, Timothy. 1989. "Garbage by the Bag: Perkasie Acts on Solid Waste." *The Margin* (September/October), p. 17.

Tresch, Richard W. 1981. *Public Finance: A Normative Theory.* Plano, TX: Business Publications.

Tuxen, Linda. 1993. "EPA's IRIS Data Base: Accessing

the Science." *EPA Journal* 19 (January–March), pp. 22–23.

"Two Faces of Risk." 1993. *EPA Journal* 19 (January–March) p. 19.

United Church of Christ, Commission for Racial Justice. 1987. *Toxic Wastes and Race in the United States: A National Report on the Racial and Socio-Economic Characteristics of Communities with Hazardous Waste Sites.* New York: United Church of Christ.

United Nations. 1978. *Water Development and Management: Proceedings of the United Nations Water Conference*, 4 vols. Oxford: Pergamon Press for the United Nations.

———. 2002. "Johannesburg Summit 2002." Available at **http://www.johannesburgsummit.org/** (last updated May 17, 2002).

———. 2000. Division for Sustainable Development. *Agenda 21.* Available at **http://www.un.org/esa/sustdev/agenda21text.htm** (June 29, 2000).

United Nations, Environment Programme. 1996. *Plan for Halon Phaseout in China.* Working paper submitted to the Executive Committee of the Multilateral Fund for the Implementation of the Montreal Protocol, 20th meeting, October 16–18, 1996.

———. 1999. *Global Environment Outlook 2000.* Nairobi, Kenya: United Nations.

———. 2002. "World Summit on Sustainable Development." Available at **http://www.unep.org/wssd/** (accessed June 6, 2002).

United Nations, Framework Convention on Climate Change. 2002a. "Kyoto Protocol: Status of Ratification." Available at **http://unfccc.int/resource/kpstats.pdf** (last modified June 4, 2002).

———. 2002b. "Kyoto Protocol Thermometer." Available at **http://unfccc.int/resource/kpthermo.html** (last modified June 4, 2002).

United Nations, Population Division. 1996. *Annual Populations 1950–2050 (The 1996 Revision)* (diskette). New York: United Nations.

"Updating Previous Big Spills." 1993. *Boston Globe,* January 6, p. 6.

U.S. Census Bureau. 1999. *Statistical Abstract of the United States: 1999,* 119th ed. Washington, DC: U.S. Government Printing Office.

———. 2000. *Statistical Abstract of the United States: 2000,* 120th ed. Washington, DC: U.S. Government Printing Office.

———. 2001. *Statistical Abstract of the United States: 2001,* 121st ed. Washington, DC: U.S. Government Printing Office.

U.S. Congress. 1996. *Safe Drinking Water Act Amendments.* Public Law 104–182, 104th Congress (August 6).

———. 1999. *Regulatory Improvement Act of 1999.* Senate Bill S.746, 106th Congress. Available at **http://thomas.loc.gov/home/c106query.html.**

U.S. Congress, CBO (Congressional Budget Office). 1984. From Office of Technology Assessment, Statement of Joel S. Hirschhorn for the Hearing Record, Senate Committee on Environment and Public Works, 98th Cong., 2d sess. (September 10, 1984).

———. 1985a. *Efficient Investments in Wastewater Treatment Plants.* Washington, DC: U.S. Government Printing Office.

———. 1985b. *Hazardous Waste Management: Recent Changes and Policy Alternatives.* Washington, DC: U.S. Government Printing Office.

———. 1986. *Curbing Acid Rain: Cost, Budget, and Coal-Market Effects.* Washington, DC: U.S. Government Printing Office, June.

U.S. Congress, Congressional Research Service. 1972. *History of the Water Pollution Control Act Amendments of 1972.* Ser. 1, 93d Congress, 1st sess., p. 137.

U.S. Congress, Office of Technology Assessment (OTA). 1983. *Technologies and Management Strategies for Hazardous Waste Control.* Washington, DC: U.S. Government Printing Office.

———. 1984. *Protecting the Nation's Groundwater from Contamination,* v. 1. Washington, DC: U.S. Government Printing Office, October.

———. 1985. *Superfund Strategy.* Washington, DC: U.S. Government Printing Office.

———. 1987. *Wastes in Marine Environments.* Washington, DC: U.S. Government Printing Office.

———. 1988. "Water Supply: The Hydrologic Cycle." In David H. Speidel, Lon C. Ruedisili, and Allen F. Agnew, eds., *Perspectives on Water: Uses and Abuses.* New York: Oxford University Press.

———. 1989. *Facing America's Trash: What Next for Municipal Solid Waste?* Washington, DC: U.S. Government Printing Office.

———. 1992. *Green Products by Design: Choices for a Cleaner Environment.* Washington, DC: U.S. Government Printing Office, October.

U.S. Department of Agriculture and U.S. Environmental Protection Agency. 1998. *Clean Water Action Plan: Restoring and Protecting America's Waters.* EPA-840-R-98-001. Washington, DC.

U.S. Department of Commerce, Bureau of Economic Analysis. Various years. "Pollution Abatement and Control Expenditures." Published periodically in *Survey of Current Business.* Washington, DC: Department of Commerce.

———. 1998. "Summary National Income and Product Series, 1929–97." *Survey of Current Business.* August.

U.S. Department of Commerce, Bureau of the Census. *Government Finances.* Various years. Washington, DC: U.S. Government Printing Office.

———. Various years. *Pollution Abatement Costs and Expenditures.* Current Industrial Reports. Washington, DC: U.S. Government Printing Office.

U.S. Department of Energy, Energy Information Administration. 1997. *International Energy Outlook 1997.* Washington, DC: U.S. Department of Energy, April.

———. 1999. *International Energy Annual 1999.* Available at **http://tonto.eia.doe.gov/FTPROOT/international/021999.pdf**

———. 2001. *Electric Power Industry Monthly Overview.* Industry Summary Statistics.

U.S. Department of Energy and U.S. Environmental Protection Agency. 2002. *Model Year 2002 Fuel Economy Guide.* Available at **http://www.fueleconomy.gov.**

U.S. Department of State. 2002. *U.S. Climate Action Report 2002.* Washington, DC: U.S. Government Printing Office, May.

U.S. DOI (Department of the Interior), Fish and Wildlife Service. 2002. *Threatened and Endangered Species.* Washington, DC: July 31.

———, U.S. Geological Survey. 1998. *Estimated Water Use in the United States in 1995.* Reston, VA: U.S. Geological Survey.

U.S. EPA (Environmental Protection Agency). n.d. *Effluent Trading in Watersheds.* Reinvention Activity Fact Sheets. Washington, DC.

———. 1974 (November). *Review of the Municipal Wastewater Treatment Works Program.* Washington, DC.

———. 1975. *Economic Report: Alternative Methods of Financing Wastewater Treatment.* Washington, DC.

———. 1984. *Final Report: The Cost of Clean Air and Water.* Washington, DC.

———. 1987 (March). *Site Discovery Methods.* EPA Contract 68-01-6888. Washington, DC.

———. 1987. (August). *The New Superfund: What It Is, How It Works.* Washington, DC.

———. 1988 (August). *Environmental Progress and Challenges: EPA's Update.* Washington, DC.

———. 1989 (August). *Comparing Risks and Setting Environmental Priorities.* Washington, DC.

———. 1990 (December). *Meeting the Environmental Challenge: EPA's Review of Progress and New Directions in Environmental Protection.* Washington, DC.

———. 1991. *Superfund NPL Characterization Project: National Results.* Washington, DC.

———. 1991. (March). *Bottled Water Fact Sheet.* Washington, DC.

———. 1994a. *1994 EPA Allowance Auction Results.* Washington, DC.

———. 1994b. *1994 EPA SO_2 Allowance Auction.* Washington, DC.

———. 2000. "Clinton-Gore Administration Acts to Eliminate MTBE, Boost Ethanol." Headquarters Press Release. Washington DC; March 20.

———. 2001. "September 18: Whitman Details Ongoing Agency Efforts to Monitor Disaster Sites, Contribute to Cleanup Efforts." Available at **http://www.epa.gov/wtc/stories/headline_091801.htm** (September 18, 2001).

———. 2002 (January). "President Bush to Commit More Than $20 Million for Watershed Protection." Headquarters Press Release. Washington, DC, January 25, 2002.

———. 2002a (September). "Benchmarks, Standards, and Guidelines Established to Protect Public Health." Available at **http://www.epa.gov/wtc/benchmarks.htm** (accessed September 24, 2002).

———. 2002b (September). "Daily Environmental Monitoring Summary." Available at **http://www.epa.gov/wtc/data_summary.htm** (accessed September 24, 2002).

———. 2002 (March 21). "EPA Administrator Whitman Announces 729 of the Nation's Top Energy Performing Buildings." Headquarters Press Release. Washington, DC, March 21, 2002.

———. 2002 (March 26). "Whitman Recognizes 36 Organizations for Energy Efficiency Accomplishments." Headquarters Press Release. Washington, DC, March 26, 2002.

———. 2002 (May). "U.S.–Mexico Border Program." Available at **http://www.epa.gov/usmexicoborder/** (last updated May 10, 2002).

U.S. EPA, Environmental Criteria and Assessment Office. 1993. *IRIS Data Base.* Research Triangle Park, NC.

U.S. EPA, Office of Air and Radiation. 1990 (November 15). *The Clean Air Act Amendments of 1990: Summary Materials.* Washington, DC.

———. 1990 (November 30). *Clean Air Act Amendments of 1990: Detailed Summary of Titles.* Washington, DC.

———. 1997 (October). *The Benefits and Costs of the Clean Air Act, 1970 to 1990.* Washington, DC.

———. 1999 (June). *Summary of Decision: American Trucking Associations Inc. v. USEPA, May 14, 1999.* Washington, DC: June 28.

———. 1999 (November). *The Benefits and Costs of the Clean Air Act, 1990 to 2010.* Washington, DC.

———. 2000 (June). *Air Quality Index: A Guide to Air Quality and Your Health.* Washington, DC.

———. 2002 (January). "Number of Days with an Air Quality Index over 100 by City." Available at **http://www.epa.gov/airtrends/** (accessed January 16, 2002).

———. 2002 (February). "Ozone Science: The Facts Behind the Phaseout." Available at **http://www.epa.gov/ozone/science/sc_fact.html** (last updated February 5, 2002).

———, Clean Air Markets Division. 2001 (May). *2000 OTC NO_X Budget Program Compliance Report.* Washington, DC, May 9.

———. 2002a (January). "SO_2 Allowance Market Analysis: Monthly Average Price of Sulfur Dioxide Allowances." Available at **http://www.epa.gov/**

airmarkets/trading/so2market/prices.html (last updated January 29, 2002).

———. 2002b (January). "SO_2 Allowance Market Analysis: SO_2 Trading Activity Breakdown." Available at **http://www.epa.gov/airmarkets/trading/so2market/qlyupd.html** (last updated January 29, 2002).

———. 2002 (July). "SO_2 Allowance Market Analysis: Monthly Average Price of Sulfur Dioxide Allowances." Available at **http://www.epa.gov/airmarkets/trading/So2market/pricetbl.html** (last updated July 1, 2002).

U.S. EPA, Office of Air and Radiation, Global Warming. 2002 (March). "Emissions." Available at **http://yosemite.epa.gov/oar/globalwarming.nsf/content/emissions/html** (last updated March 7, 2002).

U.S. EPA, Office of Air Quality Planning and Standards. 1992 (October). *National Air Quality and Emissions Trends Report, 1991.* Research Triangle Park, NC.

———. 1995 (October). *National Air Quality and Emissions Trends Report, 1994.* Research Triangle Park, NC.

———. 1997 (July). *Regulatory Impact Analyses for the Particulate Matter and Ozone National Ambient Air Quality Standards and Proposed Regional Haze Rule.* Research Triangle Park, NC, July 17.

———. 1999 (July). *Smog—Who Does It Hurt? What You Need to Know About Ozone and Your Health.* Washington, DC.

———. 2001 (March). *National Air Quality and Emissions Trends Report, 1999.* Washington, DC.

———. 2001 (June 8). "USA Air Quality Nonattainment Areas." Available at **http://www.epa.gov/airs/nonattn.html** (effective June 8, 2001).

———. 2001 (June 22). *Fact Sheet: Background Paper for the 90-Day Evaluation of the Environmental Protection Agency's New Source Review Program.* Research Triangle Park, NC, June 22.

———. 2001 (September). *Latest Findings on National Air Quality: 2000 Status and Trends.* Research Triangle Park, NC.

———. 2002. "National Ambient Air Quality Standards (NAAQS)." Available at **http://www.epa.gov/airs/criteria.html** (accessed March 29, 2002).

U.S. EPA, Office of Atmospheric Programs, Global Programs Division. 1998 (February). "Significant New Alternatives Policy (SNAP) Final Rule Summary." Available at **http://www.epa.gov/ozone/snap/fact.html** (last updated February 4, 1998).

———. 1998 (June). "HCFC Phaseout Schedule." Available at **http://www.epa.gov/ozone/title6/phaseout/hcfc.html** (last updated June 17, 1998).

———. 2001 (November). "The Accelerated Phaseout of Class I Ozone-Depleting Substances." Available at **http://www.epa.gov/ozone/title6/phaseout/accfact.html** (last updated November 1, 2001).

U.S. EPA, Office of Communications and Public Affairs. 1989 (December). *Glossary of Environmental Terms and Acronym List.* Washington, DC.

———. 1991 (May). "EPA Tightens Standards for Lead in Drinking Water." *Environmental News,* May 7.

U.S. EPA, Office of Communications, Education, and Media Relations. 2001 (February). "Supreme Court Upholds EPA Position on Smog, Particulate Rules." *Environmental News,* February 27.

U.S. EPA, Office of Communications, Education, and Public Affairs 1992 (April). *Securing Our Legacy: An EPA Progress Report 1989–1991.* Washington, DC.

———. 1992 (September). *Terms of Environment: Glossary, Abbreviations, and Acronyms.* Washington, DC.

———. 1994a (February). "Federal Actions Address Environmental Justice." *EPA Activities Update,* February 22.

———. 1994b (February). "Major Changes Recommended in the Clean Water Act." *EPA Activities Update,* February 22.

———. 1994 (March). "EPA, Chicago Board of Trade Announce Results of Second Acid Rain Auction." *Environmental News,* March 29.

———. 1998 (September). "Secondhand Smoke." Press Release. Washington, DC, September 15.

U.S. EPA, Office of Drinking Water. n.d. *Bottled Water: Helpful Facts and Information.* Washington, DC.

U.S. EPA, Office of Emergency and Remedial Response. 1992 (June). *Superfund Progress—Aficionado's Version.* Washington, DC.

———. 2000 (April). "Superfund Hazardous Waste Site Query." Internet accessible database, available at **http://ceg.eh.doe.gov/nepa/reports/statistics/tab8x9.html** (last updated April 2000).

U.S. EPA, Office of Mobile Sources. 1993 (February). *Guidance for the Implementation of Accelerated Retirement of Vehicles Programs.* Ann Arbor, MI.

———. 1997 (December). *Environmental Fact Sheet: Accelerated Vehicle Retirement Programs.* Ann Arbor, MI.

U.S. EPA, Office of Pesticide Programs. 2000 (December). "Pesticide Reregistration Facts." Available at **http://www.epa.gov/oppfead1/trac/factshee.htm** (last updated December 13, 2000).

———. 2002 (April). "Pesticide Environmental Stewardship Program." Available at **http://www.epa.gov/oppbppd1/PESP/** (last updated April 22), 2002.

U.S. EPA, Office of Pesticides and Toxic Substances. 1988 (December). *Highlights of the 1988 Pesticide Law: The Federal Insecticide, Fungicide, and Rodenticide Act Amendments of 1988.* Washington, DC.

———. 1990 (September). *Pesticides in Drinking-Water Wells.* Washington, DC.

———. 1991 (June). *For Your Information: Pesticide Reregistration.* Washington, DC.

———. 1991 (October). *Pesticides and Ground-Water Strategy.* Washington, DC.

U.S. EPA, Office of Policy Analysis. 1985 (February). *Costs and Benefits of Reducing Lead in Gasoline: Final Regulatory Impact Analysis.* Report EPA-230-05-85-006. Washington, DC.

———, Office of Policy, Planning, and Evaluation. 1987 (February). *Unfinished Business: A Comparative Assessment of Environmental Problems.* Washington, DC: U.S. Government Printing Office.

———. 1987 (August). *EPA's Use of Benefit-Cost Analysis, 1981–1986.* EPA Report 230-05-87-028. Washington, DC.

U.S. EPA, Office of Policy, Economics, and Innovation, Office of the Administrator. 2001 (January). *The United States Experience with Economic Incentives for Protecting the Environment.* Washington, DC.

U.S. EPA, Office of Policy, Planning, and Evaluation. 1990 (December). *Environmental Investments: The Cost of a Clean Environment—A Summary.* Washington, DC.

———. 1991 (March). *Economic Incentives: Options for Environmental Protection.* Washington, DC.

———. 1992 (June). *Environmental Equity: Reducing Risk for All Communities.* Washington, DC.

———. 1992 (June 29). *State Implementation of Nonpoint Source Programs.* Draft report. Washington, DC.

———. 1992 (July). *The United States Experience with Economic Incentives to Control Environmental Pollution.* Washington, DC.

———, Office of Research and Development. 1989 (December). *The Potential Effects of Global Climate Change on the United States.* Washington, DC.

U.S. EPA, Office of Pollution Prevention. 1991 (October). *Pollution Prevention 1991: Progress on Reducing Industrial Pollutants.* Washington, DC.

U.S. EPA, Office of Pollution Prevention and Toxics. 1993 (October). *Design for the Environment: Environmental Accounting and Capital Budgeting Project Update #1.* Washington, DC.

———. 1997 (March). *Chemistry Assistance Manual for Premanufacture Notification Submitters.* Washington, DC.

———. 1997 (April). *1995 Toxics Release Inventory.* Washington, DC.

———. 2001 (August). *Green Chemistry Program Fact Sheet.* Washington, DC.

———. 2001a (October). "Design for the Environment." Available at **http://www.epa.gov/dfe/about/about.htm** (last updated October 10, 2001).

———. 2001b (October). "Integrated Environmental Management Systems (IEMS)." Available at **http://www.epa.gov/opptintr/dfe/tools/iems.htm** (last updated October 10, 2001).

———. 2001c (October). "What Is Green Chemistry?" Available at **http://www.epa.gov/greenchemistry/whats_gc.html** (last updated October 21, 2001).

———. 2001 (December). "EPA's Green Chemistry Program." Available at **http://www.epa.gov/opptintr/greenchemistry/** (last updated December 26, 2001).

———. 2002 (May). "EPA's Chemical Testing and Information Home Page." Available at **http://www.epa.gov/opptintr/chemtest/** (last revised May 9, 2002).

U.S. EPA, Office of Prevention, Pesticides, and Toxic Substances. 1999 (December). *Office of Pesticide Programs: Biennial Report for FY 1998 and 1999.* Washington, DC.

———. 2001 (May). "Summary of Accomplishments." Available at **http://www.epa.gov/opptintr/newchms/accomplishments.htm** (last revised May 16, 2001).

———. 2002 (March). *Design for the Environment Projects.* Washington, DC.

U.S. EPA, Office of Research and Development, National Center for Environmental Assessment. 2001 (September). "Dioxin and Related Compounds." Available at **http://cfpub.epa.gov/ncea/cfm/dioxin.cfm** (last revised September 19, 2001).

U.S. EPA, Office of Solid Waste. 1986. *1986 National Screening Survey of Hazardous Waste Treatment, Storage, Disposal, and Recycling Facilities:* Summary of Results for TSDR Facilities Active in 1985. Washington, DC.

———. 1986 (November). *Solving the Hazardous Waste Problem: EPA's RCRA Program.* Washington, DC.

———. 1989 (January). *Recycling Works! State and Local Solutions to Solid Waste Management Problems.* Washington, DC.

———. 2001 (August). "What Is Product Stewardship?" Available at **http://www.epa.gov/epr/about/index.htm** (last updated August 29, 2001).

U.S. EPA, Office of Solid Waste and Emergency Response. 1985 (October). *The New RCRA Fact Book.* Washington, DC.

———. 1989 (November). *Decision-Makers Guide to Solid Waste Management.* Washington, DC.

———. 1990 (February). *EPA's Report to Congress on Methods to Manage and Control Plastic Wastes.* Washington, DC.

———. 1991 (October). *Summary of Markets for Scrap Tires.* Washington, DC.

———. 1992 (January). *Leaking Underground Storage Tanks and Health: Understanding Health Risks from Petroleum Contamination.* Washington, DC.

———. 1992 (spring). "The Wisdom of Waste Reduction at McDonald's." *Reusable News*, pp. 1, 3.

———. 1992 (May). *Superfund Progress, Spring 1992.* Washington, DC.

———. 1992 (fall). "FTC Announces Environmental Marketing Guidelines for Industry." *Reusable News*, pp. 1, 8.

———. 1993 (October). *WasteWise: EPA's Voluntary Program for Reducing Business Solid Waste.* Washington, DC.

———. 1997 (April). *Archival of CERCLIS Sites.* Washington, DC.

———. 1997a (May). *Characterization of Municipal Solid Waste in the United States: 1996 Update.* Washington, DC.

———. 1997b (May). *WasteWise Update: Remanufactured Products—Good as New.* Washington, DC.

———. 1998 (October). *WasteWise Update: Extended Product Responsibility.* Washington, DC.

———. 1998 (December). *Extended Product Responsibility: A Strategic Framework for Sustainable Products.* Washington, DC.

———. 2000. *Municipal Solid Waste in the United States, 1999 Final Report.* Washington, DC: U.S. EPA.

———. 2000 (March). *WasteWise Update: Moving Toward Sustainability.* Washington, DC.

———. 2000 (October). "Superfund Environmental Indicators: Cleanup of Hazardous Waste Sites." Available at **http://www.epa.gov/superfund/accomp/ei/npl.htm** (last updated October 3, 2000).

———. 2000 (December). *Superfund: 20 Years of Protecting Human Health and the Environment.* Washington, DC, December 11.

———. 2001 (June). *The National Biennial RCRA Hazardous Waste Report: National Analysis (Based on 1999 Data).* Washington, DC.

———. 2001 (July). *Municipal Solid Waste in the United States: 1999 Facts and Figures.* Washington, DC.

———. 2002 (May). "Municipal Solid Waste: Tires." Available at **http://www.epa.gov/epaoswer/non-hw/muncpl/tires.htm** (last updated May 9, 2002).

———, Office of Underground Storage Tanks. 2002 (February). "Leaking Underground Storage Tank Trust Fund." Available at **http://www.epa/swerust1/ltffacts.htm** (last updated February 14, 2002).

———. 2002 (March). "Overview of the Federal UST Program." Available at **http://www.epa.gov/swerust1/overview.htm** (last updated March 25, 2002).

U.S. EPA, Office of the Chief Financial Officer. 2000 (September). *EPA Strategic Plan.* Washington, DC.

U.S. EPA, Office of Toxic Substances. 1987 (June). *The Layman's Guide to the Toxic Substances Control Act.* Washington, DC.

U.S. EPA, Office of Transportation and Air Quality. 1994a (August). "Clean Fuels: An Overview." Fact Sheet OMS-6. Ann Arbor, MI.

———. 1994b (August). "Milestones in Auto Emissions Control." Fact Sheet OMS-12. Ann Arbor, MI.

———. 1994c (August). "Vehicle Fuels and the 1990 Clean Air Act." Fact Sheet OMS-13. Ann Arbor, MI.

U.S. EPA, Office of Water. 1988 (July). *Lead Contamination Control Act.* Washington, DC.

———. 1989 (February). *Marine and Estuarine Protection: Programs and Activities.* Washington, DC.

———. 1991 (December). *Is Your Drinking Water Safe?* Washington, DC.

———. 1992 (January). *Managing Nonpoint Source Pollution.* Washington, DC.

———. 1992a (April). *Drinking Water Regulations and Health Advisories.* Washington, DC.

———. 1992b (April). *National Water Quality Inventory: 1990 Report to Congress.* Washington, DC.

———. 1992 (May). *Phase V Rule: Fact Sheet.* Washington, DC.

———. 1994. *Clean Water Act Initiative: Analysis of Costs and Benefits.* Washington, DC.

———. 1994 (March). *National Water Quality Inventory: 1992 Report to Congress.* Washington, DC.

———. 1995 (January). *The Clean Water State Revolving Fund: Financing America's Environmental Infrastructure—A Report of Progress.* Washington, DC.

———. 1995 (December). *National Water Quality Inventory: 1994 Report to Congress.* Washington, DC.

———. 1997 (July). *Water on Tap: A Consumer's Guide to the Nation's Drinking Water.* Washington, DC.

———. 2000 (June). *National Water Quality Inventory: 1998 Report to Congress.* Washington, DC.

———. 2001 (May). *Financing America's Clean Water Since 1987: A Report of Progress and Innovation.* Washington, DC.

———. 2001 (June). *Protecting the Nation's Waters Through Effective NPDES Permits: A Strategic Plan FY 2001 and Beyond.* Washington, DC.

———. 2002 (February). *Section 319 Success Stories*, vol. 3, *The Successful Implementation of the Clean Water Act's Section 319 Nonpoint Source Pollution Program.* Washington, DC.

———. 2002 (June). "Effluent Trading in Watersheds Policy Statement." Available at **http://www.epa.gov/owow/watershed/tradetbl.html** (last revised June 15, 2002).

———. 2002 (June). "Financial Support and Flexibility." Available at **http://www.epa.gov/ow=owm.html/ewfinance/construction.htm** (last modified June 28, 2002).

———. 2002 (July). "State-EPA NPS Partnership."

Available at **http://www.epa.gov/owow/nps/partnership.html** (last revised July 24, 2002).
———. 2002 (May). "Effluent Guidelines: Effluent Guidelines Task Force Mission Statement." Available at **http://www.epa.gov/waterscience/guide/taskforce/mission.html** (last updated May 3, 2002).
———, Office of Ground Water and Drinking Water. 1991 (May). *Fact Sheet: National Primary Drinking Water Regulations for Lead and Copper.* Washington, DC.
———. 1997 (January). *Community Water Systems Survey,* vol. 1. Washington, DC.
———. 2000 (December). *Arsenic in Drinking Water Rule: Economic Analysis.* Developed by Abt Associates Inc., Bethesda, Maryland. Washington, DC.
———. 2002 (April). "Current Drinking Water Standards." Available at **http://www.epa.gov/safewater/mcl.html** (last updated April 17, 2002).
———. 2002 (May). "Drinking Water State Revolving Fund." Available at **http://www.epa.gov/ogwdw/dwsrf.html** (last revised May 8, 2002).
U.S. EPA, Office of Water, Office of Pesticides and Toxic Substances. 1992 (winter). *National Pesticide Survey: Update and Summary of Phase II Results.* Washington, DC.
U.S. EPA, Office of Water Regulations and Standards. 1988 (September). *Introduction to Water Quality Standards.* Washington, DC.
U.S. EPA, Region 2. 2000 (December). "EPA Proposes Comprehensive Plan to Clean Up Hudson River PCBs." Press release. December 6.
U.S. EPA, Region 5. 1991 (May). *A Risk Analysis of Twenty-Six Environmental Problems.* Washington, DC.
U.S. EPA, Science Advisory Board. 1990 (September). *Reducing Risk: Setting Priorities and Strategies for Environmental Protection.* Washington, DC.
———. 2002 (April). *Industrial Ecology: A Commentary by the EPA Science Advisory Board.* Washington, DC, April 4.
U.S. EPA, Stratospheric Protection Program, Office of Air and Radiation. 1987 (December). *Regulatory Impact Analysis,* vols. 1–3. Washington, DC.
———. 1999 (June). *Report on the Supply and Demand of CFC-12 in the United States 1999.* Washington, DC, June 9.
U.S. EPA and Science Applications International Corporation. 2001. *LCAccess-LCA 101.* Available at **http://www.epa.gov/ORD/NRMRL/lcaccess/lca101.htm.**
U.S. EPA and U.S. Department of Energy. 2002a. "About Energy Star." Available at **http://www.epa.gov/nrgystar/about.html** (accessed April 24, 2002).
———. 2002b. "Energy Star History." Available at **http://www.epa.gov/nrgystar/newsroom/news_estarhistory.htm** (accessed April 24, 2002).
———. 2002c. "February Newsletter." Available at **http://www.epa.gov/nrgystar/newsroom/news_headline_new.htm** (accessed April 24, 2002).
U.S. EPA, U.S. Department of Health and Human Services, and U.S. Public Health Service. 1992. *A Citizen's Guide to Radon: The Guide to Protecting Yourself and Your Family from Radon,* (3d Ed. Washington, DC, September.
U.S. GAO (General Accounting Office). 1983. *Siting of Hazardous Waste Landfills and Their Correlation with Racial and Economic Status of Surrounding Communities.* Washington, DC.
———. 1985. *Illegal Disposal of Hazardous Waste: Difficult to Detect or Deter.* Washington, DC, February 22.
———. 1987. *SUPERFUND: Extent of Nation's Potential Hazardous Waste Problem Still Unknown.* Washington, DC, December.
———. 1989. *Water Pollution: More EPA Action Needed to Improve the Quality of Heavily Polluted Waters.* Washington, DC, January.
———. 1990 (June). *Drinking Water: Compliance Problems Undermine EPA Program as New Challenges Emerge.* Washington, DC.
———. 1990 (October). *Water Pollution: Greater EPA Leadership Needed to Reduce Nonpoint Source Pollution.* Washington, DC.
———. 1991 (March). *Pesticides: EPA's Use of Benefit Assessment in Regulating Pesticides.* Washington, DC.
———. 1991 (July). *Water Pollution: Stronger Efforts Needed by EPA to Control Toxic Water Pollution.* Washington, DC.
U.S. Government, White House. 1997. "President Clinton's Proposal on Global Climate Change." Available at **http://www.epa.gov/globalwarming/publications/actions/clinton/index.html** (October 22, 1997).
———. 2001a. "The President's Energy Legislative Agenda." Press release. Available at **http://www.whitehouse.gov/news/releases/2001/06/print/energyinit.html.**
———. 2001b. "Climate Change Review—Initial Report." Available at **http://www.whitehouse.gov/news/releases/2001/06/climatechange.pdf** (June 2001).
———. 2002a. *Global Climate Change Policy Book.* Washington, DC, February.
———. 2002b. *Executive Summary—The Clear Skies Initiative.* Press release. Available at **http://www.whitehouse.gov/news/releases/2002/02/clearskies.htm** (February 14, 2002).
U.S. Office of Management and Budget. 1992. *Guidelines*

and Discount Rates for Benefit-Cost Analysis of Federal Programs. Circular A-94, Revised. Washington, DC, October 29.

U.S. Senate, Staff of the Subcommittee on Air and Water Pollution of the Committee on Public Works. 1973. *The Impact of Auto Emission Standards.* Washington, DC: U.S. Government Printing Office, October.

U.S. Trade Representative. 2002. "Encouraging Environmental Protection." *In NAFTA 5 Years Report.* Available at **http://www.ustr.gov/naftareport/encouraging2.htm** (accessed June 10, 2002).

———, WTO and Multilateral Affairs. 2000a. "The WTO and U.S. Economic Growth." Available at **http://www.ustr.gov/wto/wtofact2.html** (March 2, 2000).

———. 2000b. "What Is the World Trade Organization (WTO)?" Available at **http://www.ustr.gov/wto/wtofact3.html** (April 12, 2000).

U.S. Trade Representative, World Regions. 2002. "NAFTA Overview." Available at **http://www.ustr.gov/regions/whemisphere/overview.shtml** (accessed June 6, 2002).

Van Putten, Mark C., and Bradley D. Jackson. 1986. "The Dilution of the Clean Water Act." *Journal of Law Reform* 19 (summer), pp. 863–901.

Vaughan, William J., and Clifford S. Russell. 1982. "Valuing a Fishing Day: An Application of a Systematic Varying Parameter Model." *Land Economics* 58 (November), pp. 450–63.

Vaupel, James W. 1978. "Truth or Consequences: Some Roles for Scientists and Analysts in Environmental Decisionmaking, pp. 71–92." In Wesley A. Magat, ed., *Reform of Environmental Regulation.* Cambridge, MA: Ballinger.

Vogan, Christine R. 1996. "Pollution Abatement and Control Expenditures, 1972–94." *Survey of Current Business* 76 (September), pp. 48–67.

Wald, Matthew L. 2002. "Court Says Agency Can Tighten Smog Rules." *New York Times,* March 27, p. A18.

Wartzman, Rick. 1993. "Administration Alters Proposal for Energy Tax." *Wall Street Journal,* April 2, pp. A2, A4.

Welch, David, and Lorraine Woellert. 2000. "The Eco-Car." *Business Week,* August 14, pp. 62–68.

Wernette, D. R., and L. A. Nieves. 1992. "Breathing Polluted Air." *EPA Journal* 18 (March/April), pp. 16–17.

West, Patrick C. 1992a. "Health Concerns for Fish-Eating Tribes? Government Assumptions Are Much Too Low." *EPA Journal* 18 (March/April), pp. 15–16.

———. 1992b. "Invitation to Poison? Detroit Minorities and Toxic Fish Consumption from the Detroit River." In Bunyan Bryant and Paul Mohai, eds. A Time For Discourse. *Race and the Incidence of Environmental Hazards* Boulder, CO: Westview Press.

"What Is Environmental Technology?" 1994. *EPA Journal* 20 (fall), p. 8.

White, Gilbert W. 1988. "Water Resource Adequacy: Illusion and Reality." In David H. Speidel, Lon C. Ruedisili, and Allen F. Agnew, eds., *Perspectives on Water: Uses and Abuses,* chap. 2. New York: Oxford University Press.

"White House to Seek Reduced Use of Pesticides." *Boston Globe,* September 21, 1993, p. 7.

White, Lawrence J. 1982. *The Regulation of Air Pollutant Emissions from Motor Vehicles.* Washington, DC: American Enterprise Institute for Public Policy Research.

Whiteman, Lily. 1992. "Trades to Remember: The Lead Phasedown." *EPA Journal* 18 (May/June), pp. 38–39.

"Whitman Rejects Clean Air Plan." 2002. *New York Times,* February 20.

Winestock, Geoff. 2001. "EU Brings Global-Warming Treaty Back to Life Without U.S. Support." *Wall Street Journal* (interactive edition), July 24.

Wirth, Timothy E., and John Heinz. 1991. *Project 88: Round II Incentives for Action—Designing Market-Based Environmental Strategies.* Washington, DC, May.

Wolf, Sidney M. 1988. *Pollution Law Handbook: A Guide to Federal Environmental Laws.* New York: Quorom Books.

World Bank. *World Development Indicators.* CD-ROM annual. Washington, DC: IBRD, World Bank.

World Resources Institute. 1992a. *The 1992 Information Please Environmental Almanac.* Boston, MA: Houghton Mifflin.

———. 1992b. *World Resources 1992–93: A Guide to the Global Environment.* New York: Oxford University Press.

———. 1994. *World Resources 1994–95: A Guide to the Global Environment.* New York: Oxford University Press.

———. 1998. *World Resources 1998–99: A Guide to the Global Environment.* New York: Oxford University Press.

World Trade Organization. 2002a. "The WTO and Its Committee on Trade and Environment." Available at **http://www.wto.org/english/tratop_e/envir_e/issu1_e.htm** (accessed June 6, 2002).

———. 2002b. "The WTO in Brief." Available at **http://www.wto.org/english/thewto_e/whatis_e/inbrief_e/inbr00_e.htm** (accessed June 4, 2002).

Yang, Dori Jones, William C. Symonds, and Lisa Driscoll. 1991. "Recycling Is Rewriting the Rules of

Papermaking." *Business Week,* April 22, pp. 100H–101H.

York, Michael. 1993. "President Wins One on NAFTA; Court Reverses Order to Study Environment." *Washington Post,* September 25, p. A1.

Younger, Joseph D. 1992. "Cleaner, 'Greener' Gasoline?" *AAA World* (November/December), pp. 14–15.

Zwick, David, and Marcy Benstock. 1971. *Water Wasteland: Ralph Nader's Study Group Report on Water Pollution.* New York: Grossman.

Glossary

A

abatement equipment subsidy: A payment aimed at lowering the cost of abatement technology.

"acceptable" risk: The amount of risk determined to be tolerable for society.

acidic deposition: Arises when sulfuric and nitric acids mix with other airborne particles and fall to the earth as dry or wet deposits.

acid rain: Arises when sulfuric and nitric acids mix with other airborne particles and fall to the earth as precipitation.

action level: The manner in which MCLs are expressed, generally in milligrams per liter.

Air Quality Control Region (AQCR): A geographic area designated by the federal government within which common air pollution problems are shared by several communities.

Air Quality Index (AQI): An index that signifies the worst daily air quality in an urban area over some time period.

allocative efficiency: Requires that resources be appropriated such that the additional benefits to society are equal to the additional costs incurred.

allocatively efficient standards: Standards set such that the associated marginal social cost (*MSC*) of abatement equals the marginal social benefit (*MSB*) of abatement.

allowance market for ozone-depleting chemicals: A system that allows firms to produce or import ozone depleters only if they hold an appropriate number of tradeable allowances.

ambient standard: A standard that designates the quality of the environment to be achieved, typically expressed as a maximum allowable pollutant concentration.

anthropogenic pollutants: Contaminants associated with human activity, including polluting residuals from consumption and production.

averting expenditure method (AEM): Estimates benefits as the change in spending on goods that are *substitutes* for a cleaner environment.

B

back-end or waste-end charge: A fee implemented at the time of disposal based on the quantity of waste generated.

bag-and-tag approach: A unit pricing scheme implemented by selling tags to be applied to waste receptacles of various sizes.

behavioral linkage approach: Estimates benefits using observations of behavior in actual markets or survey responses about hypothetical markets.

benefit-based decision rule: A guideline to improve society's well-being with no allowance for a balancing of the associated costs.

benefit-based standard: A standard set to improve society's well-being with no consideration for the associated costs.

benefit-cost analysis: A strategy that compares the *MSB* of a risk reduction policy to the associated *MSC*.

benefit-cost ratio: The ratio of *PVB* to *PVC* used to determine the feasibility of a policy option if its magnitude exceeds unity.

best available technology (BAT): A treatment technology that makes attainment of the MCL feasible, taking cost considerations into account.

best management practices (BMP): Strategies other than effluent limitations to reduce pollution from nonpoint sources.

biodiversity: The variety of distinct species, their genetic variability, and the variety of ecosystems they inhabit.

brownfields: Abandoned or underutilized industrial sites where redevelopment is discouraged by actual or perceived contamination.

Btu tax: A per unit charge based on the energy content of fuel, measured in British thermal units (Btu).

bubble policy: Allows a plant to measure its emissions as an average of all emission points emanating from that plant.

C

capital costs: Fixed expenditures for plant, equipment, construction in progress, and production process changes associated with abatement.

carbon sinks: Natural absorbers of CO_2, such as forests and oceans.

carbon tax: A per unit charge based on the carbon content of fuel.

changes in manufacturing processes: The use of alternative production methods to generate less hazardous by-products.

characteristic wastes: Hazardous wastes identified as those exhibiting certain characteristics that imply a substantial risk.

chlorofluorocarbons (CFCs): A family of chemicals that scientists believe contributes to ozone depletion.

circular flow model: Illustrates the real and monetary flows of economic activity through the factor market and the output market.

clean alternative fuels: Fuels, such as methanol, ethanol, or other alcohols, or power sources, such as electricity, used in a clean fuel vehicle.

cleaner production: A preventive environmental strategy applied to products and processes to improve efficiency and reduce risk.

clean fuel vehicle: A vehicle certified to meet stringent emission standards.

Coase Theorem: Assignment of property rights, even in the presence of externalities, will allow bargaining such that an efficient solution can be obtained.

command-and-control approach: A policy that directly regulates polluters through the use of rules or standards.

common property resources: Those resources for which property rights are shared.

comparative risk analysis: An evaluation of relative risk.

competitive equilibrium: The point where marginal private benefit (MPB) equals marginal private cost (MPC), or where marginal profit ($M\pi$)=0.

Comprehensive Environmental Response, Compensation, and Liability Information System (CERCLIS): A national inventory of hazardous waste site data.

consumer surplus: The net benefit to buyers estimated by the excess of the marginal benefit (MB) of consumption over market price (P), aggregated over all units purchased.

contingent valuation method (CVM): Uses surveys to elicit responses about WTP for environmental quality based on hypothetical market conditions.

conventional pollutant: An identified pollutant that is well understood by scientists.

corrective tax: A tax aimed at rectifying a market failure and improving resource allocation.

cost-effective abatement criterion: Allocation of abatement across polluting sources such that the *MAC*s for each source are equal.

cost-effectiveness: Requires that the least amount of resources be used to achieve an objective.

cost-effective policy: A policy that meets an objective using the least amount of economic resources.

"cradle-to-grave" management system: A command-and-control approach to regulating hazardous solid wastes through every stage of the waste stream.

criteria documents: Reports that present scientific evidence on the properties and effects of known or suspected pollutants.

criteria pollutants: Substances known to be hazardous to health and welfare, characterized as harmful by criteria documents.

cyclical flow of materials: Assumes that materials run in a circular pattern in a closed system, allowing residuals to be returned to the production process.

D

damage function method: Models the relationship between a contaminant and its observed effects as a way to estimate damage reductions arising from policy.

deadweight loss to society: The net loss of consumer and producer surplus due to an allocatively inefficient market event.

declining block pricing structure: A pricing structure in which the per unit price of different blocks or quantities of water declines as usage increases.

deflating: Converts a nominal value into its real value.

demand: The quantities of a good the consumer is willing and able to purchase at a set of prices during some discrete time period, *c.p.*

***de minimis* risk**: A negligible level of risk such that reducing it further would not justify the costs of doing so.

deposit/refund system: A market instrument that imposes an up-front charge to pay for potential damages and refunds it for returning a product for proper disposal or recycling.

Design for the Environment (DfE): An initiative that promotes the use of environmental considerations along with cost and performance in product design and development.

direct user value: Benefit derived from directly consuming services provided by an environmental good.

discount factor: The term $1/(1+r)^t$, where r is the discount rate, and t is the number of periods.

dose-response relationship: A quantitative relationship between doses of a contaminant and the corresponding reactions.

Drinking Water State Revolving Fund (DWSRF): Authorizes $1 billion per year from 1994 to 2003 to finance infrastructure improvements.

E

Economic Analysis (EA): A requirement under Executive Order 12866 that called for information on the benefits and costs of a "significant regulatory action."

efficient equilibrium: The point where marginal social benefit (MSB) equals marginal social cost (MSC), or where marginal profit ($M\pi$)=marginal external cost (MEC).

effluent allowances: Tradeable permits issued up-front that give a polluter the right to release effluents in the future.

effluent reduction credits: Tradeable permits issued if a polluter discharges a lower level of effluents than what is allowed by law.

effluent reduction trading policy: Establishes an abatement objective for a watershed and allows sources to negotiate trades for rights to pollute.

emission or effluent charge: A fee imposed directly on the actual discharge of pollution.

emissions banking: Allows a source to accumulate emission reduction credits if it reduces emissions more than required by law and to deposit these through a banking program.

engineering approach: Estimates abatement expenditures based on least-cost available technology.

environmental economics: A field of study concerned with the flow of residuals from economic activity back to nature.

environmental equity: Concerned with the fairness of the environmental risk burden across segments of society or geographic regions.

environmental Kuznets curve: Models an inverted U-shaped relationship between economic growth and environmental degradation.

environmental literacy: Awareness of the risks of pollution and natural resource depletion.

environmental quality: A reduction in anthropogenic contamination to a level that is "acceptable" to society.

environmental risk: The probability that damage will occur due to exposure to an environmental hazard.

equilibrium price and quantity: The market-clearing price (P_E) associated with the equilibrium quantity (Q_E), where $Q_D=Q_S$.

excise tax on ozone depleters: An escalating tax on the production of ozone-depleting substances.

existence value: Benefit received from the continuance of an environmental good.

existing chemical: A substance listed in the TSCA inventory.

existing stationary source: A source already present when regulations are published.

explicit costs: Administrative, monitoring, and enforcement expenses paid by the public sector plus compliance costs incurred by all sectors.

exposure: The pathways between the source of the damage and the affected population or resource.

exposure analysis: Characterizes the sources of an environmental hazard, concentration levels at that point, pathways, and any sensitivities.

Extended Product Responsibility (EPR): A commitment by all participants in the product cycle to reduce any life cycle environmental impacts of products.

externality: A spillover effect associated with production or consumption that extends to a third party outside the market.

F

federal grant program: Provided major funding from the federal government for a share of the construction costs of POTWs.

feedstock taxes: Taxes levied on raw materials used as productive inputs.

first law of thermodynamics: Matter and energy can neither be created nor destroyed.

fishable-swimmable goal: An interim U.S. objective requiring that surface waters be capable of supporting recreational activities and the propagation of fish and wildlife.

fixed fee or flat fee pricing system: Pricing MSW services independent of the quantity of waste generated.

flat fee pricing scheme: Pricing water supplies such that the fee is independent of water use.

flat rate pricing: A unit pricing scheme that charges the same price for each additional unit of waste.

free-ridership: Recognition by a rational consumer that the benefits of consumption are accessible without paying for them.

front-end or retail disposal charge: A fee levied on a product at the point of sale designed to encourage source reduction.

G

gasoline tax: A per unit tax levied on each gallon of gasoline consumed.

global pollution: Environmental effects that are widespread with global implications, such as global warming and ozone depletion.

global warming: Caused by sunlight hitting the earth's surface and radiating back into the atmosphere where its absorption by GHGs heats the atmosphere and warms the earth's surface.

global warming potential (GWP): Measures the global warming effect of a unit of any GHG relative to a unit of CO_2 over some time period.

Green Chemistry Program: An initiative that promotes the development and application of innovative chemical technologies to achieve pollution prevention.

greenhouse gases (GHGs): Gases collectively responsible for the absorption process that naturally warms the earth.

greenhouse gas (GHG) intensity: The ratio of GHG emissions to economic output.

groundwater: Fresh water beneath the earth's surface, generally in aquifers.

H

halons: A major group of ozone depleters with long atmospheric lifetimes.

hazard: The source of the environmental damage.

hazard identification: Scientific analysis to determine whether a causal relationship exists between a pollutant and any adverse effects.

hazardous air pollutants: Noncriteria pollutants that may cause or contribute to irreversible illness or increased mortality.

hazardous solid wastes: Any unwanted materials or refuse capable of posing a substantial threat to health or the ecology.

hedonic price method (HPM): Uses the estimated hedonic price of an environmental attribute to value a policy-driven improvement.

hydrologic cycle: The natural movement of water from the atmosphere to the surface, beneath the ground, and back into the atmosphere.

I

implicit costs: The value of any nonmonetary effects that negatively influence society's well-being.

increasing block pricing structure: A pricing structure in which the per unit price of different blocks of water increases as water use increases.

incremental benefits: The reduction in health, ecological, and property damages associated with an environmental policy initiative.

incremental costs: The change in costs arising from an environmental policy initiative.

indirect user value: Benefit derived from indirect consumption of an environmental good.

industrial ecology: A multidisciplinary systems approach to the flow of materials and energy between industrial processes and the environment.

industrial ecosystem: A closed system of manufacturing whereby the wastes of one process are reused as inputs in another.

inflation correction: Adjusts for movements in the general price level over time.

Integrated Pest Management (IPM): A combination of control methods aimed at encouraging more selective use of pesticides and greater reliance on natural deterrents.

integrated waste management system: An EPA initiative to guide state MSW plans that promotes using source reduction, recycling, combustion, and land disposal.

international externality: A spillover effect associated with production or consumption that extends to a third party outside the market in another nation.

involuntary risk: A risk beyond one's control and not the result of a willful decision.

ISO 14000 standards: Voluntary international standards for environmental management.

J

joint and several liability: The legal standard that identifies a single party as responsible for all damages even if that party's contribution to the damages is minimal.

L

Law of Demand: There is an inverse relationship between price and quantity demanded of a good, *c.p.*

Law of Supply: There is a direct relationship between price and quantity supplied of a good, *c.p.*

life cycle assessment (LCA): Examines the environmental impact of a product or process by evaluating all its stages from raw materials extraction to disposal.

linear flow of materials: Assumes that materials run in one direction, entering an economic system as inputs and leaving as wastes or residuals.

listed wastes: Hazardous wastes that have been preidentified by government as having met specific criteria.

local pollution: Environmental damage that does not extend far from the polluting source, such as urban smog.

M

management strategies: Methods that address existing environmental problems and attempt to reduce the damage from the residual flow.

manifest: A document used to identify hazardous waste materials and all parties responsible for its movement from generation to disposal.

marginal abatement cost (*MAC*): The change in costs associated with increasing abatement, using the least-cost method.

marginal cost of enforcement (*MCE*): Added costs incurred by government associated with monitoring and enforcing abatement activities.

marginal social benefit (*MSB*): The sum of marginal private benefit (*MPB*) and marginal external benefit (*MEB*).

marginal social benefit (*MSB*) of abatement: A measure of the additional gains accruing to society as pollution abatement increases.

marginal social cost (*MSC*): The sum of the marginal private cost (*MPC*) and the marginal external cost (*MEC*).

marginal social cost (*MSC*) of abatement: The sum of all polluters' marginal abatement costs plus government's marginal cost of monitoring and enforcing these activities.

market: The interaction between consumers and producers to exchange a well-defined commodity.

market approach: An incentive-based policy that encourages conservation practices or pollution reduction strategies.

market demand for a private good: The decisions of all consumers willing and able to purchase a good, derived by *horizontally* summing the individual demands.

market demand for a public good: The aggregate demand of all consumers in the market, derived by *vertically* summing their individual demands.

market failure: The result of an inefficient market condition.

market-level marginal abatement cost (MAC_{mkt}): The horizontal sum of all polluters' *MAC* functions.

market supply of a private good: The combined decisions of all producers in a given industry, derived by *horizontally* summing the individual supplies.

materials balance model: Positions the circular flow within a larger schematic to show the connections between economic decision making and the natural environment.

materials groups: Categories in the MSW stream identified as paper and paperboard, yard trimmings, glass, metals, plastics, textiles, rubber and leather, wood, and other miscellaneous wastes.

maximize the present value of net benefits (*PVNB*): A decision rule to achieve allocative efficiency by selecting the policy option that yields greatest excess benefits after adjusting for time effects.

maximum achievable control technology (MACT): The control technology that achieves the degree of reduction to be accomplished by the NESHAP.

maximum contaminant level (MCL): A component of an NPDWR that states the highest permissible level of a contaminant in water delivered to any user of a public system.

maximum contaminant level goal (MCLG): A component of an NPDWR that defines the level of a pollutant at which no known or expected adverse health effects occur, allowing for a margin of safety.

minimize the present value of costs (*PVC*): A decision rule to achieve cost-effectiveness by selecting the least-cost policy option that achieves a preestablished objective.

mobile source: Any nonstationary polluting source, including all transport vehicles.

multimedia, multichemical approaches: Proposed regulatory procedures that would replace existing single chemical regulatory actions.

municipal solid waste (MSW): Nonhazardous wastes disposed of by local communities.

N

National Action Plan (NAP): A nation's policy to control GHGs and a statement of its target emissions level for the future.

National Ambient Air Quality Standards (NAAQS): Maximum allowable concentrations of criteria air pollutants.

National Emission Standards for Hazardous Air Pollutants (NESHAP): Set to protect public health and the environment, applicable to every major source of any identified hazardous air pollutant.

National Pollutant Discharge Elimination System (NPDES): A federally mandated permit system used to control effluent releases from direct industrial dischargers and POTWs.

National Primary Drinking Water Regulations (NPDWRs): Health standards for public drinking water supplies that are implemented uniformly.

National Priorities List (NPL): A classification of hazardous waste sites posing the greatest threat to health and the ecology.

natural pollutants: Contaminants that come about through nonartificial processes in nature.

natural resource economics: A field of study concerned with the flow of resources from nature to economic activity.

negative externality: An external effect that generates costs to a third party.

netting: Developed for PSD areas to allow emissions trading among points within a source for the same type of pollutant, such that any emissions increase due to a modification is matched by a reduction from another point within that same source.

new chemical: Any substance not listed in the TSCA inventory of existing chemicals.

new or modified stationary source: A source for which construction or modification follows the publication of regulations.

New Source Performance Standards (NSPS): Technology-based emissions limits established for new stationary sources.

nominal value: A magnitude stated in terms of the current period.

nonattainment area: An AQCR not in compliance with the NAAQS.

nonconventional pollutant: A default category for pollutants not identified as toxic or conventional.

nonexcludability: The characteristic that makes it impossible to prevent others from sharing in the benefits of consumption.

nonpoint source: A source that cannot be identified accurately and degrades the environment in a diffuse, indirect way over a relatively broad area.

Nonpoint Source Management Program: A three-stage, state-implemented plan aimed at nonpoint source pollution.

nonrevelation of preferences: An outcome that arises when a rational consumer does not volunteer a willingness to pay because of the lack of a market incentive to do so.

nonrivalness: The characteristic of indivisible benefits of consumption such that one person's consumption does not preclude that of another.

no toxics in toxic amounts goal: A U.S. goal prohibiting the release of toxic substances in toxic amounts into all water resources.

O

offset plan: Developed for nonattainment areas to allow emissions trading between new or modified sources and existing facilities such that releases from the new or modified source are more than countered by reductions achieved by existing sources.

operating costs: Variable expenditures incurred in the operation and maintenance of abatement processes.

oxygenated fuel: Formulations with enhanced oxygen content to allow for more complete combustion and hence a reduction in CO emissions.

ozone depletion: Thinning of the ozone layer, originally observed as an ozone hole over Antarctica.

ozone depletion potential (ODP): A numerical score that signifies a substance's potential for destroying stratospheric ozone relative to CFC-11.

ozone layer: Ozone present in the stratosphere that protects the earth from ultraviolet radiation.

P

performance-based standard: A standard that specifies a pollution limit to be achieved but does not stipulate the technology.

permitting system: A control approach that authorizes the activities of TSDFs according to predefined standards.

per unit subsidy on pollution reduction: A payment for every unit of pollution removed below some predetermined level.

pesticide registration: A formal listing of a pesticide with the EPA that must be approved, based on a risk-benefit analysis, before it can be sold or distributed.

pesticide reregistration: A formal reevaluation of a previously licensed pesticide already on the market.

pesticide tolerances: Legal limits on the amount of pesticide remaining as a residue on raw agricultural products or in processed foods.

photochemical smog: A type of smog caused by pollutants that chemically react in sunlight to form new substances.

physical linkage approach: Estimates benefits based on a technical relationship between an environmental resource and the user of that resource.

Pigouvian subsidy: A per unit payment on a good whose consumption generates a positive externality such that the payment equals the *MEB* at Q_E.

Pigouvian tax: A unit charge on a good whose production generates a negative externality such that the charge equals the *MEC* at Q_E.

point source: Any single identifiable source of pollution from which pollutants are released, such as a factory smokestack, a pipe, or a ship.

pollutant-based effluent fee: A fee based on the degree of harm associated with the contaminant being released.

pollution: The presence of matter or energy whose nature, location, or quantity has undesired effects on the environment.

pollution allowances: Tradeable permits that indicate the maximum level of pollution that may be released.

pollution charge: A fee that varies with the amount of pollutants released.

pollution credits: Tradeable permits issued for emitting below an established standard.

pollution permit trading system: A market instrument that establishes a market for rights to pollute by issuing tradeable pollution credits or allowances.

pollution prevention (P2): A long-term strategy aimed at reducing the amount or toxicity of residuals released to nature.

positive externality: An external effect that generates benefits to a third party.

potentially responsible parties (PRPs): Any current or former owner or operator of a hazardous waste facility and all those involved in the disposal, treatment, or transport of hazardous substances to a contaminated site.

premanufacture notice (PMN): An official notification made to the EPA by a chemical producer about its intent to produce or import a new chemical.

present value determination: A procedure that discounts a future value (*FV*) into its present value (*PV*) by accounting for the opportunity cost of money.

present value of benefits (*PVB*): The time-adjusted magnitude of incremental benefits associated with an environmental policy change.

present value of costs (*PVC*): The time-adjusted magnitude of incremental costs associated with an environmental policy change.

present value of net benefits (*PVNB*): The differential of ($PVB - PVC$) used to determine the feasibility of a policy option if its magnitude exceeds zero.

prevention of significant deterioration (PSD) area: An AQCR meeting or exceeding the NAAQS.

primary environmental benefit: A damage-reducing effect that is a direct consequence of implementing environmental policy.

primary NAAQS: Set to protect public health from air pollution, with some margin of safety.

priority contaminants: Pollutants for which drinking water standards are to be established based on specific criteria.

private good: A commodity that has two characteristics, rivalry in consumption and excludability.

producer surplus: The net gain to sellers of a good estimated by the excess of market price (*P*) over marginal cost (*MC*), aggregated over all units sold.

product charge: A fee added to the price of a pollution-generating product based on its quantity or some attribute responsible for pollution.

product groups: Categories in the MSW stream identified as durable goods, nondurable goods, containers and packaging, and other wastes.

product substitution: The selection of environmentally safe commodities in place of potentially polluting products.

profit maximization: Achieved at the output level where $MR = MC$ or where $M\pi = 0$.

property rights: The set of valid claims to a good or resource that permits its use and the transfer of its ownership through sale.

protectionism: Fostering trade barriers, such as tariffs or quotas, to protect a domestic economy from foreign competition.

public good: A commodity that is nonrival in consumption and yields benefits that are nonexcludable.

R

raw materials substitution: The use of productive inputs that generate little or no hazardous waste.

real value: A magnitude adjusted for the effects of inflation.

receiving water quality standards: State-established standards defined by use designation and water quality criteria.

reformulated gasoline: Newly developed fuels that emit less hydrocarbons, carbon monoxide, and toxics than conventional gasoline.

regional pollution: Degradation that extends well beyond the polluting source, such as acidic deposition.

Regulatory Impact Analysis (RIA): A requirement under Executive Order 12291 that called for information about the potential benefits and costs of a major federal regulation.

remanufacturing: Collection, disassembly, reconditioning, and reselling of the same product.

remedial actions: Official responses to a hazardous substance release aimed at achieving a more permanent solution.

removal actions: Official responses to a hazardous substance release aimed at restoring immediate control.

residual: The amount of a pollutant remaining in the environment after a natural or technological process has occurred.

risk: The chance of something bad happening.

risk assessment: Qualitative and quantitative evaluation of the risk posed to health or the ecology by an environmental hazard.

risk-benefit analysis: An assessment of risks of a hazard along with the benefits to society of not regulating that hazard.

risk characterization: Description of risk based upon an assessment of a hazard and exposure to that hazard.

risk management: The decision-making process of evaluating and choosing from alternative responses to environmental risk.

S

secondary environmental benefit: An indirect gain to society that may arise from a stimulative effect of primary benefits or from a demand-induced effect to implement policy.

secondary maximum contaminant levels (SMCLs): National standards for drinking water that serve as guidelines to protect public welfare.

secondary NAAQS: Set to protect public welfare from any adverse, nonhealth effects of air pollution.

second law of thermodynamics: Nature's capacity to convert matter and energy is not without bound.

shortage: Excess demand of a commodity equal to $(Q_D - Q_S)$ that arises if price is *below* its equilibrium level.

social costs: Expenditures needed to compensate society for resources used so that its utility level is maintained.

social discount rate: Discount rate used for public policy initiatives based on the social opportunity cost of funds.

society's welfare: The sum of consumer surplus and producer surplus.

sole-source aquifers: Underground geological formations containing groundwater that are the only supply of drinking water for a given area.

source reduction: Preventive strategies to reduce the quantity of any hazardous substance, pollutant, or contaminant released to the environment at the point of generation.

source segregation: A procedure that keeps hazardous waste from coming in contact with nonhazardous waste.

State Implementation Plan (SIP): A procedure outlining how a state intends to implement, monitor, and enforce the NAAQS and the NESHAP.

State Revolving Fund (SRF) program: Establishes state lending programs to support POTW construction and other projects.

stationary source: A fixed-site producer of pollution, such as a building or manufacturing plant.

stewardship: The sense of obligation to preserve the environment for future generations.

strict liability: The legal standard that identifies individuals as responsible for damages even if negligence is not proven.

Superfund cleanup process: A series of steps used to determine and implement the appropriate response to threats posed by the release of a hazardous substance.

supply: The quantities of a good the producer is willing and able to bring to market at a given set of prices during some discrete time period, *c.p.*

surface water: Bodies of water open to the earth's atmosphere as well as springs, wells, or other collectors directly influenced by surface water.

surplus: Excess supply of a commodity equal to $(Q_S - Q_D)$ that arises if price is *above* its equilibrium level.

survey approach: Polls a sample of firms and public facilities to obtain estimated abatement expenditures.

sustainable development: Management of the earth's resources such that their long-term quality and abundance is ensured for future generations.

T

technical efficiency: Production decisions that generate maximum output given some stock of resources.

technology-based effluent limitations: Standards to control discharges from point sources based primarily on technological capability.

technology-based standard: A standard that designates the equipment or method to be used to achieve some abatement level.

technology transfer: The advancement and application of technologies and strategies on a global scale.

threshold: The level of exposure to a hazard up to which no response exists.

tipping fees: Prices charged for disposing of wastes in a facility such as a landfill.

total profit: Total profit (π)=Total revenue (TR) − Total costs (TC).

toxic chemical use substitution: The use of less harmful chemicals in place of more hazardous substances.

toxic pollutant: A contaminant which, on exposure, will cause death, disease, abnormalities, or physiological malfunctions.

Toxics Release Inventory (TRI): A national database that gives information about hazardous substances released into the environment.

tradeable effluent permit market: The exchange of rights to pollute among water-polluting sources.

tradeable permit system for GHG emissions: Based on the issuance of marketable permits, where each allows the release of some amount of GHGs.

tradeable SO_2 emission allowances: Permits issued to stationary sources, each allowing the release of one ton of SO_2, which can be either held or sold through a transfer program.

travel cost method (TCM): Values benefits by using the *complementary* relationship between the quality of a natural resource and its recreational use value.

TSCA inventory: A database of all chemicals commercially produced or processed in the United States.

U

uniform rate (or flat rate) pricing structure: Pricing water supplies to charge more for higher water usage at a constant rate.

unit pricing scheme: A common designation for the use of a waste-end charge.

use designation: A component of receiving water quality standards that identifies the intended purposes of a water body.

user value: Benefit derived from physical use or access to an environmental good.

use-support status: A classification of a water body based on a state's assessment of its present condition relative to what is needed to maintain its designated uses.

V

variable rate pricing: A unit pricing scheme that charges a different price for each additional unit of waste.

vicarious consumption: The utility associated with knowing that others derive benefits from an environmental good.

volume-based effluent fee: A fee based upon the quantity of pollution discharged.

voluntary risk: A risk that is deliberately assumed at an individual level.

W

waste-end charge: A fee implemented at the time of disposal based on the quantity of waste generated.

waste management: Control strategies to reduce the quantity and toxicity of hazardous wastes at every stage of the waste stream.

waste stream: A series of events starting with waste generation and including transportation, storage, treatment, and disposal of solid wastes.

water quality criteria: A component of receiving water quality standards that gives the biological and chemical attributes necessary to sustain or achieve designated uses.

water quality–related limitations: Modified effluent limits to be met if the desired water quality level is not being achieved, even if polluters are satisfying the technology-based limits.

Z

zero discharge goal: A U.S. objective calling for the elimination of all polluting effluents into navigable waters.

Index